International Conference on Civil Engineering and Infrastructural Issues in Emerging Economics (ICCIEE 2017)

AA006259

IOP Conference Series: Earth and Environmental Science
Volume 80

Thanjavur, India
17 – 18 March 2017

ISBN: 978-1-5108-4718-7
ISSN: 1755-1307

Printed from e-media with permission by:

Curran Associates, Inc.
57 Morehouse Lane
Red Hook, NY 12571

Some format issues inherent in the e-media version may also appear in this print version.

Printed by Curran Associates, Inc. (2017)

For permission requests, please contact the Institute of Physics
at the address below.

Institute of Physics
Dirac House, Temple Back
Bristol BS1 6BE UK

Phone: 44 1 17 929 7481
Fax: 44 1 17 920 0979

techtracking@iop.org

Additional copies of this publication are available from:

Curran Associates, Inc.
57 Morehouse Lane
Red Hook, NY 12571 USA
Phone: 845-758-0400
Fax: 845-758-2633
Email: curran@proceedings.com
Web: www.proceedings.com

Table of contents

Volume 80

International Conference on Civil Engineering and Infrastructural Issues in Emerging Economies (ICCIEE 2017)

17–18 March 2017, Tirumalaisamudram, Thanjavur, India

Accepted papers received: 20 July 2017
Published online: 9 August 2017

Preface

Papers

IOP Conf. Series: Earth and Environmental Science **80** (2017) 011001 doi:10.1088/1755-1315/80/1/011001

International Conference on Civil Engineering and Infrastructural Issues in Emerging Economies (ICCIEE 2017)

17 – 18 March 2017

School of Civil Engineering, SASTRA UNIVERSITY, Tirumalaisamudram,Thanjavur - 613 401, INDIA

Preface/Introduction:

The phenomenal rise in urbanization and industrialization has created a greater need for infrastructure development. In a developing economy, resource utilization and management is the need of the hour. This creates a need for innovation in materials, methods and management with greater emphasis on limited economic solutions. This has opened avenues for experimenting and implementing novel approaches with economic viability. There is also a need for harmony between rapid development and ecological balance to ensure sustainable development. This conference jointly organized by the School of Civil Engineering, SASTRA University, Thanjavur and IGS, Thanjavur Chapter aims to provide a platform for academicians, researchers and the industry to explore the possibilities of the use of nonconventional materials and their economic viability as alternative to conventional construction material, design and construction alternatives for economic and ergonomic structures, water management, sustainable development issues in environmental, geotechnical and transportation engineering and application of geomatics in hazard prediction and management.

SASTRA
UNIVERSITY
2017

SASTRA
UNIVERSITY
Thank You
Sponsors
Build beautiful
UltraTech
CUB

GEOTROPIK
EFFECT OF GEOLOGICAL STRUCTURE ON FLYROCK PREDICTION IN CONSTRUCTION BLASTING

SASTRA UNIVERSITY
Keynote Speakers

IOP Conf. Series: Earth and Environmental Science **80** (2017) 011002 doi:10.1088/1755-1315/80/1/011002

SASTRA
UNIVERSITY
International Conference on
Civil Engineering &
Infrastructural issues in Emerging Economies
(ICCIEE - 2017)
17 & 18, 2017

IOP Conf. Series: Earth and Environmental Science **80** (2017) 011003 doi:10.1088/1755-1315/80/1/011003

International Conference on Civil Engineering and Infrastructural Issues in Emerging Economies (ICCIEE 2017)

17 – 18 March 2017

School of Civil Engineering, SASTRA UNIVERSITY, Tirumalaisamudram,Thanjavur - 613 401, INDIA

List of Sponsors:

1. Department of Science & Technology, SERB Division, New Delhi, India.
2. Indian Geotechnical Society (IGS), Thanjavur Chapter, Thanjavur, India.
3. UltraTech Cement, Chennai, India.
4. P.Srinivasan Associates, Ondipudur, Coimbatore, India.
5. MKR Constructions, Kumbakonam, India.
6. City Union Bank Limited, Thanjavur, India.
7. Lucky Borewell, Thanjavur, India.
8. KGP Traders, Vallam, Thanjavur, India.
9. NRM Constructions, Thanjavur, India.
10. Sun Builders, Thanjavur, India.
11. Shalom Pest Control, Thanjavur, India.
12. S.Sahayaraj, Fabricators, Trichy, India.

IOP Conf. Series: Earth and Environmental Science **80** (2017) 011004 doi:10.1088/1755-1315/80/1/011004

International Conference on Civil Engineering and Infrastructural Issues in Emerging Economies (ICCIEE 2017)
17 – 18 March 2017
School of Civil Engineering, SASTRA UNIVERSITY, Tirumalaisamudram,Thanjavur - 613 401, INDIA

List of Committees:

1. Organizing Committee
2. International Steering Committee
3. Advisory Committee
4. Technical Committee
5. Finance Committee
6. Registration & Reception Committee
7. Travel & Accommodation Committee
8. Catering Committee
9. Hospitality Committee
10. Public Relation Committee

IOP Conf. Series: Earth and Environmental Science **80** (2017) 011005 doi:10.1088/1755-1315/80/1/011005

Peer review statement

All papers published in this volume of *IOP Conference Series: Earth and Environmental Science* have been peer reviewed through processes administered by the proceedings Editors. Reviews were conducted by expert referees to the professional and scientific standards expected of a proceedings journal published by IOP Publishing.

Bond characteristics of reinforcing steel embedded in geopolymer concrete

Parthiban Kathirvel*, Manju Thangavelu, Rashmi Gopalan and Saravana Raja Mohan Kaliyaperumal

School of Civil Engineering, SASTRA University, Thanjavur – 613 401, India

*Email: parthiban@civil.sastra.edu

Abstract. The force transferring between reinforcing steel and the surrounding concrete in reinforced concrete is influenced by several factors. Whereas, the study on bond behaviour of geopolymer concrete (GPC) is lagging. In this paper, an experimental attempt has been made to evaluate the geopolymer concrete bond with reinforcing steel of different diameter and embedded length using standard pull out test. The geopolymer concrete is made of ground granulated blast furnace slag (GGBFS) as geopolymer source material (GSM). The tests were conducted to evaluate the development of bond between steel and concrete of grade M40 and M50 with 12 and 16 mm diameter reinforcing steel for geopolymer and cement concrete mixes and to develop a relation between bond strength and compressive strength. From the experimental results, it has been observed that the bond strength of the geopolymer concrete mixes was more compared to the cement concrete mixes and increases with the reduction in the diameter of the bar.

Key words: *Geopolymer Concrete, Ground Granulated Blast Furnace Slag, Ambient Curing, Pull-Out Test and Compressive Strength.*

1. Introduction

Over the past two decades, geopolymer concretes have developed as novel construction materials with a possibility to become a significant component in construction industry and environmentally sustainable construction [1, 2]. Geopolymer concrete is resulting from the reaction between aluminosilicates present in the GSM and alkalis to develop an inorganic polymer as binder. The materials of geological origin and industrial byproducts such as fly ash, GGBFS, etc. [3, 4] which are rich in Silica and Alumina can be used as a GSM in the development of geopolymer concrete. In spite of their diverse chemical composition and reactions [5, 6] geopolymer concretes shows a number of attributes similar to conventional cement concrete whereas, a least number of works were attempted in the past to evaluate the bond behaviour steel reinforcement in geopolymer concrete. Hence, in this investigation, the bond behaviour of GPC with deformed reinforcing steel was evaluated using standard pull out test for 12 and 16 mm diameter bars as reinforcement for the grades of M40 and M50. Since very few literatures are available to study the bond behaviour of GPC made of GGBFS, this study was carried out by employing stress-slip relation and compressive-bond strength relation. The results were compared with the performance of ordinary Portland cement (OPC) based concrete. Considering the GPC is appropriate for precast applications, this work was focused to evaluate the

IOP Conf. Series: Earth and Environmental Science **80** (2017) 012001 doi:10.1088/1755-1315/80/1/012001

bond strength of GPC under ambient temperature condition to assess its ability under cast in-situ conditions.

2. Materials and methods

2.1 Materials

Ordinary Portland Cement (OPC) of grade 53 confirming to IS 12269 was utilized in the study. The specific gravity of the OPC sample was found to be 3.12 with an initial setting time of 40 minutes and a standard consistence of 28 % with its chemical composition is given in Table 1. GGBFS a byproduct of iron and steel industry is one of the most commonly used mineral binder as it continues to gain strength over time and is chemically stable. Another interesting property about GGBFS is, when incorporated it shows latent hydraulic properties. In this experimental study with a specific gravity of 2.90, fineness 438 m^2/kg and bulk density of 1231 kg/m^3 was used as geopolymeric source material.

Table 1. Chemical composition of OPC and GGBFS.

Oxide	CaO	SiO_2	Al_2O_3	MgO	SO_3	Fe_2O_3	Na_2O	K_2O
OPC (%)	63.12	24.52	6.88	2.16	1.43	3.51	0.44	0.63
GGBFS (%)	36.77	30.97	17.41	9.01	1.82	1.03	0.69	0.46

A mixture of Sodium hydroxide and Sodium Silicate with silica to alkali molar ratio (SiO_2/Na_2O) of 2.5 was as used for the alkali activation. Commercial grade Sodium Hydroxide in the form of flakes of almost 99% purity and sodium silicate solution having 28% SiO_2, 11.2% Na_2O and 60.8% H_2O by mass were taken. Locally available river sand was used as fine aggregates. The grading of the aggregate was categorized using sieve analysis test, and the results are detailed in Table 2 and accordingly the aggregate was categorized as graded under Zone III.

Table 2. Properties of aggregates used in the mixes.

Properties	Fine Aggregate	Coarse Aggregate
Specific gravity	2.58	2.72
Fineness modulus	2.56	6.97
Bulk density, kg/m^3	1696	1485
Water absorption, %	1.37	0.77

Coarse aggregates from locally available quarry and crushed blue granite chips with a nominal size of 16 mm were used in this study. The aggregates were tested and the results obtained are detailed in Table 2. The steel bars of 12 and 16 mm high yield strength deformed bars were used as reinforcement in the cube specimens.

2.2 Methodology

The cement concrete and geopolymer concrete mixes were prepared for the grade of M40 and M50 with the concentration of NaOH solution used in geopolymer mixes were taken as 14 M and an alkaline ratio of 2.0 and the detailed mix proportioning is given in Table 3, with a liquid-binder ratio of 0.40. The cube of size 150 mm x 150 mm x 150 mm were cast with the bar was placed at the centre with a projection of 25 mm at the bottom. The cement concrete specimens were water cured, whereas the geopolymer mixes were cured under ambient temperature condition.

Table 3. Mix proportioning of the concrete mixes used in the study.

Mix	Cement	GGBFS	Sand	Aggregate	Water	NaoH Solution	Na_2SiO_3 Solution
CC40	400	---	660	1168	160	---	---
CC50	425	---	648	1087	170	---	---
GP40	---	387	432	1368	---	106.5	106.5
GP50	---	400	432	1368	---	66.7	133.3

For the compressive strength test, cubes of size 150 mm x 150 mm x 150 mm were tested in compression testing machine of 3000 kN capacity at the age of 7 and 28 days ambient curing condition in accordance with IS 516-1959. Pull out test was carried out as per the guidelines specified by IS 2770-1. The test specimens are mounted in a Universal testing machine in such a manner that the bar is pulled axially from the cube and a mechanical dial gauge was used to measure the slip rate of the reinforcing steel with the load increment from the surface of the cube specimen. The cube specimens were reinforced with 12 and 16 diameter bars to study the bond behaviour of geopolymer and cement concrete mixes with no confining reinforcement.

3. Results and discussion

The variation in the compressive strength of the cement concrete and geopolymer concrete mixes at the age of 7 and 28 days are shown in Figure 1. The 7 days compressive strength of the cement concrete mixes were found to be around 60-70% of its 28 days strength, whereas for geopolymer concrete mixes, it was found to be more than 70% of its 28 days strength. Also, the compressive strength results of the geopolymer concrete mixes were found to be superior to the cement concrete mixes.

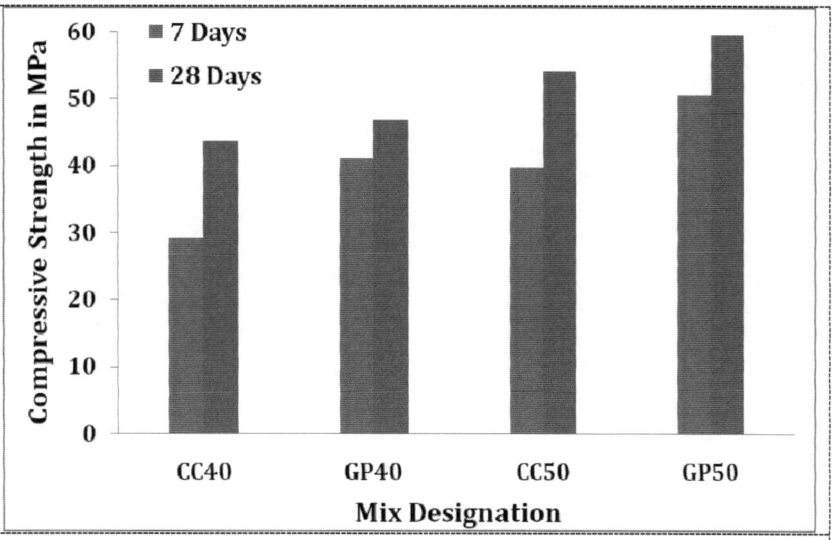

Figure 1. Compressive Strength of the mixes at 7 and 28 Days curing.

Table 4 shows the bond-slip behaviour of cement concrete and geopolymer concrete mixes at various stages of testing.

Table 4. Bond-slip results at various stages of the concrete mixes used in the study.

Mix No.	Diameter of Bar (mm)	@ 0.25 mm Slip Load (kN)	@ 0.25 mm Slip T_{bd} (MPa)	@ 2.5 mm Slip Load (kN)	@ 2.5 mm Slip T_{bd} (MPa)	@ Ultimate Level Load (kN)	@ Ultimate Level Slip (mm)	@ Ultimate Level T_{bd} (MPa)
CC40	12	1.11	0.20	7.13	1.26	50.80	6.97	8.99
	16	1.03	0.14	4.93	0.65	47.70	6.39	6.33
CC50	12	1.18	0.21	5.10	0.90	66.50	9.13	11.77
	16	1.31	0.17	7.90	1.05	61.55	8.13	8.17
GP40	12	1.46	0.26	9.00	1.59	59.20	8.04	10.47
	16	1.46	0.19	8.85	1.17	55.95	6.78	7.42
GP50	12	1.84	0.33	9.48	1.68	84.80	7.89	15.00
	16	1.44	0.19	10.28	1.36	74.40	9.30	9.87

Table 4 details the load-bond-slip behaviour of geopolymer and cement concrete mixes at various stages of loading. The load at 0.25 mm and 2.5 mm slip and at ultimate level was observed.

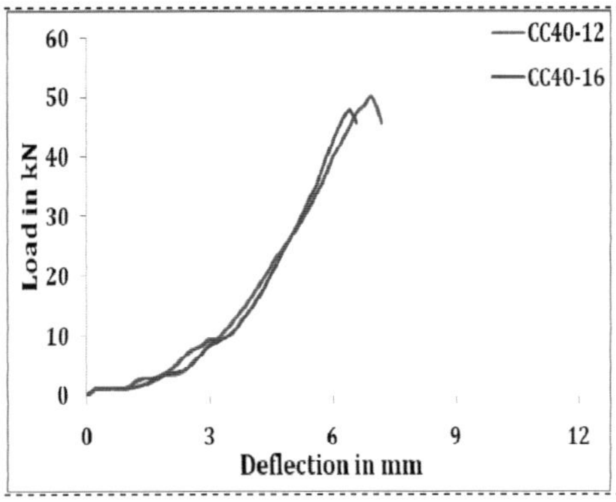

Figure 2a. Load-Slip behaviour of M40 grade cement concrete mixes.

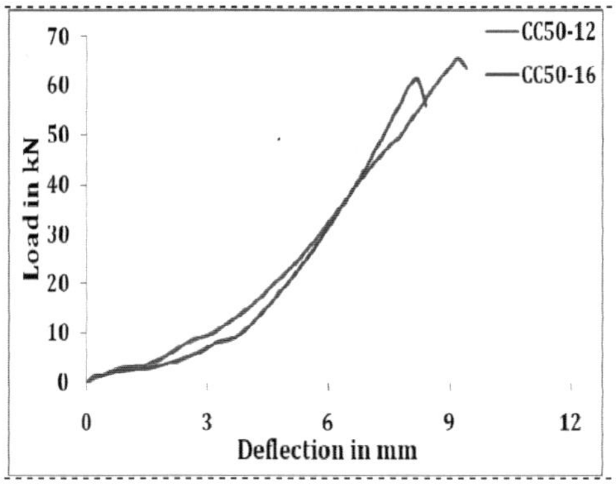

Figure 2b. Load-Slip behaviour of M50 grade cement concrete mixes.

Figure 2a and 2b, shows the load-deflection behaviour of cement concrete mixes for M40 and M50 grade mixes. The behaviour was found to be similar irrespective of grade of concrete and diameter of the reinforcement. The mixes with 12 mm diameter bars show slightly higher energy absorption capacity than the mixes with 16 mm diameter bars.

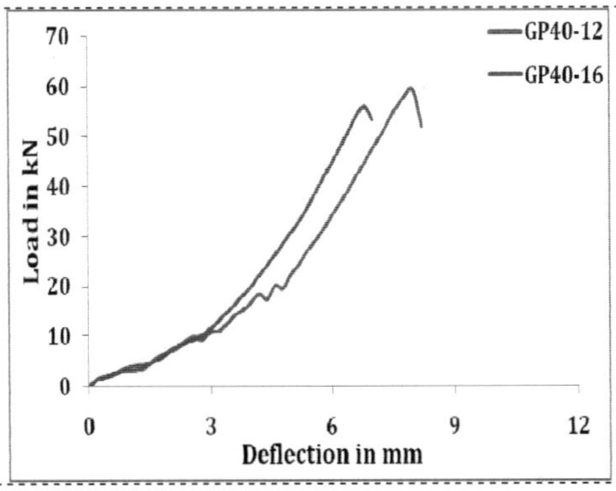

Figure 3a. Load-Slip behaviour of M40 grade geopolymer concrete mixes.

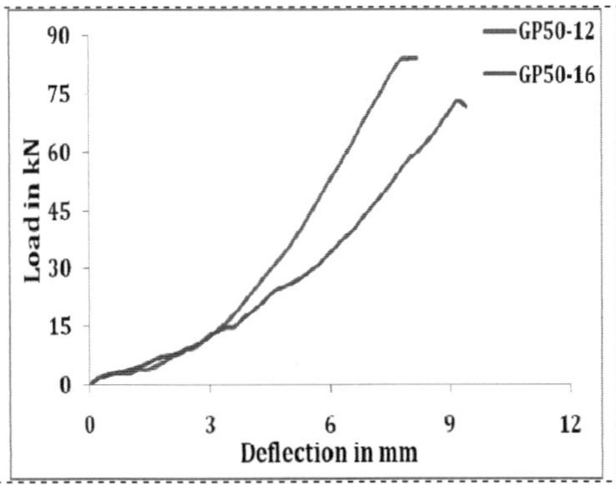

Figure 3b. Load-Slip behaviour of M50 grade geopolymer concrete mixes.

Figure 3a and 3b shows the load-deformation behaviour of geopolymer concrete mixes for M40 and M50 grade mixes reinforced with 12 mm and 16 mm diameter bars. There was a slight deviation in the curve for the mix with M40 grade in which the mix with 12 mm diameter bar experiences slightly higher slip than with the mix reinforced with 16 mm diameter bar.

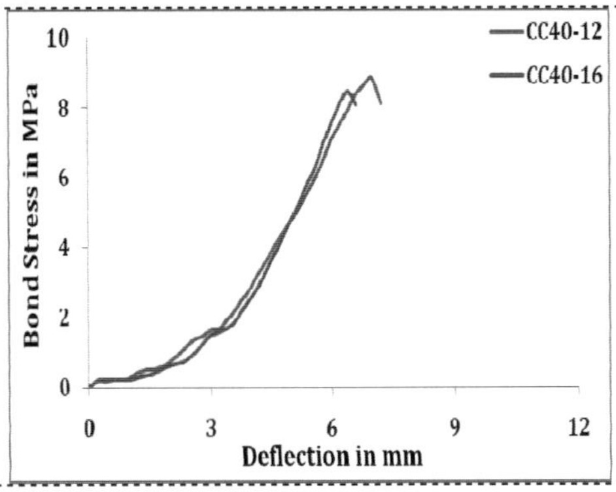

Figure 4a. Bond-Slip behaviour of M40 grade cement concrete mixes.

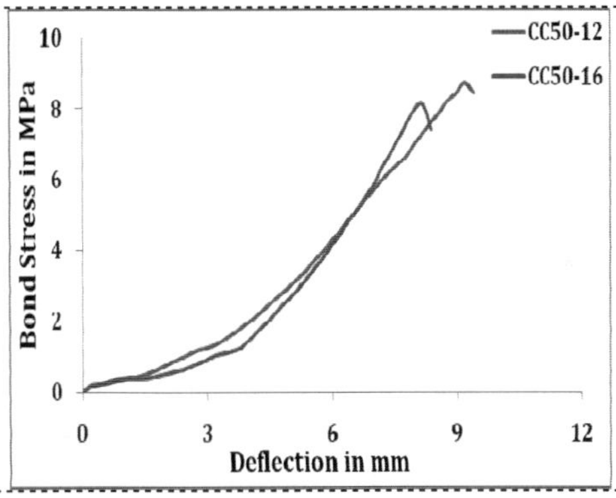

Figure 4b. Bond-Slip behaviour of M50 grade cement concrete mixes.

Figure 4 and Figure 5, shows the bond-slip relation for cement concrete and geopolymer concrete mixes respectively for M40 and M50 grade concrete. The behaviour was found to be similar to the load-slip behaviour.

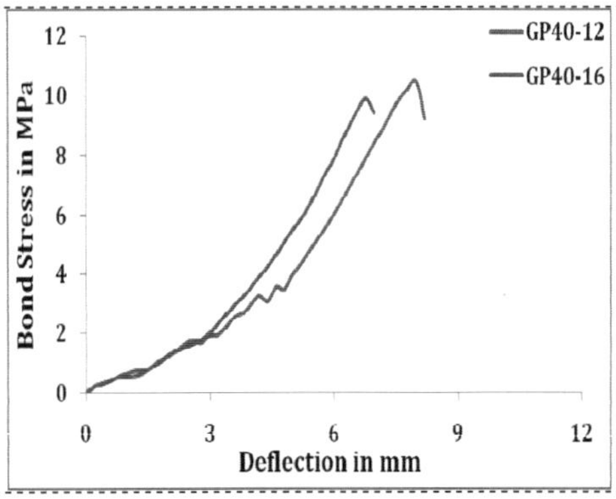

Figure 5a. Bond-Slip behaviour of M40 grade geopolymer concrete mixes.

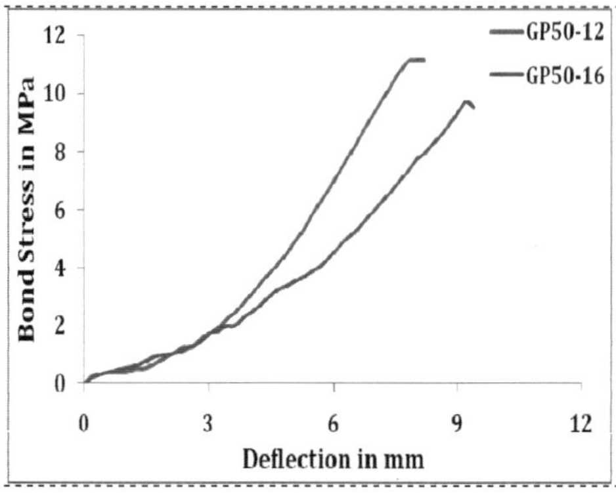

Figure 5b. Bond-Slip behaviour of M50 grade geopolymer concrete mixes.

The bond characteristics of geopolymer shows superior behaviour than cement concrete mixes irrespective of grade of concrete mix and diameter of reinforcement used, which is similar to the results observed from other studies [7,8]. This is mainly due to the strong adhesion between the aggregates and activator solution. The reason behind this improved performance is mainly attributed to the strong interfacial transition zone between aggregate and geopolymer matrix in geopolymer concrete leads to denser matrix results in higher bond strength that cement concrete mixes of similar grade [7]. It was also observed that the bond strength increase with the decrease in the diameter of the reinforcement. Similar behaviour was also observed [9]

4. Conclusion

The use of ambient temperature curing in geopolymer mixes result its ability to use in cast in-situ conditions. The compressive strength of the geopolymer mixes was found to be higher than the cement concrete mixes. There was a superior bonding between the geopolymer concrete and the reinforcement compared with the cement concrete mixes. The bond strength was found to reduce with the increase in the diameter of the reinforcement. The failure of all the specimens was observed to be splitting rather than delamination of the surface concrete.

Acknowledgements

The authors acknowledge the Vice Chancellor, SASTRA University, Thanjavur, India for providing the facilities to carry out the work and the encouragement in completing this work.

References

[1] Duxson P, Fernández-Jiménez A, Provis JL, Lukey G C, Palomo A and Van Deventer J S J 2007 Geopolymer technology: the current state of the art. *J. Mater. Sci.* **42**(9) 2917-33.

[2] Aly T and Sanjayan J G 2010 Effect of pore-size distribution on shrinkage of concretes. *J. Mater. Civ. Eng.* **22**(5) 525-32.

[3] Parthiban K, Saravanarajamohan K, Shobana S, and Anchal Bhaskar A 2013 Effect of replacement of slag on the mechanical properties of flyash based geopolymer concrete. *Int. J. Eng. Tech.* **5**(3) 2555-9.

[4] Parthiban K and Saravana Raja Mohan K 2014 Effect of sodium hydroxide concentration and alkaline ratio on the compressive strength of slag based geopolymer concrete. *Int. J. ChemTech Res.* **6**(4) 2446-50.

[5] Fernandez-Jimenez A M, Palomo A and Lopez-Hombrados C 2006 Engineering properties of alkali-activated fly ash concrete. *ACI Mater. J.* **103**(2) 106-12.

[6] Sofi M, Van Deventer J S J, Mendis P A and Lukey G C 2007 Bond performance of reinforcing bars in inorganic polymer concretes (IPCs). *J. Mater. Sci.* **42**(9) 3107-16.
[7] Sarker P K 2011 Bond strength of reinforcing steel embedded in fly ash-based geopolymer concrete. *Mater. Struct.* **44** 1021-30.
[8] Kim J S and Park J H 2014 An experimental evaluation of development length of reinforcements embedded in geopolymer concrete. *Appl. Mech. Mater.* **578-579** 441-4.
[9] Sarker P K 2010 Bond strengths of geopolymer and cement concretes. *Adv. Sci. Tech.* **69** 143-51.

Study on compressive strength of self compacting mortar cubes under normal & electric oven curing methods

G J Prasanna Venkatesh, S S Vivek* and G Dhinakaran

School of Civil Engineering, SASTRA University, Thanjavur-613401, India

*Email: vivek@civil.sastra.edu

Abstract. In the majority of civil engineering applications, the basic building blocks were the masonry units. Those masonry units were developed as a monolithic structure by plastering process with the help of binding agents namely mud, lime, cement and their combinations. In recent advancements, the mortar study plays an important role in crack repairs, structural rehabilitation, retrofitting, pointing and plastering operations. The rheology of mortar includes flowable, passing and filling properties which were analogous with the behaviour of self compacting concrete. In self compacting (SC) mortar cubes, the cement was replaced by mineral admixtures namely silica fume (SF) from 5% to 20% (with an increment of 5%), metakaolin (MK) from 10% to 30% (with an increment of 10%) and ground granulated blast furnace slag (GGBS) from 25% to 75% (with an increment of 25%). The ratio between cement and fine aggregate was kept constant as 1: 2 for all normal and self compacting mortar mixes. The accelerated curing namely electric oven curing with the differential temperature of 128°C for the period of 4 hours was adopted. It was found that the compressive strength obtained from the normal and electric oven method of curing was higher for self compacting mortar cubes than normal mortar cube. The cement replacement by 15% SF, 20% MK and 25%GGBS obtained higher strength under both curing conditions.
Key words: Self compacting mortar, silica fume, metakaolin, ground granulated blast furnace slag, compressive strength, oven curing.

1. Introduction

The appearance of any structure was made brittle, shiny, ornamental and aesthetic through the process namely plastering. The type of plastering constitutes materials such as lime, cement, cement-lime and stucco plastering. Strength of mortar should not be more than the strength of masonry unit and higher thickness could also be avoided. Once the failure occurs on the masonry unit, the plastering surface would sacrifice itself by formulation of cracks and propagating into the load bearing masonry wall. Thus it was regarded as non-structural cracks and could be repaired through maintenance. On recent developments of concrete and construction materials technology, the study on plastering (with different materials) plays a major role to cater the issues on crack repairs, damp proofing and rehabilitation issues of the structures. Self Compacting (SC) mortar was a highly fluid type which has the tendency to pass, fill and flow on the desired form to repair and rehabilitate the structures. The compressive strength of concrete cubes by different curing methods was studied and found that ponding method of curing obtained highest compressive strength than others [1].

The compressive strength of OPC mortar shown improvement (ranges from 57.5 MPa to 76 MPa at 7 and 90 days) and reduction was studied from the water curing period of 6 hours, air cured and after 20 hours of oven curing (by supplying 60 DC) [2]. An experimental investigation was made on lime-fly ash-gypsum mixtures for the masonry blocks by normal and accelerated curing methods.. It was found that steam curing for the duration of 24 hours at 80°C attained highest strength with the optimum content of gypsum as 2%. For normal curing conditions, the ratio of lime- fly ash of 0.75 obtained more compressive strength [3]. The humidity effects on mechanical properties of Portland cement mortars by formulating cube and prism specimens subjected to different curing conditions for simulating the effects of climatic conditions of Turkey regions was studied [4].

Study on self compacting concrete by replacing cement with fly ash, silica fume and by their ternary combinations were developed and found that cement replacement by 15% silica fume and 30% fly ash got highest compressive strength at 28 days of normal curing [5]. The compressive strength of cement mortars subjected to heated water curing for the duration of 20 hours and then air curing was resulted in highest compressive strength when compared with ASTM, BS and Iraq standards [6]. An experimental investigation on self compacting concrete using shrinkage reducing admixtures concluded that, the age of curing increases the pore size refinement and reduces shrinkage at early ages [7]. The research on self compacting mortar (SCM) containing mix proportions of 1:1 and 1:3 with w/c ratio of 0.34 and 0.5; with the influence of self curing chemicals namely Polyethylene Glycol 4000 & 200 enhance fresh properties, strength and durability [8].

Study on OPC replacement in binary/ ternary blend system using fly ash and spent fluid catalytic cracking catalyst improved the compressive strength up to 106 MPa at the age of 90 days [9]. The influence of zeolite and metakaolin on compressive strength of lime mortars under different humidity conditions of curing was investigated and the effect of pozzolanas cause reduced shrinkage on lime mortars was concluded [10]. For developing precast products, 50% of cement was replaced by fly ash, ground granulated blast furnace slag and rice husk ash under different curing methods and found that mortar specimens cured under hot water resulted on highest compressive strength (Evi et al., 2016) [11]. By simulating the climatic conditions of spain under curing conditions reveal that the micro structure C-S-H link was stable at higher temperatures and thus improved the mechanical properties of self-compacting concrete was studied [12]. SCC flow properties and compressive strength using silica fume as cement replacement was studied [13]. The research on metakaolin based SCC by determining fresh properties and compressive strength [14].

The simple way of SCC mix design was developed [15]. SCC mix was designed using metakaolin as cement replacement [16]. The mix design of SCC and its fresh properties test was conducted according to EFNARC guidelines [17 & 18]. In present study, the self compacting mortar (SC) cube was developed with replacing cement by three mineral admixtures namely, metakaolin (MK) from 10% to 30% (with an increment of 10%), silica fume (SF) from 5%to 20% (with an increment of 5%) and ground granulated blast furnace slag (GGBS) from 25% to 75% (with an increment of 25%) respectively. Finally the comparison of compressive strength was made between conventional cement mortar and self compacting mortar cubes in terms of normal water and electric oven curing methods.

2. Material used

Cement: OPC 53 grade of ASTM Type I [20] was used.

Fine aggregate: Natural river sand free from impurities (< 4.75 mm) grading Zone III of ASTM was used. The physical properties were determined as per ASTM 127 and their particle size distributions confirmed to the requirements of ASTMC33 [21].

IOP Conf. Series: Earth and Environmental Science **80** (2017) 012002 doi:10.1088/1755-1315/80/1/012002

Mineral admixtures: silica fume (SF) from 5% to 20%, metakaolin (MK) from 10% to 30% and ground granulated blast furnace slag (GGBS) from 25% to 75% was used as cement replacement as per ASTM C 1240-99 [19 & 22].
Chemical admixtures: High range super plasticizer (1.25% by weight of cement) and stabilizer (0.1% by weight of cement) were used for making flow of mortar and to control segregation of mix.
Water: water free from acid and base was used.

3. Experimental Programme

3.1 Basic test on materials

The constituent materials of the experimental programme where undergone the following tests namely, consistency and initial setting time of cement, specific gravity of cement, SF, MK and GGBS was determined. In order to study the properties of fine aggregate, the test on fineness modulus was conducted.

3.2 Mix proportions

For conventional mortar cubes (based on consistency test results water-cement ratio was achieved. But for self compacting mortar cubes, the water-powder ratio was fixed by adjusting the dosages of water content, super plasticizer and stabilizer (VMA). Hence the cement mortar mix proportion of 1:2 with water- cement ratio of 0.38 was used for both.

3.3 Fresh Properties test - Slump test

For ascertaining the flow behaviour of self compacting mortars by different mineral admixtures replacement, slump flow test was conducted and spread of mortar flow diameter was measured. It was done by pouring of mortar mixes on Abram's cone of dimensions, top and bottom diameter 10 cm and 20 cm with a height of 30 cm. After filling of mortar up to the top level of cone without compaction, lifted up and the resulting flow measured by spread diameter was studied.

3.4 Preparation of mould & de-moulding

The timber mould was prepared with inner dimensions of 100 mm x 100 mm x 100 mm and properly dressed with application of oil coat on its inner surfaces. Then prepared ordinary cement mortar (CM) was poured and well compacted. But for self compacting mortars, the fresh mortar was poured on the mould without any compaction (Figure 1a & 1b). After 24 hours of casting, the cubes were removed from the mould and allowed for curing (Figure 2a, 2b & 2c).

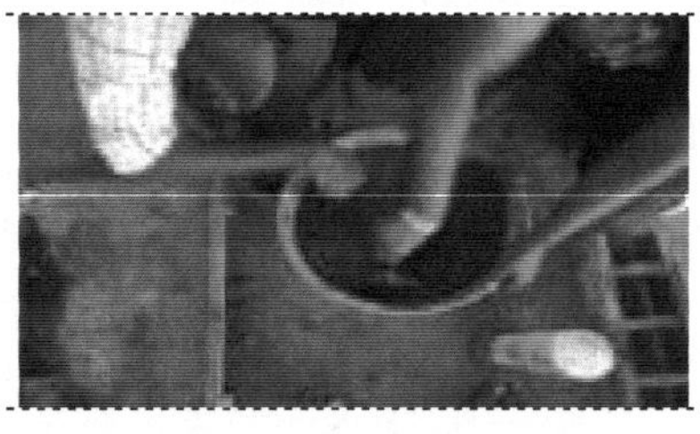

Figure 1a. Mixing of fresh SC mortar

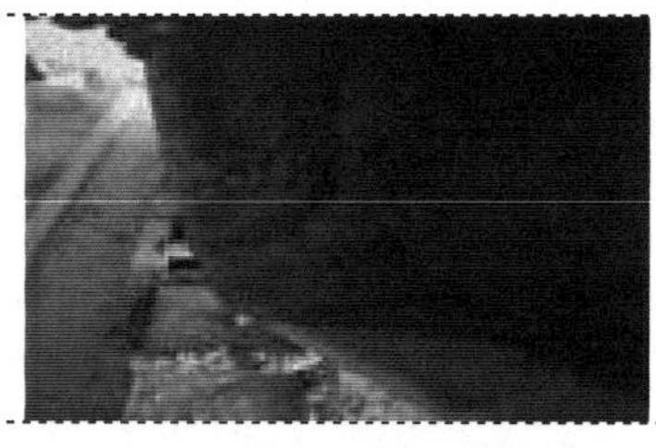

Figure 1b. Pouring of fresh SC mortar

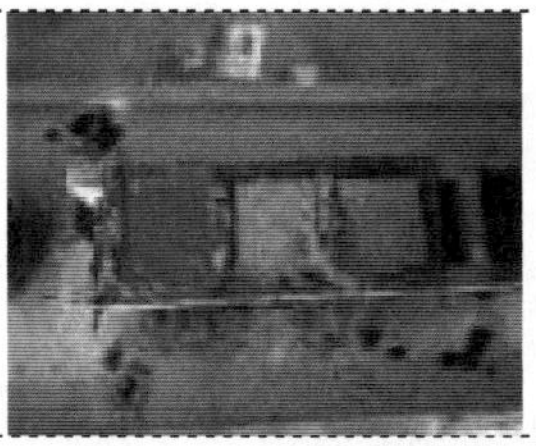

Figure 2a. Casting of SC mortar

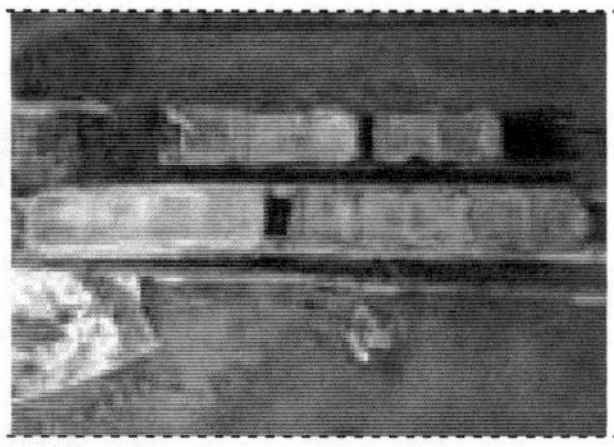

Figure 2b. Hardened state SC mortar

Figure 2c. State of cube specimens after de-moulding

3.5 Curing methods

The curing of cube specimens were undergone two series namely, series 1 consists of normal water curing for the age of 3, 7, 14 and 28 days respectively and the series 2 comprise of an accelerated curing method namely electric oven curing method (Figure 3) was adopted by maintaining initial temperature of 30°C and dried with differential temperature of 128°C for the duration of 4 hours.

Figure 3. Cube placements in Electric oven

3.6 Compressive strength

Once curing was completed by two methods, the compressive strength between conventional and self compacting mortar cubes was determined by Automatic Compression Testing Machine (ACTM) (Figure 4). The specimen was loaded up to the state of failure and fracture took place.

Figure 4. Test on Compressive strength of cube specimen

4. Results

The results obtained from the experimental programme illustrates mechanical properties of conventional and self compacting mortars by cement replacement with different mineral admixtures (SF,MK & GGBS) under normal and electric oven curing methods were discussed as follows:

Table 1. Basic test on materials

Fine aggregate	Fineness modulus:3.16
	Specific gravity:2.55
Cement(OPC)	Initial setting time:31 minutes
	Std. consistency:31%
	Specific gravity:3.13
Metakaolin (MK)	Specific gravity: 2.5
Silica fume(SF)	Specific gravity:2.1
GGBS	Specific gravity:2.81

Table 2. Self compacting mortar fresh property test

Self compacting mortar mix	Slump flow values (mm)
SF 5%	680
SF 10%	710
SF 15%	725
SF 20%	665
MK 10%	695
MK 20%	750
MK 30%	660
GGBS 25%	655
GGBS 50%	780
GGBS 75%	705

Table 3. Compressive strength on Cube mortars (in MPa)

Mix	Compressive strength				Compressive strength of Electric oven curing at 128°C for 4 hrs
	3Days	7 Days	14 Days	28 Days	
CM	19.68	28.71	42.54	47.60	52.54
SF 5%	18.73	27.32	43.54	45.08	54.25
SF 10%	19.18	29.23	42.67	49.11	53.98
SF 15%	19.86	29.94	41.92	50.00	56.72
SF 20%	20.07	28.42	44.27	45.76	52.94
MK 10%	22.32	30.65	44.23	49.65	51.64
MK 20%	21.45	29.98	45.1	48.87	55.17
MK 30%	22.75	31.23	42.79	50.75	53.84
GGBS 25%	23.69	32.87	49.72	53.25	54.63
GGBS 50%	21.56	30.23	41.17	49.27	52.19
GGBS 75%	17.21	27.23	39.99	44.25	50.07

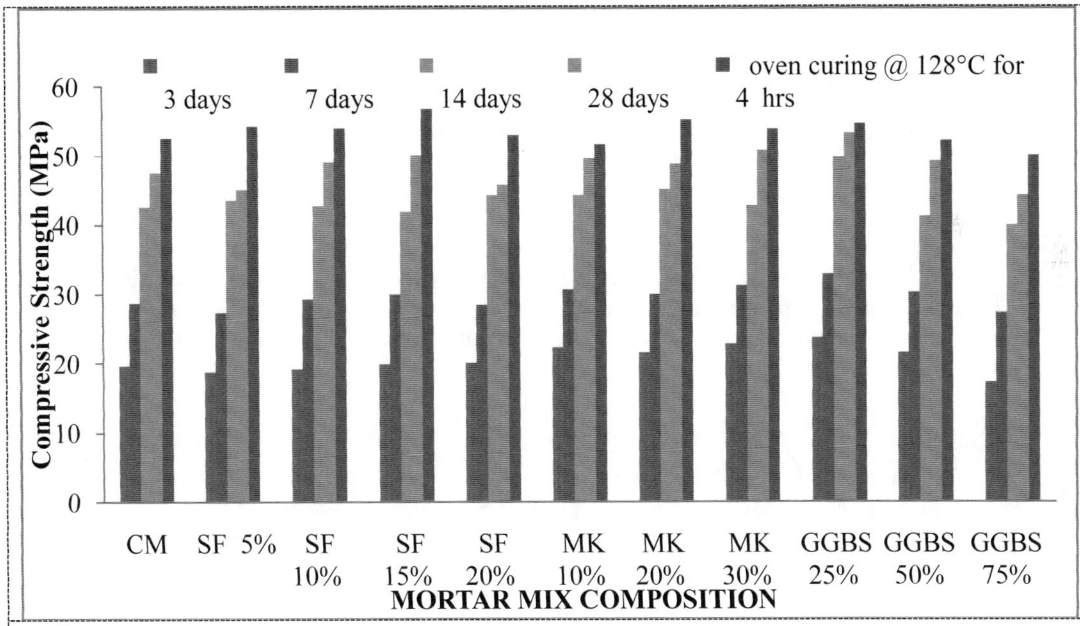

Figure 5 Compressive strength (MPa) Vs Mortar mix

4.1 Discussion

4.1.1 SC mortar by Slump flow test. This study was similar to slump test conducted in laboratory. Self compacting (SC) mortar which was replaced by three mineral admixtures was considered (Table 2) . Within SF mixes, the SF 15% made highest slump flow value compared to SF 5%, 10% & 20% respectively. It was due to optimum powder content causing influence on flow ability of SC mortar. The slump flow was made possible by fixing optimum water-powder ratio content and the optimum dosages of super plasticizer and stabilizer. In cement replacement by MK 10% to 30%, MK 20%

shown better results in comparison with other MK percentages. The mortar mix containing MK more than 20%, made the restriction on flow behaviour was noted by the lesser slump value of MK 30%. As higher the powder content the flow and the rheology of mortar gets modified. When cement mortar was replaced by GGBS from 25% to 75%, 50% replacement had shown better results, this was due to refinement of pore space causing flow ability of mortar. For the above slump flow values, the recommended range of values were taken from European Federation of Nationalised Association Representing Concrete (EFNARC Guidelines). From the overall slump flow results, GGBS 50% having spread diameter of 780 mm was higher than SF and MK mixes was due to the homogeneous behaviour between cement and mineral admixtures.

4.1.2 Mortar Compressive strength by normal curing. The cube mortar specimens prepared by conventional cement mortar and self compacting mortar containing mineral admixtures were undergone normal water curing for 28 days (Table 3). It was tested for 7 days compressive strength and obtained minimum 60% of strength achieved by self compacting mortars. The maximum compressive strength obtained by SC mortar- GGBS 25% was 10.61% higher than conventional cement mortar was due to the formation of denser C-S-H gel because of addition of slag as cement replacement at the age of 28 days. Similar trend was obtained for SF15% and MK 30% addition as partial cement replacement.

4.1.3 Mortar Compressive strength by oven curing. It was possible to obtain more than 100% of compressive strength by accelerated curing methods for the total duration of 7 hours. In present work, the cube specimens were undergone electric oven curing from the initial temperature of 30°C to the final temperature of 158°C for the duration of 4 hours after 24 hours of casting. This study resembles the precast applications where accelerated technique namely steam curing was adopted. Hence for model specimens like cubes, etc., electric oven curing was feasible and for prototype the electric furnaces were applicable. Hence the fineness of powder content governs the early strength gain. As the fineness of powder content more than 40 microns leads to later age strength gain and below 7 microns leads to early age strength gain. Here the powder content of SF 15%, GGBS 25% and MK 20% as individual cement replacement improved the refinement of pore space and also the fineness of SC mortar (Table 3). Hence within the 4 hours of electric oven curing , the early gain of compressive strength was obtained and it was higher than the normal curing by 14% (SF 15% & MK 20%) respectively.

5. Conclusion

From the above experimental programme and with the results obtained, the following conclusions were drawn.The flow behaviour of self compacting (SC) mortar was achieved by optimum dosages of chemical admixtures and water-powder ratio by keeping cement mortar proportion of 1:2. The slump flow of SC mortar was well within the limit of EFNARC guidelines. The highest compressive strength of SC mortar by replacing cement with SF 15%, MK 30% and GGBS 25% was obtained under normal water curing conditions. The obtained compressive strength values of electric oven curing was higher than normal water curing method.In SC mortar, the fineness of powder content improved early age strength in oven curing and it was obtained for SF 15%, MK 20% and GGBS 25%. The present study can also be applicable for pre-cast applications by changing the oven curing method into steam curing. Hence binary blend of SC mortar was considered for present study, further developments can also be prepared by SC of the ternary blend type.

Acknowledgements
The authors would like to thank the Vice Chancellor of SASTRA UNIVERSITY for having provided experimental facilities in the School of Civil Engineering to do this research work and also for the continuous support and encouragement given throughout this research work.

References

[1] James T, Malachi A, Gadzama E W and Anametemfiok V 2011 Effect of curing methods on the compressive strength of concrete. *Nigerian J. Tech.* **30**(3) 14-20.

[2] Fathollah S and Hashim A R 2011 Effects of curing regimes and cement fineness on the compressive strength of ordinary Portland cement mortars. *Constr. Build. Mater.* **25** 2036-45.

[3] Venkataramareddy B V and Gourav K 2011 Strength of lime–fly ash compacts using different curing techniques and gypsum additive. *Mater. Struct.* **44** 1793-1808.

[4] Hayri U N and Bulent B 2011 The effect of curing temperature and relative humidity on the strength development of Portland cement mortar. *Sci. Res. Essays* **6**(12) 2504-11.

[5] Heba A M 2011 Effect of fly ash and silica fume on compressive strength of self-compacting concrete under different curing conditions. *Ain Shams Eng. J.* **2** 79–86.

[6] Mohamed J M 2013 Effect of curing method and insoluble residue in cement on the compressive strength of portland cement mortar. *Al-Qadisiya J. Eng. Sci.* **6**(1) 74-80.

[7] Miguel J O, Bettencourt R and Fernando G B 2015 Curing effect in the shrinkage of a lower strength self-compacting concrete. *Constr. Build. Mater.* **93** 1206–15.

[8] Sri R C M, Swamynaga R G P, Rathish K P, Rajesh K G and Raveena C 2016 Effect of self curing chemicals in self compacting mortars. *Constr. Build. Mater.* **107** 356–64.

[9] Soriano L, Paya J, Monz J, Borrachero M V and Tashima M M 2016 High strength mortars using ordinary Portland cement–fly ash–fluid catalytic cracking catalyst residue ternary system (OPC/FA/FCC). *Constr. Build. Mater.* **106** 228–35.

[10] Pavlík V and Uzˇáková M 2016 Effect of curing conditions on the properties of lime, lime–metakaolin and lime–zeolite mortars. *Constr. Build. Mater.* **102** 14–25.

[11] Evi A, Payam S, Rodiah Z and Zahiruddin F A H 2016 Introducing an effective curing method for mortar containing high volume cementitious materials. *Constr. Build. Mater.* **107** 365–77.

[12] Aparicio S, Martínez R, Ranz J, Fuente J V and Hernández M G 2016 Microstructural and mechanical properties study of the curing process of self-compacting concrete. *Mater. Des.* **94** 479–86.

[13] Vivek S S and Dhinakaran G 2015 Effect of silica fume in flow properties and compressive strength of self compacting concrete. *Int. J. ChemTech Res.* **8**(1) 1-5.

[14] Anjali D, Vivek S S and Dhinakaran G 2015 Compressive strength of metakaolin based self-compacting concrete. *Int. J. ChemTech Res.* **8**(2) 622-5.

[15] Nan S, Kung C H and His W C 2001 A simple mix design method for self-compacting concrete. *Cem. Concr. Res.* **31** 1799–807.

[16] Dinakar P and Manu S N 2014 Concrete mix design for high strength self-compacting concrete using metakaolin. *Mater. Des.* **60** 661–8.

[17] *European Federation of Nationalised Association Representing Concrete (EFNARC)* guidelines February 2002 Specification and Guidelines for self-compacting concrete.

[18] *European Federation of Nationalised Association Representing Concrete (EFNARC)* guidelines May 2005 The European guidelines for self-compacting concrete (Specification, Production, Use).

[19] American Society for Testing and Materials (ASTM). Standard specification for coal fly ash and raw or calcined natural pozzolan for use in concrete. *ASTM C 618,* 2006, ASTM International, West Conshohocken, PA.

[20] American Society for Testing and Materials (ASTM). Standard specification for Portland Cement. *ASTM C150/C150M-12*, ASTM International, West Conshohocken, PA DOI: 10.1520/C0150_C0150M-12.
[21] American Society for Testing and Materials (ASTM). Standard specification for concrete aggregate. *ASTM C33*, ASTM International, West Conshohocken, PA.
[22] American Society for Testing and Materials (ASTM). Standard specifications for silica fume used in cementitious mixtures. *ASTM C 1240-99*, ASTM International, West Conshohocken, PA.

Strength characteristics of light weight concrete blocks using mineral admixtures

P Bhuvaneshwari*, U Priyadharshini, S Gurucharan and B Mithunram

School of Civil Engineering, SASTRA University, Thanjavur – 613401. India.

Email: bhuvaneswari@civil.sastra.edu

Abstract. This paper presents an experimental study to investigate the characteristics of light weight concrete blocks. Cement was partially replaced with mineral admixtures like Fly ash (FA), limestone powder waste (LPW), Rice husk ash (RHA), sugarcane fiber waste (SCW) and Chrysopogonzizanioides (CZ). The maximum replacement level achieved was 25% by weight of cement and sand. Total of 56 cubes (150 mm x 150 mm x150 mm) and 18 cylinders (100mmϕ and 50mm depth) were cast. The specimens being (FA, RHA, SCW, LPW, CZ, (FA-RHA), (FA-LPW), (FA-CZ), (LPW-CZ), (FA-SCW), (RHA-SCW)).Among the different combination, FA,FA-SCW,CZ,FA-CZ showed enhanced strength and durability, apart from achieving less density.

Keywords: Mineral admixtures, light weight concrete, wet density, dry density, compressive strength, water absorption

1. Introduction

In order to reduce environmental effects, pollution caused in manufacturing of cement and also to achieve light weight building blocks, different admixture were used as partial replacement of cement. Earlier researchers suggested replacing various mineral admixtures to design a structural lightweight high strength concrete (SLWHSC). They also found that silica fume produce good compressive strength and light weight concrete. It was found that structural lightweight concrete can reduce the dead weight and increase the strength of the structural components [1].The strength and durability of lightweight concrete was concentrated elaborately [2]. Strength characteristics were concentrated for different replacement level of light weight cinder aggregate based cement concrete [3].Development of high strength was concentrated on scoria aggregate based light weight concrete [4].

The strength of light weight was least sensitive to lack of initial curing [5]. Partial replacement of cement with various mineral admixtures for producing a lightweight concrete block as a building material was investigated. The maximum replacement level achieved was 25% of weight of cement [6]. Sludge ash aggregates could be used to develop lightweight concrete. It was found to have less thermal conductivity and good fire resistance [7].The ductility of pumice lightweight aggregate concrete replaced with steel and various fibers was studied [8]. It was found that apart from reduction in dead load, ductility was improved in all modes of loading.Flexural behaviour of lightweight concrete panel with GFRP barswas concentrated [9]. It was confirmed that light weight concrete performed well under flexural loading.

1.1*Research significance*

The objective is the partial replacement of cement with Fly ash (FA), lime stone powder waste (LPW), Rice husk ash (RHA), sugarcane fibre waste (SCW) and Chrysopogonzizanioides (CZ) to achieve light weight building blocks. The replacement would result in reduction of the dead load of the building. Parametric analysis of strength and durability characteristics of light weight concrete blocks, with that of normal concrete blocks was carried out.

2. Materials and methods

2.1 Materials used

The following materials were used to study the strength characteristics of the light weight concrete. Cement used for the research work was ordinary Portland cement of 53 grade confirming [10]. Granular river sand passing through 4.75mm sieve (grading zone III) with specific gravity 2.6 was used as fine aggregate confirming [11]. Class F fly ash with specific gravity 2.3 was obtained from thermal power plant. Rice husk ash burned in suspension at a temperature 700°C with specific gravity 2.3 and bulk density 98 kg/m^3 was used.The sugarcane waste fibers were dried and soaked in a 10% NaOH (Sodium Hydroxide) solution for about 2 to 6 hours with a temperature of about 60-70°C and are made to dry. Treated fibers were cut into pieces of 3cm length. Similar treatment was followed for Chrysopogonzizanioides. Lime stone powder waste with specific gravity of 2.61 and density of 2.45(g/cm^3) was used. The mix proportion and the specimen details were shown in table1 and table 2.

Table 1. Mix proportion of light weight concrete

Specimen	OPC	GRS	FA	LPW	RHA	SCW	CZ
Control	50	50	0	0	0	0	0
FA	25	50	25	0	0	0	0
LPW	25	50	0	25	0	0	0
CZ	46.5	50	0	0	0	0	3.5
RHA	25	50	0	0	25	0	0
SCW	46.5	50	0	0	0	3.5	0
FA-RHA	16.7	50	16.7	0	16.7	0	0
FA-SCW	29.8	25	16.7	0	0	3.5	0
RHA-SCW	29.8	50	0	0	16.7	3.5	0
FA-LPW	25	25	25	25	0	0	0
FA- CZ	29.8	50	16.7	0	0	0	3.5
LPW-CZ	29.8	50	0	16.7	0	0	3.5

Table 2. Specimen details

Specimen details (mm)	Density	Water absorption	Sorptivity	Compressive strength
Cube 150x150x150	12	12	0	12
Cylinder 100ϕ x 50 H	0	0	18	0

IOP Conf. Series: Earth and Environmental Science **80** (2017) 012003 doi:10.1088/1755-1315/80/1/012003

2.2 Methods

The consistency and setting time tests were carried out for Ordinary Portland cement according to [12].The consistency value was 33% and initial setting time was 33min. Fresh density as per [13] and dry density as per [14] were calculated for each combinations. Water absorption test was conducted to find the amount of water absorped. After 24 hours of curing, the cubes were heated in oven, cooled, weighted and kept for water absorption. Compressive strength test was conducted as per [15].From the figure 1(a) and figure 1(b) shows the cubes subjected to curing and testing. Sorptivity measures the rate of ingress of water through capillary action. Specimens for sorptivity were prepared in accordance with [16] (figure 2).

Figure 1. Specimens under sorptivity test

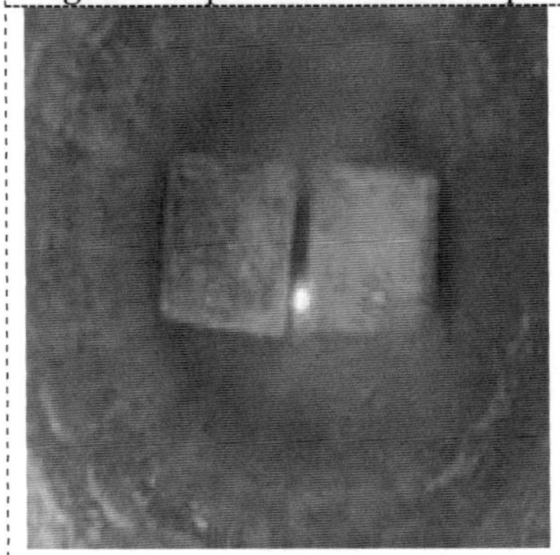

Figure 2. Specimens for water absorption

Figure 3. Specimen under compression

3. Results and discussion

3.1 Bulk density

Replacement of cement and sand with 25% of mineral admixtures showed the value of wet and dry weight of cast specimens as shown in the table 3. Compared to control specimens, reduction in fresh density achieved was 17.89, 20.75%, 24.61%, 24.10%, 19.03%, 22.9%, 24.10%, 30.05%, 14.28%, 23.07% and 30.62% respectively. Corresponding reduction in dry density was 8.76%, 15.53%, 9.94%, 27.63%, 10.76%, 22.11%, 25.62%, 31.90%, 4.32%, 11.64%and 18.43%respectively. Similar results were reported [6]. Bulk densities for the specimens are compared in figure3.

Table 3.Wet weight and dry weight of cube specimen

S.No.	Specimen	Area (mm^2)	Wet weight(kg)	Dry weight (kg)
1	Control	150 x 150	8.960	7.960
2	FA	150 x 150	7.60	7.320
3	LPW	150 x 150	7.420	6.890
4	CZ	150 x 150	7.190	7.240
5	RHA	150 x 150	6.800	5.790
6	SCW	150 x 150	8.240	7.006
7	FA-RHA	150 x 150	6.900	6.200
8	FA-SCW	150 x 150	6.840	5.920
9	RHA - SCW	150 x 150	6.240	5.420
10	FA -LPW	150 x 150	7.840	7.630
11	FA - CZ	150 x 150	7.280	7.130
12	LPW - CZ	150 x 150	6.86	6.721

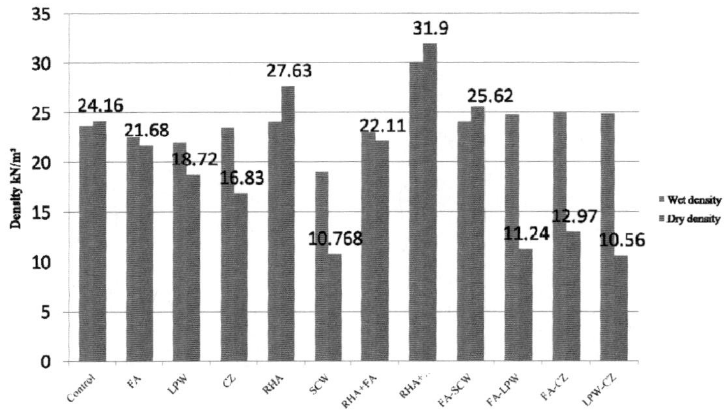

Figure 4. Wet and dry density of specimens

3.2 Water absorption

Percentage of water absorption are shown in figure 4.As compared to control specimen, RHA, (FA-RHA), (FA-SCW), (RHA-SCW) achieved the highest percentage of water absorption. As compared to control specimens, FA, LPW, CZ, SCW, (FA-LPW), (FA-CZ), (LPW-CZ) showed decrease in percentage of water absorption. It has been compared from the earlier studies, percentage increase in water absorption was gradual by using various admixtures [6].

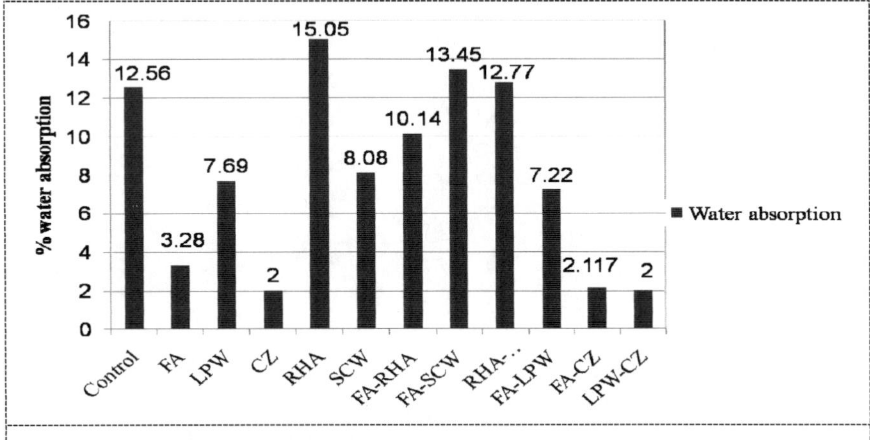

Figure 5. Percentage of water absorption for specimens

3.3 Sorptivity

The sorptivity in specimens were carried out in accordance with [16]. Behaviour of FA, CZ, (FA-CZ), (FA-SCW) was similar to control specimen. As compared to control specimen, light weight concrete specimens RHA, SCW, (FA-RHA), (RHA-SCW) showed higher value of sorptivity. Similar behaviour was noticed in light weight concrete specimens [17].

Table 4. Sorptivity for the specimens

√t	C	FA	FA-CZ	CZ	RHA	SCW	RHA-FA	RHA-SCW	FA-SCW
0	0	0	0	0	0	0	0	0	0
3.166	0.032	0.032	0.030	0.098	0.415	0.42	0.6358	0.382	0.254
5.472	0.017	0.017	0.018	0.01	0.411	0.45	0.7643	0.382	0.382
7.745	0.012	0.012	0.013	0.012	0.309	0.25	1.0191	0.509	0.382
13.41	0.007	0.007	0.007	0.007	0.508	0.65	1.1464	0.636	0.382
15.49	0.006	0.006	0.006	0.006	1.006	0.94	1.273	0.764	0.509
17.32	0.005	0.005	0.103	0.005	1.106	1.28	1.5286	0.891	0.509
18.97	0.005	0.005	0.103	0.005	1.205	1.20	1.5286	1.019	0.509

IOP Conf. Series: Earth and Environmental Science **80** (2017) 012003 doi:10.1088/1755-1315/80/1/012003

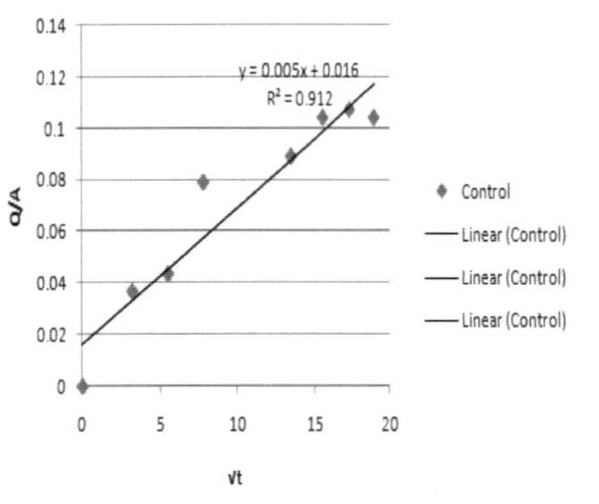

Figure 6 (a) Control

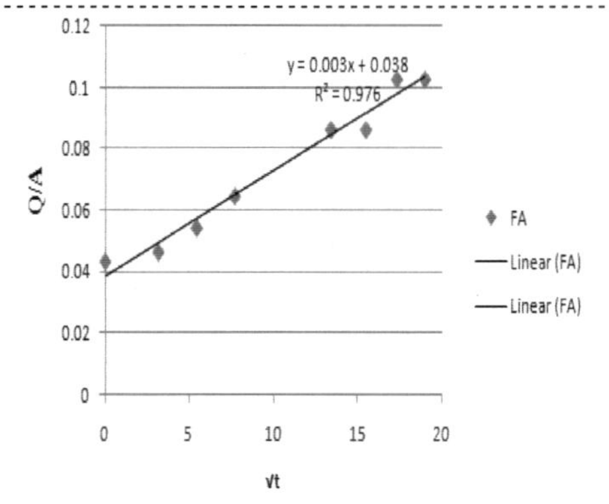

Figure 6 (b) Fly ash

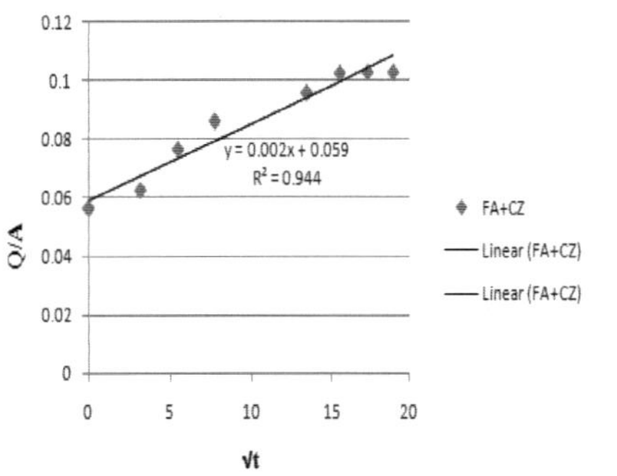

Figure 6 (c)FA-CZ

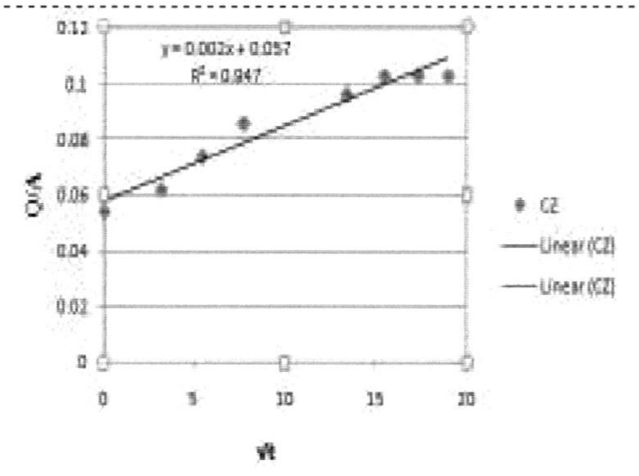

Figure 6 (d)CZ

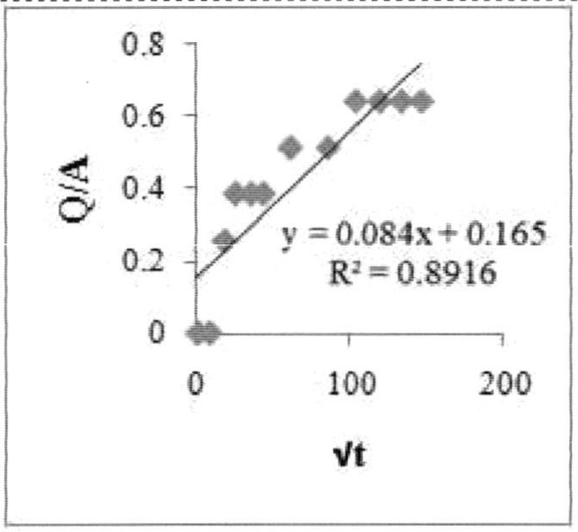

Figure 6 (e)RHA

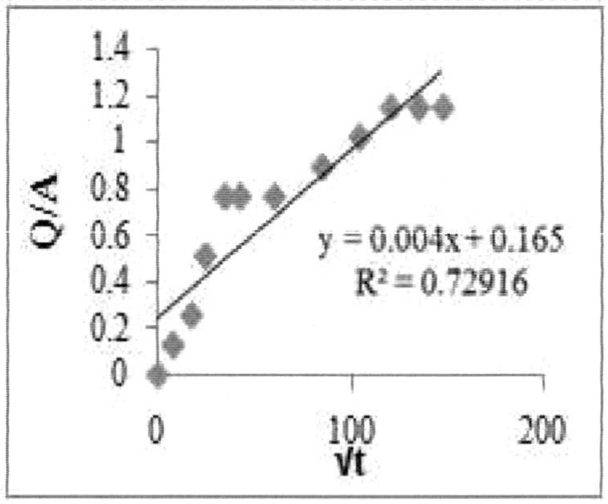

Figure 6 (f)SCW

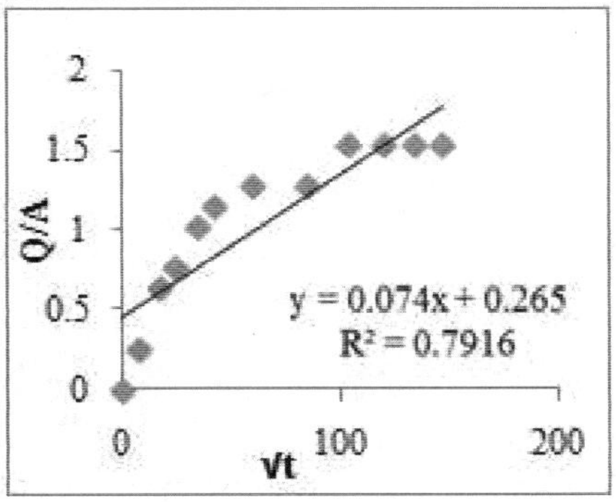

Figure 6 (g)FA-RHA

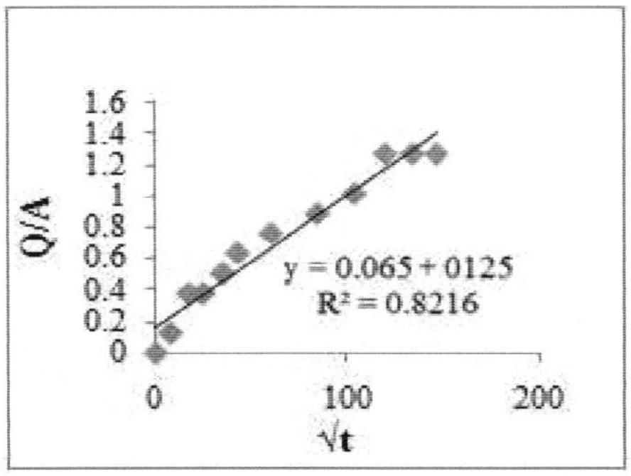

Figure 6 (h)HA-SCW

Figure 6. Sorptivity index for specimens

3.4 Compressive strength

Compressive test was achieved in CTM of loading capacity 3000 kN. After 28days of moisture curing the specimens were surface dried and tested to failure. Comparison of compressive strength are listed in the table 6, percentage of difference in strength are shown in the figure7.

Table 5. Comparison of compressive strength

S.No	Specimen	Area (mm^2)	Compressive load (kN)	Compressive stress (N/mm^2)
1	Control	150 x 150	562	24.99
2	FA	150 x 150	597	26.53
3	LPW	150 x 150	405	18.01
4	CZ	150 x 150	545	24.20
5	RHA	150 x 150	578	25.68
6	SCW	150 x 150	552	24.53
7	RHA-FA	150 x 150	375	16.66
8	RHA-SCW	150 x 150	397	17.64
9	FA-SCW	150 x 150	425	18.88
10	FA-LPW	150 x 150	582	25.86
11	FA-CZ	150 x 150	573	25.46
12	LPW-CZ	150 x 150	520	23.11

The influence of various mineral admixtures on compressive strength of concrete specimens was concentrated [1].Light density for concrete have achieved without compromising the compressive strength.

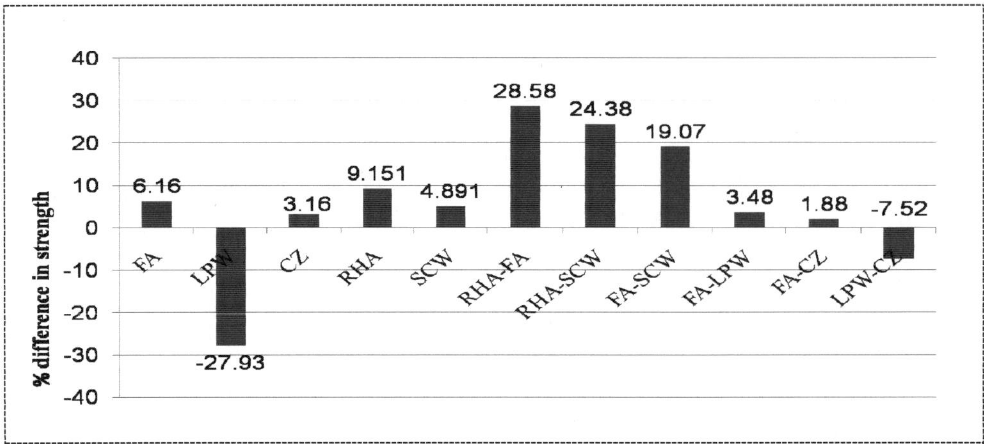

Figure 7. Percentage difference in compressive strength

From the comparison it was inferred that LPW and LPW-CZ showed decrease in compressive strength. Compared to control specimen, the other combinations FA, CZ, RHA, SCW, RHA-FA, RHA-SCW, FA-SCW, FA-LPW and FA- CZ showed enhanced results for compressive strength.

4. Conclusion

Mineral admixtures were used as partial replacement for cement to achieve light weight concrete blocks and conclusion were arrived

- FA showed less reduction in bulk density without considerable reduction in compressive strength.
- Replacement with LPW showed better reduction in density but failed to achieve the strength.
- FA and (FA-SCW) showed less reduction in the density without substantial reduction in the compressive strength.
- The results observed that (FA-LPW) specimens had higher compressive strength compared to the control specimen.
- Replacement with CZ resulted in better reduction in water absorption and density of the sample. Sample achieved good strength.
- By the addition of RHA, water absorption and bulk density of the samples considerably increased.
- Combination of (FA-CZ) sample gave reduction in density and water absorption. The samples achieve good strength.
- Combination of (LPW-CZ) sample gave reduction in density and water absorption. The samples achieved good strength.
- Sorptivity of CZ gradually decreased the sorptivity value as compared to control specimens.
- Sorptivity of FA, FA-CZ sample achieved lowest value of sorptivity as compared to control specimens.
- Study for strength and durability showed that cement replaced with FA and (FA-SCW) were effective compared to other combinations. As compared to all other specimens (FA-CZ) and (LPW-CZ) achieved light weight concrete blocks. These blocks showed good compressive strength and also performed better under durability conditions.

IOP Conf. Series: Earth and Environmental Science **80** (2017) 012003 doi:10.1088/1755-1315/80/1/012003

References

[1] Alaeetinkilic and Ceniz Duran Atis 2003 High-strength lightweight concrete made with scoria aggregate containing mineral admixtures. *Cem. Concr. Res.* **33** 1595–9.

[2] Al-Khaiat H and Haque MN 1998 Effect of initial curing on early strength and physical properties of a lightweight concrete. *Cem. Concr. Res.* **28**(6) 859–66.

[3] Bhaskar Desai V and Sathyam A 2014 Some studies on strength properties of light weight cinder aggregate concrete. *Int. J. Sci. Res. Pub.* **4**(2) 1-13.

[4] ErgulYasar and Hasan Gulsen 2003 Strength properties of lightweight concrete made with basaltic pumice and fly ash. *Mater. Let.* **57** 2267-70.

[5] Husain Al-Khaiat and Naseer Haque 2002 Strength and durability of lightweight concrete *J. Mater. Civ. Eng. ASCE.* **11**(3) 231-5.

[6] JavadTorkaman, AlirezaAshori and Ali sadrmomtazi 2014 Using wood fiber waste, rice husk ash, and limestone powder waste as cement replacement materials for lightweight concrete blocks. *Constr. Build. Mater.* **50** 432-6.

[7] Joo- Hwa Tay and Woon-Kwong Yip 1989 Sludge ash as light weight concrete material. *J. Env. Eng.* **115**(1) 56-64.

[8] Nicolas Ali Libre and Mohammad Shekarchi 2010 Mechanical properties of hybrid fiber reinforced lightweight aggregate concrete made with natural pumice. *Constr. Build. Mater.* **25** 2458-66.

[9] Ruifen Liu T and Bradon Besser T 2012 One Way Shear Behaviour of Lightweight Concrete Panels Reinforced with GFRP Bars. *J. Compos. Constr.* **16**(1) 2-9.

[10] IS 12269 2013 Ordinary Portland Cement-53 Grade-Specification: BIS, New Delhi.

[11] IS 2386 -1 1963 Methods of test for Aggregate for concrete: BIS, New Delhi.

[12] IS 4031-1 1996 Methods of physical test of cement: Part 1-BIS, New Delhi.

[13] ASTM C138 2009 Standard test method for density, yield and air content of concrete.

[14] ASTM C642 1997 Standard test method for density, water absorption and voids in hardened concrete.

[15] IS 516 1959 Methods of Tests for strength of concrete: BIS, New Delhi.

[16] ASTM C1585 2004 Standard test method for measurement of rate of Absorption (sorptivity) of water by Hydraulic-Cement Concretes.

[17] Santhosh kumar T, Balaji K V and Rajasekar K 2016 Assessmeent of sorptivity and water absorption of concrete with partial replacement of cement by silica fume and sugarcane bagasse ash. *Int. J. Appl. Eng. Res.* **85** 747-52.

Influence of Glass Fiber on Fresh and Hardened Properties of Self Compacting Concrete

R Bharathi Murugan[1], M K Haridharan[2]*, C Natarajan[3] and R Jayasankar[3]

[1] Department of Civil Engineering, Kalaslingam University, Krishnakovil, India
[2]School of Civil Engineering, SASTRA University, Thanjavore-613 401. India.
[3]Department of Civil Engineering, National Institute of Technology, Trichy. India.

*Email: haridharan@civil.sastra.edu.

Abstract. The practical need of self-compacting concrete (SCC) is increasing due to increase in the infrastructure competence all over the world. The effective way of increasing the strength of concrete and enhance the behaviour under extreme loading (fire) is the keen interest. Glass fibers were added for five different of volume fractions (0%, 0.1%, 0.3%, 0.5% and 0.6%) to determine the optimum percentage of glass fiber without compensating the fresh properties and enhanced hardened properties of SCC concrete. The fresh state of concrete is characterized by slump flow, T-50cm slump flow, and V-funnel and L- box tests. The results obtained in fresh state are compared with the acceptance criteria of EFNARC specification. Concrete specimens were casted to evaluate the hardened properties such as compressive strength, split tensile strength, flexural strength and modulus of elasticity. Incorporation the glass fiber into SCC reduces the workability but within the standard specification. The hardened properties of SCC glass fiber reinforced concrete were enhanced, due to bridging the pre-existing micro cracks in concrete by glass fiber addition.

1. Introduction

Professor Okamura of Kochi university of Technology, Japan in 1986 insisted the need of SCC. Before the first prototype innovation, multi-particle physics was chiefly investigated to clarify the dynamic segregation of the liquid matrix - particle suspension system [1 & 2]. The fresh SCC considered as multi-sized particle and understood by T. C. Powers [3]. The development SCC improved the superiority of concrete and opened up its application in all the areas of construction industry [4]. The cost reduction and environment friendly SCC can be obtained by using the industrial wastes which has binding properties [5–10]. In another study, the cement was replaced with low calcium F fly ash from 40 to 60% in SCC. The characteristic compressive strength varied between 26 and 48 MPa. The possibility of developing economical SCC mixes was by using Industrial waste which is rich in binding properties [11 & 12]. The Interfacial transition zone (ITZ) plays a very important role to ensure quality and strength aspect of the concrete. Generally the ITZ consist of water film, calcium hydroxide layer and a porous cement matrix which affects the mechanical and transport properties hence the reduction in the ITZ results in good strength between aggregate to binder. [13-15].

The aim of current research is to examine the performance of Self compacting concrete with the addition of glass fiber (SCFC). The study focused on influence of various percentage of glass fiber on the fresh and hardened properties of SCC.

2. Experimental Program

2.1 Materials and their Proportions

The cement used in this study were ordinary Portland cement of 53 grade conforming to BIS 12269 [16] and Class F fly ash was taken from Vijayawada Thermal Power Station, Andhra Pradesh. The fly-ash collected was tested for physical and chemical and the test results were in compliance with ASTM C 618 – 78 [17], BIS 3812 [18] and satisfies all the properties of Class F fly-ash. River sand was used as fine aggregates and the properties was confirming to Zone II specifications [19]and 12.5 mm downgraded metal was used as coarse aggregate. Coarse aggregate of size 12.5 mm downgraded was used in a mix proportion. The coarse aggregate with specific gravity of 2.6, fineness modulus of 6.23 and moisture content of 0.4% were used. Table 1 provides the various mix proportions considered in the study. The fly ash was used in this study for partial replacement of cement by 30% of its weight. A sulphonated naphthalene polymer based super-plasticizer was used. The glass fiber used in the study has a Specific Gravity 2.67, aspect ratio of 859.0, Elastic modulus 71GPA and Anti – Crack HD fibers.

Table 1 Mix proportions of SCC (per m^3)

Mix	Glass Fiber in %	Water/binder ratio	Cement (kg)	FA (kg)	Sand (kg)	Aggregate (kg) 12.5 mm	VMA %	Super – Plasticizer (%)
SCC-0.0	0.0%							
SCC -0.1	0.1%							
SCC-0.3	0.3%	0.40	350	142	707.7	1138.2	0.5	2.15
SCC-0.5	0.5%							
SCC-0.6	0.6%							

3. Testing Methods and Methodology

The fresh properties of SCC with and without glass fiber was carried out as per EFNARC specification [20,21]. To assess filling ability, reheology factors of the SCC slump flow test are performed. The slump flow value has to be in the range of 650 to 800 mm. T 50cm slump flow test is used along with slump flow test to assess the flow ability. V-Funnel test is used to determine the filling ability, flow ability and segregation resistance. The flow of the concrete in the congested junctions like beam column joint was assessed by the L-box test

According to BIS 516 [22] the compressive strength test, split tensile strength test, and Young's modulus tests were performed at concrete specimen after 7 and 28 days of curing. Similarly the flexural strength test was performed based on BIS 1199-1959 [23]. The cube specimens were tested upto failure in compressive strength using uniaxial compression testing machine of capacity 2,000 kN at constant loading rate of 140 kg/cm^2/min. A flexural strength test was carried out using a prism specimen of size 150 x 150x 700 mm and were tested under universal testing machine of capacity 1,000 kN at a constant loading rate 180 kg/min. Split tensile strength and Young's modulus were tested using cylindrical specimens of size 150 mm x 300 mm under uniaxial compression testing machine of capacity 3,000 kN.

4. Results and Discussion

4.1 Fresh Properties of SCC

Table 2 shows the various flow test done for varying percentage of glass fibre and the results are well within the permissible limits specified in EFNARC [20 & 21]. The table clearly depict that as the percentage of fiber content is increased the various fresh properties were slowly reaching the permissible limit.

The slump flow was varying from 692 mm for SCC to 661 mm for SCC – 0.6, it states values were well within the permissible limit of 600mm to 800mm. The passing ability of SCC through reinforcing bars was indicated by relation between the J-ring test to slump flow [20]. The passing ability of SCC was tested by using L-box, U-box, and J-ring as recommended by international standards [20, 24]. The devices were designed to assess the passing ability of SCC, in consideration with common reinforcing patterns used in the construction industry and for aggregate size of 20 mm for SCC [25]. The test result of J-Ring was found to decrease in the flow of concrete as the percentage of fiber content was increased but the results are well within the permissible value of the standards. Similar observation was observed with L-Box test, the values from 0.81 to 0.99 for SCC-0.0 and SCC-0.6 respectively. The filling ability of the SCC was measured using V-funnel test and the detail apparatus set up was made as per details of EFNARC[20]. The flow time required for SCC is 9 seconds, in the case of SCC 0.6 the flow time required was 11 seconds. It was clear from results that the flow time was increasing as the percentage of fiber content was increased. The research study consists of respondents in the construction companies specifically in South India.

Table 2. Fresh properties of SCC with and without glass fiber.

Specimen ID	Slump Flow (mm)	J-Ring (mm)	Slump for 500 mm Spread (s)	V-funnel time (s)	L- Box ratio
SCC-0.0	692	632	2.92	9	0.81
SCC-0.1	685	624	3.28	9	0.84
SCC-0.3	679	609	3.73	10	0.89
SCC-0.5	675	596	4.57	10	0.96
SCC-0.6	662	581	5.89	11	0.99

4.2 Hardened Properties of SCC

The hardened properties of concrete were studied for various percentage (0.1%, 0.3%, 0.5% and 0.6%) of glass fiber of the cement content. In this study, the values presented were average of six samples for each series of experiments.

4.2.1 Compressive Strength. Fig 1 shows the compressive strength test results of SCC with and without fiber. It is interesting to note that the mix containing fiber content resulted in the increasing rate of the strength. The percentage increase in strength was 0.80 %, 3.82%, 5.41% and 6.53% for SCC-01, SCC-03, SCC-05 and SCC-06 respectively.

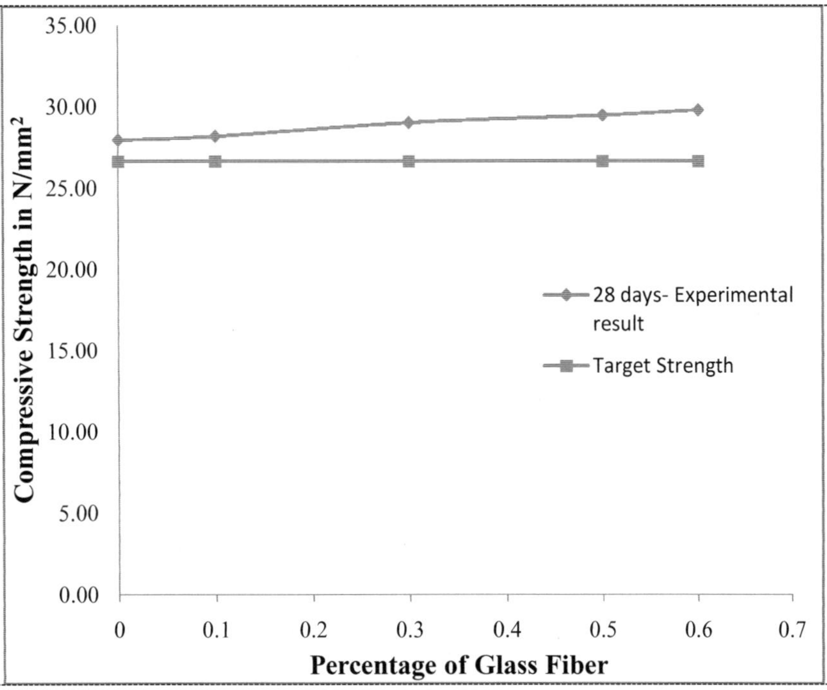

Figure 1. Variation in Compressive Strength of SCC with and without Glass Fiber

The increase in the strength may be due to bridging of pre-existing crack in ITZ and also presence of fiber improved interface between aggregate and hardened paste [26]. Similar observations were reported by other researchers [26, 27 & 28]

4.2.2 Split Tensile Strength. The average split tensile found to increase with increase in the content of glass fiber. Figure 2 shows the split tensile strength of concrete for various percentage of glass fiber and also compared with the model proposed by Felekoglu et.al [29]. The increase in split tensile strength when compared with SCC were 8.33%, 9.37%, 11.46% and 11.11% for SCC-01, SCC-03, SCC-05 and SCC-06 respectively., The increase in split tensile strength was also reported by other researchers [11, 27, 28].

4.2.3 Flexural Strength. The modulus of rupture value increases with increase glass fiber percentage. The flexural strength were 5.33 N/mm^2, 5.51 N/mm^2, 5.69 N/mm^2, 5.87 N/mm^2 and 6.04 N/mm^{2}for SCC, SCC-01, SCC-03, SCC-05 and SCC-06 respectively., the addition increase in strength is may be due to proper orientation of fiber, presence of fiber at the cracking zone which arrest theformation of micro cracks and interestingly failure of the specimens shows ductile failure when compared with SCC. The Fig 3 shows the comparison of experimentally value with the various empirical formulas developed by various codes for concrete.

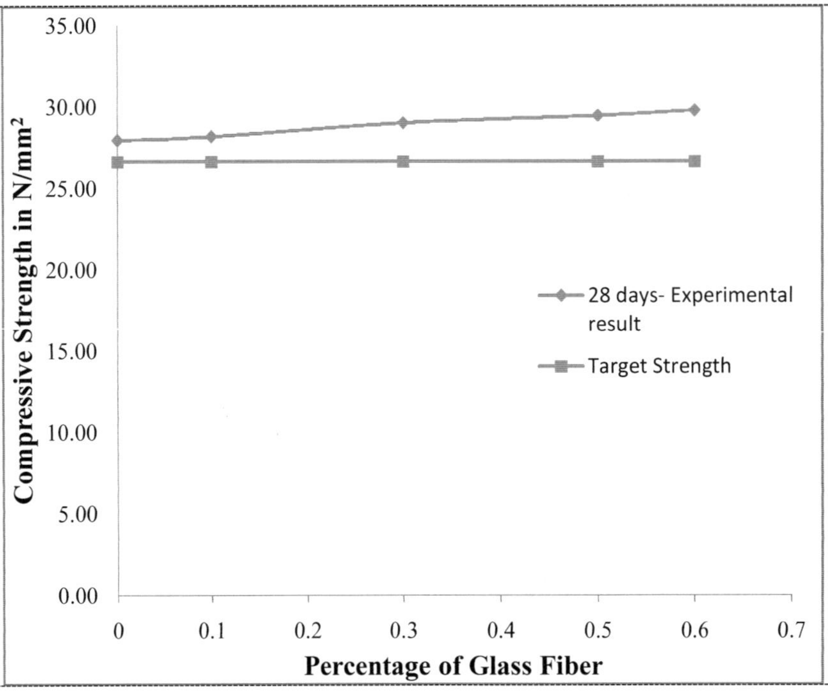

Figure 2. Variation in Split Tensile Strength of SCC with and without Glass Fiber

The SCC itself found to have modulus of rupture value greater then empirical relation and simultaneously the addition of glass fiber increases the modulus of rupture. The empirical relation provided in various codes under estimates by large margin. Based on the experimental results a new empirical relation was proposed to calculate the modulus of rupture value and the expression proposed gives a conservative prediction.

$$f_r = 0.42 * (fck)^{(3/4)} \quad (1)$$

4.2.4 Modulus of elasticity

Figure 4 shows the variation in the instantaneous elastic modulus for various percentage of glass fiber [32]. Some of the previous researchers concluded that the SFR SCC has elastic modulus greaters than the plain concrete [33-35]. Similarly other researchers reported that the variation in elastic modulus between SFR SCC and Plain SCC is insignificant[27, 28, 34 – 36].

Generally, the elastic modulus of concrete depends upon the elastic modules of aggregate and hardened cement paste [33].and also the ratio between binder content to coarse aggregate. The Fig 4 shows the comparison of experimentally value with the various empirical formulas developed by various codes for concrete. The empirical relation provided by Indian code and ACI code found to underestimates the actual modulus of elasticity, where the margin is large in the case of ACI code than India code. In considering Eurocode it over estimates the value marginally.

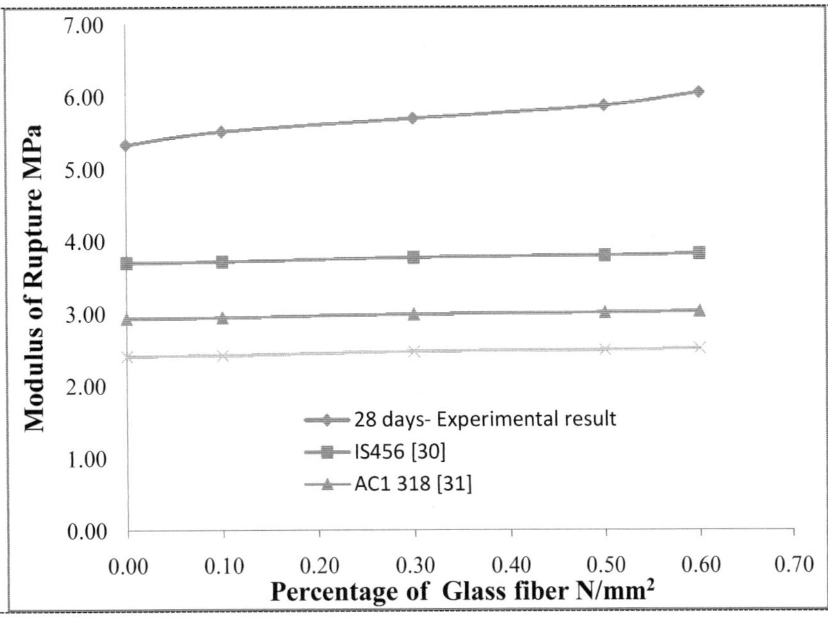

Figure 3. Variation in Modulus of rupture Value for Different percentage of Glass Fiber and Comparison with Empirical Relationship of IS456, ACI318and EC2

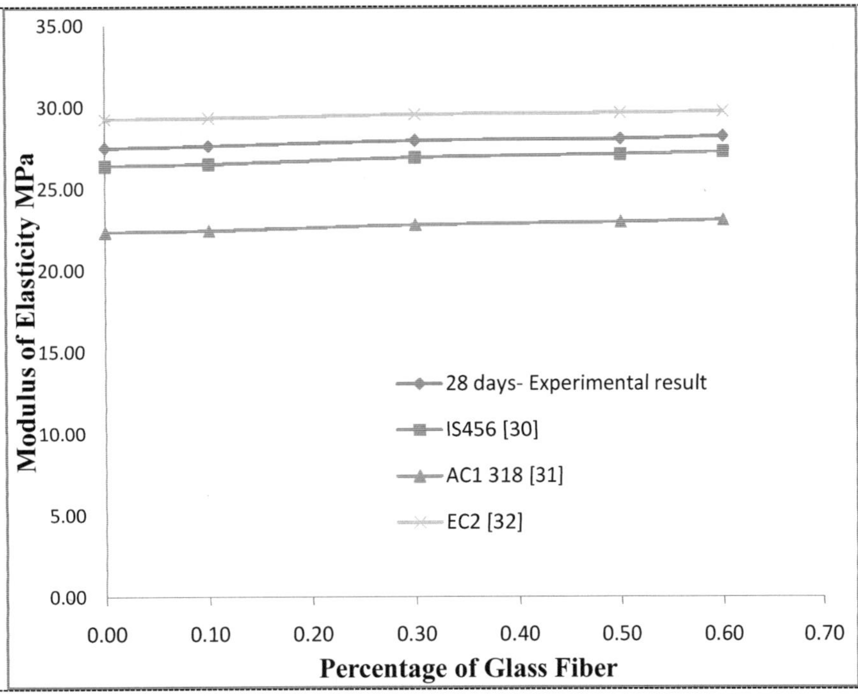

Figure 4. Variation in Modulus of Elasticity Value for Different percentage of Glass Fiber and Comparison with Empirical Relationship of IS456, ACI318and EC2

5. Conclusion

The following conclusion may be deduced from the experimental investigation

- Addition of glass fiber to SCC reduces the workability but it is within the permissible limit of EFNARC.
- The strength properties were increased with increase addition of glass fiberupto 0.6 %.
- The fiber addition enhances the strength by suppressing the development of micro crack to macro level crack; increase the mortar strength, proper orientation of the fiber in case of flexural loading and the effect of bridging between the cement mortar and coarse aggregate.
- A new empirical relation was proposed to calculate the modulus of rupture value while the existing empirical values in various code found to under estimating
- It was observed that upto 0.6% of glass fiber can be added to the SCC.

References

[1] Nanayakkara A Gunatilaka D, Ozawa K and Maekawa K1988 Flow and segregation of fresh concrete in tapered pipes - Two-phase computational model *ASME* FED **75**12.
[2] Ozawa K, Nanayakkara A and Maekawa K Flow and segregation of fresh concrete around bifurcation in pipe lines, *ASME*, FED **75**, 12.
[3] Powers T C 1968 *Properties of Fresh Concrete* (New York - MaGrow-Hill)
[4] Domone P L 2005 Self compacting concrete: an analysis of case studies *Cem. Concr. Compos.* **28** 197–208.
[5] Hearn N, Skalny J and Bajza A 1997 Microstructures and bond in cementitious solids. *Build. Res. J.* 1997 **45**(1) 103–18.
[6] Khayat K H 1999 Workability, testing, and performance of self-consolidating concrete *ACI Mater. J.* **96**(3) 346–53.
[7] Lachemi M, Hossain KMA, Lambros V and Bouzoubaa N 2003 Development of cost effective self-compacting concrete incorporating fly ash, slag cement or viscosity modifying admixtures. *ACI Mater. J.* **100**(5) 419–25.
[8] Hassan AAA, Hossain KM A and Lachemi M 2008 Behavior of full-scale self-consolidating concrete beams in shear. *Cem. Concr. Compos.* **30**(7) 588–96.
[9] Hassan AAA, Hossain KM A and Lachemi M 2010 Strength, cracking and deflection performance of large- scale self-consolidating concrete beams subjected to shear failure. *Eng. Struct.* **32**(5) 1262–71.
[10] Uysal M and Yilmaz K2011 Effect of mineral admixtures on properties of self compacting concrete. *Cem. Concr. Compos.* **33**(7)771–6.
[11] Bouzoubaa N and Lachemi M2001 Self-compacting concrete incorporating high volumes of class F fly ash: Preliminary results. *Cem. Concr. Res.* **31**(3) 413-420.
[12] Vivek SS and Dinakaran G 2017 Fresh and Mechanical Properties of Metakaolin based High Strength SCC. *Jordan J. Civ. Eng.* **11**(2)325–33.
[13] Mindess S, Young J F and Darwin D 2003*Concrete. 2nd ed.* (New Jersey - Prentice Hall)
[14] Ollivier JP, Maso J C and Bourdette B 1995 Interfacial transition zone in concrete. *Adv. Cem. Bas. Mater.* **2** 30–8.
[15] Hearn N, Skalny J and Bajza A 1997 Microstructures and bond in cementitious solids. *Build. Res. J.* **45**(1) 103–18.
[16] Bureau of Indian Standards (BIS) 1987 Code of practice for 43 Grade Ordinary Portland Cement New Delhi BIS Standard No. IS-12269.
[17] American society for testing and materials (ASTM) 2014 Standard Specification for fly ash and raw or calcined natural pozzolana for use as a mineral admixture in Portland cement concreteAmerica ASTM Standard No. ASTM C618 – 78
[18] Bureau of Indian Standards (BIS) 2003 Code of practice for pulverized fuel ash – specification, Part – I, for use as pozzolana in cement, cement mortar and cement concrete New Delhi BIS Standard No. IS3812– Part I.

[19] Bureau of Indian Standards (BIS) 1970 Code of Practice – Specification for Coarse and Fine Aggregate from Natural Sources for Concrete New Delhi BIS Standard No IS 383: 2002
[20] EFNARC. 2002 Specification and guidelines for self-compacting concrete European federation of producers and applicators of specialist products for structures.
[21] EFNARC. 2005 The European guidelines for self-compacting concrete speficication, production and use European federation of procedures and applicators of specialist products for structures.
[22] Bureau of Indian Standards (BIS) 1959 Code of Practice – Methods of test for strength of concrete New Delhi BISStandard No IS 516.
[23] Bureau of Indian Standards (BIS) 1959 Code of Practice – Methods of sampling and analysis of concreteNew DelhiBISStandard No IS 1199.
[24] Ferrara L, Park Y D and Shah S P 2007 A method for mix-design of fiber-reinforced self compacting concrete. *Cem. Concr. Res.* **37** 957–71.
[25] Madandoust R, Ranjbar M M, Ghavidel R and Shahabi S F 2015 Assessment of factors influencing mechanical properties of steel fiber reinforced self-compacting concrete *Mater. Des.* **83** 284–94.
[26] Umar A and Al-Tamimi A 2011 A critical study of the effect of viscosity modifying admixture and glass fiber on the properties of self-compacting concrete (SCC). *J. Struct. Eng.* **38**(2) 153–62.
[27] Aslani F and Nejadi S 2013 Self-compacting concrete incorporating steel and polypropylene fibers: compressive strength, tensile strength, moduli of elasticity and rupture, compressive stress- strain curve and energy dissipated under compression *Composite Part B: Eng.* **53** 121–33.
[28] Ahmad S, Umar A and Masood A 2017 Properties of Normal concrete, self-compacting concrete and glass fiber-reinforced self-compacting concrete: An experimental study. *Procedia Eng.* **173** 807–13.
[29] Felekoglu B, Turkel S and Baradan B 2007 Effect of water / cement ratio on the fresh and hardened properties of self-compacting concrete *Build. Environ.* **42**(4) 1795–802.
[30] Bureau of Indian Standards (BIS) 2000 Code of Practice for plani and reinforced concrete for general building construction New Delhi BIS Standard No IS 456.
[31] American concrete Institute 2000 Building code requirements for structural concrete and commentaryFarmington HillsMichiganACI Committee: 318-2000.
[32] BS EN 1992, Eurocode 2: Design of concrete structures. European Committee for Standardization.
[33] Ashtiani MS, Scott A N and Dhakal RP 2013 Mechanical and fresh properties of high-strength self-compacting concrete containing class C fly ash. *Constr. Build. Mater.* **47** 1217-24.
[34] Dinakar P, Babu KG and Santhanam M 2008 Mechanical properties of high-volume fly ash self-compacting concrete mixtures. *Struct. Concr.* **9**(2) 109-16.
[35] AL-Ameeri A 2013 The effect of steel fiber on some mechanical properties of self compacting concrete. *Am. J. Civ. Eng.* **1**(3) 102–10.
[36] Atis C D and Karahan O 2009 Properties of steel fiber reinforced fly ash concrete. *Constr. Build. Mater.* **23** 392–99
[37] Gencel O, Brostow W, Datashvili T and Thedford M 2011 Workability and mechanical performance of steel fiber-reinforced self-compacting concrete with fly ash. *Compos. Interfaces* **18** 169–84.

Characteristics of Cement Concrete with Nano Alumina Particles

P Jaishankar* and C Karthikeyan

School of Civil Engineering, SASTRA University, Thanjavur – 613 401 India.

*Email: jai@civil.sastra.edu

Abstract. Concrete can be nano-engineered by incorporating nano sized building blocks or objects (e.g., nano particles and nano tubes) to control material behaviour and add novel properties. In this work an attempt has been made to study the effect of nano alumina on the properties of concrete composite. In order to investigate the effect of nano-alumina on the mechanical strength of cement composite, the specimens with different volume percentages (0%, 0.5%, 0.75% and 1%) nano-alumina, in each proportion three sample were cast totally making 58 specimens and after 28 days of curing they were tested. Based on experiment, the compressive strength of concrete cubes was all increased by incorporating nano alumina into matrix, when the fraction of nano alumina was 1% of the cement by weight the compressive strength of composites increased by 33.14% at 28 days. The test results showed that addition of nano alumina enhanced the compressive strength and reduced the initial setting time of concrete composite. Micro analysis was carried out by Scanning Electron Microscope (SEM) and Energy Dispersive Spectroscopy (EDS).The studies indicated that nano particle was uniformly distributed by improving the microstructure of concrete.

1. Introduction

Nanotechnology is perhaps the buzzword of all buzzwords. There is something intrinsically interesting about the study of phenomena at a scale so small that lengths, diameters and circumferences can be measured relatively as small number of atoms. Addison of nano silica with in nano scale –size particles can bring about significantly enhanced properties and same synthetic synthesis [1]. It is incorporated with customary building materials fundamental main impetus in the development to design be that as it may, now it is utilization is expanded durability is have in a few components such as micro-cracking of concrete, stiffness and compression strength, firmness and pressure quality and contrasted and the instance of thick structure network for matrix polymer [2-4]. Since bond lattice has free structure between the concrete and total there are nano measured air voids which may have critical impact and inorganic materials including dynamic composite (SiO_2,Al_2,CaO), such as, slag, zeolite and coal powder were turned out to be crucial element of some high-quality concrete and some examination comes about, it is important for enhancing mechanical properties, solidify defrost sturdiness, the vibration damping limit, abrasion resistance, bond strength, corrosion resistance [5-6]. Disappointment is recognized in appearance of splits. It ought to be noted most auxiliary materials, outside anxieties. Since the concrete as of now has breaks exhibit in it the purpose behind utilizing nano material is to fill these splits which help in increment of toughness, quality of the concrete by increment of compressive quality, creep quality. The expansion of nano alumina lessens miniaturized scale breaking and increment compressive quality.

2. Materials

2.1 cement

Ordinary Portland cement of 53 grade confirming to IS 8112:1989of locally available RAMCO cement which comprises good quality. The chemical configuration of cement was found using X-ray fluorescence analysis and has the following properties are given in Table1.

Table 1. Shown in properties of cements

Description	**Composition**
Physical Properties	
Color	Grey
Specific gravity	3.15
Specific surface area (cm^2/g)	3540
Chemical Composition	
CaO (%)	62.8
SiO_2 (%)	20.3
Al_2O_3 (%)	5.4
MgO (%)	2.7
Na_2O (%)	0.14
K_2O (%)	62.8

2.2 Fine Aggregate

River sand was utilized as fine aggregate modulus aggregate is 3.32 and it has a place with coarse sand class which can be utilized for concrete mixing. The specific gravity of was noted as 2.64. Table 2 appeared in of fine aggregate.

Table 2. Fineness modulus of Fine aggregate.

IS Sieve (mm)	Weight retained (Kg)	Percentage Retained (%)	Cumulative percentage (%)	Cumulative percentage finer (%)
4.75	4.75	8	0.53	0.53
2.36	2.36	48	3.2	3.73
1.18	1.18	4	0.26	3.99
0.600	0.6	454	30.26	34.26
0.420	0.43	136	9.06	43.32
0.300	0.3	260	17.33	60.65
0.150	0.15	480	32	92.65
0.075	0.075	12	0.8	93.45
Pan	-	96	6.4	99.85

2.3 Coarse Aggregate

The table 3 aggregate passing through 16 mm sifter and held on 12.5 mm sieve was utilized as coarse total in the concrete mixture. The specific gravity of the coarse aggregate was noted as 2.65. The fineness modulus of the coarse total is acquired as 6.8.

Table 3. Properties of Coarse aggregate

Properties	Observed values
Fineness modulus	6.80
Specific gravity	2.75
Bulk density	1530 kg/m^3
Loose density	1445 kg/m^3
Aggregate crushing value	29%
Aggregate impact value	32%
Maximum size of aggregate	12.5 mm
Water absorption	0.15%

2.4 Nano Alumina (Al_2O_3)

Pure Nano- Alumina (Al_2O_3) was brought and its properties are tested in manufacture company give by table 4 as shown and SEM figure 1.

Table 4. Properties of Nano- Alumina (Al_2O_3)

Test item	Standard Requirement	Test result
Specific surface area(M2/G)	200-210	>202 m^2/g
Crystal structure and type	Alpha	yes
Particle shape	spherical	yes
pH value	3.7-8.5	6
Loss on drying @105^0 c (5)	<1.5	0.47
Loss of ignition @1000^{0}c (%)	<2.0	0.66
SiO_2content	>99.8	99.88
Carbon content	<0.15	0.10

Figure 1. SEM of Nano Alumina (Al_2O_3)

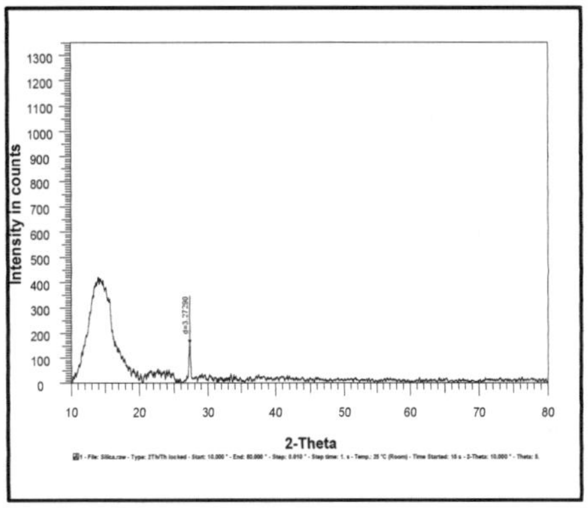

Figure 2. TEM Analysis

2.5 Super plasticizer
Conplast SP430 (G) is utilized where a high level of workability and its maintenance are required, here postponements in transportation or setting are likely or when high surrounding temperatures cause quick droop misfortune. It encourages generation of excellent cement.

2.6 Water-cement ratio
The water cement ratio was kept at 0.33, as the rate of silica smoke expanded; the necessity of water required likewise expanded.

3. Experimental program
3.1 TEM test on Nano- Alumina (Al_2O_3)
Transmission electron microscopy (TEM) is a microscopy procedure in which a light emission is transmitted through a ultra-thin example, associating with the example as it goes through it [5-7]. A picture is shaped from the cooperation of the electrons transmitted through the level of picture is amplified and centered onto an imaging gadget, for fluorescent screen on a layer of photographic film or to be recognized by a sensor as CCD camera. TEMs are fit for imaging at a fundamentally higher determination than light magnifying lens. the instrument's client to look at fine detail—even as little as

a solitary section of particles, which is a huge number of times littler than the littlest resolvable protest in a light magnifying lens [8]. TEM shapes a noteworthy examination strategy in a scope of logical fields, in physical, synthetic and organic sciences. The specimen TEM result appeared in Figure 3.

3.2 Concrete Mix design

The mix design can be characterized as the way toward choosing the reasonable elements of cement and deciding their relative extent with the question of creating cement of certain base quality and sturdiness as monetarily as would be prudent. The reason for outlining the blend is to accomplish the stipulated least quality and to make the solid in the most efficient way.

3.3 Casting Details

Concrete cubes of size 100 mm x 100 mm x 100 mm were thrown to concentrate the compressive strengths of shapes supplanted with Nano-alumina subsequent to subjecting to curing in ordinary condition. Quality trial of solid shapes was performed in programmed pressure testing machine with a limit 3000 kN.

3.4. Cylinders

Cylinders of size 300 mm x 100 mm were casted study the effect of Split tension and Modulus of elasticity in Compression testing machine.

4 Test procedures

4.1 compressive strength test

The compressive strength test is the most widely recognized test directed on the grounds that the vast majority of the alluring trademark properties of cement and the auxiliary outline reason for existing are subjectively identified with compressive strength. Test was conducted in compression testing machine of 3000kN capacity for different ages of concrete viz.7, 14 and 28 days as per the specifications given in IS 5816: 1999 under normal room temperature and figure 3 and 4 shown in experimental setup of compression, spilt tension.

Figure 3. Test setup for Compressive

Figure 4. Spilt tension

4.2 Split tensile strength test

This is an indirect test to decide the tensile strength of cylindrical specimens. Part elasticity tests were completed at 28 days for the concrete cylinder specimens of size 150 mm length across and 300 mm utilizing pressure testing machine of 3000kN limit according to IS:5816-1999. The load was applied gradually till the specimen splits and readings were noted.

IOP Conf. Series: Earth and Environmental Science **80** (2017) 012005 doi:10.1088/1755-1315/80/1/012005

4.3 SEM and EDX Results

An examining electron magnifying instrument (SEM) is a sort of electron magnifying lens that produces pictures of an example by filtering it with an engaged light emission. The electrons connect with atoms in the sample, creating different signs that contain data about the specimen's surface geography and arrangement. The electron pillar is for the most part examined in a raster filter design, and the shaft's position is joined with the recognized flag to deliver a picture [9].

SEM can accomplish determination superior to 1 nanometer. Specimens can be seen in high vacuum, in low vacuum, in wet conditions (in ecological SEM), and at an extensive variety of cryogenic or lifted temperatures. SEM test is done on 2% and 4% Nano-silica in part supplanting cement.EDS depends on a connection of some wellspring of X-beam excitation and a specimen. Its portrayal capacities are expected in vast part to the central rule that every component has a remarkable nuclear structure permitting extraordinary arrangement of crests on its X-beam emanation range.

5. Results and Discussions

5.1 compressive strength tests

Three examples of size 100 mm x 100 mm x 100 mm were utilized for compression testing for each group of blend. Perfect and surface dried examples were put in the testing machine. The platen was brought down and touched the top surface of the specimens the heap was connected step by step and most extreme load was recorded. Figure 4 and 5 as appeared in compressive and spilt tensile test of nano alumina. Three specimens of size 100 mm x 100 mm x 100 mm were used for compression testing for each batch of mix the test procedure follows as IS code.

Table 5 Compressive strength of Nano-alumina

Nano Alumina Content (%)	Compressive Strength Sample 1 (MPa)	Compressive Strength Sample 2 (MPa)	Compressive Strength Sample 3 (MPa)
0	26.2	25.9	26.6
0.5	28	28.4	27.2
0.75	33.6	32.3	31.8
1	38.4	36.8	37.8

5.2 Split Tensile strength Test

Split tensile test was conducted on cylinders of size 150mm diameter and 300mm height. Clean and surface dried specimens were placed in the testing machine. The platen was lowered and was allowed to touch the top surface of the specimen. The force was applied and maximum load at which the specimen failed was recorded.

Table 6 Split tensile strength of Nano-alumina

Nano Alumina Content (%)	split tensile (Mpa) sample1	split tensile (Mpa) sample2	split tensile (Mpa) sample3
0	2.98	3.13	3.06
0.5	3.54	3.44	3.47
0.75	3.72	3.76	3.69
1	3.94	3.90	3.84

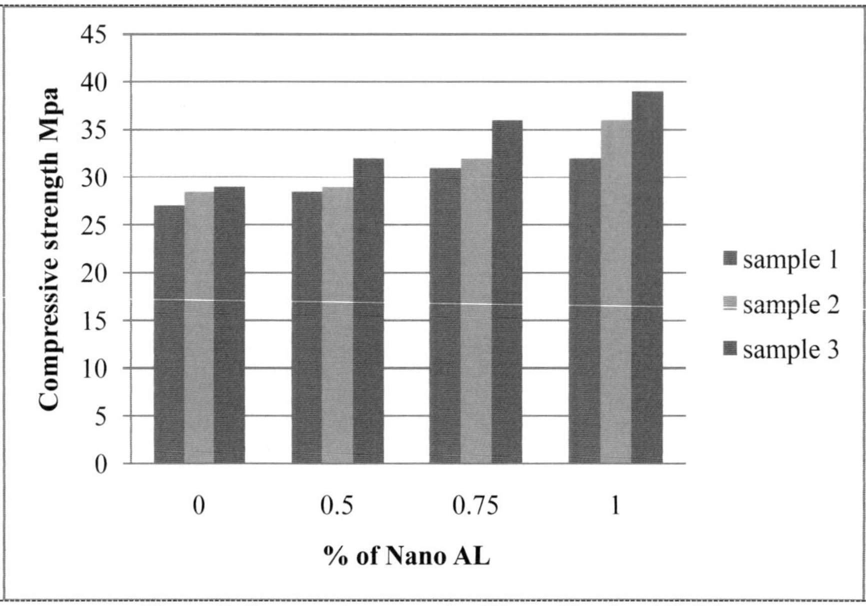

Figure 5. Compressive strength of Nano Alumina (Al_2O_3)

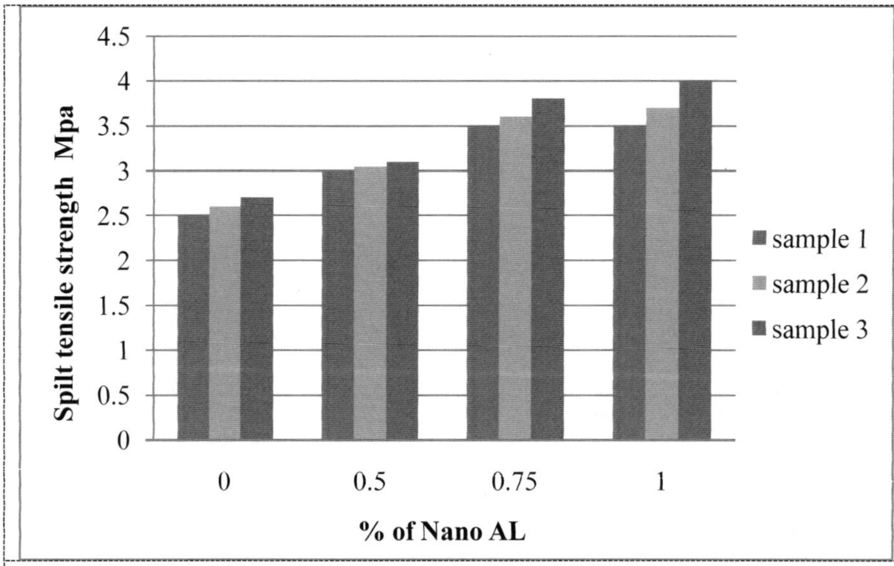

Figure 6. Split tensile strength of Nano Alumina (Al_2O_3)

5.3 SEM and EDS Results

The SEM picture and EDS range of concrete demonstrated that there exists noteworthy contrast between the 0% test and 1% nano-alumina concrete. Around the total there exist breaks under wraps test while there is a thick blame zone around the total in ITZ of 1% expansion mortars [10-12], and Figure 7 to 9 as appeared if variant of alumina present specifically region in that segment, There is more measure of aluminum and alumina around the total of the mortars with 1% alumina than the check test as appeared in Figures. Hence, amid the development of C–H–S (C_CaO, S_SiO_2), and

(H_H_2O) gel, nano-alumina topped off the pores of free net structure around the sand molecule, and diminished the porosity and expanded the thickness level of ITZ [11-12].

Figure 7. SEM image of concrete hydrated for 28 days with 0%& 0.5% nano-alumina

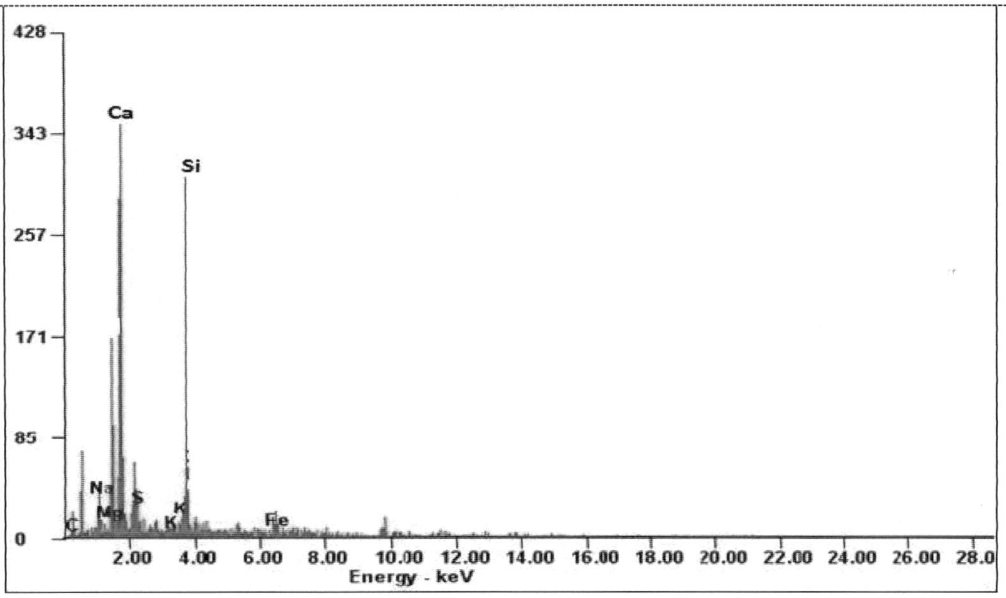

Figure 8. EDS images of 0% Nano alumina sample

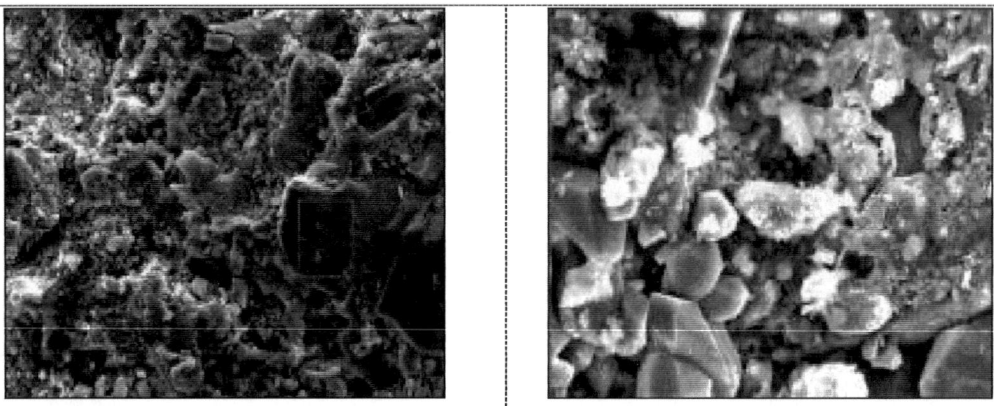

Figure 9. SEM image of concrete hydrated for 28 days with 0.75 % & 1 % nano-alumina

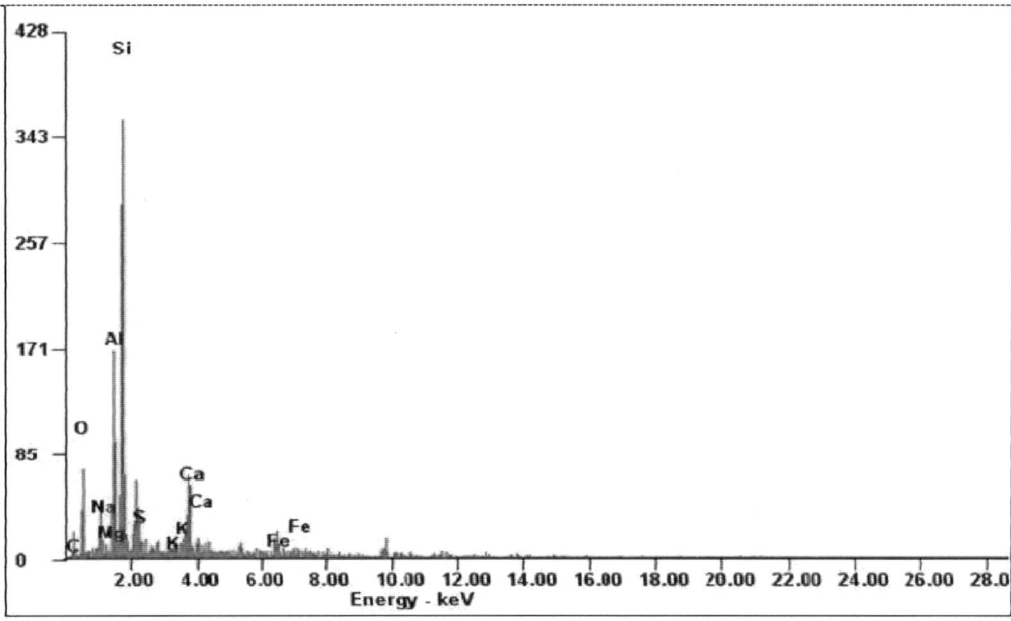

Figure 10. EDS images of 1 % Nano alumina sample

6. Conclusions

The accompanying were the conclusions touched base from the analyses directed on specimens in Standard condition: Based on the outcome, it was found that for the expansion in rate of nano alumina, mechanical properties demonstrated reliably enhanced results. The compressive strength expanded significantly, the split tensile strength increased marginally. However the split tensile strength of the concrete could be enhanced by utilizing more reasonable reinforcements with nano particles. The SEM contemplates demonstrated that the nano particles are consistently conveyed and there by microstructure of cement is progressed. The SEM perceptions likewise uncovered that the nano-particles were going about as filler, as well as an activator to elevate hydration demonstrates and to enhance the microstructure of the concrete if the nano-particles were consistently scattered. The void investigation done utilizing image J programming uncovered that the rates void brought down when the extent on nano alumina expanded. The rate voids if there should be an occurrence of 1% nano alumina content cement is relatively lesser than that of 0% nano alumina concrete. Additionally ponder toward this path is suggested, since more number of perceptions is required for getting the ideal amount of nano-particles.

References

[1] Shannag M J 2000 High strength concrete containing natural pozzolan and silica fume. *Cem. Concr. Compos.* **22** 399-406

[2] Bhikshma V, Nitturkarband K and Venkatesham Y 2009 Investigations on mechanical properties of high strength silica fume concrete. *Asian J. Civ. Eng.* **3** 335-46

[3] Nazai A and Riahia S 2010 Effect of Water-Cement Ratio on Abrasive Strength Porosity and Permeability of Nano-Silica Concrete. *World Appl. Sci. J.* **17** 929-33

[4] Sololev K and M F Gutiérrez 2012 How nanotechnology can change the concrete world. *Am. Ceram. Soc. Bull.* **84** pp 14-20

[5] Berra M Carassiti F, Mangialardi T, Paolini A E and Sebastini M 2012 Effect of Nano-Silica addition on workability and compressive strength of Portland cement pastes. *Constr. Build. Mater.* **35** 666-75.

[6] Sanchez F and Sololev K 2010 Nano Technology in concrete - a review. *Constr. Build. Mater.* **24** 2060-71

[7] Shekari A and Razzaghi M 2011 Influence of Nano particles on Durability and Mechanical properties of high performance concrete. *Procedia Eng.* **14** 3036-41

[8] Bybyung Wan Jo, Chang Hyun Kim and Jae Hoon Lim 2006 Investigations on the Development of Powder Concrete with Nano-SiO_2 Particles. *Constr. Build. Mater.* **21** 1351-5

[9] Denise A Silva, Paulo Denise Paulo and Monteiro 2006 The influence of polymers on the hydration of Portland cement phases analyzed 1046 by soft X-ray transmission microscopy. *Cem. Concr. Res.* **36** 1501-7

[10] Gengying Li 2004 Properties of high-volume fly ash concrete incorporating nano- SiO_2, *Cem. Concr. Res.* **34** 1043-9

[11] Hui Li, Hui-gang Xiao and Jinping Ou 2004 Microstructure of cement mortar with nano-particles. *Harbin Inst. Technol.* **35** 185-9.

[12] Luciano Senff, Joao Labrincha and Victor Ferreira 2009 Effect of nano-silica on rheology and fresh properties of cement pastes and mortars. *Constr. Build. Mater.* **23** 2487-91.

Study of Bond Characteristics of ReinforcedWasteGlassAggregate Concrete

P Rajagopalan, V Balaji, N Unnikrishnan, T JainulHaq and P Bhuvaneshwari*

School of Civil Engineering, SASTRA University,Thanjavur-613401. India.

*Email: bhuvaneswari@civil.sastra.edu

Abstract. The conformity of properties of waste glass aggregate with conventional aggregate was found out. Nine cubes (150mm x 150mm x 150mm) were cast out of which three were used for control concrete, three were fully replaced with waste glass as coarse aggregate, three were partially replaced(50%) with waste glass as fine aggregate. Six cylinders (150mm x 300mm) were cast out of which two for control concrete, two cylinders with coarse aggregate fully replaced with waste glass aggregate(WGA) and remaining two cylinders with partially replaced (50%) fine aggregate with waste glass aggregate. Cured specimens were subjected to compression and split-tensile test to ascertain the characteristic compressive strength and split tensile strength. Since the surface of the coarse aggregate plays a significant role in bonding of the rebar in reinforced concrete, pull-out test on both control and Waste Glass Aggregate (WGA) cube specimens(150mmx150mm with 20mm diameter steel rods) were conducted. Scanning Electron Microscopy (SEM) analysis has been done for better understanding of bonding properties in waste glass fine aggregate(WGFA) and waste glass coarse aggregate(WGCA) concrete.Comparison of the results with that of control specimens showed that waste glass could be effectively used as aggregates in reinforced concrete construction.

Keywords: Reinforced concrete beam, glass aggregate, pull-out test, bond strength

1. Introduction

The serviceability of a building depends on the quality of the materials used for its construction. The usage of cement,sand and aggregate in concrete construction give great ecological impact. To bring down the usage of these materials researchers have concentrated on potential use of waste material as an alternate concrete construction. The different waste material concentrated were fly ash, plastics, glass etc. Among the different waste materials, disposal of waste glass is still a major problem existing worldwide. Recycle of waste glass also has its limitation to ensure the quality of the manufactured new glass products. Earlier studies concentrated on suitability of utilizing the waste glass in production of construction concrete.Replacement of coarse and fine aggregate with waste glass is the recent research. The different types of waste glass generated from various industries and the problem of their disposal was concentrated[1].Past studies were concentrated on the use of waste glass as aggregate in concrete [2,3,4]. Retarders were used to suppress the alkali-silicate reaction in waste glass aggregate concrete. It was also confirmed that alkali silica reaction was predominant in the micro cracks of crushed coarse glass particles rather than finer glass particles. Even the absence of retarders gave good bonding effect for mortar with finer glass aggregates.The impurities present in glass, combination of different colors and cost of disposed glass have compelled for a development of marketing of mixed types of waste glass[5].

The usage of glass aggregate as a substitute for natural aggregate in both ordinary cement mortar and water glass activated flyash mortar were concentrated [6]. It was found that the alkali silica reaction of fly ash based mortar was less even for full replacement of natural aggregate. The optimum value of replacement of recycled coarse aggregate with natural coarse aggregate for achieving required compressive and bond strength was determined[7].Past studies confirm the effective utilization of waste glass for partial or complete replacement of natural aggregate in production of cement mortar and concrete.

1.1 Research significance
Disposing waste glass anddepletion of natural aggregates were arising problems. The study was concentrated in replacing natural coarse aggregate with waste glass. Workability and strength characteristics of WGA concrete were confirmed. Since the surface of aggregate play a role in binding the concrete, bond characteristics of WGA concrete was carried out. Bond strength of control and WGA concrete was compared.

2. Experimental procedure

2.1 Materials and methods
Ordinary Portland cementof53 gradewas used in present study.Specific gravity of the cement was found to be 3.12[8]. The maximum size of coarse aggregate and waste glass was limited to 20 mm. Natural sand and waste glasspassing 4.75mm sieve were used as fine aggregate [9]. Potable water was used to cast concrete. Waste glass was procured from bus depot located on the way from Thanjavur to Kumbakonam.Design mix was carriedoutas per [10] for M30 grade along with water cement ratio as 0.45. Fineness modulus test has been conducted for the materials used in the study [11]. The slump test (Fig.1) was carried out for both control and WGA concrete to check the workability.

Figure 1. Slump Test

2.2 Casting and Curing
Specimens were cast andkept in the mould for 24hours. The specimens were demoulded and kept immersed in clear fresh water until taken out for further test. The cast cube, cylinder and pull out specimens were shown in Fig.2, Fig.3 and Fig.4 respectively.

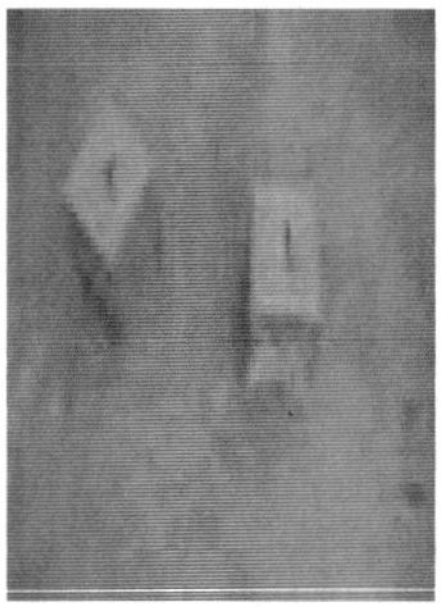

Figure 2. Cube Specimens

Figure3. Cylinder specimens

Figure 4. Specimens forPull Out test

2.3 Methods

Both compression and split tensile strength were carried out in CTM of 3000kN capacity as shown in Figure 5(a) and 5(b) respectively. The pull out test as per[12]were conducted in UTM of 1000kN capacity as shown in Fig.5(c)

(a) Cube under compression

(b) Split Tensile Test

(c) Pull-out Test

Figure5. Specimen under test

3. Results and Discussions

3.1Material property

Fineness moduli of the materials are given in Table 1. It was found that glass aggregate and coarse aggregate values were almost in the allowable range which finally denotes that glass aggregate can be effectively used in the partially replacement with the coarse aggregate in concrete.

Table 1. Fineness modulus for aggregates

Type of aggregate	Fineness modulus
coarse aggregate	7.29
fine aggregate	2.46
glass aggregate	6.04

3.2 Slump for different mix

The slump values for both control and WGA concrete are given in Table 2. Comparison of the results showed that both control and glass aggregate concrete produced similar slump values. Hence it was inferred that medium workability was achieved by control and waste glass concrete.

Table 2. Slump value for specimens

S.No.	Type of specimen	Slump (mm)
1	Control concrete	60
2	Waste glass as coarse aggregate	80
3	Waste glass as fine aggregate	70

3.3 Compressive strength

Compressive stress values of the specimens are displayed in Table 3. From the table, average peak stress for control and WGFA concrete showed minor differences and further studies are required to improve the strength of WGA concrete.

Table 3. Compressive strength for cube specimens

S.No	Type of cube specimen	Average peak load(kN)	Compressive stress(N/mm^2)
1	Control concrete	720	32
2	Glass as coarse aggregate	554	24.6
3	Glass as fine aggregate	677	30

3.4 Split Tensile strength

Tensile stress of the specimens are given in Table 4. The test values for WGFA concrete denotes that the peak load and stresses were almost nearer to that of control concrete.

Table 4. Tensile strength for cylinders

S.No	Type of specimen	Average Load(kN)	Tensile stress (N/mm^2)
1.	Control concrete	219	3.1
2.	Glass as coarse aggregate	152	2.2
3.	Glass as fine aggregate	206	2.9

3.5 Pull-out strength

The pull-out force for the specimens is displayed in Table 5. Test results inferred that the maximum force for WGFA concrete value was greater than control concrete, which showedgood bond strength between the reinforced bar and WGA concrete. The failure has been occurred in the enclosing concrete as per [12].

Table 5. Bond strength of cube specimens

S.No	Type of specimen	Maximum force(kN)
1.	Control concrete	42.5
2.	Glass as coarse aggregate	31.4
3.	Glass as fine aggregate	45

3.6 Scanning Electron Microscopy Analysis

Scanning electron microscope analysis was conducted on waste glass coarse aggregate sample (WGCA) and fine aggregate sample (WGFA).

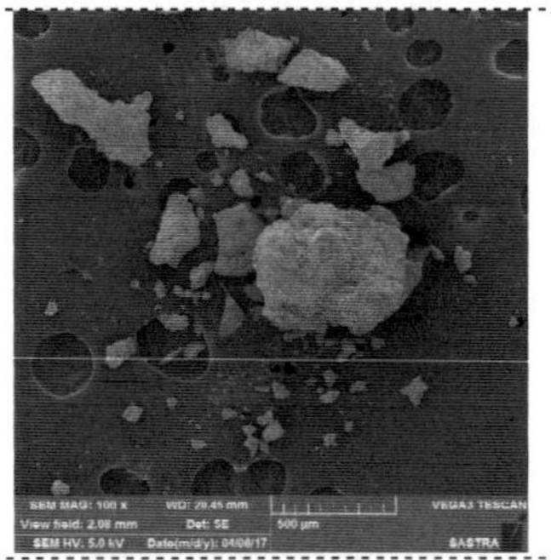

(a) WGFA **(a)**WGCA

(b)WGFA **(b)**WGCA

Figure 6. (a) 100x Magnification, (b) 500x Magnification

The surface of the WGFA sample was found to be smoother than that of WGCA sample.The silica present in glass particles has bonded better in fine aggregate than in coarse aggregate samples. The enhancement of bond strength of fine aggregate concrete might be due to the silica particles.

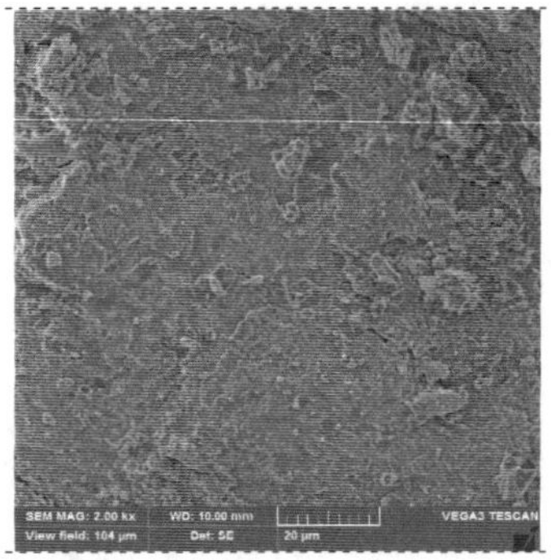

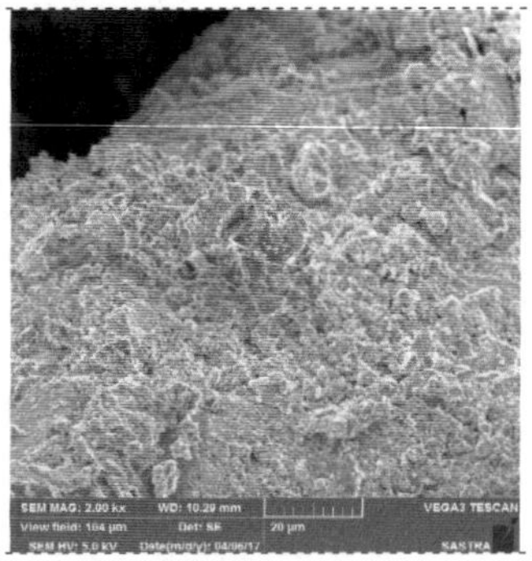

(a)WGFA **(b)**WGCA

Figure 7. 2000x Magnification

4. Conclusion

Strength characteristics of concrete produced with full and partial replacement of natural coarse and fine aggregate with waste glass was carried out and following conclusions were arrived.

- Slump test for the waste glass aggregate concrete were conducted and medium workability condition has been achieved.
- Compressive stress of WGFA concrete showed a percentage reduction of 5%, compared to control concrete
- Compressive stress of WGCA concrete showed a percentage reduction of 20%, compared to control concrete.
- Compared to control concrete, split tensile strength of WGFA concrete showed a reduction of 5%.
- Compared to control concrete, split tensile strength of WGCA concrete showed a reduction of 30%.
- Failure of pull out specimens occurred due to failure of the concrete enclosing the rebar. Pull-out test on specimens showed that the bond strength of WGFA concrete increased by 5%, compared to control concrete.
- Pull-out test on specimens showed that the bond strength of WGCA concrete decreased by 25%, compared to control concrete.
- The surface of the fine aggregate sample was found to be smoother than that of coarse aggregate sample.
- The silica present in glass particles has bonded better in fine aggregate sample than in coarse aggregate sample. Due to better bonding of fine aggregate glass in concrete,the bond strength of fine aggregate concrete was better than control concrete.
- Partial replacement of waste glass as fine aggregate gave more desirable results than replacement of glass as coarse aggregate.

References

[1] Rita Nemes and ZsuzsannaJózsa2006 Strength of Lightweight Glass Aggregate Concrete.*J. Mater. Civ. Eng.ASCE*. **18**(5) 710-4.

[2] Bashar Taha andGhassanNounu2009Utilizing Waste Recycled Glass as Sand/Cement Replacement in Concrete. *J. Mater. Civ. Eng.ASCE***21**(12) 709-21.

[3] Caijun Shi 2009 Corrosion of glasses and expansion mechanism of concrete containing waste glasses as aggregates. *J. Mater. Civ. Eng.ASCE***21**(10) 529-34.

[4] Farshad Rajabipyor, Hamad Maraghechi and Gregor Fisher 2010 Investigating the Alkali-Silica reaction of recycled glass aggregates in concrete.*J. Mater. Civ. Eng. ASCE***22**(12) 1201-8.

[5] Jian-xin Lu, Zhen-hua Duan and Chi Sun Poon 2016 Fresh properties of cement pastes or mortars incorporating waste glass powder and cullet.*Constr. Build. Mater.***131**(1) 793-9.

[6] Meyer C, Egosi N and Andela C 2001 Concrete with Waste Glass as Aggregate in Recycling and Re-use of Glass Cullet *Proc. Int.Symp.Concr.Technol.*ASCE.

[7] Mounir M Kamal, Zeinab A Etman C, Mohamed R Afify and Mahmoud M Salem 2015 Bond strength of concrete containing different recycled coarse aggregates.*Concr. Res.* **6**(2) 93-111.

[8] IS12269 2013 Methods of Physical test for hydraulic cement.*Bureau of Indian Standards* (BIS) New Delhi.

[9] IS383 1970Specification for coarse and fine aggregate from naturalsources for concrete – *Bureau of Indian Standards* (BIS) New Delhi.

[10] IS10262 2009 Indian Standard code of practice for recommended guidelines for concrete mix design – *Bureau of Indian Standards* (BIS) New Delhi.

[11] IS2386 1963 Methods of tests for aggregates for concrete – *Bureau ofIndian Standards* (BIS) New Delhi

[12] IS 2770 1 1967 Methods of testing bond in reinforced concrete– *Bureau of Indian Standards* (BIS) New Delhi.

Utilisation of Waste Marble Dust as Fine Aggregate in Concrete

G V Vigneshpandian*, E Aparna Shruthi, C Venkatasubramanian and D Muthu

School of Civil Engineering, SASTRA University, Thanjavur -613401. India.

*Email: vaithiyanathanvignesh@gmail.com

Abstract: Concrete is the important construction material and it is used in the construction industry due to its high compressive strength and its durability. Now a day's various studies have been conducted to make concrete with waste material with the intention of reducing cost and unavailability of conventional materials. This paper investigates the strength properties of concrete specimens cast using waste marble dust as replacement of fine aggregate. The marble pieces are finely crushed to powdered and the gradation is compared with conventional fine aggregate. Concrete specimen were cast using wmd in the laboratory with different proportion (25%, 50% and 100%) by weight of cement and from the studies it reveals that addition of waste marble dust as a replacement of fine aggregate marginally improves compressive, tensile and flexural strength in concrete.
Keywords: WMD-Waste Marble Dust, Concrete, Strength, Reuse, Replacement

1. Introduction

Concrete is an extensively used building item and consists of cement, coarse aggregate, fine aggregate and deliberate amount of water. Natural sand is usually used as fine aggregate. For appropriate rapid growth in structural activities, accessible natural sand supply is exhausting. In addition, there are times when it is necessary to transport high-class sand from a long distance, and because of the expense of construction, it is essential to partially or completely substitute the ordinary sand in the concrete with a substitute matter without weakening the superiority of the concrete. To substitute sand as fine aggregate, waste marble dust can be used. This research aims to exploit the usage of waste marble powder in concrete instead of sand as fine aggregate.

2. Literature Survey

Based on the extensive review of literature, following critical observation was made:

- The focus of the study so far, has understood the effect of WMD on concrete, with a view to understand / identify the mechanisms responsible for in properties of the concrete.
- Very few studies have reported on the effect of WMD on concrete and the studies were carried out under room conditions.
- Most reported studies on the effect of WMD have been carried out laboratory conditions and experimental studies for M30 grade of concrete were rarely attempted. However, such studies are essential for better understanding the effect of WMD for M30 grade of concrete.
- The effect of WMD on the concrete characteristics sophisticated method like split tensile and compression test.

3. Methodology
Conventional concrete with a water cement proportion of 0.42 was produced. The cement used was ordinary Portland cement of 53 grade using WMD. The substitution ratio of Waste Marble Dust(WMD) was 25%, 50, 100% of the weight of cement. Sieve aggregates with sizes of 20 mm and 10 mm, fine sand with a fine modulus (FM) of 1.94 were used. In order to obtain sufficient workability, the relative ratio of coarse aggregate and sand was determined for the standard weight concrete as a reference.

4. Materials Used
4.1 Cement
The cement exploited for the study is 53 grade ordinary Portland cement of IS: 8112-1989. The specific gravity of cement is 3.15

4.2 Waste Marble Dust
Wastes generated in the marble stone industry is proclaimed and used as Waste Marble Dust.

4.3 Graded Fine Aggregate
Sand grains passing through a 4.75 mm sieve are called fine aggregates. Natural sand is used as fine aggregate. Regionally available river sand compliant with zone II of IS: 383-19707 was used as fine aggregate having a specific gravity of 2.54.

4.4 Graded Coarse Aggregate
Commonly sized, locally available well-graded aggregates larger than 4.75 mm and smaller than 12.5 mm are used as coarse aggregate. ACI Mix Design Procedure: crushed stone of 20 mm and 10 mm size, taken from a local quarry with a specific gravity of 2.68.

4.5 Water
Portable water available at the university campus was used to prepare concrete in the entire project.

5. Mix Proposition
The concrete mix was designed for M30 grade concrete according to IS 10262-2009 and the mixing ratio was 1: 1.87: 3.39 with a W / C ratio of 0.42. We prepared 12 types of concrete cubes and 4 kinds of concrete cylinders and With Waste Marble Dust 4 concrete prisms of various replacements were used in total. The amounts of aggregate, water content, additives and cement are listed in the Table 1.

Table 1. Mix Proportion

S. No.	Description	Mass (kg/m^3)
1	Cement	380
2	Fine Aggregate	710
3	Coarse Aggregate	1290
	20 mm	645
	10 mm	645
4	Water	160
5	Water-Cement Ratio	0.42

6. Specimens for the Experiment
6.1 Standard Cube Specimens
To test the compressive strength of concrete, totally twelve standard cube specimens of dimension 15cm x 15cm x 15cm were cast. Three specimens were cast for each mix with 0%, 25%, 50% and 100% replacement of Waste Marble Dust with fine aggregate.

IOP Conf. Series: Earth and Environmental Science **80** (2017) 012007 doi:10.1088/1755-1315/80/1/012007

6.2 Standard Cylinder Specimens
Four-cylinder sample of diameter 100mm and altitude 200mm were prepared for each replacement mix to test the split tensile strength of the concrete.

6.3 Standard Prism Specimens
To test the flexural strength of concrete, four standard prism specimens per each fraction of replacement were cast

7. Tests Performed

7.1 Test on Physical Properties of Materials

7.1.1 Specific Gravity Test. Specific gravity is usually defined as the ratio between the weight of a given volume of material and the volume of water. The specific gravity of cement and WMD is obtained by using a density bottle with kerosene as a standard substance. Determine the specific gravity of fine aggregate and coarse aggregate using a hydrometer and a wire basket. The specific gravity of the materials used for this project work is shown in Table 2.

Table 2. Specific Gravity

Materials	Specific Gravity
Cement	3.15
Marble	2.67

7.1.2 Sieve Analysis. The sieve analysis is a practice or procedure used to evaluate the particle size distribution of the particulate material.Size distribution is often very important to the way the material performs during use. Sieve analysis is a precise method of all sorts of non-organic or organic particulate matter including sand, crushed rock, clay, granite, feldspar, coal, soil, a wide range of manufactured powders, cereals and seeds. Probably the most common as it is a simple technique like particle sizing. Fineness modules of given marble dust is 2.505.

Table 3. Sieve analysis of WMD

S. No.	IS Sieve	Particle Size (mm)	Wt retained (g)	% retained	Cumulative % retained	% finer
1	4.75	4.75	0.014	1.4	0	0
2	2.36	2.36	0.013	1.3	2.7	97.3
3	1.18	1.18	0.148	1.48	17.8	82.8
4	600micron	0.60	0.256	2.56	43.1	86.9
5	300micron	0.30	0.428	4.28	85.9	14.1
6	150micron	0.15	0.141	1.41	100	0
7	Pan	<0.75	0	0	0	0

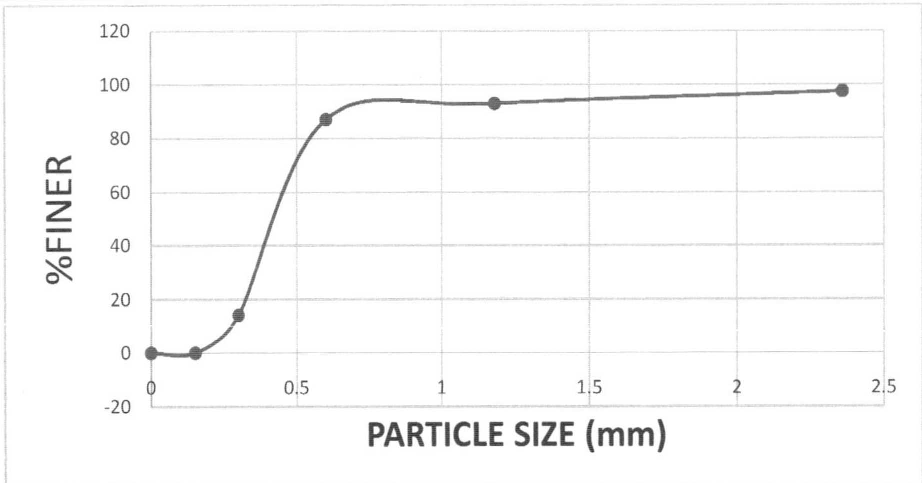

Figure 1. Sieve Analysis Graph for WMD

7.2 Test on Mechanical Properties of Concrete

7.2.1 Compressive Strength of Concrete. For the cubic compression test of concrete, a cube of 150 mm × 150 mm × 150 mm size was used. All cubes were tested in the ambient curing process. For the different proportions, cubes were tested using a 3000 kN volume compression tester 7 days and 28 days after curing. After placing the test piece in the center of the testing machine, the test was conducted with a uniform stress of 10 kg / cm 2 / min. Loading continued until the readings were reversed from the incremented value. Reversal of the reading indicates that the test piece has failed. The machine stopped, and it turned out that reading at that moment was the ultimate load. The value obtained by isolating the final load by the cross-sectional area of the sample is equal to the final cubic compressive strength.

7.2.2 Split Tensile Strength of Concrete. This is an indirect test to determine the tensile strength of a cylindrical specimen. In the divided tensile strength test, a cylinder test piece having a span of 100 mm and a length of 200 mm was subjected to curing for 28 days using a 3000KN capacity compression tester. To avoid direct loading on the test specimen, a cylindrical specimen was placed under the woody specimen. The test piece was cracked and the load was gradually added until the measured value was recorded.

7.2.3 Flexural Strength of Concrete. Flexural strength familiarized as the modulus of rupture, or flexural strength or transverse breaking strength is a property of material identified as the stress of the material before the bending test. The transverse bending test is most commonly adopted and uses a three-point bending test technique to bend a specimen either rectangular or circularin cross section to rupture or yield. Flexural strength signifies the maximum stress experienced in the material in that instant. Four-point bending is considered. In rectangular sample below a load in a four-point bending setup where the loading span is one-third of the support span.

8. Results and Discussion

The test results of specimens with varied fractions of WMD replacement are compared with the results given by the control check specimens. The table gives the comparison of 28 day split tensile, compressive and flexural strength results.

Table 4. Strength Characteristics

Percentage of Sand Replaced	Strength (N/mm^2)		
	Compressive Strength	Split Tensile Strength	Flexural Strength
0%	31.42	2.51	3.923
25%	33.02	2.069	4.022
50%	40.22	1.737	4.444
100%	29.28	1.512	3.7877

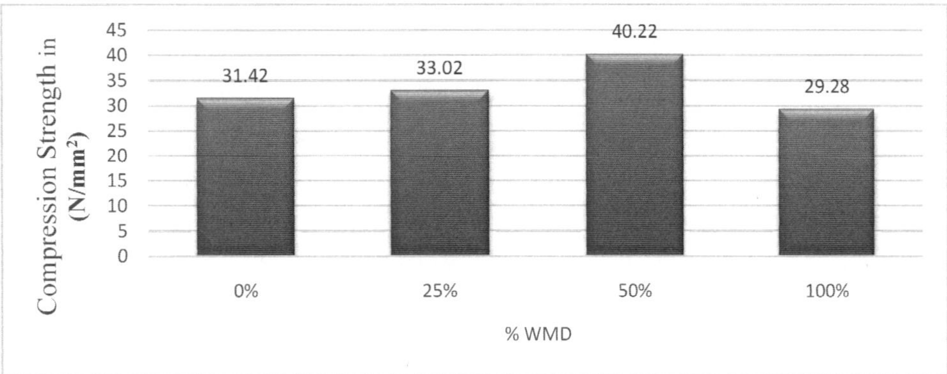

Figure 2. 28th Day Compression Strength

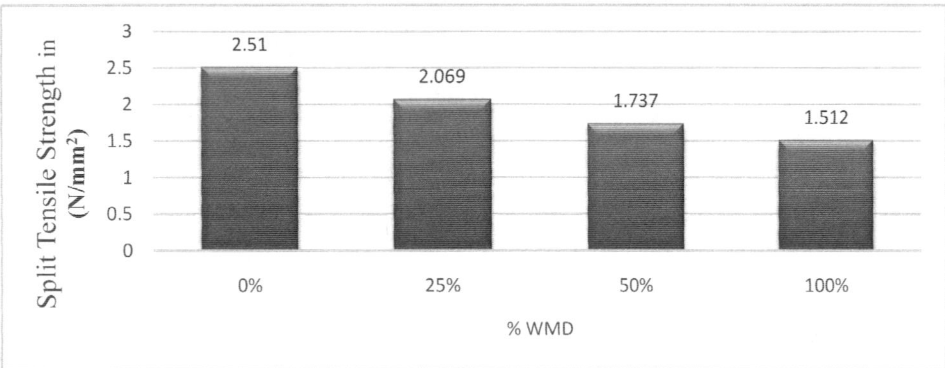

Figure 3. 28th Day Split Tensile Strength

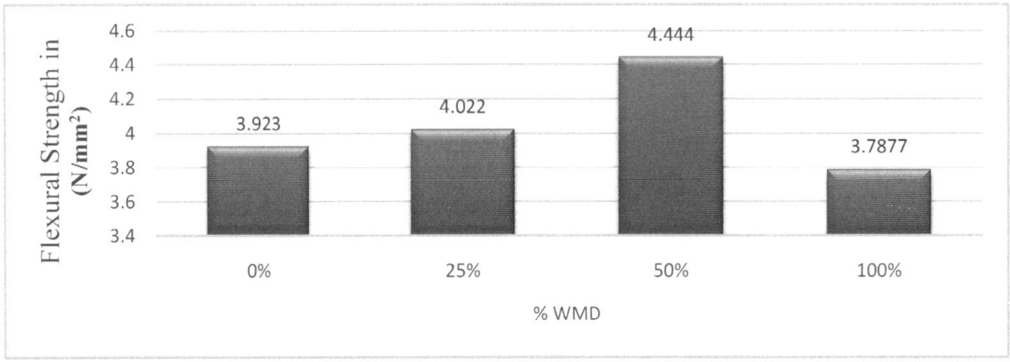

Figure 4. 28th Day Flexural Strength

8.1Discussions

- Including waste marble powder, the compressive strength of concrete progressively increases to a certain limit, but steadily decreases.
- Including marbled powder up to 50% increases the compressive strength inclusive of 28 days curing.
- Ultimate tensile strength is found with 0% replacement of marble powder and the strength gradually decreases by addition of WMD.
- Flexural strength of prism concrete increases upto 50% and then gets decreasing.
- Therefore, it was found that the optimum replacement rate by marble powder to sand in concretewas almost 50%.

9. Conclusion

This project work is intended to analyse the feasibilities of using waste marble dust as replacement with fine aggregate. It offers unique advantages of being abundance, easily accessible and cost efficient. The test result shows that the use of these Waste Marble Dust have the capability of improving the performance of the hardened concrete. From the above results, the use of Waste marble dust up to 50% replacement with the fine aggregate is recommendable. The result of flexural strength test shows the familiar behavior with the results of the compression strength test.

References

[1] Aalok D Sakalkale 2014 Experimental Study on the Use of Marble Dust in Concrete. *Int. Eng. Res. Appl. J.* **4** 44-50.
[2] Bahar Demirel Effect of fine sand marble dust on the mechanical properties of concrete. *J. Int. Phys. Soc.* **5**(9) 1372-80.
[3] Gameiro F, de Brito J and Correia da Silva D 2014 Durability of concrete containing fine aggregate from waste generated by marble quarrying. *Eng. Struct.* **59** 654-62
[4] Ramindar Singh, Manish, Bhutani and Tarnsiah 2015 Strength evaluation of concrete using marble powder and trash crushed tile. *Int. Sci. J. Latest Trend* **20**(1) 18-28
[5] Devesh Meena 2015 Study on Behavior of Large Diameter Dust in Concrete Placing. *Int. Res. J. Eng. Technol.* **2**

IOP Conf. Series: Earth and Environmental Science **80** (2017) 012008 doi:10.1088/1755-1315/80/1/012008

Experimental and Theoretical Investigations on Bond Strength of GFRP Rebars in Normal and High Strength Concrete

P Eswanth and G Dhinakaran*

School of Civil Engineering, SASTRA University, Thanjavur 613401, India

*Email: gd@civil.sastra.edu

Abstract. Bond behavior between GFRP bars and concrete is the most important parameter for constructing corrosion free structures by implementing the material. Serviceability of reinforced concrete structures are controlled by bond behavior. GFRP materials behave differently from reinforcing steel in terms of bond. They are of non-homogeneous and an-isotropic. Due to this outstanding behavior, there is a difference in transfer of loads between GFRP bars and concrete which made it as an idealized choice of a material. In the present work, the bond strength of GFRP bars in normal and high strength concrete was studied. In total, 12 specimens containing 12 mm, 16 mm diameter rebars which were embedded in 150 mm x 150 mm x 150 mm cubes were investigated. The specimens were subjected to direct tension pull out test in accordance with IS 2770 part 1. The comparison of bond properties of GFRP rebar in normal and high strength concrete showed that pull out load of non-metallic rebar fell well within the range.

1. Introduction

Fiber-reinforced polymer (FRP) reinforcement is used in structural construction as steel reinforcement is poor corrosion resistance. In structural members such as marine structures due to corrosion it leads to premature failure and serviceability of structures will be reduced. In case of corrosive conditions GFRP rebar are perfectly used as reinforcement in structures. It is important in case where design is mainly concentrated on durability of RC members.

Due to difference in properties of two materials steel and GFRP a direct substitution between these two materials is not possible. The major difference: Steel is having less tensile strength compared to that of GFRP bars and the elastic modulus is greater for steel compared to that of GFRP bars. Some FRP's have low elastic modulus and yield point is not perfectly defined in their stress strain curves. Due to outstanding behavior, GFRP materials are referred as idealized choice of material. The use of GFRP as reinforcement has been reduced due to many reasons, as it includes high cost and lack of awareness about new technology. Currently, many issues are affecting the structural members with GFRP as reinforcement such as resistance to loading and aggressive environments.

Many research studies were done using GFRP bars as reinforcement and on bond performance in concrete. The bond strength was governed by number of parameters such as bar diameter, embedment length and environmental criteria. Numerous works were carried out on these areas using

GFRP rebars [1-6]. Conventional pullout tests were performed for low to moderate concrete strengths, with bond strength in the range of 2-14MPa. Larger diameter bars will give lesser bond strengths compared to that of small diameter bars. Many a times bond strength was governed by shear criteria. It was found that bond stress decreases are lesser for larger embedment lengths than that of smaller embedment lengths [7]. Also pullout test was conducted and bond strength was measured in the range of 14MPa with the cover of ($5D_b$) mobilizing the failure and absorbed ductile response [8]. Conventional and eccentric pullout test in the range of 5-22MPa with various parameters were conducted. Curves were ductile in nature [9].

In this paper, it focuses on bond strength of steel and GFRP reinforcing bars in normal and high strength concrete using direct tension pullout test. A total of 12 specimens were tested. The experimental results are presented in detail. This research benefit engineers, professionals and provide data for code writers since GFRP technology is an important aspect and has practical significance.

2. Experimental program

2.1 Materials

2.1.1 Concrete. According to IS standards, normal strength and high strength concrete was prepared, with the composition as given in table-1. OPC 53 grade cement was used. Coarse aggregate maximum size was 20mm, and fine aggregate size interval was in the range of 0-4.75mm. Around the rebar, plywood was used of required dimension to cast the concrete cube.

Table 1. Composition and characteristics of concrete

Concrete mix	M40	M60
Water (kg/m^3)	147	190
Cement (kg/m^3)	350	633.33
Coarse aggregate(kg/m^3)	1102.5	1279.57
Fine aggregate (kg/m^3)	787.5	352.986
Superplastizer (%)	0.5	0.5

2.1.2 Fiber reinforced polymer bars. In this study GFRP bars were used which were supplied by Go Green Products. Diameters of 12mm, 16mm were used. Properties of materials are mentioned in Table.2. These are of longitudinal fibers with a thermosetting vinyl ester resin manufactured by the process called pultrusion. Due to ribs and undulations bond improvement can be achieved. The GFRP bars behave linearly up to failure in stress-stain curve.

Table 2. Mechanical properties of GFRP rebar

Material	Elastic Modulus (MPa)	Poisson ratio	Yield strength (MPa)	Ultimate tensile strength (MPa)
GFRP	0.32×10^5	0.25	446	446

2.1.3 Apparatus. Size of the mould for bond test will be of 150 mm x 150 mm x 150 mm cubes as the diameter of bars are varying from 12mm to 20mm size of cube should be 150mm. Dial micrometer was used to measure the slip of the bar from the concrete cube. Dial micrometer mounted to read in 0.0025 mm at the unloaded end and having a range of greater than 2.5 mm shall be used. At loaded end, dial micrometers mounted in 0.025 mm will be used. Universal Testing Machine was used for testing the bond strength. Tamping rod which was round and straight of 0.6m in length and 15 mm in diameter was used.

2.1.4 Specimen. To determine local bond slip relation experimentally between rebar and concrete, rebar placed at center in concrete are used for direct pullout tests. This method was everywhere used for comparing the relative bond between different materials. Pullout specimens depend upon different parameters such as concrete type, bar diameter, and bar type. The length was provided in such a way that the bond stresses should be uniformly distributed while loading is done. According to IS 2770.1.1967 direct tension test was conducted on 150 mm concrete cube, with a single rebar inserted vertically at the center of each cube specimen. The bar should extend 10mm outside the cube even after casting from the bottom face and shall provide 850mm at the top face for sufficient length of bar to extend through the blocks and to provide an adequate length to be gripped for application of load. According to IS 516-1959 three concrete were made and tested for the average compressive strength with the range of 200 to 300 kg/cm^2 at the time of making pullout tests. All the test specimens shall be cured for 28days.

2.1.5 Preparation of Test Specimen. Bars should be free from rust, grease, paint. Suitable solution must be applied so as to remove oil or grease. At the end of the bar, surface should be smooth normal to the axes of the bars. Mixing concrete and curing was done according to IS: 516-1959. After the top layer has been filled, extra quantity should be removed and it should be finished with a trowel. Bars of 12mm, 16mm were inserted in cube as shown in Figure 1. Each bar diameter of three specimens was tested. GFRP bars and steel bars were inserted in concrete as per IS standards and concrete was mixed properly, kept for curing for desired period and was kept in room temperature for dry curing to evaporate water from pores and specimens. Then pullout test was performed for that specimen.

2.1.6 Test procedure. Test was conducted in such a way that bar was subjected to axial pull from the cube in testing machine as shown in Figure 2. The distance between the loaded end of bar and concrete face in the testing apparatus was measured as slip. Load was applied at a rate not greater than 2250 kg/min or it should not greater than 1.25 mm/min. Slip was measured by using dial-micrometers. Calculation of bond strength was based on this load and results were to be recorded at maximum load at failure of the specimen as shown in Figure 3.

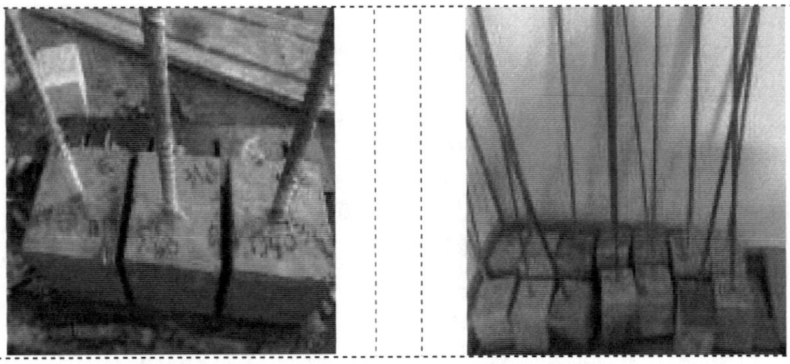

Figure 1. Cube specimens cast for pull out test

Figure 2. Testing of specimen

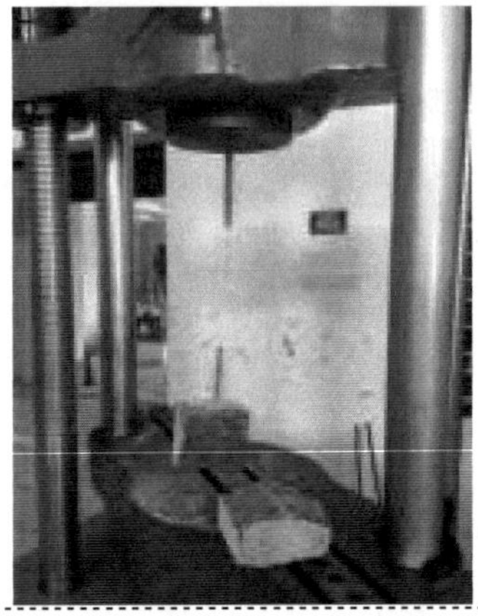

Figure 3. Failure mode of bond

3. Results and Discussion

Experimental results of bond strength of GFRP bars in concrete are mentioned in table 3 and table 4 for normal and high strength concrete and theoretical results are also mentioned in detail. Slip was high for normal strength concrete compared to that of high strength concrete for both 12 mm and 16 mm diameter of bars. Bond stress was high for small diameter compared to that of larger diameter. Mode of failure was splitting as they are of high grade concrete.

Table 3. Results of bond strength of GFRP rebar in M40 grade concrete

Diameter (mm)	Length (mm)	Embed ment length (mm)	Load at bond failure (kN)*	Load @0.125 mm slip (kN)	Gauge length (mm)	Slip (mm)	Avg. bond strength at bond failure (N/mm^2)
12	850	150	32.1	0.9185	150	12.76	5.68
12	850	150	34.85	1.125	150	13.39	6.16
12	850	150	36.59	1.125	150	13.78	6.47
16	850	150	39.05	1.00	150	14.8	5.18
16	850	150	43.62	0.985	150	13.25	5.78
16	850	150	41	1.05	150	13.00	5.44

*mode of failure is splitting

Table 4. Results of bond strength of GFRP rebar in M60 grade concrete

Diameter (mm)	Length (mm)	Embed ment length (mm)	Load at bond failure (kN)*	Load @0.125 mm slip (kN)	Gauge length (mm)	Slip (mm)	Avg. bond strength at bond failure (N/mm^2)
12	850	150	35.7	1.025	150	14.26	6.31
12	850	150	33.25	1.275	150	13.72	5.88
12	850	150	37.9	1.145	150	14.9	6.7
16	850	150	43.3	0.925	150	13.01	5.74
16	850	150	47.5	1.025	150	13.85	6.3
16	850	150	45.1	1.105	150	12.96	5.98

*mode of failure is splitting.

Ultimate bond using the following formula: $\sigma_b = \frac{P_b}{\pi * d * l}$ (1)

where, σ = ultimate bond stress, P_b= Bond failure load, d= diameter of bar, l= gauge length.

Average bond strength at failure (N/mm^2) = $\frac{P_2}{\pi * d * l}$ (2)

4. Theoretical calculation

As per IS 456-2000, design bond stress can be calculated by using following formula:

$$\tau_{bd} = \frac{\emptyset * \sigma_s}{4 * l_d} \quad (3)$$

σ_s =Stress in bar at the section considered at design load ($0.87 f_y$),τ_{bd} =design bond stress,$\emptyset$ =diameter of the bar,l_d =development length.

As per code for bond stress in case of deformed bars conforming to IS 1786 the value will be increased by 60 percent.

So for deformed bars:

$$\tau_{bd} = \frac{\emptyset * \sigma_s}{4 * 1.6 * l_d} \quad (4)$$

Results obtained based on theoretical calculations are tabulated in the table 5 for better understanding.

Table.5 Results of theoretical calculations

$\emptyset$ (mm)	f_y (N/mm^2)	l_d (mm)	$\sigma_s = 0.87 f_y$ (N/mm^2)	$\tau_{bd} = \dfrac{\emptyset * \sigma_s}{4 * 1.6 * l_d}$ (N/mm^2)
12	500	150	435	5.44
16	500	150	435	7.25

5. Conclusions

Following are the conclusions arrived from limited number of experiments conducted on bond performance GFRP in concrete. Average bond stress relationships and failure mode for each of the tested bars were experimentally determined. The results were explained as follows:

1. Failure of the bond was mostly due to length of embedment and bar surface.
2. Splitting failure occurs due to longer length of embedment. Increase in diameter reduced average bond strength.
3. Pullout load of normal strength concrete was found to be lower than that of high strength concrete. Bar diameter and embedment length are directly proportional to slip at peak bond stress.
4. Concrete strength is directly proportional to peak bond stress.

References

[1] Roman Okelo and Robert L Yuan 2005 Bond Strength of Fiber Reinforced Polymer Rebars in Normal Strength Concrete. *J. Compos. Constr.* **9** 203-13.

[2] Tastani S P and Pantazopoulou S J 2006 Bond of GFRP Bars in Concrete: Experimental Study and Analytical Interpretation *J. Compos. Constr.* **10** 381-91.

[3] Breccolotti M and Materazzi A L 2010 Structural reliability of eccentrically-loaded sections in RC columns made of recycled aggregate concrete. *Eng. Struct.* **32** 3704-12.

[4] Wambeke B W and Shield C K 2006 Development length of glass fibre-reinforced polymer bars in concrete. *ACI Str. J* **103** 11-7.

[5] Ehsani M R, Saadatmanesh H and Tao S 1995 Bond of hooked glass fibre reinforced plastic (GFRP) reinforcing bars to concrete *ACI Mater. J.* **92**(4) 391-400.

[6] Fei Yan and Zhibin lin 2016 Bond mechanism and bond strength of GFRP bars to concrete. *Compos. Part B* **98** 56-69.

[7] Achillides Z and Pilakoutas K 2004 Bond behaviour of fiber reinforced polymer bars under direct pullout conditions. *J. Compos. Constr.* **82** 173-81

[8] Cosenza E, Manfredi G, and Realfonzo R 1997 Behaviour and modelling of bond of FRP rebars to concrete.*J. Compos. Constr.* **1** 40-51.

[9] Baena M, Torres L, Turon A and Barris C 2009 Experimental study of bond behaviour between concrete and FRP bars using a pull-out test. *Compos. Part B* **40** 784-97.

Impact of Micro Silca on the mechanical properties of high volume Fly Ash Concrete

R Sripagadeesh[1], K Ramakrishnan[2*], G Pugazhmani[3], S Ramasundram[2], D Muthu[2] and C Venkatasubramanian[2]

[1]Civil Engineering Department, TKR Engineering College, Hyderabad - 500097, India.
[2]School of Civil Engineering, SASTRA University, Thanjavur– 613401, India.
[3]Civil Engineering Department, Roever Engineering College, Perambalur, India.

*E mail: ramakrishnan@civil.satra.edu

Abstract. In the current situation, to overcome the difficulties of feasible construction, concrete made with various mixtures of Ordinary Portland Cement (OPC) and diverse mineral admixtures, is the wise choice for engineering construction. Mineral admixtures viz. Ground Granulated Blast Furnace Slag (GGBS), Meta kaolin (MK), Fly Ash (FA) and Silica Fume (SF) etc. are used as Supplementary Cementitious Materials (SCM) in binary and ternary blend cement system to enhance the mechanical and durability properties. Investigation on the effect of different replacement levels of OPC in M25 grade with FA + SF in ternary cement blend on the strength characteristics and beam behavior was studied. The OPC was partially replaced (by weight) with different combinations of SF (5%, 10%, 15%, 20% and 25%) and FA as 50% (High Volume Fly Ash - HVFA). The amount of FA addition is kept constant at 50% for all combinations. The compressive strength and tensile strength tests on cube and cylinder specimens, at 7 and 28 days were carried out. Based on the compressive strength results, optimum mix proportion was found out and flexural behaviour was studied for the optimum mix. It was found that all the mixes (FA + SF) showed improvement in compressive strength over that of the control mix and the mix with 50% FA + 10% SF has 20% increase over the control mix. The tensile strength was also increased over the control mix. Flexural behaviour also showed a significant improvement in the mix with FA and SF over the control mix.

1. Introduction

In the current scenario to bulldoze the difficulties of feasible construction, concrete made with various mixtures of Ordinary Portland Cement (OPC) and diverse mineral admixtures, is the wise choice for engineering and construction. Mineral admixtures like Ground Granulated Blast Furnace Slag (GGBS), Meta kaolin (MK),Fly Ash (FA) and silica fume (SF) etc. are generally used Supplementary Cementitious Materials (SCMs). Any concrete containing Fly Ash (FA) content as 50 percent and above by mass of the total cementitious materials is considered as High Volume Fly Ash (HVFA) concrete. The concrete containing high volumes of class F fly ash was investigated for its mix proportion and its mechanical properties [1]. Past study on this subject reported that the characteristics of concrete containing HVFA ash up to 50% level of cement replacement in concrete proved efficient for reinforced concrete construction. The tests results of HVFA concrete showed that remarkable strength gain from 7 to 28 days and from 28 to 56, 91, 182, and 365 days. The durability properties of

IOP Conf. Series: Earth and Environmental Science **80** (2017) 012009 doi:10.1088/1755-1315/80/1/012009

high performance concrete viz., drying shrinkage, water sorptivity were studied to ascertain the effect of fly ash. Also the chloride ion penetration of high strength concrete having Class F fly as was investigated [2]. The OPC concrete was compared with concrete containing 30% and 40% fly ash after thorough investigation. The abrasion resistance of concrete was influenced by the presence of High-Volumes of Class F Fly Ash and its effect was investigated [3]. From the test results, it was understood that, irrespective of fly ash content, there was an increase in abrasion resistance with the increase in age for all mixtures. The effect of compressive strength and abrasion resistance of HVFA containing 70% FA and remaining SF was investigated for the influence of SF in HVFA [4]. In this work the concrete contained 70% HVFA with modification of 10% and 20% of FA by SF, GGBS and their equal combinations. At all ages, the compressive strength and abrasion resistance showed highest value due to the incorporation of 20% of silica and slag [4]. The possible use of SF along with slag in HVFA concrete exposed to elevated temperatures was studied. The weight loss and residual strength properties of Plain Cement Concrete (PCC), HVFA concrete, and HVFA concretes incorporating SF and slag after being exposed to temperatures from 400°C to1000°C in step of 200°C was investigated. The mix combinations in which a proportion part of PCC was partially replaced with the mineral admixtures viz., binary (PCC+FA), ternary (PCC+FA+SF, PCC+FA+GGBS) and quaternary (PCC+FA+SF +GGBS) cementitious blends were incorporated and investigated [5]. For the SF modified high-volume fly ash concrete, the strength and durability studies was investigated [6] The effect of variation of the cement replacement by SF in HVFA concrete on the mechanical and durability properties of concrete was studied by conducting various tests in the laboratory. The various properties of HVFA concrete incorporating Nano Silica (NS) was investigated for its durability [7]. Due to the formation of refined microstructure and smaller pores, the chloride ion penetration was reduced by the addition of SF in combination with NS in concrete mixture. By the addition of NS with HVFA or GGBS, reduction in setting time and increase in early gain of strength of concrete was achieved [8]. The alteration in compressive strength value of HVFA mortars and concrete due to the addition of NS was studied [9].

In this research work, the influence of partial replacement of OPC by FA (50%) and SF (5%, 10% and 15%) by weight in M25 grade concrete with Super Plasticizer (SP) was studied. Compressive strength at the end of 7 and 28 days was carried out on cubes of various mix proportions of SF and FA as partial substitution of cement with Water-Binder (w/b) ratio of 0.35. Beams were cast for the optimum mix proportion to determine flexural behaviour. It was found that all the mixes with the inclusion of FA and SF had improvement in strength over that of the control mix. From the experiment it was found that the mix (FA+SF)-2 had remarkable increase in strength by 20% over the control mix for M25 grade. Thus it became very important to improve the desired characteristics in HVFA by appropriate addition of mineral admixtures.

2. Experimental Program

2.1. Materials

Ordinary Portland Cement of 43 grade confirming to IS 8112:2013 [10] was used for this experimental work. Crushed granite of size ranging 20 mm – 10 mm was used and specific gravity was found to be 2.74 confirming to IS 2386 (part III) [11]. Natural river sand passing through IS sieve 4.75 mm of specific gravity 2.65 confirming to gradation Zone-III was used. Potable tap water available in the laboratory was used for mixing of concrete and curing. Conplast SP430 which is a chemical admixture complies with IS 9103 [12] and ASTM-C 496 [13] Type 'F' as a high range water reducing admixture was used to improve the workability and reduce the water content. SF and FA which are used as SCMs were obtained from ASTRRA chemicals, Chennai and Neyveli Lignite Corporation (NLC), Tamilnadu respectively. The physical and chemical properties of OPC, FA and SF are given in Table 1.

Table 1. Physical properties and Chemical composition of OPC, FA and SF

Description	OPC	FA	SF
Color	Grey	Light Grey	Light Grey
Specific gravity	3.1	2.7	2.63
CaO (%)	62.8	1.07	0.52
SiO_2 (%)	20.3	92.83	99.886
Al_2O_3 (%)	5.4	0.53	0.043
Fe_2O_3 (%)	3.9	3.97	0.040
MgO (%)	2.7	0.42	0.000
Na_2O (%)	0.14	0.16	0.003
K_2O (%)	0.53	1.02	0.001

2.2. Mix proportions

The proportion of materials for the M25 grade of concrete was ascertained by design mix and the proportion was 1:1.71:2.87 (by weight) respectively [14,15]. To study the effect of FA and SF blend as a partial substitution of cement, specimens were cast for reference and other mixes with different replacement levels of FA and SF (by weight) as given in Table 2 with w/b ratio of 0.35.

Table 2. Mix Proportions with various replacement levels of FA and SF

Mix	Replacement level (%)	
	FA	SF
Control (C)	0	0
(FA+SF)-1	50	5
(FA+SF)-2	50	10
(FA+SF)-3	50	15
(FA+SF)-4	50	20
(FA+SF)-5	50	25

2.3. Specimen casting and curing

To examine the consequence of addition of FA and SF combination (as partial replacement of cement), 150 mm cubes and 150 mm diameter, 300 mm height cylinders were cast for control mix and the other mixes comprising different mix combinations of FA and SF. Conplast SP 430 is added in all the mixes as it gives better results and good workability. Specimens were placed in the well lubricated mould and compacted and the specimens were left at room temperature for 24 hrs and after that specimens were placed in curing tank till the end of designated curing period (7 days and 28 days).

2.4. Specimen testing

The compressive strength of different mixes were found out after 7 and 28 days curing and the tensile strength were found out after 28 days curing in a compression testing machine of capacity 3000 kN confirming to IS 516:1959 [16]. The load rate adopted was 10 kN/min. Based on the test results of compressive and tensile strength, 100 mm × 150 mm × 1200 mm size beam specimens were cast for optimum mix proportion obtained for M25 grade of concrete.

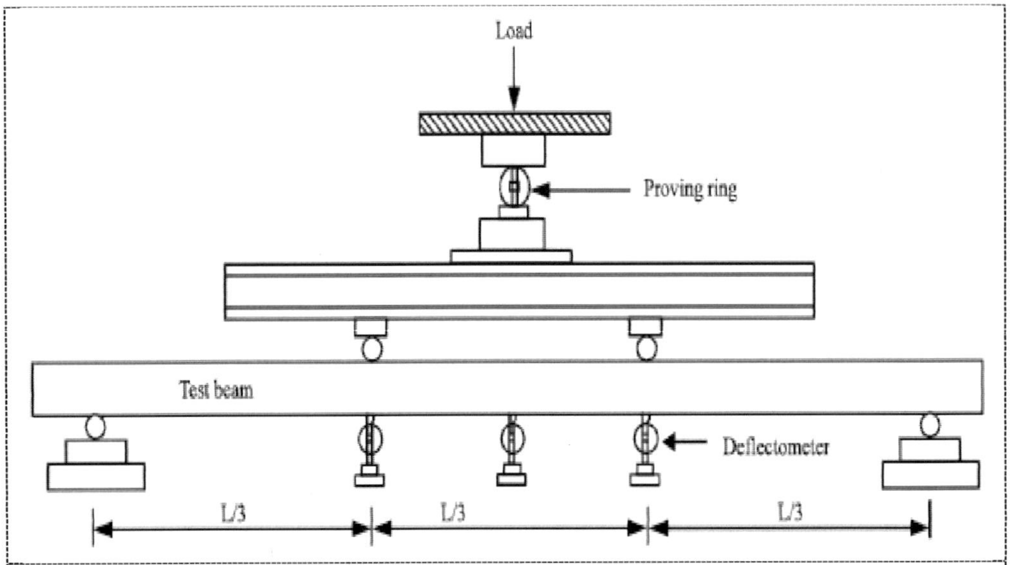

Figure 1. Test set up for flexure

3. Results and Discussion

3.1. Compressive strength

From the test results of compressive strength for concrete, it was found that the mix (FA+SF)-2 ie., 50% FA and 10% SF had higher compressive strength than that of the other mixes both in seven days and twenty eight days curing. When SF was added to concrete, initially it remained inert and then started hydrating, which produced two chemical compounds: Calcium Silicate Hydrate (CSH), formed as crystalline structure, which was responsible for the strength development and Calcium Hydroxide (CH), a by-product also called as free lime which was responsible for lining the available pores within concrete as a filler.

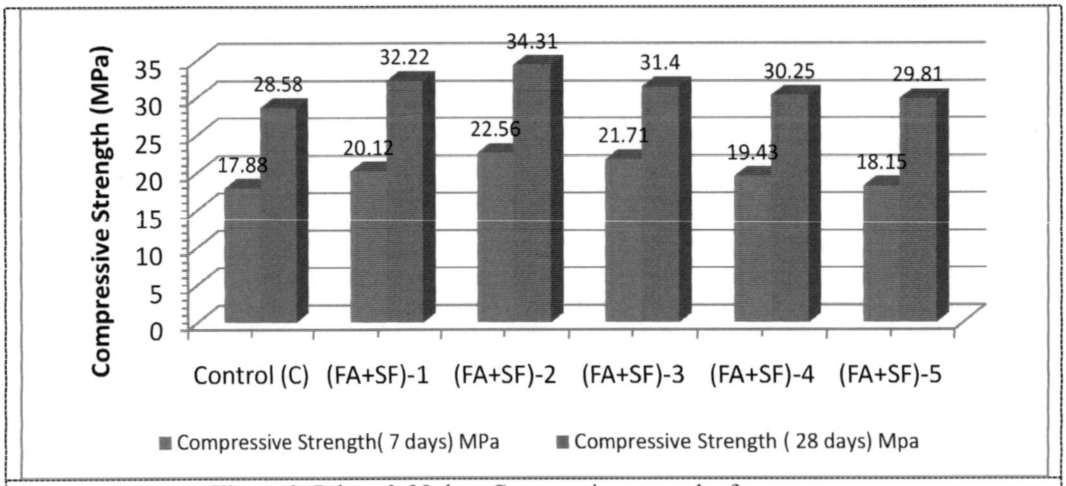

Figure 2. 7 days & 28 days Compressive strength of concrete

Pozzolanic reaction occurred between SF and the CH, producing additional CSH in many of the voids around hydrated cement particles which induced much denser matrix which was in turn improved the compressive, flexural and bond strength. Hence, less CH with increasing binding (cemen) compounds produced good strength concrete. However, in concrete with SF and FA, the interfacial zone became stronger, more homogeneous and denser. Hence, the cracks usually traverse around the coarse aggregates during the test on controlled specimens.

3.2. Split tensile strength
The findings in tensile strength gave the similar results as that of compressive strength. For FA and SF replacements there was s significant increase the split tensile strength with that of control specimen. It was noticed that the rate of mean increase of tensile strength (6.2%) was found to be low when compared with that of compressive strength (14.06%). There was an increase in tensile strength for the mix (FA+SF)-2 by 9.8% when compared with that of control mix.

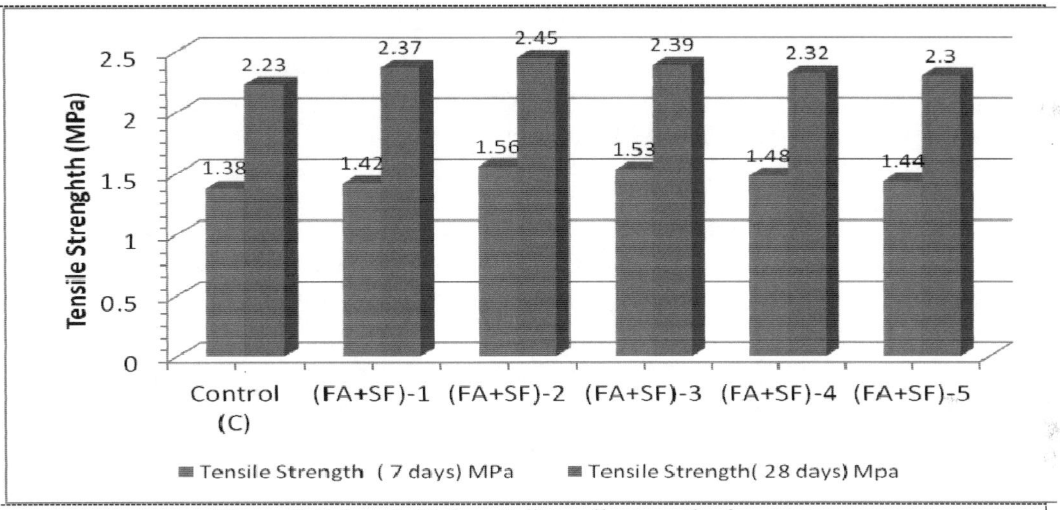

Figure 3. 7 days & 28 days Split tensile strength of concrete

3.3. Flexural behaviour
3.3.1. Ultimate load. The ultimate load for beams of different mix proportions were presented in Table 6. It can be seen that ultimate load of 66.15 kN has been achieved for mix (FA+SF)-2 which was 9.70% higher than M25 control beams respectively.

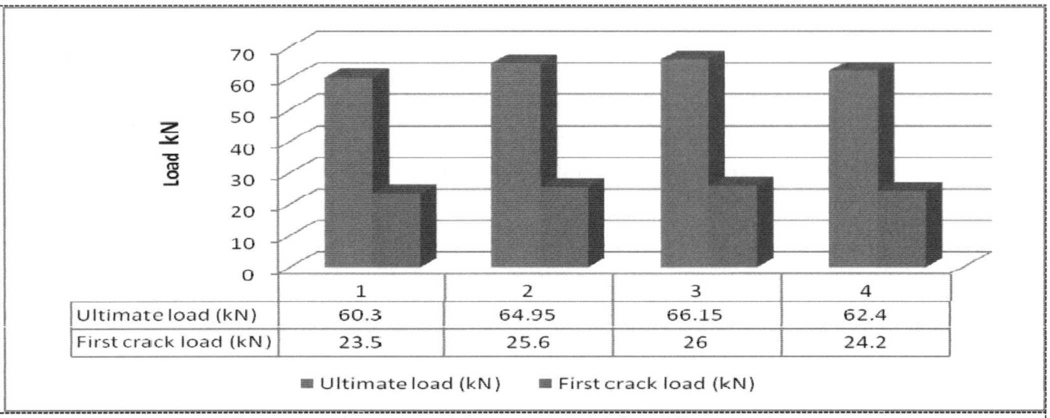

Figure 4. Ultimate and first crack load for different mixes

3.3.2. Load – Deflection behaviour. At every load increment, it was noted that the beam with optimum mix with Super Plasticizer has taken higher deflection values in comparison to that of control beam as shown in Figure 5. This showed that the replacement of cement by SF and FA leads to ductile behaviour.

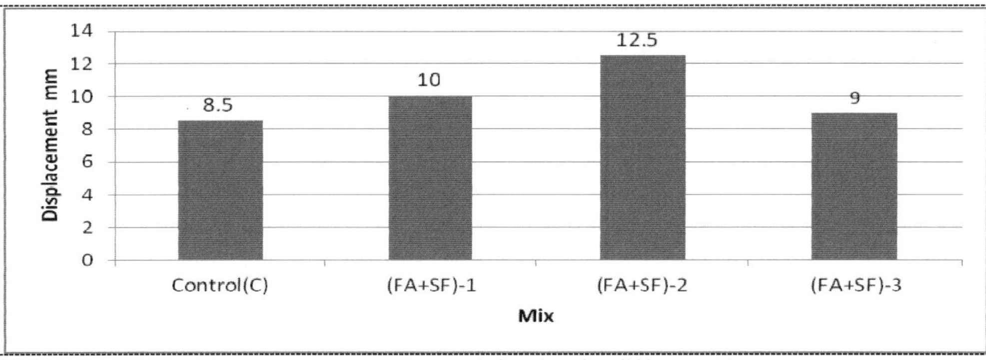

Figure 5. Maximum deflection for different mixes

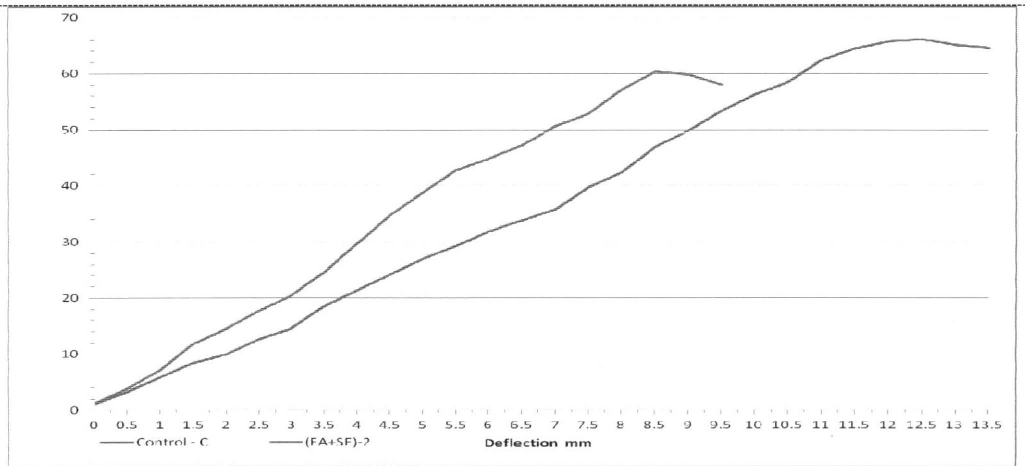

Figure 6. Load – Deflection behavior of beam

The load deflection graph (Figure 6) was plotted for the beam with control mix and for the beam with optimum mix of HVFA concrete and SF. The maximum displacement observed in control mix is 8.5 mm whereas in SF combined HVFA beam, it is 12.5 mm. From the Figure 6, it was understood that the addition of SF increased the load carrying capacity of the beam.

3.3.3. Crack pattern and failure mode. In this experimental work, the crack propagation was observed, measured and marked at every 5 kN load increment up to the failure, as shown in the Figure 7(a) and 7(b) below.

7(a). Crack pattern for M25 Control beam

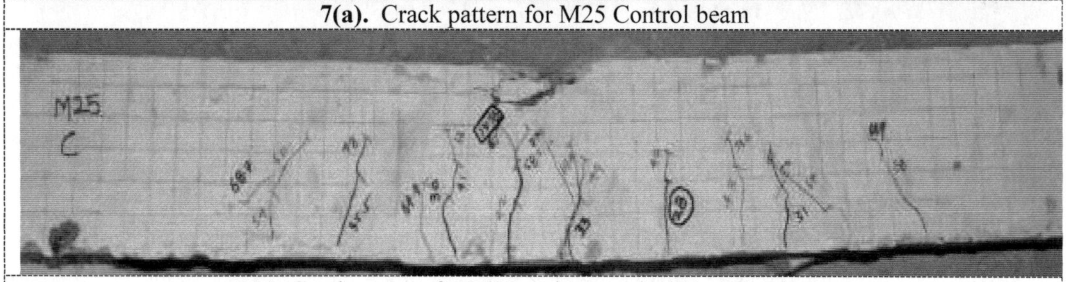

7(b). Crack pattern for M25 optimum mix (FA+SF)-2 beam

Figure 7. Crack pattern and failure mode for tested beam specimens

In the constant moment zone, flexural cracks were formed first and these cracks were extended vertically upwards and developed gradually wide as the load increased. As the load increased, the extreme fibre stresses in bending increased until the tensile strength of concrete was reached. This caused flexural cracking initially in the constant moment region. The final failure of the beam was described by large strains in the steel reinforcement and considerable deflection near collapse followed by extensive cracking.

4. Conclusions

From the results of experimental work carried out, the following conclusions could be drawn:

- ✓ The optimum dose of FA and SF in combination was found to be 50% and 10% (by weight) respectively as a partial replacement of binder (cement) for M25 grade of concrete for a w/b ratio of 0.35 to achieve high compressive strength at the end of seven and twenty eight days.
- ✓ The inclusion of FA and SF resulted in faster strength development of concrete at early age of curing. The average increase in Compressive strength at 7days was obtained to be 14.06% .There was an gradual increase of 20% in 28 days compressive strength for optimum mix (FA+SF)-2 over the control mix.
- ✓ The split tensile strength of concrete for all mixes using FA and SF combination as partial replacement of cement was higher than control mix.
- ✓ The ultimate load and first crack load for the beams with optimum mix was higher than that of control beam for both grades due to the immediate filler effect, and the acceleration of cement

hydration.

- ✓ Maximum deflection for the beam with mix (FA+SF)-2 was increased by 7.5% over the beam with control mix.
- ✓ These results encouraged the use of FA and SF, as pozzolanic material for partial cement replacement of binder, in producing concrete with superior engineering properties.
- ✓ This led to the reduction in pollution to the environment. Also the technical and economic issues caused by cement production could be sorted out.

Acknowledgement
The authors thank the SASTRA University for providing excellent computing facilities. The authors also thank the anonymous reviewers for their valuable suggestions, which resulted in a more accurate and complete presentation of the present work.

References

[1] Chung-Ho Huang Shu-Ken Lin, Chao-shun Chang, How – Ji Shen 2013 Mix proportions and mechanical properties of concrete containing very high-volume of Class F fly ash. *Constr. Build. Mater.* **46** 71-8.

[2] Nath and Sarker 2011 Effect of Fly Ash on the Durability Properties of High Strength Concrete. *Procedia Eng.* **14** 1149-56.

[3] Siddique Rafat, William Prince, Siham Kamali 2007 Influence of utilization of high-volume of Class F fly ash on the abrasion resistance of concrete. *Leonardo Electron. J. Pract. Technol.* **10** 13-28.

[4] Alaa M Rashad, H EI-D H Seleem and A F Shaheen 2014 Effect of silica fume and slag on compressive strength and abrasion resistance of HVFA Concrete, *Int. J. Concr. Struct. Mater. (IJCSM)* **8**(1) 69-81.

[5] Alaa M.Rashad 2015 Potential Use of Silica Fume Coupled with Slag in HVFA Concrete Exposed to Elevated Temperatures, *ASCE J.Mater. Civ. Eng.* Online publication.

[6] Nazeer M, Anupama P S 2014 Strength and Durability Studies on Silica Fume Modified High-Volume Fly Ash Concrete. *Int. J. Adv. Res. Eng. Technol.* **5**(10) 55-68.

[7] Steve Supit , Faiz Uddin Ahmed Shaikh 2015 Durability Properties of High Volume Fly Ash Concrete containing Nano – silica. *Mater. Struct.* **48** 2431-45.

[8] Min-Hong Zhang, Jahidul Islam, Muta Harah Zakaria 2012 Use of Nano silica to reduce setting time and increase early strength of concretes with high volumes of fly ash or slag. *Constr. Build. Mater.* **29** 573-80.

[9] Shaikh F U A, Supit S W M, and Sarker P K 2014 A study on the effect of Nano silica on compressive strength of high volume fly ash mortars and concretes. *Mater. Des.* ***60*** 433-42.

[10] IS 8112-2013, Indian standard Ordinary Portland Cement, 43 grade – specification, *Bureau of Indian Standards*, New Delhi, India.

[11] IS 2386 (Part III) – 1963, Indian Standard Methods of Test for aggregates for Concrete (Part III) Specific gravity, Density, Voids, Absorption and Bulking, *Bureau of Indian Standards*, New Delhi, India.

[12] IS 9103:1999, Specification for Concrete Admixtures, *Bureau of Indian Standards*, New Delhi, India.

[13] ASTM C496 / C496M - 11 Standard Test Method for Splitting Tensile Strength of Cylindrical Concrete Specimens.

[14] IS 10262- 2009, Indian Standard for Concrete Mix Proportioning – Guidelines, *Bureau of Indian Standards*, New Delhi, India.

[15] IS 456 – 2000, Indian Standard Plain and Reinforced Concrete - Code of Practice, *Bureau of Indian Standards*, New Delhi, India.

[16] IS 516 – 1959, Indian Standard Methods of Tests for Strength of Concrete, *Bureau of Indian Standards*, New Delhi, India.

IOP Conf. Series: Earth and Environmental Science **80** (2017) 012010 doi:10.1088/1755-1315/80/1/012010

Study on the strength characteristics of High strength concrete with Micro steel fibers

K Gowdham , A Sumathi* and K Saravana Raja Mohan

School of Civil Engineering, SASTRA University, Thanjavur, 613 401

*Email: sumathi@civil.sastra.edu

Abstract. The study of High Strength Concrete (HSC) has become interesting as concrete structures grow taller and larger. The usage of HSC in structures has been increased worldwide and has begun to make an impact in India. Ordinary cementitious materials are weak under tensile loads and fiber reinforced cementitious composites (FRCCs) have been developed to improve this weak point. High Strength concrete containing Alccofine as mineral admixture and reinforced with micro steel fibers were cast and tested to study the mechanical properties. The concrete were designed to have compressive strength of 60 MPa. Mixtures containing 0% and 10% replacement of cement by Alccofine and with 1%, 2% and 3% of micro steel fibers by weight of concrete were prepared. Mixtures incorporating Alccofine with fibers developed marginal increase in strength properties at all curing days when compared to control concrete.
Key words: HSC, Alccofine, Micro steel fibers, Fiber content, Compressive strength, split tensile strength.

1. Introduction

The study of high strength concrete has become interesting as concrete structures grow taller and larger. The usage of high strength concrete in structures has been increasing worldwide and has begun to make an impact in India. A few years ago, a characteristics compressive strength 40 MPa would have been considered high in India but now it has become normal phenomena. Presently, concrete with a 28 days curing and characteristics cube strength of 60 MPa and above will be considered as HSC. The achievement of such HSC has been made possible primarily through the introduction of supplementary materials viz. silica fume, slag, Alccofine, fly ash etc. The addition of micro steel fibers to concrete is expected to arrest cracks, improved ductility and post cracking performance of concrete and improve bond and corrosion resistance due to superior performance.Today Engineers have shown keen interest to introduce ingredients in concrete containing mineral and chemical admixtures. The purpose of adding Alccofine and steel fibers to concrete is to improve the strength, crack resistance, strain capacity and toughness parameter of concrete. Shahid Iqbal et.al [1] found out that by varying in the addition of micro steel fibers form 0 to 1.25% the concrete increases its compressive strength in the beginning and then reduces but the split tensile strength is gradually increased till final. Saurav and Ashok Kumar Gupta [2] found that by addition of a supplementary cementatious material alccofine the compressive strength can be increased by varying the percentages from 0 to 18% the optimum was found to be 13% for M50 grade. Jianming Gao et.al [3] found that addition of steel fibre to light weight improved its strength characteristics improving the compressive strength to some extent and by increasing the split tensile and flexural strength largely. Thus they came to a conclusion that by varying the fiber content in light weight concrete by 1-1.5% is extremely effective in improving the strength and toughness of concrete. Song et. al [4] found that for a high strength concrete by addition of 1.5% fibre showed maximum strength and above that percentage showed to decrease in strength but the split tensile strength showed increasing in strength by increasing in fibre volume fractions. Fouzia and Yatin et. al [5&6] found that by varying the percentages in the replacement of cement by metakaolin showed improved strength but the maximum strength was achieved when the cement was replaced by 10% of alccofine. Zoubir et.al [7] found that by addition of sulfuric acid the changes in weight and compressive strength was checked for 30, 60, 90,120 and 180 days

2. Experimental Programme

Experimental work was carried out to study the basic strength properties of the alccofine blended high strength concrete of M60 grade prepared with micro steel fibers at the percentages of 1,2 and 3 percent by weight of cement. A design mix was done as per ACI method [8] 1:0.56:1.89 with water- binder ratio of 0.33, and aggregate binder ratio of 2.0 was adopted in this investigation. In the case of super-plasticizer, the water-binder ratio was reduced from 0.33 to 0.3 and then super-plasticizer was added between 0.8 to 1.2 per cent depending upon the quantity of cement used.

Table 1.Materials and its properties

Property of Cement	Value	Grade/Type/ Source
Specific Gravity	3.12	OPC-53
Initial and Final setting time	60 and 450 minutes	
Property of Fine aggregate		
Specific Gravity	2.65	Locally available river sand passing through 4.75 mm and retained on 2.36 mm sieve was used .
Fineness Modulus	3	
Coarse aggregate		
Specific Gravity	2.7	Locally available 20 mm size was used
Fineness modulus	6.5	
Water Absorption (%)	3	
Water	Ordinary portable water was used	-
Alccofine		As per Manufacturer's Manual
Specific Gravity	2.9	
Bulk density	600-700 kg/m^3	
CaO	31-33%	
SiO_2	33-35%	
Al_2O_3	23-25%	
Micro steel fiber		Go Green Technologies, Chennai
Aspect ratio (l/d)	33-35	
Yield strength	2100 MPa	As per Manufacturer's Manual
Elastic Modulus	2.1 x 10^5 MPa	
Super plasticizer(HRWRA)	Specific gravity- 1.220-1.225 pH $\geq$6 Aspect Light Brown Liquid Relative Density 1.08 $\pm$ 0.01 at 25°c	Master Glenium Sky 8233, BASF Chemical Company Limited, Mumbai. As per Manufacturer's Manual

During the preparation of concrete mix, the water quantity was divided approximately into three parts and the admixture was added with the third part so that the interaction of admixture with the water would be more effective than adding it at the beginning itself. Subsequently, after this 10% of alccofine were added as cement replacement. Afterwards, micro steel fiber was added to this by 1, 2 and 3% by weight of concrete. Cubes of size 100 mm × 100 mm × 100 mm, cylinders of 100 mm diameter and 200mm long, 100mm diameter and 50mm long were cast. All the specimens were removed from the mould after giving a minimum period of 24 hours as setting time and curing was done using the curing tank. After 28 days the specimen was then taken from the tank for testing and allowed for surface drying for a minimum period of 5-6 hours. It was placed in ASTM Compressive Testing Machine and allowed to transfer the load at the rate of 2.9 kN/sec. The tests were conducted on concrete specimens for 7 and 28 days of curing. Cubes were tested to evaluate compressive strength of concrete as per IS516 [9], cylinders were tested as per ASTMC496 [10] to determine split tensile strength.

3. Results and Discussion

3.1 Effect of Accofine and Micro steel fiber on Compressive strength

The compressive strength of concrete cubes made with various weight fraction of steel fiber was plotted against different days of curing are shown in figure 1. Three specimens were tested to assess the compressive strength for control and 10% alccofine with steel fiber content (0%, 1%, 2% and 3%) in weight fractions. From results it is clear that the maximum compressive strength is attained for mixes containing 10% alccofine along with 2% fibers at 7 and 28 days of curing and the strength has enhanced for all fiber specimens compared to control concrete. From figure 2, the percentage improvements in compressive strength compared to control concrete are 7.35%, 23.93%, 32.51%, 25.13% and 7.62%, 24.48%, 33.56%, 28.29% at 7 and 28 days respectively.

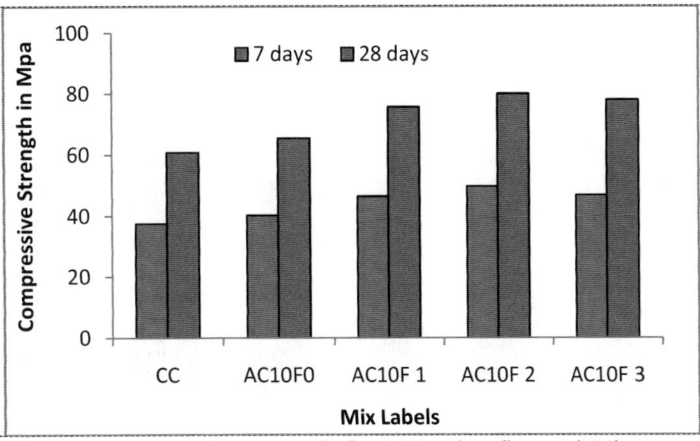

Figure 1. Compressive strength for 10% Alccofine and Micro steel fiber (%)

IOP Conf. Series: Earth and Environmental Science **80** (2017) 012010 doi:10.1088/1755-1315/80/1/012010

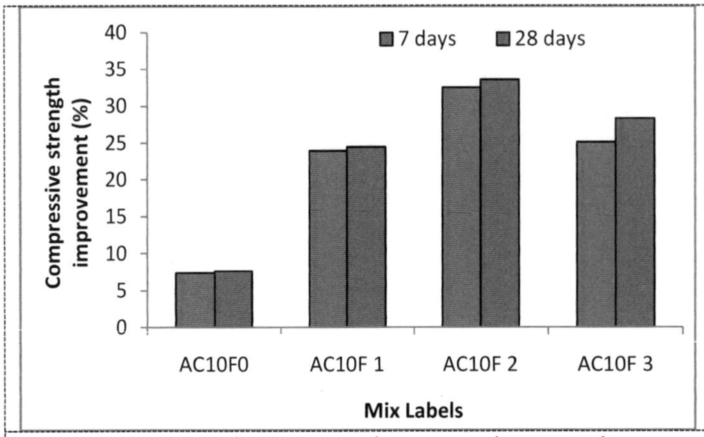

Figure 2. Percentage improvement in compressive strength

3.2 Effect of Alccofine and Micro steel fiber on Split tensile strength

Three specimens were tested to assess the split tensile strength with steel fiber content (0%, 1%, 2% and 3%) in weight fractions. From figure 3 there is an increase in split tensile strength of concrete with the increase in fiber content. The addition of steel fiber to HSC in weight fractions of 3% caused an maximum increase in strength compared to control concrete 7 and 28 days of curing. From figure 4, the percentage improvements in split tensile strength compared to control concrete are 12.12%, 27.88%, 46.46%, 60.8% and 34.29%, 41.76%, 65.57%, 73% at 7 and 28 days of curing respectively.

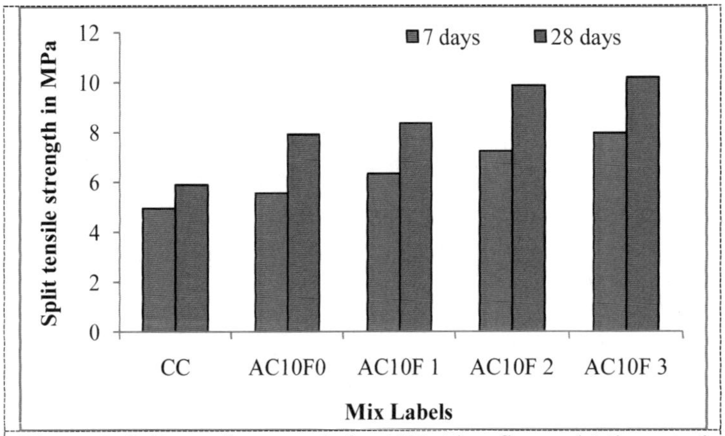

Figure 3. Split tensile strength for 10% Alccofine and Micro steel fiber(%)

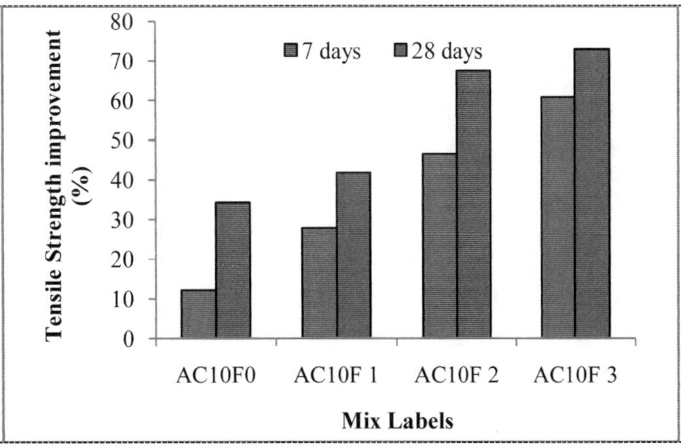

Figure 4. Percentage improvement in split tensile strength

4. Conclusions

From the result of this experimental investigation, the following conclusions were made on the addition of Alccofine and micro steel fibers.

1. The super-plasticizer for this high strength concrete was found to be 0.8 to 1.2 percent to maintain the adequate workability for Alccofine concrete and Alccofine with micro steel fiber concrete.
2. The compressive strength increases for 10% of Alccofine compared with concrete at 7 and 28 days of curing.
3. The compressive and split tensile strength of concrete increases with the addition of steel fiber in Alccofine concrete. The strength improvement was calculated for 7 and 28 days, at all fiber content exhibit superior strength improvement for split tensile than compressive strength.

Acknowledgements

The authors extend their sincere thanks to Vice Chancellor of SASTRA University for providing us laboratory facilities in school of civil engineering to successfully complete this work

References

[1] Shahid Iqbal, Klaus Holschemacher and Thomas A Bier 2015 Effect of change in micro steel fiber content on properties of High strength Steel fiber reinforced Lightweight Self-Compacting Concrete. *Procedia Eng.* **122**(3) 88-94.

[2] Saurav and Ashok Kumar Gupta. 2014 Experimental study of strength relationship of concrete cube and concrete cylinder using ultrafine slag Alccofine. *Int. J. Sci.Eng. Res.* **5**(5) 102-7.

[3] Jianming Gao,Wei Sun and Keiji Morino 1997 Mechanical Properties of Steel Fiber-reinforced, High strength, Lightweight Concrete. *Cem. Concr. Compos.* **19**(4) 307-13.

[4] Song P S and Hwang S 2004 Mechanical properties of high-strength steel fiber-reinforced concrete, *Constr. Build. Mater.* **18**(9) 669-73

[5] Fouzia Shaheen and Mohammed Samiuddin Fazil 2015Effect of metakaolin and alccofine on strength of concrete. *IRACST – Eng. Sci. Tech. Int. J.* **5**(4) 94-100.

[6] Yatin H Patel 2013 Study on durability of high performance concrete with Alccofine and fly Ash. *Int. J. Adv. Eng. Res. Struct.* **2**(3) 154-7.

[7] Zoubir Makhloufi, Madani Bederina, Mohamed Bouhicha and El-Hadi Kadri 2014 Effect of mineral admixtures on resistance to sulfuric acid solution of mortars with quaternary binders, *Phy. Procedia* **55** 329-35

[8] ACI 211.1-91, Standard practice for Selecting proportions for Normal, Heavy weight, and Mass concrete, Reapproved 2002

[9] IS 516. Indian Standard Methods of Tests for Strength of Concrete.2004, *Bureau of Indian Standards.*
[10] ASTM C496. Standard Test Method for Splitting Tensile Strength of Cylindrical Concrete Specimens 2011, *ASTM International.*
[11] IS 12269-1997. Specifications for 53 grade Ordinary Portland Cement. *Bureau of Indian Standards*, New Delhi
[12] IS 383-1970- Specification of coarse and fine aggregates from natural sources for concrete- *Bureau of Indian Standards (BIS)*: New Delhi
[13] IS 2386-1986. Methods of test for aggregates for concrete. *Bureau of Indian standards*, New Delhi.

IOP Conf. Series: Earth and Environmental Science **80** (2017) 012011 doi:10.1088/1755-1315/80/1/012011

Effect of Molarity of Sodium Hydroxide and Curing Method on the Compressive Strength of Ternary Blend Geopolymer Concrete

V Sathish Kumar*, N Ganesan and P V Indira

Department of Civil Engineering, NIT Calicut, Kerala, India

*Email: sathishkumarcivil@yahoo.co.in

Abstract. Concrete plays a vital role in the development of infrastructure and buildings all over the world. Geopolymer based cement-less concrete is one of the current findings in the construction industry which leads to a green environment. This research paper deals with the results of the use of Fly ash (FA), Ground Granulated Blast Furnace Slag (GGBS) and Metakaolin (MK) as a ternary blend source material in Geopolymer concrete (GPC). The aspects that govern the compressive strength of GPC like the proportion of source material, Molarity of Sodium Hydroxide (NaOH) and Curing methods were investigated. The purpose of this research is to optimise the local waste material and use them effectively as a ternary blend in GPC. Seven combinations of binder were made in this study with replacement of FA with GGBS and MK by 35%, 30%, 25%, 20%, 15%, 10%, 5% and 5%, 10%, 15%, 20%, 25%, 30%, 35% respectively. The molarity of NaOH solution was varied by 12M, 14M and 16M and two types of curing method were adopted, viz. Hot air oven curing and closed steam curing for 24 hours at 60°C (140°F). The samples were kept at ambient temperature till testing. The compressive strength was obtained after 7 days and 28 days for the GPC cubes. The test data reveals that the ternary blend GPC with molarity 14M cured by hot air oven produces the maximum compressive strength. It was also observed that the compressive strength of the oven cured GPC is approximately 10% higher than the steam cured GPC using the ternary blend.
Key words: *Ternary Blend, Geopolymer concrete, Compressive strength, Curing method, Alkaline activators.*

1. Introduction

Cement is the only material used in the concrete which is manufactured in the factory, and it is the key ingredient which binds all other materials in the concrete. The production of cement leads to many environmental effects due to the emission of a large amount of carbon dioxide (CO_2) during the manufacturing process. It is mandatory to develop an alternate binder for the concrete to reduce the effect of CO_2 from the cement industry for a sustainable environment. GPC is considered as one of the best alternatives as it results in cementless concrete which leads to an Eco-friendly construction material. It can also take advantage of using the Industrial waste in the concrete without any harmful effects to the environment.

GPC results from the source material that is rich in Silica and Alumina with an alkaline activator solution by the polymerisation process. The properties of the GPC are mainly depending on the composition of the source materials, type of curing and the alkaline activator. The compressive strength of the dry-cured GPC is approximately 15% larger than that of steam-cured fly ash based GPC [1]. It was reported that using GGBS with the addition of Metakaolinite affects the setting time of geopolymer mix by curing temperature, composition of source material and the type of alkaline liquid [2].

Among the source materials, Fly ash draws the attention of many researchers because of its availability and other dumping issues related to it. Metakaolin is formed from kaolinite clay which requires very low temperature and releases less Co_2 than Portland cement. It was reported that increase in GGBS content in GPC will improve the compressive strength [3]. The aluminous and siliceous structures in FA, GGBS and MK will have a higher dissolution rate in the alkaline solution. The concrete developed from these source materials will have better properties than the conventional concrete and also possess low carbon footprint.

2. Materials

2.1 Fly ash

Low-calcium Class F FA confirms to IS 3812:2003 [4] collected from Mettur Thermal Power Station, Tamil Nadu was used as a primary material for the ternary blend. The colour of the FA is dark grey, and the average particle size is 75 microns. It has a specific gravity of 2.30 and mainly consists of 55.36% 0f SiO_2 and 27.75% of Al_2O_3.

2.2 GGBS

GGBS, confirming to BS 6699:1992 [5] was used in this work. The colour of the GGBS is 0ff-white, and the mean size of the particle is 30 microns. The specific gravity of the GGBS is 2.88. It is mainly composed of 77.36% of combined CaO, MgO and SiO_2.

2.3 Metakaolin

MK was also used in the ternary blend which contains more than 92% of combined SiO_2, Al_2O_3 and Fe_2O_3 elements. It has an appearance of Creamish Ivory Powder and the average particle size is 2-3 microns. The MK used in this work is having a specific gravity of 2.56.

2.4 Fine aggregate

M-sand passing through 4.75mm IS sieve confirming to zone II of IS 383:1970 (reaffirmed 2002) [6] was used as a fine aggregate. The fineness modulus and the specific gravity of the fine aggregate were found to be 2.92 and 2.39 respectively.

2.5 Coarse aggregate

Locally available granite stone with a maximum size of 12.5mm and a specific gravity of 2.78 was used. The coarse aggregate was having a fineness modulus of 6.92.

2.6 Alkaline liquid

A combination of Sodium Silicate (Na_2SiO_3) and Sodium Hydroxide (NaOH) was used as an alkaline liquid. The Na_2SiO_3 solution consists 8% of Na_2O, 28% of SiO_2 and 64% of water by mass. NaOH was in pellets form with 99% purity.

2.7 Superplasticizer

Naphthalene based superplasticizer (SP), Conplast SP 430 with a specific gravity of 1.2 was used for the better workability of the GPC.

IOP Conf. Series: Earth and Environmental Science **80** (2017) 012011 doi:10.1088/1755-1315/80/1/012011

3. Mixing of GPC

The mix proportion is based on the recommendations given by Prof.B V Rangan [4]. The ratio of Na_2SiO_3 to NaOH was kept constant as 2.5 and the alkaline liquid to binder ratio was also chosen as 0.3. Seven combinations of the ternary binder were selected and mixed based on volume proportion. The NaOH solution was prepared by mixing water with NaOH pellets. To prepare a 12M concentration of NaOH, 480 grams (Molarity x Molecular weight) of NaOH pellets was dissolved in water and makeup to 1 litre. It is recommended to mix the NaOH solution with Na_2SiO_3 solution 24 hours before mixing the concrete to ensure the reactivity of the solution. The mass of combined aggregates was taken as 77% of the total mass of GPC mix. SP was taken as 1.5% by the weight of binder and the water to binder ratio was selected as 0.2 for all the GPC mix for better workability. Table 1 shows the summary of the ternary blend GPC mix proportion.

Table 1. Ternary Blend GPC Mixture Proportions adopted in this work.

Binder proportion (%)	FA	GGBS	MK	Coarse agg.	Fine agg.	NaOH sol.	Na_2SiO_3	SP	Water
	kg/m^3								
F60-G35-M05	234.15	169.25	21.21	1293.60	554.4	36.4	90.99	6.37	84.92
F60-G30-M10	235.80	146.10	42.72	1293.60	554.4	36.4	90.99	6.37	84.92
F60-G25-M15	237.47	122.61	64.53	1293.60	554.4	36.4	90.99	6.37	84.92
F60-G20-M20	239.17	98.79	86.66	1293.60	554.4	36.4	90.99	6.37	84.92
F60-G15-M25	240.89	74.62	109.10	1293.60	554.4	36.4	90.99	6.37	84.92
F60-G10-M30	242.64	50.11	131.87	1293.60	554.4	36.4	90.99	6.37	84.92
F60-G05-M35	244.41	25.24	154.97	1293.60	554.4	36.4	90.99	6.37	84.92

4. Casting and Curing

The dry materials were mixed in the horizontal concrete mixer for about 3-4 minutes. The Alkaline activator and the SP along with water were added to the mix and mixed for another 4 minutes. A total number of 126 ternary blend GPC cubes of size 100mm were cast and cured in a Hot air oven at 60°C for 24 hours. Another set of 42 numbers of ternary blend GPC cubes of size 100mm were cast with different proportion of source material with 14M concentration of NaOH (Maximum compressive strength) and cured in closed steam chamber at 60°C for 24 hours. 6 cubes were cast in each proportion and 3 cubes were tested for each 7days and 28days compressive strength. The GPC concrete cubes moulds were vibrated using table vibrator and covered with polythene film to reduce the water loss during the curing process. After curing the GPC cubes were removed from the mould and left to cool to room temperature until testing. Figure 1 and 2 shows the curing of GPC cubes in a Hot air oven and closed steam chamber respectively.

Figure 1. Hot air oven used for curing.

Figure 2. Closed steam chamber used for curing

5. Results and Discussion

The GPC cubes were tested in a compressive testing machine and the mean compressive strength of three cubes was tabulated for each proportion. Table 2 shows the results of the compressive strength for different mix proportion of GPC cubes kept in Hot air oven curing and closed steam chamber curing. The maximum compressive strength of GPC was achieved by Hot air oven curing and with a molarity of 14M. The study shows that the improvement in the compressive strength of ternary blend GPC cubes by Hot air oven curing over closed chamber curing was not affected significantly by the different proportions of the source material.

Table 2. Compressive Strength Results for Different Mixtures of Ternary Blend GPC cubes.

Binder proportion (%)	Molarity of NaOH	Average Compressive Strength (N/mm^2)			
		Hot air oven curing		Closed steam Curing	
		7 days	28 days	7 days	28 days
	12	43.83	52.30	-	-
F60-G35-M05	14	50.17	56.87	44.78	51.16
	16	49.52	53.12	-	-
	12	52.63	56.39	-	-
F60-G30-M10	14	52.46	59.82	47.56	53.12
	16	52.46	56.55	-	-
	12	54.59	58.35	-	-
F60-G25-M15	14	59.33	63.74	52.63	57.37
	16	54.75	59.66	-	-
	12	42.33	48.54	-	-
F60-G20-M20	14	45.60	50.50	41.84	45.93
	16	41.35	46.74	-	-
	12	35.14	42.33	-	-
F60-G15-M25	14	31.54	36.93	28.93	33.51
	16	29.26	32.20	-	-
	12	28.18	33.34	-	-
F60-G10-M30	14	30.24	34.16	27.46	31.22
	16	20.27	24.18	-	-
	12	20.10	25.66	-	-
F60-G05-M35	14	23.21	27.14	21.08	24.52
	16	14.22	19.78	-	-

It can be noted that with an increase in the molarity of NaOH there is an increase in compressive strength up to 14M and then a further increase in molarity decreases the compressive strength. The maximum compressive strength was obtained with the ternary blend contains 60% FA, 25% GGBS and 15% MK. It can be observed that increase in MK content up to 15% increases the compressive strength and further results in degradation in the strength. The density of the GPC was found to be in the range of 2400 to 2500 kg/m^3 and it was also noted that curing method does not affect the density of the samples. The lowest compressive strength was found to be 19.78 MPa which contains 60% FA, 05% GGBS and 35% MK with NaOH molarity of 16M.

6. Conclusion

The compressive strength of ternary blend GPC by Hot air oven and closed steam chamber curing was obtained. The following findings were drawn from the test results with different molarity of NaOH and the curing methods.

1. The compressive strength of ternary blend GPC increased with the increase in GGBS content up to 25% and further addition of GGBS decreases the strength.
2. In both types of curing, 80-90% of the 28 days compressive strength was achieved within 7 days in which the samples were cured for 24 hours at 60°C and then kept in ambient temperature till testing.
3. The compressive strength achieved by hot air oven curing was 10% higher than the closed steam curing for the ternary blend GPC.
4. The maximum compressive strength was obtained with 14M molarity of NaOH with ternary blend of 60% FA, 25% GGBS and 15% MK.
5. It can be concluded that the ternary blend GPC using FA, GGBS and MK can be used effectively as an alternate material to concrete for a sustainable development.

References

[1] Rangan B V 2010 Fly ash based geopolymer concrete. *Geopolymer cement and concrete Int. workshop*, India, 68-106.

[2] Cheng T W and Chiu J P 2003 Fire-resistant gopolymer produced by granulated blast furnace slag. *Miner. Eng.* **16** 205-10.

[3] Rajamane N P, Nataraja M C and Jeyalakshmi R 2014 Pozzolanic industrial waste based Geopolymer concretes with low carbon footprint. *The Indian Concr. J.* **88**(7) 49-68.

[4] IS 3812 2003 Pulvarized fuel ash – Specification. *Bureau of Indian Standards*.

[5] BS 6699 1992 Ground granulated blast furnace slag for use with Portland cement - Specification.

[6] IS 383 1970 (reaffirmed 2002) Indian standard code of practice for specification for coarse and fine aggregate from natural sources for concrete. *Bureau of Indian Standards*.

[7] Davidovits J 1999 Chemistry of Geopolymeric Systems, Terminology. *Proc. Geopolymer'99 Int. Conf.* France.

[8] Davidovits J 1991 Geopolymers: Inorganic polymeric new materials. *J. Therm. Anal.* **37**(8) 1633-56.

[9] Ganesan N, Indira P V and Anjanasanthakumar 2015 Effect of Steel Fibres on Shear Strength of Geopolymer Concrete Beams. *J. Stru. Eng.* **41**(6) 575-82.

[10] Krishna Rao M V and Rathish Kumar P 2016 Effect of Sodium Hydroxide concentration and curing type on the properties of self-compacting Geoolymer concrete. *Indian Concr. Inst. J.* **3** 17-24.

[11] Rangan B V and Hardjito 2005 Development and properties of Low Calcium Fly ash based Geopolymer concrete. *Research Report GC1*, Faculty of Engineering, Curtin University of Technology, Perth.

[12] Rangan B V and Wallah S E 2006 Low calcium based geopolymer concrete long term properties. *Research Report GC2*, Faculty of Engineering, Curtin University of Technology, Perth.

[13] Rangan B V and Sumajouw M D J 2006 Low Calcium fly ash based geopolymer concrete : Reinforced beams and Columns. *Research Report GC3*, Faculty of Engineering, Curtin University of Technology, Perth.

[14] Ranjbar N, Mehali M, Behnia A, Alengaram U J and Jumaat M Z 2014 Compressive strength and micro structural analysis of Fly ash/Palm oil fuel ash based geopolymer mortar. *Mater Des.* **59** 532-9.

[15] Sanjay Kumar and Rakesh Kumar 2014 Geopolymer: Cement for low carbon economy. *The Indian Concr. J.* **88**(7) 29-37.

[16] Shi Caijun, Krivenkopavel V and Roy della 2006 Alkali-Activated Cements and Concretes. *Taylor & Francis*, New York.

[17] Yunshengzhang, Wei sun, Li zingjin and Xiangmingzhou 2009 Geopolymer extruded composites with incorporated fly ash and polyvinyl alcohol short fiber. *ACI Mater. J.* **106**(1) 3-10.

[18] van Jaarsveld J G S, van Deventer J S J and Lukey G C 2002 The effect of composition and temperature on the properties of Fly ash and kaolinite based geopolymers. *The Chem. Eng. J.* **89**(1-3) 63-73.

[19] Zhang hai-yan, Kodurvenkatesh, Wu bo, Cao liang and Qi shu-liang 2015 Effect of Carbon Fibers on Thermal and Mechanical Properties of Metakaolin Fly-Ash-Based Geopolymers. *ACI Mater. J.* **112**(3) 375-84.

[20] Zhang hai-yan, Kodurvenkatesh, Wu bo, Cao liang and Qi shu-liang 2015 Comparative Thermal and Mechanical Performance of Geopolymers derived from Metakaolin and Fly Ash. *J. Mater. Civil Eng.* **28**(2) 1-12.

IOP Conf. Series: Earth and Environmental Science **80** (2017) 012012 doi:10.1088/1755-1315/80/1/012012

Experimental studies on effect of cow dung ash (pozzolanic binder) and coconut fiber on strengthproperties of concrete

C Venkatasubramanian*, D Muthu, G Aswini, G Nandhini and K Muhilini

School of Civil Engineering,SASTRA University, Thanjavur – 613 401. India.

*Email: cv@civil.sastra.edu

Abstract. The studies on durability of concrete have attracted attention in the recent years and its long term strength depends on quality of ingredients used in production of concrete. Now a days, the availability of ingredients is limited and in order to overcome this problem, research studies focuses on some alternate materials in the concrete production process. Also, Incorporation of waste materials consumes less energy leading to reduction of emission of green house gases. The application of fly ash and cow dung ash as a pozzolanic binder instead of cement and coir fibers finds extensive application in the manufacturing process of building materials. In this project an attempt has been made to utilize cow dung ash and coconut fiber as a replacement material of cement in the production of concrete. The cement is partially replaced with cow dung ash by about 2.5, 3 & 3.5 % by weight and with 1% of coconut fiber.The Compressive and Tensile strengths of concrete were found at different curing periods (7,14 & 28 days).From this study, it is inferred that these replacements will have a reasonable improvement in the strength properties of concrete by about 55-70%.The substitution of CDA, CF is economical in terms of cost and this usage eliminates the problem of landfills, reducing the environmental risk, maintaining the ecological balance, which is very much required for our nation.

1. Introduction

Concrete is the second largest material used by mankind for making the best shelter and taller structures. Due to increase in population, Industrialization, Infrastructural development, etc. there is a demand for both green and cost effective buildings.Concrete is one of the best construction material and has more advantages than other materials in terms of strength, thermal resistance, cost effectiveness. Cement is required for production of concrete which act as a very good binder. But, its production requires more energy and it also lead to pollution. It has been estimated by World Health Organization (WHO) that about 5% of total CO_2 released into the atmosphere is from cement industry. Besides being an environmental pollutant, the amount of heat energy required for production of cement is about 1500 °C which requires more energy. Hence, it is necessary to focus on a material which is of best quality and also involves less energy for its production. Many naturally available materials like coconut fiber, cow dung ash, straw, rice husk ash, etc. are used as binders by partially replacing the cement in concrete. This project work is dealt with incorporation of coconut fiber and cow dung ash in concrete. Coconut fiber is an agricultural waste obtained from fibrous husk of coconut. When it is added to concrete, the workability of concrete decreased, whereas its cohesiveness increased.It was found that coconut fiber incorporation in concrete can decrease about 2 percentage of steel reinforcement decreasing the cost to certain extent. Cow Dung ash is obtained by drying and burning of dried cowdung pats and has large content of Nitrogen, Potassium, Calcium, Carbon and Phosphorous.It has a low thermal conductivity and requires a maximum of 400°C of heat energy.This method also serve as a good method for reuse of coconut fiber,which is an agricultural and food waste. Incorporation of coconut fiber and cow dung ash in certain amount in concrete proves to improve the strength of concrete. But increase in amount of either coconut fiber or cow dung ash above a certain value may decrease the strength of concrete conversely. So, it is necessary to find out the optimum value of these materials to be added in the concrete which maintains the strength with reduction in cement content. This project deals with the replacement of cement in concrete using cow dung ash and coconut fiber to achieve reduction in cement content with optimum amount of inclusion of these two materials for better strength performance of concrete. This may, in future can lead to use of waste as construction materials at large scale after various research works carried out at various scales.

CDA and RHA with admixtures and plasticizers increased workability and strength of concrete for a range of 0% to 25 %.Compressive strength of concrete increased 30% for 5% replacement with CDA and RHA [1]Coconut fiber enhanced toughness ,torsion ,cracking and spalling resistance & tensile strength of concrete.A maximum of 3% of Coconut fiber can be used after which strength decreased [2].Up to 2% of steel reinforcement reduction can be achieved if sizable coconut fibers are used.CFA contains silica, alumina, iron oxides that were found in OPC.Workability decreased with increase in Coconut Fiber ash[3].With increase in percentage of CDA,Workability ,Initial & Final setting time of concrete decreased.CDA has low thermal conductivity & suitable for Light weight structures [4].Compressive strength of concrete decreased if Cow-dung ash of more than 10% by weight of cement is added.CDA needed more water when added to concrete [5].A maximum of 5% of Cattle Waste Ash(CWA) can be used in small rural housings which reduces cost.

2. Materials and Methodology

In this study, we used CDA and CF as a replacement of cement in concrete, CDA is a good binding material reducing voids in concrete and CF can make the concrete more tough.

2.1 Coconut fiber

Dried Coconut Fibers that were finely chopped to a length less than 1 cm as shown in Figure 1 was used.

Figure 1. Coconut fiber.

Figure 2. Cow dung ash.

2.2 Cement

Cement used in this experimental work was (PPC) Portland Pozzolanic Cement with a specific gravity of 3.20 which passed through 90 μm sieve with a Fineness of 7.1% and Consistency of 30, 30.7, 30.9 and 43.7 % for 0, 2.5, 3 and 3.5 % of CDA respectively.

2.3 Cow dung ash

Cow dung ash (Figure 2) that has been obtained from villages are dried under sunlight, burnt at a temperature of 450 to 500 °C and cooled. After cooling it was crushed to powder form, sieved under 300 micron sieve was stored in an air tight container preventing moisture ingress. Its properties were Specific gravity of ash - 2.55, pH value-9.5, Loss on ignition-12.28%, Blaine's fineness - 338.

Table1. Chemical Composition of Cow Dung Ash (CDA) vs Ordinary Portland Cement (OPC)

Oxide/Oxide Percentage	CDA	OPC
SiO_2	65.7655	20.26
Al_2O_3	4.4575	6.30
Fe_2O_3	3.1625	3.26
CaO	12.98975	65.51
MgO	2.01775	0.96
SO_3	0.9405	0.69
P_2O_3	2.83375	0.88
Na_2O	0.511	0.89
P_2O_5	1.38275	0.25
Mn_2O_3	0.59925	0.21
$SiO_2+Al_2O_3+Fe_2O_3$	73.3855	-
$CaCO_3$	23.18225	-
TiO_2	0.36875	0.24

2.4 Coarse Aggregate
Crushed granite that passed through 20mm sieve and retained on 4.75mm sieve was used as coarse aggregate.

2.5 Water
The water used for this experimental purpose was obtained from underground source through pump that was free from bacteria and visible sediments.

2.6 Fine Aggregate
Fine aggregate used for testing satisfied the required properties of fine aggregate conforming to zone 1 as per IS 383:1970.

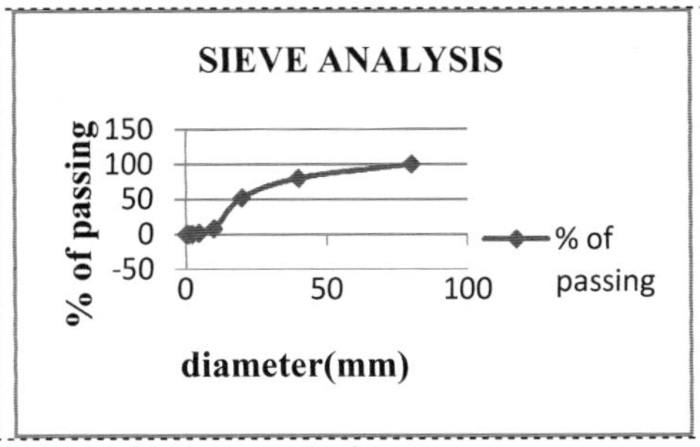

Figure 3. Sieve analysis - Coarse aggregate.

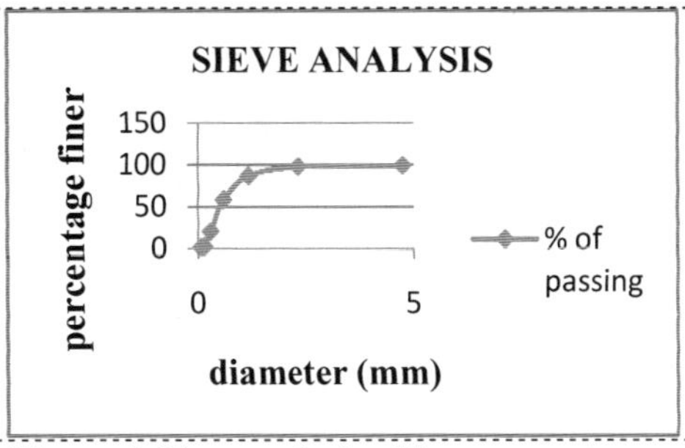

Figure 4. Sieve analysis - Fine aggregate.

3. Concrete Mix Design
The Concrete of M-20 grade was made using Binders, Fine and Coarse aggregate with a mix proportion of 1:1.7:3.3 and Water-Cement ratio of 0.45.

4. Casting of Samples

Initially, Cement and Sand were mixed up and added with CDA and CF.Coarse aggregate and Water were added after procuring and weighing them properly and finally were casted to cubes of size 150 mm and cylinders of diameter 150 mm and height 300 mm. After curing, the cubes and cylinders were tested at 7,14 and 28 days respectively under Universal Testing Machine.

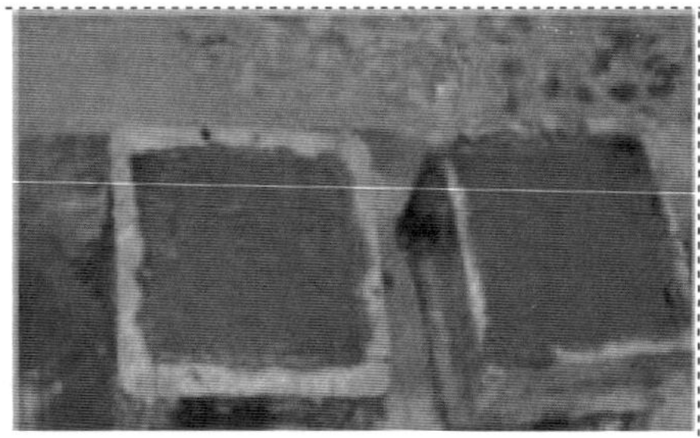

Figure 5. Compaction of cube.

Figure 6. Testing cubes under UTM.

5. Results and Discussion

Table 2. Tensile strength (Cylinder).

S. No.	% of CDA added	Tensile Strength(MPa)
1.	2.5	2.54
2.	3	2.63
3.	3.5	2.92

Table 3. Compressive strength(cube)

S.No	% of CDA	Days Of Curing	Weight after Curing (kg)	Compressive Strength (MPa)
1.	2.5	7	8.54	19.43
		14	8.60	34.56
		28	8.56	35.44
2.	3	7	8.48	17.22
		14	8.41	27.34
		28	8.46	30.10
3.	3.5	7	8.44	18.16
		14	8.36	34.79
		28	8.23	35.11

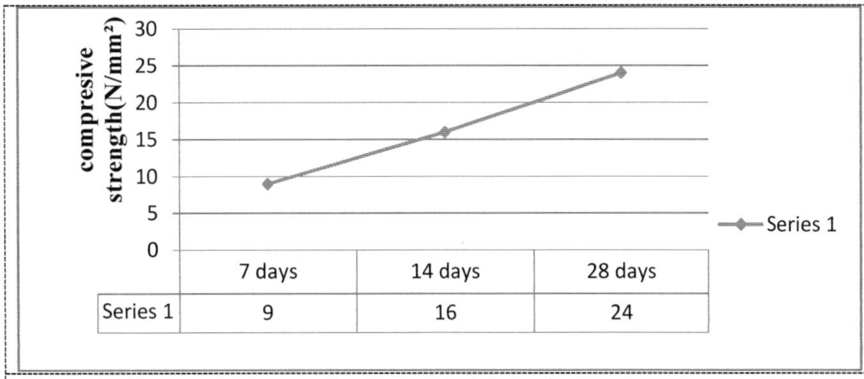

Figure 7. Compressive strength of concrete (Control).

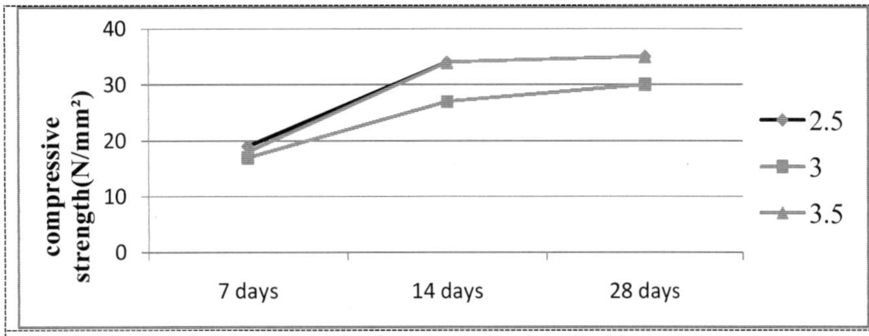

Figure 8. Compressive strength of concrete (replaced with CDA AND CFA).

From the results, it was found that the Concrete's compressive strength increased by about 77.2% and 75.5% for 2.5% and 3.5% replacement of cement using CDA and CFA. The Tensile strength increased by about 46% for 3.5% replacement of cement using CDA and CFA. So, it can be inferred that there was a significant increase in both the Compressive and Tensile strength of concrete for 3.5% replacement of Cement using cow dung ash and coconut fiber.

6. Conclusion

At present, the process of manufacturing of cement involves a larger emission of harmful gases which poses a threat to environment in many ways. Many research works are being carried out to replace either partially or wholly, the fine aggregate, coarse aggregate and cement in concrete. In this project, an attempt has been made to use cow dung ash and coconut fiber as a replacement material of cement. The addition of these materials marginally improved the structural properties of concrete.It also lead to reduction of environmental impact caused by land filling or dumping of these materials. Replacement using these materials can bring down the original cost of production of concrete and it also lead to reuse of waste materials. The above tested methods for production of concrete can be applied in the construction of compound walls,non-load bearing partition walls,lightly loaded precast members-shelf slabs, sill slabs, cut lintels and sunshades, kerb walls and medians of road.

Acknowledgement

We are extremely thankful to our Management and Civil engineering department for providing us with the necessary equipments required for our project work.We thank all the teaching and non-teaching staffs and labors who helped us in each and every aspect for making our project a successful one.I also thank almighty, my parents and friends for their constant encouragement without which our project would not have been possible.

References

[1] Inderveer Singh Gurjar and Gautam Bhadouriya 2015 A study on use of Cow-Dung Ash & Ricehusk ash in concrete. *Int. J. Res. Eng. Technol.* **4**(11) 101-9.
[2] Abhijeet R Agarwal, Sanket S Dhase and Kautuk S Agarwal 2014 Coconut Fiber in concrete to enhance its strength and making light-weight concrete. *Int. J. Eng. Res. Dev.* **9**(8) 64-7.
[3] Ojedokun O Y, Adeniran A A, Raheem S B and Aderinto S J 2014 Cow dung ash (CDA) as partial replacement of cementing material in the production of concrete. *British J. Appl. Sci. Technol.* **4**(24) 3445-54.
[4] Jitender Kumar Dhaka and Surendra Roy 2015 Utilization of fly ash and cowdung ash as partial replacement of cement in concrete. *Int. J. Civ. Stru. Eng.* **6**(1) 323-46.
[5] Pavan Kumar V S R, Polu Raj 2012 Incorporation of cowdung ash to mortar and concrete. *Int. J. Eng. Res. Appl.* **2**(3) 580-5.

Optimization and influence of parameter affecting the compressive strength of geopolymer concrete containing recycled concrete aggregate: using full factorial design approach

Thulasirajan Krishnan* and Revathi Purushothaman

Department of Civil Engineering, Pondicherry Engineering College, Puducherry-605014,India.

*Email:kthulasirajan@gmail.com

Abstract. There are several parameters that influence the properties of geopolymer concrete, which contains recycled concrete aggregate as the coarse aggregate. In the present study, the vital parameters affecting the compressive strength of geopolymer concrete containing recycled concrete aggregate are analyzedby varying four parameters with two levels using full factorial design in statistical software Minitab® 17. The objective of the present work is to gain an idea on the optimization, main parameter effects, their interactions and the predicted response of the model generated using factorial design. The parameters such as molarity of sodium hydroxide (8M and 12M), curing time (6hrs and 24 hrs), curing temperature (60°C and 90°C) and percentage of recycled concrete aggregate (0% and 100%) are considered. The results show that the curing time, molarity of sodium hydroxide and curing temperature were the orderly significant parameters and the percentage of Recycled concrete aggregate (RCA) was statistically insignificant in the production of geopolymer concrete. Thus, it may be noticeable that the RCA content had negligible effect on the compressive strength of geopolymer concrete. The expected responses from the generated model showed a satisfactory and rational agreement to the experimental data with the R^2 value of 97.70%. Thus, geopolymer concrete comprising recycled concrete aggregate can solve the major social and environmental concerns such as the depletion of the naturally available aggregate sources and disposal of construction and demolition waste into the landfill.
Keywords: *Geopolymer Concrete, Recycled concrete aggregate, Factorial design, curing condition, molarity of sodium hydroxide and Compressive Strength.*

1. Introduction

In 1978, Davidovits suggested that an alkaline solution could be employed to react with the silicon (Si) and the aluminum (Al) to form a solid matrix. The polymerization action incorporates a significant accelerated chemical reaction under alkaline medium on Si-Al minerals, emerging in a three dimensional polymeric group and ring framework consisting of Si-O-Al-O bond [1]. The polymerisation process of geopolymer concrete are controlled by various parameters such as curing temperature, curing time, molarity of sodium hydroxide,ratio of alkaline liquid-to-fly ash, ratio of Si to Al, water content of the mixture, method of mixing the ingredients and method of curing [2-4].In the past, many studies were conducted on geopolymer concrete comprising natural coarse aggregate(NCA),but studies ongeopolymer concrete by full replacement of natural coarse aggregate with recycled concrete aggregate (RCA) is reported minimum. The parameters varied in the previous studies are the molarity of sodium hydroxide [5-8], percentage RCA content [9-13], curing temperature [7] and method of curing (Ambient and heat curing) [5]. It was reported that strength of geopolymer concrete containing RCA declines with the addition of RCA [9-13]. However, variation in the curing time and combined analysis of the parameters such as molarity of sodium hydroxide, curing time, curing temperature and percentage of RCA content with optimization for a particular compressive strength value using full factorial design of experiments has been not yet reported.

2. Experimental Program

2.1 Production of Geopolymer Concrete

Geopolymer concrete was prepared using Class-F low calcium fly ash obtained from Udupi thermal power plant, sodium silicate, sodium hydroxide (NaOH), river sand, NCA and RCA. Crushed concrete samples from the concrete laboratory of original compressive strength ranging from 25 – 35 MPa was used as a source of RCA and further crushed using lab model jaw crusher. The aggregates passing 10 mm sieve size and retained on 4.75 mm sieve size were used as the Coarse Aggregate. The properties of NCA and RCA are given in the Table 1.

Table 1. Properties of Coarse Aggregate

S. No.	Properties	NCA	RCA
1	Specific gravity	2.85	2.65
2	Water absorption(%)	0.33	1.33
3	Percentage adhered mortar(%)	0	33.33
4	Bulk density(kg/m^3)	1452	1434
5	Abrasion value(%)	26	30.5
6	Impact value(%)	32.82	30.69

The composition of Na_2SiO_3 was Na_2O= 15.2%, SiO_2 = 29.7% and H_20 =55.1%. The density was found to be 1.53gm/cm^3. The NaOH solution was taken in the form of flakes of 98% purity which was procured from the local market. The NaOH solution of specific molarity was produced 24 hours earlier to use, as the addition of NaOH flakes and tap water, liberates large amount of heat. The Na_2SiO_3 and NaOH solutions were blended half an hour before the casting of the specimens. Conplast SP 430 based on Sulphonated Naphthalene polymer was used as high range water reducing Superplastizer with a dosage of 2% by the weight of fly ash. The quantity of materials required per cubic meter are tabulated in Table 2. Conventional mixing procedure was followed for producing geopolymer concrete. The fresh concrete was then casted in cube moulds of 100mm size. The cubes along with the mould were cured in hot air drying oven and tested after placing in a room temperature of 24 hours.

Table 2. Quantity of Materials required (kg/m^3)

S. No	RCA %	Coarse aggregate NCA	Coarse aggregate RCA	Fine aggregate	Fly ash	Sodium silicate	NaOH	Superplastizer
1	0	1201.2	----	646.8	380.7	122.4	48.95	7.614
2	100	----	1201.2	646.8	380.7	122.4	48.95	7.614

2.2. Factorial design of experiments

In this study, 2^4 full factorial design analysis (four parameters and two control levels) was performed with 16 trials of experiments using Minitab® 17 statistical software. The major advantage of the factorial design was the precise estimation of the main effects of each factor and their interactions in fewer experiments [14]. Parameters and the levels used in 2^4 full factorial design is shown in Table 3. The geopolymer synthesis parameter such as molarity of sodium hydroxide (8M and 12M), curing time (6hrs and 24 hrs), curing temperature (60°C and 90°C) and percentage of recycled concrete aggregate (0% and 100%) and the effects of interactions between those parameters were examined. The experimental data were analyzed using second ordered interaction and the regression model form is shown in equation (1).

$$Y= \beta_0 + \beta_1 A + \beta_2 B + \beta_3 C + \beta_4 D + \beta_5 AB + \beta_6 AC + \beta_7 AD + \beta_8 BC + \beta_9 BD + \beta_{10} CD \quad (1)$$

Where Y is the value of the response, β_0 is a constant coefficient and β terms specify the regression coefficient estimates of main and interaction effects. The A, B, C and D are notations for the parameters as indicated in Table 3.

Table 3. Parameters and the levels used in 2^4 full factorial design

Notations	Parameters	Type	Parameter Level	
			Low level (-1)	High level (1)
A	Molarity of sodium Hydroxide(M)	Numerical	8	12
B	Curing time(Hours)	Numerical	6	24
C	Curing temperature(°C)	Numerical	60	90
D	RCA content (%)	Numerical	0	100

3. Results and discussion

3.1 Effect of parameters on the strength of geopolymer concrete using analysis of variance (ANOVA)

The significant parameters affecting the strength of geopolymer concrete were obtained by performing analysis of variance (ANOVA). It is a statistical approach which separates the total variance in an entire data into various parts linked to definite sources of variations for the objective to test the hypotheses on the factors used in the model [14]. ANOVA results for the full factorial model are presented in Table 4. The P-values lesser than 0.05 indicates the parameters are statistically significant. Thus from the Table 4, the parameters namely the molarity of sodium hydroxide, curing time and curing temperature in the one way interaction and the combination of factors, namely molarity of Sodium hydroxide * RCA content and curing time * curing temperature in the two way interaction are statistically significant indicating their major effects on the strength of geopolymer concrete. These results are further justified by the Pareto charts and normal plot of the standardized effect as shown in Figure 1.

Table 4. Analysis of variance (ANOVA)

Source	DF	Adj SS	Adj MS	F-Value	P-Value
Model	10	2953.65	295.37	20.93	0.002
Linear	4	2669.91	667.48	47.29	0.000
A	1	991.78	991.78	70.27	0.000
B	1	1582.65	1582.65	112.13	0.000
C	1	93.94	93.94	6.66	0.049
D	1	1.54	1.54	0.11	0.754
2-Way Interactions	6	283.74	47.29	3.35	0.103
A*B	1	3.16	3.16	0.22	0.656
A*C	1	24.53	24.53	1.74	0.245
A*D	1	102.57	102.57	7.27	0.043
B*C	1	152.71	152.71	10.82	0.022
B*D	1	0.78	0.78	0.06	0.824
C*D	1	0.00	0.00	0.00	0.991
Error	5	70.57	14.11		
Total	15	3024.22			

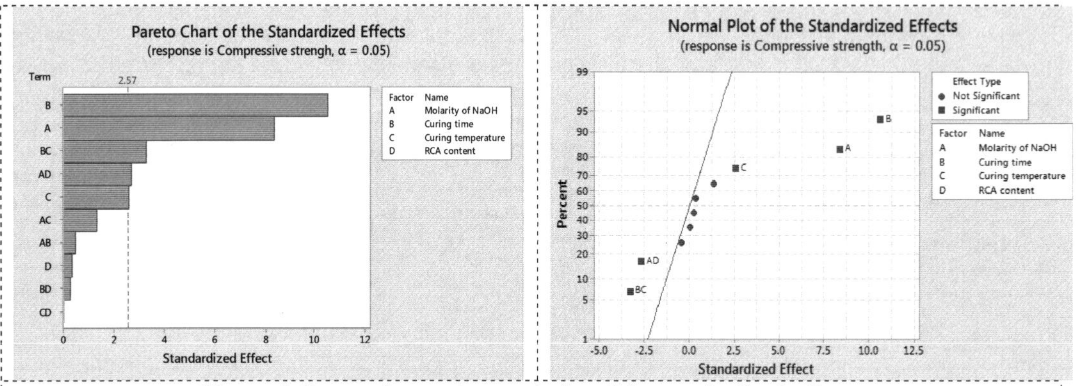

Figure 1. Pareto chart and normal plot of the standardized effects

3.1.1 Main effect. The main effect shows variations of the mean among the high and low values of each factor. The magnitude of slope represents the intensity of the effects that each factor exerts. When the slope is positive, the response increases for higher levels of that factor and vice-versa [15]. The main effect plot for the mean strength of geopolymer concrete is shown in Figure 2. It is seen that all the factors shows a slope of positive response increase to the higher level. The mean strength of geopolymer concrete reduces gradually in the order of curing time > molarity of NaOH> curing temperature > RCA content. As compared to other parameters, curing time parameter with steeper slope plays a major role in strength enhancement of geopolymer concrete. Thus, prolonged curing for 24 hours may boost up the geopolymeric reaction (polymerization reaction) and forms a stronger bond between alumina (Al) and silica (Si) compounds. The molarity of sodium hydroxide is the second parameter which played a vital part in the disintegration (leaching) of Al and Si atoms. By the dissolution process, the free Al and Si atoms are released in the solution, enhancing the reaction mechanism pertaining to the strength increase at higher molarity. The curing temperature ranks third in influencing the strength of geopolymer concrete. The strength variation at higher temperature is not as identical as compared to the curing time and molarity of NaOH solution. The possible reasons may be, at higher temperature the breakdown of the Al and Si bonds occurs with the evaporation of the water (OH ions). The last parameter, namely RCA has negligible influence on the strength of geopolymer concrete with minimum slope line. But, this result is contradictory to the previous research work predominantly by the decrease in strength, as the RCA content is increased to 100% [9-13]. The possible reasons are the use of good quality RCA as tabulated in Table 1 with the properties of the RCA are almost similar to NCA.

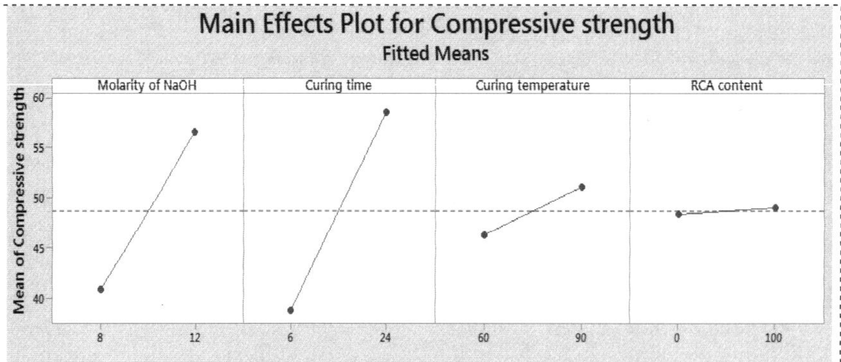

Figure 2. Main effect plot of compressive strength

3.1.2 Interaction effect. The interaction effect indicates the combined interactive effect of two parameters, when a response of one parameter (low to high) depends on the level of a second parameter. Graphically, two parallel lines of parameters indicate no interaction between them, whereas non-parallel lines suggest that the two parameters interact together. From the Figure 3, it is noted that the molarity of Sodium hydroxide * RCA content and curing time * curing temperature interacts with each other showing non parallel lines. Further, the mean strength for all the interaction parameters increase from lower level to higher level.

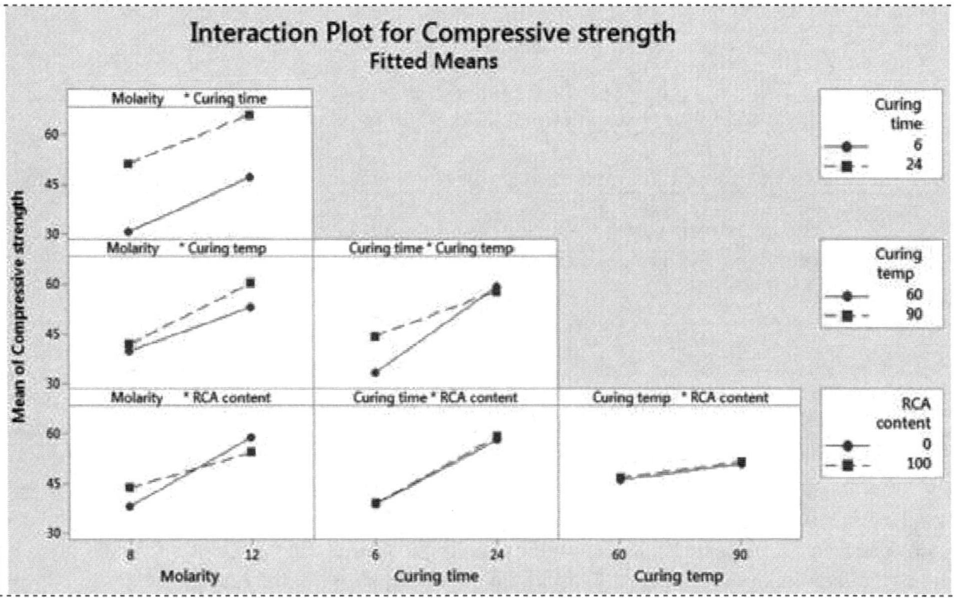

Figure 3. Interaction effect for the mean compressive strength

3.2 Full factorial model and optimization of the parameters

An empirical regression model equation (second-order) was generated using the experimental data as shown in equation (2) and the strength of geopolymer concrete was predicted with an R^2 value of 97.7%. The fitted line plot between the experimental and predicted values of the geopolymer concrete strength is shown in Figure 4. A response optimization tool from Minitab software was used to optimize the parameters for the goals of maximum and target strengths of geopolymer concrete as shown in the Table 5.

$$\text{Compressive strength} = -30.4 + 2.48\,A + 3.044\,B + 0.091C + 0.251\,D - 0.0247\,AB + 0.0413\,AC - 0.02532\,AD - 0.02288\,BC + 0.00049\,BD + 0.00001\,CD \quad (2)$$

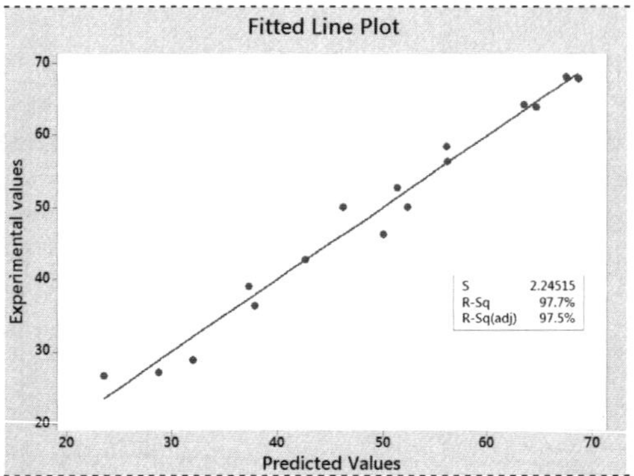

Figure 4. Experimental strength vs predicted strength

Table 5. Optimization of parameters for target and maximum compressive strength

S. No	RCA content (%)	Molarity of NaOH (M)	Curing time (Hrs)	Curing temperature (°C)	Compressive strength (MPa)	Goals
1	0	9.35	6	60	30.0	Target
2	100	8.55	6	60	30.0	Target
3	0	12	24	90	68.6	Maximum
4	100	12	24	90	64.6	Maximum

3.2.1 Goodness of model fit. Figure 5 shows the goodness of model fit by checking the normality and equal variance assumptions. The normal probability plot is a graphical description for concluding either the data are allocated normally indicated by the closer accumulation of points [16]. Figure 5a shows the data are assignednormally, as all the points are closer to the straight line.

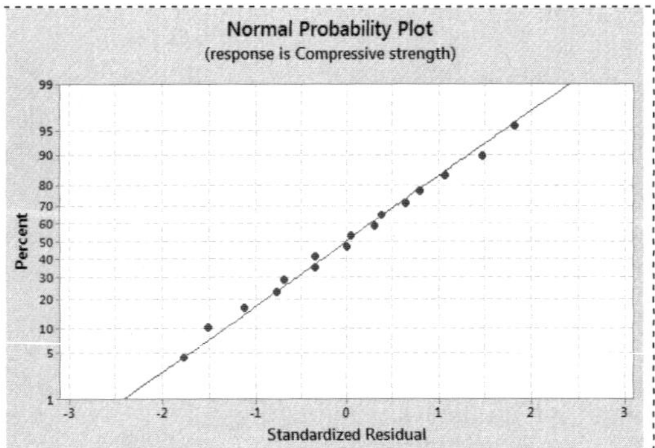

Figure 5a. Normal probability plot

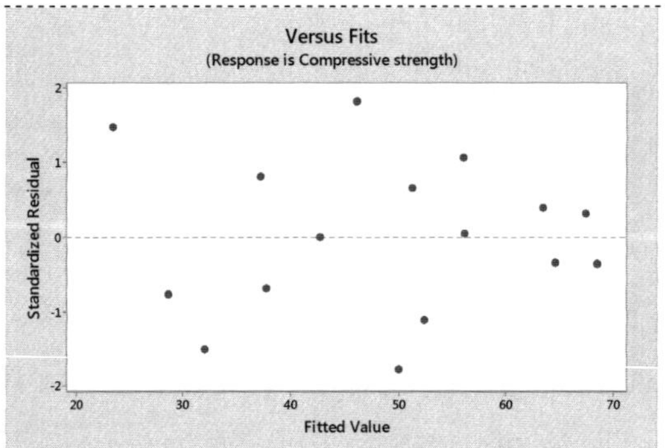

Figure 5b. Residuals versus fitted values plot

Figure 5. Check for goodness of model fit

The residual values describes the changes among the predicted values (model) and the experimental values. Standardized residuals greater than +2 and less than -2 are usually considered as outliers. Figure 5b shows the Spread of residual values in the plots are within this limits and on both sides of 0 indicating no outliers in the data. Thus the predicted model specified an insignificant divergence in the fitted value from the experimental value [17].

3.3 Validation of full factorial model

The factorial model generated for the particular set of parameters with the predicted response are checked with the new intermediate level of a particular parameter, namely the molarity of sodium hydroxide (10M). Other parameters such as curing time, curing temperature and RCA content were kept constant. The predicted compressive strength using the generated model is compared with the experimental value and the percentage deviations are tabulated in table 6. It is noted that the deviation of the predicted compressive strengths from the experimental value are within the acceptable range of (± 6 %), thereby validates the generated model.

Table 6. Validation of the full factorial model

S. No	Molarity of NaOH (M)	Curing time (Hrs)	Curing time (°C)	RCA Content (%)	Compressive strength (MPa)		Deviation (%)
					Predicted Value	Experimental Value	
1	10	24	90	0	57.41	60.12	-4.72
2	10	24	90	100	58.50	62.08	-6.12
3	10	24	60	0	58.80	57.3	2.55
4	10	24	60	100	59.80	61.69	-3.16
5	10	6	90	0	44.13	52.81	-6.07
6	10	6	90	100	44.34	41.57	6.25
7	10	6	60	0	33.13	40.4	-3.83
8	10	6	60	100	33.30	31.83	4.41

4. Conclusion

The 2^4 full factorial design of experiment was successfully used to generate prediction model for geopolymer concrete. The expected responses of the generated model showed a satisfactory and rational agreement to the experimental data with the R^2 value of 97.70%. The analysis of variance (ANOVA) implies that the curing time and the molarity of the sodium hydroxide were the most significant parameters affecting the strength of geopolymer concrete. The percentage of RCA content was statistically insignificant and had a negligible effect on the compressive strength of geopolymer concrete. The set of parameters required to develop the maximum strength of geopolymer concrete containing 100% RCA content is as follows: Molarity of NaOH = 12M; Curing time = 24 Hrs; Curing temperature = 90°C. Thus, it may be inferred that the 100% replacement of natural coarse aggregate with recycled concrete aggregate does not impact the strength of geopolymer concrete. Overall, geopolymer concrete comprising recycled concrete aggregate can solve the major social and environmental concerns such as the depletion of the naturally available aggregate sources and disposal of construction and demolition waste into the landfill.

Acknowledgements

This work has been carried out with the support of the funding provided by UGC through the Projecttitled 'Recycled Aggregate Geo-Polymer Concrete for Pre-Cast Reinforced Concrete Elements'.

References

[1] Davidovits J 1991 Geopolymers: Inorganic Polymeric New Materials. *J. Therm. Anal.* **37** 1633-56.

[2] Hardjito D, Wallah SE, Sumajouw DMJ and Rangan BV 2004 on the development of fly ash-based geopolymer concrete. *ACI Mater. J.* **101**(6) 467-72.

[3] Fernandez-Jimenez A, Palomo JG and Puertas F 1999 Alkali activated slag mortars:Mechanical strength behaviour. *Cem. Concr. Res.***29**(8) 1313-21.

[4] Palomo A, Grutzeck MW and Blanco MT 1999 Alkali-Activated Fly Ashes: A Cement for the Future. *Cem. Concr. Res.* **29**(8) 1323-29.

[5] TawatchaiTho-In, VanchaiSata, Trinh Cao and PrinyaChindaprasirt 2017 Use of Recycled Concrete Aggregate in High-Calcium Fly Ash Geopolymer Concrete. *Key Eng. Mater.* **718** 163-8.

[6] Peem Nuaklong, Vanchai Sata and Prinya Chindaprasirt 2016 Influence of recycled aggregate on fly ash geopolymer concrete properties. *J. Cleaner Prod.* **112**(4) 2300-7.

[7] Kostas Komnitsas, Dimitra Zaharaki, Antigoni Vlachou, Georgios Bartzas and Michalis Galetakis 2015 Effect of synthesis parameters on the quality of construction and demolition wastes (CDW) geopolymers. *Adv. Powd. Technol.* **26**(2) 368-76.

[8] Preethy K Thomas, Binu M Issac and Deepak John Peter 2015 Assessment of demolished concrete as coarse aggregate in geopolymer concrete *Int. J. Adv. Res. Sci. Eng.* **4**(1) 607-19.

[9] Shi X S, Wang Q Y, Zhao X L and Collins F 2012 Discussion on properties and microstructure of geopolymer concrete containing fly ash and recycled aggregate. *Adv. Mater. Res.***450-451** 1577-83.

[10] Shi X S, Collins F G, Zhao X L and Wang Q Y 2011 Experimental Study on Geopolymeric Recycled Concrete Using as Sustainable Construction Material. *Proc. Int. Conf. Adv. Constru. Mater. Sci. Eng.* Hong Kong 186-94.

[11] Shi X S, Collins F G, Zhao X L and Wang Q Y 2012 Mechanical properties and microstructure analysis of fly ash geopolymeric recycled concrete. *J. Haz. Mater.* **237–238** 20-9.

[12] Faiz Uddin Ahmed Shaikh 2016 Mechanical and durability properties of fly ash geopolymer concrete containing recycled coarse aggregates. *Int. J. Sustainable Built Environ.* **5**(2) 277-87.

[13] Benjamin Galvin and Natalie Lloyd 2011 Fly Ash Based Geopolymer Concrete with Recycled Concrete Aggregate Curtin University of Technology Perth Australia.

[14] DC Montgomery 1991 Design and Analysis of Experiments 3rd edition Wiley New York.

[15] Saima Farooq, Abubakr Saeed, Muhammad Sharif, JavidHussaina, FazalMabood and Maryam Iftekhar 2017 Process optimization studies of crystal violet dye adsorption onto novel, mixed metal $Ni_{0.5}Co_{0.5}Fe_2O_4$ ferrospinel nanoparticles using factorial design. *J. of Water Process Eng.* **16** 132-41.

[16] Pokhrel D and Viraraghvan T 2006 Arsenic removal from aqueous solution by iron oxide coated fungal biomass: factorial design analysis. *Water Air Soil Pollut.***173** 195-208.

[17] Rathinam A, Rao JR and Nair BU 2011 Adsorption of phenol onto activated carbon from seaweed: determination of the optimal experimental parameters using factorial design *J. Taiwan Inst. of Chem. Eng.***42**(6) 952-56.

IOP Conf. Series: Earth and Environmental Science **80** (2017) 012014 doi:10.1088/1755-1315/80/1/012014

Influence of Waste Tyre Crumb Rubber on Compressive Strength, Static Modulus of Elasticity and Flexural Strength of Concrete

M K Haridharan[1], R Bharathi Murugan[2]*, C Natarajan[3] and M Muthukannan[2]

[1]School of Civil Engineering, SASTRA University, Thanjavore – 613 401. India.
[2]Department of Civil Engineering, Kalaslingam University, Krishnakovil. India.
[3]Department of Civil Engineering, National Institute of Technology, Trichy. India.

*Email: rbmmecivil@gmail.com.

Abstract. In this paper, the experimental investigations was carried out to find the compressive strength, static modulus of elasticity and flexural strength of concrete mixtures, in which natural sand was partially replaced with Waste Tyre Crumb Rubber (WTCR). River sand was replaced with five different percentages (5%, 10%, 15%, 20% and 25%) of WTCR by volume. The main objective of the experimental investigation is to find the relationship between static modulus of elasticity and flexural strength with compressive strength of concrete with WTCR. The experimentally obtainedstatic modulus of elasticity and flexural strength results comparing with the theoretical values (various country codes recommendations).

1. Introduction

Recent years the rapid development is noticed in the construction industry due to urbanization. Concrete is one of the widely used materials in the construction industry. In the concrete preparation 70% to 80% occupying natural aggregates like crushed rock (coarse aggregate) and river sand (fine aggregate). The commonly used coarse aggregate and fine aggregate is likely becoming huge demand and scared it is today. Now construction people need the alternatives to natural aggregates. Therefore, finding alternatives to naturally available materials is important to sustaining construction industry. On the other hand a large quantity of waste materials is being produced by various industries and the governments are seeking ways to reduce the problem of disposal and to control the health hazard from the accumulation of waste materials. Some of the waste materials, such as coal fly ash, bottom ash, glass granules, plastic granules, copper slag, and crushed rock dust were used in the concrete. So such waste materials are utilized to alter the mechanical and durability characteristics of concrete so as to make it fit for any condition, this would also give some more benefits in terms of cost reduction, promoting environmental sustainability,energy savings, and reduction in the use of natural resources [1-3].

In this paper, waste tyre crumb rubber partially replaces the fine aggregate by volume with different percentage (5%, 10%, 15%, 20% and 25%) and study to the main characteristics of the concrete made of crumb rubber and to compare the test results with the normal concrete (without rubber). The compressive strength, flexural strength and static modulus of elasticity are the main properties of concrete. In fact, these are very important properties for designing and analysing the concrete structures. Design codes of various countries are recommended the empirical relationship between static modulus of elasticity; flexural strength and compressive strength of concrete at 28 days are shown in Table 1.

Table 1. Empirical relationship between static modulus of elasticity; flexural strength and compressive strength of concrete at 28 days

S. No	Empirical relationship between static modulus of elasticity and compressive strength	Empirical relationship between flexural strength and compressive strength	Code details
1	$E_C = 5000\sqrt{f_c}$	$f_r = 0.7\sqrt{f_c}$	IS 456 [4]
2	$E_C = 4734\sqrt{f_c}$	$f_r = 0.62\sqrt{f_c^{'}}$	ACI 318 [5]
3	$E_C = 3320\sqrt{f_c^{'}} + 6900$	$f_r = 0.6\sqrt{f_c^{'}}$	NZS 3101 [6]
4	$E_C = 9500(f_c^{'} + 8)^{0.33}$	$f_r = 0.3(f_c)^{0.67}$	EC 02 [7]
5	$E_C = 4500\sqrt{f_c^{'}}$	$f_r = 0.6\sqrt{f_c^{'}}$	CSA A23.3-04 [8]

Where, E_c is the static modulus of elasticity at 28 days in MPa, f_r is flexural strength of concrete in MPa, f_c is cube compressive strength of concrete at 28 days in MPa and $f_c^{'}$ is cylinder compressive strength.

The above empirical relationship is only for conventional concrete. Therefore, in this paper the tests were made to establish a new empirical relationship between static modulus of elasticity, flexural strength based on the compressive strength of concrete incorporating waste tyre crumb rubber.

2. Experimental Program

2.1 Material properties

Ordinary Portland cement 53 grade (OPC 53) confirming to IS 12269 [9] was used in this study. The properties of cement presented in Table 2. Properties of cement test results satisfied the requirements.

Table 2. Properties of Cement

S. No.	Property		Results	Requirements as per IS 12269
1	Specific surface (m^2/kg)		320	225 (Min.)
2	Normal consistency		31%	---
3	Specific gravity		3.14	---
4	Setting time	Initial	65 min	Not less than 30 min
		Final	280 min	Not more than 600 min
5	Compressive strength (N/mm^2)	7 days	38.60	37.00
		28 days	56.96	53.00

Locally available river sand was used as a fine aggregate. The properties are confirmed as per IS 383. River sand confirms grading Zone II with specific gravity of 2.65 and fineness modulus of 2.45 were used.Crushed granite stone having a maximum size of 20mm and down with specific gravity of 2.63 and fineness modulus of 7.2 were used. The properties are confirmed to IS 383

[10].Shredding of waste tyres were to produce crumb rubber having a maximum size of 4.75mm down with specific gravity of 0.689.

2.2 Mix Proportion

The mix designed was designed for M40 grade concrete as per IS 10262 [11] which yielded proportions of 1:2.22:2.66 with water- cement ratio of 0.5. The dosage of super-plasticizer 1% (by weight of cement) was added to improve the workability. Replacing 5%, 10%, 15%, 20% and 25% of fine aggregate with waste tyre crumb rubber by volume and re-proportioning the conventional concrete mixes gave the waste tyre crumb rubber concrete mixes. Water – cement ratio and dosage super plasticizer was kept constant for all the mixtures. The details of mix proportions designed for both conventional and waste tyre crumb rubber based concrete is shown in Table 3.

Table 3. Mix proportions

S. No.	Mix ID	Details of replacement	Ratio C: S: CA: CR
1	R0	Conventional concrete	1: 2.2:2.66:0.00
2	R5	5% fine aggregate replaced by waste tyre crumb rubber	1:2.09:2.66:0.09
3	R10	10% fine aggregate replaced by waste tyre crumb rubber	1:1.98:2.66:0.18
4	R15	15% fine aggregate replaced by waste tyre crumb rubber	1:1.87:2.66:0.27
5	R20	20% fine aggregate replaced by waste tyre crumb rubber	1:1.76:2.66:0.36
6	R25	25% fine aggregate replaced by waste tyre crumb rubber	1:1.65:2.66:0.45

Note: C – Cement, S – sand, CA – Coarse aggregate and CR – Crumb rubber.

2.3 Casting of Test Specimens

Cube specimens of size 150 mm× 150 mm× 150 mm were cast for cube compressive strength. Cylinder specimens of 200mm length and 100 mm diameter were cast for cylinder compressive strength and static modulus of elasticity. Prism of size 500 mm× 100 mm× 100 mm was cast for flexural strength.

3. Results and Discussion

3.1 Compressive Strength

The concrete cube and cylinder compressive strength test results as shown in figure 1. The test result shows that the concrete cube and cylinder compressive strength were decreased with increasing crumb rubber content in the concrete mix. The reduction in the compressive strength of cubes with the addition of 5%, 10%, 15%, 20% and 25% of WTCR was observed 4.09%, 7.73%, 13.72%, 23.64% and 33.85% for concrete cubes 4.26%, 9.85%, 16.22%, 27.20% and 36.02% for concrete cylinders at 28 days respectively as compared to normal concrete (without WTCR).

The major cause for the drop in the strength is the softness of the rubber particles are much softer due to this while compressive load applied to the specimen the crack propagation was very fast around the rubber particles which will affect the rubber-cement matrix and leads to the failure [12-13].

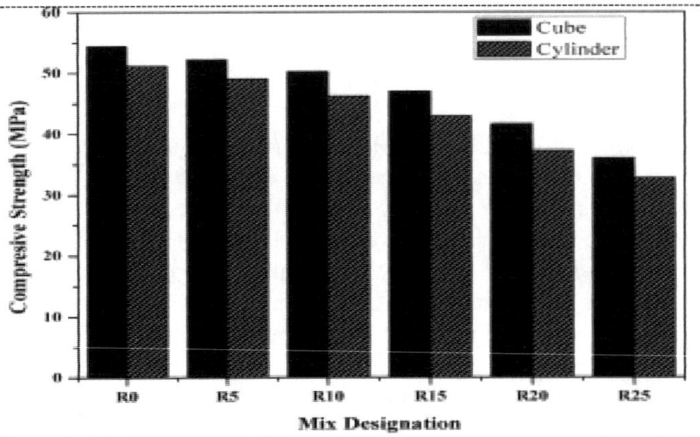

Figure 1. Compressive strength of concrete cubes and cylinders

3.2 Flexural Strength

Figure 2 represents the comparison of experimental and theoretical (Various country codes, recommendations) obtained flexural strength for both normal concrete and WTCR mixed concrete is presented in Figures 2.

Two point load was applied to the specimens and the breaking load for the specimens were measured. From the breaking load the flexural strength was calculated. The test results show that the flexural strength was increased up to 15% WTCR replacement. This might be because of the improved tensile load carrying capacity of the rubber particles [14].

Figure 2 shows the experimentally obtained flexural strength value is higher than comparable to other codal provision. The experimentally obtained flexural strength of concrete higher than 4.35 to 13.74 percent for IS 456, 17.87 to 27.16 percent for ACI 318 , 19.12 to 32.01 percent for EC 02 and 20.52 to 29.51 percent for NZS 3101 and CSA A23.3.04 codes for all concrete mixes.

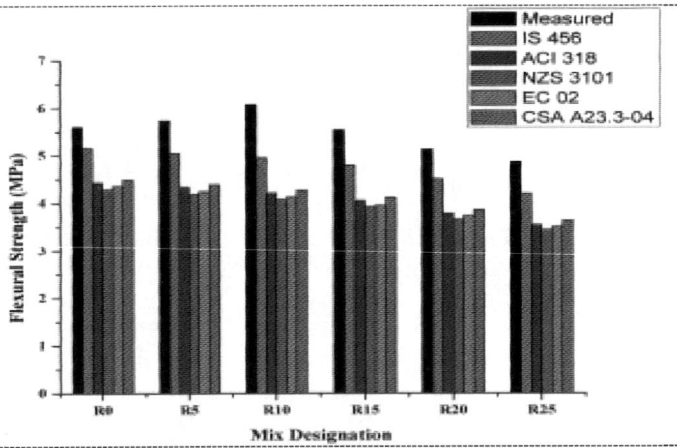

Figure 2. Comparison of experimental and theoretically obtained flexural strength for controland WTCR mixed concrete

3.3 Modulus of Elasticity

Figure 3 clearly shows that the experimentally obtained modulus of elasticity at 28 days and calculated values from the empirical equation given in the various country codes for both normal and rubberized concrete. The figure clearly shows that the measured modulus of elasticity is lower than the theoretical one.

The modulus of test procedure was followed the guidelines of ASTM C 469-02 [15]. The Static modulus of normal and rubberized concrete was determined as the secant modulus measured at a stress level equal to 40 percent of the average compressive strength of concrete cylinders. The gauge length of the cylinder was marked on the central height of specimens. Longitudinal extensometer was attached to the specimen such that gauge length is 200mm, placed the cylinder with a longitudinal extensometer on the compression testing machine and adjusted the deflect meter reading to zero. The load is increased in increment up to 40% of the cylinder compressive strength of concrete. The deflect meter readings are tabulated and the static modulus of elasticity was found. The reading of one unit in the extensometer is equal to 1/100 mm.

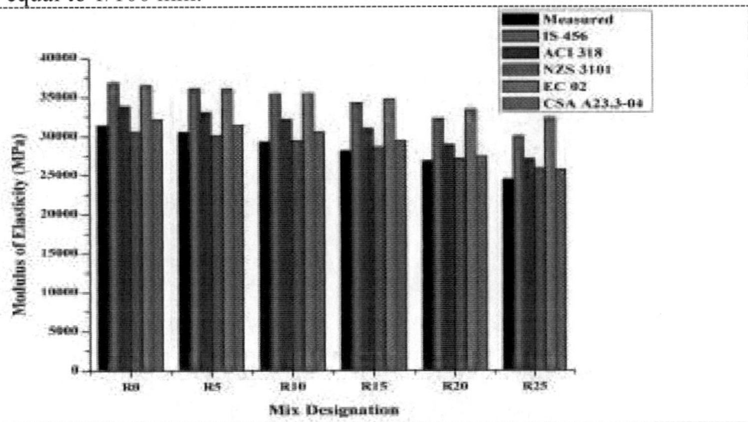

Figure 3. Comparison of experimental and theoretically obtained for modulus of elasticity for normal and rubberized concrete

The figure 3 shows the modulus of elasticity calculated by EC 02 and IS 456is higher than those predicted by NZS 3101, ACI 318 and CSA A23.3.04. The figure 3 also shows the measured modulus of elasticity is lower than the theoretical values. The difference between experimentally obtained modulus of elasticity and calculated modulus of elasticity increased as the strength of concrete decreased with increasing the percentage of WTCR content.

4. Conclusion

- The concrete cubes and cylinder compressive strength were decreased with increasing in WTCR content..
- Flexural strength of concrete was increased up to 15% of WTCR replacement. It the percentage of WTCR replacement increases over 15% the flexural strength was decreased. The measured flexural strength was more than the predicted flexural strength by various country codes for all the percentage replacements of sand by WTCR.
- Static modulus of elasticity of WTCR mixed concrete was less than control concrete, but there was withstanding large deformationand displacement due to the properties of rubber. Normally rubber having more ductile or flexible in that it has the ability to withstand larger deformation. The experimentally obtained modulus of elasticity of control and WTCR mixed concrete was less than predicted from the theoretically obtained values..

References

[1] Azmi N J, Mohammed B S and Al-Mattarneh H M 2008 Engineering properties of concrete containing recycled tire rubber. *ICCB B* **34** 373–82.

[2] El-Gammal A, Abdel Gawad A K, El-sherbini Y and Shalaby A 2010 Compressive strength of concrete utilizing waste tire rubber. *J. Emerg. Appl. Sci.* **1(1)** 96–9.

[3] Ganesan N, Bharathi Raja J and Shashikala A P 2013 Flexural Fatigue Behavior of Self Compacting Rubberized Concrete. *Constr. Build. Mater.* **44** 7–14.

[4] Indian standard code for plain and reinforced concrete for general building constructions IS: 456-2000, *Bureau of Indian Standard*, New Delhi.

[5] Building code requirements for structural concrete and commentary, ACI Committee: 318-2000, *American concrete Institute*, Farmington Hills, Michigan.

[6] The design of concrete structures (part-1) and commentary on the design of concrete structures (part2), NZS: 1995, Wellington, New Zealand.

[7] BS EN 1992, Eurocode 2: Design of concrete structures. *European Committee for Standardization.*

[8] Design of concrete structures, CSA A 23.3:1994, Technical Committee, *Canadian Standards Association*, Rexdale, Ontario.

[9] Indian Standard Specification for 53 Grade Ordinary Portland cement, IS 12269:1987. *Bureau of Indian Standards*, New Delhi.

[10] Indian Standard Specification for coarse and fine aggregates from natural sources for concrete IS: 383-1970, *Bureau of Indian Standard*, New Delhi.

[11] Recommended guidelines for concrete mix design, IS: 10262-1982, *Bureau of Indian Standard*, New Delhi.

[12] Gunneyisi E, Gesoglu M and Ozturan T 2004 Properties of rubberized concrete containing silica fume. *Cem. Concr. Res.* **34** 2309–14.

[13] Issa C A and Salem G 2013 Utilization of recycled crumb rubber as fine aggregate in concrete mix design. *Constr. Build. Mater.* **42** 48–52.

[14] Ozbay E, Lachemi M and Sevim UK 2011 Compressive strength, abrasion resistance and energy absorption capacity of rubberized concretes with and without slag. *Mater. Struct.* **44(7)** 1297–307.

[15] Standard test method for static modulus of elasticity and Poisson's ratio of concrete in compression, ASTMC: 469 – 2002, *ASTM Standards*, United states.

Experimental studies on effect of Date Seed Ash (DSA) on strength properties of cement sand mortar

G I Gunarani* and S P Chakkravarthy

School of Civil Engineering, SASTRA University, Thanjavur – 613 401. India.

*E-mail: gunarani@civil.sastra.edu

Abstract. The need for alternative material for the cement is arising and being compromised by many engineering researchers. However, the growing demand and surging prices of raw materials challenges the constructional field. India, being one of the largest agricultural economy, produces a quantitative volume of agro-waste that is being dumped. In the conventional concrete production, coarse aggregate (CA) plays an important filler material. The initial study on date seed as a replacement for CA was not successful. This study primarily focuses on Date seed ash as a replacement material for ordinary Portland cement. OPC was replaced by Date Palm Seed Ash (DPSA) in the ratio up to 10% in terms of 2% interval. The main objective of this paper was to study the variation of strength properties of mortar by DPSA in specified ratio along with curing period of 3,7,14 and 28 days. The stress strain behavior has indicated a significant improvement. The overall results indicated the increase in replacing ratio, decreases the strength properties. However the physical, chemical and mechanical properties increased gradually in strength in minimal ratio.
Key words: Cement Mortar, Date Palm Seed Ash, Cement replacement and Strength properties.

1. Introduction

The world is leaping forward towards the modern living standards through building up a sustainable infrastructure. The huge need and demand for buildings has created a direct increase in the consumption as well the production of cement. Due to the need for supplementary material for the ordinary Portland cement many researches are been conducted. Fly ash, one of the hazardous air pollutants, is presently used as a major partial raw material for the Portland Pozolonic Cement (PPC). Pioneers of civil engineering research and many engineering students have taken this in their concern and because of which many scholarly articles stating the usage of Rice husk ash, Sunflower seed ash, Palm kernel ash, Coconut shell ash, etc., are being used as a partial replacement material for Ordinary Portland Cement. For a sustained growth in the construction industry, the necessity for a alternative or a supportive material is needed for cement [1-7].This has to be researched without compromising the strength as well as the environment which can be made possible only with recyclable natural waste with pozzolanic reactivity. [8-13] Recycling industrial or agricultural waste is vital for the sustainable development which plays a crucial role in pollution control. The current society has immense interest in the use of innovative materials, which reduces cost with enhanced quality, in place of domestic products, which also aimed at reducing the environmental impact. The high cost of construction materials like cement has led to increased cost of construction. Mortar is a material having tiny spaces through which liquidness or air may pass. The durability of mortar depends largely on movement of mortar.

Date, Phoenix dactylifera, is an edible fruit in the palm family. This is widely grown in temperate and arid zones which has a high fibrous strength. The research was carried out on date seed if it would find an application in building construction. The use of date seed in concrete mixtures will not only reduce the pile up of the same in the dump yards but will greatly help in cost cutting in construction industry. Adding colour to the ongoing research in utilization of waste as solution to the senior high cost of construction materials, this research was initially planned to investigate its suitableness as light weight aggregate material in concrete production since little or no work exist on it.

2. Materials and Methods

2.1 Cement

Ordinary Portland cement of 43 grade conforming to IS: 8122-1989 was used.

Table 1. Properties of Ordinary Portland cement (OPC)

Description	Composition
Specific gravity	3.05
Initial setting time	30 minutes
Final setting time	600 minutes

2.2 Date Palm Seed Ash (DPSA)

Dates seed were dried in open air for about two months and burnt in open air until the seed loses its internal moisture. Then these burnt seeds were grinded using conventional method and then sieved to less than 90 microns manually to meet the size of cement.

Table 2. Chemical Properties of DPSA

S. No.	Element Name	Symbol	Normal DPSA (in %)	Calcinated DPSA (in %)
1	Potassium	K	43.02	37.06
2	Oxygen	O	28.85	31.21
3	Phosphorous	P	7.40	8.55
4	Chlorine	Cl	5.54	4.74
5	Calcium	Ca	5.52	3.82
6	Magnesium	Mg	3.50	3.67
7	Sulphur	S	2.77	2.61
8	Silicon	Si	0.89	1.90
9	Iron(Ferrous)	Fe	0.70	3.61

2.3 Aggregates

Coarse aggregate (Granite) of maximum size 20 mm were used. River sand was used as fine aggregate.

Table 3. Properties of coarse and fine aggregates

Properties	Coarse aggregate	Fine aggregate
Specific gravity	2.68	2.39
water absorption	1.56	0.80

3. Experimental investigation

Specimens were casted for a replacement of DPSA by 0%, 2%, 4%, 6%, 8% and 10% with cement. The cube specimens of size 150 × 150 × 150 mm were cast for compressive strength, and cylindrical specimens of diameter 150 mm and height of 100 mm for sorptivity test. The broken specimen powder from compressive strength was taken in a beaker dissolved in pH neutral water to test for alkalinity.

4. Results and Discussion

4.1 Effect of Compression

The compression test was carried using a compression testing machine and the results in MPa for three samples are given below. Each sample was arranged to face the load in X, Y and Z axis and tabulated below in the same order.

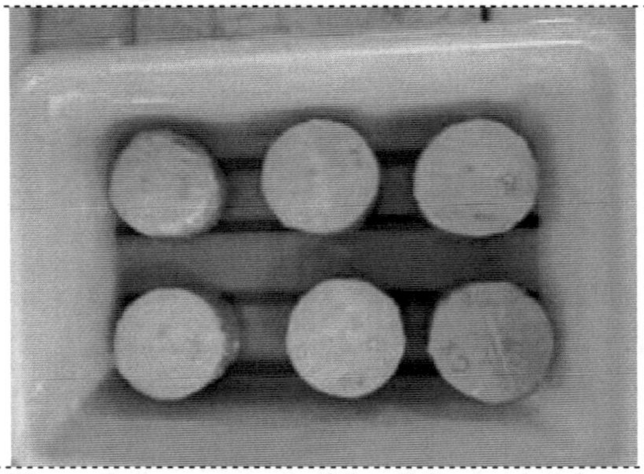

Figure 1. Arrangement of Sorptivity test in lab

Figure 2. Arrangement for Compression Test

Table 4. Effect of DSA replacement in strength at the end of 3,7,14 and 28 days.

Mix ratio / Days	3 days	7 days	14 days	28 days
	Strength in MPa			
	9.85	10.02	12.18	8.59
0	8.81	10.52	12.11	8.33
	7.76	11.02	12.04	8.07
	22.36	20.30	21.84	22.02
2	23.00	19.31	21.57	21.38
	23.64	18.31	21.30	20.73
	10.66	20.65	25.27	23.36
4	10.07	19.39	23.43	21.80
	9.47	18.12	21.58	20.23
	13.00	18.61	16.03	15.89
6	11.63	16.20	16.16	17.19
	10.26	13.78	16.28	14.50
	9.73	12.70	22.10	16.22
8	8.50	11.88	17.04	15.56
	7.27	11.06	11.98	16.68
	5.18	6.01	5.89	6.96
10	6.33	6.45	5.94	6.72
	7.47	6.89	5.99	6.66

The average of each replacing ratio against the strength is plotted in the graph below

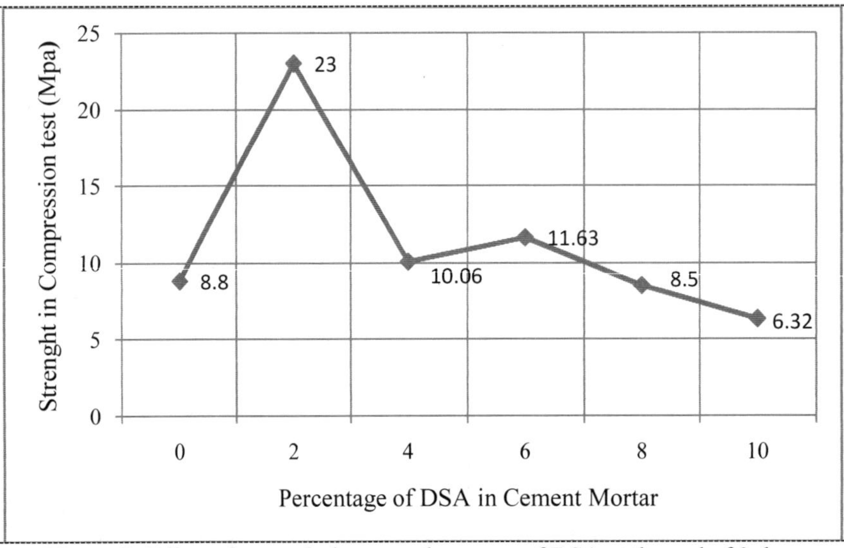

Figure 3. Effect of strength due to replacement of DSA at the end of 3 days

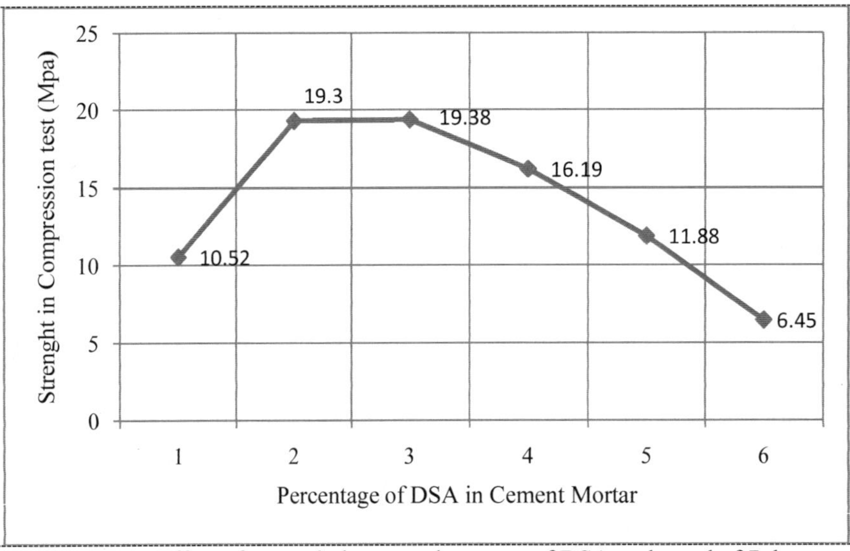

Figure 4. Effect of strength due to replacement of DSA at the end of 7 days

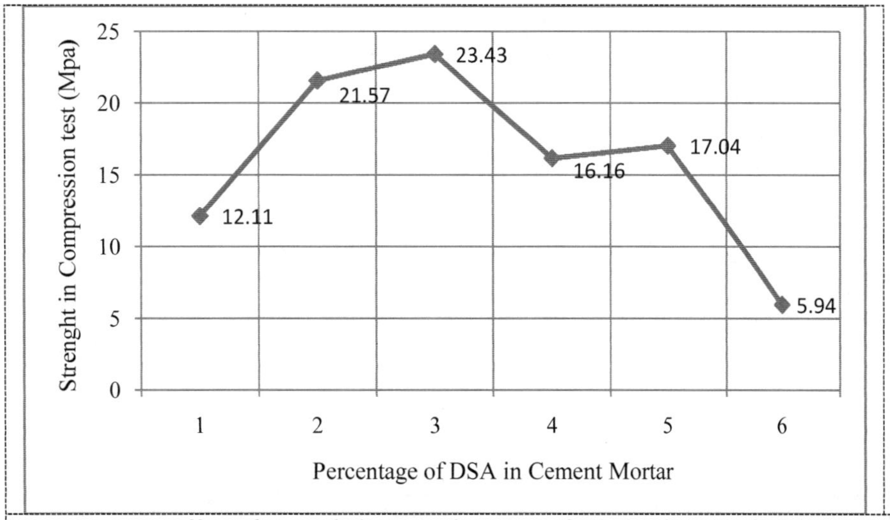

Figure 5. Effect of strength due to replacement of DSA at the end of 14 days

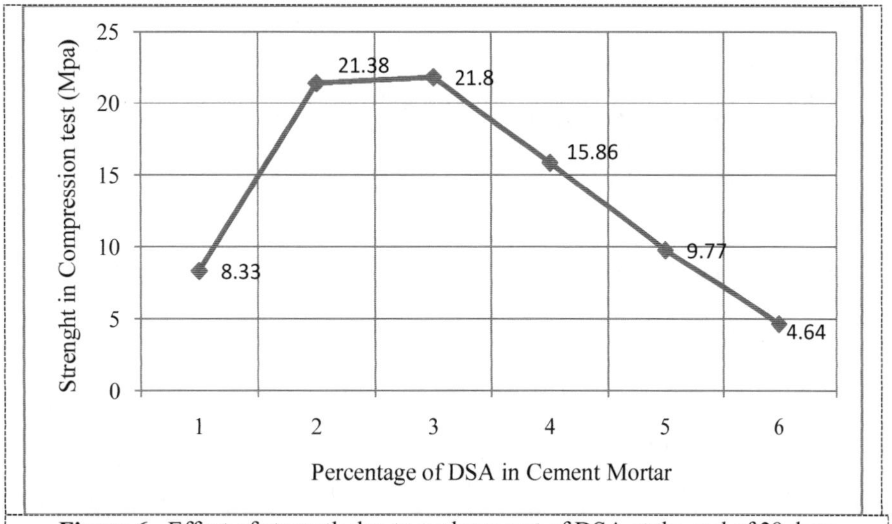

Figure 6. Effect of strength due to replacement of DSA at the end of 28 days

From the above observations, we could see that the initial compressive strength was poor for 4% ratio for 3 days curing, but it increased during the period of curing. However, 4% DPSA ratio had a sustained performance towards the end.

4.2 Water Absorption Test

The mortar cube after casting was immersed in water for 28 days curing. These specimens were then over dried for 24 hours at the temperature 85°C until the mass became constant and again weighed. The weight was noted as the dry weight of the specimen. After that the specimen was kept in water at 85°C for 24 hours. Then this weight was noted as the dry weight of the specimen.

Table 5. Water absorption test result for mortar cubes

Ratio in %	W1 in grams	W2 in grams	Water absorption in %
0	700	736	5.14
2	700	732	4.57
4	700	719	2.7
6	700	724	3.42
8	700	742	6.00
10	700	746	6.57

Here 4% DPSA ratio was giving a better water absorption index; however the other ratio were comparatively higher.

4.3 Sorptivity Test

The samples were placed in a recipient in contact with level of water capable to submerge them about 5 mm as shown in the Fig.1 The weight increased with time intervals was plotted the graph between cumulative volumes of water per unit surface area to the square root of time. This determines the rate of capillary rise absorption by a concrete cube which rest on a small support in a manner such that only lowest 2 to 5 mm of the cube is submerge. The specimen in contact with water was kept around 5 hours. In the test, several measurements are taken over a period of up to five hours and a straight line is fitted to the plot of the increase in mass or rise of water front, versus the square root of time as given below.

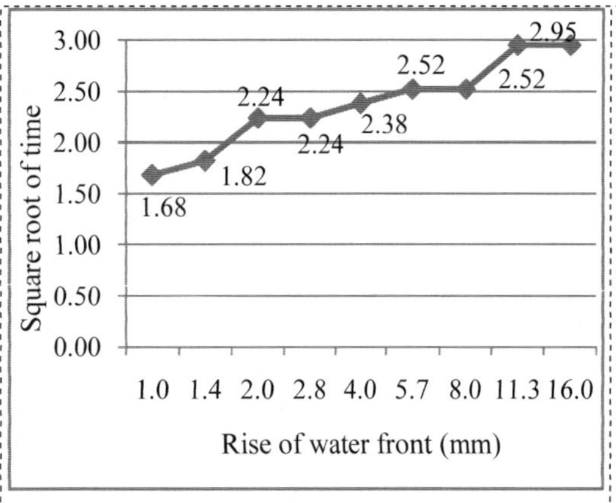

Figure 7. Sorptivity results for 0% DPSA ratio

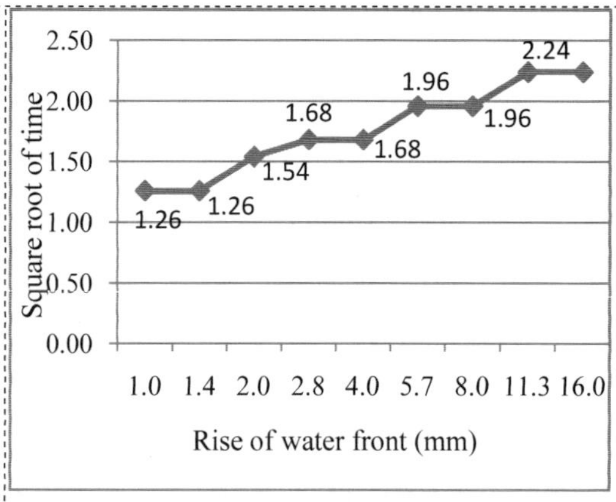

Figure 8. Sorptivity results for 2% DPSA ratio

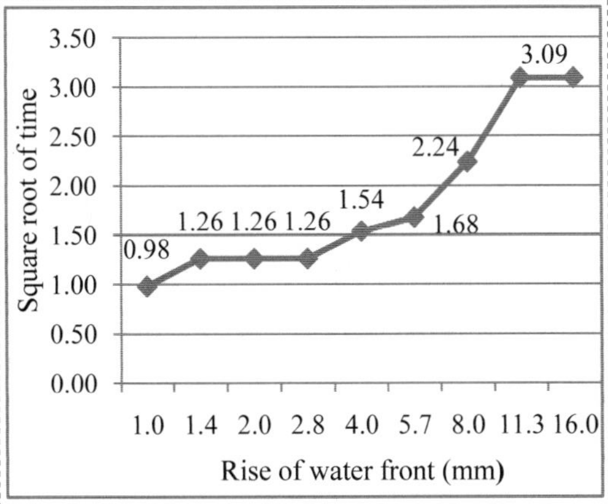

Figure 9. Sorptivity results for 4% DPSA ratio

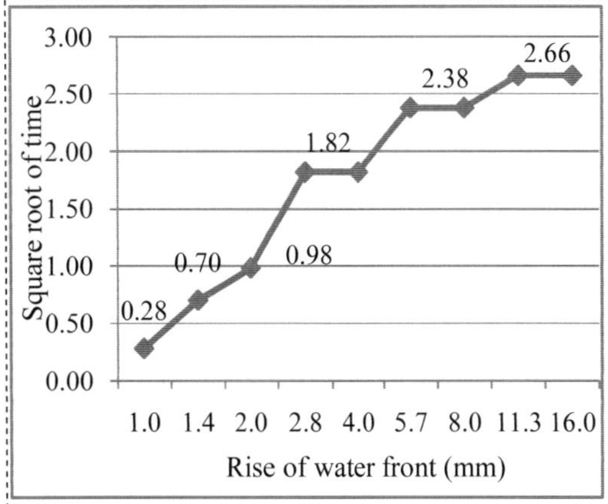

Figure 10. Sorptivity results for 6% DPSA ratio

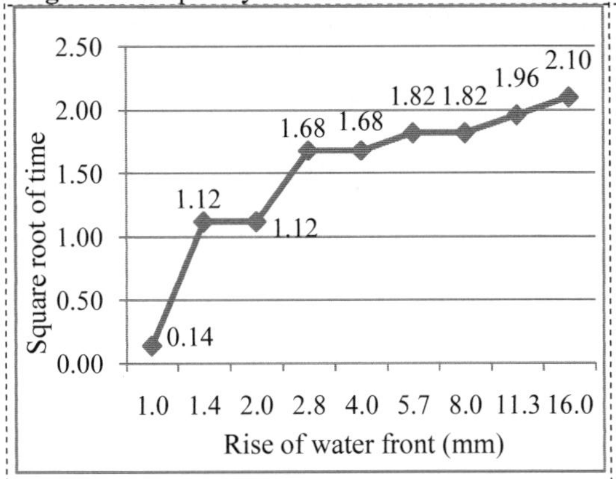

Figure 11. Sorptivity results for 8% DPSA ratio

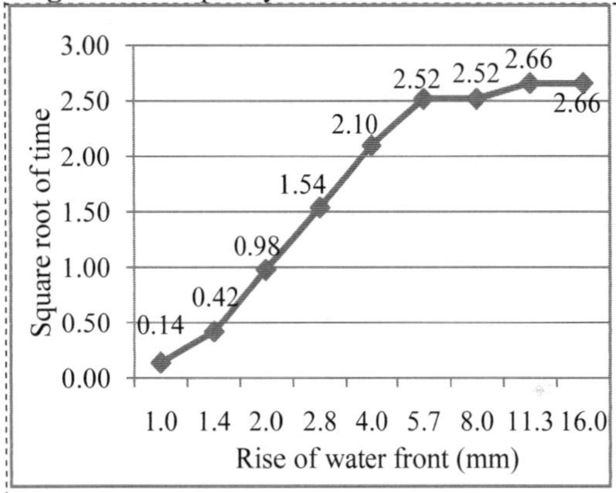

Figure 12. Sorptivity results for 10% DPSA ratio

From the sorptivity test, we can arrive that 4% ratio of DPSA gives a better sorptivity coefficient about comparing to other ratio of DPSA

4.4 Alkalinity Test

The powder from approximately middle of the broken cube was weighed for 200 grams and then was dissolved in the distilled water. This setup was left undisturbed for 72 hours, so that the powder completely dissolved in the water and the same was then taken in a beaker. The pH electrodes were immersed to it to understand the alkaline effect of the concrete. The results that were obtained are tabulated below.

Table 6. Change in pH concentration due to replacement of DSA

S. No.	Percentage of Replacement	pH
1	2	10.18
2	4	10.19
3	6	10.22
4	8	10.23

From the table, As the DPSA ratio was increased, the pH kept increasing however we limited to 8% so as to compromise along with the compressive strength.

5. Conclusion

From the above experiments and studies that were made, the below conclusions were suggested against the usage of date seed ash as a replacement for OPC in the ratio.

- 4% DPSA ratio would be an optimum amount of replacement of OPC with DPSA
- 8% replacement shows an enhanced alkaline property. Thus it can be adopted in placed where acid resistance plastering is done
- 2% replacement can be adopted in places where higher bond strength is needed. Since this gives twice the compressive strength comparative to OPC
- Calcinated DPSA do not show a good bonding strength because the calcites are being oxidized during the calcination process.

References

[1] Mehta P K 2004 High-Performance, High Volume Fly Ash Concrete for Sustainable Development. *Proc. Int. Workshop on Sustainable Develop. Concr. Technol.* 3-14.

[2] Mehta P K 1994 *Proc., of an Eng. Found. Conf.* University of New Hampshire **3** 243-56.

[3] Ramezanianpour A, Gafarpour F and Majedi M H 1995 The use of rice husk ash in the building industry. *Build. Hous. Res. Cent.* (BHRC) winter **4** 9.

[4] Ramezanianpour A Bina G and Rahimi H 2000 the role of rice husk ash in production of light-weight structural panels. *Proc, 3rd Int. Conf. Concr.* **12** 13-22.

[5] Ramezanianpour A A and Ramezanianpour A M 2005 Chloride diffusion in silica fume concrete mixtures. *Int. Symp. Concr.* **22** 456

[6] Nehdi M, Duquette J and El Damatty A 2003 Performance of rice husk ash produced using a new technology as a mineral admixture in concrete. *Cem. Concr. Res.* **33** 1203–10.

[7] Zhang M-H and Malhotra V M 1996 High-Performance Concrete Incorporating rice Husk Ash as a Supplementary Cementing Material. *ACI Mater. J.* **93**(6) 629-36.

[8] Ramezanianpour A Gafarpour F and Majedi M H 1997 Use of rice husk in the production of masonry cements. *Build. Hous. Res. Cent.* (BHRC) **7**(2).

[9] Khoddam Razavi S R 2005 Rice husk ash a mineral admixture for high performance concrete. *MSc Thesis* Amirkabir University of Technology Press

[10] Vosugh Sh 2001 Mechanical properties and Durability of Concretes Containing of Rice Husk Ash. *MSc Thesis* Amirkabir University of Technology Press

[11] Bui D D 2001 Rice husk ash a mineral admixture for high performance concrete. *PhD Thesis* Delft University Press

[12] Bui D D, Hu J and Stroeven P 2005 Particle size effect on the strength of rice husk ash blended gap-graded Portland cement concrete *Cem. Concr. Compos.* **27** 3357-66

Study on Mechanical Properties of Concrete Using Plastic Waste as an Aggregate

B Jaivignesh* and A Sofi

School of Civil and Chemical Engineering, VIT University, Vellore-632014. India.

*Email: jai_vignesh@yahoo.com

Abstract. Disposal of large quantity of plastic causes land, water and air pollution etc.., so a study is conducted to recycle the plastic in concrete. This work investigates about the replacement of natural aggregate with non-biodegradable plastic aggregate made up of mixed plastic waste in concrete. Several tests are conducted such as compressive strength of cube, split tensile strength of cylinder, flexural strength test of prism to identify the properties and behavior of concrete using plastic aggregate. Replacement of fine aggregate weight by 10%, 15%, 20% with Plastic fine (PF) aggregate and for each replacement of fine aggregate 15%, 20%, 25% of coarse aggregate replacement also conducted with Plastic Coarse(PC) aggregate. In literatures reported that the addition of plastic aggregate in concrete causes the reduction of strength in concrete due to poor bonding between concrete and plastic aggregate, so addition of 0.3% of steel fiber by weight of cement in concrete is done to improve the concrete strength. Totally 60 cubes, 60 cylinders and 40 prisms are casted to identify the compressive strength, split tensile strength and flexural strength respectively. Casted specimens are tested at 7 and 28 days. The identified results from concrete using plastic aggregate are compared with conventional concrete. Result shows that reduction in mechanical properties of plastic aggregate added concrete. This reduction in strength is mainly due to poor bond strength between cement and plastic aggregate.
Keywords: *Plastic aggregate, Bond strength, Concrete, Mechanical properties.*

1. Introduction

In civil engineering construction use of shredded plastics has increased drastically as a partial replacement of aggregates. As it is an added advantage in terms of environmental and potential economic consideration incorporation of waste in concrete increases[1-4]. Plastic can be incinerated with energy recovery, if material recycling is not feasible. Air pollution may cause and results in acid rain because of waste incineration[6,9]. Past investigations suggest that partial replacement of aggregate of concrete with waste plastic can improve properties such as abrasion resistance, impact resistance, ductility, shock absorption and thermal conductivity[10]. It also shows that addition of plastic to concrete causes some reduction in mechanical properties such as compressive strength, split tensile strength, flexural strength[5,7,8]. Literature reviews suggest that addition of steel fiber in normal concrete improves these mechanical properties. Taking the advantages of this, steel fiber is added to plastic waste aggregate concrete to make steel fiber reinforced waste plastic mix concrete. [12,13]This paper reports the strength characteristics of plastic waste aggregate mix concrete with the addition of steel fiber for specimens of strength 30 MPa. Tests were performed for cube and cylinder and prisms for compressive strength, split tensile strength and flexural strength respectively.

2. Material Properties

2.1. Cement

Cement used as a binder material in concrete. The main characteristics of cement in the concrete mix design are developing the compressive strength within period. In this study 53 grade of ordinary Portland cement properties are confirming to code IS:12269-1987. The specific gravity of cement found in the laboratory is 3.13.

2.2. Aggregate

Fine aggregate passing through 4.75 mm IS sieve conforming to grading zone II of IS 383:1970 and having specific gravity of 2.6 was used in this study. Crushed aggregate available from local sources with maximum size of 20 mm and conforming to IS2386:1963 (part I, II and III) was used as coarse aggregate in this study. Specific gravity of coarse aggregate is found out as 2.7.

2.3. Water

Potable water was used in the experimental work for both mixing and curing purposes

2.4. Plastic waste

Waste plastic represents the discarded waste from plastic containers that were collected from plastic manufacturing plants located in Vellore. After collection of waste plastic, Plastic aggregate is obtained by crushing the plastic waste [11].

3. Mix Design

The M25 grade concrete mix proportions were arrived as per the Indian Standard IS: 10262-2009. Replacement of fine aggregate weight by 10%, 15%, 20% with Plastic Fine (PF) aggregate and for each replacement of fine aggregate 15%, 20%, 25% of coarse aggregate replacement also conducted with Plastic Coarse (PC) aggregate.0.3% of steel fiber is added by weight of cement.

4. Mix Proportion and Quantities

M-25 grade of concrete

Cement	:	395 kg/m^3
Fine aggregate	:	706kg/ m^3
Coarse aggregate	:	1151 kg/ m^3
Water	:	197kg/ m^3
Mix ratio	:	1:1.79:2.92:0.45

5. Experimental Work

The experimental work is carried out in order to find out the mechanical properties of the concrete. The cubes of size 100mm × 100mm × 100mm are casted for all the mixes to find out the compressive strength. The cylinders of 100mm Ø× 200mm deep are casted to measure the split tensile strength. The prisms of size 500×100×100 are casted to measure flexural strength. The specimens are casted, cured and tested as per IS:516-1987. Testing was done for the same samples after 7 and 28 days of curing. The Universal Testing Machine (UTM) was used for testing.

6. Test Results and Discussion

6.1 Compressive Strength

Concrete specimens are taken out from the curing tank and specimens are allowed to dry the surface moisture and kept in the compression testing machine. The load applied on the specimen until the specimen were failed and found the ultimate load of each specimen. The compressive strength of concrete was calculated and the average of three specimen compressive strength was estimated at different curing age of all the concrete mixtures. By using the result plotted the graph and shown in Figure 1 and Figure 2.

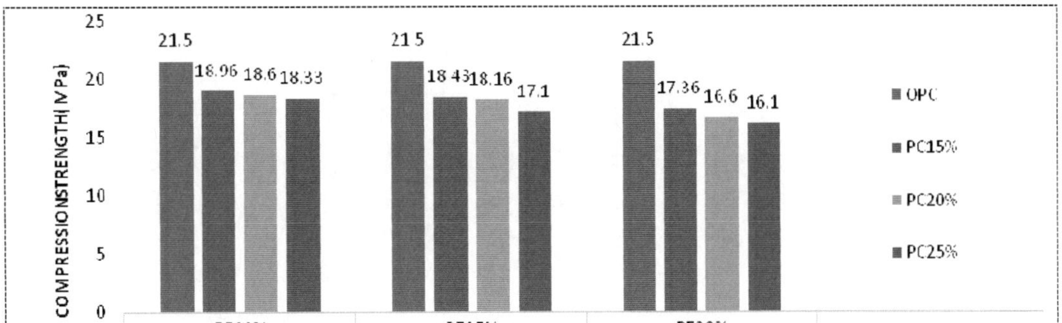

	PF10%	PF15%	PF20%
OPC	21.5	21.5	21.5
PC15%	18.96	18.43	17.36
PC20%	18.6	18.16	16.6
PC25%	18.33	17.1	16.1

Figure 1. Compressive Strength (7th day).

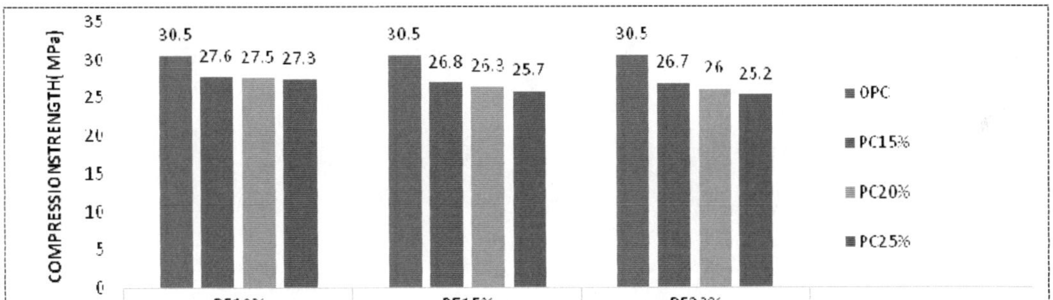

	PF10%	PF15%	PF20%
OPC	30.5	30.5	30.5
PC15%	27.6	26.8	26.7
PC20%	27.5	26.3	26
PC25%	27.3	25.7	25.2

Figure 2. Compressive Strength (28th day).

6.1.1 Discussion on compressive strength test results. The compressive strength of waste plastic used concrete was lowered by the addition of plastic. The reduction is being in the range of 9 to 17 %. This loss may be attributed to the poor bond strength between plastic particles and concrete.

6.2 Split tensile Strength

The concrete is weak in tension. There is no direct method to find out the splitting tensile strength of concrete. So the cylinder of 100mm Ø and 200mm depth were casted and cured into portable water for 7 days and 28 days. The specimens are allowed to dry the surface moisture and kept longitudinally in the compression testing machine. Load is applied at the longitudinal surface of the specimen and the ultimate load is found at which the specimen may crack. From the results obtained graph has been plotted as shown in Figure 3 and Figure 4.

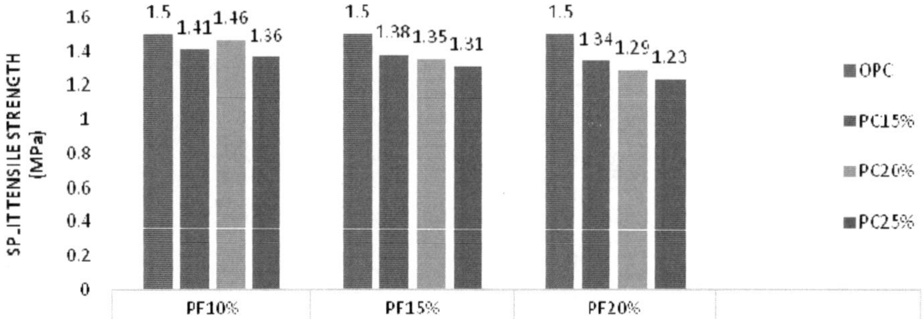

	PF10%	PF15%	PF20%
OPC	1.5	1.5	1.5
PC15%	1.41	1.38	1.34
PC20%	1.46	1.35	1.29
PC25%	1.36	1.31	1.23

Figure 3. Split Tensile Strength (7th day).

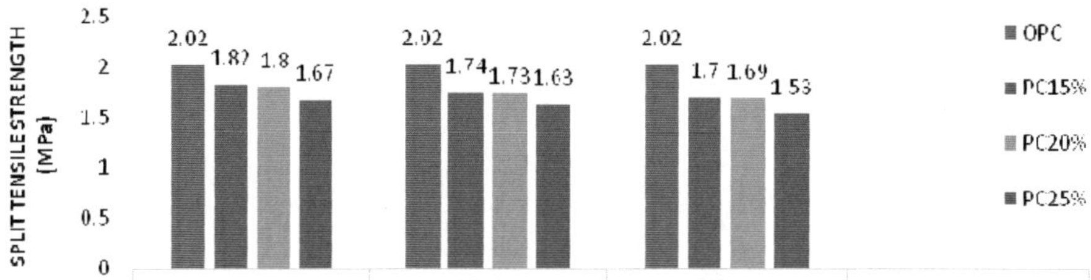

	PF10%	PF15%	PF20%
OPC	2.02	2.02	2.02
PC15%	1.82	1.74	1.7
PC20%	1.8	1.73	1.69
PC25%	1.67	1.63	1.53

Figure 4. Split Tensile Strength (28th day).

6.2.1 Discussion on split tensile strength test results. The split tensile strength of waste plastic used concrete was lowered by the addition of plastic. The reduction is being in the range of 10 to 24 %.

6.3 Flexural strength

The flexural strength was conducted on prisms of size 100 × 100 × 500 mm at the age of 7 days and 28 days to find out the behavior of beams. The specimen was mounted on the universal testing machine and two point loading was applied hydraulically which was increased until failure. From the results obtained graph has been plotted as shown in Figure 5 and Figure 6.

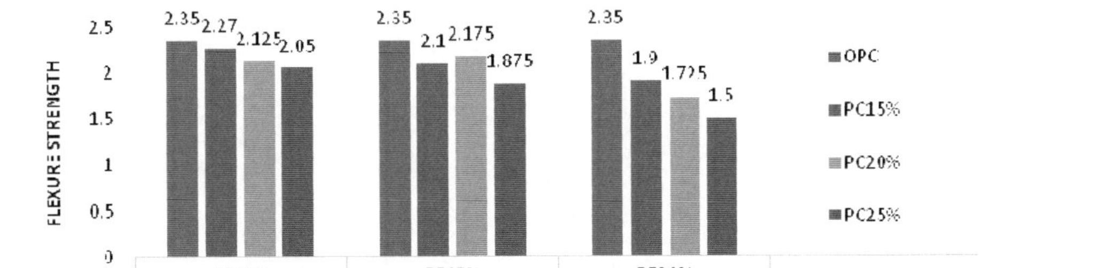

	PF10%	PF15%	PF20%
OPC	2.35	2.35	2.35
PC15%	2.27	2.1	1.9
PC20%	2.125	2.175	1.725
PC25%	2.05	1.875	1.5

Figure 5. Flexural Strength (7th day).

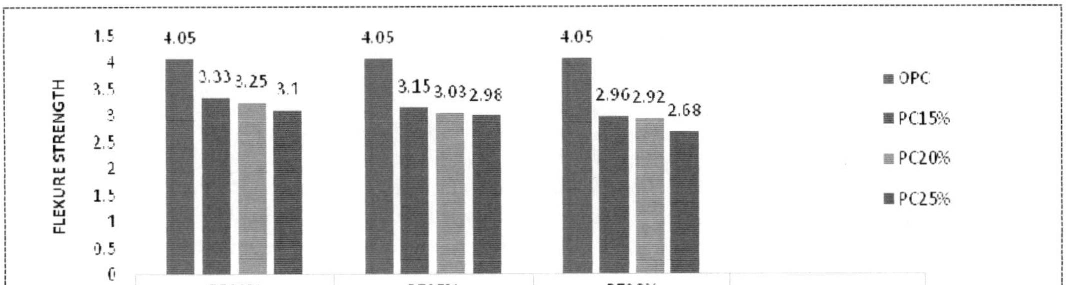

	PF10%	PF15%	PF20%
OPC	4.05	4.05	4.05
PC15%	3.33	3.15	2.96
PC20%	3.25	3.03	2.92
PC25%	3.1	2.98	2.68

Figure 6. Flexural Strength (28th day).

6.3.1 Discussion on flexure strength test results. The flexure strength of waste plastic used concrete was lowered by the addition of plastic. The reduction is being in the range of 20 to 30 %.

7. Conclusion

At all curing ages the compressive strength decreases when compared to control concrete. The decrease is mainly due to adhesive strength between the surface of the waste plastic and cement paste. The hydration of cement is also restricted by means of using waste plastic which it is a hydrophobic material. In addition waste plastic is hydrophobic material which may restrict the hydration of cement. Due to presence of plastic aggregate the reduction in split tensile strength and flexural strength were relatively less when compared with compressive strength. The use of waste plastic aggregate can be effectively used as aconservation point of view, cost-effective and energy-conserving benefits.

Acknowledgements

We would like to express our sincere thanks for VIT University Management for providing facilities in the laboratory for performing experiments.

References

[1] Osei D Y 2014 Experimental Investigation on Recycled Plastics as Aggregate in Concrete. *Int. J. Struct. Civ. Eng. Res.* **3**(2) 168-74.

[2] Praveen Mathew, Shibi Varghese, Thomas paul and Eldho Varghese 2013 Recycled Plastic as Coarse Aggregate for Structural Concrete. *Int. J. Innovat. Res. Sci. Eng. Technol.* **2**(3), 687-90.

[3] Liu F, Yan Y and Lan C 2015 Performance of Recycled Plastic-Based Concrete. *J. Mater. Civ. Eng.* **27**(2)

[4] Zainab Z. Ismail, Enas A. AL Hashmi, October 2007,"Use of waste plastic in concrete mixture as aggregate replacement", *Departmentof Environmental Engineering, college of Engineering, University of Baghdad, Iraq.*

[5] Rebeiz K S, Rosett J W and Craft A P 1996 Strength Properties of Polyester Mortar using PET and Fly Ash Wastes. *J. Energ. Eng.* **122**(1) 11052.

[6] Bhogayata A, Shah K D, Vyas B A and Arora N K 2012 Performance of concrete by using Non-Recyclable plastic wastes as concrete constituent. *Int. J. Eng. Res. Technol.* **1**(4) 1-3.

[7] Elzafraney M, Soroushian P and Deru M 2005 Development of energy Efficient Concrete Buildings Using Recycled Plastic Aggregate. *J. Archit. Eng.* **11**(4) 122-30.

[8] Patil P S, Mali J R, Tapkire G V and Kumavat H R 2014 Innovative Techniques of Waste Plastic Used in Concrete Mixture. Int. J. Res. Eng. Technol. **3**(9) 29-32.

[9] Sofi A 2016 Sustainable concrete with waste tyre rubber - An overview. *J. Chem. Pharmaceut. Sci.* **9**(2) 848-53.

[10] Sofi A and Phanikumar B R 2015 Effect of pond ash and steel fibre on engineering properties of concrete. *Ain Shams Eng. J.* **7**(1) 89-99.

[11] Sofi A and Phanikumar B R 2015 An experimental investigation on flexural behaviour of fibre-reinforced pond ash-modified concrete. *Ain Shams Eng. J.* **6**(4) 1133-42.

[12] Sofi A and Phanikumar B R 2015 Durability properties of fibre-reinforced pond ash-modified concrete. *J. Eng. Sci. Technol.* **11**(10) 1385-402.

Papercrete brick as an alternate building material to control Environmental Pollution

J S Sudarsan*,S Ramesh, M Jothilingam, Vishalatchi Ramasamy and Rajitha J Rajan
Department of Civil Engineering, SRM University, KattanChennai – 603203. India.

*Email: sudarsan.j@ktr.srmuniv.ac.in

Abstract. Utilization of concrete in the construction industry is increasing day by day. The increasing demand for concrete in the future is the major issue, for which an alternate option is to find out at a reduced or no additional cost and to reduce the environmental impact due to increase of cement industries that are important ingredient to economic development. It turns out urgent to find out alternate for the partial replacement of concrete and cement, as natural sources of aggregates are becoming exhausted. As large quantity of paper waste is generated from different countries all over the world which causes serious environmental problems, So in this present study abandoned paper waste was used as a partial replacement material in concrete,. Study indicates that 80% of the construction cost of a building was contributed by building material and still millions of people in developing countries like India are not able to afford the cost of construction of house. This study is based on potential use of light weight composite brick as a building material and potential use of paper waste for producing at low-cost. Experimental investigation was carried out to analyse optimization of mix for papercrete bricks depending upon the water absorption, compressive strength and unit weight. Papercrete bricks were prepared out of waste paper, and quarry dust with partial replacement of cement by another industrial by-product Fly Ash in varying proportions of 25%, 40% and 55%. The properties like mechanical strength, standard quality comparisons with the conventional bricks through standard tests like hardness, soundness, fire resistance and Cost-Benefit Analysis were performed and studied. The specimens of dimension 230mm x 110mm x 80mm were subjected to 7 Days and 28 days air curing and sun drying before tests were performed on them. Based on the study it was found that for non-load bearing walls papercrete bricks are best suited.
Keywords Low Cost Material, Waste Paper, Light Weight Material, Papercrete Hollow Block, Compressive Strength

1. Introduction

Large amount of paper is used for different activities and four hundred and fifty (450) million tons of paper is produced across the world. It involves chopping off trees and deforestation that poses a major environmental problem faced by our society in the current scenario, as 42% of all global wood harvest is used to produce paper [1]. This causes contamination, which causes release of chlorine-based bleaches, which is used during production, and the methane gas generated as a result of rotting of paper [2]. The 3rd largest industrial polluter of environment is the pulp and paper industry and it is difficult to identify landfill sites to deposit them [1]. In Fast Growing countries, this causes a threat to municipal solid waste management. In 2015, approximately 62 million tons of municipal solid wastes were generated in India. Only 12 percent of this was processed and safely disposed. The staggering 88 percent of India's solid waste was dumped either in unsanitary landfills or in public areas indiscriminately [3]. It contains plastic waste, electronic and medical waste, construction and demolition waste etc. The raw materials required become deficit, when the waste produced reaches their maximum level.

A large demand being placed on the materials used for construction , this study aims in taking the best out of the paper waste, by utilizing it as a building material. The thrust on the construction material increases in proportion to the generation of waste paper. This inconsistency can be managed with the use of papercrete bricks. It not only leads to the natural resources conservations but will also harvest better ways of tackling residuals and by-products. The purpose of this research is to take advantage of the waste materials like paper and to replace the costly and rare conventional building materials. Papercrete material has been found many years ago but rediscovered recently [4]. Papercrete is a material consisting of fly ash, waste paper, quarry dust and cement. To be used as a long-lasting building material, the substances are mixed with water, which can then be placed into a mould and allowed to dry.

2. Objective of the Research

The objective of the study is to investigate the properties of Papercrete bricks that were prepared out of waste paper, quarry dust and Fly Ash in varying proportions of 25%, 40% and 55%. The following properties were studied and a comparative study with conventional bricks was done:

- Mechanical properties
- Weight comparison
- Quality Standards like hardness, soundness and fire resistance
- Environmental compatibility

3. Experimental Procedure

3.1 Specific Gravity Test for Fine Aggregate and Paper Pulp

- A clean and dry pycnometer with its cap is taken and weighed in grams (W1).About 200g of dry sample (quarry dust or paper pulp) passing through it was put in it and gross weight of the pycnometer and the sand was again taken (W2).
- The pycnometerwas then filled with distilled water up to the hole in the conical cap and shaken gently until all the air bubbles escaped. Then the weight of the pycnometer with the sand and water was taken in grams (W3).
- The pycnometerwas then emptied and cleaned thoroughly. The clean pycnometerwas then filled with distilled water up to the hole in the conical cap and weighed in grams (W4).

3.2 Sieve Analysis

About 1 kg of dry sample of fine aggregate was taken and sieved on IS sieve number: 470, 240, 60, 30 and 15. This operation was continued until number particles passed through the sieves finally. Then the material retained in each sieve was collected and weighted. The results were tabulated and the percentage of fine aggregate of varying size that passes through each sieve was calculated and recorded. The value obtained was compared with grading limit chart for fine aggregate [7] and the actual zone, to which particular fine aggregate comes under, was determined.

3.3 Determination of Bulk Density

To find, the bulk density of quarry dust, the following procedure was employed. The empty weight of the container was taken as W_c. The container was filled with aggregates sample for about one third height and was tamped evenly with 25 blows. Similarly same quantity of aggregates were added as second layer and was tamped with 25 blows. A third layer of aggregate was added until it overflows and was tamped with 25 blows. Using tamping rod as a straight edge, the surplus aggregate was struck off. Measure the weight W_1. The container was emptied and it was filled again until it over flows using a shovel, aggregates being poured from a height not exceding 5 cm above the top of the container. Surface was leveled and the weight was taken as W_2.

3.4 Specific Gravity of Fly-Ash and Cement

The procedure to find the Specific Gravity of Fly-Ash/Cement is as follows . Having taken a clean and dry Le Chatlier Flask with its stopper, its weight was measured as W1. Filling half the flask with test specimen (fly ash/cement) the weight was measured as W2. Following which kerosene was added until it is partially filled. The mixture was mixed using a glass rod to remove entrapped air. Stirring was continued and more kerosene was added up to the graduation mark. The weight W3 was measured. Emptying the flask and refilling it with clean kerosene up to the graduation mark the weight W4 was measured.

Specific gravity of fly-ash/cement was then calculated by using the formula given in Eq. (1):

$$\text{Specific gravity} = W3 - W1/\left[(W4 - W2) - (W3 - W1) \times 0.79\right] \quad (1)$$

Here 0.79 is the specific gravity of kerosene

3.5 Quality Standard Tests on Papercrete bricks

This involves the study of various qualities of Papercrete bricks. The presence of soluble salts was determined by efflorescence test. In which the fibrous concrete brick was immersed in water for 24 hours. Following which they are kept in shade to dry. Having taken all the three proportion bricks, the hardness of the brick was determined by making a scratch on the brick surface. To study the soundness property of the brick, two bricks of same proportion were struck with each other. By setting a few bricks to fire, it enabled the study on the fire resistive nature of the bricks. In projects we prefer a simple and easy test that can be conducted in a very short span of time for certain specifications and to control the quality. The very thought is that we can make use of rapid chloride permeability test (RCPT). By observing the quantity of flow of electrical current through a 50 mm thick by 100 mm in diameter sample for duration of 6 hours, the rapid chloride permeability test was performed. The sample was cut as a slice of a core or cylinder. Throughout the test a potential difference of 60V DC was sustained across the electrodes. One lead was immersed in a 0.3 M caustic soda (NaOH) solution and the other in a 3.0% salt (NaCl) solution. A conditional ranking was made for the concrete's permeability based on the charge that passes through the sample. Finally, in compression testing machine the brick was tested for cracking pattern under different loads.

4. Results and Discussions

As per the procedure given, to study the properties of the brick various tests were conducted. On performing the Specific Gravity Test on Fine Aggregates for three different samples, the specific gravity was calculated as 2.52. The summary of the entire result is shown in Figure 1. Secondly, Sieve Analysis test performed on quarry dust revealed that it has a fineness modulus of 2.8 and the data obtained is tabulated in Table 01. From the results obtained in the table, a graph was plotted between sieve sizes and the percentage passing as shown in Figure 2. The third test being the determination of bulk density of quarry dust, helped in calculating the value of bulk density of compacted quarry dust as 1.7kg/l and that of loaded quarry dust was 43%. By Le Chatelier method, the specific gravity of Fly-Ash and Cement was determined. Specific gravity of cement was 3.13 and fly ash was 2.33. Specific Gravity Tests on Paper Pulp was done and the results are shown in Figure 3.

Among the Quality Standard Tests, Efflorescence Test on Papercrete Bricks proved no white or grey surface deposits on the bricks. Hence, the bricks are free from soluble salts (Figure 4). The result of Hardness Test proved that the brick is sufficiently hard with a less impression due to the scratch. The clear ringing sound produced during the Soundness Test proves that the bricks are good. From Fire Resistance test, it was observed that in an open flame the Papercrete bricks did not burn. They burnt slowly like charcoal. However flaming for several hours, these bricks would be reduced to ashes as shown in Figures 5 and 6. If the interior plaster and exterior stucco are provided in papercrete bricks, the bricks will not burn. In addition, if plaster is applied without any holes or leakage in the bricks, due to absence of oxygen it will not get burnt inside. As we know corrosion of the reinforced steel which is held within the concrete as a result of chloride intrusion is a frequent environmental attacks that guides to the downturn of the concrete structures.

IOP Conf. Series: Earth and Environmental Science **80** (2017) 012017 doi:10.1088/1755-1315/80/1/012017

This durability issue has received prevalent awareness in modern years due to its regular occasion and the linked high cost of repairs. Rapid Chloride Permeability Test (RCPT) results indicated Chloride permeability as per ASTM C1202 as HIGH . Conventional Bricks are highly brittle in nature and fail as soon as load is applied. They usually crack from middle and break into multiple pieces. However, papercrete bricks never failed catastrophically, they just are compressed like squeezing rubber. Therefore, the test was conducted under full compression. Even at higher load, the structure experiences outer face cracks and peels out without structural fail, whereas the papercrete brick failed. This is mainly due to the higher elastic nature at failure which compresses like rubber. Since the bricks experiences compression under heavy loading, the roof may get collapsed resulting its application for non-load bearing partition walls alone.

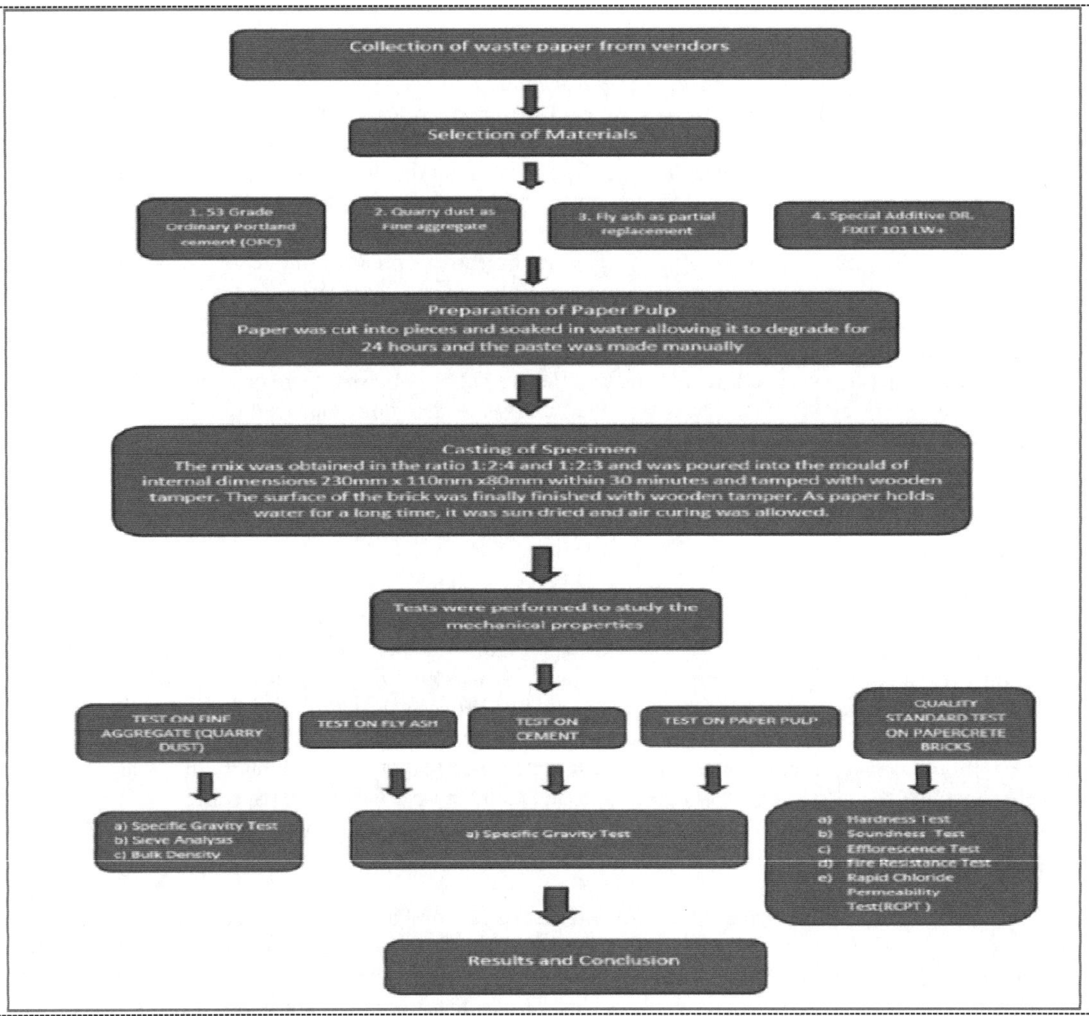

Figure 1. Flow Chart Showing Experimental Procedure.

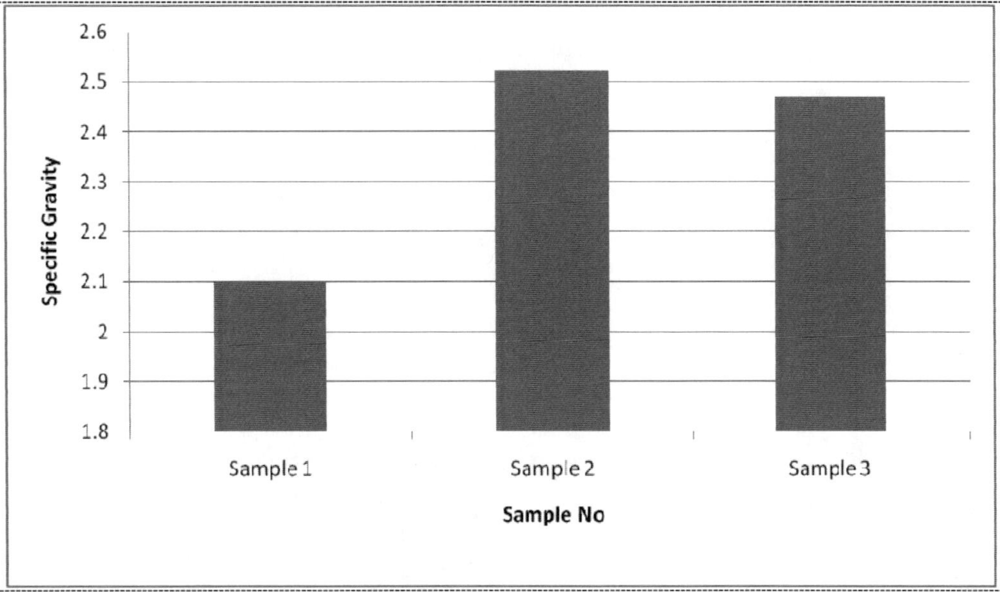

Figure 2. Specific gravity test of fine aggregates.

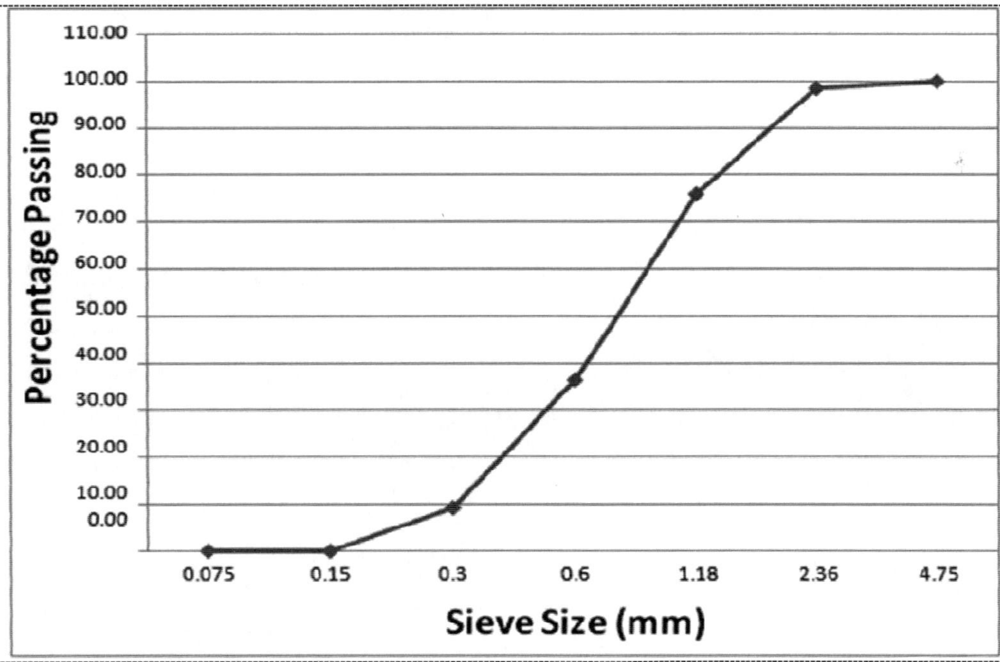

Figure 3. Sieve Analysis of Fine Aggregate.

Table 1 Sieve Analysis of Fine Aggregate

Sieve Size	Weight Retained	% Weight Retained	Cumulative % Weight Retained	% Passing
4.75mm	0	0	0	100.00
2.36mm	15	1.5	1.5	98.50
1.18mm	225	22.5	24	76.00
600 μ	396	39.6	63.6	36.40
150 μ	92	9.2	100	0
75 μ	0	0	-	-
Pan	-	-	-	-
			$\sum = 279.9$	

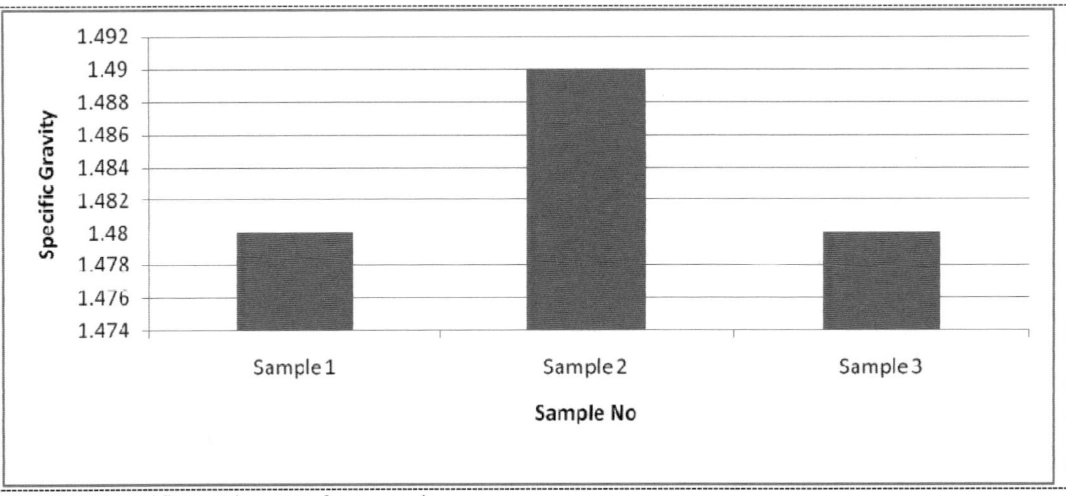

Figure 4. Specific gravity test of paper pulp.

Figure 5. Efflorescence Test of Papercrete bricks.

Figure 6. Papercrete brick during fire test.

Figure 7. Papercrete brick after fire test.

5. Conclusion

Papercrete bricks satisfy the basic characteristics of conventional bricks as various tests results proved in the study. The basic qualities of conventional brick are: they should be of uniform colour, should have even surfaces, free from cracks and should have sharp and defined edges. They should be hard; so that no impression is left when scratched with fingernails and should produce clear ringing sound when struck with each other. The bricks when tested in accordance with the procedure laid down, the rating of efflorescence shall be 'Nil'. Average weight should be 30N to 35N.And the specific gravity ranges from 2.6 to 2.8. The bricks should have low thermal conductivity and should have percentage of water absorption by weight less than 20%, when soaked in cold water for 24 hours.

The above characteristics are clearly observed in our study and they are special in the case of fire resistive nature. Thus, papecrete bricks prove to be one of the emerging solutions for eco-friendly buildings.

References

[1] Rohit Kumar Arya and Rajeev Kansal 2016 Utilization of Waste Papers to Produce Ecofriendly Bricks *Int. J. Sci. Res.* **5**(8) 92-96

[2] Fuller B J, Fafitis A and Santamaria J L 2006 The Paper Alternative. *Civ. Eng. Mag. Arch.* **76**(5) 72-7.

[3] Joo-Hong Chung, Byoung-Hoon Kim, Hyun-ki Choi and Chang-Sik Choi 2015 Development of Papercrete due to Paper mixing ratio. *Int. Conf. Sustain. Build. Asia* 317-20.

[4] Shewit Birhane, Mikyas Mesfin and Werku Koshe 2017 Experimental Study on Some Mechanical Properties of Papercrete Concrete. *Adv. Mater.* **6**(1) 1-6

[5] Anandaraju K, Jose Ravindra Raj and VijayaSarathy R 2015 Experimental Investigation of Papercrete Brick. *Int. J. Machine Constr. Eng.* **2**(2).

[6] Ahmad S and Malik M 2013 Study of Concrete Involving Use of Waste Paper Sludge Ash as Partial Replacement of Cement. *IOSR J. Eng.* **3**(11) 6-15.

[7] IS: 383-1970 Specification for coarse and fine aggregates from natural sources for concrete. *Bureau of Indian Standards*. New Delhi.

[8] IS: 2386 (3) 1963 Methods of test for aggregates for concrete. *Bureau of Indian Standards*. New Delhi.

[9] IS: 1077-1992 and IS: 3495-1992 Common burnt clay building bricks – specification. *Bureau of Indian Standards*. New Delhi.

[10] IS: 3495 1992 Methods of tests of burnt clay building bricks. *Bureau of Indian Standards*. New Delhi.

[11] IS: 4031(1) 1996 Method of physical tests for hydraulic cement. *Bureau of Indian Standards*. New Delhi.

[12] IS:8112 1989 Ordinary Portland Cement, 43 grade specification. *Bureau of Indian Standards*. New Delhi.

[13] Cusidó J, Cremades L, Soriano C, Devant M 2015 Incorporation of paper sludge in clay brick formulation: Ten years of industrial experience. *Appl. Clay Sci.* **108** 191-8.
[14] Raut S, Sedmake R, Dhunde S, Ralegaonkar R and Mandavgane S. 2012 Reuse of recycle paper mill waste in energy absorbing light weight bricks. *Constr. Build. Mater.* **27**(1) 247-51.
[15] MonteiroSVieira C 2014 On the production of fired clay bricks from waste materials: A critical update. *Constr. Build. Mater.* **68** 599-610.

IOP Conf. Series: Earth and Environmental Science **80** (2017) 012018 doi:10.1088/1755-1315/80/1/012018

Strength and durability studies on concrete with partial replacement over burnt brick bat waste

S Kanchidurai*, G Bharani and K Saravana Raja Mohan

School of Civil Engineering, SASTRA University, Thanjavur, Tamil Nadu, India.

*Email: kanchidurai@civil.sastra.edu

Abstract. This paper presents the partial and complete replacement of over burnt brick bat (OBB) 20-30mm as coarse aggregate in the concrete. OBB are formed at extreme heating to a temperature not less than 1600 degree Celsius. The burnt bricks change from red to blue-black ceramics color. The series of tests are conducted to study the effect of 0%, 25%, 50%, 75% and 100% replacement of coarse aggregate with over burnt bricks. Totally 36numbers of 150mm concrete cube with 5 different percentage replacement mix are cast and tested and three numbers of the flexural beam. In durability aspects, water absorption and sorptivity were tested. Experimental results found 25-50% of overburnt brick bat wastes can be replaced with the normal and mass concrete without quality compromisation.
Keywords: Brick bats, Compressive strength, Flexural strength and sorptivity

1. Introduction

Concrete is the base material for construction industries. It is strong in compression and weak in tension, the main constituent of the concrete in cement, sand, coarse aggregate and water, replacing some of these material makes significant changes in cost as well as performance. Coarse aggregate filled almost 70% of volume in concrete [1], the cost of coarse aggregate rapidly increasing also the availability of the aggregate is getting reduced. The major cost of the concrete is belonged to the aggregate [2]. The over burnt brick bat (OBB) wastes 20-30mm size having available in the brick manufacturing industries. In recent research, the OBB wastes were replaced with concrete. This OBB maintains strength and performance to the concrete also reduce the weight of the concrete [3[4[5]. Strongly OBB replaced with concrete can be performed in the mass concrete filling area [6]. In environmental aspects replacement of OBB in concrete is reduce the conservation in the natural resources [7].Raw material utilization can be diminished which at last spare time and vitality. These will diminish the measure of ozone-harming substance era [8]. The blocks which are close to the fire in the oven subjected to high warmth more than 1000 degree centigrade it will shrink and changes in shape, the shading ends up noticeably ruddy and its appearance like rosy to blackish inclination stone[9][10]. This over consumed block fills in as waste in the development business and needs to amass some place during the time spent reusing.

The primary objective of this research is to study the influence of partial and complete replacement of coarse aggregate with brick bats, and to compare it with the compressive and flexural strength of ordinary M20 concrete. Also trying to find the percentage of brick bats replaced in concrete that makes the strength of the concrete maximum. Brick bats are usually considered as waste material. So, by replacing coarse aggregate with brick bats, proposing a method that can be of great use in reducing pollution to a great extent.

2. Properties of Material

Ordinary Portland Cement (OPC) 53 grade cement conforming to IS 10262 – 2009 [11] was used. The physical properties of the used cement, sand, coarse aggregate and brick bat aretested provision given by Indian standard Table 1 shows the physical properties of raw materials.

3. Mix design

Mix proportion of the available raw material is derived as per IS456:2007 [12] as specified in Table2. Mix Ratio1:1.57:2.75.

Table 1. Physical properties

No	Material	Description	Value
1	Ordinary Portland Cement	Specific gravity	3.145
2	Ordinary Portland Cement	Standard consistency (%)	28%
3	Ordinary Portland Cement	Initial setting time	32 min
4	Ordinary Portland Cement	Compressive strength	54.25 MPa
5	River sand (FA)	Specific Gravity	2.60
6	Coarse Aggregate	Specific Gravity	2.71
7	Coarse Aggregate	Water absorption	0.50
8	Over burnt bricks	Specific Gravity	2.17
9	Over burnt bricks	Water absorption	4.4

Table 2. Mix proportion for M20 grade concrete

Cement (kg/m^3)	Fine aggregate (kg/m^3)	Coarse aggregate (kg/m^3)	Water (l/m^3)	Water cement ratio
399.2	642.8	1097.2	191.6	0.48

4. Experimental investigation

4.1 Compressive strength

Totally 36 numbers of cubes and 3 number of beams were casted, out of which 9 cubes were conventional and other 27 cubes were replaced with 25%, 50%, 75% and 100% for coarse aggregate by OBB, this was formed five sets of concrete cube compressive strength tests. Figure 1 shows the concrete mould 150 x150 x150mm placed M20 grade different replaced concrete [14].

4.2 Flexural strength

OBB replaced 25 and 50% of coarse aggregate, conventional beam casted for comparison. The specimen size was maintained 100 x 150 x 100 as specified in Indian standard [13][14]. The beams are designed as the singly reinforced beam with 2 numbers of 10mm diameter bar in tension zone and two numbers of 10mm diameter bar at compression zone. For shear reinforcement, 8mm diameter bars were used at 100mm spacing. The beams are identified as (CB) conventional beam and for beams with brick bats as

(BBB) Brick bat beam. The specimens are tested using two point loading by 100T loading frame. Figure 2 shows the flexural beam mould placed by different percentage

4.3 Water absorption
As per guidelineof ASTM C642 [15] water absorption of partially mixed OBB concrete was tested. Table 5 shows the percentage of water absorption of OBB mixed concrete. Initially, the sample dry weight was measured in cool temperature, after 110 degree Celsius and 24hours kept in the oven specimen.Similarly, different masswas taken and compared natural temperature as per ASTM procedure.

4.4 Sorptivity
This strategy is planned to decide the helplessness of an unsaturated cement to the entrance of water. At 50'C for 3 days the 100mm diameter and 50mm height test specimen were placed hot oven, after oven curing, specimen placed normal temperature in for 15 days.The complete circumference of the specimen was sealed and top and bottom portion kept open.Mass of dried specimen was noted and the procedure as per ASTM C1585 [16].The absorption I (mm) = Δm/ (a x d) (a-exposed area, Δm- change in mass in g and d-density of water in g/cu.m).

An experimental test was performed in the compressive testing machine (CTM) which is having 3000 kN capacity. The flexural strength of beams istested by universal loading machine (UTM). The ultimate load carrying capacity of beam was found by subjecting the beam to two-point loading using 1000kN loading frame. The deflection and flexural strength of beam is measured by placing LVDT at the mid-span and quarter of span of the beam. In this two point loading method, the load is divided into two parts as, P = P/2+P/2, here 'P' is the total load given on the beam through the loading frame and is divided into two separate loads by placing two I-sectioned rollers below the beam and the load is transferred to beam. The load at first crack is noted when the cracking was developed in the concrete and the maximum load is indicated by LVDT. From this test it is found out there is no much difference in load carrying capacity between the conventional beam and beam with replaced brick bat.

Figure 1. Concrete specimens for compressive strength and durability test

Figure 2. Flexural member concrete placement for 25% OBB

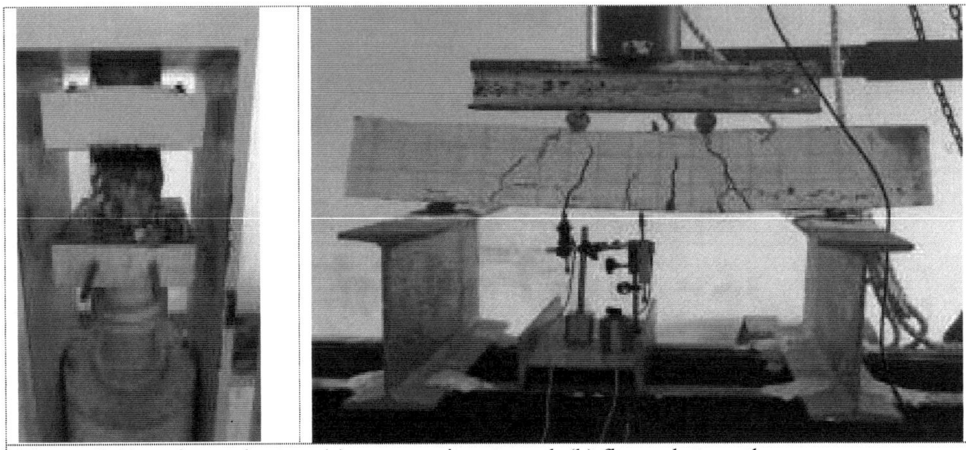

Figure 3. Experimental set-up (a) compressive strength (b) flexural strength

5. Results

Table 3. Compressive strength of cubes

Replacement / Curing	7 Days	14 Days	28 Days
Conventional	21.06 MPa	29.15 MPa	30.18 MPa
25% Replacement	14.57 MPa	17.3 MPa	24.6 MPa
50% Replacement	8.46 MPa	12.08 MPa	16.4 MPa
75% Replacement	6.31 MPa	11.40 MPa	12.06 MPa
100% Replacement	1.92 MPa	3.99 MPa	4.98 MPa

Table 4. Flexural Strength of beam

Description	Conventional	25% Replacement	50% Replacement
Peak Load(kN)	63.1	58	46.9
First Crack Load(kN)	30.1	24.2	21.5
Flexural Strength(MPa)	19.63	18.04	14.59
Maximum Deflection(mm)	9.86	9.45	9.28
Deflection under loading(mm)	7.25	6.24	6.56

Table 5. Sorptivity test values

Replacement	I = Δm/(a x d)											
	5min	10min	15min	30min	1hr	2hr	3hr	4hr	5hr	6hr	24hr	48hr
0%	0.25	0.51	0.71	1.02	1.02	1.53	1.78	2.03	2.29	2.55	2.55	2.55
25%	0.25	1.02	0.53	1.78	2.29	2.55	2.55	2.29	2.55	2.80	2.80	2.80
50%	0.25	0.71	0.27	1.78	2.80	4.07	4.83	5.09	5.34	5.53	7.09	7.09
75%	0.51	0.51	1.02	1.53	2.29	3.31	4.07	4.33	4.84	5.34	9.31	9.31
100%	0.25	0.51	0.51	1.27	12.22	12.73	3.31	3.56	4.33	5.09	11.14	11.14

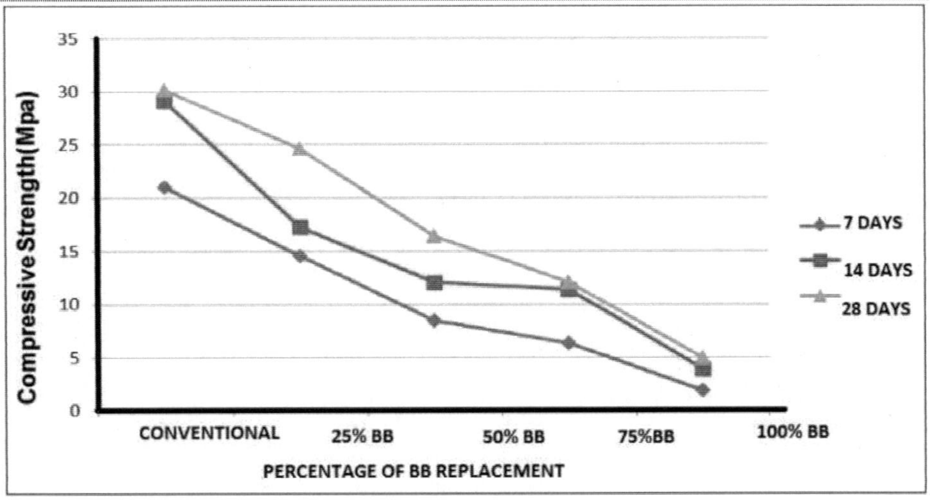

Figure 4. Graphical representation of compressive strength of cubes

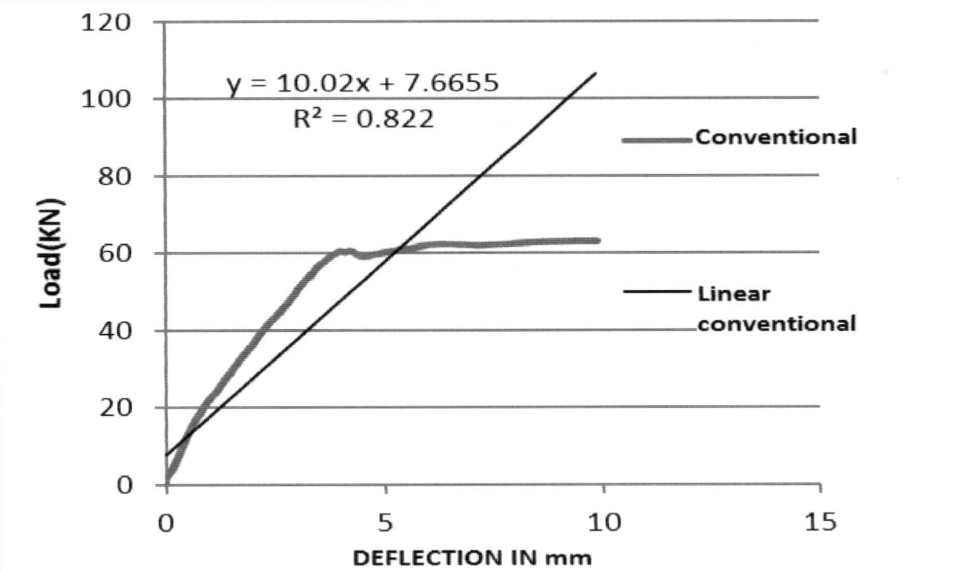

Figure 5. Graphical representation of flexural strength of conventional beam

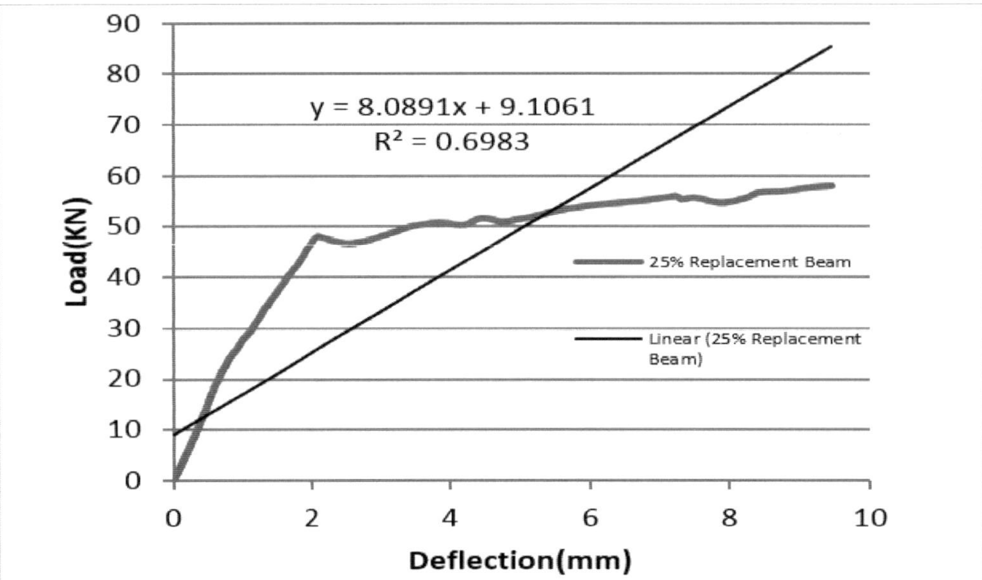

Figure 6. Graphical representation of flexural strength of 25% replaced beam

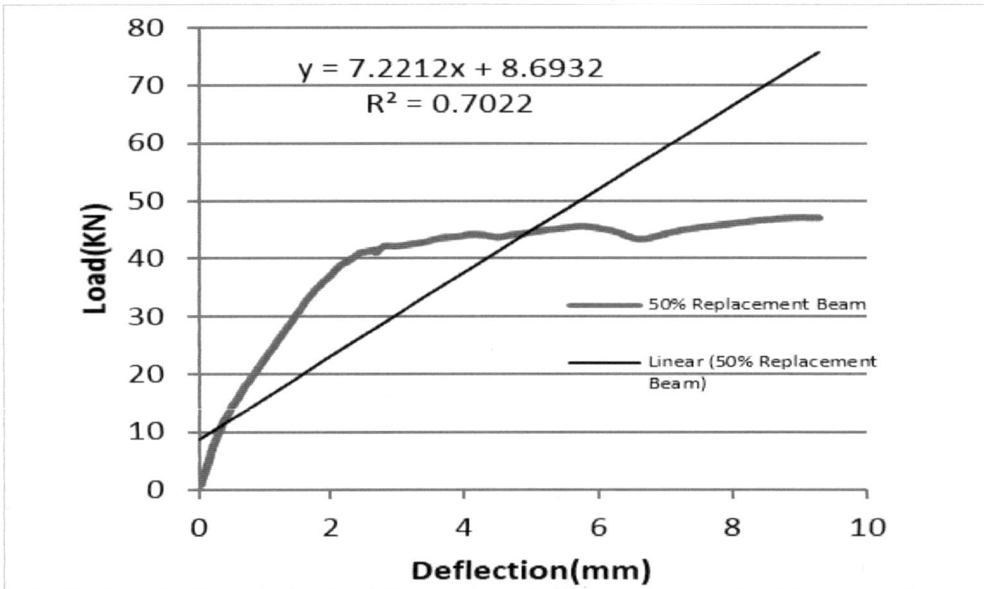

Figure 7. Graphical representation of flexural strength of 50% replaced beam

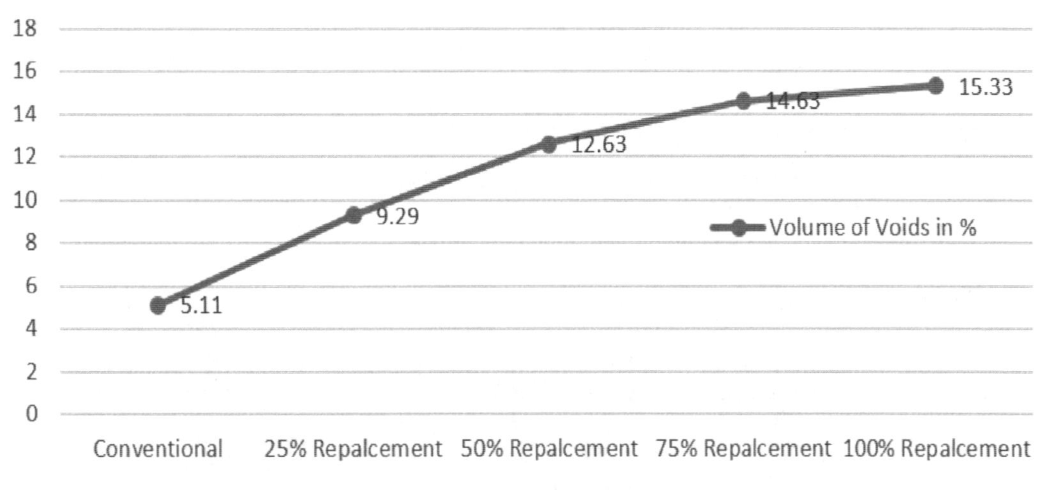

Figure 8. durability - volume of voids in %

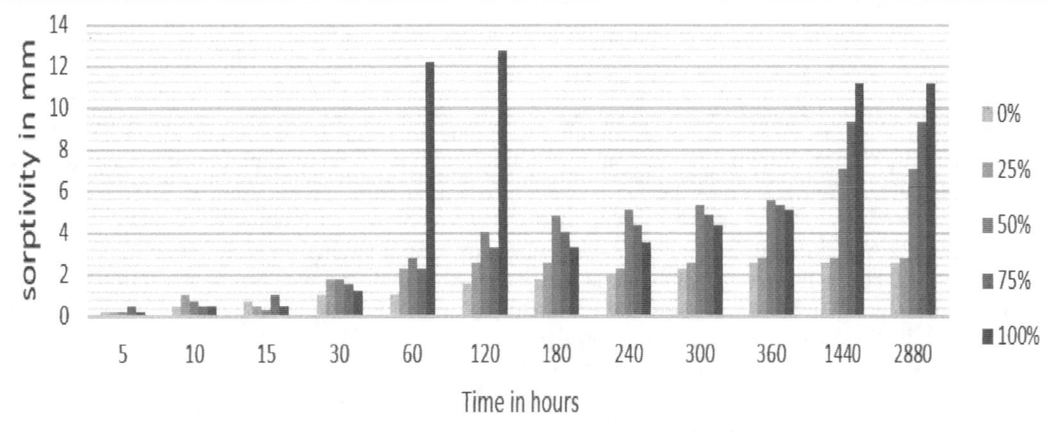

Figure 9. durability - Sorptivity I value

Table 6. Water absorption test values [17]

Description	Formula	Replacement %				
		0%	**25%**	**50%**	**75%**	**100%**
Mass of oven dry sample in air (g)	A	926	882	810	826	810
Mass of Surface-dry sample in air after immersion (g)	B	962	942	876	900	882
Mass of surface-dry sample in air after immersion and boiling(g)	C	956	930	866	882	868
Apparent mass of sample in water after immersion and boiling(g)	D	400	400	500	500	400
Absorption after immersion (%)	$[(B-A)/A]\times 100$	3.89	6.80	8.15	8.98	8.89
Absorption after immersion and boiling (%)	$[(C-A)/A]\times 100$	3.24	5.44	6.91	6.78	7.16
Bulk density, dry (g/cc)	$[A/(C-D)]\times\rho = g_1$	1.67	1.66	2.21	2.16	1.73
Bulk density after immersion (g/cc)	$[B/(C-D)]\times\rho$	1.05	1.58	2.39	2.36	1.89
Bulk Density after immersion and boiling (g/cc)	$[C/(C-D)]\times\rho$	1.72	1.05	2.37	2.31	1.86
Apparent density (g/cc)	$[A/(A-D)]\times\rho = g_2$	1.76	1.83	2.61	2.53	1.98
Volume of voids (%)	$[(g_2-g_1)/g_2]\times 100$	5.11	9.29	12.63	14.63	15.33

6. Discussion and Conclusion

Table 3 shows the decrement in the compressive due to replacing OBB but at the same time the replacement for 25% satisfies the strength for M20 grade, also the OBB replacement reduces the concrete weight slightly. Figure 4 shows the continuous decrement in compressive strength. In Table 4 shown the flexural performance, the flexural strength of 25% and 50% OBB replaced concrete attains nearly conventional flexural resistance -8% to -30% reduced flexural strength noticed, Figure 5,6 and 7 shows curves for load Vs deflection. Table 6 and figure 8 shows the durability – water absorption test results in the percentage of a volume of voids, 25% of OBB replacement in the concrete gives the nearest value of voids content. Table 5 and Figure 9 shows the results for durability in sorptivity aspect, results gives 'I' (in mm) almost near to the 25% OBB replacement in concrete.

- From the present study, it is found that brick batcan be used as a replacement of coarse aggregate by various percentages.
- The strength of the replaced concrete is less than conventional concrete and the workability is little poor than conventional concrete. This can be easily overcome by adding water reducing admixtures.
- From the compressive strength of cubes, it is found that the optimum replacement level of brick bats is 25% volume of coarse aggregate.
- Flexural strength of concrete was studied for 3 beams (Conventional, 25% Replacement and 50% Replacement). Some durability parameters such as sorptivity and water absorption were studied for all various percentages. From the results 25% and 50% replacement of brick bats are found to be suitable for concrete.
- Durability aspects water absorption and sorptivity results hold good for 25% OBB replacement.

Acknowledgement
Authors express heartfelt gratitude to Vice-Chancellor, SASTRA University for given me the permission to carry out works at SASTRA.

References

[1] Adamso M, Razmjoo A and Poursaee A 2015 Durability of concrete incorporating crushed brick as coarse aggregate. *Constr. Build. Mater.* **94** 426-32.
[2] Apebo N S, Iorwua M B and Agunwamba J C 2013 Comparative analysis of the compressive Strength of concrete with gravel and crushed over burnt bricks as coarse aggregate. *Nigerian J. Technol.* **32**(1) 7 - 12.
[3] Attaullah Shah, Irfan U Jan, Raza U Khan and Ehsan U Qazi 2013 Experimental investigation on the use of recycled aggregate in producing concrete. *Struct. Eng. Mech. An Int. J.* **47**.
[4] Ezeokonkwo J C, Apebo N S and Agunwamba J C 2014 The Suitability of crushed over burnt bricks as coarse aggregates for concrete. *Int. J. Eng. Sci. Innov. Technol.* **3**(1) 315-21.
[5] Saravanakumar P and Dhinakaran G 2013 Durability characteristics of recycled aggregate concrete. *Struct. Eng. Mech.* **47**(5).
[6] Paulo B Cachim 2009 Mechanical properties of brick aggregate concrete. *Constr. Build. Mater.* **23** 1392-7.
[7] Subhash C, Yaragal, Dumpati C Teja and Mohammed Shaffi 2016 Performance studies on Concrete with recycled coarse aggregate. *Adv. Concr. Constru.* **4**(4).
[8] Subramani T and Kumaran S 2015 Experimental investigation of using concrete waste and brickwaste as a coarse aggregate. *Int. J. Appl. Innov. Eng. Manage.* **4**(5) 294-303.
[9] Tanvir Hossain, Md. Abdus Salam and Mohiuddin Abdul Kader 2012 Pervious concrete using brick chips as coarse aggregate: An experimental study. *J. Civ. Eng. (IEB)* **40**(2) 125-37.
[10] Zhi Ge, Hongya Yue and Renjuan Sun 2015 Properties of mortar produced with recycled claybrick aggregate and PET. *Constr. Build. Mater.* **93** 851-6.
[11] IS 10262: 2009, Indian Standard Recommended Guidelines for Concrete Mix Design. *Bureau of Indian Standards.*
[12] IS 456: 2000, Indian Standard Code of Practice for Plain and Reinforced Concrete. *Bureau of Indian Standards.*
[13] SP16: 1980, Design Aids for Reinforced Concrete.
[14] IS 516: 1959, Method of Tests for Strength of concrete. *Bureau of Indian Standards.*
[15] ASTM C642 -2006, Standard Test Method for Density, Absorption and Voids in Hardened Concrete.
[16] ASTM C1585- 2004, Standard Test Method for Measurement of Rate of Absorption of Water by Hydraulic Cement Concrete.
[17] Sharmila P and Dhinakaran G 2016 Compressive strength, porosity and sorptivity of ultra-fine slag based high strength concrete. *Constr. Build. Mater.* **120** 48-53.

An Experimental Evaluation of Mechanical Properties of Hybrid Reinforcements

A Sai Kumar*, G Ganesan and K Karthikeyan

School of Mechanical and Building Sciences, VIT University, Chennai. India.

*Email: sai2103.sk21@gmail.com

Abstract. This paper presents the mechanical properties of unidirectional hybrid reinforcements formed from continuous fibres impregnated with a fibre binding material which are used for reinforcing the concrete. Recently FRP (Fibre Reinforced Polymer) manufacturers and suppliers have been increased all over the world because of the superior performance of FRP products in the construction industry. Its non-corrosive nature has turned the attention of many researchers to make several studies on different type of FRP products. Through a vast research, several standards also have been formulated. In this regard a new combination of FRP materials is tried in this paper and its properties have been derived. Carbon fibre and glass fibres fuse in this study to form a new hybrid rebar. The design properties such as tensile strength, tensile modulus, and compressive strength have been studied as per ASTM standards and it has been identified that the Hybrid rebar show a superior performance in comparison with GFRP (Glass FRP) and Steel rebars. This extraordinary performance of hybrid composite material increases the extensive engineering applications such as transport industry, aeronautics, naval, automotive industries.
Keywords: Hybrid, FRP, Mechanical properties, tensile property, compressive property

1. Introduction

For the past two decades, the hybrid composites have been successfully substituted universally which possess superior properties than their parental composites such as Glass, Carbon, Aramid and Basalt fibre composites. The performance of composite materials can be improved based on the fusion of two or more fibers in a single polymeric matrix which leads to the advanced material called hybrid composite with a great diversity of material properties. The main purpose of the combination of two or more materials is to bypass the demerits of those materials. Thus it is mandatory to investigate the properties of hybrid composites in order to substitute the conventional steel reinforcements.

It is well known that the concrete structures suffers deterioration due to external loading and severe environmental conditions such as corrosion of steel rebar, alkali reaction, de-icing salt, and freeze-thaw among which the corrosion dominates. As a consequence the serious structural damage and, structural failure takes place. Several studies are going on to find a solution to this corrosion problem [1-24] and in this regard the present study has been carried out with carbon-glass hybrid composite reinforcements.

In 1994, the tensile behavior of hybrid which is a combination of FRP skin and a steel inner core has been evaluated [6].It has been observed that the failure of hybrid rods occurred in the FRP Skin and then to the inner core. Vinylon has been introduced to increase the strength as protection. Harry G.Harris and Frank K.Ko (1998) investigated a new ductile hybrid fibre reinforced polymer bar of carbon core with epoxy and used as reinforcement in new or repaired concrete structure.
Due to the anisotropic nature of the specimen being tested in this analysis, the bending failure may be caused by tensile, compressive, shear or a combination of these stresses.

The mechanical properties like bending, fatigue, stiffness and strength of carbon-glass/epoxy hybrid laminates were reported by the researchers [7,8].Jones *et al.* have performed experiments on the combined composites of carbon/carbon, glass/glass, aramid/carbon, glass/carbon fibers, and then obtained pseudo-ductile behaviors when small amounts of fibers with low stretch were well distributed. In these cases, however, there was a problem of increasing the unit price of products.
From all these literature studies, it has been observed that the efficiency of composite reinforcement depends on the adhesion between matrix and fiber; hence it is needed to determine their mechanical properties. The present work depicts the mechanical properties such as the tensile properties and compressive properties of Carbon-glass Hybrid composite reinforcements as per ASTM standards.

2. Materials and methods

Hybrid reinforcements used in this study were manufactured by Meena Fiberglass Industries, Pondicherry, India. **Using Pultrusion process,** the reinforcements were manufactured with a volume of 60% of E-glass fibre and 40% carbon fibre and are reinforced with epoxy resins. The epoxy resin, hardener Tri Ethylene Tetra Amine (TETA) and Catalyst (Methyl ethyl Kethone Peroxide) were used. The glass fiber of bi-directional woven mat with 200 gm and the density of 2.5 gm/cc were used. The carbon fiber of bi-directional woven mat with 200 gm and the density of 1.78 gm/cc were used. The glass fiber and carbon fiber in the form of rovings used in the fabrication of hybrid fiber reinforced composites are shown in Figure 1.

Figure 1. Raw Carbon and Glass fibres

The weight fraction of fibers and epoxy matrix materials were determined by considering the density, specific gravity and mass. Initially, the fabrication of the composite was done at room temperature by Pultrusion technique. Pultrusion process is used to pull any type of fibre and resin , through a shaped die having a constant cross section. The resin matrix gives form and protection from the external environment to the fibres. Chemical, thermal, and electrical performance can be affected by the choice of matrix resin. As the resin rich fiber exits the resin impregnation system, the un-cured composite material is guided through a series of tooling which helps to arrange the fibre into the correct shape, and the excess resin is squeezed out known as debulking. The die is heated to a constant temperature of 145°, which will cure the thermosetting resin. The profile that exits the die is now a cured pultruded Hybrid Fiber Reinforced Plastic (H- FRP) composite and the hybrid rod manufactured for the present study is shown in Figure 2.

Figure 2. Hybrid Rebar

The manufactured hybrid rods are subjected to tensile test and compression test, according to ASTM standards.

3. Experimental Programme

This paper mainly presents the tensile properties and compressive properties of the Hybrid rod. A minimum of three identical test specimens were prepared for each test. The density which is considered as an important physical property was at first studied and observed that Hybrid and GFRP rods possess lighter densities such as 2502.43 and 2122.33 kg/m^3 which is almost three times lesser than that of steel rods (7480.58 kg/m^3).

3.1 Uniaxial Tensile Test

As per ASTM D638 rods a total of 15 specimens i.e., 5 specimens in each variety (GFRP, Hybrid and Steel rods) were cut into pieces of 600 mm. The specimen was fixed in universal testing machine (UTM) between the adjustable grips. The load was applied perfectly on the rod till the failure of the specimen. The test was repeated for five times on the five identical specimens. The tensile strength was calculated using the following Equation (1) and the average of the five values was taken as the ultimate tensile strength of the specimen.

$$\text{Tensile strength}=\frac{\text{Maximum applied Load}}{\text{Original Cross sectional area}} \quad (1)$$

Figure 3. Test set up and rupture of rebars

Table 1. Tensile properties of reinforcements

Type of specimen	Diameter (mm)	Area (mm^2)	Initial modulus (MPa)	Tensile strength (MPa)
GFRP	10	78.54	1.83×10^4	553.85
Hybrid	10	78.54	6.47×10^3	1082.25
Steel	10	78.54	2.08×10^5	496.56

The ultimate tensile modulus can be determined by the stress strain graph. Since the hybrid is brittle in nature ,it exhibits a linear behavior until it fails.The test setup and the rod failure are shown in Figure.3.The tensile properties of reinforcements are shown in Table.1.The stress-strain relationship of a Hybrid rod sample is shown in Fig.4 showing strain along X axis and stress along Y axis. The ultimate strain was range between 4.1 to 4.7%.

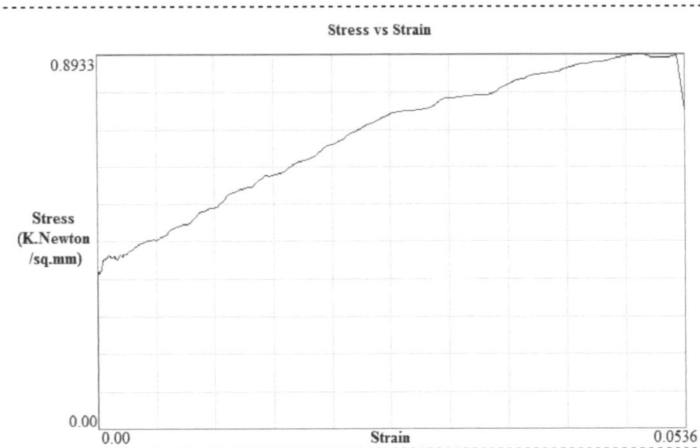

Figure 4. Stress Vs Strain Graph for various types of reinforcement

It was found that the FRP rods display a linear elastic behaviour up to failure with a modulus of elasticity lower than that of steel. The elastic modulus is mainly dependent on fibre type and volume percentage. The applied load is transferred to the core through the bond between core and skin. Unlike steel rods the hybrid and GFRP rods exhibit the rupture of rods along the longitudinal direction without yielding at the breaking stage.

3.2 Compressive Strength

According to ASTM D695 the compressive strength of Hybrid rods were determined in UTM. Usually due to the anisotropic nature the compressive property of composite rods is not recommended. But a thorough knowledge of strength characteristics is important and hence the load carrying capacity in the opposite direction has been studied.The specimens were fixed in between the platens of UTM.The length of the specuimen is taken as 50 mm . The load was applied at an uniform rate till the specimen fails and the failure load was noted from the dial indicator, the load dial gauge was adjusted to read zero before the load application. Fig.5 shows the Experimental setup for compression and the failure of the specimen and Table.2 depicts the Compressive properties of various reinforcements.

Figure 5. Experimental setup for compression and the failure of the specimen

Table 2. Compressive properties of various reinforcements

Specimen	Diameter (mm)	Area (mm^2)	Compressive load (kN)	Compressive stress (N/mm^2)
GFRP	10	78.5	35.0	445.86
Hybrid	10	78.5	37.8	481.28
Steel	10	78.5	30.0	381.97

4. Discussion of Results

The test results can be used for material specifications, quality control and assurance, structural design and analysis, and research and development. From the tensile test, it has been noted that the hybrid rods perform superior than the other specimens. The stress strain curve shows that the hybrid and GFRP rods behave bilinearly. All the FRP and hybrid rods clearly exhibited a point at which the steel rebar started yielding. The modulus of elasticity of GFRP rods is improved due to the addition of carbon core. If the volume fraction of carbon increases, the ductility will improve. Since the carbon is a good conductor of electricity its volume fraction is kept low in this study. Anyhow, the ultimate tensile strength of hybrid is found to be 49% greater than gfrp and 54% greater than steel reinforcements.

The compression modes of failure reflects the transverse tensile failure, fibre micro buckling, or shear failure. The modes of failure depend on the type of fiber, the fiber-volume fraction, and the type of resin. In general, compressive strengths are higher for rods having higher tensile strengths. From the present study it has been observed that hybrid rods show 7.4% and 20% higher compressive strength than GFRP and steel rebars respectively.

5. Conclusion

It has been observed from the tests that the combination of Glass and Carbon FRP has a predictable performance based on the law of mixtures. The present study peeps through the excellent mechanical properties of hybrid rods over the GFRP and conventional rods. Pultrusion technique has been used to fabricate the hybrid rods. The fusion of glass and carbon fibre in a volume fraction 60:40 enhanced the ultimate tensile strength, yield strength and peak load of the composite. The ductility of hybrid rebarsis higher than GFRP composite rebars. The glass is used as outer core to prevent the corrosion as well as to prevent conduction of electricity as we know the carbon is a good conductor of electricity. The present investigation can be extended in the future for different mix proportions of two different FRP materials using different manufacturing technique to increase the strengths in all aspects.

Acknowledgments
The authors wish to express their gratitude and sincere thanks to the Meena Fiberglass Industries, Pondicherry, India for manufacturing and testing the Hybrid and GFRP Reinforcements used in the present study.

References
[1] ACI 440.1R-06 2006 Guide for the Design and Construction of Structural Concrete Reinforced with FRP Bars *ACI Committee 440, Am. Concr. Inst.*
[2] ACI 440.5-08 2008 Specification for Construction with Fiber-Reinforced Polymer Reinforcing Bars, *ACI Committee 440, Am. Concr. Inst.*
[3] ACI 440.6-08 2008 Specification for Carbon and Glass Fiber-Reinforced Polymer Bar Materials for Concrete Reinforcement, *ACI Committee 440, Am. Concr. Inst.*
[4] ACI 440R-07 2007 Report on Fiber-Reinforced Polymer (FRP) Reinforcement for Concrete Structures, *ACI Committee 440, Am. Concr.*
[5] Aiello M A and Ombres L 2000 Environmental Effects On the Mechanical Properties Of Glass-FRP And Aramid-FRP Rebars. *Mech. Compos. Mater.* **36**(5) 395-398.
[6] Antonio Nanni, Markus J Henneke and Tadashi Okamoto 1994 Tensile properties of hybrid rods for concrete reinforcement. *Constr. Build. Mater.* **8**(1) 27-34.
[7] Belingardi G and Cavatorta M P 2006 Bending fatigue stiffness and strength degradation in carbon-glass/epoxy hybrid laminates: Cross-ply vs. angle-ply specimens. *Int. J. Fatig.* **28** 815-25.
[8] Chensong Dong and Ian J Davies 2012 Flexural properties of glass and carbon fiber reinforced epoxy hybrid composites. *J. Mater. Design Appl.* **227**(4) 308-17.
[9] Harry G.Harris, Win Somboonsong and Frank K Ko1998 New Ductile Hybrid FRP Reinforcing Bar for concrete structures. *J. Compos. Constr.* **2** 28-37.
[10] Jawaid M, Abdul Khalil H P S, Abu Bakar A and NoorunnisaKhanam P 2011 Chemical resistance, void content and tensile properties of oil palm/jute fiber reinforced polymer hybrid composites. *Mater. Des.* **12** 1014-9.
[11] Jones K D and di Benedetto A T 1994Fiber Fracture in Hybrid Composite Systems. *Compos. Sci. Technol.* **51** 53-62.
[12] Jong-pil Won and chan-gi Park 2006 Effect of Environmental Exposure on the Mechanical and Bonding Properties of Hybrid FRP Reinforcing Bars for Concrete Structures. *J. Compos. Mater.* **40**(12) 1063-76.
[13] Kailash C Jajam and Hareesh V Tippur 2012 Quasi-static and dynamic fracture behavior of particulate polymer composites: A study ofnano- vs. micro-size filler and loading-rate effects. *Compos. Part B- Eng.* **43**(8) 3467-81.
[14] Kocaoza S, Samaranayakeb V A and Nannia A 2005 Tensile characterization of glass FRP bars. *Compos. Part B* **36** 127-34.
[15] Mathieu Robert and Brahim Benmokrane 2010 Behaviour of GFRP Reinforcing Bars Subjected to Extreme Temperatures. *J. Compos. Constr.* **14** 353-60.
[16] Pazhanivel K, Ramadoss N, Poyyathappan K, Anandan P and Bhaskar G B 2013 Flexural analysis on GFRP composites subjected to cyclic gradual load and cyclic impact. *Adv. Mater. Res.* **685** 35-9.
[17] Radhouane Masmoudi, Abdelmonem Masmoudi, Mongi Ben Ouezdou and Atef Daoud 2011 Long term bond performance of GFRP bars in concrete under temperature ranging from 20°C to 80°C. *Constr. Build. Mater.***25** 486-93.
[18] Rahmanian S, Thean K S, Suraya A R, Shazed M A, MohdSalleh M A and Yusoff H M 2013 Carbon and glass hierarchical fibers: Influence of carbon nanotubes on tensile, flexural and impact properties of short fiber reinforced composites. *Mater. Des.* **43** 10-6.
[19] Sakthivel R and Rajendran D 2014 Experimental Investigation and Analysis a Mechanical Properties of Hybrid Polymer Composite Plates. *Int. J. Eng. Trends Technol.* **9**(8) 407-14.

[20] Saleh Alsayed, Yousef Al-Salloum, Tarek Almusallam, Sherif El-Gamal and Mohammed Aqel 2012 Performance of glass fiber reinforced polymer bars under elevated temperatures. *Compos. Part B* **43** 2265-71.

[21] Wang Y C and Kodur V 2005 Variation of Strength and Stiffness of Fibre Reinforced Polymer Reinforcing bars with Temperature. *Cem. Concr. Compos.* **27** 864-74.

[22] Wang Y C, Wong P M H and Kodur V 2007 An experimental study of the mechanical properties of fibre reinforced polymer (FRP) and steel reinforcing bars at elevated temperatures. *Compos. Struct.* **80** 131-40.

[23] Young-Jun You, Jang-Ho Jay Kim, Sung-Jae Kim and Young-Hwan Park 2015 Methods to enhance the guaranteed tensile strength of GFRP rebar to 900 Mpa with general fiber volume fraction. *Constr. Build. Mater.* **75** 54-62.

[24] Yun fu Ou and Deju Zhu 2015 Tensile behavior of glass fiber reinforced composite at different strain rates and temperatures. *Constr. Build. Mater.* **96** 648-56.

IOP Conf. Series: Earth and Environmental Science **80** (2017) 012020 doi:10.1088/1755-1315/80/1/012020

Properties of concrete modified with waste Low Density Polyethylene and saw dust ash

P Srimanikandan and S Sreenath*

School of Civil Engineering, SASTRA University, Thanjavur – 613 401. India.

*Email: sreenath@civil.sastra.edu

Abstract. The increase in industrialization creates need for disposal of large quantity of by-products. To overcome the difficulty of disposal, these by-products can be used as a replacement for raw material. In this concern, non-conventional industrial wastes such as plastic bags, PET bottles, pulverized waste Low Density Polyethylene (LDPE) and biological waste such as saw-dust ash, coconut coir were used as a replacement in concrete. In this project, saw-dust ash and pulverized waste LDPE were introduced as the partial replacement for cement and fine aggregates respectively. 0%, 5%, 10%, 15% and 20% of sand by volume was replaced with LDPE and 0%, 1%, 3%, 5% and 10% of cement by volume was replaced with saw dust ash. Standard cube, cylinder and prism specimens were cast to assess the compressive strength, split tensile strength and flexural strength of modified concrete after 28 days of curing. Optimum percentage of replacement was found by comparing the test results. The mix with 5% of LDPE and 3% of saw dust ash showed a better result among the other mixes.

1. Introduction

Concrete is one of the major materials in construction industry. It has an important role in the economy development of a country due to its consumption in large volumes. Concrete is a composite material which consists of different constituents such as binding materials, water, admixtures and aggregates [1]. Among these, aggregate plays a crucial role, which occupies 60-75% of total concrete volume. Over the past century, concrete based on Portland cement became the product manufactured in highest volume around the world. This concrete possesses highly reliable performance, low cost and versatility. Presently the manufacturing industry facing serious consequences as concrete matrix materials became unsustainable. Huge amount of raw material and energy is consumed for large scale production of Ordinary Portland cement (OPC), which is commonly used as the binder in the manufacture of concrete. Its production not only consumes significant amount of natural resources and energy but also release greenhouse gases. The sources of fine aggregate are natural river sand and quarries which are also not sustainable. Recent researchers studied the possibilities of replacement of conventional constituents in concrete with recycled materials and industrial by-products. This will lead to a sustainable production of concrete with less consumption of natural resources [1-6]. In this study, an attempt was made to produce sustainable concrete by replacing the fine aggregate and cement partially with pulverised waste LDPE and saw dust ash respectively. The optimum percentages of replacement of fine aggregate and cement were found by comparing the rest results.

2. Experimental Investigation

2.1 Specifications of Materials

Ordinary Portland Cement (OPC) of 53 grade was used to prepare the concrete mixes. The physical and chemical properties of OPC are given in the Table 1 and 2 respectively.

Table 1. Chemical composition of OPC

Calcium oxide	62.68%
Silica	20.36%
Alumina	6.87%
Iron oxide	3.67%
Magnesia	1.77%
Sulphuric anhydride	2.24%
Tricalcium silicate	42.57%
Dicalcium silicate	26.33%
Tricalcium aluminate	12%
Tetra calcium alumino-ferrite	2.24%

Table 2. Physical Properties of Cement

Specific gravity of cement	3.13
Consistency	35%
Initial setting time	28 min
Final setting time	60 min

The Coarse aggregate (CA) of hard blue granite of size 20mm aggregates was used to manufacture the concrete. The aggregates were in angular shape. The properties of coarse aggregate are given in the Table 3.

Table 3. Physical Properties of Coarse Aggregate

Specific gravity	2.75
Water absorption	0.9%
Moisture Content	10%
Bulk density	1.66g/cc
Fineness modulus	7.4

Locally available natural river sand was used to manufacture the concrete. The physical and chemical properties of fine aggregate (FA) are given in the Table 4.

Table 4. Physical Properties of Fine Aggregate

Specific gravity	2.68
Water absorption	2.9%
Moisture Content	3%
Bulk density	1.86g/cc
Fineness modulus	3.05

Normal potable water without any impurities is used for preparing of concrete mixes. Saw dust ash (SA) was a heterogeneous mixture of different sized particles generally angular in shape. These particles basically consist of partially burned or unburned wood and bark. Physical and chemical properties of wood ash are given in the Table 5 and 6 respectively.

Table 5. Physical Properties of saw dust ash

Specific Gravity	2.57
Fineness	6.0%
Colour	Grey
Density	1472 kg/m^3

Table 6. Chemical composition of saw dust ash

Silicon dioxide (SiO_2) %	31.00
Aluminum oxide (Al_2O_3)%	14.40
Iron oxide (Fe_2O_3) %	6.90
Calcium oxide (CaO) %	12.60
Magnesium oxide (MgO) %	0.69
Potassium Oxide (K_2O) %	1.57
Loss of Ignition	34.30
Moisture content %	1.60
Alkalis %	0.89

Pulverized waste LDPE was used as a substitute for fine aggregate. The physical properties of LDPE are given in the Table 7.

Table 7. Physical Properties of LDPE

Tensile Strength	0.20 - 0.40 N/mm^2
Notched Impact Strength	No break
Thermal Coefficient of Expansion	100 - 220 x 10^{-6}
Max. Continued Use Temperature	65 °C (149 °F)
Melting Point	110 °C (230 °F)
Glass Transition Temperature	-125 °C (-193 °F)
Density	0.910 - 0.940 kg/m^3

2.2 Tests conducted

Workability of the freshly prepared mix was assessed by conducting slump cone test. The variation in density of the concrete was assessed by weighing the hardened concrete specimens of each mixes. The strength characteristics of the mixes were assessed by conducting compressive strength, split tensile strength and flexural strength tests.

2.3 Mix Proportions

In this research, a concrete mix of grade M325 with the characteristic compressive strength of 25 N/mm^2 was used. The proportion of the constituents was designed to conform to Indian standards [10,11] and obtained a ratio of 1:1:2 with water to cement ratio 0.50. The conventional mix was then modified by adding various fractions of LDPE and saw dust ash. 25 mixes were prepared and the details of mixes are given in the Table 8.

Table 8. Mix Proportions

Mix id	LDPE (%)	S.A (%)	F.A (kg/m^3)	CEMENT (kg/m^3)	C.A (kg/m^3)	LDPE (kg/m^3)	SA (kg/m^3)
0-0	0	0	493	493	1030.14	0	0
0-1	0	1	493	488.07	1030.14	0	4.38
0-3	0	3	493	478.21	1030.14	0	13.16
0-5	0	5	493	468.35	1030.14	0	21.93
0-10	0	10	493	443.7	1030.14	0	43.87
5-0	5	0	468.35	493	1030.14	14.29	0
5-1	5	1	468.35	488.07	1030.14	14.29	4.38
5-3	5	3	468.35	478.21	1030.14	14.29	13.16
5-5	5	5	468.35	468.35	1030.14	14.29	21.93
5-10	5	10	468.35	443.7	1030.14	14.29	43.87
10-0	10	0	443.7	493	1030.14	28.59	0
10-1	10	1	443.7	488.07	1030.14	28.59	4.38
10-3	10	3	443.7	478.21	1030.14	28.59	13.16
10-5	10	5	443.7	468.35	1030.14	28.59	21.93
10-10	10	10	443.7	443.7	1030.14	28.59	43.87
15-0	15	0	419.05	493	1030.14	42.89	0
15-1	15	1	419.05	488.07	1030.14	42.89	4.38
15-3	15	3	419.05	478.21	1030.14	42.89	13.16
15-5	15	5	419.05	468.35	1030.14	42.89	21.93
15-10	15	10	419.05	443.7	1030.14	42.89	43.87
20-0	20	0	394.4	493	1030.14	57.18	0
20-1	20	1	394.4	488.07	1030.14	57.18	4.38
20-3	20	3	394.4	478.21	1030.14	57.18	13.16
20-5	20	5	394.4	468.35	1030.14	57.18	21.93
20-10	20	10	394.4	443.7	1030.14	57.18	43.87

3. Results and discussions

3.1 Workability of mixes

Slump test was conducted for each mix and the results are tabulated in the Table 9. From the test results it was observed that, the slump value is getting reduced as the percentage of LDPE is increased in the concrete. The addition of saw dust ash also leads to the reduction in slump. This shows the decrease in workability of modified concrete mixes compared to the conventional mix. This is due to the hydrophobic nature of the LDPE and saw dust ash.

3.2 Density of hardened concrete

The hardened cube specimens of each mixes were weighed to calculate the density of modified mixes. The results are tabulated in the Table 9. It was observed that the density was constantly reduced by the addition of LDPE. It was further reduced due to the addition of saw dust ash. This is also attributed to the hydrophobic nature of LDPE and saw dust ash.

3.3 Compressive strength

Standard cube specimens of size 150mm×150mm×150mm was tested conforming to Indian standards [6], in the compression testing machine and the results are tabulated in the Table 9.

Table 9. Test results

Mix ID	Slump (cm)	Density (kg/m^3)	Compressive strength (N/mm^2)	Split tensile strength (N/mm^2)	Flexural strength (N/mm^2)
0-0	10	2464	34.82	3.52	4
0-1	11.5	2428	32.45	3.56	5.25
0-3	12	2448	33.93	3.35	5
0-5	12	2384	33.63	3.5	5
0-10	12	2440	33.36	3.30	5.2
5-0	15	2400	38.06	3.42	6.5
5-1	10	2312	38.28	3.20	4.75
5-3	7.5	2372	37.53	3.28	5.0
5-5	7.5	2292	38.38	3.05	2.75
5-10	7.5	2336	34.12	3.00	2.5
10-0	9.0	2288	35.27	3.33	6.5
10-1	6.5	2360	34.04	3.19	6.5
10-3	5.5	2268	29.99	3.16	2.5
10-5	4.5	2332	28.06	2.67	3.75
10-10	4.5	2288	24.03	2.69	3.5
15-0	4.25	2204	27.47	2.43	5.5
15-1	3.5	2188	22.04	3.2	4.75
15-3	3.5	2268	26.76	2.37	5.5
15-5	2	2236	27.74	2.65	3.25
15-10	1.5	2216	22.20	2.52	2.25
20-0	3.5	2203	25.22	2.33	5.0
20-1	3	2132	27.95	3.08	4.75
20-3	2.75	2240	22.05	2.34	4.25
20-5	2.75	2204	21.03	2.59	2.75
20-10	2.75	2200	27.91	3.26	2.5

It was observed that the compressive strength of the mixes was getting reduced compared to the conventional mix by the addition of LDPE and it was further reduced by the addition of saw dust ash. This also attributed to the hydrophobic nature of LDPE and saw dust ash because of which matrix formed by the binder and fine aggregate will be weak. The weaker bond between the binder and aggregates lead to the decrease in the compressive strength of mix. The variation of compressive strength is represented in the figures 1,2,3,4 and 5.

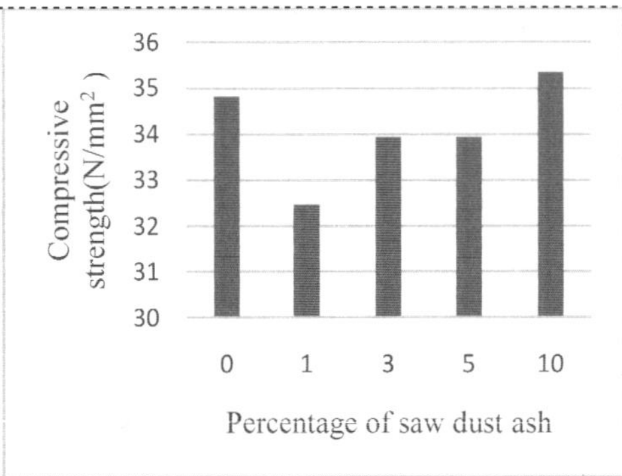

Figure 1. Variation of compressive strength of mix with 0% LDPE

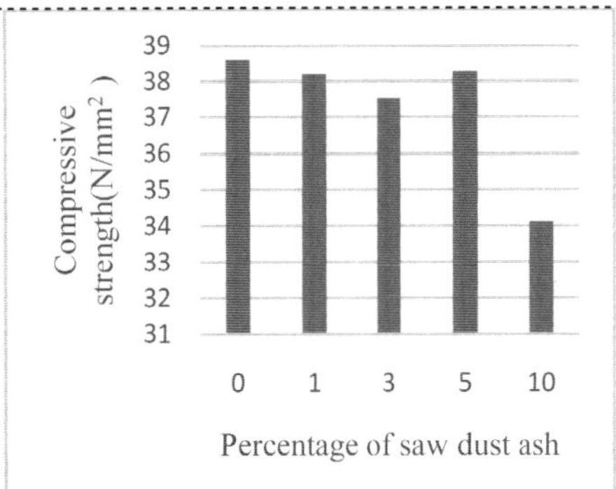

Figure 2. Variation of compressive strength of mix with 5% LDPE

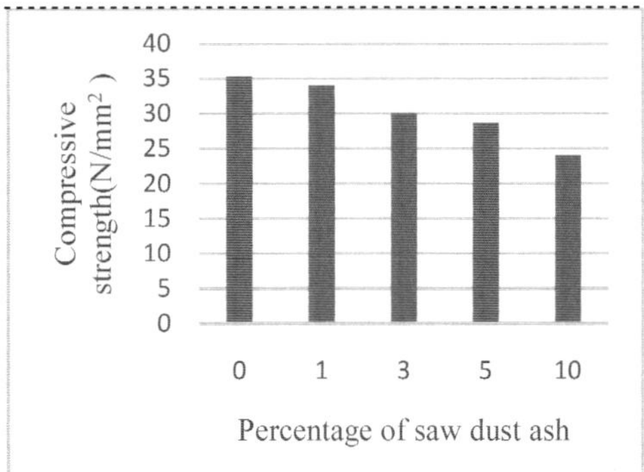

Figure 3. Variation of compressive strength of mix with 10% LDPE

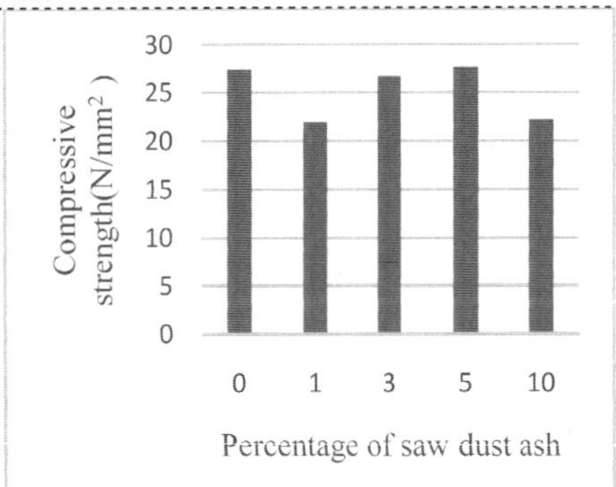

Figure 4. Variation of compressive strength of mix with 15% LDPE

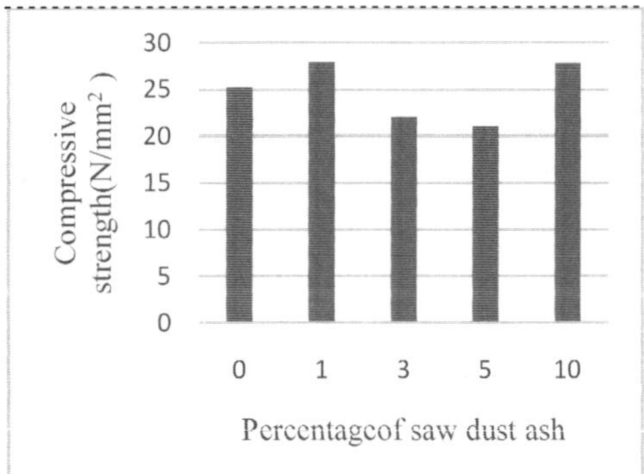

Figure 5. Variation of compressive strength of mix with 20% LDPE

3.4 Split tensile strength test

To assess the tensile nature of the modified mixes, standard cylinder specimens were tested conforming to the Indian standards [7]. The results are tabulated in the table 8. It was observed that, split tensile strength was also reduced compared to the conventional mix, as the percentage of substitutes was increased. This effect is also attributed to the hydrophobic nature of LDPE and saw dust ash. The variation of split tensile strength is represented in the figures 6,7,8,9 and 10.

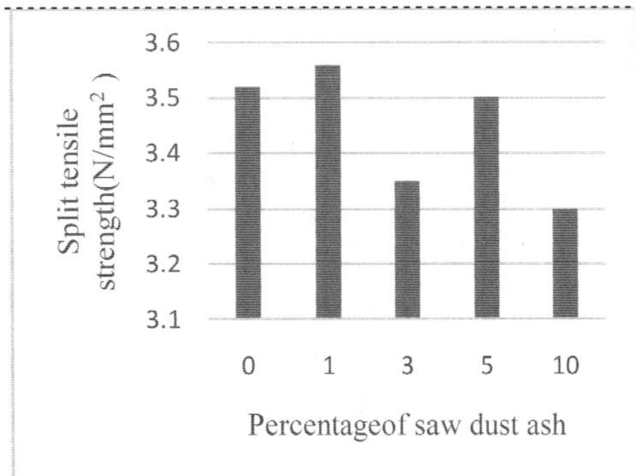

Figure 6. Variation of Split tensile strength of mix with 0% LDPE

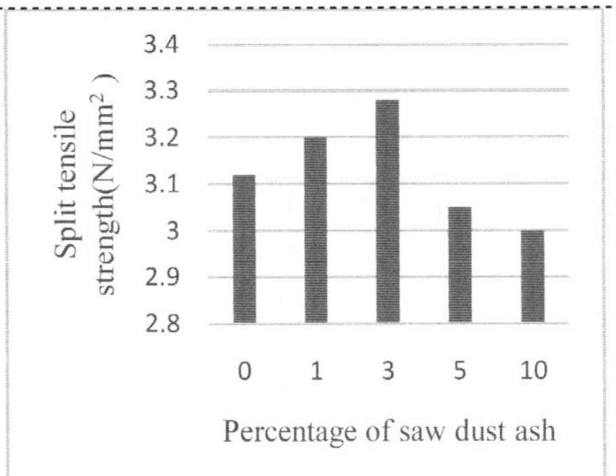

Figure 7. Variation of Split tensile strength of mix with 5% LDPE

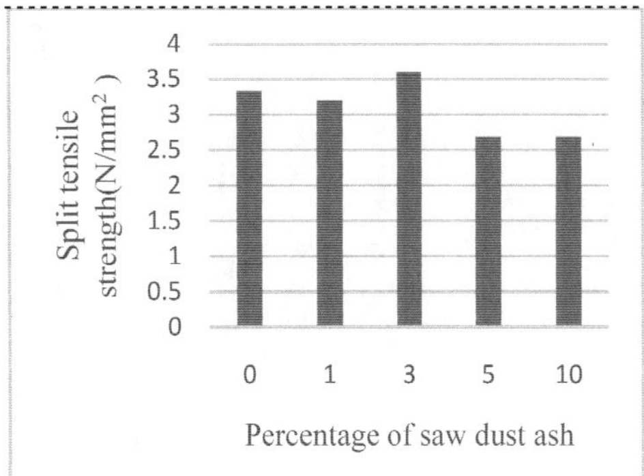

Figure 8. Variation of Split tensile strength of mix with 10% LDPE

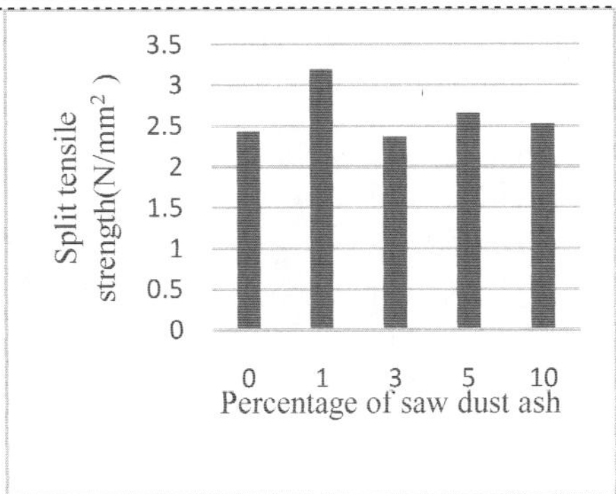

Figure 9. Variation of Split tensile strength of mix with 15% LDPE

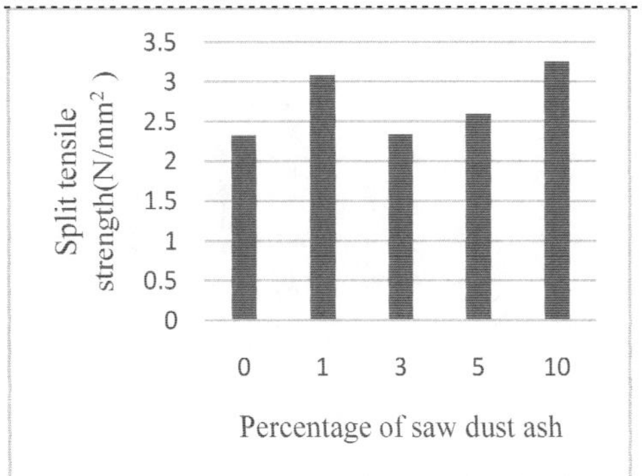

Figure 10. Variation of Split tensile strength of mix with 20% LDPE

3.5 Flexural strength

Standard prism specimens were tested for flexure conforming to the Indian standards [6], to assess the flexural strength of the modified mixes. Reduction in the flexural strength compared to the conventional mix was observed with increase in the percentage of LDPE and it was further decreased due to the addition of saw dust ash. The variations of flexural strength with increase in the percentages of LDPE and saw dust ash are given in the figures 11,12,13,14 and 15. The reduction in flexural strength is also attributed to the hydrophobic nature of the substitutes.

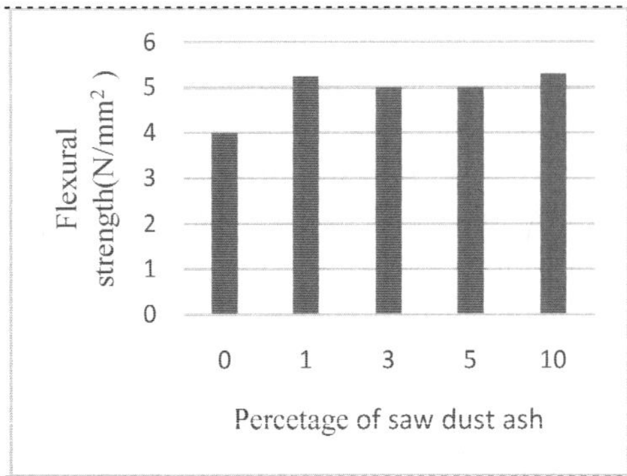

Figure 11. Variation of Flexural tensile strength of mix with 0% LDPE

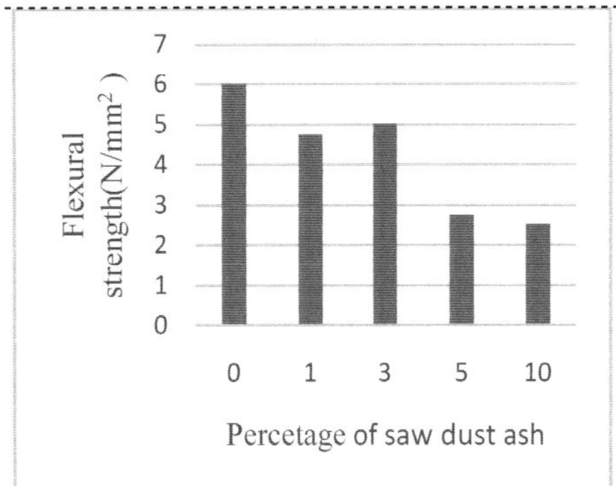

Figure 12. Variation of Flexural tensile strength of mix with 5% LDPE

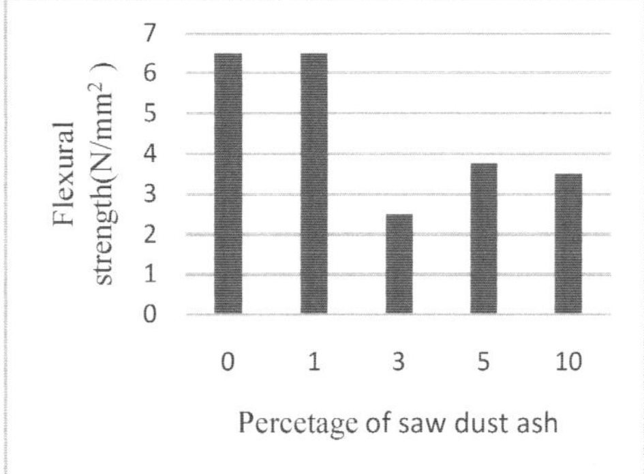

Figure 13. Variation of Flexural tensile strength of mix with 10% LDPE

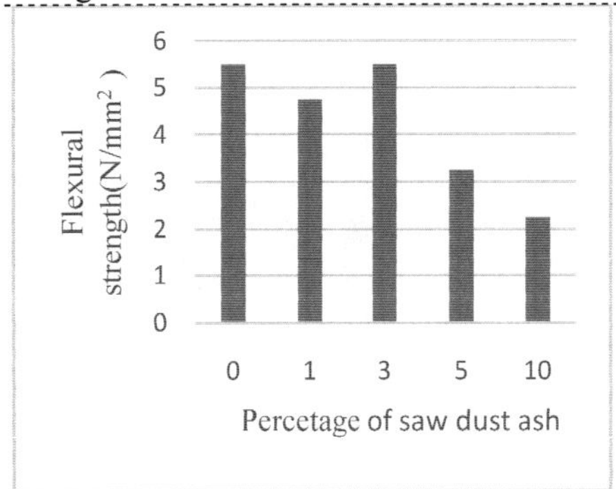

Figure 14. Variation of Flexural tensile strength of mix with 15% LDPE

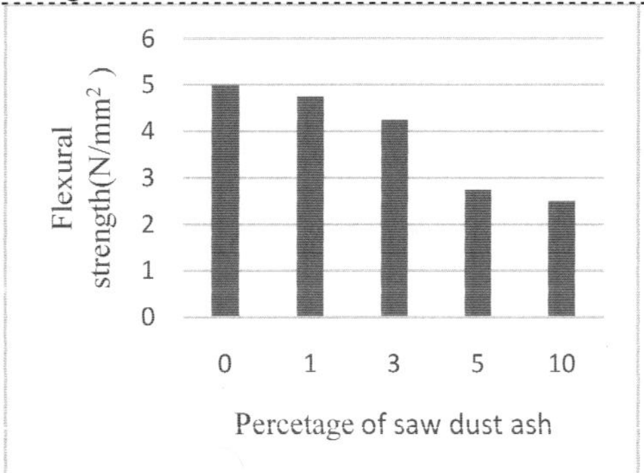

Figure 15. Variation of Flexural tensile strength of mix with 20% LDPE

4. Conclusions

- From the test results, it was observed that, the workability of the modified mixes were decreasing as the percentage of LDPE and saw dust ash increased in the mix. Due to th hydrophobic nature of LDPE and saw dust ash, less water will be absorbed for the preparation of matrix which will lead to the formation of dry mixes.
- It was observed that the density of modified mixes was less compared to the conventional mix. This is due to the addition of low density materials as substitutes for the conventional materials.

- The compressive strength, split tensile strength and flexural strength of the mixes were getting reduced compared to the conventional mix by the addition of LDPE and those were further reduced by the addition of saw dust ash. This also attributed to the hydrophobic nature of LDPE and saw dust ash because of which matrix formed by the binder and fine aggregate will be weak. The weaker bond between the binder and aggregates lead to the decrease in the strength of mix.
- The mix with 5% of LDPE and 3% of saw dust ash showed a better result among the other mixes. This can be considered as the optimum ratio of LDPE and saw dust ash to achieve the targeted properties.

References

[1] Patil P S, Mali J R, Tapkire G V and Kumavat H R 2014 Innovative Techniques of Waste Plastic Used in Concrete Mixture. *Int. J. Res. Eng. Technol.* **3** 29-32.

[2] Chowdhury S, Mishra M and Suganya O 2015 The incorporation of wood waste ash as a partial cement replacement material for making structural grade concrete. *Ain Shams Eng. J.* **6** 429–37.

[3] Suganthy P, Chandrasekar D and Sathish Kumar P K 2013 Utilization of Pulverized Plastic in Cement Concrete as Fine Aggregate. *Int. J. Res. Eng. Technol.* **2** 1015-19.

[4] Raghatate Atul M 2012 Use of plastic in a concrete to improve its properties *Int. J. Adv. Eng. Res. Stud.* **1** 109-10.

[5] Praveen Mathew, Shibi Varghese, Thomas Paul and Eldho Varghese 2013 Recycled Plastic as Coarse Aggregate for Structural Concrete *Int. J. Innov. Res. Sci. Eng. Technol.* **2** 687-90.

[6] Bureau of Indian Standards (BIS) 1959 *Method of Tests for Strength of Concrete* (IS 516-1959) New Delhi India.

[7] Bureau of Indian Standards (BIS) 1999 *Splitting Tensile Strength of Concrete, Method of Test* (IS 5816-1999(R2004)) New Delhi India.

Experimental study on the strength parameter of Quarry Dust mixed Coconut Shell Concrete adding Coconut Fibre

Victor Matangulu Shrestha, S Anandh* and S Sindhu Nachiar
Department of Civil Engineering, SRM University, Kattankulathur. India

*Email: anandhsekarcivil@gmail.com

Abstract. Concrete is a heterogeneous mixture constitute of cement as the main ingredient with a different mix of fine and coarse aggregate. The massive use of conventional concrete has a shortfall in its key ingredients, natural sand and coarse aggregate, due to increased industrialisation and globalisation. To overcome the shortage of material, an alternate material with similar mechanical properties and composition has to be studied, as replacement of conventional concrete. Coconut shell concrete is a prime option as replacement of key ingredients of conventional concrete as coconut is produced in massive quantity in south East Asia. Coconut shell concrete is lightweight concrete and different research is still ongoing concerning about its mix design and composition in the construction industry. Concrete is weak in tension as compared to compression, hence the fibre is used to refrain the crack in the concrete. Coconut fibre is one of many fibres which can be used in concrete. The main aim of this project is to analyse the use of natural by-products in the construction industry, make light weight concrete and eco-friendly construction. This project concerns with the comparison of the mechanical properties of coconut shell concrete and conventional concrete, replacing fine aggregate with quarry dust using coconut fibre. M25 grade of concrete was adopted and testing of concrete was done at the age of 3, 7 and 28 days. In this concrete mix, sand was replaced completely in volumetric measurement by quarry dust. The result was analysed and compared with addition of coconut fibre at varying percentage of 1%, 2%, 3%, 4% and 5%. From the test conducted, coconut shell concrete with quarry dust has the maximum value at 4% of coconut fibre while conventional concrete showed the maximum value at 2% of coconut fibre.
Keywords: conventional concrete, light weight concrete, coconut shell, coconut fibre, quarry dust

1. Introduction
Concrete is an important construction material used in the construction field. The increase in its use and its limited resource has challenged the engineers and scientist in making of lightweight concrete and replace the ingredients used for making a concrete mix. The research has been going on for the use of natural ingredients as replacement of fine and coarse aggregate to make lightweight concrete. Generally, when the density of concrete is lower than 2000 kg/m^3, it is categorized as Light Weight Concrete(Gunasekaran et al). Different waste material such as fly ash, glass, rubber, palm shell, coconut shell, and much other agricultural wastes has been researched as an alternative for replacing coarse aggregates. Similarly, the quarry dust is another waste product formed after crushing the stones. Initially, these were used in road pavements, but after finding the properties and behaviour similar to sand, it is proposed for replacement of sand in concrete(Balamurugan & Perumal). The study shows

good strength possessed by Quarry dust during partial or fully replacement of fine aggregate, as an alternate of sand(Balamurugan & Perumal 2013a).

As we all know that Concrete is good in compression than tensile force. To sustain the tensile force different research has been carried out where the use of fibre has so far sustained the impact load and retard the concrete from early failure due to crack and sudden force (Abbass 2015),(Kumar 2012). There are many fibres used such as steel fibre, glass fibre, synthetic fibre and natural fibre. The coconut fibre serves as an alternate of other costly fibre and is light weighted compared to other.

Therefore, use of natural by-products in concrete production helps to make construction eco-friendly and lightweight concrete. These material are not only eco-friendly but also fulfils the requirement of construction material, reduce the material cost and exhibit resistance against impact.

2. Material Properties

The materials used in this project were collected from nearby market and quarry site. The concrete mix of M-25 was designed as per IS10262-1989. This project is for investigation of mechanical properties of lightweight concrete using coconut shell, quarry dust and coconut fibre. Even thou there are numerous design mix codes and procedure carried out for different grade of concrete, there is no definite method available for lightweight concrete. Hence the mix design for lightweight concrete is established by regular trial mix. The trial mix used in this research were selected from previous research work (Gunasekaran et al. 2011). The selected trial mix are given below.

Grade of concrete = M25
Normal concrete = 1:2.22:3.66:0.55
Coconut Shell Concrete = 1:1.47:0.65:0.42

For these concrete mix sand was replaced by quarry dust by volume and coconut fibre were used in the proportion of 1%, 2%, 3%, 4% and 5% by volume of concrete.

2.1 Cement

The cement used in this project was 53-grade OPC cement. The properties of cement are tabulated in Table 1.

Table 1. Properties of Cement

S. No.	Description	Value
1.	Normal Consistency (%)	31.000
2.	Initial setting time, minutes	32.000
3.	Specific Gravity	3.075

2.2 Coconut Shell

The coconut shell was collected from nearby coconut oil mill industry. These were crushed by a mini crusher and only the shells passing through 12.5 mm sieve were used. The surface was properly cleaned with fairly smooth on the inner side and rough on the outer side. Due to the porosity of CS, high water absorption is expected, so coconut shell was dipped in water for 24 hours before its use. These CS were saturated surface-dry (SSD) condition before used in concrete mix.The coconut shell as coarse aggregate has properties (Balamurugan & Perumal 2013b) as shown in Table 2.

Table 2. Properties of Coconut Shell

S. No.	Description	Value
1.	Maximum size used (mm)	12.500
2.	Water Absorption (%)	24.000
3.	Sp. Gravity	1.100
4.	Fineness modulus	6.260
5.	Crushing value (%)	2.580
6.	Shell thickness (mm)	2-4

2.3 Fine aggregate
Quarry dust was used as a complete replacement of Sand as fine aggregate, passing through 4.75 mm sieve, for Conventional concrete and CSC. The sand was air dried and free from any foreign material. The various properties of fine aggregate are tabulated in Table 3.

Table 3. Properties of Fine Aggregate

S. No.	Description	Sand	Quarry dust
1.	Specific Gravity	2.639	2.487
2.	Fineness Modulus	2.730	3.900

2.4 Coconut fibre
Coconut fibre was derived from the coconut seed of 0.2-0.5 mm diameter which is cut into 50 mm in length. The fibre was mixed with Conventional Concrete with Quarry dust (CCQ) and CS Concrete at 1%, 2%, 3%, 4%, and 5% by volume of concrete. The fibre was selected to improve the impact loading in the concrete because it has a high energy absorption capability compared to other natural fibres.

3. Test Result and Discussion
The overall test was conducted to determine the mechanical properties of the specimen. The test was performed on each material before casting and these specimens were tested on the standard compressive machine. The properties determined by the test are shown in tables and results are discussed.

3.1 Mechanical Properties
A different test is carried out to determine the characteristics strength of concrete. This test defines the properties of Conventional concrete and coconut shell concrete, replacing fine aggregate, with an addition of fibre. The ultimate percentage of fibre is determined to make lightweight concrete and use of natural by-products.

3.1.1 Compression test. Compression test is a measure used to determine the properties of concrete formed by mixing of various ingredients. The specimen were cast on 100×100×100mm steel moulds and was tested at the age of 3 days, 7 days, and 28 days, after water pond curing and surface dried. The test result shows that CC with quarry dust has a greater value of 38.567 Mpa at 2% of fibre while CSC with Quarry dust has 25.50 Mpa at 4% of fibre. The mean compressive strength of the specimen is shown in Table 4 and Figure 1.

Table 4. Compressive strength test (in MPa) on CC/ CSC with Quarry Dust on % addition of Coconut Fibre

Mix	Fibre %	3rd day	7th day	28th day
Conventional Concrete with Quarry dust	0	20.267	27.067	33.033
	1	20.200	26.567	36.533
	2	22.833	28.933	38.567
	3	17.400	24.467	34.533
	4	16.900	24.033	32.767
	5	13.667	17.100	28.733
Coconut shell Concrete with Quarry dust	0	10.667	14.900	19.300
	1	13.133	14.867	20.033
	2	13.733	16.500	21.533
	3	14.533	19.133	23.967
	4	17.833	21.533	25.500
	5	13.667	17.333	23.267

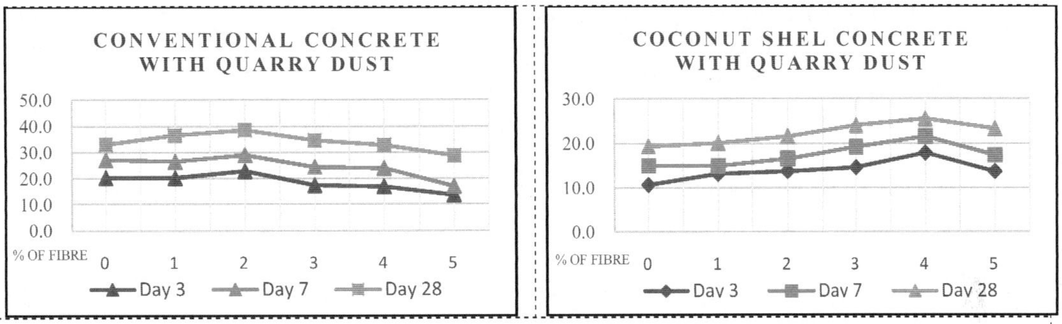

Figure 1. Line chart of CC & CSC with Quarry dust with% of Fibre,

3.1.2 Split tensile test. A steel cylindrical mould of 100×200mm was used for a split tensile test. In this test, a load is applied to a cylindrical specimen, placed horizontally, in a compression test machine until a vertical crack is formed. A strip of wooden plank placed between the testing machine and specimen reduce the magnitude of high compression stress. The value is about 0.05-0.12 times greater than direct tensile strength. The test results are given in Table 5 and their variations are shown in Figure 2.

Table 5. Comparison of Split tensile strength (in MPa) of CCQF with CC and CSCQF with CSC

M25	Fibre %	3 Day	7 Day	28 Day
CC	0	2.22	3.55	3.96
CCQ	2	2.28	3.68	4.04
CSC	0	1.11	2.17	2.56
CSCQ	4	2.04	2.55	2.62

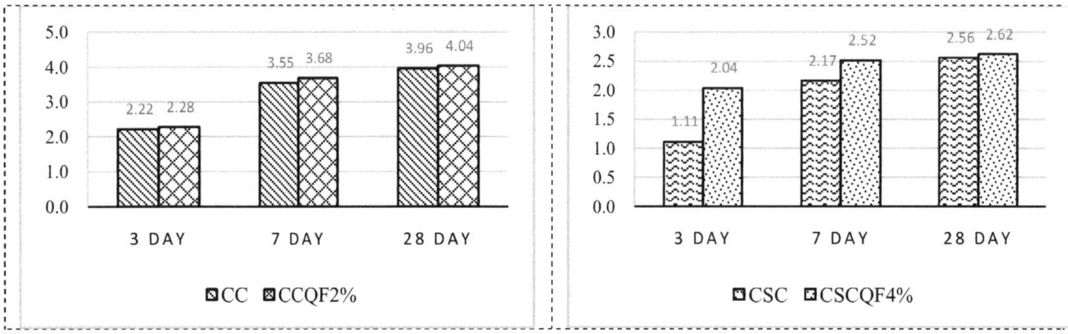

Figure 2. Comparison of Split tensile strength of CCQF with CC and CSCQF with CSC,

3.1.3 Flexure test. The flexure test of the beam (100×100×500mm) was done by four point loading method. The load was applied continuously increasing until the specimen failed. The load is divided equally between the two loading rollers without subjecting the specimen to any torsional stresses. The flexural strength of specimen is expressed as the modulus of rupture f_b. The test result is shown in Table 6 and Figure 3.

Table 6. Comparison of Flexure strength (in MPa) of CCQF with CC and CSCQF with CSC

M25	Fibre %	3 Day	7 Day	28 Day
CC	0	2.13	3.33	4.13
CCQ	2	2.27	3.47	4.67
CSC	0	2.00	2.67	3.73
CSCQ	4	2.13	3.07	4.00

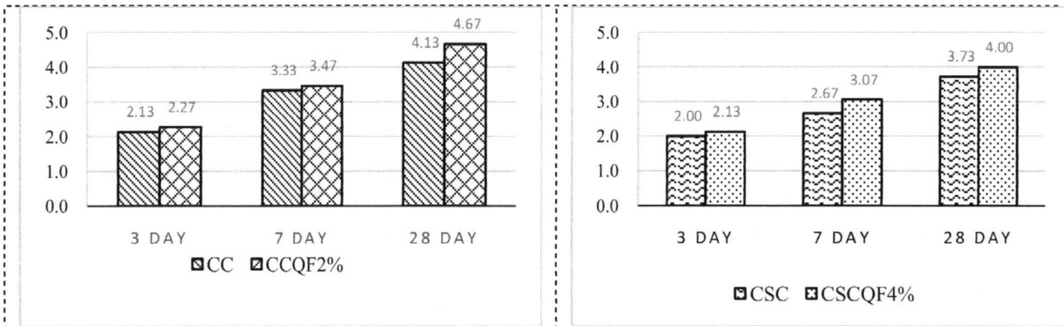

Figure 3. Comparison of Flexure strength of CCQF with CC and CSCQF with CSC,

3.1.4 Impact test. The impact test defines the resistance of the specimen on sudden loading. This test shows the energy resist by the specimen during a number of blows from initial crack until it is broken down. The specimen of 65×150mm disc was used to analyse the impact test. The impact value is tabulated in Table 7.

Table 7. Impact value (in MPa) on CCQ and CSCQ

M25	Fibre %	No of blows in 28 Day	
		Initial Crack	Fractured crack
CCQ	2	256	268
CSCQ	4	182	232

3.2 Discussion

3.2.1 Compressive strength. The compressive strength of coconut shell concrete with quarry dust using coconut fibre has shown the concrete properties gaining the maximum value of 25.50 Mpa at 4% of fibre. Similarly, CC with quarry dust has its maximum value of 38.567 Mpa at 2% of fibre. This shows that the quarry dust and coconut fibre can be used as an alternative material for replacing the fine aggregate in the construction industry.

3.2.2 Split tensile strength. The specimen with quarry dust and coconut fibre has shown the improved value than its conventional concrete. Table 5 shows the comparative chart between the conventional concrete and coconut shell concrete replaced with quarry dust and coconut fibre. Even thou the value are in marginal difference, it shows that quarry dust and coconut fibre are good in tensile strength. The CSCQF4% has the strength of 2.62 Mpa and CCQF2% has the strength of 4.04 Mpa at 28 days. The tensile property was improved due to the use of coconut fibre, which has shown the good tensile property (Zia et al. 2015).

3.2.3 Flexural strength. The flexural strength of CCQF2% (4.67 Mpa- 13.07% increase) and CSCQF4% (4.0 Mpa- 7.23% increase) are better than its conventional concrete in aflexural test at 28 days test. The modulus of rupture or bending strength shows that these materials have the ability to withstand the fracture strength, better than CC & CSC.

3.2.4 Impact test. The impact test is resistance to sudden load. The test performed on CCQF2% & CSCQF4%, as shown in table 7, has good ability to withstand the impact force as compared to conventional concrete. The study shows that the use of fibre improves the properties such as toughness, flexural strength, fatigue and stress. The energy was transformed in the formation of cracks with fibre holding the particles from dissipation rather than spalling in CC.

4. Conclusions

From the test following conclusion were made.

- The compressive strength of CCQF & CSCQF has shown the significant mechanical properties compared to CC and CSC specimen.
- The flexure and split tensile test shows the increased strength in fracture and tensile properties at thereplacement of fine aggregate with quarry dust and theaddition of fibre.
- The impact value shows its resistance against impact and load transfer through the fibre, but the use of fibre should be limited up to 5% where further increasing decreases the strength of the specimen.
- Coconut shell and coconut fibre are natural products and have more water absorption value compared to normal aggregates, hence, increase on these natural aggregates and fibre decrease the workability and form true slump.
- Results showed that quarry dust and coconut fibre produced workable concrete with satisfactory strength, useful for low cost and lightweight building construction, as an alternate to future construction material.

References

[1] Abinesh R and Gunasekaran K 2015 Study on Flexural behaviour of Coconut Shell Concrete using Quarry Dust. *Int. J. Innov. Res. Sci. Technol.* **1**(11) 426-32.

[2] Abdullah A, Jamaludin S A, Noor M M and Hussain K 2011 Composite cement reinforced coconut fibre: Physical and mechanical properties and fracture behaviour. *Australian J. Basic Appl. Sci.* **5**(7) 1228-40.

IOP Conf. Series: Earth and Environmental Science **80** (2017) 012021 doi:10.1088/1755-1315/80/1/012021

[3] Reddy B D, Jyothy S A and Shaik F 2014 Experimental analysis of the use of coconut shell as coarse aggregate. *IOSR J. Mech. Civ. Eng.* **10**(6) 6-13.

[4] Balamurugan G and Perumal P 2013 Behaviour of concrete on the use of Quarry dust to replace sand- and Experimental study. *IRACST Eng. Sci. Technol.* **3**(6) 776-81.

[5] Ruben J S and Baskar G 2014 Experimental study of coir fibre as concrete reinforcement material incement based composites. *Int. J. Eng. Res. Appl.* **4**(1) 128-31.

[6] Gunasekaran K and Kumar P S 2008 Lightweight concrete using Coconut shells as aggregate. *Proc. Int. Conf. Adv. Concr. Constr.* (ICACC- 2008). Hyderabad, India 450-9.

[7] Gunasekaran K, Kumar P S and Lakshmipathy M 2011 Mechanical and bond properties of coconut shell concrete. *Constr. Build. Mater.* **25**(1) 92-8.

[8] Shetty M S 1999 Concrete technology- theory and practice. *S. Chand & Co. Ltd.*

[9] Abbass M 2015 Coconut fibre as fibre reinforcement: A Review. *SSRG Int. J. Civ. Eng.* 62-5.

[10] Kumar N 2014 Strength properties of coconut shell as Coarse Aggregate in concrete. **3**(5)

[11] Zia S I 2015 Use of coconut fibre concrete. *Int. Res. J. Eng. Technol.* **2**(6) 235-7.

[12] Wang W and Chouw N 2014 An experimental study of coconut fibre reinforced concrete under impact load. *Proc. NZSEE Conf.* **P34** 1-7.

IOP Conf. Series: Earth and Environmental Science **80** (2017) 012022 doi:10.1088/1755-1315/80/1/012022

Study on Strength and Durability Characteristics of Concrete with Ternary Blend

C Nissi Joy, K Ramakrishnan*, M Snega, S Ramasundram, C Venkatasubramanian and D Muthu

School of Civil Engineering, SASTRA University, Thanjavur – 613401, India.

*Email: ramakrishnan@civil.sastra.edu

Abstract. In the present scenario to fulfill the demands of sustainable construction, concrete made with multi-blended cement system of Ordinary Portland Cement (OPC) and different mineral admixtures is the wise choice for the construction industry. In this research work, M20 grade mix of concrete (with water - binder ratio as 0.48) is adopted with glass powder (GP) and SugarCane Bagasse Ash (SCBA) as partial replacement of cement. GP is an inert material, they occupy the landfill space for considerable amount of time unless there is a potential for recycling. Such glass wastes in the crushed form have a good potential in the infrastructure industry. Replacement of cement by GP from 30% to 0% by weight of cement in step of 5% and by SCBA from 0% to 30% in step of 5% respectively was adopted. In total, seven different combinations of mixes were studied at two different ages of concrete namely 7 and 28 days. Compressive strength of cubes for various percentage of replacement were investigated and compared with conventional concrete to find out the maximum mix ratio. Flexural strength of concrete for the maximum mix ratio was found out and durability parameters viz., water absorption and sorptivity were studied. From the experimental study, 20% SCBA and 10% GP combination was found to be the maximum mix ratio.

1. Introduction

1.1. General

Now a day the world is witnessing the construction of very challenging and difficult structures, concrete being the most important and widely used structural material is called upon to possess very high strength. The main ingredient in the conventional concrete is the Ordinary Portland Cement (OPC). The amount of cement production emits approximately equal amount of carbon dioxide into the atmosphere. Cement production is consuming significant amount of natural resources. To overcome the above ill effects, the advent of newer material and construction techniques and in this drive, admixture has taken newer things with various ingredients has become a necessity. The addition of pozzolanic materials with OPC a century old practice is an alternative in the construction industry.

IOP Conf. Series: Earth and Environmental Science **80** (2017) 012022 doi:10.1088/1755-1315/80/1/012022

1.2. Sugarcane bagasse ash (SCBA)

Bagasse is the fibrous residue of sugarcane after crushing and extraction of juice.Sugarcane Bagasse Ash is a by-product of sugar factories found after burning sugarcane bagasse which itself is found after the extraction of all economical sugar from sugarcane [12].Mostly, bagasse produced is burnt for energy needed for sugar processing The disposal of this material is already causing environmental problems around the sugar factories [2]. The use of bagasse ash as partial replacement of cement at 10%, 20% and 30% and there was an increase in strength at 10% and low permeability was investigated [13]. 5 to 30% Ground Bagasse Ash (GBA) in partial replacement of cement in concrete and found that there is an increase in early and later compressive strength at 20% and decrease in water permeability [7] [11]. The use of SCBA as sand replacement material to improves quality and reduce the cost of construction material [8].The main composition of bagasse ash is siliceous oxide that react with free lime from cement during hydration process.

1.3. Glass powder (GP)

Glass is one the most versatile substances on Earth, used in many applications and in a wide variety of forms. When manufactured, glass is a mixture of silica, soda, and lime, the amount of waste glass produced has gradually increased over the recent years due to an ever growing use of glass products. Most waste glass is being dumped into landfill sites. The land filling of waste glass is undesirable because waste glass is non biodegradable which makes the environment harmful. There is huge potential for using waste glass in the concrete construction sector. When waste glasses are reused in making concrete products, the production cost of concrete will go down. This move will serve two purposes; first, it will be environment friendly; second, it will utilize waste in place of precious and relatively costlier natural resources [10].Waste glass when ground to a very fine powder shows some pozzolanic properties as it contains high SiO_2 and therefore to some extent it replaces the cement and contributes for strength development.The use of finely ground glass powder as partial replacement of cement or aggregate upto 20%. Compressive strength was decreased when cement was partially replaced with glass powder when compared with glass aggregate [9].The use of ground waste glass powder as partial replacement of cement at 0%,10% and 20% was studied [1].Due to high alkali content, Alkali Silica Reaction (ASR) expansion was drastically reduced during the pozzolanic reaction. Partial replacement of cement by glass powder at 10%,20% and 30% was investigated.There was an increase in strength at 20% when the size of glass powder was less than 90 microns [4].

2. Materials

2.1. Cement

OPC 53 grade cement conforming to IS 10262 – 2009 was used in this work. The specific gravity was 3.15.

2.2. Fine aggregate (FA)

The locally available river sand with specific gravity 2.60 was used as fine aggregate in this investigation. The sand is free from clay, silt, and organic impurities. Fine aggregate were found to be less than 4.75 mm size which confirms to IS: 383-1970 (Zone: II).

2.3. Coarse aggregate (CA)

Crushed granite metal obtained from a local source with specific gravity 2.60 was used as a coarse aggregate that passed through 20 mm sieve and retained in 12 mm sieve.

2.4. Water

The quantity of water in the mix plays vital role on the strength of the concrete. Hardening occurs because of the chemical reaction, called hydration, between the cementitious materials and water.Water available in the campus conforming to the requirements of water for concreting and curing as per IS: 456-2009.

2.5. SugarCane Bagasse Ash (SCBA)

SCBA ash was collected from Dhanalakshmi sugar factory, Perambalur, Tamil Nadu, India. The value of Specific gravity was 2.17. The particle size of SCBA was found to be 0.840 microns.

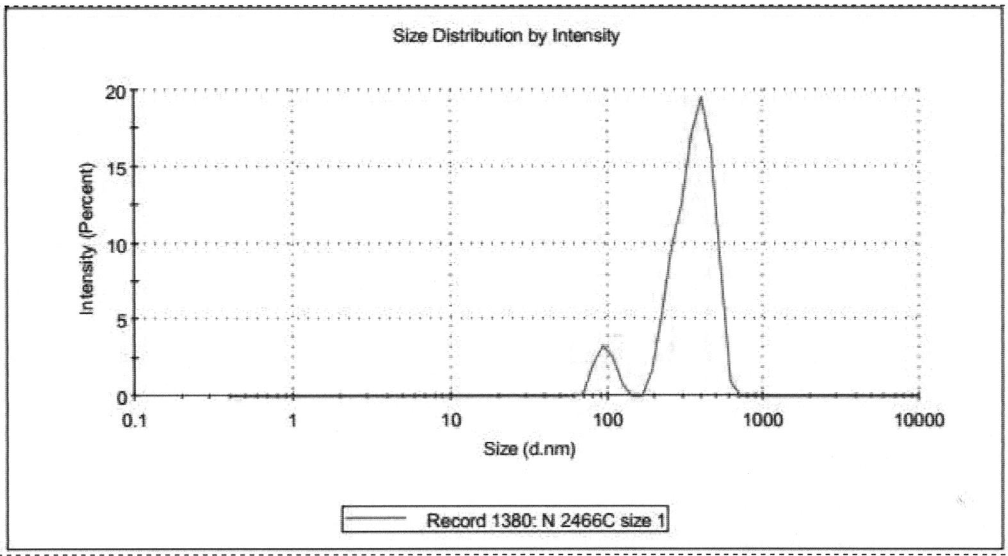

Figure 1. Particle size distribution of SCBA by intensity

2.6. Glass Powder (GP)

Glass powder was prepared from waste glasses. The value of Specific gravity was 2.42.The particle size of GP was found to be 1.168 microns.

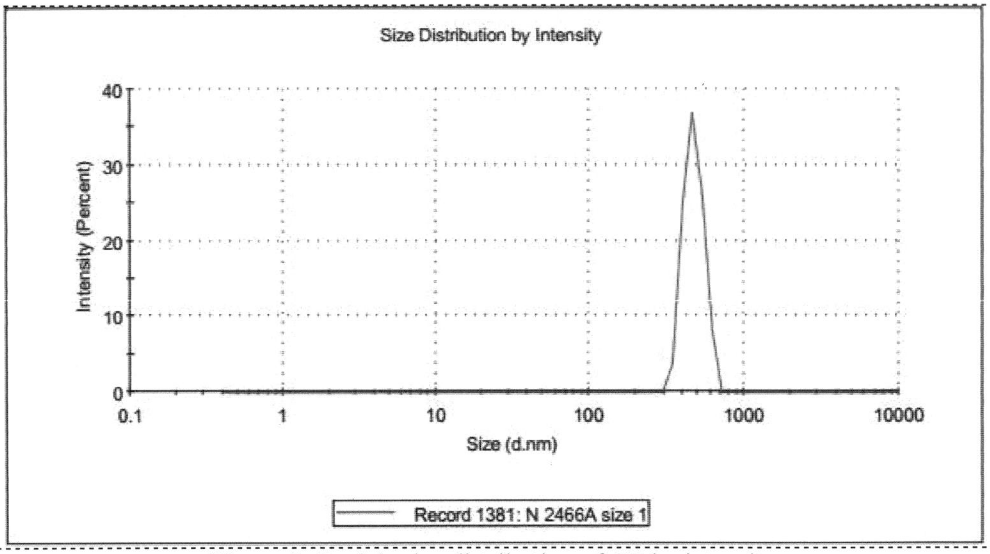

Figure 2. Particle size distribution of GP by intensity

Table 1. Chemical composition of SCBA and GP

Components	SCBA	GP
Silica (SiO_2)	79%	70%
Alumina (Al_2O_3)	9.52%	1.50%
Lime (CaO)	2.60%	11.24%
Iron Oxide (Fe_2O_3)	3.86%	2.65%
Magnesia (MgO)	2.50%	2.50%
Sodium Oxide (Na_2O)	0.31%	12.20%
Chromium Oxide (Cr_2O_3)	1.00%	1.00%
Sulphur Trioxide (SO_3)	0.4%	0.4%
Loss in Ignition	0.32	0.32

3. Methodology and Mix proportions

M20 Grade of concrete is used. The percentage of binder (cement) was kept constant at 70% by weight of the mix. Replacement of cement by SCBA was from 0% to 30% in step of 5% and GP with range of replacement from 30% to 0% in step of 5% respectively. In total, seven different combinations of mixes were studied at two different ages of concrete namely 7 and 28 days. The mix ratio which had maximum value in compression at 7 and 28 days age was selected from different mix ratios. Beams were cast for conventional and mix ratio having maximum compression value. Then the beams were tested for the flexural behaviour. Durability parameters viz., Water absorption and Sorptivity were studied for all different mix ratios. The specimens of size 150 mm × 150 mm × 150 mm were cast by partially replacing the binder (cement) with GP and SCBA for compressive strength estimation. The same mix was followed for casting the beam of size 100 mm × 150 mm × 1000 mm. For durability test, 100 mm diameter and 50 mm height cylinder specimens were cast for sorptivity test and water absorption test.

Table 2. Mix design and Proportions

M20 Mix proportion	
Cement (Kg/m^3)	399.2
Fine aggregate (Kg/m^3)	672.8
Coarse aggregate (Kg/m^3)	1097.2
Water (Litres/m^3)	191.6
Water cement ratio	0.48
Mix Ratio	1:1.68:2.75

Table 3. Mix Proportion Percentages

Mix designation	Cement	SCBA	GP	Fine Aggregate (FA)	Coarse Aggregate (CA)
S	100%	0%	0%	100%	100%
S1	70%	0%	30%	100%	100%
S2	70%	5%	25%	100%	100%
S3	70%	10%	20%	100%	100%
S4	70%	15%	15%	100%	100%
S5	70%	20%	10%	100%	100%
S6	70%	25%	5%	100%	100%
S7	70%	30%	0%	100%	100%

4. Experimental Investigations

4.1. Compressive strength

In this experimental work, totally 48 number of cubes with the standard size of cube 150 mm × 150 mm × 150 mm were cast, 6 cubes were conventional and other 42 cubes were made with partial replacement of binder by GP from 30% to 0% in step of 5% and SCBA with range of replacement from 0% to 30% in step of 5% respectively. The test was conducted in compression testing machine (CTM) of 3000 kN capacity for different ages of concrete viz., 7 and 28 days as per the specifications given in IS 516: 1959 under normal room temperature.

Figure 3. Compression test on cube specimen in-progress

4.2. Flexural Strength

All the beams are of same dimensions, 100 mm x 150 mm x 1000 mm. The beams were designed as singly reinforced beam with 2 nos. of 8 mm diameter bar in tension zone and 2 nos. of 8 mm diameter bar at compression zone. For shear reinforcement 6 mm diameter bars were used at 100 mm spacing. The specimens were tested using two point loading by 100 T loading frame. The deflection and flexural strength of beam was calculated by placing LVDT at the mid span and quarter of span of the beam. In this two point loading method, the load was divided into two parts as, P = P/2+P/2. Here 'P' is the total load given on the beam through the loading frame and was divided into two separate loads by placing two I-sectioned rollers below the beam and the load was transferred to beam. The load at first crack was noted when the cracking was developed in the concrete and the maximum load was indicated by LVDT.

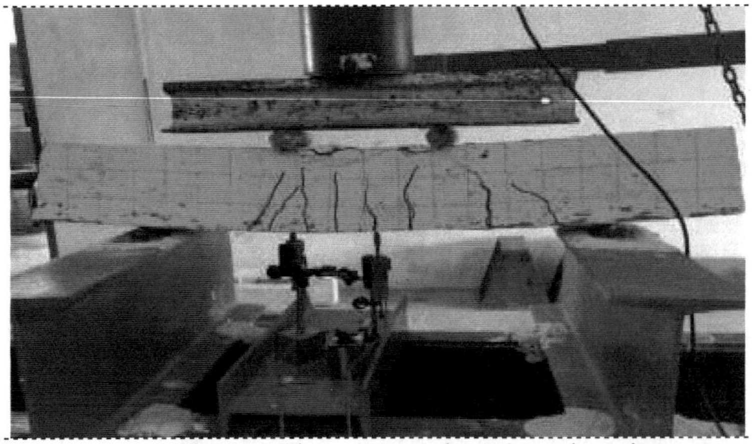

Figure 4. Flexure test on beam1 specimen in-progress

4.3. Durability tests
Total number of 32 specimens of size 50 mm height and 100 mm diameter were cast to study the durability properties such as water absorption in accordance with ASTM C 642 and sorptivity in accordance with ASTM C 1585.

4.3.1. Water absorption. Water absorption specimens were dried in the oven at a temperature of 100 to 110 °C for not less than 24 hours. Then the specimens were removed from the oven, allowed it to cool in dry air (preferably in desiccators) to a temperature of 20 to 25 °C and the oven dry mass was determined and the value was designated as A. Then the specimens were immersed in water at approximately 21 °C for not less than 48 hours. Saturated mass after immersion was determined and designated as B. The specimens were immersed in the tub water and allowed it to boil for 5 hours .Then the specimens were allowed to cool by natural loss of heat for not less than 14 hours to a final temperature of 20 to 25 °C. Surface moisture was removed with a towel and the mass of the specimen was determined. The soaked, boiled, surface-dried mass was designated as C. Finally immersed apparent mass was determined and the values were designated as D.

4.3.2. Sorptivity test. This method is intended to determine the susceptibility of an unsaturated concrete to the penetration of water. Test specimen was placed in the hot air oven at a temperature of 50 °C for 3 days. After the 3 days oven curing, specimen was kept in a room temperature for 15 days .The specimen's side and top surfaces were sealed with sealing material by keeping bottom side of the specimen open. Bottom surface was exposed to water. Then the mass of the specimen was noted at different intervals of time. Mass of dried specimen was noted and the procedure as per ASTM C1585 was followed.

5. Results and Discussion

5.1 Compressive strength
The seven different combination mixes were tested and the results were shown in Table 4.

Table 4. Compressive test results

S. No.	Designation	Mix ratio	7 DAYS (MPa)	28 DAYS (MPa)
1.	S	Conventional	14.24	22.28
2.	S1	0% SCBA + 30% GP	12.25	21.69
3.	S2	5% SCBA + 25% GP	15.92	23.65
4.	S3	10% SCBA + 20% GP	16.39	26.72
5.	S4	15% SCBA + 15% GP	17.08	28.81
6.	S5	20% SCBA + 10% GP	18.63	29.18
7.	S6	25% SCBA + 5% GP	16.87	27.21
8.	S7	30% SCBA + 0% GP	15.81	26.12

From the experimental investigations, conventional mix had compressive strength of 22.28 MPa. The compressive strength increased from S2 mix (5% SCBA+ 25% GP) and maximum compressive strength was obtained with S5 mix (20% SCBA+10% GP). The compressive strength of S5 mix was 30.96% more than that of S (conventional mix). This increase was attributed to the pozzolanic property of Supplementary Cementitious Materials (SCM) added viz., SCBA and GP.

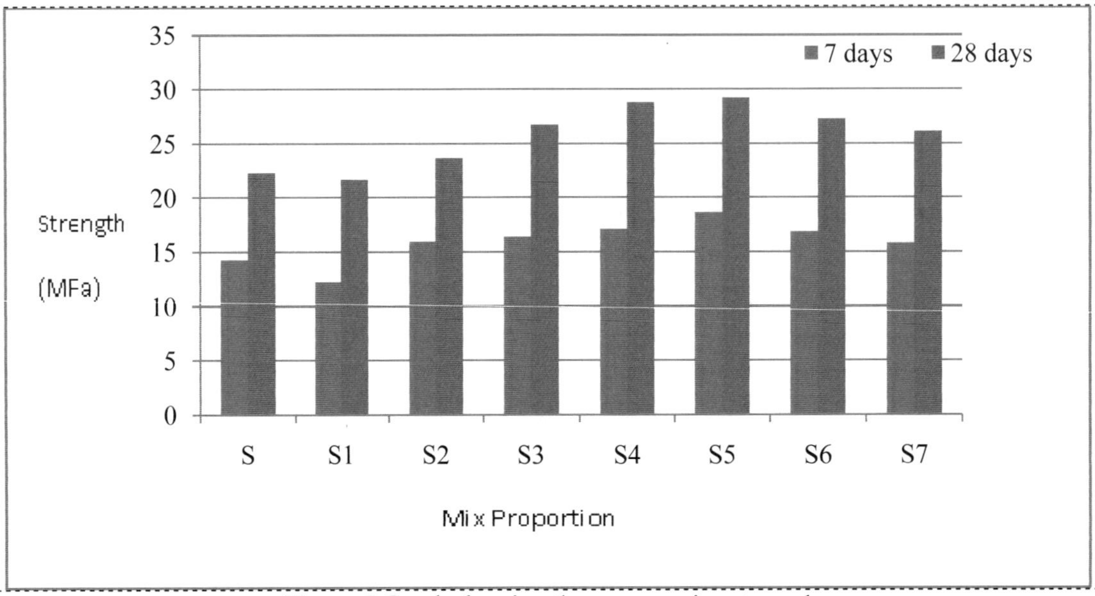

Figure 5 Graph showing the compressive strength

5.2. Flexural strength

The beam prepared with the mix ratio S5 (20% SCBA and 10% GP) produced slightly higher flexural strength compared with that of conventional concrete. The flexural strength values are shown in Table 5.

Table 5. Flexural strength values

Description	Conventional mix	S5 mix (20% SCBA 10% GP)
Peak load (kN)	46.9	49.60
First crack load (kN)	23.3	24.00
Maximum deflection in mm	9.28	9.14
Deflection under loads in mm	6.56	4.26
Flexural strength (MPa)	14.59	15.43

From the flexure test, the peak load carrying capacity of S5 mix (20% SCBA 10% GP) had increase in strength of 5.75% in comparison with that of conventional mix and the load at first crack was also higher than that of the conventional mix by 3%. The load-deflection curve for the conventional beam and the beam made with S5 mix are shown in the Figure 6.

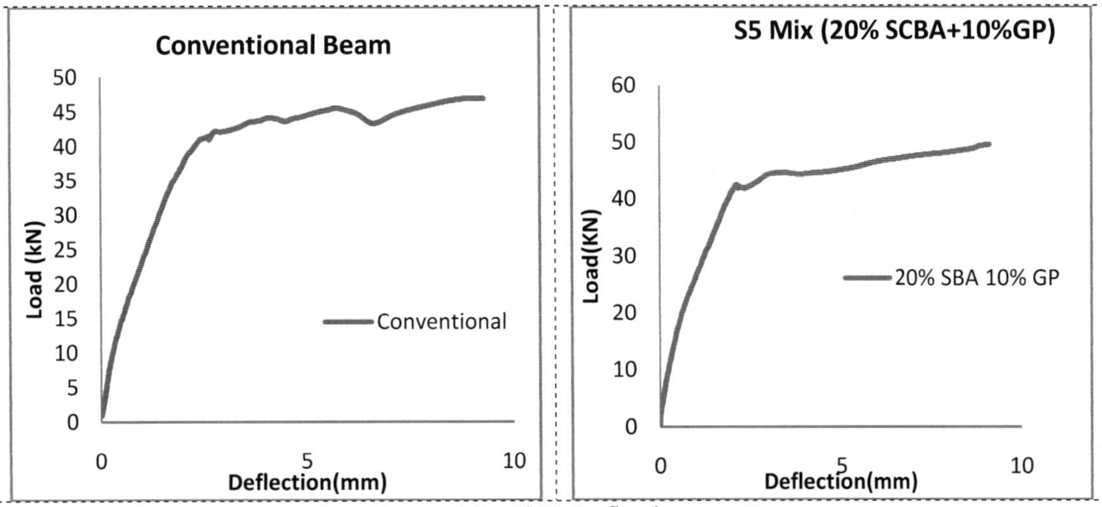

Figure 6. Load Vs Deflection curve

5.3. Water absorption and sorptivity

From water absorption and sorptivity results, S5 mix (20% SCBA+ 10% GP) showed 6.97% and 4.08% decrease in the amount of water absorped respectively compared with that of S mix (conventional mix). The decrease in water absorption of S5 mix (20%SCBA+10% GP) is due to the filling up of micro pores in the concrete matrix by GP and SCBA leading to the dense compact and impervious concrete.

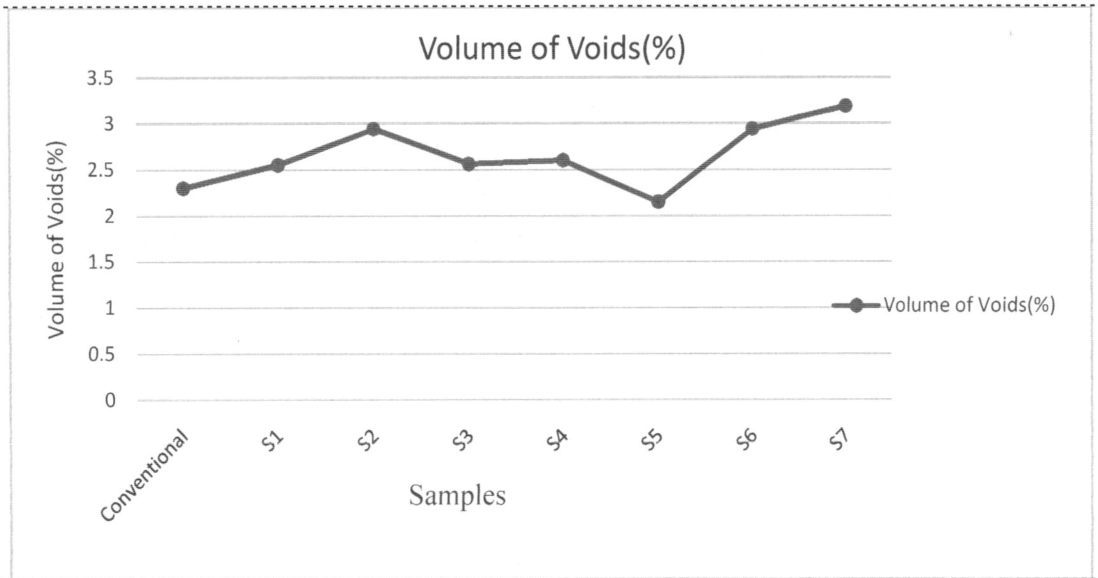

Figure 7. Water absorption values

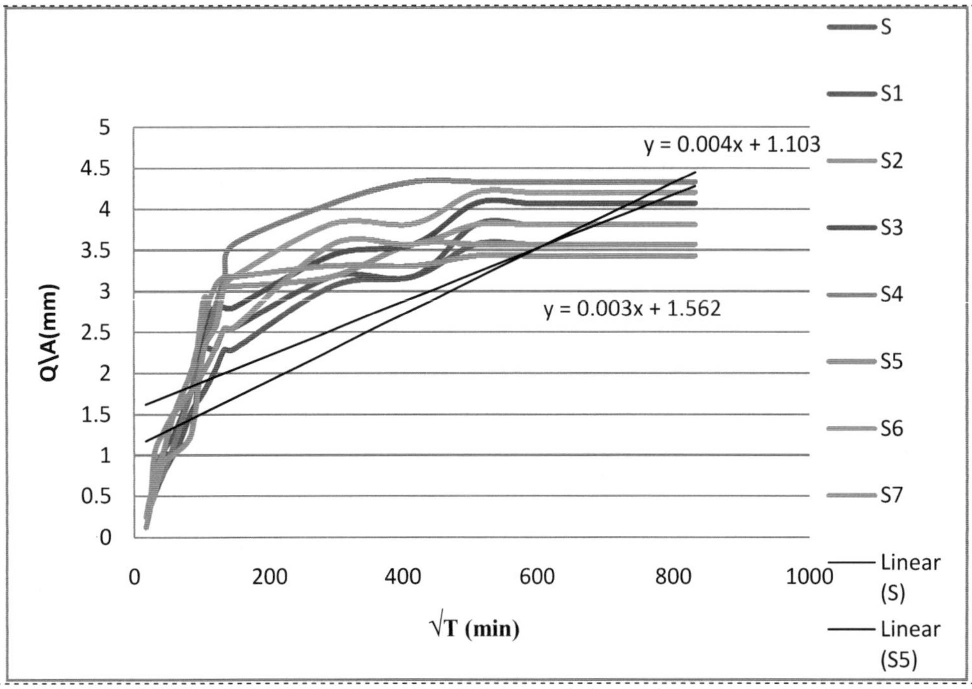

Figure 8. Sorptivity Values

6. Conclusions

From the experimental results, the maximum mix ratio percentage was found to be S5 mix (20% SCBA + 10% GP). In this research, Out of the total weight requirement of the binder (Cement), 30% by weight of total binder was fulfilled with SCM viz., combination of SCBA as 0% to 30% in step of 5% and GP as 30% to 0% in step of 5%. The remaining 70% of the total requirement of binder was Cement i.e., cement percentage as binder was kept constant at 70% for all mix ratio. Totally seven different combinations of mix ratios were studied for a curing period of 7days and 28 days.

- In the mix ratio S5 (20% SCBA+10% GP) the compressive strength was increased compared with other mix ratios (including conventional concrete) i.e.,18.63 MPa and 29.18 MPa for 7days and 28 days respectively. There was 30.96% increase in compressive strength compared with that of conventional.
- The peak load for the conventional beam was 46.9 kN and for S5 (20% SCBA+10% GP) mix ratio was 49.6 kN. S5 mix showed increase in strength of 5.75% in comparison with that of conventional mix and the load at first crack was also higher than that of the conventional mix by 3%.
- The fineness of GP and SCBA helped to fill up the micro pores in the concrete matrix which in turn produced a compact dense and impervious mass. This helped in achieving better resistance to compression and flexure.
- From the water absorption and Sorptivity test results, S5 mix ratio (20% SCBA+10% GP) showed decreased in water absorption by 6.97% and 4.08% respectively compared with that of S mix (conventional mix ratio).

- Overall the compressive strength and flexural strength were increased compared with conventional without adding any chemicals or superplasticizer.
- Due to the pozzolanic properties of both SCBA and GP, there was an increase in strength in some mechanical properties like compressive strength and flexure etc. and some durability properties viz., water absorption and sorptivity..Therefore these waste materials can successfully enhance the durability and strength which further contribute to sustainability in construction.

Acknowledgement
The authors thank the SASTRA University for providing excellent computing facilities. The authors also thank the anonymous reviewers for their valuable suggestions, which resulted in a more accurate and complete presentation of the present work.

References
[1] Ana Mafalda Matos and Joana Sousa-Coutinho 2012 Durability of mortar using waste glass powder as cement replacement. *Constr. Build. Mater.* **36** 205-15.
[2] Aukkadet Rerkpiboon, Weerachart Tangchirapat and Chai Jaturapitakkul 2015 Strength, chloride resistance and expansion of concretes containing ground bagasse Ash. *Constr. Build. Mater.* **101** 983-9.
[3] Cordeiro G C, Toledo Filho R D, Tavares L M and Fairbairn E M R 2012 Experimental characterization of binary and ternary blended-cement concretes containing ultrafine residual rice husk and sugar cane bagasse ashes. *Constr. Build. Mater.* **29** 641-6.
[4] Dhanaraj Mohan Patil and Keshav Sangle K 2013 Experimental investigation of waste glass powder as partial replacement of cement in concrete. *Int. J. Adv. Technol. Civ. Eng.* **2**(1) 112-7.
[5] Dhinakaran G, Revanth Kumar K, Vijayarakhavan S and Avinash M 2017, Strength and durability characteristics of ternary blend and lightweight HPC. *Constr. Build. Mater.* **134** 727-36.
[6] Ettu L O, Ezeh J C, Anya U C, Nwachukwu K C and Njoku K O 2013 Strength of ternary blended cement concrete containing Afikpo rice husk ash and saw dust ash. *Int. J. Eng. Sci. Invention* **2**(4) 38-42.
[7] Ganesan K, Rajagopal K and Thangavel K 2007 Evaluation of bagasse ash as supplementary cementitious material. *Cem. Concr. Compos.* **29** 515-24.
[8] Juliana Moretti P, Almir Sales, Fernando Almeida, Mariana Rezende A M and Pedro Gromboni P 2016 Joint use of construction waste (CW) and sugarcane bagasse ash sand (SBAS) in concrete. *Constr. Build. Mater.* **113** 317-23.
[9] Kaveh Afshinnia and Prasada Rao Rangaraju 2016 Impact of combined use of ground glass powder and crushed glass aggregate on selected properties of Portland cement concrete. *Constr. Build. Mater.* **117** 263-72.
[10] Mahsa Kamali and Ali Ghahremaninezhad 2015 Effect of glass powders on the mechanical and durability properties of cementitious materials. *Constr. Build. Mater.* **98** 407-16.
[11] Noorul Amin 2016 Use of bagasse ash in Concrete and its impact on the strength and chloride resistivity. *J. Mater. Civ. Eng.* ASCE **23**(5) 717-20.

[12] Nuntachai Chusilp, Chai Jaturapitakkul and Kraiwood Kiattikomol 2009 Utilization of bagasse ash as a pozzolanic material in concrete. *Constr. Build. Mater.* **23** 3352-8.
[13] Sumrerng Rukzon and Prinya Chindaprasirt 2012 Utilization of bagasse ash in high strength concrete. *Mater. Des.* **34** 45-50.
[14] Obilade I O 2014 Strength of ternary blended cement concrete containing rice husk ash and saw dust ash. *The Int. J. Eng. Sci.* **3**(8) 22-7.
[15] Vijaya Bhaskar Reddy S and Srinivasa Rao P 2016 Experimental studies on compressive strength of ternary blended concretes at different levels of micro silica and GGBS. *Mater. Today: Proc.* **3** 3752-60.
[16] ASTM C 1585 2004 Standard test method for Measurement of Rate of Absorption of water by Hydraulic cement concretes.
[17] ASTM C 642 2006 Standard test method for Density, Absorption and Voids in Hardened Concrete.
[18] IS 10262 2009 Indian Standard for Concrete Mix Proportioning – Guidelines, *Bureau of Indian Standards*, New Delhi, India.
[19] IS 456 2000 Indian Standard Plain and Reinforced Concrete - Code of Practice, *Bureau of Indian Standards*, New Delhi, India.
[20] IS 516 1959 Indian Standard Methods of Tests for Strength of Concrete, *Bureau of Indian Standards*, New Delhi, India.

Analysis of Flexural Fatigue Strength of Self Compacting Fibre Reinforced Concrete Beams

G Murali*, J P Arul Sudar Celestina, N Subhashini, M Vigneshwari

School of Civil Engineering, SASTRA University, Thanjavur-613 401. India

*Email: murali_220984@yahoo.co.in

Abstract. This study presents the extensive statistical investigation ofvariations in flexural fatigue life of self-compacting Fibrous Concrete (FC) beams. For this purpose, the experimental data of earlier researchers were examined by two parameter Weibull distribution.Two methods namely Graphical and moment wereused to analyse the variations in experimental data and the results have been presented in the form of probability of survival. The Weibull parameters values obtained from graphical and method of moments are precise. At 0.7 stress level, the fatigue life shows 59861 cyclesfor areliability of 90%.
Keywords: Weibull parameters, Fibre, Concrete, fatigue, beams.

1. Introduction

According to ACI [1]self-compacting concrete is an innovative technique and is characterised by its ability toconsolidate under its own weight without anymechanical consolidation [2]. Addition of fibres in self-compacting concrete (SCC) improves its mechanical properties when compared to conventional concrete [2-6]. The immanent benefits of self-compacting fibrous concrete (SCFC) makes it suitable for wide range of application, especially, in road pavements, off shore platforms under deep sea to produce oil, and bridges where the flexural fatigue loading is predominant [7]. The phenomenon of progressive permanent internal structural change is known as fatigue failure and is caused by repetitive stresses[8-9]. Numerous studies have been carried out in the past few decades to evaluate the fatigue behaviour of concrete and its failureprobability. The fatigue life of concrete depends on various factors viz., the typical stress level [10], stress ratio [11], loading frequency [12] and presence of fibre reinforcement [13].

Substantial researches has been accomplishedto investigate the flexural fatigue life of SCFC. However, the understanding on the flexural fatigue life in terms of probability of survival, is still lackingalthough it is important in design of concrete structures and therefore, one is encouraged to work in this direction.Thus, the present investigation deals with the analysis of flexural fatigue life data of an earlier researcher [7]using Weibull distribution and assessment of probability distributions of SFRC.

2. Weibull distribution

The concrete is not homogeneous material, although the level of stress is the same in flexural fatigue test, the experimental data are very scattered. Hence, it is essential to adopt a suitable method to analyse the testing data.Weibull, in 1939,pioneered a probability density function to analyse the fatigue testdata [14-15].The further study exhibited that the Weibull distribution has convincing backgrounds and may be sufficient to describe the behaviour of concrete under fatigue loading [16-17]. Therefore, in this paper, the Weibull distribution was selected to examine the fatigue performance of SFRC is shown in Table 1.

Table 1. Fatigue life (N) of SCFC with 0.5% fibre dosage [7].

S. No	Stress level 'S'				
	0.9	0.85	0.8	0.75	0.7
1	1114	5685	17414	96251	282014
2	1240	6106	19650	119562	380132
3	1327	6832	23237	131198	472329
4	1522	7229	24363	150947	518928
5	1617	7987	31927	178217	635492
6	1726	8921	34256	191326	724120
7	1810	9542	37236	211319	812596
8	1918	10146	42035	235282	893415
9	2103	13240	51089	346603	1022520
10	2685	14538	63549	394352	1698642
11	2948	15831	72632	468581	1882857
Mean ($\bar{N}$)	1114	5685	17414	96251	282014
Standard Deviation (σ)	577	3473	18043	121727	517441

The cumulative distribution function can be expressed as follows [7,13,18].

$$L_N(n) = exp\left[\frac{n}{u}\right]^{\propto} \quad (1)$$

where 'n' represents the fatigue life data N; 'α' is shape parameter at different stress level; 'u' is fatigue life at particular stress level. The fatigue failure life N_R is based on probability of survival (R) can be written as.

$$N_R = u(-\ln (R)^{\frac{1}{\propto}}) \quad (2)$$

2.1 Graphical method

The natural logarithm transformation twice for both sides of equation (1) gives

$$ln\left[ln\left(\frac{1}{L_N}\right)\right] = \propto ln\, n - \propto \ln u \quad (3)$$

The equation (3) can be written as follows

$$Y = \propto X - \beta \quad (4)$$

Where, Y = ln[ln(1/L_N)], X = α ln(n), and β = α ln u.

Initially, the flexural fatigue failure data should be placed in ascending order and the empirical survivorship function $L_N$of each fatigue failure data is obtainedfrom the following expression as [19-20]:

$$L_N(n) = 1 - \frac{i}{k+1} \quad (5)$$

where 'i' is the failure order number: and 'k' denotes the number of fatigue data.

$$u = exp^{-(\frac{slope}{intercept})} \quad (6)$$

2.2 Method of Moments (MOM)

A prominent technique called MOM is commonly used in the engineering research area for assessing Weibull parameters. Itdepends on the numerical iteration of average fatigue life data and its standard deviation. The 'α' and 'u'can be determined as follows [21]:

$$\alpha = \left(\frac{0.9874}{\frac{\sigma}{\bar{N}}}\right)^{-1.086} \tag{7}$$

$$\bar{N} = u\Gamma\,(1 + 1/\alpha) \tag{8}$$

The gamma function is defined by

$$\Gamma(x) = \int_0^{\infty} t^{x-1}\exp(-t)\,dt \tag{9}$$

3. Results and discussions

Using the graphical method, the Weibull parameter value can be assessed from the ln[ln(1/L_N(n))] vs ln(n) graph in Fig. 1. The slope of line gives the values of 'α' and 'u' can be calculated from equation (6). It can be seen from Figure. 1that the coefficient of determination (R^2) for all stress levels are hitting the value greater than 0.90, indicating the linear relationship of ln[ln(1/L_N(n))] and ln(N), thus proving that the mixesfollow the two parameter Weibull distribution. In the case ofmethod of moments, the parameter values attained are α = 3.48 and u = 2022.34 at a stress level S = 0.9. The Weibull parameters values obtained at other stress levels i.e., S = 0.85 to 0.7 are given in Table 2.

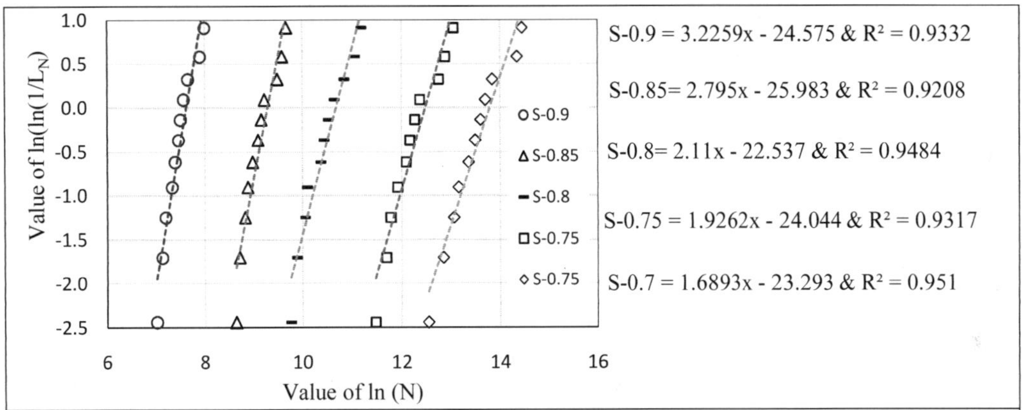

Figure 1. Graphical analysis of fatigue life data of SCFRC with 0.5% steel fibres.

Table 2. Weibull parameters.

Stress level	Parameters	GM	MOM	Average
S-0.9	α	3.23	3.48	3.35
	u	2034.55	2022.34	2028.44
S-0.85	α	2.80	3.03	2.91
	u	10897.01	10793.18	10845.09
S-0.80	α	2.11	2.23	2.17
	u	43522.91	42845.94	43184.42
S-0.75	α	1.93	1.98	1.95
	u	263710.78	258824.05	261267.41
S-0.7	α	1.69	1.70	1.69
	u	973401.07	949741.76	961571.41

The shape and scale parameter values estimated using the graphical method and method of moments arereported in Table 2. It canbe noticedfrom the Table 2,that the average 'α'value for the fatigue life data of SCFC decreasescorresponding to the decrease in stress level, thereby indicating lesser deviationsat higher stress levels[19]. This lesser deviation in the fatigue life distribution of SCFC might be due to the elimination of conventional vibration techniques that possess the influence of inherent deficiencies and material defects due to bleeding or segregation. Thus, the SCFC is highly homogenous and predominantly increasedfatigue life. Also, the orientation of steel fibresalong the direction of concrete flow, enhances the flexural fatigue properties of SCFCsignificantly when compared to SCC, as reported by few researchers [22-23].

By using Weibull parameters, α and u, the fatigue life of SCFCat different reliability levels (i.e. R= 0.99, 0.9 to 0.1 and 0.01) corresponding to a stress level S,can be determined usingequation (2) and it is shown in Table 3.The fatigue life of SCFC at 0.7 stress level, shows that the cycles of loading decreases to 59861 for90%reliability (R=0.9). Similarly, for each specific stress level, the fatigue life decreases with increasing reliability level.

Table 3. Flexural fatigue life data (N_R) in form of reliability.

R	S-0.9	S-0.85	S-0.80	S-0.75	S-0.70
0.01	2785	17158	91623	616332	2616448
0.1	1393	8579	45811	308166	1308224
0.2	973	5996	32021	215399	914409
0.3	728	4486	23954	161133	684043
0.4	554	3414	18230	122632	520595
0.5	419	2582	13791	92767	393815
0.6	309	1903	10163	68366	290228
0.7	216	1329	7096	47736	202646
0.8	135	831	4440	29864	126780
0.9	64	393	2096	14101	59861
0.99	6	37	200	1345	5710

4. Conclusions

Based on the statistical analysis the following conclusions can be drawn,

- Using the two parameter Weibull distribution, thefatigue life of SCFC at a particular stress level can be modelled precisely with the correlation coefficientvalue higher than 0.9.
- The shape parameter of SFRC decreases with decrease in stress level (S-0.9 to S-0.7), thereby denoting lesser deviation in the fatigue life distribution of SCFC at higher stress levels.
- The Weibull parameter values obtained from graphical and method of moments yields almost similar results.
- Fatigue life distribution for SFRC in forms of probability of survival have been obtained by using Weibull distribution function. The fatigue life at 0.7 stress level (considering 59861 cycles of loading) has a reliability of 90%. The survival probability (%) of fatigue life can be found easily corresponding to any stress level. Safe design life is considered much important for brittle structured composites (Concrete). These methods of statistical analysis can be considered as reliablewith higher safety limits in identification of the fatigue life under any stress level.

Acknowledgements

The authors would like to gratefully acknowledge the support of SASTRA University.

References

[1] ACI Committee 237 2007 Self-consolidating concrete. *ACI Emerging. Technol. Series* 1–30. 237R-07.
[2] Yehia S, DoubaA, AbdullahiO and Farrag S2016 Mechanical and durability evaluation of fiber-reinforced self-compacting concrete. *Constr. Build. Mater.* **121** 120-33.

[3] Cunha V M C F, Barros J A O and Sena-Cruz J M 2010 Pullout behaviour of steel fibers in self-compacting concrete. *J. Mater. Civil Eng.-ASCE* **22** 1-9.
[4] Khaliq W and Kodur V.K.R2011 Effect of high temperature on tensile strength of different types of high-strength concrete. *ACI Mater. J.* **108**(4) 394-402.
[5] Akcay B and Tasdemir M A 2012 Mechanical behaviour and fibre dispersion of hybrid steel fibre reinforced self-compacting concrete. *Constr. Build. Mater.* **28**(1) 287–93.
[6] Sideres K K andManita P2013 Residual mechanical characteristics and spalling resistance of fiber reinforced self-compacting concretes exposed to elevated temperatures. *Constr. Build. Mater.* **41**(4) 296–302.
[7] Goel S, Singh S P and Singh P 2012 Residual mechanical characteristics and spalling resistance of fiber reinforced self-compacting concretes exposed to elevated temperatures. *Eng. Struct.* **40** 131–40.
[8] ACI 215R-74 1997 Considerations for design of concrete structures subjected to fatigue loading, *ACI Committee Report*.
[9] ZandiY and Akpinar M V 2012 Evaluation of internal resistance in asphalt concretes. *Int. J. Concr. Struct. Mater.* **6**(4) 247-50.
[10] Lee M K and Barr B I G 2004 An overview of the fatigue behavior of plain and fibre reinforced concrete. *Cem. Concr. Compos.* **26**(4) 299–305.
[11] Tepfers R and Kutti T 1979 Fatigue strength of plain, ordinary, and lightweight concrete. *ACI J. Proceeding* **76**(5) 635–52.
[12] Saucedo L, Yu R C, Medeiros A, Zhang X and Ruiz G 2013A probabilistic fatigue model based on the initial distribution to consider frequency effect in plain and fiber reinforced concrete. *Int. J. Fatig.* **48** 308–18.
[13] Mohammadi Y and Kaushik S K2005 Flexural fatigue-life distributions of plain and fibrous concrete at various stress levels. *J. Mater. Civ. Eng.* **17**(6) 650–8.
[14] Kaitai F and Jianlun X 1987 Statistical distribution. Beijing: Science Press.
[15] Gao Zhentong 1986 Fatigue application statistics. Beijing: National Defence Industry Press.
[16] Oh B H 1986 Fatigue analysis of plain concrete in flexure. *J. Struct. Eng.* **112**(2).
[17] Oh B H 1991 Fatigue-life distributions of concrete for various stress levels. *ACI Mater. J.*
[18] Gumble E J 1963 Parameters in distribution of fatigue life. *J. Eng. Mech. ASCE* 45–63.
[19] Ganesan N, Bharati Raj J and Shashikala A P 2013 Flexural fatigue behavior of self-compacting rubberized concrete. *Constr.Build. Mater.* **44** 7–14.
[20] Arora S and Singh S P 2016 Analysis of flexural fatigue failure of concrete made with 100% Coarse Recycled Concrete Aggregates. *Constr. Build. Mater.* **102** 782–91.
[21] Kumar K S P and Gaddada S 2015Statistical scrutiny of Weibull parameters for wind energy potential appraisal in the area of northern Ethiopia. *Renewables* **2**(14)1-15.
[22] Grunewald S and Walraven J C2001 Parameter-study on the influence of steel fibers and coarse aggregate content on the fresh properties of self-compacting concrete. *Cem. Concr. Res.* **31**(12) 1793–8.
[23] Goel S, Singh S P and Singh P 2012 Fatigue analysis of plain and fiber reinforced self-consolidating concrete. *ACI Mater. J.* **109**(5) 573–82.

Failure analysis of lattice tower like structures

RaghavanRamalingam

Civil Engineering Department, SASTRA University, Thanjavur, Tamil Nadu, India

*Email: raghavan.ramalingam@gmail.com

Abstract. The Experimental investigations have revealed significant mismatches between analytical estimates and experimentally measured deflections of tower structures. Conductor and groundwire tension and sag are influenced by the peak and cross-arm deflections and hence accurate estimates of the tower deflections at service loads are of key interest. This paper presents a nonlinear formulation for analysis of tower structures in an attempt to close the mismatch. The analysis includes geometric nonlinear behaviour of the tower, material nonlinearity as well as leg member buckling in compression. The nonlinear analysis reveals that the ultimate collapse loads of the towers tested to failure can be predicted. However the, deflections continue to have a mismatch despite nonlinear analysis. This demonstrates the need for inclusion of other parameters to be able to reduce the error of analytical deflection estimates.

1. Introduction

Lattice tower like structures are commonly encountered in the form of communication tower masts and transmission line towers. The property of tall heights compared with their base widths, while having limited bending action due to the lattice or truss form, make these structures unique in behaviour. The various loading conditions on towers cause specific deflection shapes and failure locations which are further dependent on properties of both members and connections. Towers are designed traditionally for reliability and security requirements by ensuring the strength limit states for individual members and connections. The designs however cannot ensure the safety against ultimate collapse of the tower as a whole. The ultimate collapse of towers are mandatorily studied through full-scale tests by clients as specified in the testing codes (IEC 60652: 2002 and IS 802:1995 – part 3). Active research on the ultimate collapse and deflection performance of transmission line towers has aimed to address several mismatches existing between test observations, analytical procedures and design provisions.

Initial research on the behaviour of transmission line towers (Al-Bermani and Kitipornchai, 1992) employed a nonlinear formulation that considered the members as thin-walled beam column elements to capture the geometric nonlinear effects. Satisfactory predictions of collapse loads were not matched by satisfactory deflection predictions, which the authors attributed due to the joints. A similar procedure was proposed to replicate and reduce the need for full-scale tests (Al-Bermani et al, 2009). The effect of hip bracing patterns on the failure of transmission towers was studied analytically with additional suggestions for appropriate bracings to adjust slenderness of leg members in different types of towers (Rao et al, 2010). A common discrepancy highlighted in the above studies is the inability to match deflection predictions with values measured from the full-scale tests. The primary cause is stated to be the occurrence of bolt slippage in the splices provided for joints of leg members. Two bolt slippage models were suggested – sudden and continuous slip (Kitipornchai et al, 1994). A detailed thesis on the analysis of tower structures with consideration of bolt slippage was presented by Kroeker (2000). Component level studies of bolt slippage in joints are also present in literature for further reference (Ungkurapinan et al, 2003). Modified analytically estimated deflections have also been compared with full-scale test measurements of deflections (Rao et al, 2012). Provisions for experimentally obtaining the bolt slip portion of overall deflections are present in the testing codal standards.

2. Tower modelling and design

In this paper, as truss elements were adopted for the tower models, the secondary bracings were omitted and checked for 1.5-2% of the axial force in the main leg members. This was to avoid joint instability which occurs at the joints of secondary bracing members (Fig. 1) due to no out-of plane restraint with respect to the plane of the secondary bracing. The leg members and main bracings should comply with the permissible stresses and clauses present in the design codes (IS 802:Part 1 – Section 2; ASCE Manual 52 – Guide for design of steel transmission towers; IEC 60826, 2003). Criteria for deflections of towers are not specified in either of the IEC codes or the Indian Standards. It is reported that erstwhile USSR codes contained provisions to limit tower deflections. This clause has subsequently been made obsolete given that deflections estimated by structural analysis do not match actual full-scale test deflections. Thus it would be erroneous for a designer to set deflection limits in reference to such a clause. This sharply highlights the necessity for improvements in analysis procedures of transmission line or lattice towers in general. This study aims to attempt to close this mismatch in deflections at the service load levels using nonlinear analysis of towers with inclusion of member yielding and buckling.

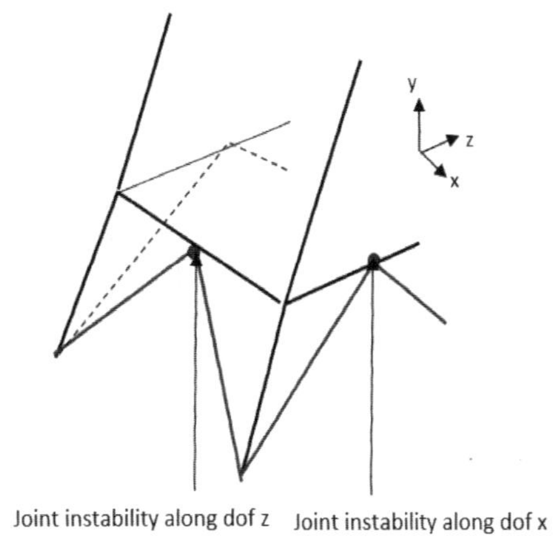

Figure 1. Joint instability at secondary bracing joints

3. Nonlinear analysis of lattice towers

In this study a formulation based on the corotated (CR) approach (Mattiasson, 1983) was employed within the updated Lagrangian form of nonlinear finite element equations. This formulation was used owing to the fact that, for truss elements in a CR approach, the local tangent stiffness matrix is 1×1 instead of 6×6. This is a consequence of the separation of strain causing deformations from rigid body motions of the total deformation in the CR approach. After applying transformations from local to global coordinate system, the global tangent stiffness matrix is given by

$$\boldsymbol{K_G} = \boldsymbol{A^T}\left(\boldsymbol{E^T K'E} + \boldsymbol{R_{x2}B}\right)\boldsymbol{A}\delta\boldsymbol{p} \tag{1}$$

where $\boldsymbol{A^T}$, $\boldsymbol{B}$ and $\boldsymbol{E}$ are transformations matrices described in reference (Mattiasson, 1983), $\boldsymbol{R_{x2}}$ is the element force at the single degree of freedom after applying corotation and $\delta\boldsymbol{p}$ is the change in global displacement vector. The local tangent stiffness $\boldsymbol{K'}$ (in the corotated coordinate system) is obtained using nonlinear finite element equations in the updated Lagrangian framework. But the displacement function in this equation contains just one displacement due to the application of corotation (instead of displacements at two nodes in traditional displacement function of truss elements).

The local tangent stiffness, $\boldsymbol{K'}$ can be written as

$$\boldsymbol{K'} = \frac{A}{L}[\boldsymbol{C} + \boldsymbol{\sigma}] \tag{2}$$

where$\boldsymbol{K'}$ is the element local tangent stiffness and is a single value matrix due to the use of the corotated approach and $\boldsymbol{\sigma}$ is the element stress vector.

The above equations handle geometric nonlinear analysis. Inclusion of material nonlinearity and member buckling require modifications to the stiffness given by Eq (2). Member buckling is considered by including an initial imperfection as a small fraction of the member length (0.1%) to estimate the reduced stiffness of buckled members. This is achieved through the arc length method (Crisfield, 2000) which traces the load-deflection of members undergoing buckling. The tangent stiffness expression then includes the form lateral displacement component due to buckling (Jayachandran et al, 2004) as shown below

$$\boldsymbol{K'} = \frac{1}{\left[\left(\frac{1}{\frac{A}{L}(\boldsymbol{C}+\boldsymbol{\sigma})}\right) + \frac{F_\Delta L}{\boldsymbol{R_{x2}}}\right]} \tag{3}$$

where, with δ_c as the maximum lateral deflection

$$F_\Delta = \left[1 - \frac{1}{1 + \frac{2}{3}\left(\frac{\delta_c}{L}\right)^2}\right] \tag{4}$$

Material nonlinearity is taken as the inelastic behaviour of the members and a mixed hardening model is considered, though for static analysis a flat yielding curve would be sufficient. The state determination procedure (Bhatti, 2000) and von-Mises yield criterion for mixed hardening modify the stress component $\boldsymbol{\sigma}$ in Eq (2) to account for material nonlinearity.

$$\begin{gathered} \delta\boldsymbol{\sigma}_{\mathbf{y}} = MH\delta\mathbf{E}'^{\boldsymbol{p}} \\ \delta\boldsymbol{\alpha} = (1-M)H\delta\mathbf{E}'^{\boldsymbol{p}} \\ f = (\boldsymbol{\sigma} - \boldsymbol{\alpha}) - \boldsymbol{\sigma}_{\mathbf{y}} \end{gathered} \tag{5}$$

In the above equations, $\delta\boldsymbol{\sigma}_{\mathbf{y}}$and $\delta\boldsymbol{\alpha}$ denote the change in subsequent yield stress due to hardening and the stress shift vector respectively. $M$is the mixed hardening parameter (Axelsson and Samuelsson, 1979), H is the hardening slope and $\delta\mathbf{E}'^{\boldsymbol{p}}$ is the increment of plastic strain which is evaluated within the formulation. This combined geometric and material nonlinear analysis procedure is incremental-iterative. The results obtained using the formulation with elastic analysis and no member buckling is compared with results from ABAQUS to verify the formulation. However, modelling member buckling in commercial FE software is tedious and this is an advantage of the formulation.

4. Results and discussion

Three examples of towers from literature are included in this study. The first example is the double circuit tower shown in Fig 2 (units in kN,mm). The tower was modelled and analysed in ABAQUS (Fig 2) considering only elasticity and this result was used to validate the formulation in the present study for nonlinear analysis of tower structures. The estimates of deflections at the groundwire (GW) and bottom panel (BP) joints on the leeward side and the comparisons of deflections with ABAQUS results are shown in Fig 3(a) and Fig 3(b). Also, the stiffness shown by the current study is 0.93 times that of commercial FE software.

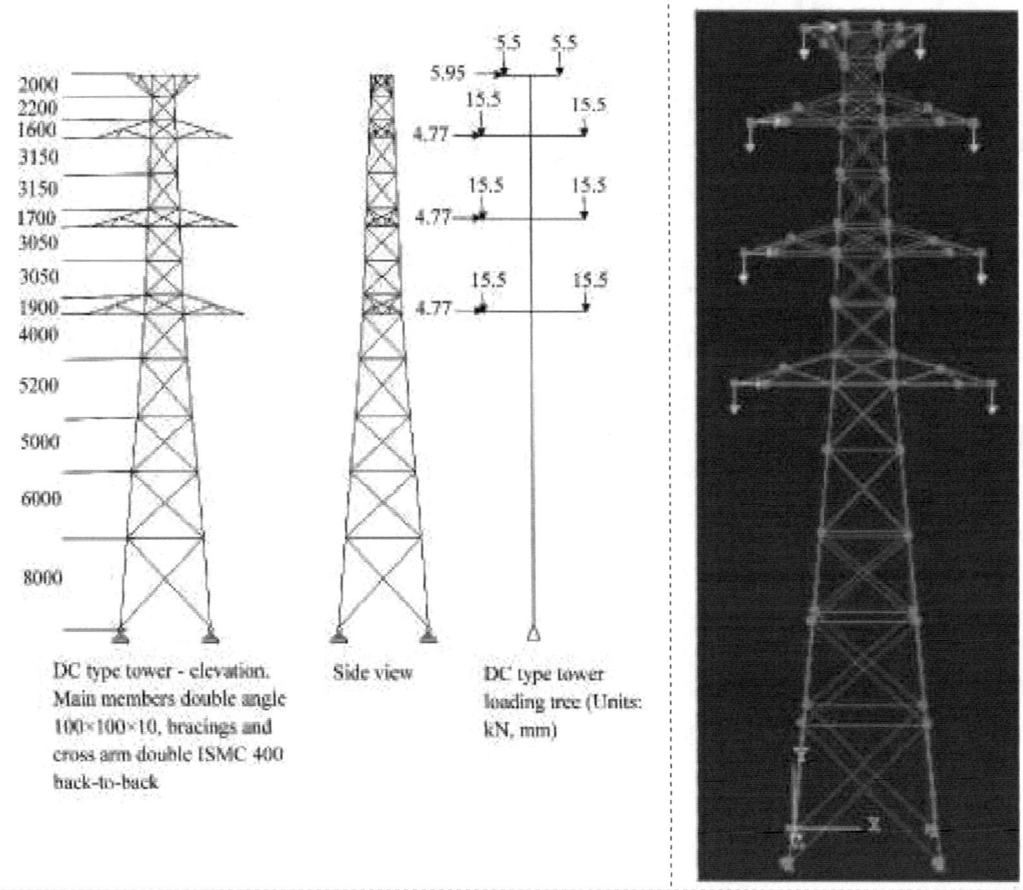

Figure 2. Double circuit tower

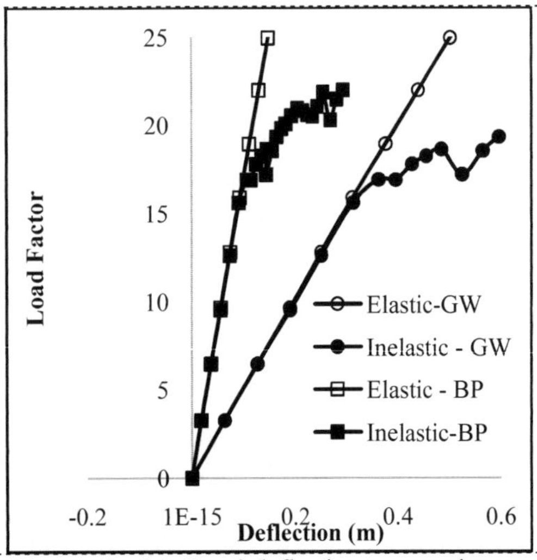

Figure 3. (a) Load-deflection comparisons of double circuit tower at groundwire and bottom panel

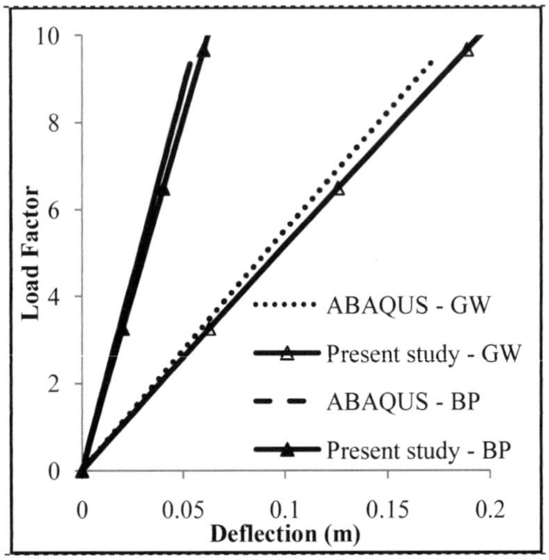

Figure 3. (b) Comparison of present study with ABAQUS results

The second example is the communication mast (Tanaka et al, 1985) in Fig 4(a) studied under vertical loads on the top nodes and an imperfection load of 0.1% of any one vertical load along the transverse direction. The axial (lb) and flexural rigidities (lb-in^2) for the members are: 7.08×10^6 and 2.16×10^6 for longerons, 2.7×10^5 and 6.43×10^4 for diagonals and battens, 1.65×10^8 and 2.2×10^8 for short longerons, 1.37×10^8 and 1.52×10^8 for plan diagonals. Fig 4(b) shows the load-deflection comparison with consideration of both member inelasticity and buckling. The softening part of the curve of elastic analysis in the study is due to the imperfection load which makes the overall tower resemble a slender column and this is not reflected in the reference curve. Member buckling is governs the ultimate collapse of the tower.

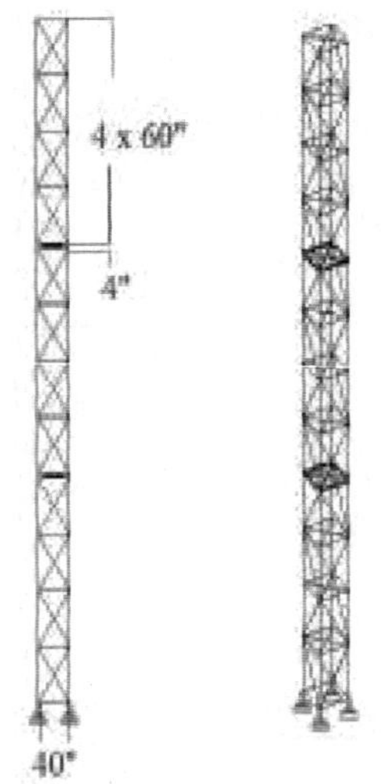

Figure 4. (a) Communication mast

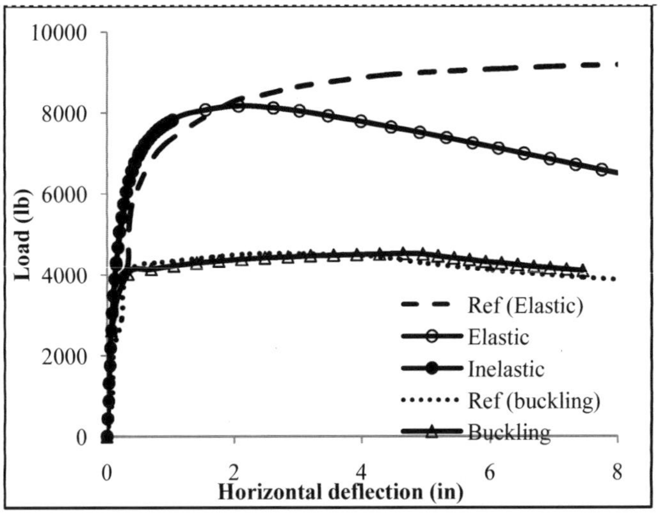

Figure 4. (b) load-deflection of communication mast

The final example is the 33 kV AT type tower (Rao et al, 2012) shown in Fig 5. The linear analysis from the reference and the ABAQUS results do not capture the ultimate collapse load level which occurs at a load factor of above 1.0 (GW – ground wire peak, C1 to C3 – top cross arm to bottom cross arm) from the current study in Fig 6. This is predicted at load factor 1.4 in analysis by buckling of bottom leg members (at 1.15 load factor in experimental test). Also, the stiffness shown by the current study is 0.87 times that of commercial FE software. The measured deflection at GW peak reported in the paper was 350mm, while that obtained from analysis is 257 mm (an error of 26%). This shows that despite nonlinear analysis, the error in deflection estimates is not bridged.

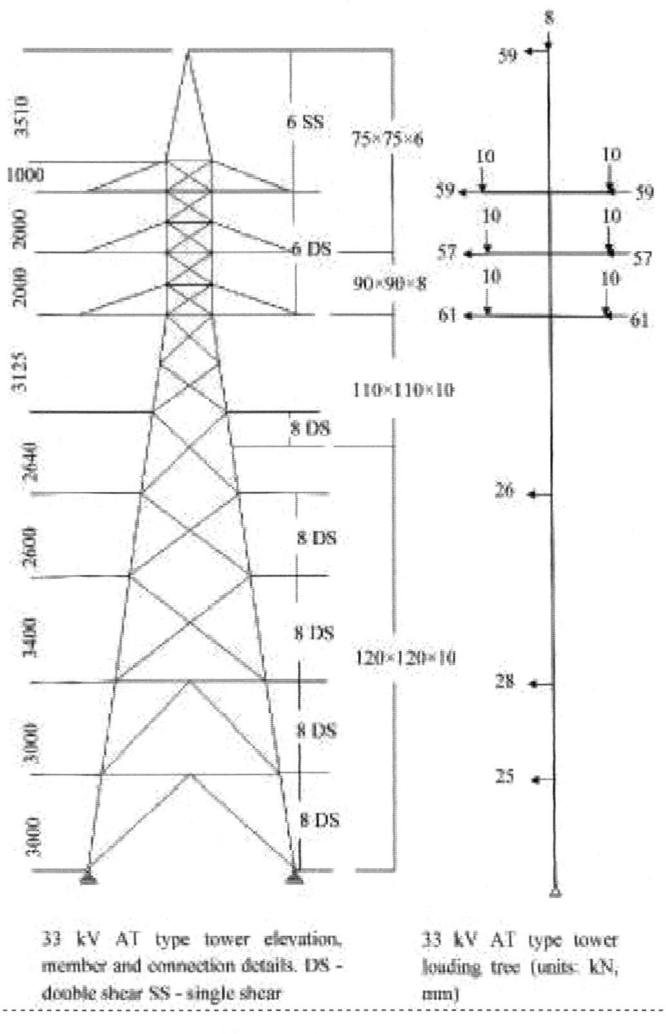

Figure 5. 33 kV AT type tower.

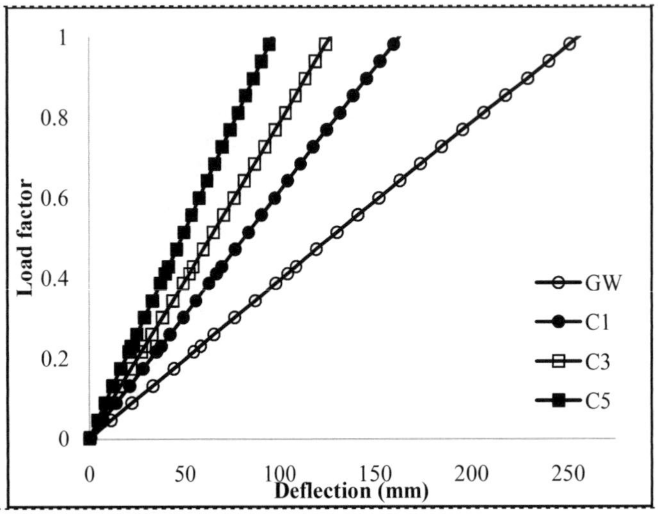

Figure 6 Load-deflection of 33 kV AT type tower

5. Conclusions

The study presents an efficient formulation for the nonlinear analysis of lattice towers based on the corotated approach and is verified through comparisons with an elastic analysis of the towers in commercial FE software ABAQUS. Determination of ultimate collapse loads of the towers is enabled by inclusion of material nonlinearity and reduced stiffness of buckled members. Experimentally measured values of ultimate load and deflections in literature showed that the ultimate loads are predicted with reasonable accuracy through nonlinear analysis. Also the cause and location of failure are predicted from the analysis. However, the analytically estimated deflections are at considerable error in relation to the test values inspite of nonlinear analysis. Thus, other effects like bolt slip and connection stiffness are necessary to enable accurate estimates of deflections through analysis. This necessitates component level studies for joint effects which can be incorporated into the overall analysis of the tower.

References

[1] Al-Bermani FGA and Kitipornchai S 1992 *Eng. Struct.***14(3)**139-51.
[2] Albermani F, Kitipornchai S, Chan RWK 2009*Eng. Fail. Anal.* **16**1922–8
[3] Rao NP, Knight GMS, Lakshmanan N and Iyer NR 2010*Eng. Fail. Anal.***17**1127-41.
[4] Kitipornchai S, Al-Bermani FGA and Peyrot AH 1992*J. Struct Eng ASCE***120(8)**2281-7.
[5] Kroeker D 2000*Structural analysis of transmission towers with connection slip modelling* MS thesis, University of Manitoba.
[6] Ungkurapinan N, ChandrakeerthySRDeS, Rajapakse RKND and Yue SB 2003*Eng. Struct.***25**779-88.
[7] Rao NP, Knight GMS, Lakshmanan N and Iyer NR 2012*Pract. Period. Struct. Des. Constr. ASCE.***17(2)**65-73.
[8] Mattiasson K 1983*On the corotational finite element formulation for large deformation problems*DrIng thesis, Chalmers university of technology.
[9] Jayachandran SA Kalyanaraman V and Narayanan R 2004 *Int. J. Struct. Stab. Dyn.***4(1)** 1-19.
[10] Crisfield, MA 2000 *Non-linear Finite Element Analysis of Solids and Structures, Volume 1: Essentials* John Wiley & Sons, Chichester.

[11] Bhatti MA 2006 *Fundamental finite element analysis and applications: with Mathematica and Matlab computations* John Wiley & Sons.
[12] Axelsson K and Samuelsson A1979*Int. J. Numer. Methods Eng.***14(2)** 211-25.
[13] Tanaka K, Kondoh K and Atluri SN 1985 Finite Elem. Anal. Des., **1**, 291-311.

IOP Conf. Series: Earth and Environmental Science **80** (2017) 012025 doi:10.1088/1755-1315/80/1/012025

Behaviour of wrapped cold-formed steel columns under different loading conditions

B Hari Baabu and S Sreenath*

School of Civil Engineering, SASTRA University, Thanjavur – 613 401. India.

*Email: sreenath@civil.sastra.edu

Abstract. The use of Cold Formed Steel (CFS) sections as structural members is widely accepted because of its light nature. However, the load carrying capacity of these sections will be less compared to hot rolled sections. This study is meant to analyze the possibility of strengthening cold formed members by wrapping it with Glass Fiber Reinforced Polymer (GFRP) laminates. Light gauge steel columns of cross sectional dimensions 100mm x 50mm x 3.15mm were taken for this study. The effective length of the section is about 750mm. A total of 8 specimens including the control specimen is tested under axial and eccentric loading. The columns were tested keeping both ends hinged. For both loading cases the buckling behaviour, ultimate load carrying capacity and load-deflection characteristics of the CFS columns were analyzed. The GFRP laminates were wrapped on columns in three different ways such that wrapping the outer surface of web and flange throughout the length of specimen, wrapping the outer surface of web alone throughout the length of specimen and wrapping the outer surface of web and flange for the upper half length of the specimen where the buckling is expected. For both loading cases, the results indicated that the column with wrapping at the outer surface of web and flange throughout the length of specimen provides better strength for it.
Keywords: CFS column – channel section – GFRP laminates – axial loading – eccentric loading – wrapping – buckling.

1. Introduction

The application of thin-walled section at the global scenario in the field of engineering is having a massive spike due to its low self-weight, durability, non-combustibility and cost effectiveness in case of transportation and erection and also it provides greater flexibility which makes it suitable for longer spans. It is used as a structural element in the field of engineering. It is commonly available in C-section with and without lips, I-section, hat section and tubular section. Even though it has more merits, it has some demerits also. The main drawback is that the load carrying capacity is less as compared to hot rolled steel. These thin walled sections are liable to instability showing various modes of failure such as global buckling, distortional buckling and local buckling. The global buckling is further classified into flexural buckling, torsional buckling and flexural-torsional buckling. The primary factors which influence the buckling behaviour is the slenderness ratio. The other factors that will affect the behaviour of CFS sections are the imperfections in the material, damages that occur for the section during transportation and installation, connections between the members and eccentricity of loading. These possibilities of failure of elements raises demand for strengthening techniques to meet serviceability for the designed life span. The performance of CFS members majorly depends on its material and sectional properties. As of now many studies have been done in strengthening the existing members. Some of the common methods of strengthening are by providing bracing, steel jacketing and wrapping it with fibre laminates. In this study, an attempt was made to strengthen the CFS column specimen by wrapping it with Glass Fiber Reinforced Polymer (GFRP) laminates. Previous studies carried out by Sreedhar Kalavagunta et al. provides details of optimum strengthening of columns with Carbon Fiber Reinforced Polymer [1] and P. Manikandan et al. explains the behaviour of stiffened CFS channel section [2]. Bambach et al. proposed that strengthening of column with CFRP subjected to axial loading increases the strength up to 2 times of that of control specimen [3]. Thomas H.K.Kang et al. described the different types of buckling modes for CFS column with different cross sections [4]. Hélder D. Craveiro et al. provided details about the difference in the buckling behaviour of single and built-up cross section with different end conditions [5]. Sreedhar Kalavagunta et al. gives details of the adhesive which can be used for attaching the laminate with the column section and gives a clear idea of the setting time for better binding [6]. M.A. El Aghoury et al. carried out a study on battened column subjected to eccentric loading which shows method of provision of eccentricity for loading in detail [7].

In this study channel section without lips is used and GFRP is used as the strengthening material. The strength mainly depends on the bond between fiber and the member hence proper bond is provided with epoxy resin. The main objective of this study is to determine the ultimate load carrying capacity and to differentiate the buckling behaviour of CFS members with and without wrapping.

2. Materials

2.1. Cold Formed Steel

The light gauge steel sections are cold formed. That is, these are made by rolling it in cold condition or by bending it in press brakes. These are used in the structure where the elements are subjected to light or moderate loads which make the structure economical.

2.2. GFRP Laminate

Fiber glass is a strong light weight material and it is mainly available in various forms such as bars, sheets, laminates. The GFRP laminates are typically processed in a hand lay technique, where the sheet of fibers are placed and brushed with a resin at room temperature.

3. Experimental Setup

3.1. Test Specimens

In this study total of 8 numbers of CFS channel section of 100mm x 50mm x 3.15mm cross sectional dimension as shown in Figure 1 with an effective length of 750mm is tested as columns. Specimens are grouped in to two based on the loading conditions. Each group is having 4 specimens in which one is the control specimen. The other three specimens are wrapped in 3 different patterns such as wrapping the outer surface of web and flange throughout the length of specimen, wrapping the outer surface of web alone throughout the length of specimen and wrapping the outer surface of web and flange for the upper half length of the specimen where the buckling is expected. The load is applied axially and eccentrically with both ends hinged condition. The hinged condition is provided with a setup of two plates with a ball at the center. The sectional details of CFS channel section are given in Table 1 [8].

Table 1. Sectional details

Dimensions l x b x t mm	A mm^2	C_y Mm	I_{xx} mm^4	I_{yy} mm^4	R_{xx} mm	R_{yy} mm	Z_{xx} mm^3	Z_{yy} mm^3
100 x 50 x 3.15	587	14.2	90×10^4	14.6×10^4	39.2	15.8	18×10^3	4.06×10^3

The GFRP laminates of thickness 1mm and length which is required for wrapping is taken. Since the load applied is compressive, a uni-directional GFRP laminate is chosen. That is all the fibers are arranged in a single direction. The bond between the GFRP laminate and the column section is made using epoxy based adhesive and let it to set in a week of time at room temperature. The specimen details are given in Table 2.

Table 2. Specimen details

Notations	Specimen Details	Notations	Specimen Details
AC	Axially loaded Control Specimen	EC	Eccentrically loaded Control Specimen
AWS	Axially loaded Web Strengthened specimen.	EWS	Eccentrically loaded Web Strengthened specimen.
AWFS	Axially loaded Web and Flange strengthened specimen.	EWFS	Eccentrically loaded Web and Flange strengthened specimen.
AES	Axially loaded partially (upper half length) wrapped specimen.	EES	Eccentrically loaded partially (upper half length) wrapped specimen.

A coupon test was conducted to determine the material properties of GFRP laminates. GFRP laminates did not show any yielding and the failure was observed at the ultimate load. The material properties of CFS and GFRP are listed in Table 3.

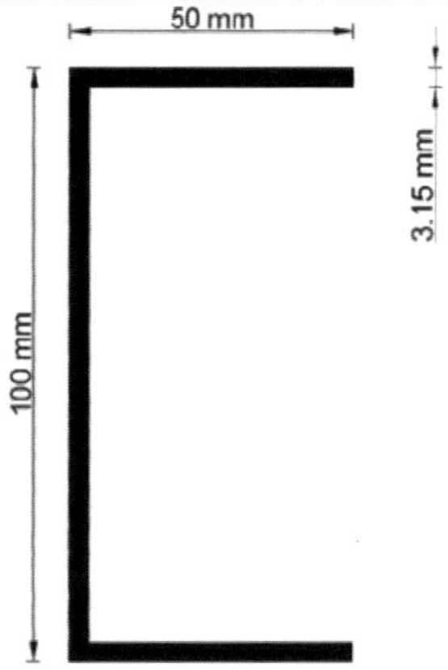

Figure 1. Cross sectional view of specimen

Table 3. Material properties

Material	Young's modulus, E N/mm^2	Poisson's ratio, n	Yield stress, S_y N/mm^2	Ultimate Tensile Strength N/mm^2
CFS	2×10^5	0.3	545	632
GFRP	0.26×10^5	0.28	530	530

3.2. Test Setup and Experimental Program

In axial loading case, the test specimens were placed in a loading frame of capacity 1000kN in such a way that the line of action of load coincides with the axis of the specimen. In eccentrically loaded columns the line of action of load coincided with the eccentric axis. The maximum eccentricity of

0.05d was considered where d is the higher lateral dimension. One LVDT is used to measure the mid height deflection in web and another is used to measure the lateral deformation of flange. The load and deflection values were acquired using the digital data acquisition system. The scheme and the test setup is shown in Figure 2.

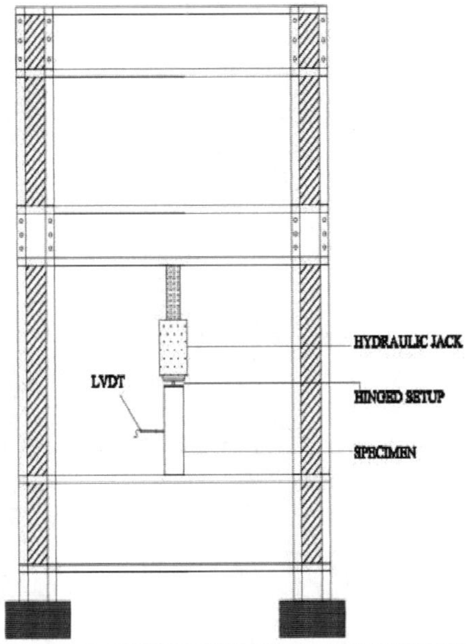

Figure 2a. Schematic diagram of the experimental setup

Figure 2b. Test setup

4. Results and Discussion

The buckling behavior, ultimate load carrying capacity and load-deflection characteristics of the wrapped and unwrapped CFS columns under axial and eccentric loading were analyzed. The ultimate load carrying capacity of the wrapped specimen is compared with that of control specimen for both the loading conditions. The ultimate load carrying capacity of specimens obtained experimentally and theoretically for axial and eccentric loading conditions is listed in Table 4.

Table 4. Ultimate load of axially and eccentrically loaded specimens

Specimens	Experimental kN	Specimens	Experimental kN
AC	82.5	EC	53.11
AWS	129.1	EWS	66.5
AWFS	148.6	EWFS	89.9
AES	117.1	EES	81.9

(a) AC (b) AWS (c) AWFS (d) AES

Figure 3. Buckling behaviour of axially loaded specimens

(a) EC (b) EWS (c) EWFS (d) EES

Figure. 4. Buckling behaviour of eccentrically loaded specimens

The comparisons between the ultimate load carrying capacity of axially and eccentrically loaded specimen are shown in Figure 5 and Figure 6 respectively. The buckling behaviour of the CFS sections were observed. It is noted that the control specimens showed distortional buckling as shown in Figure 4 and Figure 5 and the specimens which are wrapped at the web alone also shows the same behaviour but the specimens with wrapped web and flange and the specimen with partially wrapped web and flange showed local buckling as shown in Figure 4 and Figure 5.

It was observed that in both loading cases, the column with fully wrapped web and flange shows better results. The Load-Deflection behaviour of the axially and eccentrically loaded specimens is shown in Figure 7 and Figure 8 respectively. Figure 7 and Figure 8 represents that the control specimen deflects more in both cases and the wrapped specimens deflects less as compared to control specimen.

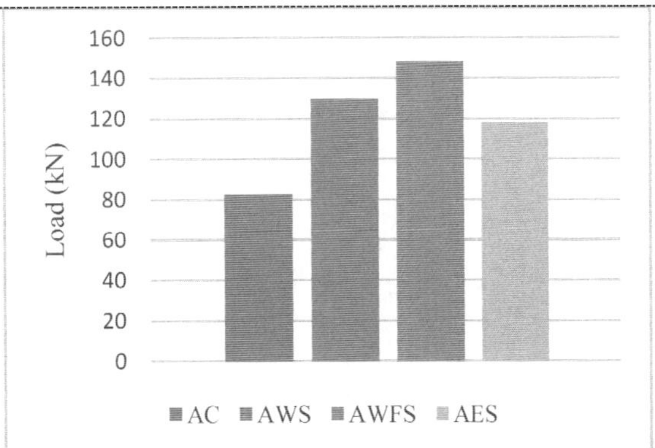

Figure 5. Ultimate load of axially loaded specimens

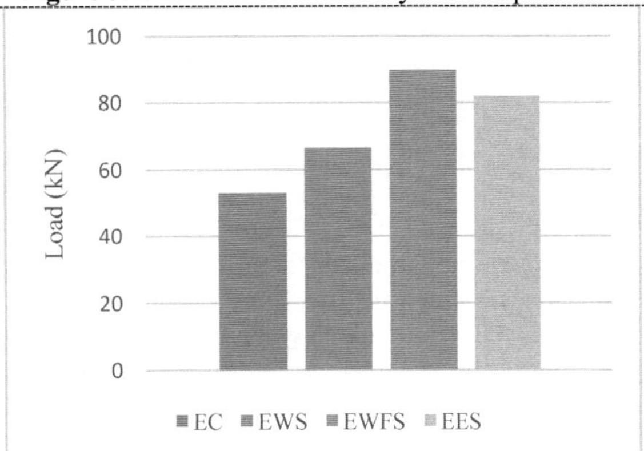

Figure 6. Ultimate load of eccentrically loaded specimens

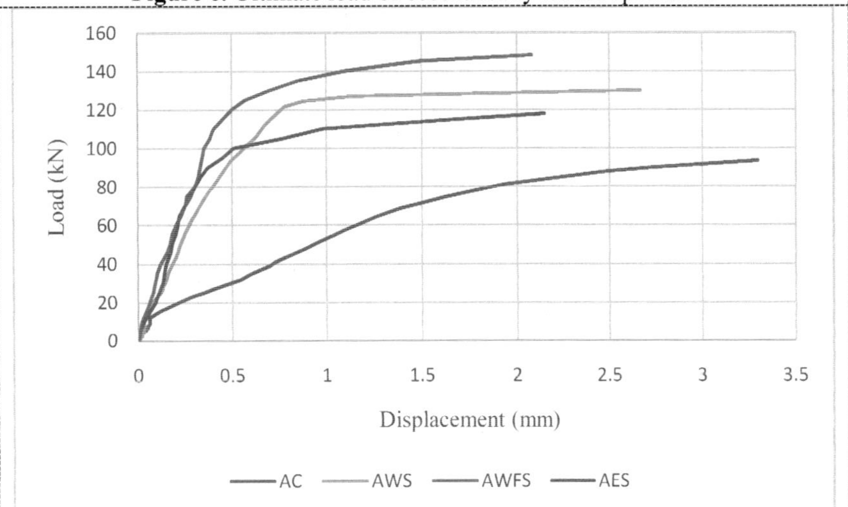

Figure 7. Load vs Deflection - axially loaded specimens

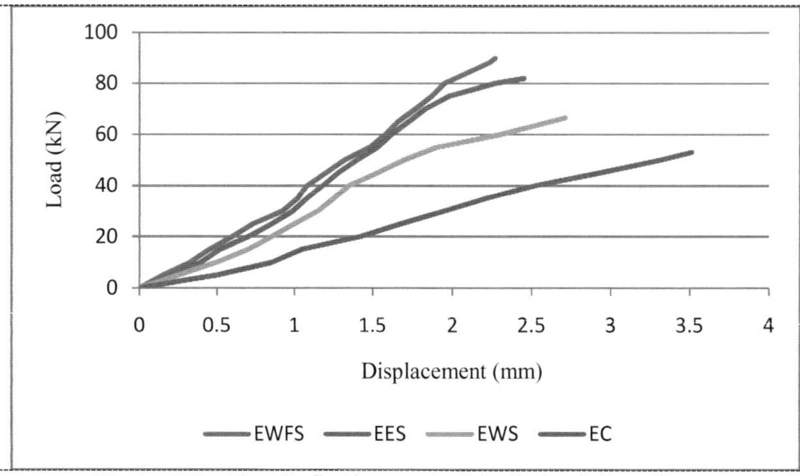

Figure 8. Load vs Deflection - eccentrically loaded specimens

5. Conclusion

- There was 79.3% and 67.7% increase in load carrying capacity of column with fully wrapped web and flange under axial and eccentric loading respectively.
- For columns wrapped at web portion alone, the load carrying capacity was increased by 57.5% and 25.2% under axial and eccentric loading respectively.
- 47.5% and 54.2% increase in load carrying capacity was observed for columns wrapped partially at the top portion of web and flange under axial and eccentric loading respectively.
- The columns which are wrapped at flange and web throughout the length of the column provides better strength than others specimens. This is due to the strengthening of column as a result of the confinement.

References

[1] Sreedhar Kalavagunta, Sivakumar Naganathan and Kamal Nasharuddin Bin Mustapha 2014 Axially Loaded Steel Columns Strengthened with CFRP. *Jordan. J. Civ. Eng.* **8** 58-69.

[2] Manikandan P, Sugumar S and Balaji T U 2014 Effective Shaping of Cold-Formed Thin-Walled Built-up Beams in Pure Bending *Arab. J. Sci. Eng.* **39** 6043–54.

[3] Bambach M R, Jama H H and Elchalakani M 2009 Axial capacity and design of thin-walled steel SHS strengthened with CFRP *Thin-Walled. Struct.* **47** 1112–1121.

[4] Thomas H K Kang, Kenneth A Biggs and Chris Ramseyer 2013 Buckling Modes of Cold-Formed Steel Columns *IACSIT Int. J. Eng. Technol.* **5** 447-51.

[5] Hélder D Craveiro, João Paulo, Rodrigues C and Luís Laím 2016 Buckling resistance of axially loaded cold-formed Steel columns *Thin-walled. Struct.* **106** 358–75.

[6] Sreedhar Kalavagunta, Sivakumar Naganathan and Kamal Nasharuddin Bin Mustapha 2013 Proposal for design rules of axially loaded CFRP Strengthened cold formed lipped channel steel sections *Thin-Walled. Struct.* **72** 14–9.

[7] M A El Aghoury, Salem A H, Hanna M T and Amoush E A 2015 Strength of cold formed battened columns Subjected to eccentric axial compressive force *J. Constr. Steel Res.* **113** 58–70.

[8] Bureau of Indian Standards (BIS) 1987 *Specifications for Cold Formed Light Gauge Structural Steel Sections* (IS 811 (1987)) New Delhi India.

Study on Flexural Behaviour of Ternary Blended Reinforced Self Compacting Concrete Beam with Conventional RCC Beam

M Marshaline Seles, R Suryanarayanan, S S Vivek* and G Dhinakaran
School of Civil Engineering, SASTRA University, Thanjavur- 613401, India

*Email: vivek@civil.sastra.edu

Abstract. The conventional concrete when used for structures having dense congested reinforcement, the problems such as external compaction and vibration needs special attention. In such case, the self compacting concrete (SCC) which has the properties like flow ability, passing and filling ability would be an obvious answer. All those SCC flow behavior was governed by EFNARC specifications. In present study, the combination type of SCC was prepared by replacing cement with silica fume (SF) and metakaolin (MK) along with optimum dosages of chemical admixtures. From the fresh property test, cube compressive strength and cylinder split tensile strength, optimum ternary mix was obtained. In order to study the flexural behavior, the optimum ternary mix was taken in which beam specimens of size 1200 mm x 100 mm x 200 mm was designed as singly reinforced section according to IS: 456-2000, Limit state method. Finally the comparative experimental analysis was made between conventional RCC and SCC beams of same grade in terms of flexural strength namely yield load & ultimate load, load- deflection curve, crack size and pattern respectively.
Keywords: Self compacting concrete, silica fume, metakaolin, flexural strength, cracks, load and deflection

1. Introduction

In recent developments of concrete technology many special concrete were developed ensuring high strength and high performance characteristics. In that scenario, self compacting concrete (SCC) was developed in 19th century in Japan to cater the durability aspects because of unskilled labour in the placement of concrete in bridge construction. It has a dynamic behaviour such as flow, passing and filling ability in the desired form. SCC had vast field applications in construction, repairs and rehabilitation of structures in countries such as Japan, Switzerland and European countries. European Federation of Nationalised Association Representing Concrete (EFNARC) made guidelines on conducting the test on SCC in fresh state and recommended the range of values for the tests such as slump flow, V-funnel, U-box and L-box test. SCC could be prepared as powder type and combination type.On SCC, the use of slag to understand the rheology was developed and found that 15% slag was optimum [1]. The research on fly ash addition in high strength SCC with reduced segregation potential enhances later age compressive strength (i.e. after 28 days) [2].The addition of nano-silica and mineral admixtures (fly ash& silica fume) improved the pore space denser and the rheology of high performance SCC [3]. In SCC, the increase of w/c ratio from 0.35 to 0.7 decreases the compressive and tensile strength by 66% and 51% respectively [4]. Research on SCC using GGBS and RHA by conducting axial compression and flexure tests reveals that the GGBS efficiency was improved from 0.7 to 1.8 at 28 days, obtained from Bolomey's empirical expression [5]. After structural assessment of corroded SCC beams on durability properties such as half-cell potential, cracks, corrosion control, etc., beams prepared by SCC mix was superior to normal concrete [6]. The analytical & experimental study on rehabilitation of RC beams using SCC jacket was found that the structure was recovered completely in terms of strength and ductility property and the good correlation exhibits between analytical and experimental results [7]. The experimental investigation on the flexural characteristics of SCC pre-tensioned bridge girders was studied at high early age compressive strength after 16 h and at later ages, SCC obtained higher strength than conventional concrete [8]. The SCC flow behaviour and mechanical properties by 10% SF as cement replacement performed better was investigated [9]. Using 15% MK as cement replacement in SCC, fresh properties test and compressive strength was determined [10]. The simplest mix design was developed for SCC mixes [11]. The research was carried out on SCC using cement replacement by MK [12]. The mix design of SCC and its fresh properties test was conducted according to EFNARC guidelines [13-14].

In present study, the combination type (mineral & chemical admixtures) of SCC was developed. The powder content of SCC consists of mineral admixtures namely Silica Fume (SF) and Metakaolin (MK). The chemical admixtures include the dosages of super plasticizer (SP) and viscosity modifying agent (VMA). The ternary blend SCC mix was prepared and studied by conducting fresh properties test, compressive strength on cubes, split tensile strength on cylinders and flexural behaviour of beams.

2. Material used

Cement: Ordinary Portland cement 53 grade ASTM Type I [16] was used.
Mineral Admixtures: Silica Fume (SF) and Metakaolin (MK) were used as per ASTM C 1240-99[15, 18].
Coarse Aggregates: For Self Compacting Concrete, the coarse aggregate was used as per ASTM C 127-12 [19]. Hence the property of fluidity was developed.
Fine aggregates: Natural river sand (< 4.75 mm) grading Zone III of ASTM was used. The physical properties were determined as per ASTM 127 and their particle size distributions confirmed to the requirements of ASTMC33 [17].
Chemical Admixtures: Super plasticizers and stabilizers play a major role in workability of concrete particularly for SCC.

3. Mix design

The mix design was developed according to conventional M60 grade concrete. By laboratory test trials, the fine and coarse aggregate proportions were adjusted and with addition of optimum water-powder ratio (0.4), dosages of super plasticizers (1.25% weight of cement) and stabilizers-VMA (0.1% weight of cement) SCC mix were prepared (Table 1). Then the prepared mix was subjected to fresh properties test according to EFNARC guidelines.

Table 1. Materials required for SCC

MIX	C Kg/m^3	SF kg/m^3	MK kg/m^3	FA kg/m^3	CA kg/m^3	Water l/ m^3	SP l/m^3	VMA l/m^3	w/p ratio
CVC	600	-	-	396	1074	198	-	-	0.33
C+10% SF+ 20%MK	420	60	120	810	660	240	7.5	0.6	0.4
C+5% SF + 20%MK	450	30	120	810	660	240	7.5	0.6	0.4
C+20%SF + 5%MK	450	120	30	810	660	240	7.5	0.6	0.4

4. Experimental Programme

The experimental programme includes conducting test on basic materials to understand their physical properties and test on SCC fresh properties. In hardened state, the mechanical properties were studied by the preparation of cube, cylinder and beam specimens at 28 days of testing.

4.1 Basic test on materials

Before undergoing mix preparation, the quality of materials was found by conducting corresponding tests such as specific gravity, fineness modulus, bulk density, etc., for cement, fine aggregate and coarse aggregate, silica fume (SF) and metakaolin (MK).

4.2 Test for fresh properties of concrete

The SCC fresh property test namely slump flow test & T500, L box test, U box test and V funnel test was conducted for each mix and done according to EFNARC guidelines.

4.3 Preparation of mould and Casting of specimens

Once fresh properties test were conducted on each of those prepared mixes, the cube mould prepared from wooden material of inner dimensions 100 mm x 100 mm x100 mm, cylinder mould of PVC material having inner diameter 100 mm and 200 mm height, SCC was poured. From the optimum results obtained from cube compressive strength and cylinder split tensile strength, beam specimen was prepared (Figure 1& 2). Two beams were cast with the wooden mould of inner dimensions 1200 mm × 100 mm × 200 mm size.

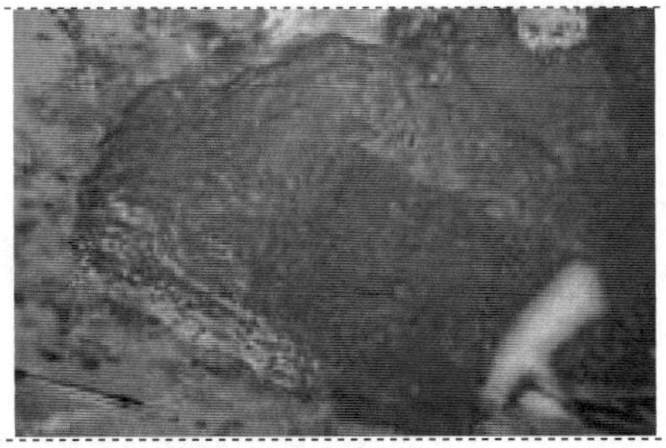

Figure 1. Mix preparation

Figure 2. Casting of beam.

4.4 Curing of specimens

The cube and cylinder specimens were removed from the mould and normal water curing was done for the duration of 28 days. But for the beams, the formwork of beams were de-moulded after 24 hours of casting and then cured in gunny bags with ponding of water for the age of 28 days. Finally the specimens were taken out from water tank after curing; its surfaces were made dry in normal atmosphere temperature for 3 hours before testing in laboratory.

4.5 Testing of specimens

The cube and cylinder specimens were tested in the Universal Testing Machine (UTM) to determine compressive and split tensile strength. For testing of beams, the self supporting frame was used (Figure 3). The load & deflection was measured using digitized recorder and LVDT's were used for measuring deflection at mid span (L/2) and under the load (L/3).

Figure 3. Experiment test set up on beam

5. Results and discussion

5.1 Results

The results obtained from the experimental programme were categorized into three stages. Stage 1 includes results from basic properties of materials and SCC fresh properties test. Stage 2 includes test on compressive and split tensile strength on cubes and cylinders. Stage 3 includes results obtained by conducting test on beams.

Table 2. Basic test on materials.

Fine aggregate	Fineness modulus:3.16
	Specific gravity:2.547
Coarse aggregate	Fineness modulus:7.02
	Specific gravity:2.74
Cement(OPC)	Initial setting time:33 minutes
	Std. consistency:31%
	Specific gravity:3.14
Metakaolin (MK)	Specific gravity: 2.5
Silica fume(SF)	Specific gravity:2.08

Table 3. Test on SCC fresh properties.

MIX	Fresh properties test (As per EFNARC guidelines)				
	Slump flow test (mm)	T 500 (s)	V-funnel test (s)	L-Box test (H2/H1)	U-Box test (H2-H1) (mm)
C+10% SF +20%MK	675	3	10	0.84	25
C+5% SF +20%MK	660	3.5	11	0.91	28
C+20%SF +5%MK	655	2.5	9	0.87	27.5
Accepted Range	650 to 800	2 to 5 s	6 to 12 s	0.8 to 1	0 to 30

Table 4. Test on cube compressive strength.

MIX	Cube Compressive Strength (MPa)		
	3 days	7 days	28 days
CVC	12.35	33.6	55.94
C+10% SF +20%MK	15.2	38.3	65.5
C+5% SF +20%MK	13.7	36.5	60.3
C+20%SF +5%MK	13.1	34.5	58.1

Table 5. Test on cylinder split tensile strength.

MIX	Cylinder Split tensile strength (MPa)	
	7 days	28 days
CVC	2.15	3.55
C+10% SF +20%MK	2.37	4.01
C+5% SF +20%MK	2.28	3.84
C+20%SF +5%MK	2.19	3.6

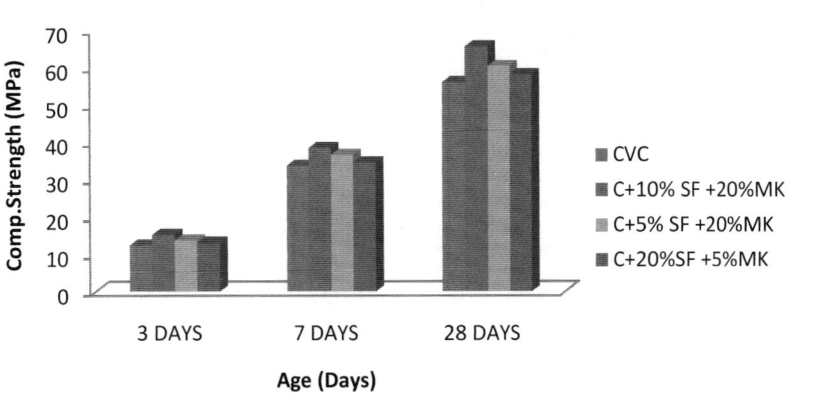

Figure 4. Cube compressive strength (MPa) Vs Age (Days).

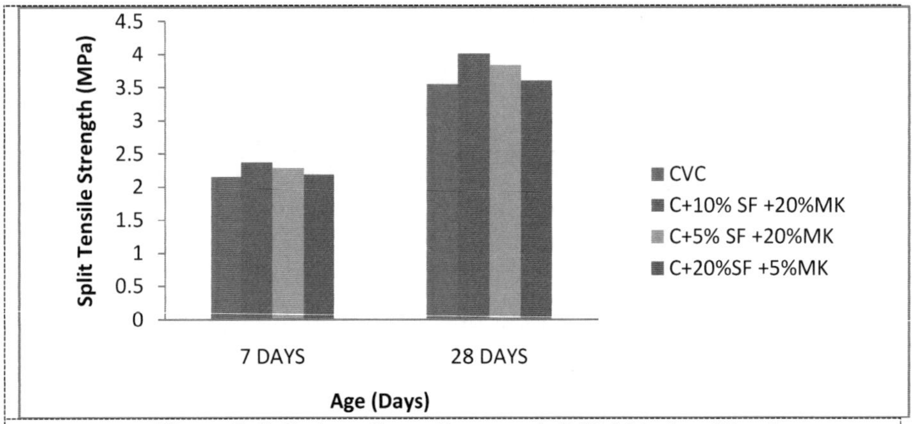

Figure 5. Cylinder split tensile strength (MPa) Vs Age (Days).

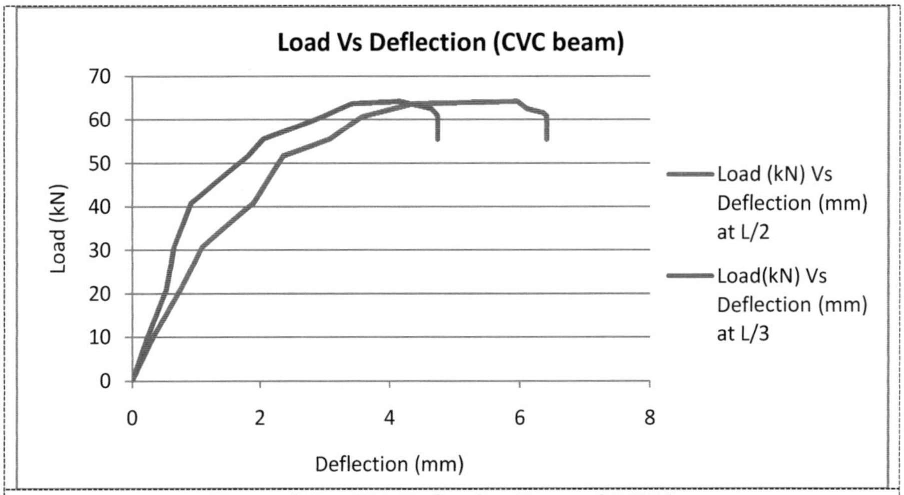

Figure 6. Load Vs Deflection Curve of CVC beam.

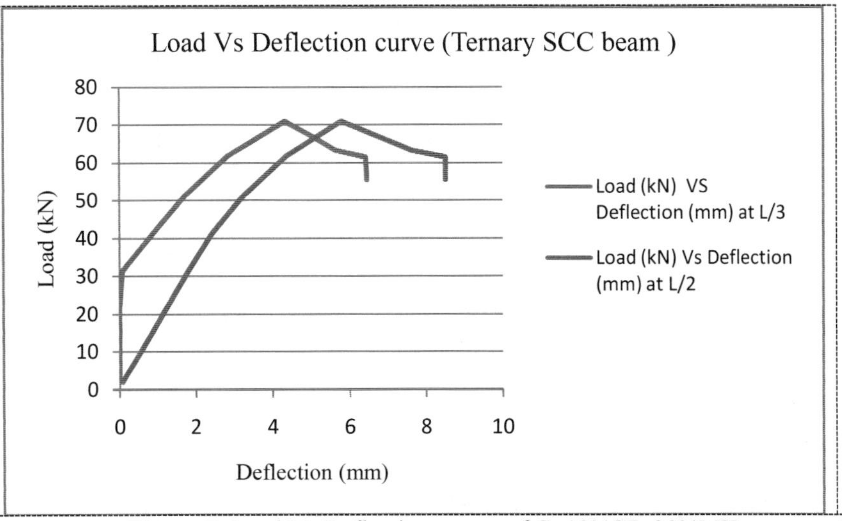

Figure 7. Load Vs Deflection curves of C+10%SF+20%MK.

5.2 Discussion

5.2.1 SCC fresh properties. SCC rheology was studied by conducting fresh properties test according to EFNARC guidelines (Table 3). In SCC ternary blend type, two different heterogeneous materials (SF & MK) were used to replace cement. The conglomeration of SCC mix was obtained by the addition of chemical admixtures namely SP and VMA. The determination of water powder ratio was done by conducting laboratory test trials. First the control over aggregate proportions was developed and on the next stage water content, dosages of SP and VMA were optimized. In such a way the flow was developed with w/p ratio 0.4, SP 1.25% weight of cement and VMA 0.1% weight of cement. It should also be noted that the over dosages of w/p ratio and chemical admixtures leads to less strength and segregation of mix. The slump flow & T500 test measures flow ability of SCC mix, all ternary SCC mixes were attained minimum spread diameter of 650 mm. The reason of SCC flow behaviour was due to addition of powder content namely MK and SF used up to 25% and 30% in the ternary mix. Similar trend was obtained for L-box, U-box and V-funnel test which determines the filling ability of SCC in the field applications such as dense congested reinforcement. Hence all SCC fresh properties test were well within the range of EFNARC guidelines and found satisfactory.

5.2.2 CVC and SCC mechanical properties. The hardened state properties of CVC and ternary SCC mixes were studied by conducting cube compressive strength and cylinder split tensile strength at the age of 28 days (Table 4 & 5). The cement replacement by two mineral admixtures namely 10%SF and 20% MK plays a significant role in improving mechanical properties was shown (Figure 4 & 5). It reveals that optimal cement replacement confines the interfacial transition zone (ITZ) and it makes denser in ternary SCC mixes. Here out of 3 ternary SCC mixes, two mixes were replaced up to 25% and one having maximum cement replacement of 30% respectively. Hence increase of fines in the form of powder content develops higher strength in the ternary form, which resembles that in binary form pozzolonic activity was less whereas in the ternary form combined reaction takes place. The above statement could be correlated with the composite behaviour of materials with fibres and matrix form. In this study, due to above highlighted reasons, SCC ternary form shown higher mechanical properties than conventional vibrated concrete (CVC).

5.2.3 Flexural behaviour of CVC and SCC beam. The structural behaviour of any concrete materials could be studied by flexure or axial compression test. In present study the beams were cast with CVC and ternary SCC type. The beam test set up (Figure 3) comprises of two point loading system in which the corresponding deflection was measured at the position of L/2 and L/3 of beam simply supported span (L). The initial crack obtained for CVC was obtained at 40kN whereas in ternary SCC (C+ 10%SF+20%MK) was 50 kN. The maximum load obtained by ternary SCC was 71 kN with a maximum displacement at L/2 was 8.52 mm was relatively a better performance with CVC beam (Figure 6 &7). The wider cracks say 1.5 mm were found in CVC beam than SCC of 1mm. The failure modes of beams were under flexure only. Hence SCC beam performed better than CVC on flexure study.

6. Conclusion

Based on experimental investigation on self compacting concrete with cement replacement by SF and MK as ternary blend type the following important conclusions were made. By laboratory trials, optimum w/p ratio of 0.4 was achieved by conducting fresh tests with the dosages of Super plasticizer (1.25% by weight of cement) and Stabilizer (0.1% by weight of cement) respectively. The SCC fresh properties tests values were well within the recommended range of EFNARC guidelines. The compressive strength and split tensile strength values of ternary SCC mix was 15% and 12% higher than CVC mix. It was found that cement replacement by 10%SF and 20% MK performed better in terms of compressive strength and split tensile strength respectively. From the load –displacement curve between CVC and SCC, SCC beam were ductile in nature compared to CVC beam was observed on flexure study. The width, spacing and propagation of cracks were within the flexure zone was observed.

Acknowledgements
The authors would like to thank the Vice Chancellor of SASTRA UNIVERSITY for having provided experimental facilities in the School of Civil Engineering to do this research work and also for the continuous support and encouragement given throughout this research work.

References

[1] Boukendakdji O, Kenai S, Kadri E H and Rouis F 2009 Effect of Slag on the Rheology of Fresh Self-compacted Concrete. *Constr. Build. Mater.* **23** 2593–8.
[2] Sri R R, Siladyi D and Adamopoulos B 2003 Development of High strength Self-compacting Concrete with Reduced Segregation Potential. *3rd International RILEM Symposium* **1** 1048
[3] Mostafa J, Alireza P, Omid F H and Davoud J 2015 Comparative study on effects of Class F fly ash, nano silica and silica fume on properties of high performance self compacting concrete. *Constr. Build. Mater.* **94** 90–104
[4] Nikbin I M, Beygi M H, Kazemi M T, Vaseghi A J, Rabbanifar S, Rahmani E and Rahimi S A 2014 Comprehensive investigation into the effect of water to cement ratio and powder content on mechanical properties of SCC. *Constr. Build. Mater.* **57** 69–80.
[5] Kiran D, Abhishek G, Vivek V and Bhupinder S 2014 Experimental Study of Self Compacting Concrete made With GGBS and RHA under Axial Compression and Flexure. *Int. J. Recent Res. Aspects* **1**(3) 118-20.
[6] Hassan A A A , Hossain K M A and Lachemi M 2010 Structural assessment of corroded self-consolidating concrete beams. *Eng. Struct.* **32** 874-5.
[7] Constantin E C, Georgia E T and Stavroula J P 2014 Behaviour of rehabilitated RC beams with self-compacting concretejacketing – Analytical model and test results. *Constr. Build. Mater.* **55** 257–73
[8] Young H K, Mary B D H and David T 2015 Flexural Behavior of High-Early-Strength Self-Consolidating Concrete Pretensioned Bridge Girders *J. Bridge Eng.* **20**(2) 1-13
[9] Vivek S S and Dhinakaran G 2015 Effect of silica fume in flow properties and compressive strength of self compacting concrete. *Int. J. ChemTech. Res.* **8**(1) 1-5
[10] Anjali D, Vivek S S and Dhinakaran G 2015 Compressive strength of metakaolin based self-compacting concrete. *Int. J. ChemTech Res.* **8**(2) 622-5

IOP Conf. Series: Earth and Environmental Science **80** (2017) 012026 doi:10.1088/1755-1315/80/1/012026

[11] Nan S, Kung C H and His WC 2001 A simple mix design method for self-compacting concrete. *Cem. Concr. Res.* **31** 1799–807
[12] Dinakar P and Manu S N 2014 Concrete mix design for high strength self-compacting concrete using metakaolin. *Mater. Des.* **60** 661–8
[13] *EFNARC* guidelines 2002 Specification and Guidelines for self-compacting concrete
[14] *EFNARC* guidelines 2005 The European guidelines for self-compacting concrete (Specification, Production, Use)
[15] American Society for Testing and Materials (ASTM) 2006 Standard specification for coal fly ash and raw or calcined natural pozzolan for use in concrete *ASTM C 618,* ASTM International, West Conshohocken, PA
[16] American Society for Testing and Materials (ASTM). Standard specification for Portland Cement *ASTM C150/C150M-12*, ASTM International, West Conshohocken, PA DOI: 10.1520/C0150_C0150M-12
[17] American Society for Testing and Materials (ASTM).Standard specification for concrete aggregate. *ASTM C 33*, ASTM International, West Conshohocken, PA
[18] American Society for Testing and Materials (ASTM). Standard specifications for silica fume used in cementitious mixtures. *ASTM C 1240-99*, ASTM International, West Conshohocken, PA
[19] American Society for Testing and Materials (ASTM). Standard Test Method for Density, Relative Density (Specific Gravity), and Absorption of Coarse Aggregate. *ASTM C 127-12*, ASTM International, West Conshohocken, PA, DOI 10.1520/C0127-12

Experimental Study On Flexural Behaviour Of Beams Reinforced With GFRP Rebars

G Naveen Kumar and Karthik Sundaravadivelu*
School of Civil Engineering, SASTRA University, Thanjavur, 613401, India.

* E-mail: karthik@civil.sastra.edu

Abstract. In saline, moisture and cold conditions corrosion of steel is inevitable and the lot of economy is used for rehabilitation works. Corrosion of steel is nothing but oxidation of iron in moisture conditions and this corrosion leads to the spalling of concrete which intern reduces the strength of the structure. To reduce this corrosion effects, new materials with resistance against corrosion have to be introduced. Many experiments are going on using Glass Fiber Reinforced Polymer (GFRP) as alternate material for steel due to its non-corrosive nature, weight of GFRP is nearly one third of steel and ultimate tensile strength is higher than steel. In this paper, six beams are casted in which three beams are casted with steel as main and shear reinforcement and another three beams are casted with GFRP as main reinforcement with steel as shear reinforcing material. All beams casted are of same dimensions with variation in reinforcement percentage. The size of the beams casted is of length 1200 mm, breadth 100 mm and depth 200 mm. The clear cover of 25 mm is provided on top and bottom of the beam. Beams are tested under two-point loading with constant aspect ratio (a/d) and comparing the flexural strength, load deflection curves and types of failures of beams reinforced with GFRP as main reinforcement and beams reinforced with conventional steel. The final experimental results are compared with numerical results. M30 grade concrete with Conplast as a superplasticizer is used for casting beams.
Key words: GFRP bars, aspect ratio, flexural strength and superplasticizer.

1. Introduction

Corrosion is one of the main problem for the steel reinforced concrete structures in moisture conditions. The steel in concrete structures corrode due to the carbonation of metal. Generally cement is very high alkaline in nature and when react with water forms hydrated calcium silicates (CSH) and calcium hydroxide. All these elements are very high alkaline in nature in fresh concrete and in ages this alkalinity is neutralized by reaction of lime with atmospheric carbon dioxide and pH value of concrete comes below 10 and corrosion starts in steel. To overcome this corrosion new material has to be introduced. Fiber Reinforced Polymer FRP materials are getting more popular due to its high strength and non-corrosive nature. There are different types of FRP in which Glass Fiber Reinforced Polymer (GFRP), Carbon Fiber Reinforced Polymer (CFRP) and Basalt Fiber Reinforced Polymer (BFRP) are more popular. In this project GFRP is adopted as reinforcing material. GFRP is popular not only for its non-corrosive property but also for its light weight, nonmagnetic and high strength properties. Manufacturing of GFRP bars are simple and it requires bonding material to hold fibers together. Manufacturing of GFRP rods is quite simple than that of Steel production, GFRP production

Process involves pultrusion process. The roving's of GFRP fiber is given a resin bath and sent in to a mould of required thickness. The fibers are treated under 160 degrees Celsius in mould and rod is constantly pulled out from mould with constant rate. These rods are kept under 140 degrees Celsius as a part of post curing process [1]. In different types of FRPs, GFRP is most economical material with less tensile strength value. Main variables that influence the flexure behavior of GFRP reinforced beams are low modulus of elasticity and rupture strain [2]. In a balanced reinforced section the tensile reinforcement reaches its ultimate strength when compression concrete reaches its maximum strain i.e. 0.0003 (ACI 2003). Using of GFRP rods as main reinforcement had reduced the ultimate load of 23% for beam prototype and failure of beam is under flexure. Replacing GFRP as main and shear reinforcement showed the reduction of 33% in ultimate load and beam showed shear failure [3]. GFRP reinforced beams fails mainly due to bond failure between GFRP rods and concrete, reduced post cracking stiffness [4]. Effect of GFRP reinforcement in transverse direction shows significant change in shear strength of beam. There is no significant increase in strength of beam with increase in concrete strength. Mode of failure changes with change in ratio of longitudinal reinforcement [5]. In direct pull out test between GFRP and concrete shows failure of GFRP rod layers and increase in diameter of bar decrease in bond strength [6]. GFRP reinforced slabs shows more deflection and more number of cracks when compared with steel reinforced concrete [7]. Concrete beams reinforced with GFRP bars shows higher strain, deflection values. Behavior of stress strain curve is linear before crack and after crack beam behaves linearly with reduced stiffness [8]. Increase in a/d ratio decreases the ultimate load carrying capacity and stirrups in GFRP reinforcement shows higher strain than in steel reinforced beams [9].

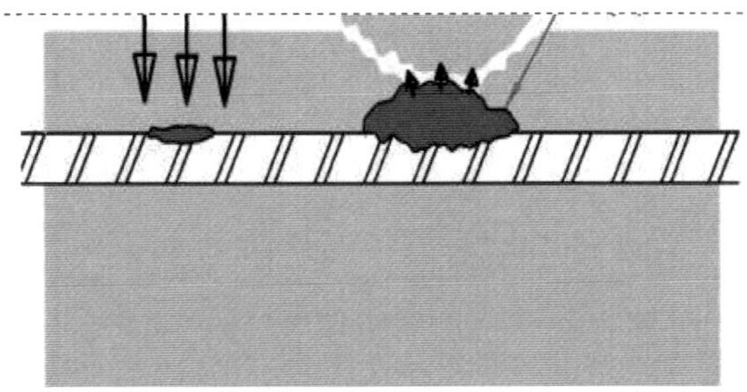

Figure 1. Rusting of Iron

3. Experimental Program

The experimental program consists of preparing and testing of six prototype beams under two point loading system. Beams are casted with M30 grade concrete, the grade of concrete is finalized after few trail mixes and the ratio is 1:1.67:2.54. Three beams casted with GFRP reinforcement with ratios 1%, 1.5%, and 2.5% along with corresponding control beams are casted with steel as main reinforcement providing same reinforcement ratios. The width, depth and length of the beam were 100 mm, 200 mm and 1200 mm respectively. All the beams were casted and cured for 28 days.

4. Material Properties

The properties of the materials used in experimental program is illustrated below

4.1. Cement

OPC 53 grade cement of locally available NAGARJUNA cement satisfying IS 12269:1987 [10] is used, which comprises of good quality. Chemical configuration of cement is given in Table 1.

4.2 Fine Aggregate For fine aggregates, uncrushed and naturally available local river sand of maximum size 4.75 mm with a fineness modulus of 3.35 and specific gravity of 2.65 using IS 2386 (Part III):1963[11] was used.

Table 1. Physical and chemical composition of ordinary Portland cement (OPC)

Description	Composition
Physical Properties	
Color	Grey
Specific gravity	3.15
Specific surface area (cm^2/g)	3540
Standard consistency (%)	29 %
Final setting time	175 minutes
Initial setting time	33 minutes
Chemical Composition	
CaO (%)	61.3
SiO2 (%)	19.8
Al2O3 (%)	5.3
Fe2O3 (%)	4.01
MgO (%)	2.85
Na2O (%)	0.13
K2O (%)	61.96

4.3 Coarse Aggregate

The coarse aggregates size used for casting ranges between 12.5 mm to 20 mm of specific gravity 2.74 using IS 2386(Part III):1963[11]. Specifically, the mixture having 70% of 20 mm and 30% of 12.5 mm by weight is used. The preliminary test results are shown in Table 2.

Table 2. Material test results of aggregate

Property	Fine aggregate	Coarse aggregate	
		20mm down	10mm down
Fineness modulus	3.35	7.56	3.20
Specific gravity	2.65	2.74	2.69
Water absorption (%)	1.20	1.83	1.35
Bulk density	1753	1741	1711

4.4 Chemical Admixture

CONPLAST is used as chemical admixture complies with IS: 9103:1999 [12] and having a specific gravity of one was used as a high range water reducing agent, 0.5% of cement is added to concrete to increase workability and strength.

4.5 Water

Ordinary Potable tap water was used for mixing and curing.

4.6 Reinforcement

Steel bars of Fe500 grade and sand coated GFRP bars were used for main reinforcement. Yield tensile stress, young modulus and yield strain of steel bars tested in Universal Testing Machine (UTM) were 512 MPa, 210 GPa, and 0.0028 respectively. As for the steel used in shear reinforcement, the values were found to be 164 GPa, 433MPa and 0.0028 as the modulus, yield stress and yield strain respectively. The tensile yield strength of GFRP bar obtained is 617 N/mm2 and the young's modulus is 45 N/mm2. Sand coating is done to increase the bond strength between GFRP and concrete. The physical properties of reinforcement used in the experiment is detailed in the Table 3.

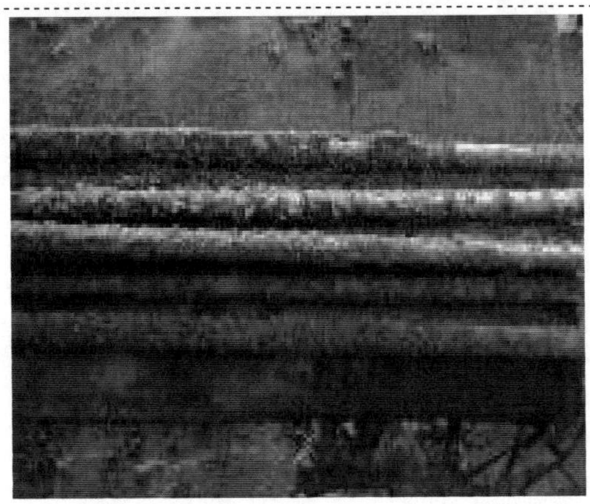

Figure 2 Sand coated GFRP bars

Figure 3 Failure of GFRP bars

Table 3 Physical properties of steel and GFRP

S. No	Description	Value	
		GFRP	Steel
1	Density	2100 kg/m3	7850 kg/m3
2	Young's modulus	41 GPa	210 GPa
3	Thermal coefficient	6 x 10-6/ °c	12 x 10-6/ °c
4	Poisson's ratio	0.2	0.3
5	Shear modulus	12 GPa	81 GPa

5. Concrete

5.1 Mix Proportions

Mix design of concrete was done by using IS 10262-2009 and IS 456-2000. Concrete grade for M30 is prepared and the ratio by weights is obtained as 1:1.87:2.43, superplasticizer as water reducing agent is added and detailed proportion is given in Table 4.

Table 4. Quantity of Material

MATERIAL	WEIGHT in kg/m^3
Cement	380
Fine aggregate	710.6
Coarse aggregate	965
Water cement ratio	0.42
Superplasticizer	0.5% by weight of cement

5.2 Specimen Casting and Testing

150 mm cubes and 150 mm diameter with 300 mm height cylinders were casted and found the average compressive strength and split tensile strength after 28 days of curing and is given in Table 5

Table 5 Physical properties of concrete

Description	Compressive strength	Tensile strength
With admixture	34.63 MPa	3.45 MPa
Without admixture	30.52 MPa	2.82 MPa

6. TEST MATRIX AND SPECIMEN PREPARATION

In total 6 prototype beams were casted with same dimensions of width 100 mm, depth 200 mm and length 1200 mm. Longitudinal reinforcement bars used are of three different diameters i.e. 10, 12 and 16 mm. Anchorage reinforcement used is of 8 mm in diameter and shear reinforcement used is 6 mm diameter bars with uniform spacing of 100 mm. Concrete with water reducing chemical admixture is used for casting beam prototypes. Based on type of longitudinal reinforcement provided beams are divided in two groups and identity is given to each beam based on reinforcement percentage. Group S consist of beams reinforced with steel and Group G consists of beams reinforced with GFRP as main reinforcement.

Table 6 Test matrix

Group	Specimen ID	Longitudinal reinforcement	Reinforcement (%)
S	M30-S1	Steel	1
	M30-S2	Steel	1.5
	M30-S3	Steel	2.5
G	M30-G1	GFRP	1
	M30-G2	GFRP	1.5
	M30-G3	GFRP	2.5

7. TEST SETUP

Two point Load is applied on prototype using single point loading unit from which load is distributed equally for both points. LVDT are placed in bottom face at mid-section and one third of beam length to note the deflection. The first crack load and crack pattern is noted on both sides of beam at regular intervals. Gauge length between load points is 300 mm and the constant aspect ratio of 150 mm is provided. All specimens are tested under same setup conditions and load applied on beam at a constant rate of 10 kN/min in 1000 kN load cell. Automatic data acquisition technique is used to record the load and deflections.

8. Results and Discussion

8.1 Ultimate Load Carrying Capacity

The ultimate load carrying capacity is the load at the failure of the specimen. The values obtained from the two point loading bending test for different specimen reinforced with GFRP and steel reinforcement is given in the table 7 and we can clearly observe the increase in the ultimate load value for M30 G1 than M30-S1.

Table 7 Ultimate loads for Beams

Sl.No.	Reinforcement Ratio	Ultimate Load (kN)	
		Steel	GFRP
1	1	88.5	126
2	1.5	124.3	74.80
3	2.5	162.9	48

8.2 Load Vs Deflection Behavior

The Load Vs Deflection behavior for different percentages is shown in Figures 4. It is clearly shown that M30-G1 beam takes maximum load than M30-S1 beam and steel reinforced beams take maximum load than GFRP reinforced beam. The deflections for GFRP reinforced beams are similar to their corresponding control beams.

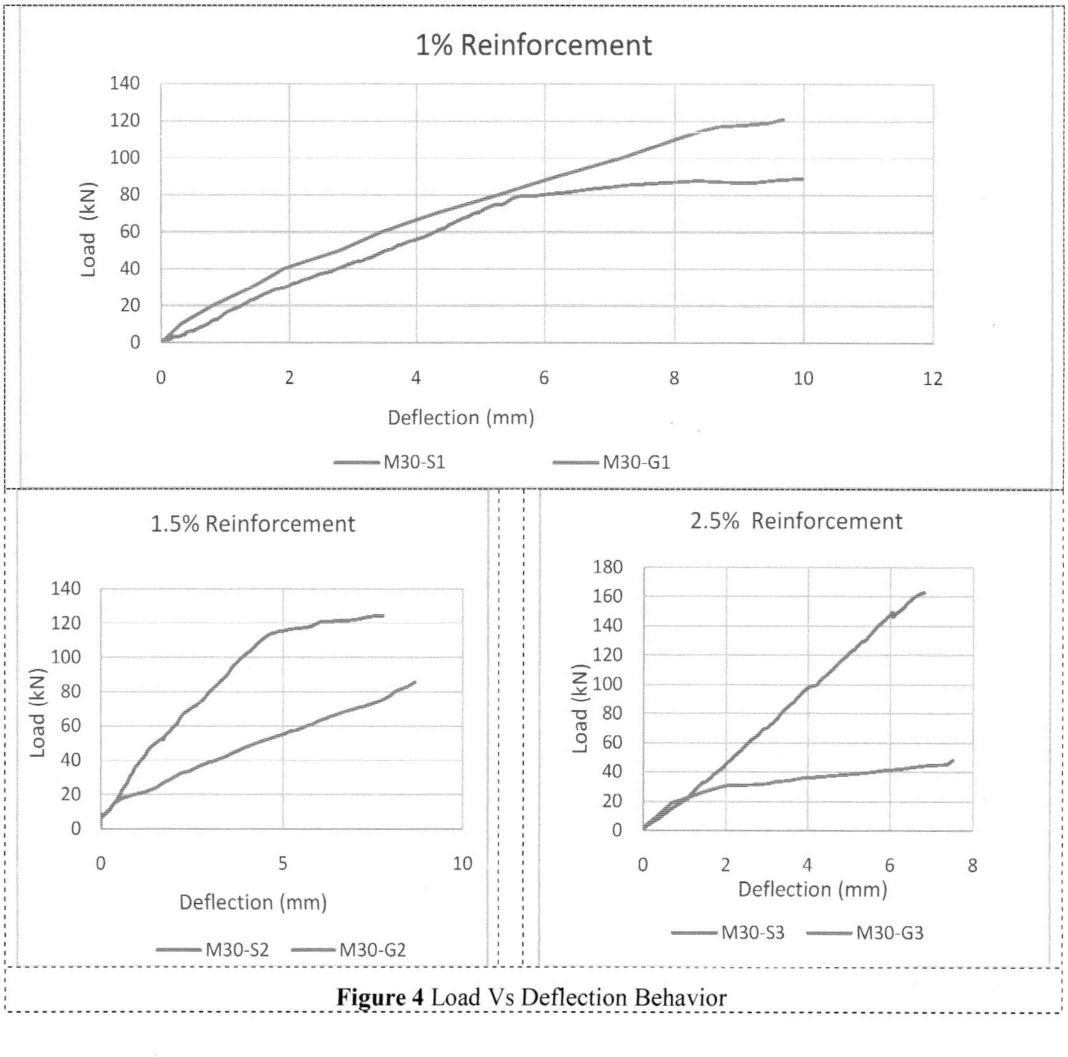

Figure 4 Load Vs Deflection Behavior

8.3 Stiffness

M30-S1 shows more stiffness than M30-G1, it shows that the GFRP reinforced beams under goes more deflection for small loads. The initial stiffness for M30-G1 and G2 beams is higher than M30G1 and nearly equal to M30-S2 and S3 respectively. Stiffness for beams with steel shear reinforcement is more and with the increase in shear reinforcement will increase stiffness. The load vs stiffness graphs for beams are compared and shown in figure.5.

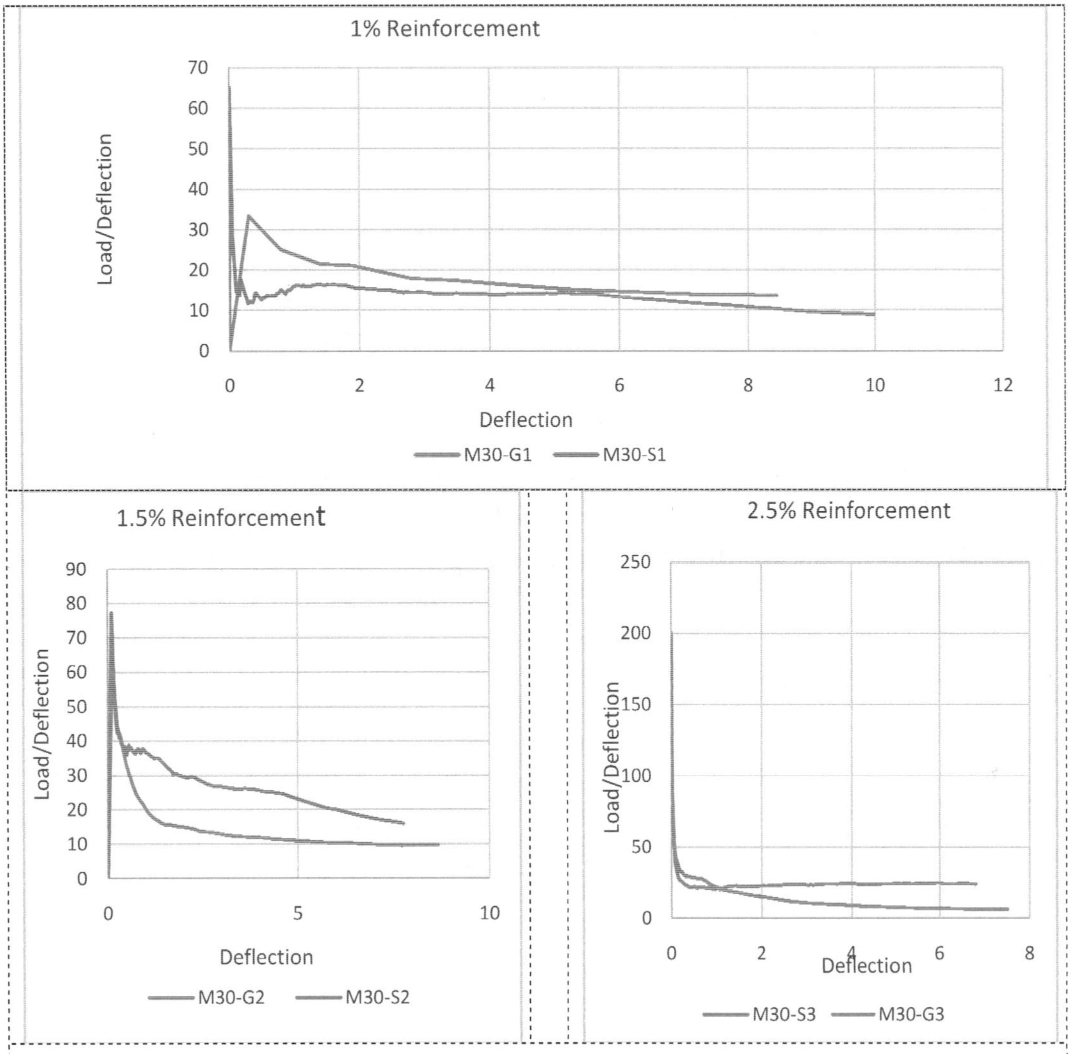

Figure 5 Stiffness

8.4 Crack Pattern

a) Increase in reinforcement ratio in controlled specimen changes the crack pattern from flexural failure to pure shear failure and GFRP reinforced beams has local and pure shear cracks. The behavior of crack failure for beams are mentioned in Table 8

Table 8 Modes of Failure of Beams

Sl.No	Reinforcement Ratio	Modes of failure of beams reinforced with STEEL	GFRP
1	1 %	Flexure	Flexure with web shear cracks
2	1.5 %	Flexure with prolonged shear cracks	Flexure
3	2.5 %	Pure shear	Flexure with brittle behavior

b) The spacing between cracks is less in GFRP reinforced beams and width of cracks is more. The crack pattern for the beams are shown schematically in figure 6.

c) The crack to crack spacing in GFRP reinforced beams are less and number of cracks is more. This is due to the lesser modulus of elasticity of GFRP rods.

d) The crack behavior of GFRP reinforced beam with 2.5% of reinforcement is very similar to the over reinforced flexure member.

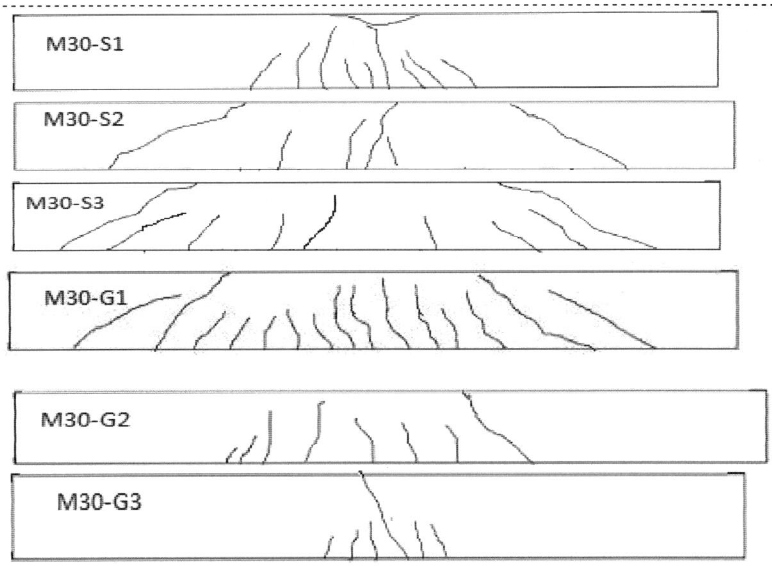

Figure 6 Crack Patterns

9. Conclusion

1) Experimental results obtained for conventional beams are satisfied with theoretical values from IS 456-2000 but GFRP reinforced beams with 1.5 % and 2.5% reinforcement disagrees with IS 456-2000.

2) With the increase in percentage of steel reinforcement there is an increase in peak load value and with increase in GFRP percentage there is decrease in peak load value.

3) The failure of GFRP reinforced beams shows that with increase in the percentage of reinforcement concrete reaches its ultimate stress first and beams fails under flexure(which is similar to a over reinforced section).

4) The peak load value for M30-G1 is 36% more than M30-S1, M30-G2 is 39% less than M30S2.

5) The initial stiffness for GFRP beams is very high when compared with corresponding steel reinforced beams but increase in load stiffness for GFRP beams reduced.

6) Number of cracks is more in GFRP reinforced beams than steel reinforced beams for same reinforcement percentage.

7) The decrease in load value with increase in GFRP reinforcement is due to the bond failure bond slip between GFRP bars and concrete.

8) Modes of failure for GFRP reinforced beams is due to pure and local flexure but in steel reinforced beams failure changes from flexure to shear with increase in reinforcement percentage.
9) The load between first crack and ultimate load is very less in GFRP reinforced beams but in steel reinforced beams takes more load between first crack and ultimate load.

Acknowledgement

The authors would like to thank to the authorities of SASTRA University, Thanjavur, for providing laboratory facilities and administrative assistance to perform the present research work in the School of Civil Engineering.

References

[1] D. H. Tavares et al. September, (2008). "Behavior of reinforced concrete beams reinforced with GFRP bars" *Volume 1, Number **3** p. 285 - 95 • ISSN 1983-4195.*
[2] American Concrete Institute (2003). ACI Committee 440.1R *Guide for the Design and Construction of Concrete Reinforced with FRP Bars.*
[3] S.Vivekanand et al. (2015). "Static Behavior of Concrete Beams Reinforced in Shear with GFRP Bars" *Int.J. ChemTech Res.**8** (2), pp 635-42.*
[4] B. Saikia et al. "Strength and serviceability performance of beams reinforced with GFRP bars in flexure" *Construction and Building Materials **21** (2007) 1709–19.*
[5] Ramadass S & Job Thoma 2010. "Flexure-shear Analysis of Concrete Beam Reinforced with GFRP bars". *The 5th International Conference on FRP Composites in Civil Engineering September 27-29, 2010.*
[6] Zenon Achillides and Kypros Pilakoutas(2004)."Bond Behavior of Fiber Reinforced Polymer under direct pullout condition" *Journal of Composites Constr., 2004, **8**(2): 173-81.*
[7] M V Venkateshwara Rao et al. (2010). "Behaviour Of Gfrp R.C. Slabs In Flexure In Local Environment - An Experimental Study" *35th Conference on OUR WORLD IN CONCRETE & STRUCTURES: 25 - 27 August 2010,Article Online Id: 100035057.*

[8] F.M. Wegian, H.A. Abdalla, "shear capacity of concrete beams reinforced with fiber reinforced polymer" *Composite Structures* ***71*** *(2005) 130–8.*

[9] Noor Azlina Abdul Hamid and et al."Experimental Investigation on the Shear Behaviour of Concrete Beams Reinforced with GFRP Reinforcement Bars".*Advanced Materials Research Vol.* ***626*** *(2013) pp 559-63.*

[10] *IS 12269:1987*, Indian standard ordinary Portland cement, 53 grade ;specification, *Bureau of IndianStandard, Manak Bhavan, 9 Bahadur Shah Zafar Marg, New Delhi 110002.*

[11] *IS 2386 (Part III) : 1963*, Indian Standard Methods of Test for aggregates for Concrete Part III Specific gravity, Density, Voids, Absorption and Bulking, *Bureau of Indian Standard, Manak Bhavan 9 Bahadur shah Zafar Marg, New Delhi 110002.*

[12] *IS 9103:1999,* Specification for Concrete Admixtures, Bureau of Indian Standard, Manak Bha van, 9 Bahadur Shah Zafar Marg, New Delhi 110002.

[13] *IS 10262: 2009*, Indian Standard for Concrete Mix Proportioning Guidelines, Bureau of Indi anStandard, Manak Bhavan, 9 Bahadur Shah Zafar Marg, New Delhi 110002.

[14] *IS 456-2000, Indian* Standard Plain And Reinforced Concrete Code of practice, Bureau of Indian Standard, Manak Bhavan, 9 Bahadur ShahZafar Marg, New Delhi 110002.

Behaviour of Nano Silica in Tension Zone of High Performance Concrete Beams

P Jaishankar* and D Vivek

School of Civil Engineering, SASTRA University, Thanjavur 613 401, India,

*Email: jai@civil.sastra.edu

Abstract. High performance concrete (HPC) is similar to High strength concrete (HSC).It is because of lowering of water to cement ratio, which is needed to attain high strength and generally improves other properties. This concrete contains one or more cementitious materials such as fly ash, Silica fume or ground granulated blast furnace slag and usually a super plasticizer. The term 'high performance' is somewhat different because the essential feature of this concrete is that it's ingredients and proportions are specifically chosen so as to have particularly appropriate properties for the expected use of the structure such as high strength and low permeability. Usage of nano scale properties such as Nano SiO_2 can result in dramatically improved properties from conventional grain size materials of same chemical composition. This project is more interested in evaluate the behaviour of nano silica in concrete for 5%, 10%, and 15% volume fraction of cement. Flexural test for beams were conducted with two point loads, at different percentage as mentioned above. From results interpolated, Nano silica with higher order replacement gives optimized results compared to control specimens.

1. Introduction

Utilization of nano scale measure particles can bring about drastically enhanced properties from regular grain estimate materials of same synthetic creation of materials. If nano particles are added with routine high rise building materials, will posses excellent properties for the construction of long span, high durable for civil engineering infrastructures development [1-3]. In late patterns utilization of nano materials in cement has expanded due to its improved properties, for such as sturdiness and mechanical properties and solidness. The mechanical conduct of concrete materials depends, as it were, of basic components that are successful utilizing on nano scale [4]. The size of calcium silicate hydrate (C-S-H) stage which goes about as an essential part assumes a noteworthy part for in charge of quality and different properties in nano run. The structure of C-S-H has affect on execution of cement in light of the fact that the structure is delicate to dampness condition [5-6].From past research comes about, nano particles is important for enhancing mechanical and sturdiness properties and improving stop defrost toughness, scraped area resistance, and the bond strength with steel bar, the concoction assault resistance and erosion resistance of steel bar [8-11].Failure of the structures sometimes identified with appearance of cracks. In concrete, strength is related to the stress required to cause fracture and is synonyms with degree of failure at which applied stress is maximum value. Since concrete already has cracks present in it the reason for using nano material is to fill these cracks which help in increase of durability and other strength parameters [12].

2. Experimental Investigation

2.1 Materials

Ordinary Portland Cement (OPC) with 53 grade which is satisfies the physical and chemical properties within the range of ASTM C150, Compressive strength of 54.5 MPa was used for complete experimentation process and density of the OPC is 3104 Kg/ m^3. Local river sand is added as Fine aggregate which specific gravity 2.65 and its fineness modulus 2.64. The properties of sand were determined by conducting tests as per IS: 2386 (Part-I). The results indicate that the sand conforms to Zone III of IS: 383-1970.

2.2 Nano SiO_2

Pure Nano-silica was brought and its properties are tested in manufacture company give by Table 1, SEM Figure as shown in 1.

Table 1. Properties of Nano-silica

Test item	Standard Requirement	Test Result
Specific surface area (M2/G)	200-210	202
pH value	3.7-4.5	4.12
Loss on drying @105^0 c (5)	<1.5	0.47
Loss of ignition @1000^{0}c (%)	<2.0	0.66
Sieve Residue (5)	<0.04	0.02
SiO_2content	>99.8	99.88
Carbon content	<0.15	0.06

Figure 1. SEM of Nano SiO_2

2.3 Specimen Preparation

The experimental program consists of preparation of reinforced high performance concrete. Beams specimens of size (100 x 150 x 1200) mm were prepared with reinforcement of 12 mm bars at tension side. Nano silica is replaced for the cement having different replacement levels ranging from 5%, 10%, and 15% for cement. These replacements are mainly concerned with the tension zone of the beam specimen and hence replaced in the bottom layer of beam. Two point loads was applied at 133.33mm from center of beam at either side. The beams were subjected to loading conditions under different replacement criteria.

2.4Concrete Mix design

For this present study design of mix has been done based on ACI code. In order to meet high performance concrete that meet the strength requirement, careful selection of material is needed. After designing the mix, various trails have been carried out by varying the ingredients to arrive at the optimum mix. The following procedure is adopted for mix proportioning of high strength concrete as per guidelines of ACI 211.4R-93

2.5Casting Details

The casting specimen required amounts of the fixings were weighed and kept prepared for blending. For simple evacuation of the specimens, before throwing of specimen oil was connected to the inward surface of the molds and on the floor on which the specimen was threw. The measure of super plasticizer utilized for the mix were measured and kept prepared for mixing. Before mixing initiates super plasticizer was mixed a large portion of the amount of water. For blending, at first 50% of the considerable number of fixings were very much mixed in dry condition in a concrete blender half of computed measure of water was added to the dry mix and mixed altogether to get a uniform mix for around 2 minutes. After that the staying half fixings additionally added to the mix and the staying half of water blended with super plasticizer was poured into the mixer and very much mixed. The concrete set in beam, which was at that point kept prepared. Concrete was filled in the shape in three equivalent layers and compacted utilizing a needle vibrator. Subsequent to filling, best layer was done to show signs of improvement surface. Control specimen was likewise arranged by a similar mix.

2.6 Curing of Specimens

The specimens were covered using wet gunny bags one hour after casting to prevent the loss of moisture from concrete, which has low water cement ratio. After 24 hours of casting, specimens were removed from the mould and again moist cured for 28 days using wet gunny bags. After the curing period was over, the specimens were white washed and kept ready for testing. The control specimens were also cured under same environment.

2.7 Test procedures

Specimens of size (100 x 150 x 1200) mm were tested in a Universal compression testing machine. In order to note down the applied load precisely, load cell of 25T capacity was used Two Linear Variable Differential Transducer (LVDT) were for measuring the longitudinal strains at the top and bottom of the beam to find out the moment curvature relationship. The longitudinal deformation at the top and bottom were measured using LVDT with a range of 5mm and a resolution of 0.001mm. The beams were supported using two rollers having 30 mm diameter and one of which was fixed and other was capable of rotation. The effective span was kept as 1000 mm. The specimens were tested under two point loadings. Two rollers each of diameter 30 mm served as load points and was kept on the beams at a distance of 333.33 mm, kept in position by plaster of paris. A rolled steel joist was used to transmit the load from the machine to the two locations through the rollers. The load was applied in stages. For every stage of loadings, the following readings have been noted for deflection at mid span of the specimen.

Figure 2 Test setup for beam specimen

3. Results and Discussion

3.1 Flexural Behaviour of beams

Table 2 below shows beams with different replacements of Nano Silica varying from 0%. 5%, 10%, 15% for cement, First crack, ultimate load, Failure load values for different specimens are studied and the values are shown in Table 3.

Table 2. Beam designation

Beam designation	Volume fraction of NS replacement % for cement
B1	0
B2	5
B3	10
B4	15

Table 3. Flexural Behaviour of Beams

Beam designation	First Crack Load (kN)	Ultimate Load (kN)	Failure Load (kN)
B1	27	50	44
B2	29.5	55	47
B3	31	58	51
B4	33	62	55

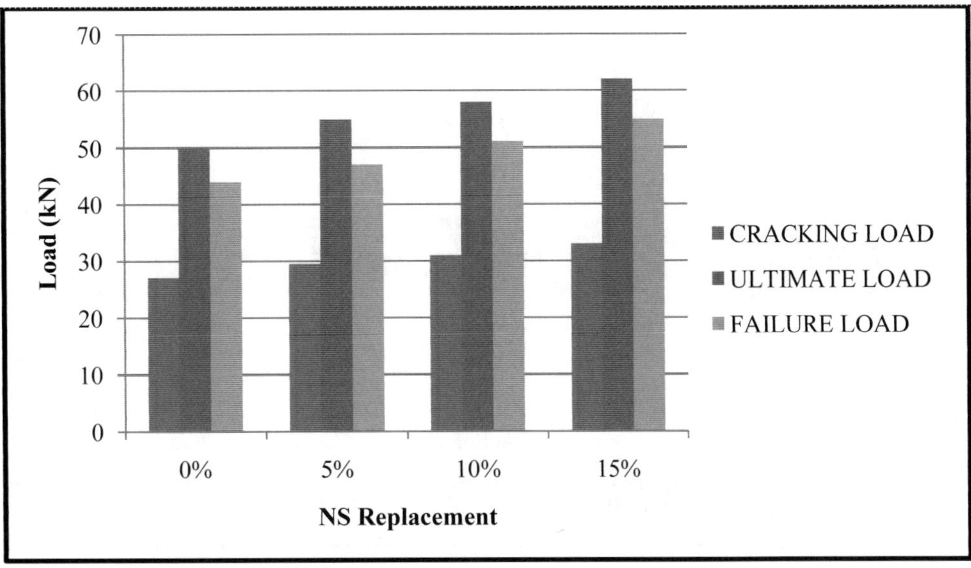

Figure 3 Flexural Strength of beam specimen

3.2 Stiffness

In Figure 4, the stiffness (k) is the resistance of the part against bending distortion. It is the capacity of elastic modulus (E), the area of moment of inertia (I) of the bar cross segment about the hub of intrigue, length of the beam and shaft limit conditions [9-11].Bending stiffness in beam is also known as flexural rigidity. Typical Stiffness versus Deflection plots are shown below for the different replacement levels of nano silica.

3.3 Loads versus Deflection Behaviour

The load versus mid span deflection response of the beam specimens under monotonic loading is discussed in Figure 5. The recorded values of load and deflection have been used to obtain load versus deflection plot. From the above graph it is found that specimen with higher order replacement of NS is practically effective in bringing about the desired modification in the deformation characteristics from beginning to failure and contribute significantly to maintain the structural integrity of the beams.

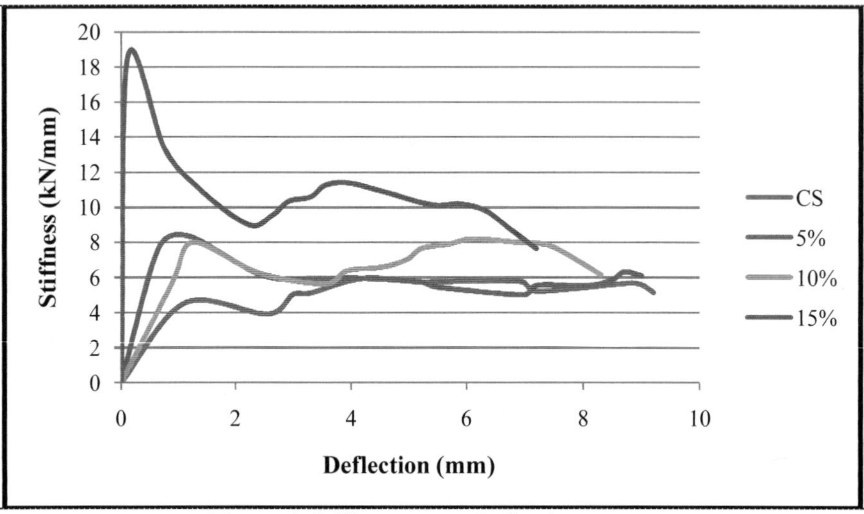

Figure 4. Stiffness versus Deflection curves

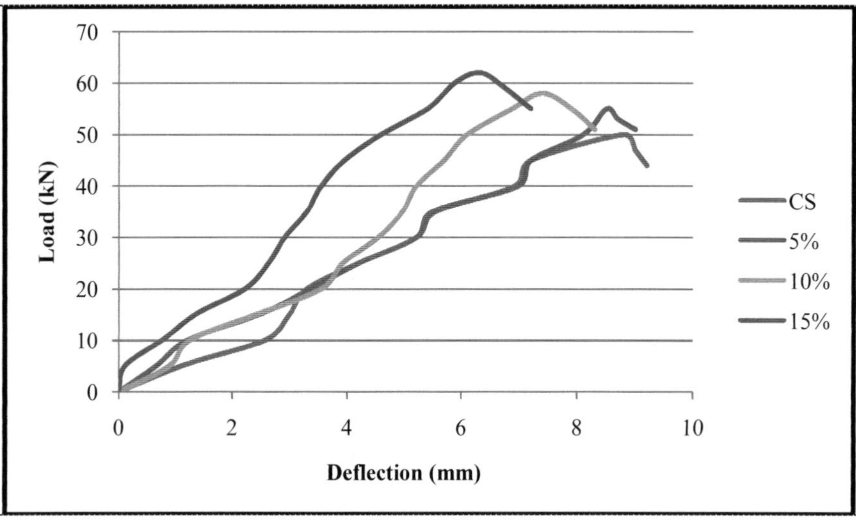

Figure 5. Load versus Deflection curves

4. Conclusions

The review of results above indicates the following general conclusions.

a. Addition of nano silica improves the consistency and workability of fresh and hardened concrete because an additional volume of fines is added to mixture.
b. Replacement of nano silica in concrete plays a major role because it acts as a key factor in modifying the micro structure of concrete consequently reducing the permeability thereby reducing the penetration of water which results in prevention of corrosion.
c. From load versus deflection graph plotted, Cement replaced with 15% of NS shows higher load carrying capacity 24 % higher comparing to control specimen also have lesser deflection comparing with other nominal replacements (5% and 10%).
d. From stiffness versus deflection graph plotted, comparing the results for various replacements, cement with 15% NS replacement shows increase in stiffness comparing other replacements.

This is mainly due to interlocking of nano particles with the concrete specimen which acts as pores filling agent.

References

[1] Jaishankar P and Saravana Raja Mohan K 2015 Influence of Nano particles in High Performance Concrete (HPC) *Int. J. ChemTech Res.* **8** 278-84

[2] Nittaya Thuadaij and Apinon Nuntiya 2008 Synthesis and Characterization of Nano silica from Rice Husk Ash Prepared by Precipitation Method. *Cem. Concr. Res.* **7** 55-9.

[3] Manoj Kumaar C, Mark U, Vivin Raj and Mahadevan D 2015 Effect of Titanium di-oxide in Pervious Concrete. *Int. J. ChemTech Res.* **8** 183-7.

[4] Priya A K and Nithya M 2016 Experimental Investigation on Developing Low Cost Concrete by Partial Replacement of Waste ludg. *Int. J. ChemTech Res.* **9** 240-7.

[5] Madhavi T, Pavan Kumar P and Jothi lingam M 2015 Effect of Copper Slag on the Mechanical Strengths of Concrete. *Int. J. ChemTech Res.* **8** 442-9.

[6] Praveen Kumar T, Sudheesh C and Sasi Kumar S 2015 Strength Characteristics of Saw Dust Ash Based Geo polymer Concrete. *Int. J. ChemTech Res.* **8** 738-45.

[7] American Society for Testing and Materials (ASTM) Standard specification for Portland Cement ASTM C150/C150M ASTM International West Conshohocken PA 2015

[8] Shekari A and Razzaghi M 2011 Influence of Nano particles on Durability and Mechanical Properties of high performance concrete *Procedia Eng.***14** 3036-41.

[9] Sumathi A and Saravana Raja Mohan K 2015 Study on the Strength and Durability. Characteristics of High Strength Concrete with Steel Fibers. *Int. J. ChemTech Res.* **8** 241-8.

[10] Jai Shankar P and Saravana Raja Mohan K 2015 Experimental investigation on nano particle In high performance concrete. *Int. J. ChemTech Res.* **8** 1666-70.

[11] Jaishankar P and Muthu siva chandru B 2016 Effect of Nano silica additions on Mechanical and Micro structural analysis of High performance concrete. *Int. J. ChemTech Res.* **9** 453-61.

[12] Ye Qing, Zhang Zenan, Kong Deyu and Chen Rongshen 2008 Influence of nano-SiO_2 addition on properties of hardened cement paste as compared with silica fume. *Mater. Des.* **21** 539-45.

Experimental investigations on steel-concrete composite columns for varying parameters

V Aparna[1], D Vivek[1], Kancharla Neelima[2] and B Karthikeyan[3*]

[1]KPR Institute of Engineering and Technology, Arasur, Coimbatore-614407, India.
[2]NVLN Constructions Pvt. Ltd., Hydrabad, India
[3] School of Civil Egineering, SASTRA University, Thanjavur-613401, India.

*Email: karthikeyan@civil.sastra.edu

Abstract. In this study, the experimental investigations on steel tubes filled with different types of concrete are presented. Steel tubes filled with fibre reinforced concrete using lathe waste and steel tube with concerned confined with steel mesh were investigated. The combinations were compared with steel tubes with conventional concrete. A total of 4 concrete filled steel tube (CFST) combinations were made with tubes of diameter 100 mm with wall thickness 1.6 mm and a height of 300 mm. Axial compression test to examine the resisting capacity of the columns and push-out test for noting the bond strength were performed. Coupon tests were also conducted to determine the mechanical properties of steel. The structural behaviour of the composite columns was evaluated from on the test results. It was observed that steel tube filled fibre reinforced possessed better bond strength and resistance to axial load.
Keywords: CFST, lathe waste, steel mesh, steel fibre, push-out test

1. Introduction

Studies on steel-concrete composite sections and encased columns have extensively been made over recent years and have also been implemented in many new constructions. Due to the benefits such as increased strength, better ductility and lower construction costs as form works are not required and they are preferred in the construction field [1]. Many research works have already shown the advantages of using concrete filled tubular columns under various categories. Works from various researchers explain about performing axial loading tests on concrete filled tubular columns [2-7]. Few research works discussed about strength and state of art research on non-linear analysis of concrete filled tubular columns [8-10]. Research work of Varma et al discusses the behaviour of steel- concrete columns under seismic load conditions [11]. Azizinamini and Schneider discussed about connection of circular CFST columns [12]. In addition to using steel tubes filled with concrete many researches were done using FRP tubes, confining steel tube with FRP [13-17]. Ehab Ellobody and Marian did an experimental investigation of eccentrically loaded fibre reinforced concrete-filled stainless steel tubular columns [18]. Though many works were done as mentioned in previous literatures using CFST, FRP tubes, confining concrete filled tubes (CFT) with FRP still a lot more space is available in this area for further improvement and implementation of new techniques in that field and this study helps to identify such innovations and improvement that can be made in this fixed composite structures.

1.1 Objective

The major objective of this experimental research work is to study the mechanical properties of CFT columns when the concrete is mixed with fibres and mesh. Since no form work is required for structural elements made with concrete filled tubes, it is difficult to check whether the concrete has occupied the tubular section fully or it has dumped over the aggregates leaving a hole in between. Use of fibres can help the concrete arresting the propagation of cracks also improving it's tensile and impact strength. Instead of choosing steel fibres or any costly fibres, lathe waste was added as a fibre to the concrete mixture. In addition to that steel mesh was also used as a fibre and the behaviour was studied. The main objective of the present work is to study the structural performance of the fibre reinforced CFST columns with lathe waste and steel mesh used as fibres and to compare them against those with the normal concrete mixes.

2. Materials

2.1 Cement

OPC 43 grade cement from the local market was used and the physical properties of the cement were tested as specified in IS: 4031(1988) [19] and IS: 12269 (1987) [20]. The specific gravity of cement is 3.15.

2.2 Aggregate

For fine Aggregate River sand was used and aggregates passing through 16 mm sieve and retained on 12.5 mm sieve were used as coarse aggregate in the concrete mix. The basic properties such as specific gravity, bulk density, and gradation and fineness modulus are tested as specified in IS: 2386 Part 1 (1963) [21]. The specific gravity of fine aggregate is 2.6.

2.3 Steel sheets

Mild steel hot rolled sheets of 1.6 mm thickness were used for our study. Steel sheets were shaped into the desired shape with the use of welding.

2.4 Lathe waste

In this research the lathe waste which otherwise would be dumped as a waste was tried using as a fiber in the concrete filled tubes and the parameters were studied. Steel fibers balance the force by transmitting tensile forces to the steel fibers which run along the cracks, as the result flexural toughness and flexural strength increases to great amount. Steel scrap fibers from lathe machines of random dimensions are used, usually of length 20-30 mm, width 1.5-2 mm and 0.3-0.6 mm thick, are used to reinforce the concrete cubes. It is a normal practice to add steel Fibers in concrete in fewer dosages to reduce cracking. Though steel bars are used in tension zone of concrete they can improve the tensile strength in one particular direction only and steel fibers can act as multidirectional reinforcement. Steel scraps obtained from the wastes of lathe machine are generally known as lathe waste. Fig.1 shows the Lathe waste used for the research work and the properties are listed out in Table 1. Since lathe fibers are actually obtained from waste, they do not have a well defined material property. So, we tried to fix its properties with reference to a commercially available steel fiber. A crimped steel fiber of length 12.5 mm and diameter 0.3 mm was taken as a reference and the lathe waste fibers were made into pieces, matching the dimensions of the crimped steel fiber. Though the fibers were a little curly, they were cut and made approximately straight.

Table 1. Properties of lathe waste

Cross section	**Length (mm)**	**Diameter**	**Aspect ratio**
Straight	25	0.3	83.33

IOP Conf. Series: Earth and Environmental Science **80** (2017) 012029 doi:10.1088/1755-1315/80/1/012029

2.5 Steel mesh
Steel mesh of 1mm thickness is used as a minimal reinforcement for the columns to check its behavior. Fig. 2 shows the steel mesh and its position before pouring concrete. It is a common practice to provide a lean reinforcement in the tubular section before pouring concrete though it is not necessary as the outer steel tube is sufficient and can act both as longitudinal and lateral reinforcement. Here instead of providing reinforcement with steel bars we tried with steel mesh.

Figure 1. Lathe waste used as fibre

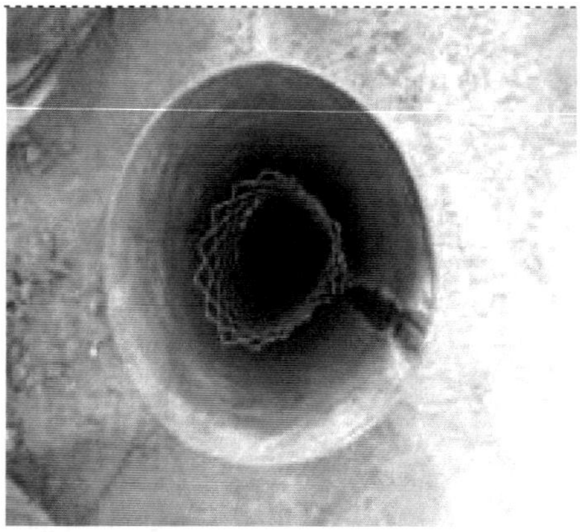

Figure 2. CFSTM before pouring concrete

3. Testing methods

In the present study, a total of 14 specimens were considered for two separate tests viz, Test I and Test II. Test I was for finding the maximum axial load bearing capability by axial compression test and the test II for finding the bond stress of the specimen by performing push out test. All the specimens were chosen according to Euro Standards [22] provided for the selection of CFST column dimensions.

The circular models having diameter 100 mm and of wall thickness of 1.6 mm (i.e. D/t proportion with 62.5) were utilized, the exact sizes were chosen from previous research works [23]. For performing axial compression empty steel tubes (HT), Concrete Filled Steel Tubes with normal concrete (CFST), with lathe waste fibre reinforced (CFSTF), and with steel mesh (CFSTM) were examined. In case of push out test, Concrete Filled Steel Tubes, Concrete filled steel tube with fibre reinforced concrete and Concrete Filled Steel Tubes with Steel mesh were considered.

3.1 Concrete mix proportions
A normal concrete mix of M25 is adopted for this work. 4 combinations were made for the present research work namely HT, CFST, CFSTF and CFSTM, where HT is a hollow tubular section, CFST representing specimens with normal concrete, CFSTF concrete with lathe waste in which the lathe waste fibre was added in 1.5% to the total volume of the concrete and CFSTM is the one with steel fibre mesh used along with concrete.

3.2 Specimen details
Steel sheets of thickness 1.6 mm thickness were cut to 300 mm length and were welded such that tubes of inner diameter 100 mm were made. The yield strength and ultimate strength of the steel tube were obtained from coupon tests. After cutting the sheets for the desired length and before making them as tubes, grooves were made inside them to enhance the gripping property of the steel tube. The specimens were subjected to curing by wetting the top and bottom concrete portions with water. The ends of the specimens were later covered with steel plates and welded to induce the condition both

ends fixed while testing it for compressive strength. All the tubes were of same size and material and the difference in behaviour due to the use of various concrete mixtures were studied.

3.3 Coupon test
The test was done by placing the specimen in the Universal Testing Machine (UTM) between the two chucks after marking the gauge lengths and extending it slowly till it fails. The elongation of the gauge section was noted later. The computerized UTM recorded the percentage of elongation; ultimate tensile strength, yield strength of the material and stress vs. strain graph were arrived.

3.4 Push out test
The steel specimens were cut for the obliged L/D ratio, and the edges were flattened well to obtain a level surface for uniform loading. In the course of preparation of the test specimens, the specimens were kept in upright position and the compaction was carried out in layers for maintaining a strategic distance from air voids in concrete. And also for performing push out test for finding bond stress, before the casting of specimens a gap of 50 mm left not filled at the top for enabling the movement of concrete within the outer shell at load. The specimens were kept inverted and load was applied on the concrete alone using a circular steel solid less than the diameter of the specimen and the 50 mm gap at the bottom got collapsed, the slip was noted up to the failure load after it attained the peak load. Fig.3 illustrates the push out test being conducted.

Figure 3. Setup for push out test

3.5 Axial compression test
The sections were fixed ended at the closures with the help of steel plates of 5mm thickness welded to the top and bottom of the tubes and the load applied exactly on the centroid of the section. A circular steel collar plates were given at the closures of specimens to exchange the concentric action with no slip. The specimens mounted with dial gauge at the compression side of the tube in order to record the amount of axial shortening for the given load at regular intervals of deflection. The tests were carried out through a 1000 kN capacity universal testing machine. In order to perform axial compression test, steel plates were made for required dimensions with a small groove exactly at the middle of the plate and steel ball of required dimensions were placed at the centre point of the two plates.

4. Results and discussion

4.1 Push out test

Push out test is conducted to find out the bond stress at the steel concrete interface. For all specimens the failure load and its corresponding axial displacement were observed, the bond strength of the specimens was shown in the Table.2. For determining the bond stress of the specimens from the failure load equation (1) was used

$$\text{Bond Stress} = \frac{P}{\pi DL} \tag{1}$$

Where P was the failure load, D was the inside diameter of the steel tube and L was the length of the concrete steel interface. According to the test results, it can be noticed that the strength of the concrete and load conveying limit of the specimen increases with the provision of mesh in CFST. A sound can be perceptible once the failure load reaches. Then solid body motion of the concrete in relation with steel tube starts with increasing displacements. The failure loads of the specimens are shown in the Table 2 .The bond failure state of the specimen is shown in Fig. 10. It is observed that the higher bond stress is found in tube with fibres mixed, since fibre is provided, bond between steel and concrete was really high as the fibres actually stitched the aggregates with concrete.

Table 2. Results of push out test

Specimen	Average ultimate compression load (kN)	Average bond-stress (N/mm^2)
CFST	50.2	0.532
CFSTF	62.8	2.09
CFSTM	55.2	1.85

4.2 Axial compression test

Graph was plotted between axially applied load and the axial deformation for the CFST segments. The axial load was noted from the universal testing machine while the axial shortening was noted from dial gauge. It was observed that for the specimens with normal concrete, at first, the axial load increased as the axial shortening increased. As the shortening increased further, the micro-cracks in the central solid concrete grew quickly and consequently the Poisson's ratio of concrete increased alarmingly and its value was more than that of steel tube. The load bearing capability of CFSTF is 16.7% more than that of CFST. The axial shortening significantly reduced in CFSTF and its value was 10.5 mm as against the value of 14.4 mm for CFST at failure loading.

4.3 Failure modes of columns

The various modes of crushing of concrete were observed in the specimens for different parameters. They are discussed as below.

4.4 Hollow tube without infill concrete (HT)

The maximum load that the HT specimens resisted was 59.5 kN and it failed shortly at an axially displacement of 4.6 mm. Deflection occurred earlier in this column than other specimens and the load deflection behavior is depicted in Fig. 4. The empty hollow tubular sections (HT) experienced local failure mechanism which can be noted from the dents made in the outer core. Fig 5 shows the failure of hollow tubular column.

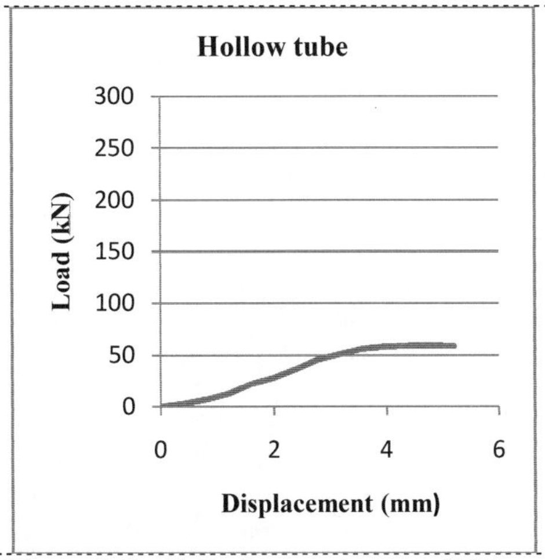

Figure 4. Load displacement curve of HT

Figure 5. Crushing failure of HT

4.5 Concrete filled steel tube (CFST)

In this tubular column, concrete of grade M25 was filled in the steel tube. Fig.6 shows the load deflection curve. On observing the results a drastic increase in strength from 59.35 kN to 245 kN can be noticed. The concrete core which is good at taking up compression load has resisted well up to 245 kN until reaching an axial displacement of 14.4 mm. The failure was mainly due to crushing of the concrete. Concrete got crushed at many local points under compression and it can be noted from the outer surface of the steel tube, many dents have formed on the steel tube section which are due to the concrete pieces disintegrated during crushing and getting dumped at certain locations which can be observed from Fig.7. Though they behaved better compared to the hollow tube, it is observed that they can offer more resistance when mixed and used with other mineral admixtures.

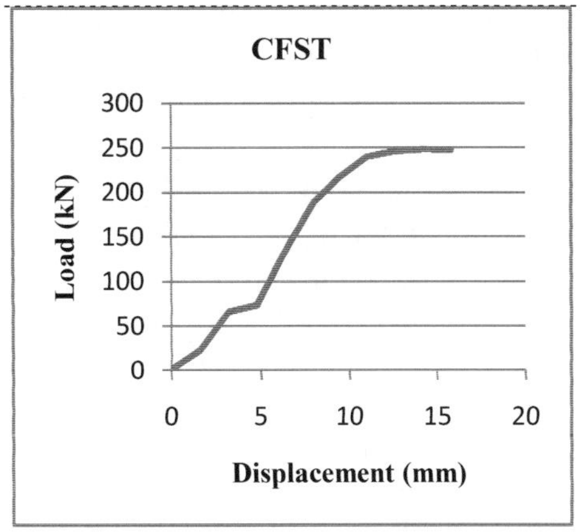

Figure 6. Load displacement curve of CFST

Figure 7. Crushing failure of CFST

4.6 Concrete filled steel tubes with fibre reinforced concrete (CFSTF)

This column failed at 284.5 kN. From the results it is clear that column strength capacity has been improved on adding steel fibres to concrete. Also, the ductile properties of CFSTF columns are also better which can be understood from the high degree of deformation capacity. Crushing of concrete at many places was reduced due to the presence of fibres as the presence of lathe waste fibres, stitched the pores, and prevented the crack propagation in the concrete core, thereby leading to an increased compressive strength. Figures 8 and 9 show load displacement curve and the crushing failure state of CFSTF.

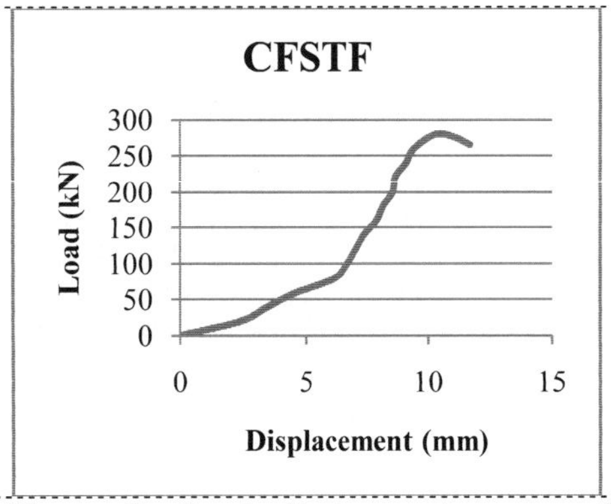

Figure 8. Load displacement curve of CFSTF

Figure 9. Crushing failure of CFSTF

4.7 Concrete filled steel tube with mesh reinforcement (CFSTM)

This column showed a moderate load carrying of 253.5 kN, which is better than the specimens cast only with concrete. Fig. 10 shows the deflection characteristics. Steel mesh was provided as minimum reinforcement for the column instead of steel bars. One of the main advantages of providing mesh is that it reduces the weight of the column. CFT with mesh reinforcement behaves well compared to all other because of the stiffer bond between the steel the concrete as a core. On observing the profile of the specimen after failure, it can be noted that the local failure point due to the crushing of concrete pieces were less. Generally the crushed concrete particles will create pores inside the concrete core and the disintegrated pieces will blocked somewhere inside creating dents on the steel section when load is applied. Such, effects were limited due to the provision of mesh and crushing of concrete was observed only in the cover regions. It can be noticed from Fig. 11.

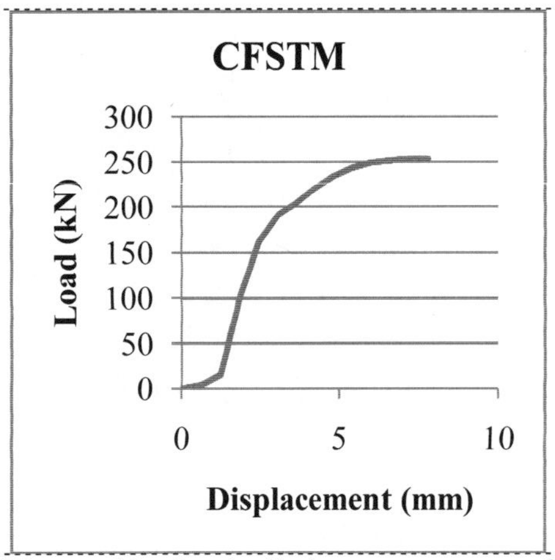

Figure 10. Load displacement curve of CFSTM

Figure 11. Crushing failure of CFSTM

5. Conclusions

On the basis of the results, the following conclusions were arrived.

- There were significant enhancements in the physical performance of CFSTF in reference to ultimate load bearing capacity, stiffness and ductility.
- The consequences from push out tests specified that bond strength of CFSTF specimens were found to be superior to the other CFSTs specimens.

  From load versus deflection curve, the proportional increase in load carrying capability of CFSTF increased in the range of 16.6%, when compared with CFST columns.
- Though, there was no drastic improvement in load carrying capacity of the axial loaded CFSTF columns, it offers more ductility than the CFST columns.
- Hence, it is once again proved that addition of fibers in concrete will have effect not only in normal, high strength, high performance concrete but can also add strength to structures with concrete filled tubular section.

References

[1] Guoqiang Li 2007 Experimental study of hybrid composite cylinders *Compos. Strut.* **78**, 170-81

[2] Schneider S P 1998 Axially loaded concrete–filled steel tubes *J. Struct. Eng. ASCE* **124**(10) 1125-38

[3] Han L H 2002 Tests on stub columns of concrete-filled RHS sections *J. Constr. steel Res.* **58**(3) 353-72

[4] Han L H, Yao G H and Zhao X L 2005 Tests and calculations of hollow structural steel (HSS) stub columns filled with self- consolidating concrete (SCC). *J. Constr. Steel Res.* **60**(9) 1241-69

[5] Giakoumelis G and Lam D 2004 Axial capacity of circular concrete-filled tube columns. *J. Constr. Steel Res.* **60**(7) 1049-68

[6] Lam D, Young B and Ellobody E 2009 Behaviour of normal high-strength concrete-filled compact steel tube circular stub columns. *J. Constr. Steel Res.* **62**(7) 706-15

[7] Liang QQ and Fragomeni S 2009 Nonlinear analysis of circular concrete-filled steel tubular short columns under axial loading. *J. Constr. Steel Res.* **65**(12), 2186-96

[8] Knowles RB and Park R 1969 Strength of concrete filled steel tubular columns. *J. Struct. Eng. ASCE* **95** 2565-87

[9] Hajjar J F and Goorley B C 1996 Representation of concrete-filled steel tube cross-section strength. *J. Struct. Eng. ASCE* **122**(11) 1327-36

[10] Spacone E and El-Tawil S 2004 Nonlinear analysis of steel-concrete composite structures: State of the art. *J. Struct. Eng. ASCE* **130**(2) 159-68

[11] Varma A H Ricles J M Sause R and Lu L W 2004 Seismic behavior and design of high-strength square concrete-filled steel tube beam columns. *J. Struct. Eng. ASCE* **130**(2) 169-79

[12] Azizinamini A and Schneider S P 2004 Moment connections to circular concrete-filled steel tube columns. *J. Struct. Eng. ASCE* **130**(2) 213-22

[13] Mirmiran A and Shahawy M 1996 A new concrete-filled hollow FRP composite column. *Compos Part-B: Engineering* **27B** 263-8

[14] Li G, Torres S, Alaywan W and Abadie C 2005 Experimental study of FRP tube-encased concrete columns. *J. Compos. Mater.* **39**(13) 1131-45

[15] Fam A Z, Flisak B and Rizkkalla S 2003 Experimental and analytical modelling of concrete-filled fiber reinforced polymer tubes subjected to combined bending and axial loads. *ACI Struct. J.* **100** 499-509

[16] Mirmiran A, Shahawy M, El Khoury C and Naguib W 2000 Large beam-column tests on concrete filled composite tubes. *ACI Struct. J.* **97** 268-76

[17] Fam A and Rizakalla S 2003 Large scale testing and analysis of hybrid concrete/composite tubes for circular beam-column applications *Constr. Build. Mater.* **17** 507-16.

[18] Ehab Ellobody and Mariam F 2007 Experimental investigation of eccentrically loaded fibre reinforced concrete-Filled stainless steel tubular columns. *J. Constr. Steel Res.* **76** 167-76

[19] IS: 4031–1988, Methods of physical tests for hydraulic cement. *Bureau of Indian Standards.*

[20] IS: 12269-1987, Specification for 53 grade ordinary Portland. *Bureau of Indian Standards.*

[21] IS 2386(Part 1):1963 Methods of test for aggregates for concrete: Part 1 Particle size and shape. *Bureau of Indian Standards.*

[22] Euro code 4- *Design of composite steel concrete structures.*

[23] Brian Uya Zhong Taoa and Lin-Hai Han 2011 Behaviour of short and slender concrete-filled stainless steel tubular columns *J Constr steel res* 360-378.

IOP Conf. Series: Earth and Environmental Science **80** (2017) 012030 doi:10.1088/1755-1315/80/1/012030

Behaviour of axially and eccentrically loaded short columns reinforced with GFRP bars

S Sreenath[1*], S Balaji[2] and K Saravana Raja Mohan[1]

[1]School of Civil Engineering, SASTRA University, Thanjavur – 613401. India.
[2]Kongu Engineering College, Thoppupalayam, Perundurai – 638052. India.

*Email: sreenath@civil.sastra.edu

Abstract. The corrosion of steel reinforcing bars is a predominant factor in limiting the life expectancy of Reinforced Cement Concrete (RCC) structures. Corrosion resistant Fibre Reinforced Polymer (FRP) bars can be an effective alternative to steel bars in this context. Recent investigations reported the flexural behaviour of RCC beams reinforced with Glass Fibre Reinforced Polymer (GFRP) bars. This study is meant to investigate the suitability of Sand Coated GFRP reinforcement bars in short square columns which when loaded axially and loaded with a minimum eccentricity. Standard tests to assess mechanical properties of GFRP bars and pullout test to quantify the bond strength between the bars and concrete were conducted. GFRP reinforced column specimens with a cross-sectional dimension of 100mm X 100mm and of length 1000mm were cast and tested under axial and eccentric loading. The assessed load carrying capacity was compared with that of conventional steel reinforced columns of the same size. The yield load and ultimate load at failure withstood by the steel reinforced columns were considerably more than that of GFRP reinforced columns. The energy absorption capacity of GFRP reinforced columns was also poor compared to steel reinforced columns. Both the columns exhibited nearly the same ductile behaviour. Hence GFRP reinforcements are not recommendable for compression members.

1. Introduction

Columns are structural members subjected to compressive forces, which will transmit the load from the slabs and beams to the lower level. Hence, these columns are the most critical elements in the structure [1]. Corrosion of steel bars is one of the threats to the strength and serviceability of Reinforced Cement Concrete (RCC) columns. The gradual process of corrosion will weaken the columns and the problem will become severe when it is exposed to aggressive environments. The difficulty in repairing these degraded elements raises the demand for a noncorrosive material as reinforcements. Fiber Reinforced Polymer (FRP) composites are popular for strength and corrosion resistance. Both cost and durability benefits can be achieved by properly employing these composites in the field of infrastructural development. Its high strength and stiffness to weight ratios, tractable thermal expansion, damping properties and electromagnetic neutrality are the additional benefits [2-4]. These benefits can bring in improved safety and life cycle with a possible reduction in production, equipment and maintenance costs. FRPs are composites made of polymer matrices reinforced with fibers. The fibers may be glass, carbon, aramid or basalt and the polymer may be epoxy, vinyl ester, phenol formaldehyde etc. [5]. Glass Fibre Reinforced Polymer (GFRP) is selected for this study. Investigators recently studied the possibilities of using GFRP bars in flexural members as longitudinal and shear reinforcements. Studies are less on the behaviour of GFRP bars in compression members. Micro-buckling of fibers is expected to occur in the bars when it is compressed and it becomes deficient to withstand compressive loads further. This is because of the non-homogeneous and anisotropic nature of the GFRP matrix. Hence the use of GFRP bars is less recommendable. American Concrete Institute (ACI) through their standard code [6] demands for further studies on the behaviour of GFRP reinforced compression members [1].

1.1 Previous Studies

Previous researchers confirmed the poor performance of GFRP reinforcements under compression. The reported compressive strength of GFRP reinforcement bars was as low as 55% of its tensile strength [7]. The tests conducted on small scale square columns of size 200 × 200 × 650mm with grid type FRP reinforcement reported a conservative axial load carrying capacity ignoring the part of FRP bars [8]. The study of compression behaviour of rectangular columns of size 450 × 250 × 1200mm with GFRP longitudinal reinforcement reported a reduction of 13% in the load carrying capacity compared to the conventional steel reinforced columns irrespective of the type of lateral ties (Steel or GFRP). An investigation done on square columns reported the improvement in the ductile behaviour of the column with increasing the reinforcement ratio [9].

1.2 Objectives of the study

The study is aimed to

- Provide a clear idea of the mechanical properties of GFRP reinforcement bars.
- Illustrate the behaviour of Sand Coated GFRP reinforced short columns under axial loading and loading with minimum eccentricity.
- Frame equations to predict the load carrying capacity of the columns and to validate the same.

2. Experimental Investigation

2.1 Specifications of Materials

In this research, a concrete mix of grade M30 with the characteristic compressive strength of 30 N/mm^2 was used. The proportion of the constituents was designed to conform to Indian standards [10,11] and obtained a ratio of 1:1.16:2 with water to cement ratio 0.50.

Ordinary Portland Cement (OPC) of grade 53 with specific gravity 3.15 was used in the mix. The fine aggregate passing through a 4.75 mm sieve and retained on 150-micron sieve having a specific gravity of 2.68 was used. The grading zone of fine aggregate was zone II as per Indian standard specification [12]. The maximum size of coarse aggregate used for this investigation is limited to 12.5 mm to avoid the chances of honeycombing in the specimen. The specific gravity of the coarse aggregate was determined based on Indian standards [12] and it was 2.78.

Potable water was used which was free from acids, oil, alkalis and other organic impurities. Fe500 grade High Yield Strength Deformed (HYSD) bars were used for steel reinforced columns. 12 mm diameter bars were used as longitudinal reinforcement. The lateral ties were about 8 mm diameter steel bars of the same grade for both steel and GFRP reinforced columns. 12 mm diameter HYSD bars were used for the column head reinforcements for both types of columns.

Table 1. Mechanical properties of steel and GFRP

Property	Steel bar	GFRP bar
Tensile Strength (in MPa)	549	965
Yield Strength (in MPa)	500	-
Compressive strength (in MPa)	520	320
Tensile Modulus of Elasticity (in GPa)	210	41
Specific gravity	7.9	1.7

Sand Coated type GFRP rebars were used. The GFRP reinforcements are shown in Figure 1. The diameter of GFRP bar used in this study was 12 mm (including the thickness of sand coating). Table 1 gives the properties of steel and GFRP bars used in this study.

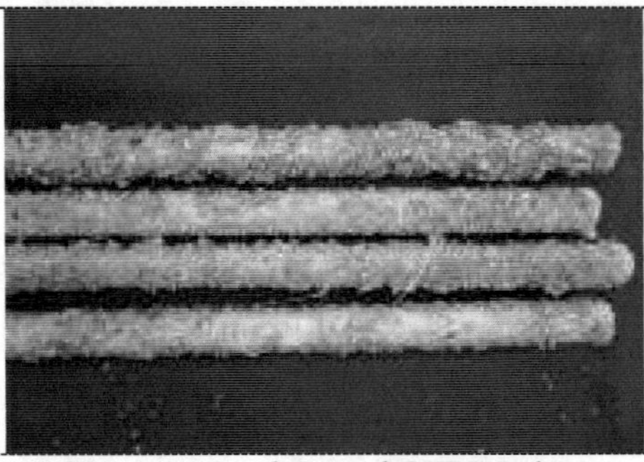

Figure 1. Sand Coated GFRP Rebar.

2.2 Details of test specimens

Totally four numbers of column specimens were cast and tested. Out of four specimens, two were reinforced with GFRP rebars and the remaining two were reinforced with steel rebars. The specimens were tested under both axial and eccentric loading. The columns were designed as short columns with a cross-sectional dimension of 100 mm x 100 mm. The length of the column is 1000 mm. All were provided with end corbels of cross-sectional dimension 200 mm x 100 mm to avoid direct crushing failure [13]. The depth of end corbels (column head) was 100 mm. The control columns were reinforced with four 12 mm diameter HYSD steel bars as longitudinal reinforcement. It was provided with steel lateral ties of 8mm diameter with a spacing of 100 mm from centre to centre at its mid height. The spacing was reduced to 50mm from centre to centre at the supports. 12mm diameter HYSD steel bars were used for column head reinforcement.

For GFRP reinforced columns the longitudinal bars were of 12mm diameter Sand Coated GFRP rebars and the lateral ties were of 8mm diameter HYSD steel bars. 12mm diameter HYSD steel bars were used for column head reinforcement. The dimensions and reinforcement detailing of the column specimen are shown in Figure 2. Figure 3 shows a typical reinforcement arrangement. The details of cast specimens are shown in Table 2.

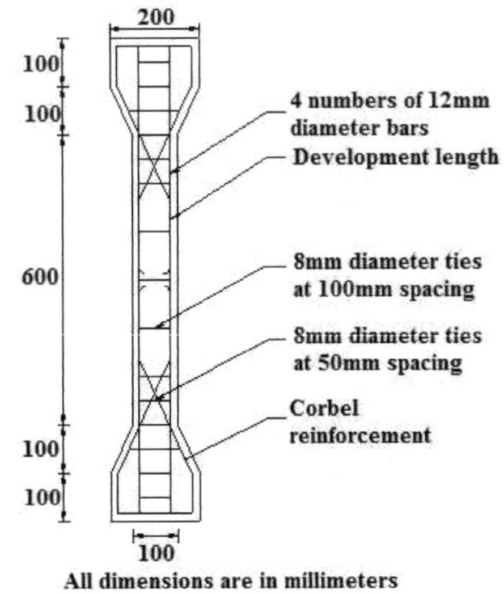

Figure 2. Dimensions and detailing of the column specimen.

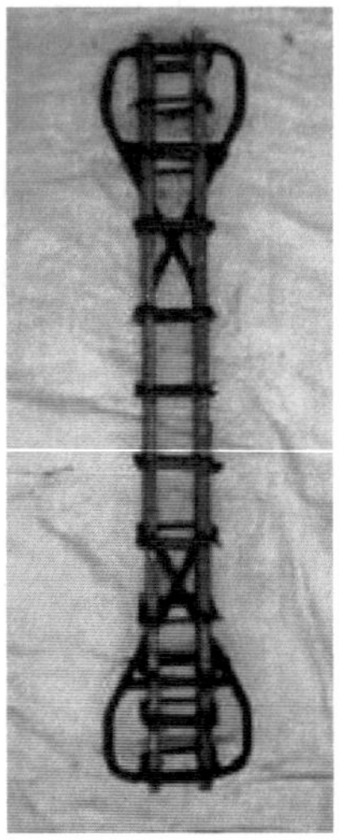

Figure 3. Typical reinforcement arrangement for the column

Table 2. Details of cast specimens

Column designation	Reinforcement type	Type of loading
CRSA	Steel	Axial loading
GRSA	GFRP	
CRSU	Steel	Eccentric loading
GRSU	GFRP	

2.3 Experimental setup and Instrumentation

The column specimens were tested in a loading frame of 2000kN capacity. The axial load was applied using a hydraulic jack of 1000 kN capacity. The load applied to the specimen was measured using an electronic load cell of 1000 kN capacity, which is connected to the data acquisition system. Axial load was transmitted to the column through the steel plate and ball setup to provide hinged end condition. Figure 4 shows the specimen with support condition. The column specimens CRSA and GRSA were tested under axial compression. The specimens were placed in such a way that the line of action of the axial load coincides with the axis of the column. The verticality of the column specimens was checked using a plumb bob and a spirit level to avoid unexpected eccentricity in loading. A mechanical strain gauge, which is demountable, having a gauge length of 200mm and least count of 0.002 mm was used to measure the axial deformation at the mid-height of the specimen. Two numbers of Linear Variable Differential Transformer (LVDT) were used to measure the mid-height lateral deflections (if any) of the short column specimens. The strain is measured in four faces at the middle of the column.

The specimens CRSU and GRSU were loaded by applying compression on a point with an eccentricity of 0.05D from the axis, where D is the lateral dimension of the column. This was performed for accounting the behaviour of column specimens under loading with accidental eccentricity. The specimens were so placed that the axis of loading passes through the eccentric point. Throughout the test setup, care was taken to ensure that the load was applied with permissible eccentricity. The plumb bob was used to check the verticality of columns.

Figure 4. Experimental setup

Initially, an axial load of 5 kN was applied to hold the specimen in its position and then the instruments were normalized and initial readings were observed. Lateral deflections were measured at mid height using Linear Variable Differential Transducers (LVDT) and the axial deformations were measured using the demountable mechanical strain gauge. The load was applied gradually and the deflections were measured at various load stages. At the same time, axial deformation values were also measured. Initiation of crack was observed and the corresponding load was noted, ultimate load to failure and mode of failure were taken. The experimental setup for testing of columns is shown in Figure 4.

3. Results and discussions

3.1 Behaviour of axially loaded columns

The test results are given in the Table 3. The columns CRSA and GRSA were tested under axial compression and axial deformation was observed. The results are tabulated in Table 4. The specimen CRSA started yielding at a load of 428kN. This is the load under which the first crack was observed in the specimen. The deformation observed under this load was 1.21 mm. The specimen failed at a loading of 547kN. The axial deformation under this ultimate load was 1.54 mm. In the case of GRSA, the first crack was observed at a load value of 312kN and the corresponding deflection was 1.12kN. The specimen failed at a loading of 407kN. The deflection at failure was 1.46kN. While comparing the first crack load and ultimate load of CRSA and GRSA, it was noted that CRSA showed a better performance under axial compression than GRSA. The yielding point load of CRSA was 37% more than that of GRSA and the ultimate load of CRSA was 34% more than that of GRSA. But the deflection of CRSA at the yielding point was 8% more than that of GRSA. The behaviour of GRSA was almost the same as stated by the previous researchers.

3.1.1 Load – axial deformation behaviour. The load- axial deformation curve was plotted for CRSA. The plotted curve was then used to compute the energy absorption and the ductility factor of the specimens. A linear variation was found until the load corresponding to the initiation of a crack. After that point, the curve was of parabolic nature. This profile of the curve revealed that the specimen started yielding from the point of initiation of a crack. The load versus axial deformation curve for

GRSA was also plotted. GRSA also showed the same load-deformation behaviour as CRSA. Figure 5 shows the comparative load-axial deformation curve for CRSA and GRSA.

Table 3. Specimen details and results of pullout test

Column designation	Yield load (kN)	Deflection at yield (mm)	Ultimate load (kN)	Deflection at failure(mm)	Energy Absorbed (kNmm)	Ductility Factor
CRSA	428.00	1.21	547.00	1.54	598.74	1.27
GRSA	312.00	1.12	407.00	1.46	426.78	1.30
CRSU	248.00	0.69	419.00	1.18	311.20	1.71
GRSU	189.00	0.67	320.00	1.12	201.40	1.67

3.1.2 Energy absorbed by specimen. The energy absorbed by the column specimens can be quantified as the area under the load-deformation curve. The energy absorbed by CRSA was 598.74kNmm while that of GRSA was 426.78kNmm. That is the energy absorbed by the steel reinforced column under the axial loading is 40% more than that of GFRP reinforced column.

3.1.3 Ductility factor. Ductility is the ability of the column to sustain inelastic deformation without significant reduction in its load carrying capacity. Ductility factor is the ratio of deformation at ultimate load to that at the onset of yielding. The ductility factor exhibited by CRSA was 1.27 while the same exhibited by GRSA was 1.30. That is the GFRP reinforced column exhibited a better ductile behavior than a steel reinforced column.

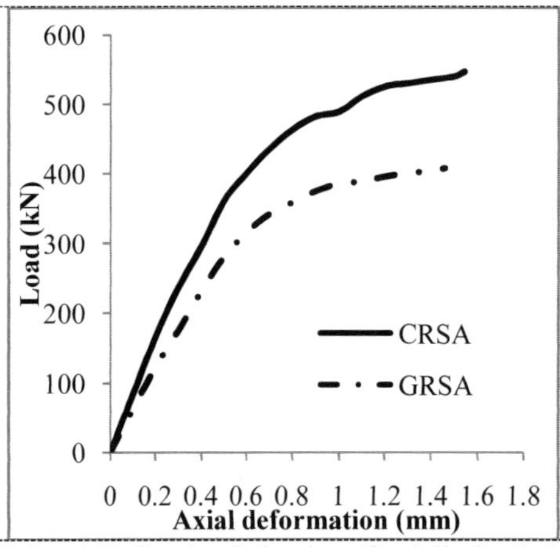

Figure 5. Load-Deflection behavior of axially loaded columns

3.2 Behaviour of axially loaded columns

The column specimens CRSU and GRSU were tested under loading with a uniaxial eccentricity of 0.05D, where D is the lateral dimension of the column. That is 5mm from the axis of the column. Load versus axial deformation curves were plotted for both CRSU and GRSU. The first crack was observed on CRSU at a loading of 248kN. This is the yield point load and the corresponding deflection was 0.69mm. CRSU failed at a load of 419kN and the corresponding deflection was 1.18mm. GRSU started yielding at a load of 189kN and the deformation under this load was 0.67mm. The specimen failed at a load of 320kN and the corresponding deflection was 1.12mm. The results indicated that the yield point load of CRSU was 31% more than that of GRSU and the deflection of CRSU at the yield point was 3% more than that of GRSU. The ultimate load of CRSU was 34% more than that of GRSU and the deflection of CRSU at failure was 5% more than that of GRSU.

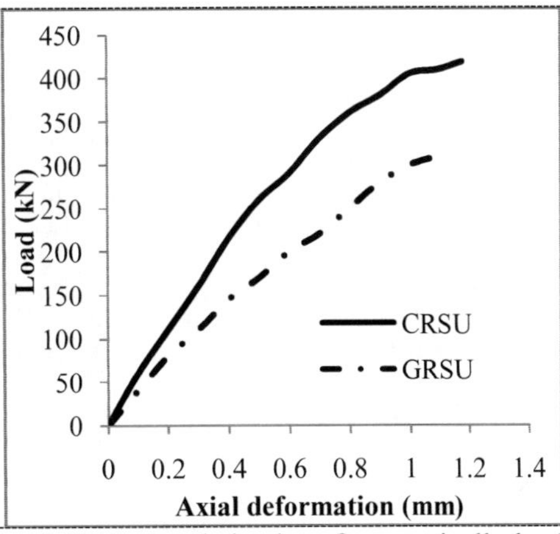

Figure 6. Load-Deflection behavior of eccentrically loaded columns

3.2.1 Load – axial deformation behaviour. The load- axial deformation curve was plotted for CRSU. The plotted curve was then used to compute the energy absorption and the ductility factor of the specimens. A linear variation was found until the load corresponding to the initiation of a crack. After that point, the curve was of parabolic nature. This profile of the curve revealed that the specimen started yielding from the point of initiation of a crack. The load versus axial deformation curve for GRSA was also plotted. GRSU also showed the same load-deformation behaviour as CRSU. Figure 6 shows the comparative load-axial deformation curve for CRSU and GRSU.

3.2.2 Energy absorbed by specimen. The energy absorbed by CRSU was 311.20kNmm and that of GRSU was 201.40kNmm. That is the energy absorption capacity of CRSU was 55% more than that of GRSU.

3.2.3 Ductility factor. The ductility factor exhibited by CRSU was 1.71 and that of GRSU was 1.67. This shows the better ductile behaviour of steel reinforced columns than that of GFRP reinforced columns.

3.3 Mode of failure

Since the load was applied at the bottom end, crushing failure was observed at the base of the specimen. Both steel reinforced and GFRP reinforced columns showed the same mode of failure due to crushing. The typical mode of failure exhibited by axially and eccentrically loaded steel and GFRP reinforced columns is shown in the Figure 7.

Figure 7. Typical failure pattern of the specimens

4. Ultimate capacity and design equations

The equation recommended by American codes for finding the ultimate load carrying capacity of conventional steel reinforced columns is as given below.

$$P_o = k_c f'_c(A_g - A_{st}) + f_y A_{st} \tag{1}$$

Where P_0 is the ultimate load carrying capacity f'_c is the compressive strength of concrete, A_g is gross sectional area of the column and A_{st} is the area of steel reinforcement.
The parameter k_c id defined as the ratio of in-place strength of concrete to the concrete cylinder strength. The value of this factor varies with respect the effect of size, shape and even with respect to the practice of concreting [1]. Previous researchers suggest a value of 0.85 for k_c [14]. Equation 2 gives the expression for ultimate load-carrying capacity with a substituted value of k_c.

$$P_o = 0.85\, f'_c(A_g - A_{st}) + f_y A_{st} \tag{2}$$

Because of the poor performance of GFRP bars under compression ACI standards does not recommend the use of it. Canadian Standards [15] permits the use of GFRP bars as longitudinal reinforcements in compression members subjected to axial load only, ignoring the contribution of the bars in the load carrying capacity of the column. Equation 3 gives the recommendation of CSA (Canadian Standards Association).

$$P_o = \propto_1 f'_c(A_g - A_f) \tag{3}$$

Where A_f is the area of GFRP reinforcement and $\alpha_1 = 0.85\text{-}0.0015f'_c \geq 0.67$.

The Equation 4 gives the ACI 318-11 [16] design equation ignoring the contribution of reinforcement in the load carrying capacity of the column.

$$P_o = 0.85\, f'_c(A_g - A_f) \tag{4}$$

Previous researchers [1] introduces a new expression for finding the theoretical load carrying capacity of GFRP reinforced column as given in the Equation 5.

$$P_o = 0.85\, f'_c(A_g - A_f) + \propto_g f_{yf} A_f \tag{5}$$

Where f_{yf} is the strength of GFRP bars and α_g is a new factor which is used to account the reduced compressive strength of GFRP bars as a function of its tensile strength. Previous researchers fix a value of 0.35 for α_g [8, 17]. The ratio of the experimental values of load carrying capacities of axially and eccentrically loaded GFRP reinforced columns to the theoretical values (that is P_o/P_{ult}), were 1.12 and 1.02 respectively. This shows the accuracy of the equation to predict the load carrying capacity of GFRP reinforced columns.

5. Conclusion

Based on the results obtained from the experimental program the following conclusions were made.

- Under the axial loading, the yield point load showed by the steel reinforced column was 37% more than that of GFRP reinforced column and the ultimate load showed by the steel reinforced column was 34% more than that of GFRP reinforced column. When loaded eccentrically, the yield load showed by steel reinforced column was 31% more that of GFRP reinforced columns and the ultimate load at failure of the steel reinforced column was 34% more than that of GFRP reinforced column. The deficiency of GFRP reinforced columns is due to the micro buckling of fibres in the bars.
- The energy absorbed by steel reinforced column was 40% more than that of GFRP reinforced column when loaded axially and it was 55% more than GFRP reinforced column when eccentrically loaded. The poor energy absorption capacity is also attributed to the micro buckling of the fibres in the bars.
- The result of the experimental program indicates the poor performance of GFRP reinforced column under both axial and eccentric loading.
- When axially loaded the GFRP reinforced column showed a better ductile behaviour. But when loaded eccentrically steel reinforced column showed better ductility.
- An analytical model for predicting the ultimate load carrying capacity was also made and the experimental results were compared with analytical values. The values given by the equation were conforming to the experimental values obtained.
- Irrespective of the non-corrosive nature of GFRP bars, its contribution in the load carrying capacity of columns is less compared to that of steel reinforcement bars. Hence GFRP bars can be used in members where the contribution of reinforcements in load carrying capacity is less.

References

[1] Mohammad Z Afifi, Hamdy M Mohamed and Brahim Benmokrane 2014 Axial Capacity of Circular Concrete Columns Reinforced with GFRP Bars and Spirals. *J. Compos. Constr.* **18** 04013017-1-11.

[2] Benmokrane B, El-Salakawy E, El-Ragaby A and Lackey T 2006 Designing and testing of concrete bridge decks reinforced with glass FRP bars. *J. Bridge Eng.* **11** 217-19.

[3] Rizkalla S, Hassan T and Hassan N 2003 Design recommendations for the use of FRP for reinforcement and strengthening of concrete structures. *Progr. Struct. Eng. Mater.* **5** 16-28.

[4] El-Salakawy E, Benmokrane B and Desgagné G 2003 FRP composite bars for the concrete deck slab of Wotton Bridge. *Can. J. Civ. Eng.* **30** 861-70.

[5] Ehsani M R 2005 *Alternative materials for the reinforcement and prestressing of concrete* J.L. Clarke United Kingdom G64 2NZ **1** 34-53.

[6] American Concrete Institute (ACI) 2006 *Guide for the design and construction of concrete reinforced with FRP bars* (ACI 440.1R-06) Farmington Hills, MI.

[7] Wu W P 1990 *Thermomechanical properties of fiber reinforced plastic (FRP) bars* Ph.D. dissertation West Virginia Univ. Morgantown, WV.

[8] Kobayashi K, and Fujisaki T 1995 Compressive behavior of FRP reinforcement in non-prestressed concrete members. *Proc. 2nd Int. RILEM Symp. on Non-Metallic (FRP) Reinforcement for Concrete Structures* E & FN Spon London 267-74.

[9] Ehab M Lotfy 2010 Behavior of reinforced concrete short columns with Fiber Reinforced polymers bars. *Int. J. Civ. Struct. Eng* **1** 545-57.

[10] Bureau of Indian Standards (BIS) 2009 *Concrete mix proportioning – Guidelines* (IS 10262-2009) New Delhi India.

[11] Bureau of Indian Standards (BIS) 2000 *Plain and reinforced concrete - Code of practice* (IS 456 - 2000) New Delhi India.
[12] Bureau of Indian Standards (BIS) 1970 *Specification for coarse and fine aggregates from natural sources for concrete* (IS 383 - 1970) New Delhi India.
[13] Muthupriya P, Subramanian K and Vishnuram B G 2011 Experimental investigation on high performance reinforced concrete column with silica fume and fly ash as admixtures *Asian J. Civ. Eng. (Building and Housing)* **12** 597-618.
[14] Lyse I and Kreidler C L 1932 Fourth progress report on column tests at Lehigh University *ACI J.* **28** 317-46.
[15] Canadian Standards Association (CSA) 2012 *Design and construction of building components with fiber reinforced polymers* (CAN/CSAS806-12) Toronto.
[16] American Concrete Institute (ACI) 2011 *Building code requirements for structural concrete* (ACI 318-11) and commentary Farmington Hills, MI.
[17] Tobbi H, Farghaly A S and Benmokrane B 2012 Concrete columns reinforced longitudinally and transversally with glass fiber-reinforced polymers bars *ACI Struct. J.* **109** 1-8.

Experimental Study on Voided Reinforced Concrete Beams with Polythene Balls

P Sivaneshan and S Harishankar*

School of Civil Engineering, SASTRA University,Thanjavur 613 401, India.

*Email: harishankar@civil.sastra.edu

Abstract. The primary component in any structure is concrete, that exist in buildings and bridges. In present situation, a serious problems faced by construction industry is exhaustive use of raw materials. Recent times, various methods are being adopted to limit the use of concrete. In structural elements like beams, polythene balls can be induced to reduce the usage of concrete. A simply supported reinforced concrete beam has two zones, one above neutral axis and other below neutral axis. The region below neutral axis is in tension and above neutral axis is in compression. As concrete is weak in tension, steel reinforcements are provided in tension zone. The concrete below the neutral axis acts as a stress transfer medium between the compression zone and tension zone. The concrete above the neutral axis takes minimum stress so that we could partially replace the concrete above neutral axis by creating air voids using recycled polythene balls. Polythene balls of varying diameters of 75 mm, 65 mm and 35 mm were partially replaced in compression zone. Hence the usage of concrete in beams and self-weight of the beams got reduced considerably. The Load carrying capacity, Deflection of beams and crack patterns were studied and compared with conventional reinforced concrete beams.

Keywords: Concrete, Polythene balls, Compression, Deflection, Crack

1. Introduction

1.1 General

Now-a-days, construction industry starts to pay a great attention to special material that is used in building construction in terms of their effects on environment. Ordinary Portland cement (OPC) is one of the main components used in casting reinforced concrete (RC) structures[1]. So the usage of concrete is high now-a-days, and hence there becomes a shortage of raw materials for its preparation. To reduce the usage of concrete many researches were carried out for alternate materials that can be used instead of concrete. Some of the researchers used local materials like copper slag, rice husk and fly ash for alternate material and they were evaluated experimentally. A simply supported reinforced concrete beam has two zones, one above the neutral axis and other below the neutral axis. The region below neutral axis is in tension and above neutral axis is in compression. As concrete is weak in tension, steel reinforcements are provided in tension zone. The concrete below the neutral axis acts as a stress transfer medium between the compression zone and tension zone. The concrete above the neutral axis takes minimum stress. By recycling scrap or waste plastic material turn them into useful products, and plastic is non-biodegradable material. So, recycling the plastic is global effort to reduce plastic in the waste stream, especially in a year approximately 8 million metric tons of waste plastic that entered into earth's ocean. By recycling the waste plastic material it can reduce the high rate of plastic pollution. By using this method the concrete is replaced from above neutral axis by creating air voids using recycled polythene balls. These balls are less in weight and when we use these balls, it will not affect geometrical shape of beam. Also it greatly does not affect the strength and stress characteristics of beam. The testing of ball is shown in fig.1.

Figure 1. Ball size and Testing of ball

From the previous studies it is evidently understood that in RC beams, concrete near the neutral axis can be replaced with lightweight materials like hollow pipes, terracotta hollow blocks, brick and polystyrene sheets etc.,. When we use these materials, they are good in the reduction of self-weight of the beam but the ultimate load carrying capacity and first crack load is very small when compared to conventional beam [2]. To overcome these drawbacks, concrete above neutral axis can be replaced by creating air voids using recycled polythene balls of diameter 75mm, 65mm and 35mm.The main objective of this study is to introduce a new method by replacing the concrete present above the neutral axis region by creating air voids using recycled polythene balls. This polythene balls were made by waste recycled plastic materials.

2. Materials and methods

The materials used for the experimental works and the design methods adopted for the paper work are included below.

2.1 Materials

OPC 53 grade cement conforming to ASTM C 150/TYPE 1 was used. The physical properties of cement is tested and calculated as followed. The specific gravity of cement is 3.15 and consistency is 29%. Zonal Sand and coarse aggregate were obtained.

2.1.1 Mix Design

Table 1.M30 Mix proportion

Material	
Cement (Kg/m^3)	380
Fine aggregate (Kg/m^3)	711
Coarse aggregate (Kg/m^3)	924
Water (Lit/m^3)	159.6
Water cement ratio	0.42
Mix Ratio	1:1.87:2.43

2.1.2 Methods. The methodology of work consists of the selection of Grade of concrete: M30and Mix design for M30 grade concrete was calculated by using the code provision IS 10262 – 2009. After the calculation of mix ratio, cubes and cylinders were cast using M30 grade concrete. The cubes and cylinder were cast with recycled polythene balls of 75mm, 65mm and 35mm. After casting they were allowed to dry for 24 hours and were under normal water curing for 7 days and 28 days. The cube is tested after 7 days curing and test results for normal M30 cube gave a value of 22.39kN and the cube with 75mm balls are gave a value of 20.31kN. After getting better compression strengths, casting of RC conventional beam and beams with 75mm, 65mm and 35mm balls were done. These beams were kept under water curing for the period of 28 days. After curing period, the beams were subjected to four point bending test using 100T loading frame. Totally 7 number of beams were casted, out of which, a couple was conventional RC beam and other 5 beams with 75mm, 65mm and 35mmrecycled polythene balls. All the beams were of same dimensions, 1200mm x 100mm x 200mm where the effective span was 1100mm. The beams were designed as singly reinforced beam with 2 numbers of 12mm diameter bar in tension zone below neutral axis and 2 numbers of 8mm diameter bar at compression zone above neutral axis [1]. For shear reinforcement 6mm diameter bars were used at 100mm spacing as shown in figure 2.

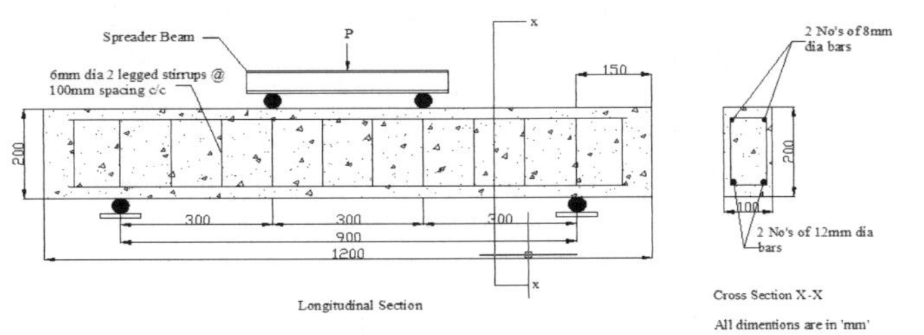

Figure 2. Reinforcement Details

The conventional beams were identified as CB and the beams with balls were identified as PBB (Polythene Ball Beam). The specimens are tested using two point loading by 100T loading frame. The neutral axis depth is calculated by considering M30 grade concrete and Fe415 steel with an effective cover of 50mm. All the beams were designed to fail at compression strain. The equilibrium of forces in bending requires a tension that is equal to compression at all times. It is derived as,

$$\text{Total tension, } T = f_{st}A_{st} \tag{1}$$

$$\text{Total compression, } C = 0.36 f_{ck} b (X_u) \tag{2}$$

Where,

f_{st}= actual tension in steel corresponding to the strain in steel.

Equating the two expressions, we obtain,

$$f_{st}A_{st} = 0.36 f_{st} \text{b} (X_u) \tag{3}$$

$$\text{i.e.} X_u = \frac{f_{st}\, A_{st}}{0.36 f_{ckb}} \tag{4}$$

For under reinforced beams, steel reaches yield stress of 0.87fy first.

Substituting this value in equation (3) and dividing both sides by the effective depth, d (IS 456 Annexure G), we get,

$$\frac{X_u}{d} = \frac{0.87 fyAst}{0.36Fck\ b} \tag{5}$$

$$X_u = \frac{0.87 \times 415 \times 226.08}{0.36 \times 30 \times 100} = 75.58mm$$

The zone above the neutral axis is selected and replaced with 75mm, 65mm and 35mm recycled polythene balls. From this test, the flexural behavior, deflection and crack patterns were studied and comparison was done between conventional reinforced concrete beams and the beams with balls.

2.1.3 Test Procedure. The flexural strength, crack pattern and ultimate load bearing capacity of the beam was found using 100T loading frame and using LVDT (Linear Variable Differential Transformer) the deflections at the centre and quarter span of the beam were calculated. The specimen had 1100mm effective span achieved through a wooden mould with proper cover blocks. The flexural strength of the beam was found using 1000kN loading frame by subjecting the beam to four-point-bending test. The test set up was made as simply supported using two I-sections at proper offsets from the edges. The LVDT was set at centre of the beam to find maximum deflection at a given load and another LVDT was placed at quarter span of the beam from its one of the edges. Both the deflection values were observed from beginning till the failure of the beam. The loading was stopped when first crack has appeared and the readings were noted, after that the ultimate load of the beam is monitored. Finally, a load vs. deflection plot was made.

2.1.4 Load Carrying Capacity. The ultimate load carrying capacity of beam was found by subjecting the beam to four-point-bending using 1000kN loading frame that works through hydraulic motor. The deflection or flexure of beam was calculated by placing LVDT at the mid span and quarter of span of the beam [3]. In this, four point bending test, the load is divided into two parts as, P = P/2+P/2. Here 'P' is the total load given on the beam through the loading frame and was transferred as separate loads by placing an I-section above two rollers above the beam where the load has to be applied. The test setup is shown in figure 3.

Figure 3. Reinforcement Details

The load at first crack was noted when the crack was developed in the concrete and the maximum load was observed and the corresponding deflection was indicated by LVDT. From this test, it was found that there existed no great difference in load carrying capacity between the conventional beams and beams with recycled polythene ball. The beams with recycled polythene balls were named, PBB1, PBB2, PBB3, PBB4 and PBB5 for easy identification. And the normal conventional beams were named, CB1 and CB2. The balls were also tested using CTM (Compression Testing Machine), and the CTM reaches its limit at certain point, until that point the ball gets pressed.

Table 2. Percentage of Concrete Replaced

S. No.	Beams	Percentage of concrete replaced by this balls
1.	*Control Beam 1*	-
2.	*Control Beam 2*	-
3.	*Beam with 75mm balls (PBB1)*	10%
4.	*Beam with 65mm balls (PBB2)*	6%
5.	*Beam with 65mm balls (PBB3)*	12%
6.	*Beam with 35mm balls (PBB4)*	2%
7.	*Beam with 35mm balls (PBB5)*	6%

3. Results and Discussion

The compression behavior of the cubes, deflection, ultimate load, load at first crack, deflections and crack patterns of the beams are discussed below.

In this study, the concrete in beam at its compression zone is replaced. Contribution from compression zone to the load carrying capacity is comparatively lesser than tension zone. The 75mm balls were used for 10% of concrete replacement, 65mm balls for 6% and 12% of concrete replacement and 35mm balls for 2% and 6% of concrete replacement from total concrete. While comparing the load carrying capacity and deflection of the beams with polythene balls andthe conventional beam, the values were observed to be nearly same. Through this method, in wide range of constructions, the usage of concrete is reduced by greater amount. In this method the composite action takes place between concrete and polythene balls. This composite action and placing of balls in between shear reinforcement does not affect the load carrying capacity of beam. The behavior of RC beam at four-point-bending test, the first crack strength, toughness and load deflection response were studied. The fig. 4 shows load vs. deflection response curves for each specimen of conventional beam and beam with 75mm, 65mm, and 35mm recycled polythene balls. From this graph, we can predict the strength of material by knowing the deflection at the respective load values. For example, a specimen concrete and reinforcement bars initially works together to resist deflection of beam. The test result for CB1 is compared to beams with balls, and conventional beam got its first crack at 58kN and corresponding deflection was 1.93mm. The ultimate breaking load for CB1 is 124.3kN and the corresponding deflection was 7.76mm.In PBB1, when load was 40kN, first crack appeared and beam got deflected (1.53mm) in mid-span. After that, load kept on going till 114.6kN and the specimen failed. The load was observed to be the same when compared to the conventional beam, and the deflection at that point was observed to be 5.14mm. PBB2 also got its first crack at 40kN and broke at 108.9kN. At that point the deflection of beam was 7.35mm. And the test result for PBB3 had similar first cracking load. Its first crack at 40kN and the corresponding deflection at mind span was 2.21mm. The ultimate load for this beam was 105kN and the corresponding deflection of the beam was 5.62mm.

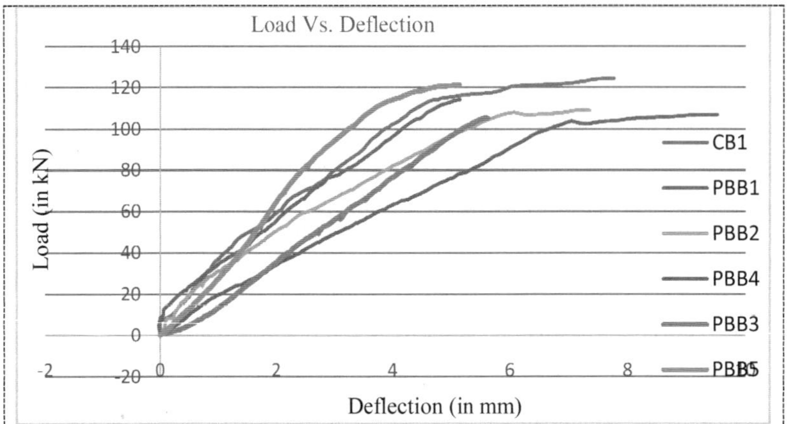

Figure 4. Reinforcement Details

The test result for PBB4 had similar first crack load compared to other two beams. It had the first crack at 50kN and the corresponding deflection at mind span was 3.06mm. The ultimate load for this beam was 107kN and the deflection of the beam was 9.52mm. This beam failed by crushing of concrete. Comparing PBB4 and PBB3, both had meagre variation but the deflection of beam was controlled in PBB3 beam. PBB5 beam had its first crack at 58kN. At that point, deflection of beam was 1.84mm, and it failed at 121kN with a deflection of 5.81mm. PBB5 had more load carrying capacity and deflection was controlled compared to PBB4.From these test results, the conventional beam and beam with balls had the similar value and crack pattern. The crack pattern for all beams is shown in figure 5.

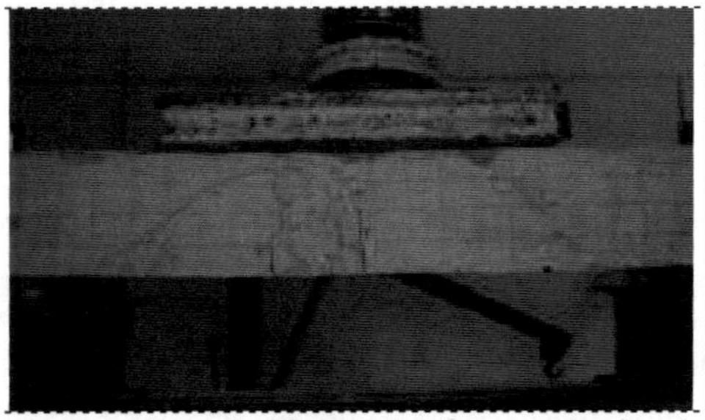

Figure 5(a). Crack pattern for conventional beam

Figure 5(b). Crack Pattern for 75mm ball beam

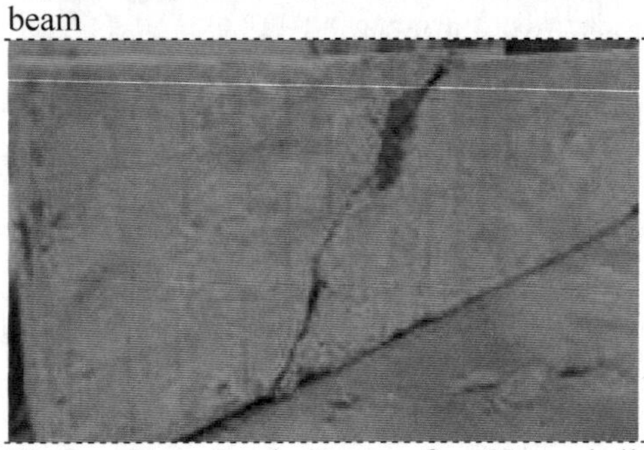

Figure 5(c). Crack Pattern for 65mm ball beam

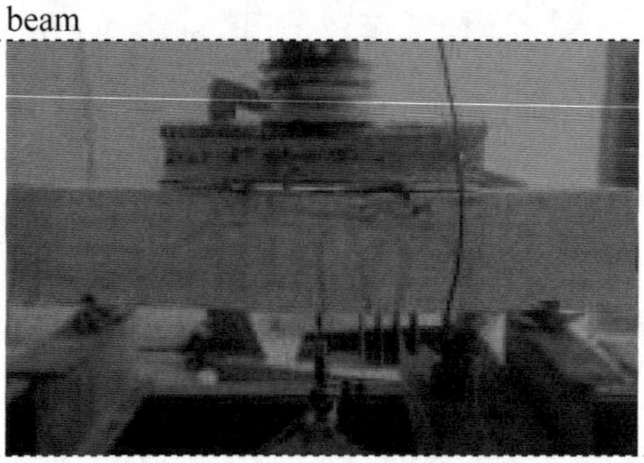

Figure 5(d). Crack Pattern for 35mm ball beam

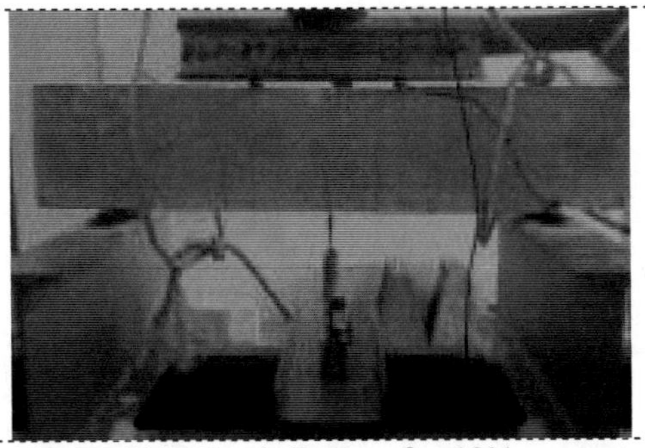

Figure 5(e). Crack Pattern for 12% concrete replacement

Figure 5(f). Crack Pattern for 6% concrete replacement

4. Conclusion

The flexural behaviour of the beam with replaced polythene balls are tested and compared with conventional beam in this study. The theoretical explanation of the test beams were carried out. From this present study the following conclusion are listed.

- Flexural behaviour of conventional beam and beam with polythene balls are marginally similar.
- Replacing the concrete by polythene balls in compression zone does not exhibit significant change in the load carrying capacity of the beams.
- In this Study compared to other polythene ball beams PBB5 had more load carrying capacity, and deflection of the beam is controlled.
- Deflection of the polythene ball beam is considerably lower than the conventional beam.
- By replacing concrete with these balls in reinforced concrete beams had no need of additional labors and time. Economically this method is effective and percentage of concrete replaced from beam is based upon count of balls. In this method huge amount of concrete reduction can be done without affecting the behaviour of the structure and it minimizes the construction cost too.
- When this method is used in long span construction considerable amount of concrete is reduced which in turn decreases the usage of raw material.

References

[1] Rami A Hawileh, Jamal A Abdalla, Fakherdine Fardmanesh, Poya Shahsana and Abdolreza Khalili 2017 Performance Of Reinforced Concrete Beams Cast With Different Percentages Of GGBS Replacement To Cement. *Arc. of Civil and Mech. Eng.* **17** 3511-9

[2] Aswathy S Kumar, Anup Joy 2013 Experimental Investigation on Partial Replacement of Concrete below Neutral Axis of Beam. *IJSR* **8** 1670-4.

[3] Adam M Knaack and Yahya C Kurama 2012 Behavior of Reinforced Concrete Beams with Recycled Concrete Coarse Aggregates. *J. Str. Eng. ASCE* **3** 1–12.

[4] Arati Shetkar and Nagesh Hanche 2015 An Experimental Study On Bubble Deck Slab System With Elliptical Balls *Pro. NCRIET & Indian J S R* **12** 21-7

[5] Amer M Ibrahim, Nazar K Ali and Wissam D. Salman 2013 Flexural capacities of reinforced concrete two-way bubble deck slabs of plastic spherical voids. *D. J. Eng. S.* **6** 9-20

[6] Ade S. Wahyuni, Hamid Nikraz, and VanissornVimonsatit 2012 Reinforced Concrete Beams With Lightweight Concrete Infill. *S. R. Eng.* **27** 2370-9.

[7] Paul Makesh A, Moses Aranganathan S and Seileysh Sivaraja S 2011 Cost Effectiveness to Residential Building using Green Building Approach. *I. J. Eng. S. T.* **3**(12).

[8] Jain Joy and Rajesh Rajeev 2014 Effect of Reinforced Concrete Beam with Hollow Neutral Axis. *I.J. Sci. Res*, **2** 1670-4

[9] Chung J H, Kim B H, Choi H K, Lee S C and Choi C S 2010 Flexural capacities of hollow slab with material properties. *P. Korea Concr. Inst.* **22** 1-4.
[10] Pathak K K, Rakesh Patel and Dubey S K 2014 Analysis of infilled beams using method of initial functions and comparison with FEM. *Eng. Sci. Technol. Int. J.* **17** 158-64.
[11] Surendar M and Ranjitham M 2016 Numerical and Experimental Study on Bubble Deck Slab *I J S R* **5** 5957-5960
[12] Mihai Bindea, Raul Zagon and Zoltan kiss 2013 Flat Slab with spherical voids. Part II: Experimental tests concerning shear strength *Civil Eng. A A* **1** 74-81.
[13] Osama A Abaza and Zaids S Hussein 2015 Flexural Behavior of Steel Fiber-Reinforced Rubberized Concrete. *J. Mater. Civ. Eng. ASCE* **10** 1-11
[14] Salman Wissam D and Ali K Nazar Ibrahim 2013 Finite Element Analysis Of Reinforced Concrete Slabs With Spherical Voids. *D. J. Eng. Sci.* **06** 415- 37
[15] Karthik S, Eramma H and Madhukaran 2014 Behaviour of Concrete Grade Variation in Tension and Compression Zones of RCC Beam. *Int. J. Adv. Tech. Eng. Sci.* **02** 4067-72.
[16] Xiaochun Fan and Mingzhong Zhang 2013 Experimental study on flexural behaviour of inorganic polymer concrete beams reinforced with basalt rebar. *S.C. Eng. Arch. Comp.* **93** 174-83.
[17] IS 10262 2009) Concrete Proportioning Guidelines (First Revision), *Bureau of Indian Standards*.
[18] IS 456 2000 Plain Reinforced Concrete Code of Practice (Fourth Revision), *Bureau of Indian Standards*.

Experimental investigation on different patterned bolted/welded structural connection in steel and GFRP plates

S Kanchidurai* and P Vivek

School of Civil Engineering, SASTRA University, Thanjavur – 613 401, India.

*Email: kanchidurai@civil.sastra.edu

Abstract: The experimental investigation is explained different structural connection like various patterned single shear lap bolted connections and slot complete penetration welded connection. Totally 12 numbers of 300 x 50 x 4 mm steel connections made by lap joint as per IS800: 2007 provisions. The patterns are linear, diamond and staggered for the bolted and slot complete penetration welded connections, 4.6 grade bolt and tungsten electrode rods are used for structural slot complete penetration welded connection conforming to IS 800: 2007. And 6 numbers of 300 x 50 x 6mm Glass fibre reinforced polymer (GFRP) plate with bolted connection was casted, shear resistance capacity is considerably higher in the staggered pattern structural connections then the linear and diamond pattern connections. The connection which is made by GFRP, failure occurred in the plate itself and the shear resistance capacity is lowered 60% then the steel plate even though GFRP plate high resistivity against aggressive environment
Keywords: lap joints, single shear failure mode, bolt, fasteners arrangement and slot complete penetration weld.

1. Introduction

A steel structure is an assemblage of beam, column and tension members. These different members have to be connected properly by means of fasteners in order to act as a single composite unit. Facilitating flow of forces and moments from one member to another is not the only function of connections but it also allows transfer of forces up to foundation level[1]. The performance of a structure depends upon behaviour of joints so before analyzing the structure it is important to analyze the joints within them .In order to avoid connection failure proper design and detailing is needed otherwise it results in a weaker connection. Safe, economical and simple are the basic goal of connection design to produce a joint so that without any difficulty it can be manufactured and assembled at site. A connection failure should not occur before a member failure because it may lead to failure of whole structure and steel member failure is more ductile when compared with connection failure. There are different failure modes in which a bolted connection may fail such as shear failure of bolt, shear failure of plate, bearing failure of bolt, bearing failure of plate, tensile failure of bolts, bending of bolts and tensile failure of plate. Any one of the above mentioned following modes controls the strength of bolted connections [2-4].

In steel structures different types of simple connections are used such as lap joint, butt joint, truss joint connections, connection at beam column junctions, tension and flange splices. For the members, which are composed of plate elements lap joints are often used to connect. In case of lap joint two members are simply overlapped or connected together by means of bolts or welds[5]. Different types of fastener patterns are available for connecting plate members such as linear, diamond and staggered or chain connection. In the case of linear pattern, fastener holes are located in a straight line and in the case of diamond pattern, fastener holes are located in diamond shape and for staggered pattern fastener holes are arranged in a zigzag pattern[6].

Heating two pieces of metal to a plastic or fluid state for the occurrence of fusion to connect them is called welding. To join various metals there are several welding processes available today. Different types of welds are available for jointing; they are groove welds, fillet welds slot or plug and slot complete penetration welds. In steel construction slot complete penetration welds are widely used nowadays. In slot complete penetration welding, weld metal is used to fill the hole completely or partially. Slot complete penetration welds are used when it is not possible for fillet welding to fill up the holes in connections[7]. The design strength of slot complete penetration weld is similar to that of fillet weld and also they assumed to fail in shear [8]. In preventing overlapping parts from buckling slot complete penetration welds are useful [9-10]. Bolts have many advantages such as simple mechanism, high bearing ability and anti-fatigue behaviour.

2. Experimental procedures

2.1 Materials

Steel: IS2062 – E415 mild steel flat section is used for the bolted and welded connections, 300x50x4mm thick flat strip used for the lap joint, similarly 300x50x6mm size manufactured glass fibre reinforced polymer flat strip connected by lap joint using E4.6 grade bolts. Table1 shows the mechanical properties of IS2062 steel and GFRP

Table 1. properties of the materials

Material	Elastic modulus (MPa)	Poisson ratio	Yield strength (MPa)	Ultimate tensile strength (MPa)	Percentage elongation (%)
Steel	2×10^5	0.30	312	423	52.05
GFRP	0.26×10^5	0.25	360	385	5.13

2.2 Specimens

Totally 18 numbers of specimens are casted to conduct experimental study. Each pattern 2nos of specimen fabricated in linear, diamond and staggered; for analysing mechanical property coupon test was conducted separately results are mentioned in Table1. Out of 18numbers of specimen 12 numbers fabricated by using steel flat strips and 6 numbers of specimen casted in GFPR flat strips. Table 2 shows the number of experiment conducts for different pattern. 6mm diameter with 4.6 grade bolts is used for all bolted connections. Fabrication done with provisions in IS800:2007. Bolted and slot complete penetration welded connections made 7.2mm diameter (d_o) bolt clearance provided as specified in IS800.

Table 2. Design of experiments

Description	pattern	Numbers of connection fabricated
Bolted connection in steel flat	linear	2
	diamond	2
	staggered	2
Slot complete penetration welded connection in steel flat	linear	2
	diamond	2
	staggered	2
GFRP with bolted connection	linear	2
	diamond	2
	staggered	2

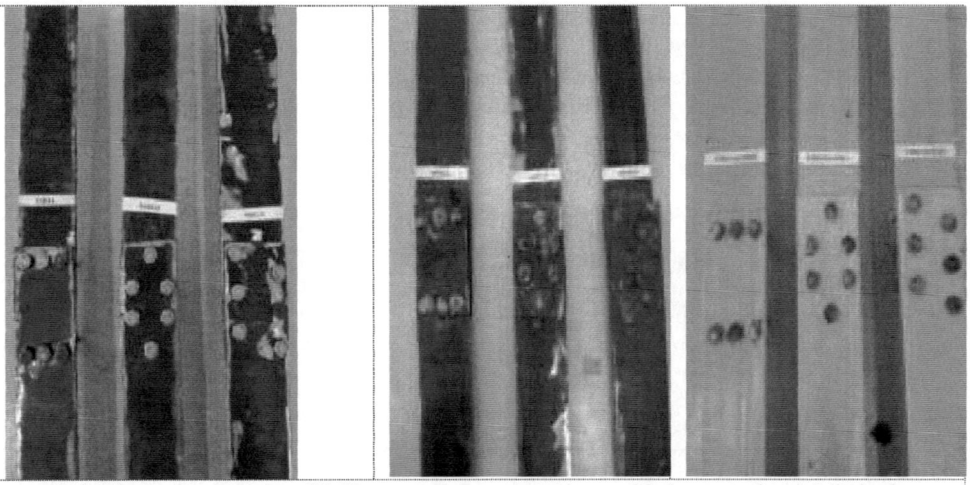

Figure 1. (a) Bolted connection (b) Slot complete penetration welded connection (c) Bolted connection in GFRP.

Figure 2. Experimental set-up

3. Results and discussion

Experimentation result shows that tensile strength of the slot-plug welded connection and bolted is provides greater results then the codal provisions, also from different pattern of lap connection, bolt value having significant variation with same number of bolt. Table 3 shows the Experimental results for carbon steel bolted lap connection, among linear, diamond and staggered connections the bolt value of staggered one gives higher resistance comparing to other two connections. But where as in the slot welded connection linear pattern gives greater than the other two in Table 4. In GFRP strips lap bolted connections gives significant strength about 40% of steel connections strength in staggered connection pattern.

Figure 3, 4 and 5 shows the comparison between theoretical tensile strength and the experimental tensile strength of bolted and welded connections , all specimens satisfies the codal requirement and tensile strength 15-30% higher than the theoretical value in that staggered pattern connection gives better value in bolted connection and linear pattern gives better value in welded connections.

Table 3 Experimental results for carbon steel bolted lap connection

S. No.	Specimen ID	Connections	Shear strength of connections		
			Codal provisions (kN)	Experimental (kN)	Variation in percentage %
1	CS-BL-01	Bolted linear pattern	25.058	42.215	68.469
2	CS-BL-02	Bolted linear pattern	25.058	40.639	62.179
3	CS-BD-01	Bolted diamond pattern	25.058	32.217	28.569
4	CS-BD-02	Bolted diamond pattern	25.058	36.715	46.52
5	CS-BS-01	Bolted staggered pattern	25.058	44.452	77.396
6	CS-BS-02	Bolted staggered pattern	25.058	45.621	82.062

Table 4. Experimental results for carbon steel welded lap connection

S. No.	Specimen ID	Connections	Shear strength of connections		
			Codal provisions (kN)	Experimental (kN)	Variation in percentage %
1	CS-WL-01	Slot complete penetration welded linear pattern	45.238	66.352	46.670
2	CS-WL-02	Slot complete penetration welded linear pattern	45.238	69.175	52.913
3	CS-WD-01	Slot complete penetration welded diamond pattern	45.238	64.850	43.353
4	CS-WD-02	Slot complete penetration welded diamond pattern	45.238	63.342	40.020
5	CS-WS-01	Slot complete penetration welded staggered pattern	45.238	61.727	36.460
6	CS-WS-02	Slot complete penetration welded staggered pattern	45.238	59.770	32.123

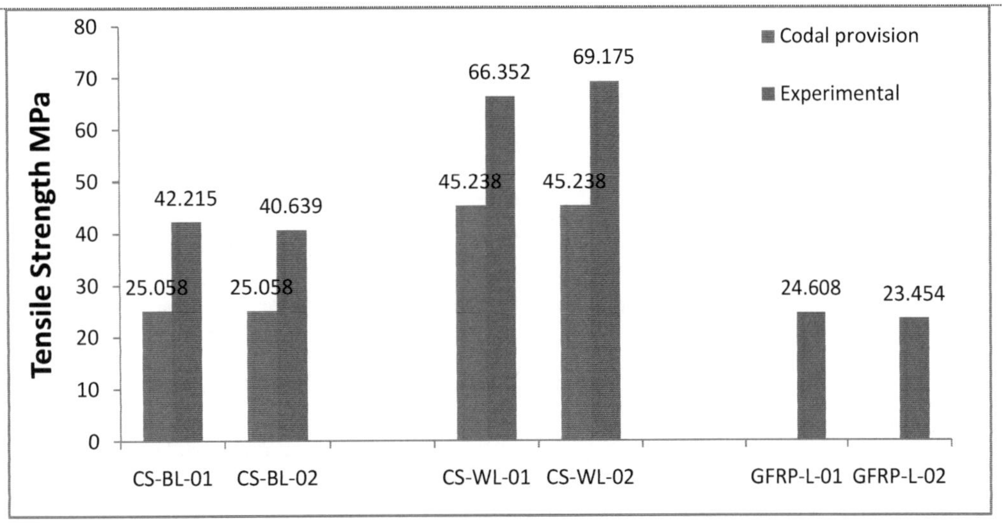

Figure 3. Comparisons of Linear pattern structural connection

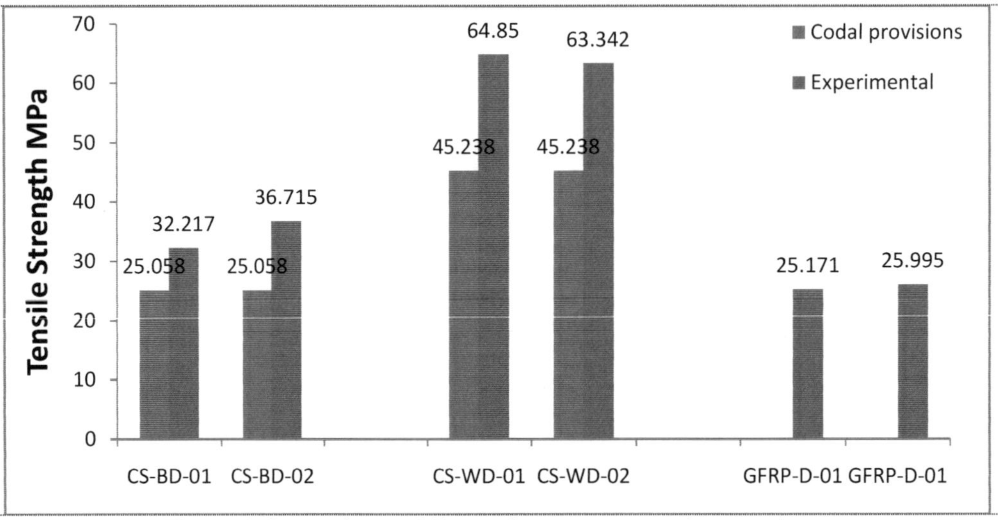

Figure 4. Comparisons of diamond pattern structural connection

Table 5. Experimental results for GFRP with bolted connections

No.	Specimen ID	Connections	Shear strength GFRP	Variation
1	GFRP-L-01	Bolted linear pattern	24.608	-1.86
2	GFRP-L-02	Bolted linear pattern	23.454	-6.838
3	GFRP-D-01	Bolted diamond pattern	25.171	+0.449
4	GFRP-D-01	Bolted diamond pattern	25.995	+3.739
5	GFRP-S-01	Bolted staggered pattern	26.854	+7.167
6	GFRP-S-02	Bolted staggered pattern	29.766	+18.788

4. Conclusion

From experimental results, linear pattern of lap joint bolted connections provides almost 65% higher strength considering provisions is made by IS800:2007, but the staggered patterned connections shows the greater shearing strength of 22.3% comparing to linear pattern in bolted connections. Comparing complete penetration of slot type welded connections shearing strength slightly lowered. GFRP plate connections are not satisfied the provisions made by IS800:2007 but the shear strength gives relatively well (-6.838 to +18.788%). GFPR connections may suitable for offshore structures, aggressive environmental conditions, freezing thawing region, low maintenance areas, low cost and where high durability required.

Acknowledgement

Author express heartfelt gratitude to Prof. R. Sethuraman, Vice Chancellor (SASTRA University) for given permission to carry out works in SASTRA university laboratory.

References

[1] Gamdanis F 2015 Tensile strength of open-hole, pin-loaded and multi-bolted single-lap Joints in woven composite plates. *Mater. Des.* **88** 702-12.

IOP Conf. Series: Earth and Environmental Science **80** (2017) 012032 doi:10.1088/1755-1315/80/1/012032

[2] Xianong G 2016 Load-bearing capacity of occlusive high-strength bolt connections. *J. Constru. Steel Res.* **127** 1-14.

[3] Yonghyun C 2016 Estimation of ultimate strength in single shear bolted connections with aluminium alloys (6061-T6). *J. Thin-walled Struct.* **101** 43-57.

[4] Ling L 2015 Effect of beam web bolt arrangement on catenary behaviour of moment connections. *J. Constru. Steel Res.* **104** 22-36.

[5] Gerami M 2011 Cyclic behaviour of bolted connections with different arrangement of bolts, *J. Constru. Steel Res.* **67** 690-705.

[6] Kim T 2007 Finite element modelling of bolted connections in thin-walled Stainless steel plates under static shear. *J. Thin-walled Struct.* **45** 407-21.

[7] Dusicka P 2010 High strength steel bolted connections with filler plates. *J. Constru. Steel Res.* **66** 75-84.

[8] Kamtekar A G 2012 On the bearing strength of bolts in clearance holes. *J. Constru. Steel Res.***79** 48-55.

[9] Moze P 2014 A complete study of bearing stress in single bolt connections. *J. Constru. Steel Res.* **95** 126-40.

[10] Draganic H 2014 Investigation of bearing failure in steel single bolt lap connections. *J. Constru. Steel Res.* **98** 59-72.

IOP Conf. Series: Earth and Environmental Science **80** (2017) 012033 doi:10.1088/1755-1315/80/1/012033

A review of inducing compressive residual stress – shot peening; on structural metal and welded connection

S Kanchidurai[1*], P A Krishanan[2], K Baskar[2] and K Saravana Raja Mohan[1]

[1]School of Civil Engineering, SASTRA University, Thanjavur, Tamil Nadu, India.
[2]National Institute of Technology, Trichy, Tamilnadu , India.

*Email: kanchidurai@civil.sastra.edu

Abstract Shot peening treatment (SPT) is a significant mechanical method to enhance the surface of the material by inducing compressive residual stress on the layer. This study provides a review of prominent improvement in fatigue life on high strength aluminium alloy, steel and welded connection by SPT. Compressive residual stress measurement and its factors data are extracted from assorted literature, optimized peening process commented in this paper, also different types of mechanical peening methods and its effectiveness are mentioned. Fatigue life improvement is focused commented to welded structural connections. The extracted results shows significant changes in the surface layer of metals, aluminium alloy 15 – 250% of fatigue life improvement, steel plain members 6-200% of fatigue life improvement, welded connections 50-75% of fatigue life improvement and significant improvement in mechanical properties like roughness reduction, wear, hardness, tensile strength, corrosion and scuffing.
Keywords: Shot peening, fatigue performance, weld, stress concentration, residual stress.

1. Introduction

Inducing residual stresses to the material will provide certain kind of resistance to the material against fatigue, impact, creep etc., in the case of fatigue life improvement many mechanical methods are available [1]. Improving fatigue strength in welded and non-welded structures is influenced by mode of fracture failure; making the delay of crack initiation and propagation can be done by inducing compressive residual stress in the steel. Residual compressive stress has a great influence on fatigue strength of metallic materials, but the tensile residual stress will reduce the fatigue performance [5]. Figure 1 shows the different notable types of residual stress methods discussed for welded structures.

1.1 Mechanism to induce residual stress

The material can be stressed either at the stage of the manufacturing process or by an external application, during manufacturing stage the material plastically deform to bring perfect shape and it's happened by means of rolling, bending, forging, drawing and

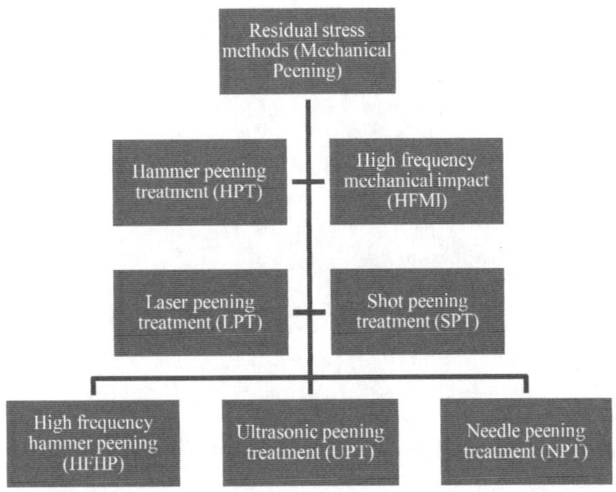

Figure 1. Classification of significant mechanical surface treatment methods [1-4]

1.2 Mechanism to induce residual stress

The material can be stressed either at the stage of the manufacturing process or by an external application, during manufacturing stage the material plastically deform to bring perfect shape and it's happened by means of rolling, bending, forging, drawing and extrusion. Also, some surface modification did during manufacturing like grinding, machining, peening and carburizing [6], due to an external application like welding, casting, quenching and phase transformation major part in inducing residual stress. Figure 2 shows the parabolic curve of tensile residual stress induced by means of tensile force.

Nomenclature

σ_y MPa	Yield stress	USPT	Ultra sonic peening treatment
$E_{L,\%}$	Elongation	LCF	Low cycle Fatigue
API	Almen Peening Intensity	Al	Aluminium Alloy
P_a	Air pressure	Ts	Tensile strength increment
S_r	Stress Ratio	SS	Stainless steel
FLI	Fatigue life increment	d_s	Mean diameter of Shot size
e_A	Strain Amplitude factor in range	MMA	manual metal arc
CSP	controlled shot peening	MIG	metal inert gas
WAl	Wrought aluminium alloy	TIG	tungsten inert gas
SCC	surface corrosion crack	ADI	Austempered ductile iron
HESP	high energy shot peening		

1.3 Residual stress measurement methods

Residual stress evaluation is quite complicated because of its degree of independency of stress to plot load versus deformation curves. Considering the energy release rate by relaxation method or in Griffith's analysis it is possible to find, many relaxation methods have been developed both general and specific types of specimen namely [6]

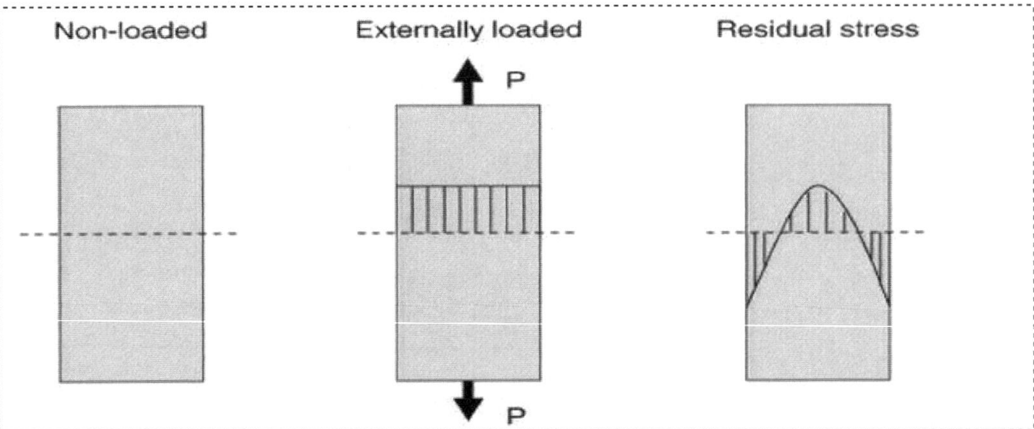

Figure 2. Schematic diagram of residual stress existence (toughened glass sheet) [6]

Relaxation methods are [7-8]		Diffraction methods are		Some other methods are	
i.	Splitting method	i.	X- ray diffraction method	i.	Magnetic residual stress measurement method
ii.	Sectioning method	ii.	Synchrotron X-ray method	ii.	Ultrasonic method
iii.	Layer removal method	iii.	Radiation diffraction method	iii.	Thermo elastic method
iv.	Hole drilling method	iv.	Neutron diffraction method	iv.	Photo elastic method
v.	Slitting method			v.	Indentation method
vi.	Contour method				

1.3.1 Classification of inducing compressive residual stress methods. Residual stress methods are discussed below

a) Hammer peening (HP)

HP is one of the most reliable methods by working in cold region operation speed can adjust based on machine availability. Operation done either Manual or machine by engines, electrical and pneumatic system is used, all tools have some curved nose or blunt nose to produce even local plastic flow. This is usually judged by eye. Frequently, special tools are made for special purposes, figure 3 shows the typical graphic diagram of the machine hammer peening tool component, spherical carbide tip has been used to produce the peening on the welded surface. This actuator controlled by CAM software and rate of peening is given by repeated frequency of the plunger movement [9].

b).High frequency hammer peening (HFHP)

Sometimes high-frequency hammer peening required to improving fatigue strength for welded and non-welded structures having high yield strength steel like 960 N/mm^2, 1100 N/mm^2 and 1300 N/mm^2. High frequency will make the modification of residual stress locally on high strength steel, greater fatigue resistance can achieve using HFHP and Figure 4 shows HFHP the typical model of peening application resulting plastic deformation [10]. HFHP produced high compressive residual stresses and the hardness increase in the upper surface layer resulting from the reduction of surface roughness like glazed surfaces [11]

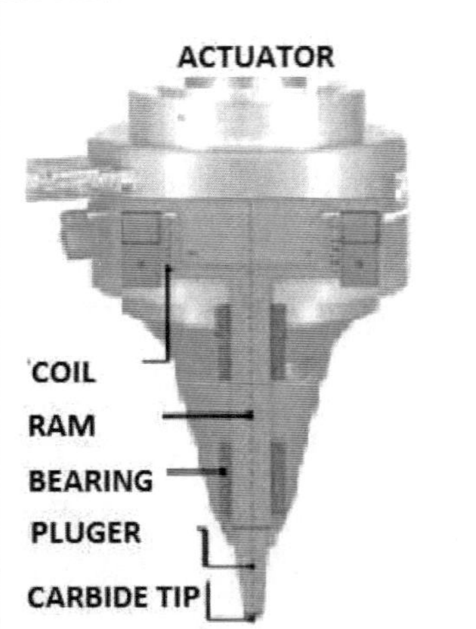

Figure 3. Typical components of Machine hammer peening tool [9]

Figure 4. Typical components of Machine hammer peening tool [10]

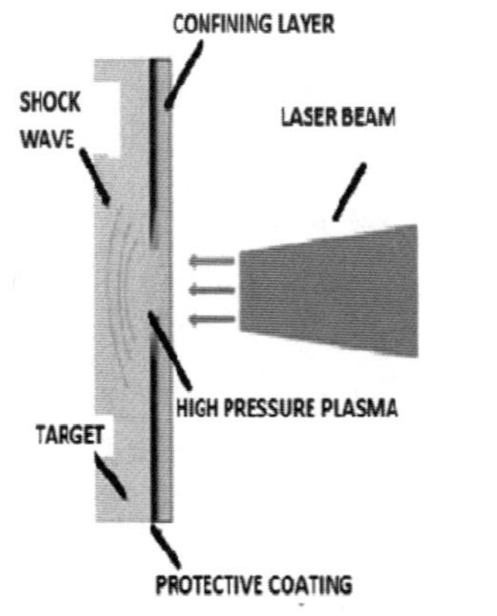

Figure 5. Laser shock peening principle[13]

c) Laser peening (LPT)

Deep compressive residual stress induced by higher level controller process by LPT, figure 5 shows the typical application principle of LPT [13]. Here high power laser beam nearly20 N.m enables irradiances between 20 MW/mm^2 and 100 20 MW / mm2. Laser spot generally 3mm to 10mm provided with plasma force builds to 10-30 kbar force in nanosecond time [12]. Laser peening improves surface quality and fatigue performance more effectively [14]. Another way using diode laser, re-melting of weld toe also improves fatigue performance in the welded structures [15]. Fatigue resistance evaluates in the hook shank application for aircraft results shows notable increment in crack initiation time by LPT [16].

d).Ultra sonic peening (UPT)

Inducing plastic deformation by means of compressive residual stress method UPT is a most promising method for enhancing fatigue life and strength, corrosion resistance and increase surface micro-hardness. The UPT process cylindrical rods or pins used for surface finishes of metallic materials [17]. UPT provides ultra long life in fatigue performance for post welded structures [19]. Ultrasonic impact treatment (UIT) gives significant improvement in the fatigue strength of highly stressed spokes of cast aluminum wheels [20]. Figure6. Shows UPT equipment components.

e).High frequency mechanical impact (HFMI)/(HiFiT)

Operating HFMI provides minimized spacing between alternate impacts, finer surface finishes and user-friendly, also HFMI device can lead in ultrasonic and pneumatic operations [21]. HFMI treatment can improve up to 26% fatigue strength and 353 N/mm2 welded fatigue strength [22]. Figure7. Shows the device commonly used for HFMI, UPT and needle peening. In lightweight design, the fatigue strength improvement is successfully implemented by HFMI treatment of longitudinal stiffeners [24][25][26].

f).Needle peening treatment (NPT)

NPT also notable treatment for welded connections, Peening process is done by rounded headed needles to prepare a surface. [27]

2. Shot Peening (SPT)

SPT is the process to produce high compressive stress at very near surfaces with very limited surface roughness reduction. It is similar like sand blasting instead garnet is replaced by steel balls, figure 8. Showing SPT process which is involved impact at high speed creates a dimple, compressive stress development and residual stress concentration area. Normally the steel ball used as 0.8mm – 0.05mm diameter size [28]. Area coverage and creation of surface nano-layer is the major advantage in SPT, also it requires less time to finish the treatment process and testing [28-29].

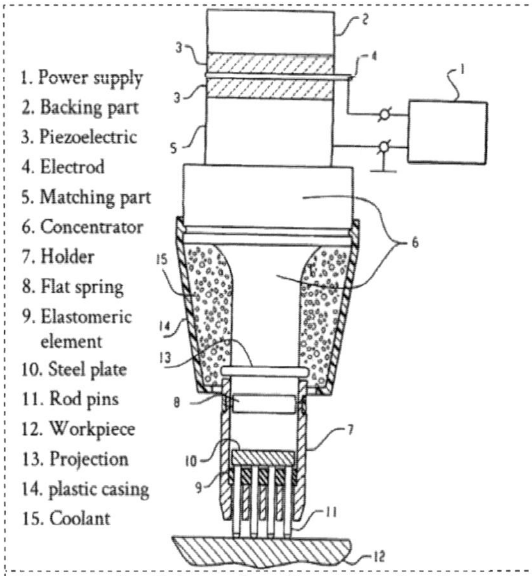

Figure 6. Schematic diagram of ultrasonic peening tool [18]

Figure7. HFMI device and its indenter [23].

Vaibhav Pandey et.al was studied about low cycle fatigue (LCF) behavior on aluminum alloy AA7075 and W.C.lie et.al was investigated about high cycle (HCF) fatigue behavior of the magnesium alloy Mg–10Gd–3Y with four different condition as - cast, cast- T6, as-extruded and extruded-T5 both experimentation shows maximum beneficial effects due to LCF and HCF[30][31]. SPT enhance the corrosion resistance property which is dealt with welded connections, 75% of corrosion can be decreased after 1000 hours of salt mist exposure [33].

Residual Stress Distribution:- With various experimental data Jinxiang Liu and Ming Pang analyzed to predict fatigue life by calculating distribution of residual stress, existing approach is to find fatigue strength is theory of critical distances[34]

$$\Delta\sigma_{av}|(\text{Area}, r = a_0) = \Delta\sigma_{max} / \{1 - (4/3\pi)\, a_o\, \beta\} \quad (1)$$

Here, $\Delta\sigma_{av}$ = $\Delta\sigma$ averaged over a semi-circular area of radius $r = a_o$
$\Delta\sigma_{max}$ = Cyclic stress at notch root ('hotspot' stress)
a_o = ElHaddad constant
β = normalized stress gradient $\{1/\sigma_{max} \times (d\sigma / dr)$.
inducing residual compressive make surface harder and reduction in grain size results high corrosion resistance to the metal [35]

3. Experimentation Results

Residual stress are induced by SPT to Wrought aluminium alloy specimens results in Table 1,2 and 3. High strength aluminium alloys is the predominant material, normal alloy metals are magnesium, manganese, silicon, copper, zinc and tin. Normally used in machineries, vehicles, aircraft and engineering structures component. WAl 2000 series are alloyed with copper, 5000 series are alloyed with magnesium, 6000 series are alloyed with magnesium and silicon and 7000 series are alloyed with zinc. T3 represents solution heat treated and cold worked, T6 represents solution heat treated and artificially aged, T8 represent solution heat treated, cold worked and artificially aged and H1 represents strain hardened without thermal treatment. Table 4 shows the fatigue life improvement in structural materials such as carbon steel, alloys, stainless steel and Austempered ductile iron. Table 5 and 6 shows improvement of mechanical behavior like roughness, hardness, tensile strength, wear and Scuffing resistance with considering API, P_a, d_s and S_r.

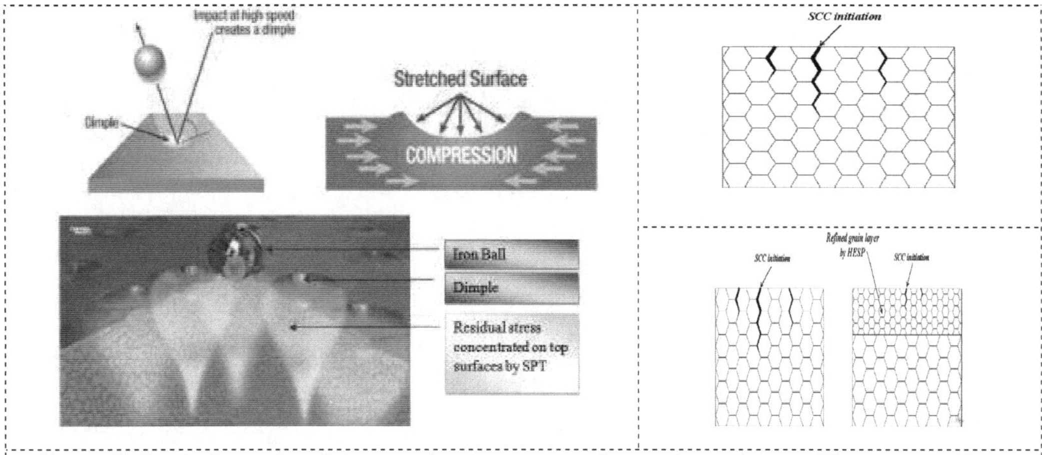

Figure 8. Typical process of shot peening treatment and corrosion resistance [Courtesy Curtiss – Wright][32][35]

Table 1. Fatigue life prediction of SPT on Wrought aluminium alloy Al

References	Type	σ_y, MPa	E_L, %	API,	P_a, bar	d_s, mm	S_r	FLI, %
[36]	Al 2195	215	9	00.008–0.012mmA	-	0.59	0.1	increased
[37]	Al 7075	462	11	0.27mmA	0.7	0.19	0.1	70
[38]	5183 H11A1	143	26	16-20 A	-	0.8	(i) 0.1	135
[39]	Al 6082	200		0.15 – 0.2mmA	-	0.6	0.1	26
[40]	Al7075–T6	520	-	-	-	-	0.1	130

[41]	Al-7075-T6	534	11	4.5N	-	0.06-0.12	-1	20
[42]	Al-7075-T6	515	15	4.5N	-	0.06-0.12	0.05	15
[43]	Al-7075	480	-	0.25-0.14mm		0.21- 0.32	0	250
[44]	Al 2024-T3	347.4	15	14A	-	0.5842	0.1	significant improvement
[45]	Al 5083	196	-	0.5& 0.9mm		0.9906	-1	75%

Table 2. Deflection control and hardness of SPT on AA5A06 [42][46][55]

References	Type	σ_y, MPa	E_L, %	deflection reduction in %	hardness increment in %	mechanical property
[46]	Al5A06	124	27	92.3	10	considerably increased
[47]	2124-T851Al	453	10	reduced	13	3.92 Pa and 0.84 dia shot ball
[48]	Al2024	451	11	-	53.5	87nm-13nm size, 180sec

Table 3. Corrosion resistance and fatigue performance by SPT on Al

References	Type	σ_y, MPa	API,A	Corrosion density, (A/cm^2)		Fatigue performance	note
[49]	Al 7075	400	12	As-machined 2.1×10^{-6}	Shot-peened 5.2×10^{-6}	ability to retard crack propagation 87%	5% of NaCl

Table 4. Fatigue life prediction of SPT on steel plain member

Referencs	Type	σ_y, MPa	E_L, %	API,	P_a, bar	d_s, mm	S_r	Ts, %	FLI, %	note
[50]	Carbon steel	770.57	-	0.3 mmA	6.8	0.36	0.1	6.04	6.18	Shot wire used
[51]	12CrNi Mo	868.1	-	8-12A		0.584	0.93	increase	Increase	Din No 1.493
[52]	ASTM A516 grade 70	262	17-21	14-16A		0.431	0.1	-	50 – 63	MMA, MIG, TIG by CSP
[53]	20CrMnTi.	-	-	0.56 mmA	-	0.8	0.1	-	38.87	
[54]	Q345 steel	391	30.5	0.3 mmA	3	0.56	0.1	95		
[57]	En 3 CS	386	29	0.008-0.01 mmA	-	0.5842	-	-	22	

Table 5. Stainless steel roughness reduction, hardness, compressive yield strength, and wear resistance

References	Type	σ_y, MPa	P_a, bar	d_s, mm	Roughness reduction in %	Hardness increase in %	Ts, in %	Wear	Remarks
[55]	17-4 SS	914	3.79	0.07-0.14	70	16	3	much lower	-

Table 6. Austempered ductile iron hardness, roughness reduction, wear and scuffing

References	Type	API, mmA	d_s, mm	Hardness increase, in %	Surface Roughness reduction in %	Wear resistance improvement, in %	Scuffing resistance
[56]	ADI	0.38	0.8382	30.84	93.5	30	higher

4. Discussion

In this section, mechanical properties such as yield strength, elongation index, hardness, tensile strength increment, surface roughness reduction, wear resistance; deflection control, corrosion and scuffing resistance are discussed for the following materials like carbon steel, stainless steel, austempered ductile iron and aluminium alloy.

From different literature, results of shot peening treated materials are collected. Basically, the data's categorized from following factors as shown in figure 9.

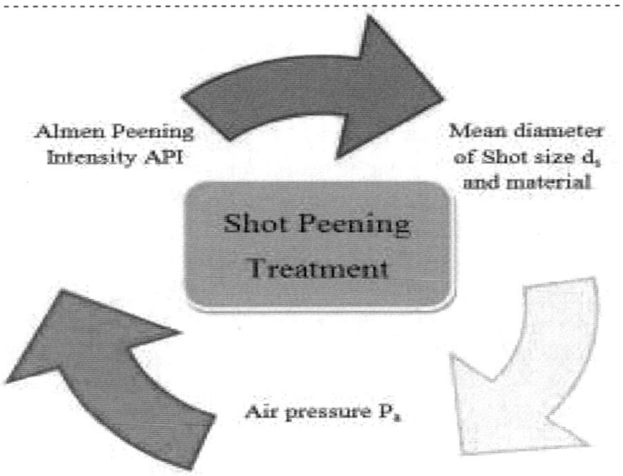

Figure 9. Important factors influencing SPT

Almen strips are used to measure the intensity of shot peening, standard strips are almen A(cast shot), almen N (glass bead peen and ceramic bead peen)and almen C (thicker strip for higher intensity application). The different diameter of shot ball used for appropriate material based on this the property improvement identified. Peening pressure also involves surface variations. Omer et.al Al 2195 with the yield of 215 MPa SPT done 0.008 – 0.012 mmA almen intensity using 0.59mmdia dS improves limited FLI but D.W. Hatamlen, N. Sihhoma et.al proves increasing almen intensity (0.27 to 0.3 mmA) at the same time ds for Al7075 from 0.7 and 0.8mm provides increases FLI 70-135%. Ivo cerny et.al fix the almen intensity in a range of 0.25-0.14mmA and ds 0.21-0.32mm for the same

Al7075 proves up to 250% of FLI hence the almen intensity and shot dia plays important role in the FLI. Some other mechanical property like hardness and deflection for the Al material gives significant improvement depends on the duration of the shot peening process [46][47][48]. Considerably thickening the surface layer will arrest the corrosion development in Al and gives the ability to stop crack propagation against fatigue loading Figure 8 shows the same [35][49]. In plain structural steel member considerably keeping API range of 0.56mmA and ds 0.8mm gives the greater value of FLI 38.87% [53]. Whereas the structural welded connection the API and ds limited in the range of 14-16A and 0.431mm to increase in the FLI 50-63%[52]. R.B. water house et.al proved the increases of hardness, surface roughness reduction, wear resistance and scuffing resistance using 0.38mmA API and 0.8382mm ds.

5. Conclusion and suggestion

The Wrought aluminium alloy (Al series) for fatigue life improvement is suggested to provide the range of 0.27 to 0.3 mmA API and 0.6-0.8mm shot peening cast steel ball will make constant improvement with 180 sec SPT. Structural plain member best API and ds is 0.38mmA – 0.56 and 0.8mm cast steel ball to improve for yield, fatigue strength, hardness, surface roughness reduction, wear resistance and scuffing resistance. In Structural welded connection 14-16A API and 0.431mm ds acquired the significant improvement in fatigue strength.

Acknowledgement

Author express heartfelt gratitude to Vice-Chancellor, SASTRA University for given me the permission to carry out works at SASTRA.

References

[1] Haagensen P J 2011 Fatigue strength improvement methods. *Norwegian University of Science and Technology*. Norway.
[2] Kasra Ghahremania 2015 Fatigue Strength Improvement of Aluminum and High Strength Steel Welded Structures using High Frequency Mechanical Impact Treatment. *Procedia Eng.* **133** 465-76.
[3] Joern Berg and Natalie Stranghoener 2014 Fatigue strength of welded ultra high strength steels improved by high frequency hammer peening. *Procedia Mater. Sci.* **3** 71-6.
[4] Haagensen P J and Maddox S J 2013 IIW Recommendations on Methods for Improving the Fatigue Strength of Welded Joints *IIW*-2142-110. *A volume in Wood head Publishing Series in Welding and Other Joining Technologies*. 29-34.
[5] Doremus L 2015 Influence of residual stresses on the fatigue crack growth from surface anomalies in a nickel-based super alloy. *Mater. Sci. Eng.* **644** 234-46.
[6] Gary S Schajer 2015 Practical residual stress measurement methods. *A John Wiley & sons, Ltd. Publication.*
[7] Lu J 1996 Hand book of measurement of residual stresses. *Fairmount press liburn USA*
[8] Schajer G S 2010 Relaxation methods for measuring residual stresses: Technique and opportunities. *Experimental mechanic.* **50** 1117-27
[9] Friedrich Bleicher 2013 Improving the tribological characteristics of tool and mould surfaces by machine hammer peening. *CIRP Annals – Manuf. Technol.* **62** 239-42.
[10] Joern Berga, Natalie Stranghoenera, Andreas Kernb and Marion Hoevel 2016 Variable amplitude fatigue tests at high frequency hammer peened welded ultra high strength steel S1100. *Procedia Struct. Integrity*. **2** 3554-61.
[11] Friedrich Bleicher 2012 Mechanism of surface modification using machine hammer peening technology. *CIRP Annals – Manuf. Technol.* **61** 375-8.
[12] Lloyd Hackela 2015 Laser Peening to Improve Fatigue Strength and Lifetime of Critical Components. *Procedia Eng.* **133** 545-55.

[13] Hao Wang 2017 Improvement of abrasion resistance in artificial seawater and corrosion Resistance in *NaCl* solution of 7075 aluminum alloy processed by laser shock peening. *Optics and Lasers in Eng.* **90** 179-85.
[14] Gao Y K 2011 Improvement of fatigue property in 7050-T7451 aluminum alloy by laser peening and shot peening. *Mater. Sci. Eng.* **528** 3823-8.
[15] Christoph Gerritsena, Sofie Vanrostenberghe and Matthew Dore 2013 Diode laser weld toe re-melting as a means of fatigue strength improvement in high strength steels. *Procedia Eng.* **66** 171-80.
[16] Leap M J 2011 Effects of laser peening on fatigue life in an arrestment hook shank application for naval aircraft. *Int. J. Fatig.* **33** 788-99.
[17] Massoud Malaki and Hongtao Ding 2015 A review of ultrasonic peening treatment. *Mater. Des.* **87** 1072-86.
[18] Prokopenko G I, Kozlov O, Kleiman J I and Micheev P P 2002 Device for ultrasonic peening of metals, *Integrity Testing Laboratory*, US Patent, No. **US6467321**-B2
[19] Danqing Yin, Dongpo Wang, Hongyang Jing and Lixing Huo 2010 The effects of ultrasonic peening treatment on the ultra-long life fatigue behavior of welded joints. *Mater. Des.* **31** 3299-307.
[20] Berg-Pollack, Voellmecke F J and Sonsino C M 2011 Fatigue strength improvement by ultrasonic impact treatment of highly stressed spokes of cast aluminium wheels. *Int. J. Fatig.* **33** 513-518.
[21] Halid Can Yildirim, Marquis and Zuheir Barsoum B 2013 Fatigue assessment of high frequency mechanical impact *(HFMI)* improved fillet welds by local approaches. *Int. J. Fatig.* **52** 57-67.
[22] Ebrahim Harati, Lars-Erik Svensson, Leif Karlsson and Mattias Widmark 2016 Effect of high frequency mechanical impact treatment on fatigue strength of welded 1300 *MPa* yield strength steel. *Int. J. Fatig.* **92** 96-106.
[23] Halid Can Yıldırım and Marquis Gary B 2014 Fatigue design of axially-loaded high frequency mechanical impact treated welds by the effective notch stress method. *Mater. Des.* ***58*** 543-50.
[24] Kasra Ghahremani, Rakesh Ranjan, Scott Walbridge and Ayhan Ince 2015 Fatigue Strength Improvement of Aluminum and High Strength Steel Welded Structures using High Frequency Mechanical Impact Treatment. *Procedia Eng.* **133** 465-76.
[25] Yıldırım H C, Marquis G and Sonsino C M 2016 Lightweight design with welded high-frequency mechanical impact treated high-strength steel joints from S700 under constant and variable amplitude loadings. *Int. J. Fatig.* **91** 466-74.
[26] Halid Can Yıldırım, Gary Marquis and Cetin Morris Sonsino2015 Lightweight Potential of Welded High-Strength Steel Joints from S700 under Constant and Variable Amplitude Loading by High- Frequency Mechanical Impact Treatment. *Procedia Eng.* **101** 467-75.
[27] http://www.sonats-et.com/page_23-needle-peening.html.
[28] Kazuyuki Oguri 2011 Fatigue life enhancement of aluminum alloy for aircraft by Fine Particle Shot Peening (FPSP). *J. Mater. Process. Technol.* **211** 1395-9.
[29] Habibi N 2012 The effect of shot peening on fatigue life of welded tubular joint in offshore structure. *Mater. Des.* **36** 250-7.
[30] Vaibhav pandey, Chattopadhyay K, Santhi Srinivas N C and Vakil Singh 2016 Low Cycle Fatigue behavior of *AA*7075 with surface gradient structure produced by Ultrasonic Shot Peening. *Procedia Struct. Integrity* **2** 3288-95.
[31] Liua b J W C Dong, Zhang P, Korsunsky A M, Song X and Dinga W J 2011 Improvement of fatigue properties by shot peening for *Mg-10Gd-3Y* alloys under different conditions. *Mater. Sci. Eng.* **528** 5935-44.
[32] Curtiss-Wright Surface Technologies https://cwst.com Paramus U.S. state of New Jersey.
[33] Barbara Nasiłowska, Zdzisław Bogdanowicz and Michał Wojucki 2015 Shot peening effect on 904 L welds corrosion resistance. *J. Constr. Steel Res.* **115** 276-82.

[34] David Taylor 1999 Geometrical effects in fatigue: a unifying theoretical model. *Int. J. Fatig.* **21** 413-20.
[35] Lu Zhiming 2015 Effect of high energy shot peening pressure on the stress corrosion cracking of the weld joint of 304 austenitic stainless steel. *Mater. Sci. Eng.* **637** 170-4.
[36] Omar Hatamleh 2009 A comprehensive investigation on the effects of laser and shot peening on fatigue crack growth in friction stir welded *AA 2195* joints. *Int. J. Fatig.* **31** 974-88.
[37] Hammond D W 1990 Crack propagation in the presence of shot peening residual stresses. *Eng. Fracture Mech.* **3** 7373-87.
[38] Sidhoma N 2005 Fatigue strength improvement of 5083 *H11 Al*-alloy T-welded joints by shot peening: experimental characterization and predictive approach. *Int. J. Fatig.* **27** 729-45.
[39] Moridi 2015 Fatigue behavior of cold spray coatings: The effect of conventional and severe shot peening as pre-post treatment. *Surface & Coatings Technol.* **283** 247-54.
[40] Majzoobi G H 2009 Modification of fretting fatigue behavior of *AL*7075–T6 alloy by the application of titanium coating using *IBED* technique and shot-peening. *Tribology Int.* **42** 121-9.
[41] Benedetti M 2016 On the combination of the critical distance theory with a multiaxial fatigue criterion for predicting the fatigue strength of notched and plain shot-peened parts. *Int. J. Fatig.* **93** 133-147.
[42] Benedetti M 2014 Multiaxial fatigue resistance of shot peened high-strength aluminum alloy. *Int. J. Fatig.* **61** 271-282.
[43] Ivo Cerny 2014 Short fatigue crack growth in an aircraft *Al*-alloy of a 7075 type after shot peening. *Surface & Coatings Technol.* **243** 20-7.
[44] Ali 2007 The effect of controlled shot peening on the fatigue behavior of 2024-T3 aluminium friction stir welds. *Int. J. Fatig.* **29** 1531-45.
[45] Zivkovic D and Anzulovic B 2005 The fatigue of 5083 aluminium alloy welds with the shot-peened crater hot-cracks. *Mater. Des.* **26** 247-50.
[46] Ya-zhang 2016 Correction of buckling distortion by ultrasonic shot peening treatment for 5A06 aluminum alloy welded structure. *Trans. Nonferrous Met. Soc. China.* **26** 1531-7.
[47] Yong-Seog Nam 2015 Enhancing surface layer properties of an aircraft aluminum alloy by shotpeening usi ng response surface methodology. *Mater. Des.* **83** 566-76.
[48] Kyun Taek Cho 2012 Surface hardening of aluminum alloy by shot peening treatment with Zn based ball. *Mater. Sci. Eng.* **543** 44-9.
[49] Zupanc U and Grum J 2010 Effect of pitting corrosion on fatigue performance of shot-peened aluminium alloy 7075-T651. *Process Technol.* **210** 1197-202.
[50] Khany S E 2015 An experimental study of the effect of shot peening on the low carbon steel and identification of optimal process parameters. *Mater. Today Proc.* **2** 3363-70.
[51] James M N 2010 Shot Peening of Steam Turbine Blades: Residual Stresses and their Modification by Fatigue Cycling. *Procedia Eng.* **2** 441-51.
[52] Lah N A C 2010 The effect of controlled shot peening on fusion welded joints. *Mater. Des.* **31** 312-24.
[53] Hongbin X 1991 The effect of shot peening on rolling contact fatigue behaviour and its crack initiation and propagation in carburized steel. *Wear* **151** 77-86.
[54] Gan J, Di Sunb, Wanga Z, Luo P and Wua W 2016 The effect of shot peening on fatigue life of Q345D T-welded joint. *J. Constr. Steel Res.* **126** 74-82.
[55] Al Mangour B and Yang J M 2016 Improving the surface quality and mechanical properties by shot-peening of 17-4 stainless steel fabricated by additive manufacturing. *Mater. Des.* **110** 914-24.
[56] Zammit 2016 The effect of shot peening on the scuffing resistance of Cu-Ni austempered ductile iron. *Surface & Coatings Technol.* **308** 213-9.
[57] Waterhouse R B and Saunders D A 1979 The effect of shot-peening on the fretting fatigue behaviour of an austenitic stainless steel and mild steel. *Wear* **53** 381-6.

Study and Behaviour of Prefabricated Composite Cladding

Sai Avinash P[1], Thiagarajan N[2] and Santhi A S[1*]

[1] Department of Structural and Geotechnical Engineering, VIT University, Vellore-632014, Tamil Nadu, India
[2]CSIR-National Aerospace Laboratories, Banglore-5600177, Karnataka, India

*Email: as_santhi@vit.ac.in

Abstract. The incessant population rise entailed for an expeditious construction at competitive prices that steered the customary path to the light weight structural components. This lead to construction of structural components using ferrocement. The load bearing structural cladding, sizing 3200x900x100 mm, is chosen for the study, which, is analyzed using the software ABAQUS 6.14 in accordance with the IS:875-87 Part1, IS:875-87 Part2, ACI 549R-97, ACI 318R-08 and NZS:3101-06 Part1 standards. The Ferrocement claddings (FCs) are fabricated to a scaled dimension of 400x115x38 mm. The light weight-high strength phenomena are corroborated by incorporating Glass Fibre Reinforced Polymer Laminates (GFRPL) of thickness 6mm, engineered with the aid of hand layup (wet layup) technique wielding epoxy resin, followed by curing under room temperature. The epoxy resin is employed for fastening ferrocement cladding with the Glass fiber reinforced polymer laminate, with the contemporary methodology. The compressive load carrying capacity of the amalgamated assembly, both in presence and absence of Glass Fibre Reinforced polymer laminates (GFRPL) on either side of Ferrocement cladding, has been experimented.
Keywords: Cladding, Ferrocement, Epoxy resin, Glass fibre reinforced polymer laminate.

1. Introduction

Construction is the strategy for development which incorporates connecting segments of a structure in an assembling or construction site, it is method of good outline with present day elite segments and quality controlled assembling methodology. This system may permit work to proceed regardless of poor climate conditions and ought to diminish any waste in time and material at the site. The prefab parts and prefab structures wipe out space and time over traditional developments. In spite of the fact that construction is utilized to a huge degree in a wide assortment of nations, in India, development industry, regardless of its extension keeps on embracing same routine strategies. Typically, post natural disaster demands precast dwellings for rapid repossession, so availability of precast houses will secure them due to its faster construction.

Ferrocement panel, kind of thin fortified cement for the most part with cement mortar strengthened together with firmly set layers having ceaseless. Metallic or other reasonable materials are utilized to build up the mesh. Ferrocement contains a high elasticity to-weight proportion and more noteworthy splitting conduct when contrasted with conventional panel. In this way, the thin ferrocement structures appear to be relatively light and watertight. Epoxy gum frameworks are concoction blends containing two primary parts, the epoxy and the curing agent (likewise called hardener). Epoxies are additionally used to bond cement to itself, to steel or two dissimilar materials. The preferences of a FRP composites panels are lightweight, high quality and elite, synthetic and erosion safe.

In this study, analysis and experimental work is carried over to find the compressive load carrying capacity of load bearing cladding. Three different cladding panels were considered, 1. R.C.C Cladding, 2. Ferrocement Cladding, 3. Ferrocement Cladding with application of GFRP. Each of 2 panels were fabricated, cured and tested after 28 days under Universal Testing Machine (UTM).

2. Literature Review

To adopt low carbon foot print, fly ash based design mixes are came in to existence. Fly ash based mix increment by 50% compared to control mix. Raise in compressive strength of panel under steam curing by 184kN, with accelerated curing of 142.5kN, steam curing has shown a rapid strength with other curing methods [1]. The precast claddings at the initial phase were made of non-composite mechanism by incorporating insulation in concrete Wythes at obligatory intervals. Idea of precast cladding lead to design as a load bearing cladding which the engineer or architects are advocate to use of these precast concrete sandwich cladding for better execution, ideal for transmitting load to the foundation from superstructure. The integrity action of precast concrete sandwich cladding most part based on type of connection between the concrete Wythes [2]. Construction industry looks after faster, rapid with light weight material with variable considerations in design. Ferrocement cladding is the one for its low cost and light weight being used till now. The tensile crack pattern will not widen further due to the application of chicken wire mesh in mortar mix [3]. Precast sandwich panels comprised of two concrete Wythes which are apart by insulation material such as polystyrene, polyurethane foam and which are joined by a shear connectors to transfer the longitudinal shear on panel. Density of the insulating material also plays a crucial role in transferring the load along the interface of the panel [4]. Under complex and expensive circumstances glass fibre reinforced polymers (GFRP) profiles which enclosure the surface of the building. Building integrity with precast components with lightweight materials tend to ameliorate the manufacturing quality and reduction of building cost [5]. Precast concrete sandwich panel compose two concrete Wythes designed as load bearing and also non load bearing wall. The connection mechanism that adopted the use of continuous shear connector to integrate two structural concrete Wythes which able to transfer adequate shear and full composite action. Load carrying capacity of the sandwich panel depends on life span of shear connector [6]. Behaviour of prefabricated RCC sandwich panels (RCSPs) under simulated seismic loading. RCC sandwich panel (RCSP) contains Expanded Polystyrene core with steel wire mesh reinforcement incorporated between layers of sprayed concrete. Typical wall thickness ranges from 15cm to 20cm based on the thickness of foam [7]. Alternative change that has been found in construction industry in using fibre reinforced plastic materials. Due to its high structural rigidity it used as a barrier. This system provides outstanding thermal performance and reduces work to labour. Thermal resistance of the building mainly depends on the R value which shown a good interest based on material thickness [8]. Retrofitted technique was applied in application of ferrocement with GFRP profiles. Lateral resisting factor increase by 25-32% compared to normal wall, energy absorption increases by 85%.lateral storey drift results shown undesirable. Energy dissipation and load transferring characteristics can be achieved if fabricated properly [9]. Highly versatile ferrocement panel made of cement mortar and inclusion of chicken mesh layers. A polymer modified ferrocement panels are fabricated shown that 12.5% increase in durability. Based on % of fly ash the development of first crack delayed and also 0-12.5% addition of polymers in fabrication. Moisture resisting capacity decreases by increase of fly ash, polymer cement [10].

3. Design of Cladding

Residential building is taken in to consideration to design the conventional wall as load bearing cladding. The maximum axial load carrying capacity of the wall is calculated as per IS: 875(part1)-1987, IS: 875(part2)-1987.The cladding dimension 3200x900mm is taken in to account as per availability in market and thickness is restricted to 100mm for cases 1, 2 and 88mm for case 3 with 6mm GFRP laminates on either side. Three different claddings were designed in ABAQUS bench work.

3.1 Stress and Deformation

Case 1: R.C.C Cladding

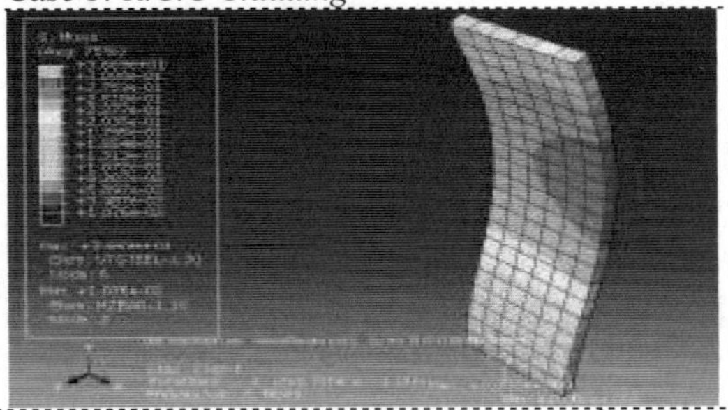

Figure 1. Max. Stress: +3.669e+01N/mm^2.

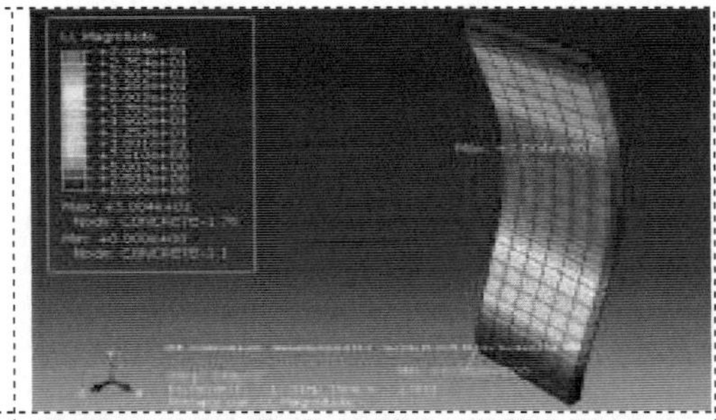

Figure 2. Max. Deformation: +3.004e+01mm.

Case 2: Ferrocement Cladding

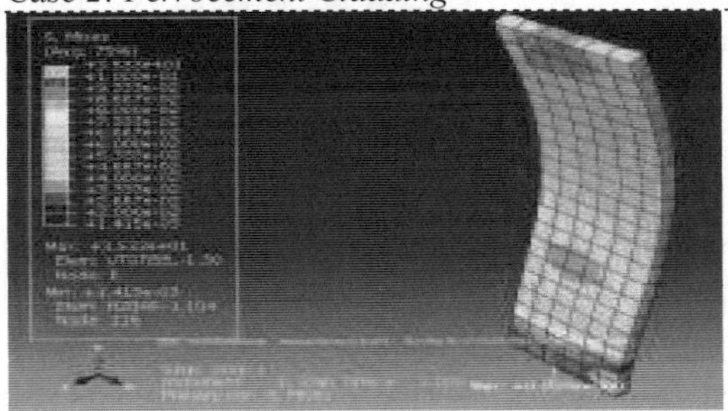

Figure 3. Max. Stress: +3.533e+01 N/mm^2.

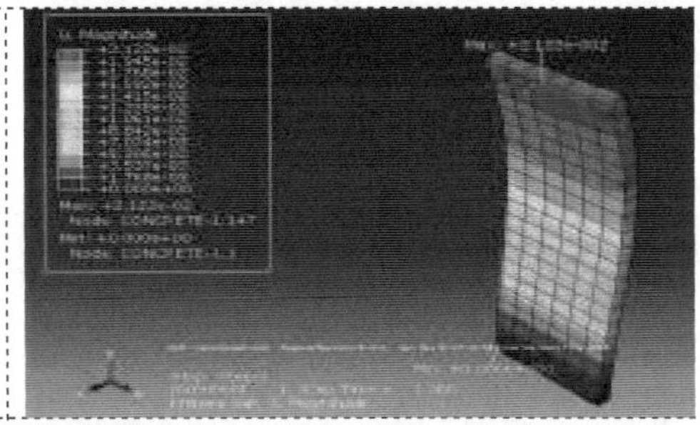

Figure 4. Max. Deformation: +2.122e+01mm.

Case 3: Ferrocement Cladding with application of GFRP Laminates

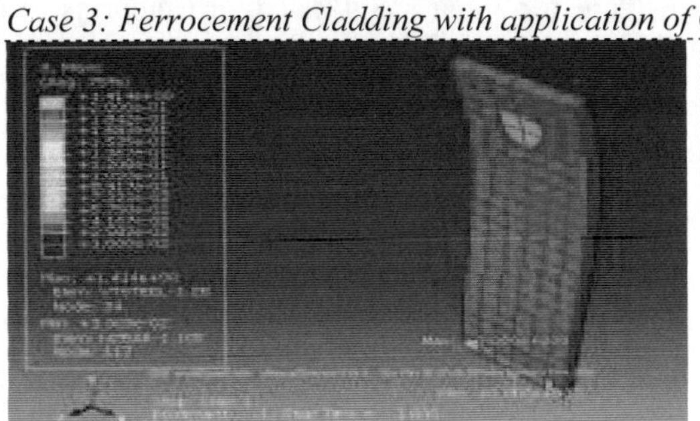

Figure 5. Max. Stress: +1.414e+00 N/mm^2.

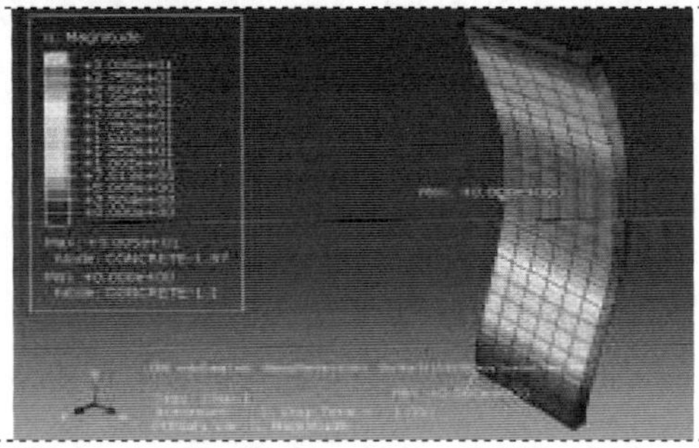

Figure 6. Max. Deformation: +3.005e+01mm.

4. Experimental Study

The prototype was scaled down to study the behaviour of cladding panels dimension 400x115 mm. Where the thickness restricted to 50mm in cases 1, 2 and 38mm in case 3 with 6mm GFRP panels on either side of cladding. Superplastizer of 0.35% of cement is added to reduce the water content, to achieve higher density. The connection mechanism that is adopted in case 3 by using Epoxy resin which is cured under room temperature. All the panels were tested under compressive load carrying capacity. Experimental considerations were shown in Table 1, 2.

Table 1. R.C.C. Cladding

Mix Design	M25
Reinforcement bars	6mm dia.
Bar Spacing	90mm
Concrete Cover	20mm
W/C ratio	0.4
Cement	975 Kg/m^3
Water	390 L/m^3
Fine Aggregate	1462 Kg/m^3
Coarse Aggregate	2924 Kg/m^3

Table 2. Ferrocement Cement Cladding

Mix Design	1:2
Mesh spacing	12.2mmx12.2mm
Mesh Diameter	1mm
Concrete Cover	20mm
W/C ratio	0.38
Cement	1584 Kg/m^3
Water	602 L/m^3
Fine Aggregate	3168 Kg/m^3
Mesh Layers	2

4.1 Casting and Curing

A total of 5 panels were casted using mix ratio of 1:1.5:3 for R.C.C and 1:2 ratio mortar mix for ferrocement panels. Conventional type of curing is done for 28 days and tested for Compressive load carrying capacity of panels. Figure 7& 8 shows the scaled model after casting and curing of panels.

IOP Conf. Series: Earth and Environmental Science **80** (2017) 012034 doi:10.1088/1755-1315/80/1/012034

Figure 7. Scaled Panel.

Figure 8. Conventional Curing.

4.2 Testing

The scaled models of wall cladding were tested for axial load carrying capacity in transverse direction using universal testing machine (UTM).prefabricated composite cladding is fabricated by joining of ferrocement cladding of thickness 38mm with glass fibre reinforced polymer laminates using epoxy resins which cured under room temperature and further achieve the high strength it is examined under 45°C temperature for 4 hours. Prefabricated composite cladding and test setup is shown under figure 9 & 10.

Figure 9. Prefabricated Composite Laminate.

Figure 10. Test Setup.

5. Results and Discussion

Ferrocement cladding of 1:2 mix proportions has shown the axial load carrying capacity of 145kN with deflection of 26.5mm at 100kN load at mid height. Whereas R.C.C cladding with 1:1.5:3 mix gain the strength of 165kN when compared to ferrocement cladding with deflection of 25mm at 100kN load at mid height. In case of ferrocement cladding of 38mm thick with aid of glass Fibre Reinforced Polymer Laminates (GFRP) sustain up to 86kN with a deflection of 28mm at 70kN load, but delamination of GFRP from ferrocement cladding occur at 75kN load, also bending failure is observed only in ferrocement cladding in application of glass fibre reinforced polymer laminates since thickness of ferrocement cladding is less than 50mm. GFRP laminates shown unique property without failure even after test.

6. Conclusions

Ferrocement cladding under axial load shows the maximum stress of 35.3N/mm^2 and deflection of 21mm at 100kN load. In case of R.C.C cladding the axial load is raised up to maximum stress of 36.6N/mm^2 and deflection of 30mm at 100kN load. Ferrocement Cladding with application of GFRP Laminates stress pattern of 1.4N/mm^2 at 70kN load. Ultimate load carrying capacity of ferrocement cladding is 145kN and for R.C.C cladding is of 165kN but on application of GFRP it resists up to 86kN.Bending failure is observed in composite cladding at 75kN but it reaches ultimate load up to 86kN.

Acknowledgment

The authors would like to thank for the extensive help provided by Chidambaram R (Technical Assistant, CSIR-National Aerospace Laboratories, Banglore) and Selvi (Lab Assistant, VIT University, Vellore).

References

[1] Akruth K, Santhi A S and Mohan Ganesh G 2015 Behaviour of precast concrete wall panels. *Int. J. Appl. Eng. Res.* **10** 10-4.

[2] Benayoune A, Samad A A A, Abang Ali A A and Trikha D N 2007 Response of pre-cast reinforced composite sandwich panels to axial loading. *Constr. Build. Mater.* 21 677–85.

[3] Mugahed Amran Y H, Raizal S M Rashid, Farzad Hejazi, NorAzizi Safiee and Abang Ali A A 2016 Response of precast foamed concrete sandwich panels to flexural loading. *J. Build. Eng.* **7** 143-58.

[4] Fabrizio Gara, Laura Ragni, Davide Roia and Luigino Dezi 2012 Experimental tests and numerial modelling of wall sandwich panels. *J. Eng. Struct.* **37** 193-204.

[5] Hassan Adbolpour, Julio Garzon-Roca, Goncalo Escusa, Jose M Sena-Cruz, Joaquim A O Barros and Isabel B Valente 2016 Development of composite prototype with GFRP profiles and sandwich panels used as a floor module of an emergency house. *J. Compos. Struct.* **153** 81-95.

[6] Mugahed Amran Y H, Abang Ali A A, Raizal S M Rashid, Farzad Hejazi and Nor Azizi Safiee 2016 Structural behaviour of axially loaded precast foamed concrete sandwich panels. *Constr. Build. Mater.* **107** 307-20.

[7] Alberto Pavese and Dionysios A Bournas 2011 Experimental assessment of the seismic performance of a prefabricated concrete structural wall system. *J. Eng. Struct.* **33** 2049–62.

[8] Ossama A Abdou, Kris Murali and Ahmed Morsi 1996 Thermal performance evaluation of a prefabricated fiber-reinforced plastic building envelope system. *Energ. Build.* **24** 77-83.

[9] Mosaad Ei-diasity, Hussein okail, Osama kamal and Mohamed Said 2015 Structural performance of confined masonary walls retrofitted using ferrocemet and GFRP in-plane cyclic loading. *J. Eng. Struct.* **94** 54-69.

[10] Bhikshma V, Ravandekishore and Srinivas R 2011 Durability of Polymer and Fly Ash Modified Ferro Cement Elements. *Proc. 12th East Asia-Pacific Conf. Struct. Eng. Constru.* **14** 2642-9.

[11] NZS 3101: part 1:2006 Concrete Structures Standard.

[12] ACI 318R-08 Guide for Precast Concrete Wall Panels.

[13] ACI 549R-97 Guide for Ferrocement.

[14] IS: 875(Part 1)-1987 Dead Loads: Code of practice for design loads (other than earthquake) for buildings and structures.

[15] IS: 875(Part 2)-1987 Imposed Loads: Code of practice for design loads (other than earthquake) for buildings and structures.

IOP Conf. Series: Earth and Environmental Science **80** (2017) 012035 doi:10.1088/1755-1315/80/1/012035

Ultimate Strength of Ferro-Geopolymer Composite Built-Up I Joist

K T Vipin*, N Ganesan and P V Indira

National Institute of Technology Calicut, India

*Email:ktvipin@gmail.com

Abstract: An experimental study was carried out to study the behaviour of ferro-geopolymer built-up I- joist with different types of mesh reinforcements under flexure. Mesh reinforcements considered in this study are square welded meshes, square woven meshes and hexagonal meshes. First crack load as well as ultimate strength of ferro-geopolymer built-up I-joist in flexure was obtained. An attempt was made to predict the first crack load and ultimate moment capacity of the specimen.
Key words: Geopolymer, Ferrocement, First Crack Load and Ultimate Strength.

1. Introduction

Ferrocement can be described as a composite, where the brittle cement mortar matrix is reinforced layers of wire mesh throughout the matrix, resulting in better performance than the individual one. Ferrocement members are very thin and usually the thickness doesnot exceed 40 mm. The product formed due to the polymerisation reaction of any source material which is rich in silicon and aluminium such as fly ash, slag, metakaolin etc. with an alkali solution can be termed as Geopolymer. These can be utilised as a substitute for cement and thereby reducing the overall production of greenhouse gases. Ferro-geopolymer composite is obtained when the geopolymer mortar matrix is reinforced with the wire meshes to form the desired structural element. Many studies have been conducted in the past on investigating the flexural strength of ferrocement elements. Collen [1] tested trough-sectioned ferrocement beams of varying the span, based on which he designed and constructed roof of flour mills upto span of 7.82 m. Balaguru [2], Desayiet. al.[3], Logan and Shah [4] also conducted studies on flexural strength of ferrocement elements. However very limited studies are available in the field of ferro-geopolymer composite. Considering the gap it is proposed to conduct an investigation on ferro-geopolymer composites.

2. Materials and methods

2.1 Materials

In the present study, three ferro-geopolymer composite built-up I joists have been cast and tested. The different types of wire meshes used and volume fraction are the variables included in the study. The specimens were tested for flexure under two-point loading. The ultimate loads as well as maximum deflection were measured and attempts have been made to develop methods for predicting the ultimate strength in flexure. The computed values have been compared with the experimental results. The materials used for this study consisted of wire meshes, HYSD reinforcement bars of 6 mm diameter, fly-ash Class-F, manufactured sand and alkaline solution. Three different types of wire mesh i.e., square welded mesh, square woven mesh and hexagonal wire mesh; details are provided in Table 1. The number of layers of wire meshes were chosen such that volume fraction along the loading direction (the equivalent volume fraction, V_f was kept constant in all the three beams.

Table 1. Wire mesh designation

Mesh type	Designation gauge	Proof Stress, f_s (MPa)
Square Welded Mesh	2/21	366
Square Woven Mesh	5/22	275
Hexagonal Mesh	C24	230

Note: 1. 2/21 is a square welded mesh of 21-gauge wire; i.e., mm wire at 12.7 mm c/c.
2. 5/22 is a square woven mesh of 22-gauge wire; i.e., 0.81 mm dia wire at 5.08 mm c/c.
3. C24 is a hexagonal chicken mesh of 24-gauge (0.56 mm dia) wire.

Fly-ash (class F) was obtained from a thermal power plant near Coimbatore, Tamil Nadu. The manufactured sand was available locally; it was carefully sieved to obtain the grading size as per table 3.1, ACI 549. The geopolymerisation reaction occurs under the presence of alkaline activators. A solution of Sodium Silicate to Sodium Hydroxide (14M) was prepared in the ratio 2.5:1. The Solution was prepared 24 hours prior to casting in order to ensure proper reactivity. The mix details are as shown in Table 2. In case of geopolymer, as there is no standard code available, the optimum mix design was arrived at based on the method suggested by Rangan.The cross sectional detail of the built-up I-joist and the reinforcement details of the channel specimen are as shown in Figure 1a and 1b respectively.

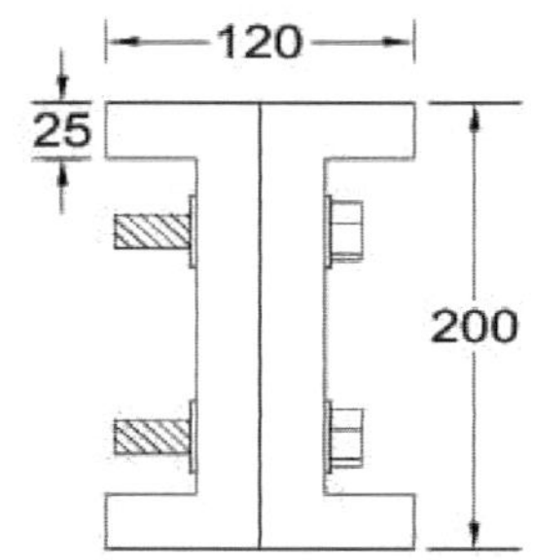

Figure 1a. Cross Sectional Details of the Specimen.

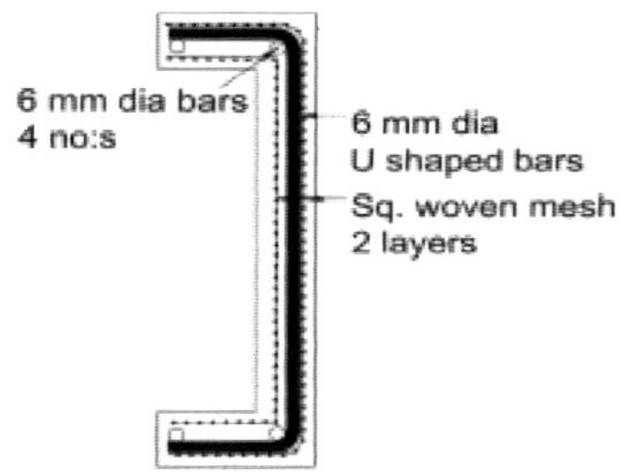

Figure 1b. Reinforcement distribution.

Table 2. Mix Design details

S. No.	Ratios	Values
1	Alkaline activator solution to Binder ratio	0.4
2	Sodium Silicate to Sodium Hydroxide ratio	2.5
3	Binder to Aggregate ratio	2.5
4	Water to Binder ratio	0.1
5	Super plasticiser to Binder ratio	0.01

The test specimens comprised of three built-up I joists having three different types of wire mesh reinforcements. Table 3 gives the details of wire mesh and bar reinforcements used in the channel specimen.Diameter of the holes was increased wherever required with the help of a driller as depicted in the Figure 2a. The bolts were designed properly before providing in the shear span. Figure 2b depicts the final built-up I-joist.

Table 3. Details of reinforcement

Specimen Designation	No. of 6mm dia bars in longitudinal direction	No. of 6mm dia U shaped bars in transverse direction	No of layers			Volume fraction in loading direction
			Square Welded Mesh	Square Woven Mesh	Hexagonal Mesh	
B1	4	14	3	0	0	0.510
B2	4	14	0	2	0	0.533
B3	4	14	2	0	3	0.535

Figure 2a. Drilling holes for provision of bolt

Figure 2b. Cross section of Built-up I-joist

All the three beams were of 2.5 m effective span. Four bars of 6 mm diameter were provided along the longitudinal direction, which acted as the skeletal reinforcement for the channel section. U shaped bars were provided along the transverse direction, spacing was reduced in the shear span so that shear failure is avoided.A brick masonry mould was prepared of the required dimension. The skeletal steel along with meshes on either side was tied securely by binding wire and placed over the mould. The mortar was mixed using rotary pan mixer having a capacity of 8 m^3/hr. Admixture in the form of plasticiser (Complast SP 430) wasadded in order to enhance workability of the mix. The mortar was applied into the meshes using trowels and floats were used to obtain a perfect finish, simultaneously mortar cubes were cast to check the compressive strength. The specimen was left for open steam curing under a pressure of 0.5 psi at 60°C for a period of 24 hours. Thick tarpaulin sheets were used to cover the whole arrangement so that the steam would be retained inside. The built-up I joist was prepared by connecting the channel sections back to back using 20 mm dia bolts in a staggered arrangement along the shear span on either side of the joist. The specimen was tested under two-point loading. 300 tonne capacity UTM was used.The deflections, stain, crack patterns were noted at regular intervals. All the three elements failed in flexure exhibiting high curvature. Table 4 gives the results of the flexure test. Figure 3 shows the load-deflection plots for all the three elements.

3. Results and Discussion

The first crack load as well as ultimate load of the members under flexure was determined analytically. Equations were adopted from previous literatures.

3.1 Cracking Moment

The cracking moment, M_{cr} is calculated using the Eq. (1) given by

$$M_{cr} = f_{cr} \times I_g / y_t \quad (1)$$

Where f_{cr} is the modulus of rupture of mesh-mortar mix computed from Eq. (2),

$$f_{cr} = 0.1962\sqrt{f_c'} \quad (2)$$

The following equation was adopted from the work carried out by Desayi et al. [4]. $f_c' = 0.8fc$, where f_c is the compressive strength of mesh-mortar mix and is calculated using the Eq. (3)

$$f_c = 0.8614f_{cu}\left[1 + 1.095\frac{V_{ft}f_s}{f_{cu}}\right] \quad (3)$$

where, V_{ft} = total volume fraction in the specimen
I_g = gross moment of inertia
y_t = distance from neutral axis to extreme fibre
f_{cu} = maximum compressive strength of the geopolymer mortar cube specimen
f_s = proof stress or yield strength of the wire mesh adopted
P_m = ratio of area of wire mesh in longitudinal direction to the gross area of the specimen.

Also the cracking load, P_{cr} is given by Eq. (4),

$$P_{cr} = M_{cr}/\left(\frac{L'}{2}\right) \quad (4)$$

Where L' is the shear span. Table 5 shows the results of the cracking load obtained and the ratio of computed to experimental values of cracking load for all the three specimens.

Table 4. Strength of Test Specimens

Specimen Designation	Type of Failure	Mortar Cube Strength (MPa)	First Crack Load (kN)	Ultimate Load (kN)
B1	Flexure	36	5.9	27.47
B2	Flexure	35	5.9	23.54
B3	Flexure	36	5.9	21.60

As the experimental first crack loads were deduced from the Load-Deflection plots, the deviation from the computed value was found to be higher for all the three beams.

Table 5. First Crack Load of Test Specimens

Specimen Designation	Experimental First Crack Load (kN)	Computed First Crack Load (kN)	Ratio of Computed Load to Experimental Load
B1	5.9	3.89	0.66
B2	5.9	3.45	0.59
B3	5.9	3.22	0.55

3.2 Ultimate Moment

Two different methods were used to determine the ultimate moment of resistance of the section, M_{UR1}. In the first method, the ultimate moment is determined using the procedure similar to that employed for reinforced concrete sections. Flexural compressive stress distribution in mortar at ultimate is assumed to be rectangular in shape with a value equaling to 68% of that of the cube strength. Tensile strength contribution of mortar is neglected. The wire meshes are assumed to be uniformly distributed over the entire depth of the section with a value equaling to the yield stress, as shown in the Figure 4. Table 6 shows the computed as well as the experimental ultimate loads of all the three specimens.

IOP Conf. Series: Earth and Environmental Science **80** (2017) 012035 doi:10.1088/1755-1315/80/1/012035

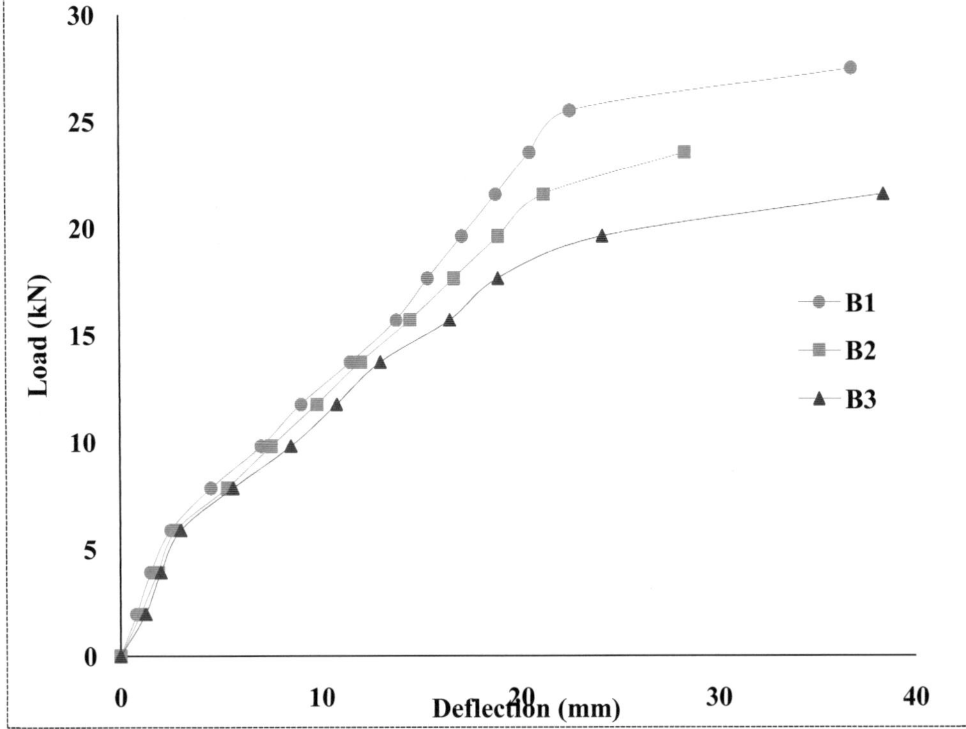

Figure 3. Load-deflection plot.

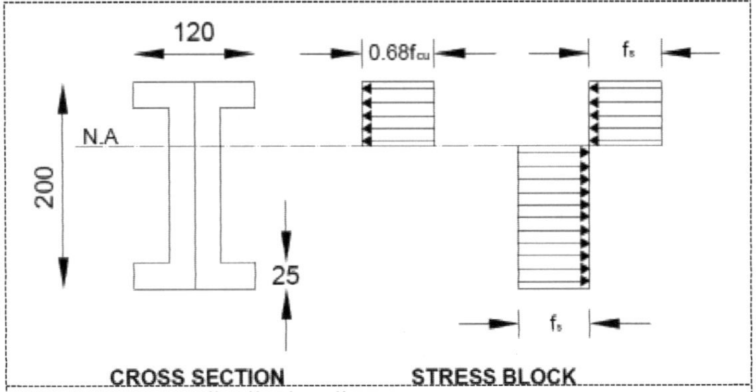

Figure 4. Stress distribution diagram.

In the second method, the ultimate resisting moment, M_{UR2} is determined using the Eqs. (5-7)

$$M_{UR2} = M_{fu} + M_{su} \tag{5}$$

$$M_{fu} = f_{cb}{}'I_g/y_t \tag{6}$$

$$M_{su} = \min\{A_{sc}\sigma_s(d-t), A_s\sigma_s(d-t)\} \tag{7}$$

where,

A_{sc} = Area of steel in compression

A_{st} = Area of steel in tension

σ_s = Yield strength of steel reinforcement bar

d = Effective depth of cross section

t = Thickness of the section

Table 6. Ultimate Loads by Method I & Method II

Specimen Designation	Experimental Ultimate Load (kN)	Computed First Crack Load (kN)		Ratio of Computed Load to Experimental Load	
		Method I	Method II	Method I	Method II
B1	27.47	28.96	24.31	1.05	0.88
B2	23.54	23.24	23.98	0.98	1.01
B3	21.60	22.87	23.95	1.06	1.10
		Average		1.03	0.99
		Coefficient of Variation		0.06	0.10

The ultimate moment as well as loads of all the three specimens computed using the second method is shown in Table 6. The results are compared with the experimental values. From both the tables it can be observed that, both the methods agreed satisfactorily with the test results.

4. Conclusions

All the three joists exhibited high ductility and failed in flexure and following conclusions are arrived at:

- The beam reinforced with square welded meshes was observed to have the maximum load carrying capacity among the three.
- Maximum ductility was exhibited by the beam reinforced with the combination of hexagonal wire mesh and welded square mesh, even though the specimen had the least load carrying capacity.
- The first crack load as well as ultimate moment capacity of the section was calculated and compared with the experiment values.
- The ultimate moment of resistance was computed using two different methods; it was found that the values predicted by Method I showed a smaller coefficient of variationwhen compared to that of Method II, hence Method I provides value closer to the experimental value.
- Computation using Method II is relatively easier than Method I, hence it can be employed for rough checks.

References

[1] Collen L D G 1960 Some Experiments in Design and Construction with Ferrocement. *Bulletin The Inst. Civ. Eng. Ireland*, 39-56.
[2] Balaguru P 1981 Prediction of Crack Widths in Ferrocement Beams. *J. Ferrocement* **11**(3) 203-14.
[3] Logan D and Shah S P 1973 Moment Capacity and Cracking Behaviour of Ferrocement in Flexure. *J. Am. Concr. Inst.* ***70***(12) 799-804.
[4] Desayi P, Viswanatha C S and Kannapan S 1982 Some Studies on Ferrocement Roofing Elements. *J. Ferrocement* **12**(3) 273-88.
[5] Rangan B V 2010 Fly Ash-Based Geopolymer Concrete. *Proc. Int. Workshop on Geopolymer Cem. Concr.* Annamalai University, Tamil Nadu, India.
[6] ACI-549R 2011 Thin Reinforced Cementitious Products & Ferrocement. *Am. Concr. Inst.*
[7] Paul B K and Ricardo P Pama 1978 Ferrocement. *IFIC, Asian Inst. Technol.*
[8] Ganesan N and Suresh Kumar S 1994 Prediction of Moment Capacity of Fibrous Ferrocement Flex- ural Members. *Proc. 2nd Pacific Symposium on Ferrocement*, Roorkee, India.
[9] Kaushik S K, Trikha D N and Kotdawala R R 1982 Ultimate Strength Behaviour of Ferrocement Beams. *J. Ferrocement* **12**(1) 1-12.

[10] Skudra A A 1994 Elastic Characteristics of Ferrocement Reinforced with Woven Meshes. *Mech. Compos. Mater.* **30**(4).

Effects of slag on flexural strength of slurry infiltrated fibrous concrete

D Elavarasi*, K Saravana Raja Mohan, P Parthasarathy and T Dinesh

School of Civil Engineering, SASTRA University, Thanjavur-613401. India.

*Email: elavarasi@civil.sastra.edu

Abstract. Slurry infiltrated fibrous concrete is one of the new advanced concrete composite which differs from method of fabrication and composition of the matrix. Extensive research is being carried out on alternative binders or supplements to cement aiming to reduce environmental impact. However, little has been published to investigate the structural behaviour of SIFCON incorporating with mineral admixtures, particularly as regards its ultra-high ductility, which may alter the mode of failure from brittle to the more desirable ductile. An experimental study was carried out to investigate the flexural behavior of SIFCON containing three different percentage of fibre content 6%,8% &10% with incorporation of optimum dosage of blast furnace slag replaced by cement. Strength characteristics such as Compressive strength and splitting tensile strength test were carry out for SIFCON incorporating 10% of fibre content and different percentage of slag (0, 15, 30, 45, 60%&75%) to optimize the replacement level. The test results found that the maximum strengths were attained at 30% of blast furnace slag replaced by cement. To study the flexural behavior of SIFCON beam of size 1.2x0.1x0.2m containing different percentages (6%, 8% & 10%) fibre content incorporating with and without the optimum percentage of slag were cast and tested .Both flexural strength and Load displacement characteristics of the specimens were studied under flexure. The outcomespresentedfromtest resultshave been compared. The test results reveals that the flexural strength, toughness, ductility and stiffness characteristics were significantly improved due to incorporation of optimum dosage of slag enhancing when compared to without mineral admixtures also compared to conventional concrete(RCC). Major conclusions were drawn from the investigations which are presented.
Keywords: SIFCON, GGBFS, Flexural strength, Ductility, Stiffness, Toughness.

1. Introduction

Slurry-infiltrated fibrous concrete (SIFCON) is a high strength composite material originated by Lankard. It possess superior mechanical properties and toughness characteristics. SIFCON can be considered as steel fibre reinforced concrete (SFRC) with high volume of fibre content which differs from the conventional SFRC in terms of volume of fibre content and method of fabrication. The matrix consists of cement mortar or slurry without containing course aggregates. In most cases, SIFCON is prepared by infiltrating the cement slurry into a pre-packed fibre bed. Even though SIFCON is a new construction material, it has found many applications in the areas of rehabilitation of pavements, prefabricated products, coatings, bridge decks and safety vaults, seismic and explosive-resistant structures, military applications such as anti-missile hangers, under-ground shelters ,sea-protective works, shielding of nuclear containment, platforms for aerospace repair and rehabilitation of air-field, solar tower, mega structures like long span structures and offshore structures etc.

It is already established by many researchers concluded that the SIFCON composites exhibit excellent performance in static loading such as shear strength and flexural strength compared to plain cement concrete(PCC), reinforced cement concrete(RCC) and fibre reinforced concrete(FRC) . Also the structures subjected to impact and dynamic loading gaining excellent ductility and energy absorption characteristics. The strength of the composite depends upon four main design factors such as slurry strength, fiber volume, fiber alignment, and fibre type and aspect ratio. The fiber content by volume fraction is based on the fiber type, aspect ratio and the vibration effort needed for proper compaction. [1] revealed the basic mechanical properties of SIFCON like load-deflection response, ultimate compression and flexural strengths, impact and abrasion resistance.[2,3] studied the effect of incorporation of high volume class C fly ash on mechanical strength of SIFCON and conclusions have been made that the flexural strength and toughness characteristics were improved with higher amount of mineral admixtures and increased percentage of fibre content by volume fraction.[4] investigated the behaviour of slurry infiltrated fibrous concrete slabs subjected to impact loading and reported that the energy absorption capacity of SIFCON exhibits excellent performance compared to fibre reinforced concrete(FRC), reinforced cement concrete (RCC) and plain cement concrete (PCC). The test result revealed that SIFCON slabs with 12 % fiber content performed well in strength and toughness characteristics compared to all other SIFCON specimens. Energy absorption capacity of SIFCON slabs were increases with increasing fiber volume and also from visual observation the level of damageis less when compared with SIFCON slabs without conventional reinforcement. [5] investigated the performance of restrained SIFCON two way slab was subjected to flexure and found that SIFCON slab containing 12% of fibre content yielding higher ultimate load with lesser deflection and became reported that the increased percentage of fibre content increases the stiffness of SIFCON slabs. And also observed the first crack load was higher than all other specimens. [6] investigated the simply supported two way SIFCON slab containing different percentage of fibre content(8%,10% &12%) under flexure and reveals that the SIFCON slabs exhibit excellent toughness characteristics and spacing of cracks have been minimized with increased volume of fibre content in SIFCON compared to Fibre reinforced concrete and plain cement concrete slabs. [7] studied the relative behaviour of SIFCON subjected to abrasion,flexure, and impact loads and also studied the toughness characteristics.The test results reveals that the SIFCON specimen exhibit superior ductility behavior and greater spalling and crack resistance. [8] the pull-out behavior was studied and indicated that increasing pull out strength depends on the fiber type, length of fibre, condition of curing, geometry of fibre and matrix strength and also increasing diameter increases the matrix-fibre bond. It was concluded that the pullout toughness increased due to increasing length of fibre and also observed that the hooked end fibre shows better bond strength when compared with straight smooth fibre.[9] investigated experimentally the triaxial compressive behavior of SIFCON with 10% of fibre content increases the higher tri-axial compressive strength and poisons ratio with increasing confining pressure and fibre content compared to high strength concrete (HSC), HPFRC. [10] investigated experimentally the shearing properties of SIFCON specimens with two different diameters and aspect ratio of fibres using cored cylindrical specimens subjected to pure torsion and concluded that the SIFCON beam shows more ductile , stable, post yielding cyclic behavior compared to reinforced cement concrete(RCC. [11] investigated the durability characteristics of ground granulated blast furnace slag (GGBFS) concretes and reinforced concrete beams were affected by corrosion under various loading conditions. The test results found that the slag was partial replacement by cement has the effects of arresting on both permeability and total charge-pass and also significant effects on corrosion rate and flexural rigidity of RCC beams. [12,13] examined the durability characteristics of alkali-activated slag(AAS) concrete subjected to acid attack and revealed that both the strength and durability characteristics of AAS shows better performance when compared to plain cement concrete using the same grade of ordinary Portland cement(OPC). [14] investigated the reported the efficiency of GGBFS in concrete observed that increasing the strength upto 80% replaced by cement for getting required design strength.[15] investigated on optimum percentage of ground granulated blast furnace slag (GGBFS) on mechanical property of concrete and the test results concluded that the presence of GGBFS in the matrix acting as filler material enhancing the strength characteristics were improved upto 55% subsequently causes reducing strength. [16] investigated experimentally studied the stress – strain characteristics SIFCON composites in tension and compression and concluded that the elastic modulus of SIFCON increases due to the fibre volume , fibre alignment and orientation, aspect ratio and method of testing etc. [17] found that the incorporation of the mineral admixtures decreases permeability of chloride ion, thus improved durability of concretes and also found that the incorporation of ground granulated blast furnace slag(GGBFS) gives better durability characteristics when compared to flyash and silica fume. [18] investigated experimentally the effect of slag on strength and durability properties of concrete and reported that the concrete containing slag exhibits lower early strength and higher later strength were attained when compared to conventional concrete.

However, literature review reveals that a very little work has been carried out on SIFCON to determine the durability characteristics. Hence there is a need to conduct experiments to study the durability characteristics of SIFCON. Further more, the utilization of higher amount of cement in slurry not only influenced the cost of production, but also has an unfavorable effects on releasing enormous amount of heat causes shrinkage problems. Substitution of ground granulated blast furnace slag by cement suggest to solve these complications.

2. Experimental Investigations

An experimental program was designed to produce a slurry infiltrated fibrous concrete made with binary cementitious blend of blast furnace slag. For optimizing the slag was replaced different percentage (0%,15%,30% ,45,60% &75%) with constant a fibre content (10%) by volume fraction on compressive strength and split tensile strength. An optimum dosage of slag is to be found experimental results.

To study the ductility, ultimate load carrying and stiffness characteristics of beams are to be evaluated on two point bending test using loading frame testing setup. It comprises of casting of six reinforced SIFCON beams (1200 x 100 x 200 mm) with varying fibre content (6%, 8% &10%) with and without incorporating optimum percentage of slag and one conventional concrete(RCC) for comparison, The materials used, mix proportions and the test methods are illustrated in the following headings.

2.1 Materials used

Grade 53 ordinary Portland cement was used to IS: 12269. The specific gravity of cement was found to be 3.15. The binary cementitious blend of slag conforming to ASTM C 989 was partially replaced by cement for improving strength characteristics and its specific gravity was found to be 2.91was used.. Locally available river sand of particle size less than 4.75mm having a specific gravity of 2.71 and fineness modulus of 2.43% was used for preparation of slurry. Hooked end steel fiber having a circular cross section of 1mm diameter, tensile strength of1056MPa and aspect ratio 30 was used. Fibers were randomly oriented in all directions. For casting and curing of the specimens a fresh portable water which is available from local source. In addition to water a high range water reducing admixture (Superplasticizer CONPLAST 430) was added to improve the workability and flow ability of slurry into the fibre bed.

Table 1. Chemical constituents of Cement and Slag

Chemical Constituents	Cement (%)	Slag (%)
Silica (SiO_2)	21.8	39.18
Alumina (Al_2O_3)	6.6	10.18
Ferric oxide (Fe_2O_3)	4.1	2.02
Calcium oxide (CaO)	60.1	32.82
Magnesium oxide (MgO)	2.1	8.52
Sodium oxide (Na_2O)	0.4	1.14
Potassium oxide (K_2O)	0.4	0.30
Sulphuric anhydride (SO_3)	2.2	-
Loss on Ignition (LOI)	2.4	1.0
Specific gravity	3.1	2.56

2.2 Mix proportions

The slurry mix consisted of partial replacement of cement with different percentage of slag (15,30, 45%, 60%&75%) for getting an optimum dosage of slag in the matrix and rich mix proportion of 1:1 with water binder ratio of 0.4 were maintained for fabrication of the SIFCON specimens containing a constant 10% of fibre content was adopted . During mixing of slurry a high range water reducing admixture (Superplasticizer) was added for improving more workable and easy infiltration capacity into the packed of fibre. The matrix was compacted well manually to verify thecomplete penetration of the slurry into the pre-packed fibre bed.

To study the flexural behavior of reinforced SIFCON beam with varying fibre content (6%, 8% &10%) and incorporation of an optimum dosage of slag (30%) for enhancing the ductility, stiffness and toughness characteristics. Reinforced cement concrete specimen of mix proportion 1:1:2.8 and water cement ratio of 0.39 was adopted for comparison. The mixratio of different mixes were presented in Table 2.

Table 2. Mix ratio of SIFCON specimens

S. No.	Nomenclature of mix	Cement content (%)	Slag (%)	Fine Aggregate (%)	Fibre content (%)	W/B ratio	SP (%)
1.	RCC	Mix proportion 1:1:2.8			-	0.39	-
2.	SIF6(S0)	100	-	100	6	0.40	2
3.	SIF6(S30)	70	30	100	6	0.40	2
4.	SIF8(S0)	100	-	100	8	0.40	2
5.	SIF8(S30)	70	30	100	8	0.40	2
6.	SIF10(S0)	100	-	100	10	0.40	2
7.	SIF10(S30)	70	30	100	10	0.40	2

2.3 Casting of specimens

For investigating the flexural behavior of SIFCON incorporating with an optimum dosage of blast furnace slagcontaining different percentage of fibrecontent(6%,8%&10%)were fabricated by adopting three layer techniques. Hand compaction have been done to assure the slurry is infiltrated into full depth of packed bed of fibres. After casting the specimens were demouldedafter 24 hours and cured in water for 28 days. The reinforcement details were shown in figure 1.

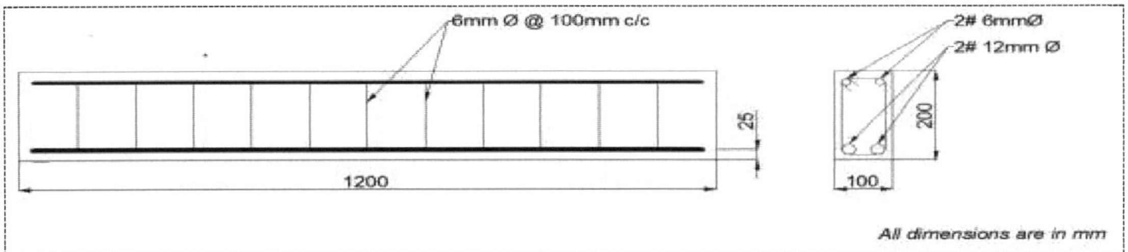

Figure 1. Reinforcement details.

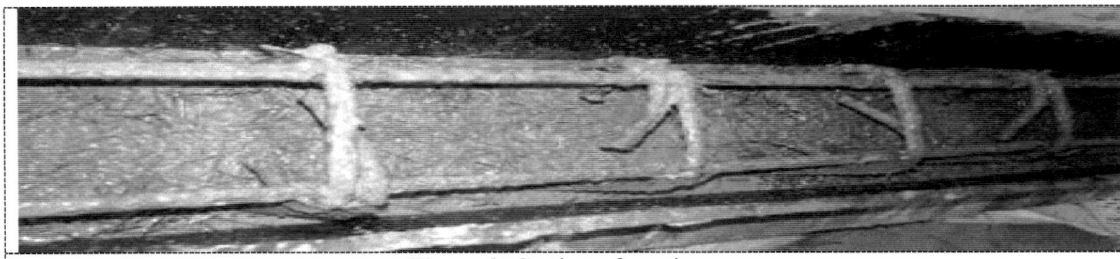

Figure 2. Casting of specimen.

2.4 Testing of specimens
The flexural performance of SIFCON containing three different percentage of fibre content with and without an optimum percentage slag were conducted on a computerized loading frame test setup as shown in figure 3. The beam specimens were erected and static load was applied gradually increasing the rate of loading till the specimens were collapsed. Mid span deflections at the initiation of first crack, yield point and ultimate stage were measured by a dial gauge with an accuracy of 0.01mm was used.To obtain an optimum dosage of blast furnace slag in SIFCON was also carried out on compression and spit tensile strength by using compression testing capacity of3000kN at constant rate of loading.

3. RESULTS AND DISCUSSIONS
3.1 Compressive and split tensile strength
The compressive and splitting tensile strength of SIFCON specimens containing 10% of fibre content were increased with increasing the replacement level of slag 30% and 45% respectively. One drawback of using slag solidification of the concrete is delayed causes strengths were attained on later stage due to incorporation of slag in SIFCON specimens.The test results were given in Table 3.

Table 3. Strength characteristics of different mixes

S. No.	Mix type	Compressive strength(N/mm^2)	Split tensile strength (N/mm^2)
1.	SIF10(CS)	65.50	11.03
2.	SIF10(S15)	65.85	12.67
3.	SIF10(S30)	66.97	13.08
4.	SIF10(S45)	64.85	13.78
5.	SIF10(S60)	61.97	12.06
6.	SIF10(S75)	59.73	11.98

3.2 Flexural strength

The flexural behaviour of SIFCON specimens containing differentfibre content of 6%,8% and 10% were tested and shown in Fig.4. The test results reveals that the flexural performance was gradually improved with increasing fibre content (SundarsanaRao et al. 2008) in the matrix with and without slag. Due to the incorporation of slag the strength was increased by 12.05%, 13.37% & 11.63% in SIFCON containing 6%,8% and 10% respectively when compared to without slag in SIFCON.

3.3 Ultimate load

The ultimate load carrying capacity and first crack loads were presented in table 4 observed that load carrying capacity was improved due to incorporation of blast furnace slag and increasing percentage of fibre content and also the displacement of slag based SIFCON specimens were decreased with increasing load. The ultimate lad carrying capacity increases also depend on the fibre alignment and orientation, mthod of production (Homrich et.al, 1987).

Table 4. Mid deflection at first crack load and at ultimate load

Mix type	First crack		Yield		Ultimate	
	Load (kN)	Deflection (mm)	Load (kN)	Deflection (mm)	Load (kN)	Deflection (mm)
RCC(CS)	29.6	3.4	51.7	4.75	73.8	6.25
SIF6(0)	52.3	3.9	85.5	6	127.6	12.5
SIF6(S30)	57.2	1.5	96.1	2.94	142.8	7.83
SIF8(0)	54.95	4	87.65	6.5	129.25	15.5
SIF8(S30)	61.1	2.1	101.8	3.24	146.5	10.1
SIF10(0)	56	7.6	90.4	7.3	133.2	16.16
SIF10(S30)	64	1.2	102.5	1.86	148.9	6.08

3.4 Stiffness

The stiffnes behaviour of differement mixes were shown in fig.5.The ultimate stiffness of slag based SIFCON mixes were increased by 78.7%,74.75% and 197% in 6%, 8% and 10% of fibre content.Whereas in SIFCON without mineral admixtures , it was decreased with the addition of fibre content. The ultimate stiffnes was increased due to increasing load carrying capacity without increasing displacement.(Gnaneswar et al. 2009)

3.5 Ductility factor

Figure 6 shows the ductility factor of various mixes. The SIFCON specimens were more ductile than RCC. Among the SIFCON without blast furnace slag shows the ductility factors were improved with increasing fibre content by volume fraction. It was observed that the incorporation of slag in SIFCON the ductility were increased upto 8% , beyond it was decreased because the load carrying capacity was increased without increasing displacement.

3.6 Toughness

The toughness characteristics found from the experimental test results were shown in figure7. It was calculated the area under load deflection curve. When compared to RCC specimen SIFCON mixes were more toughness value. Among the SIFCON beam without incorporation of slag the toughness was gradually increased with increased percentage of fibre content. Due to incorporation of slag enhancing the stiffness characteristics were improved with decreasing toughness.

3.7 Failure pattern

The first crack loads and failure loads were observed for all specimens were presented in table 3 and failure pattern was shown in figure 3(b). The failure pattern was observed that all the specimens were getting flexure failure and the cracks spacing have been decreased with addition of fibre content and also the deformation was quite lessin SIFCON incorporated with slag when compared to without slag, it became reported that the stiffness was increased due to slag which is present in the mix and also the growth of cracks were not extended into the entire depth of the beam.

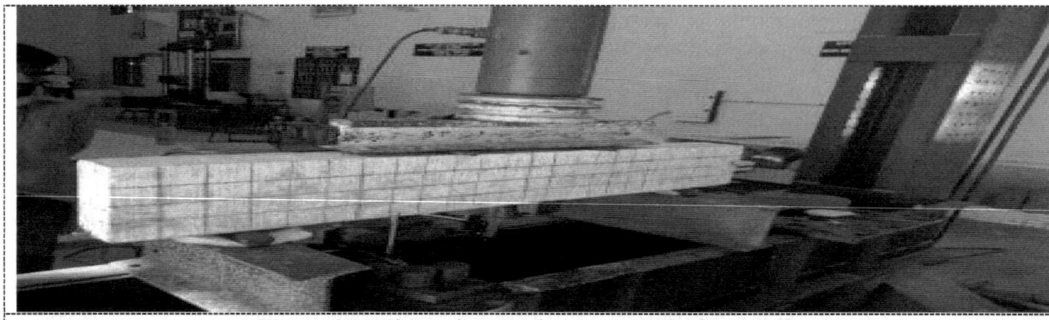

Figure 3(a) Flexural strength test setup (Before loading)

Figure 3(b) Failure pattern(After loading)

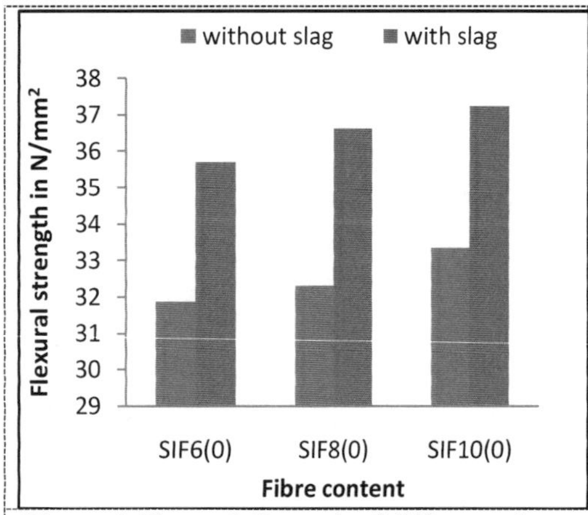

Figure 4. Flexural Strength of different mixes

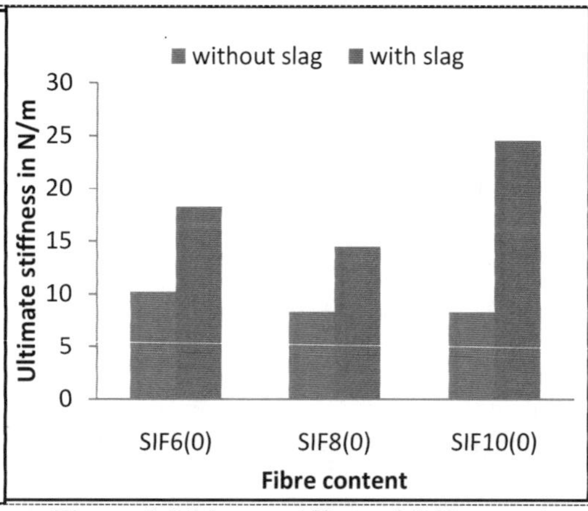

Figure 5. Ultimate stiffnessof different mixes

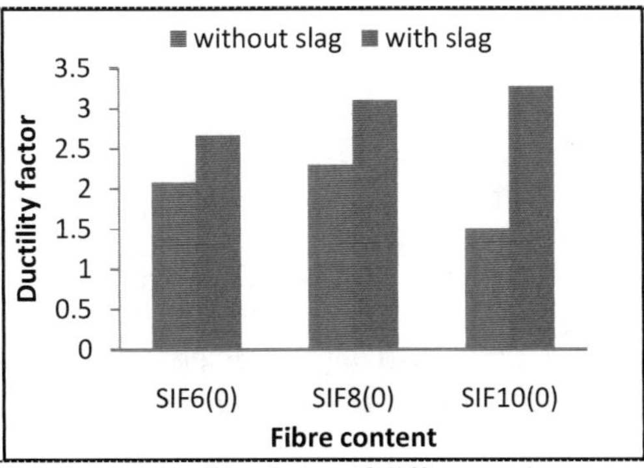

Figure 6. Ductility factorof different mixes

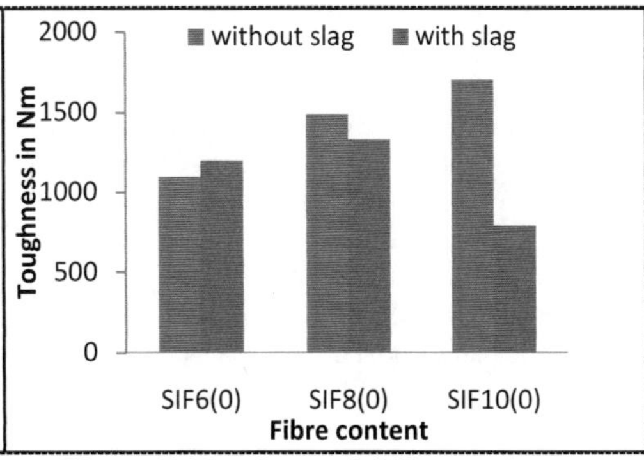

Figure 7. Toughnessof different mixes

4. Conclusions

Based on the investigations have been done experimentally the following inferences were drawn from test results:

- It is observed that the utilisation of ground granulated blast furnace slag(GGBFS)is well accepted as an alternative source for cement thereby reducing the consumption of cement which will cause saving energy, construction cost and also protect our environment from pollution.
- The impact of ground granulated blast furnace slag (GGBFS) increases the mechanical properties such compression and splitting tension were attained at 30% compared to other replacement level.
- The flexural strength and ultimate load carrying capacity of slag based SIFCON specimens were increased by 12.05%,13.37% &11.63% in SIFCON containing 6, 8 and 10% of fibre content respectively compared with SIFCON specimens without incorporation of slag.
- The load carrying capacity and ultimate stiffness of SIFCON specimens were shows better performance in SIFCON containing higher volume of fibre content (10%) with incorporation of slag when compared toSIFCON without slag.
- The toughness characteristics were also improved due to increasing fibre content in SIFCON without blast furnace slag, whereas in slag based SIFCON containing 10% was decreased the due to arresting the displacement.
- The ductility performance was improved with increased amount of fibre content containing an optimum dosage slag in SIFCON compared to without inclusion of slag .The incorporation of slag enhancing the ductility behaviour.
- It was concluded that the replacement of slag by cement shows better ultimate load carrying capacity and excellent stiffness characteristics. The addition of slag in SIFCON improving the flexural properties and also modifying the microstructure of concrete by means of reducing pores which is present in the composites arrested the deformation and also reduce its permeability thereby reducing the intrusion of water.

References

[1] Lankard D R 1984 Properties of Slurry infiltrated fibrous concrete (SIFCON). *Int. J. Appl. Eng. Technol.* **29** 44–7.

[2] Halityazıcı, serdaraydın, hüseyinyiğiter, meryardımcı and girayalptuna 2010 Improvement on SIFCON performance by fiber orientation and high-volume mineral admixtures. *J. Mater. Civ. Eng.* **49** 1093-101.

[3] HalitYazici, HuseyinYigiter, SerdarAydin and BulentBaradan 2006 Autoclaved SIFCON with high volume Class C fly ash binder phase. *Cem. Concr. Res.* **36** 481–6.

[4] Sudarsana Rao H, Vaishali G, Ghorpade, Ramana and Gnaneswar K 2010 Response of SIFCON two-way slabs under impact loading. *Int. J. Impact Eng.* **37** 452–8.

[5] Sudarsana Rao H, Ramana N V and Gnaneswar K 2009 Behaviour of restrained SIFCON two way slabs. *Asian J.Civ. Eng* **10**(4) 481-94.

[6] Sudarsana Rao H, Ramana N V and Gnaneswar K 2008 Behaviour of simply supported steel reinforced SIFCON two way slabs in punching shear. *Indian J. Eng. Mater. Sci.* **15** 326-33.

[7]. Parameswaran V S, Krishnamoorthy T S and Balasubramanian K 1990 Behaviour of high volume fiber cement mortar in flexure. *Cem. Concr. Compos.* **12(4)** 293-301.

[8] Murat Tuyan and HalitYazıcı 2012 Pull-out behaviour of single steel fiber from SIFCON matrix. *Constru. Build. Mater.* **35** 571-7.

[9] Farnam Y, Moosavi M, Shekarchi M, Babanajad S K and Bagherzadeh A 2010 Behaviour of slurry infiltrated fibre concrete under triaxial compression. *Cem. Concr. Res.* **40** 1571-81.

[10] Wang M L and Maji A K 1994 Shear properties of slurry-infiltrated fibrous concrete. *Constr. Build. Mater.* **8**(3) 161-8.

[11] An Cheng, Ran Huang, Jiann-Kuo Wu and Cheng-Hsin Chen 2005 Influence of GGBS on durability and corrosion behaviour of reinforced concrete. *Mater. Chem. Physic.* **93** 404-11.

[12] Bakharev T, Sanjayan J G and Cheng Y B 2002 Sulfate attack on alkali-activated slag concrete. *Cem. Concr. Res.* **32** 211-6.

[13] Bakharev T, Sanjayan J G and Cheng Y B 2003 Resistance of alkali-activated slag concrete to acid attack. *Cem. Concr. Res.* **33** 1607-11.

[14] Ganesh Babu K, Sree Rama Kumar V 2000 Efficiency of GGBS in concrete. *Cem. Concr. Res.* **30** 1031-6.

[15] Oner A and Akyuz S 2007 An experimental study on optimum usage of GGBS for the compressive strength of concrete. *Cem. Concr. Compos.* **29** 505-14.

[16] Homrich J R and Namman A E 1987 Stress-strain properties of SIFCON in compression, *Am. Concr. Inst.* **105** 283-304.

[17] Mehmet Gesogˇlu, ErhanGüneyisi, ErdogˇanÖzbay 2009 Properties of self-compacting concretes made with binary, ternary, and quaternary cementitious blends of fly ash, blast furnace slag, and silica fume. *Constr. Build. Mater.* **23** 1847-54.

[18] Wang Qiang, yan Peiyu, Yang ianwei and Zhang B O 2013 Influence of steel slag on mechanical properties and durability of concrete. *Constr. Build. Mater.***47** 1414-20.

IOP Conf. Series: Earth and Environmental Science **80** (2017) 012037 doi:10.1088/1755-1315/80/1/012037

A Study on Hybrid Precast walls

Rohith S, Chidambaram C R, Karthikeyan K* and Helen Santhi M

School of Mechanical and Building Sciences, VIT University, Chennai. India.

*Email:karthikeyan.kothanda@vit.ac.in

Abstract. This study proposes an innovative wall system so called as hybrid precast walls (HW). HW uses both mild steel and high strength steel angles for flexural resistance across horizontal joint. High strength steel plates are used for steel angles connection at the horizontal precast joint and also at places where angles cross which assist for the re-centering property of the HW walls. HW is designed with mild steel to yield both tension and compression whereas the steel angles provide re-centering capability and thereby reducing lateral displacements. Efficiency of the structure depends upon the performance of both reinforcements accordingly to what it is meant to be. This paper presents the seismic performance studies of HW under gradually applied lateral static load. Experimental studies were carried out which account for steel yielding, concrete cracking and crushing.
Key words: *Hybrid walls, precast walls, Experimental study and Static lateral load.*

1. Introduction

The main challenge faced by Indian construction industry is maintaining quality as well as speed of building constructions. For tackling this major challenge hybrid precast walls can be used with wide range of applicability from moderate to high seismic regions. The research works in abroad included both monolithic and hybrid precast wall which gave an overview of drafting, design, construction and codes to be followed [1, 2]. Special design and detailing of hybrid walls are required for the seismic regions [3, 4]. Perforated walls have been also used to study the behavior under lateral loads [5, 6, 7].

2. Experimental Study

2.1 Material Properties

Ordinary Portland Cement (OPC) of grade 53, fine aggregate conforming to zone II and coarse aggregate were used in the study. The material properties are shown in Table 1.

2.2Wall Specimens

Three hybrid precast wall models and one monolithic model was used in the study. All the specimens were of size 400 mm wide, 80 mm thick and 950 mm tall. These walls were casted on a footing slab of size 800 mm x 500 mm x 150 mm. The wall was provided with double steel mesh of #8 mm at 120 mm spacing, one mesh on each face of the wall. The footing slab was provided with double mesh of #12 mm at 150 mm c/c, one mesh on each face of the slab. The bottom end of vertical reinforcement of the walls was extended into footing slab for a length equal to the development length of the bar. The connection between the wall and footing slab was monolithic. M30 grade concrete and HYSD steel bars were used for the study. The precast specimens were made by introducing a joint 300 mm above the footing slab. The bottom part was cast monolithically with the footing slab. The top part was cast separately. The top part was connected to the bottom part by means of bolting and then covered by grouting. The monolithic wall (MW) is shown in Figure 1.

Table 1. Material Properties

Property	**Values**
Specific gravity of fine aggregate	2.6
Specific gravity of coarse aggregate	2.6
Specific gravity of cement	3.15
Bulk density of fine aggregate	1467 kg/m^3
Bulk density of coarse aggregate	1492 kg/m3
Bulking of sand	8%

The precast specimens had structural steel angle embedment arranged in different patterns. The angles were jointed near the horizontal joints of the walls using gusset plates to ensure safe and secure connection of the top part to the bottom part. The angles 25 x 25 x 3 were of grade E250 (Fe410) and the M10 bolts were of grade 8.8. The gusset plate of thickness 3 mm was used and it was of grade E250 (Fe410). The back to back angles placed vertically near the ends of the walls were used in hybrid wall 1 (HW1)as shown in Figure 2. In case of hybrid wall 2 (HW2), single angles were used and they were placed diagonally as shown in Figure 3. In case of hybrid wall 3 (HW3), single angles were used and they were placed diagonally and also vertically near the ends of the walls as shown in Figure 4.

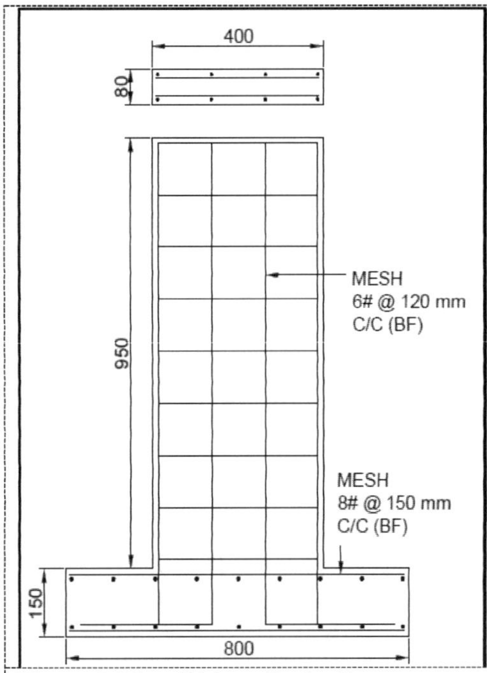

Figure 1. Monolithic wall - details

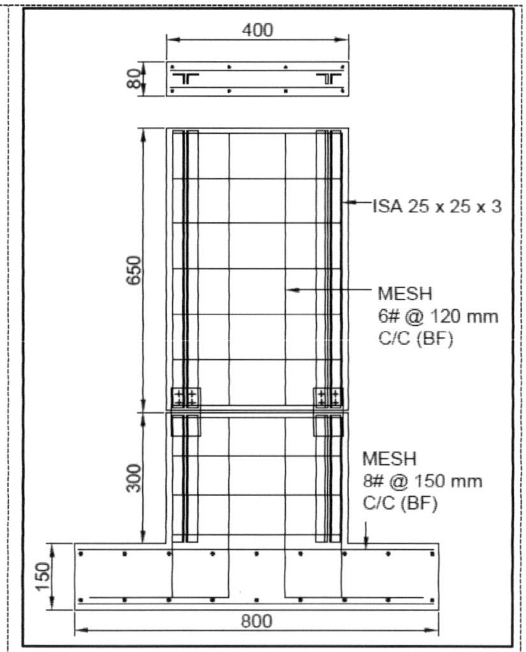

Figure 2. Hybrid wall 1 - details

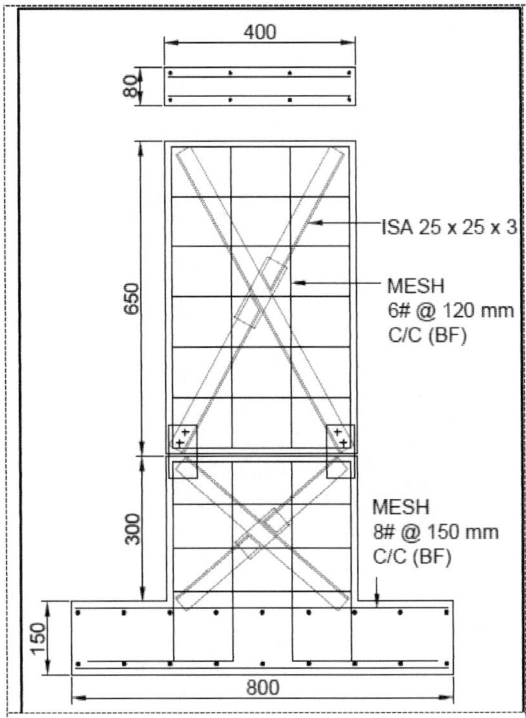

Figure 3. Hybrid wall 2 – details

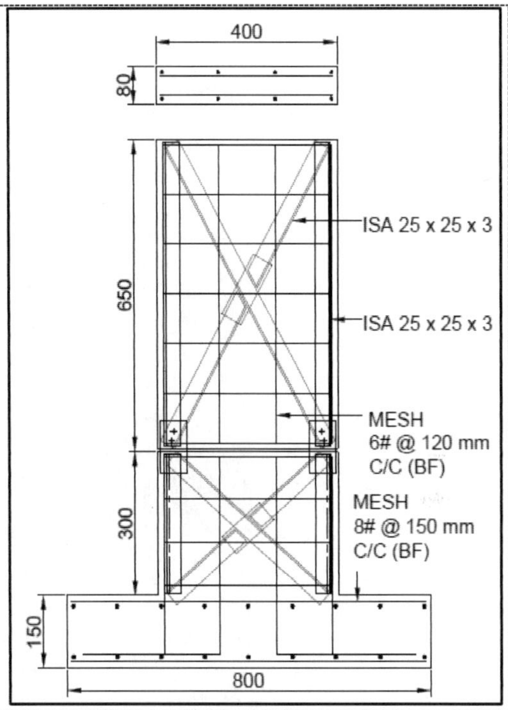

Figure 4. Hybrid wall 3 - details

2.3 Testing

For testing a static lateral load was applied to the wall specimen using a hydraulic jack-collar setup which was connected to a pumping jack that has the load reversal technique for applying load in both the lateral direction. LVDT fixed to the collar measures the lateral displacement of the wall specimen. The data was used to plot the load v/s displacement curve for all the wall specimens. The loads were increased till the failure of the structural components and the crack of the specimen was viewed for analyzing the performance of the wall under loading similar to the seismic loading. Monolithic wall was also tested in a similar way and the test results were compared with the precast walls for performance comparison. The test setup is shown in Figure 5.

Figure 5. Test setup

IOP Conf. Series: Earth and Environmental Science **80** (2017) 012037 doi:10.1088/1755-1315/80/1/012037

3. Results and discussion

3.1Crack Pattern

The cracks were found the junction of wall and footing slab in monolithic specimen. No cracks were found on the wall. The behavior was a typical cantilever action. All the precast walls had developed cracks at the joint of the wall panels. The grouted panel joints cracked. Inclined hair line cracks were also observed near the ends of the wall panels indicating crushing of toe concrete at those locations. The thickness of crack width was significant in all the precast walls. The cracking pattern is shown in Figure6 - 9.

Figure 6. Crack Pattern – MW

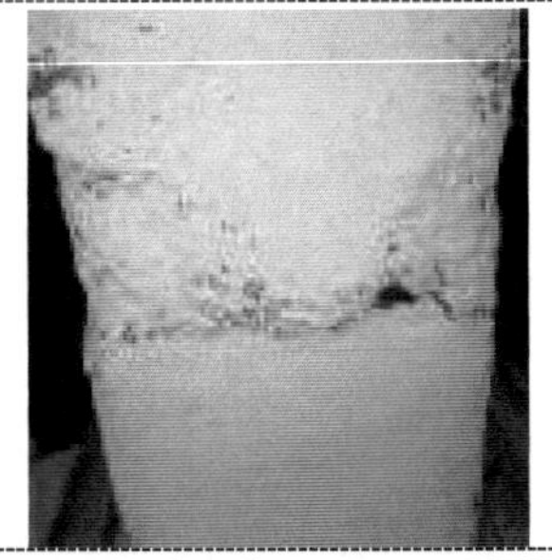

Figure 7. Crack Pattern – HW1

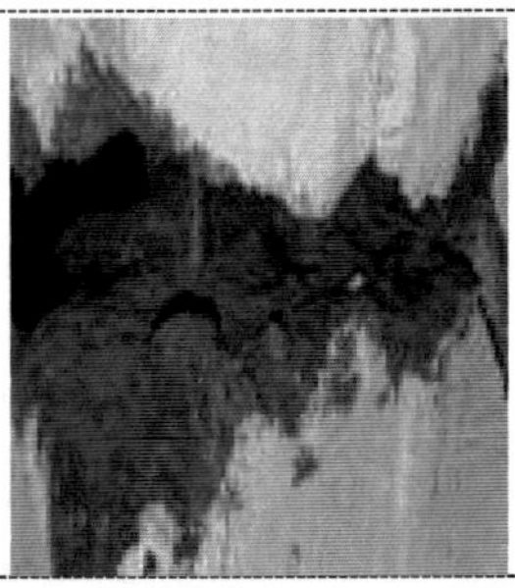

Figure 8. Crack Pattern – HW2

Figure 9. Crack Pattern – HW3

3.2 Load vs. displacement

The load vs. displacement cracks are shown in Figure 10 – 13. From the above graphs it is evident that all the walls follow the same pattern. The graphs pattern is like it follows a linear pattern till the failure load and takes a deviation downwards from the cracking load point. The lateral displacement and the corresponding failure load were being noted for all the wall specimens at the peak point. For monolithic wall the data set is (3.4mm,77.1KN), for HW1 it is (8mm,52.9KN), for HW2 it is (35.2mm,79.7KN) and for HW3 the data set is (26mm,115.7KN). Monolithic wall behaved essentially as a rigid body showing minimum lateral deflection and a medium width crack generation along the wall foundation joint which is a good result as the seismic performance point of view. HW1 wall showed some premature cracking at a lower level of failure load which was the result of poor placement of the structural members and the lateral displacement was little higher compared to the failure load. HW3 and HW2 performed well than the monolithic wall in terms of the seismic performance point of view. HW3 and HW2 walls failed at higher loads than the monolithic wall even though the corresponding lateral displacements were higher. This shows that the truss action and be-truss action placement of structural members is an excellent way for providing good seismic performance for the walls which is to be used in seismic regions. All hybrid precast walls showed high lateral displacement exhibiting ductility.

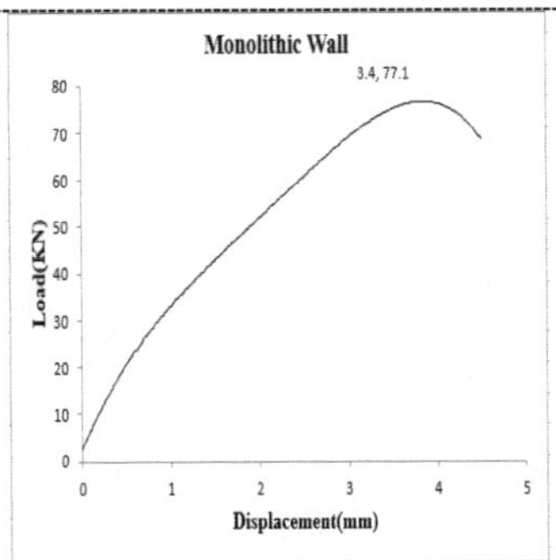

Figure 10. Load vs. deflection – MW

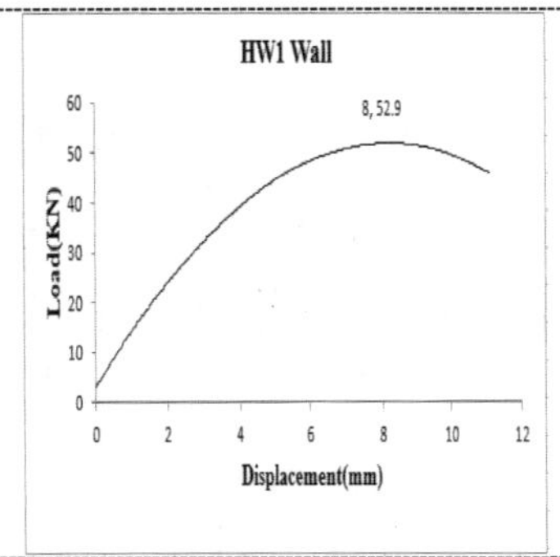

Figure 11. Load vs. deflection – HW1

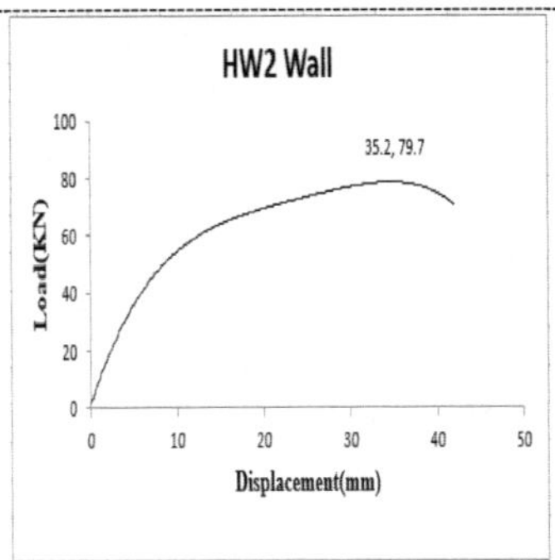

Figure 12. Load vs. deflection – HW2

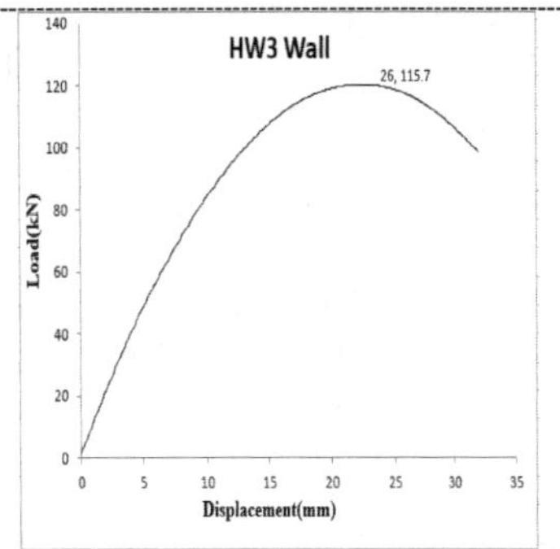

Figure 13. Load vs. deflection – HW3

4. Conclusion

The following conclusion can be made from the above study,

- All hybrid wall specimens showed less crack width than the monolithic wall which shows that the re-centering capability of the high strength angles gives an advantage for the hybrid walls over the monolithic wall in case of the cyclic lateral load applied
- The cracking load of all the hybrid walls HW2 and HW3 were more than the monolithic wall which again shows that the high strength angles along with the reinforcement mesh wrapping can sustain higher seismic load than the monolithic wall
- HW1 exhibited premature cracking at lower loads than the monolithic wall due to poor structural member placements
- Structural member placement influence the seismic performance of the hybrid walls to a larger extend
- All the test results show that the hybrid walls with correct structural member placement can be high end alternative for the normal monolithic wall in seismic regions having more chances of seismic load acting on the structures
- As the hybrid walls performs well than monolithic wall in every aspect the structural codes of all countries should include design guidelines and specifications for usage of such kind of walls in seismic region during the next revisal stage

Acknowledgement
The authors acknowledge VIT University, Chennai, Tamil Nadu,India for providing the facilities to carry out the research work and the encouragement in completing this work.

References
[1] Smith B J, Kurama Y Cand McGinnis M J 2010 Design and Measured Behavior of a Hybrid Precast Concrete Wall Specimen for Seismic Regions *J. Struct. Eng.* **137**(10) 1052-62.
[2] Smith B J, Kurama Y Cand McGinnis M J 2012 Comparison of solid and perforated hybrid precast concrete shear walls for seismic regions.*Struct. Cong.* 1529-40.
[3] Smith B J, Kurama Y Cand McGinnis M J 2013 Behavior of Precast Concrete Shear Walls for Seismic Regions: Comparison of Hybrid and Emulative Specimens.*J. Struct. Eng.* **139**(11) 1917-27.
[4] Smith B J and Kurama Y C 2013 Seismic Displacement Demands for Hybrid Precast Concrete Shear Walls.*Struct. Cong.* 2626-37.
[5] Rossley N, Aziz F N AA and ChewH C 2014 Behaviour of Precast Walls Connection Subjected to Shear Load.*J. Eng. Sci. Technol.Spl. Issue Appl. Eng. Sci.* 142–50.
[6] Sanghvi H R and Dhankot M A 2015 Analysis of Precast Shear Wall Connection - State of the Art Review. *Int. J. Res. Eng. Technol.* **4**(2) 767–70.
[7] AdemSolak, YavuzSelim Tama, SalihYılmaz and Hasan Kaplan 2015 Experimental study on behavior of anchored external shear wall panel connections.*Bull. Earthquake Eng.* **13** 3065–81.
[8] Ioannis N Psycharis, Harris P Mouzakis 2012 Shear resistance of pinned connections of precast members to monotonic and cyclic loading.*Eng. Struct.* **41** 413–27.
[9] EliyaHenin and George Morcous 2015 Non-proprietary bar splice sleeve for precast concrete construction, *Eng. Struct.* **83** 154–62.
[10] Fabio Biondini, Bruno Dal Lago and GiandomenicoToniolo 2013 Role of wall panel connections on the seismic performance of precast structures.*Bull Earthquake Eng.* **11** 1061–81.
[11] RaminVaghei, FarzadHejazi, Hafez Taheri, MohdSalehJaafar and Abang Abdullah Abang Ali 2014 Evaluate Performance of Precast Concrete wall to wall Connection.*APCBEEProcedia* **9** 285–90.
[12] Todut C, Dan D andStoianV 2014 Theoretical and experimental study on precast reinforced concrete wall panels subjected to shear force.*Eng. Struct.* **80** 323–38.
[13] Hamid N H and ManderJ B 2010 Lateral seismic performance of multicore precast hollow core walls.*J. Struct. Eng. ASCE* **136**(7) 795–804.
[14] Su-min Kang,Ook-jong Kim and Hong Gun Park 2013 Cyclic loading test for emulative precast concrete walls with partially reduced cross section.*Eng. Struct.* **56** 1645–57.
[15] Zhu Zhangfeng and GuoZhengxing 2017 Experimental Study on Emulative Hybrid Precast Concrete Shear Walls.*KSCE J. Civ. Eng.* **21**(1) 329-38.

Retrofitting of Reinforced Concrete Beams using Reactive Powder Concrete (RPC)

S Karthik and Karthik Sundaravadivelu*

School of Civil Engineering, SASTRA University, Thanjavur 613 401, India.

*Email : karthik@civil.sastra.edu

Abstract. Strengthening of existing damaged structures is one of the leading studies in civil engineering. The purpose of retrofitting is to structurally treat the member with an aim to restore the structure to its original strength. The focus of this project is to study the behaviour of damaged Reinforced Concrete beam retrofitted with Reactive Powder Concrete (RPC) Overlay. Reinforced concrete beams of length 1200 mm, width 100 mm and depth 200 mm were casted with M30 grade of concrete in the laboratory and cured for 28 days. One beam is taken as control and are tested under two point loading to find out ultimate load. Remaining beams are subjected to 90 % ultimate load of control beams. The partially damaged beams are retrofitted with Reactive Powder Concrete Overlay at the full tension face of the beam and side overlay depends upon the respectable retrofitting techniques with 10 mm and 20 mm thick layer to find optimum. Materials like steel fibres are added to enhance the ductility by eliminating coarse particle for homogeneity of the structure. Finally, the modes of failure for retrofitted beams are analysed experimentally under two point loading & compared the results with Control beam.
Key words: RPC, Retrofitting, reinforced concrete beam, steel fibre, silica fume.

1. Introduction

1.1. General

For a variety of reasons, subsist concrete structures may become deficient and found to be perform not satisfactory which leads extreme deflections and cracking, or results in not enough ultimate strength [1]. For this purpose, retrofitting is the suitable way of strengthening the structures by means of adding new features or technologies to the structural member. The main aim of retrofitting is to structurally treat the member with an aim to restore the structure to attain its original strength. Some of the commonly used repairing and retrofitting techniques are patch repair, resin injection, stitching, bonding of external reinforcement, routing and sealing, drilling and plugging. Depending upon the type of distress, the techniques used for carrying out repair of damaged structures vary from one structure to the other. Although these techniques are favourable in particular cases, but there are lot of specialized specifications according to ACI 440.2R-08 [2] and disadvantages to follow the same in field work [1, 3]. The suitable way to deal with subsist problem for retrofitting is cement-based mortars. Fibre based composites has step-up a lot of notice in strengthening the flexural members [4].
From literature studies, retrofitting by use of FRP materials have gained lot of benefits in corrosion resistance, can be adhesively bonded, and have a stiffness greater compared to that of steel. Despite its benefit over other methods, the strengthening technique using FRP is not entirely problem-free. The

cost of resins is also high and it affects the manual worker by release of toxic fumes during application pose potential hazards [5].

From the earlier literature studies, the new generation of cement-based materials called reactive powder concrete (RPC) which is also called as ultra- high performance concrete (UHPC) has gained a lot of attention in achieving maximum compressive strength in the range of 200–800 MPa due to its super-high strength, superior toughness and extreme durability [6]. Elimination of coarse aggregate for homogeneity by incorporating steel fibres to improve ductility with addition of silica fume which is a micro-cementitious materials into the concrete substance to enhance pozzolanic reaction and to reduce the water/cement ratio with the use of superplasticizers [7]. RPC proves to have higher flexural strength and ductility of about 250 times when compared to that of normal concrete. Generally production cost of RPC is high as cement dosage varies from 800–1000 kg/m^3. Due to high mechanical performance of RPC, it has some economic advantages by reducing the thickness of concrete members which effect in materials and cost savings. It is also implemented in other applications such as for special type of precast and pre-stressed concrete members [8]. In the literature studies, reported on RPC they gained importance regarding the increase in compressive strength and also finding the optimized curing conditions. But narrow studies have been described in increasing the flexural capacity of a member with various repair techniques. This study has been done to increase the flexural capacity of a RC beam as retrofit with RPC overlay in tension side of a beam and also sides of the beam where crack is applicable for economical purpose with three different repair techniques to compare and find the optimum. The main purpose of an overlay is to extend the service period of the structure and increase the structural performance and durability of a wearing surface.

In this paper, seven RC beams of M30 grade with same dimension are measured for experimental purpose. For partial loading, beams are subjected to 90% of the ultimate load of control beam. The partially loaded beams are then retrofitted with 10 and 20 mm thickness with RPC overlay and allowed for normal curing 28 days and tested for its flexural behaviour.

1.2 Significance of work
The objective of this work is to find out the effective thickness of RPC overlay in strengthening for reinforced partially loaded beams and also finding the suitable retrofitting techniques to achieve higher ultimate load.

2. Experimental Program
2.1 Reactive Powder Concrete
The mix design details of RPC are presented Table 1. The RPC mix is prepared by the following constituents: Ordinary Portland cement (C) 53 grade as per IS 12269:1987, with quantity of cement content high as 1000 Kg/m^3 than that used in common concrete with 3.15 as a specific gravity. Silica fume (SF) is used as a secondary binder with specific gravity of 2.2, to neglect the micro-particulate voids in the binder paste. Fine aggregate (FA) according to IS 650–1991 with a specific gravity of 2.65 and size as 4.75 mm, hooked micro steel fibres (ST) of 30 mm length and 0.96 mm diameter, Nitobond EP epoxy adhesive for bonding old concrete and new concrete, Super plasticizer (SP) and ordinary tap water.

2.2 Reinforced concrete beams
The mix design of M30 grade concrete are shown in Table 1. Totally 7 RC beam are casted with a dimension of 1200 mm length, 100 mm wide and 200 mm depth. The control beam details are shown in Table 2. The schematic diagram of beam are represented in Figure 1. The characteristic strength of the steel f_y is 500 N/mm^2. Three types of retrofitting techniques are considered for the study are grooving (A), meshing (B) and stitching (C).

Table 1. Mix ratio of M30 and RPC

MIX NAME	C (Kg/m^3)	FA (Kg/m^3)	CA (Kg/m^3)	SF (Kg/m^3)	QP (Kg/m^3)	WATER (Lit/m^3)	SP (by weight)	STEEL FIBRE (by volume)
M30	380	711	924	-	-	160	0.5%	-
RPC	970	1080	-	244	-	272	3%	2%

Table 2. Control beam details

Description	Main reinforcement	Ultimate load (KN)
CB	2 no's of 12 mm	123.68

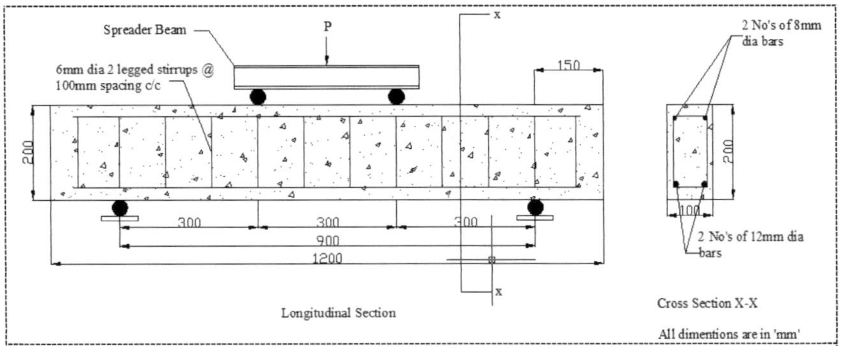

Figure 1. Schematic diagram.

Table 3 Retrofitted beams details

Group	Induced partial load	Retrofit beam	Overlay thickness (mm)
A	90% of the ultimate load of CB	A1	10
		A2	20
B	90% of the ultimate load of CB	B1	10
		B2	20
C	90% of the ultimate load of CB	C1	10
		C2	20

Two numbers of 12 mm diameter steel rod is used as bottom main reinforcement and 2 numbers of 8 mm diameter steel is used as anchorage bars and 6 mm diameter is used as stirrups to avoid shear failure and are placed at 100 mm c/c distance.

2.3 Casting of RPC overlay

The overlay thickness is chosen based on the literature study [4]. The retrofitted beams details are given in Table 3. The overlay is applied by using a Nitobond EP epoxy in the tension region of the partially loaded beam for effective bonding.

2.4 Testing

The casted control beam and retrofitted beams are tested under two-point loading. The centre to-centre distance between supports is 1100 mm. In a 1000 kN load cell unit, the load is applied under a rate of

10 kN/min. The tested beams are placed with two LVDT one at centre of the span and the other at quarter span just below the loading points. To locate the progression of the crack, the specimens are white washed and grids are drawn of size 50 mm x 50 mm on two sides of surface of beam.

3. Results and Discussion

3.1 Mechanical properties

The mechanical properties of conventional M30 grade concrete and RPC for 28 days with an average value are shown in Table 4.

Table 4. Mechanical properties of M30 concrete and RPC

Description	Compressive Strength (N/mm^2)	Split – Tensile Strength (N/mm^2)
M30	32.24	3.45
RPC	65.7	6.31

To achieve higher compressive strength than the conventional concrete the water content usage must be reduced than the optimum amount, so that the part of cementitious constituents remains anhydrate which affects the microstructure of RPC [9].

3.2 Bending behaviour of control beam

The control beam is tested to find the ultimate load. With formation of minute cracks, a slight decrease in the stiffness of the beam is seen during testing. Visible cracks are seen at 58 KN. The crack develop upwards due to complete yielding of reinforcement and a mid-span deflection of 6–8 mm with crushing of concrete can be seen in the compressive region.

3.3 Damaged beams

The ultimate load are found out by testing the control beam till failure. Remaining beams are partially loaded till 90% of ultimate load of control beam [4]. The pre damage load of RC beams is 110 kN. The partially loaded beams are retrofitted with three different retrofitting techniques – Grooving (A), Meshing (B), Stitching (C) with 10 mm and 20 mm thick overlay and its load vs deflection response is shown in Fig 2.

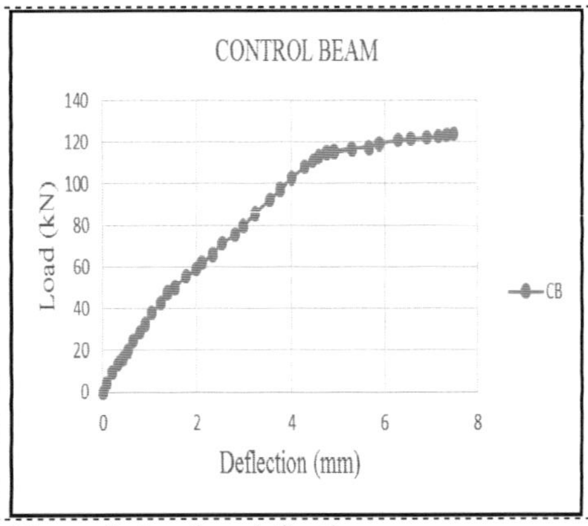

Figure 2. Load vs deflection response: (a) Control beam till failure

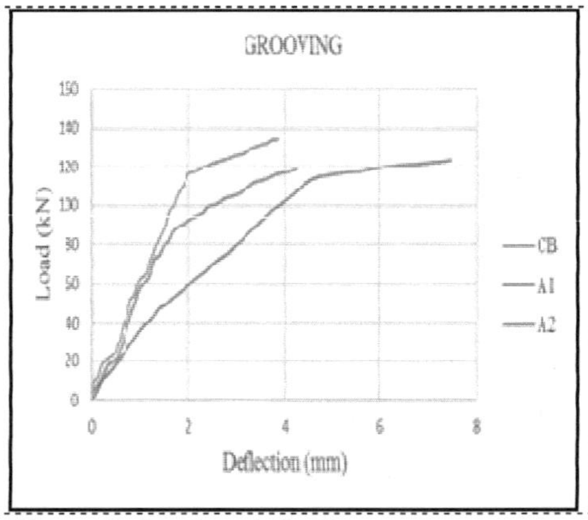

Figure 2. Load vs deflection response: (b) CB, A1-A2,

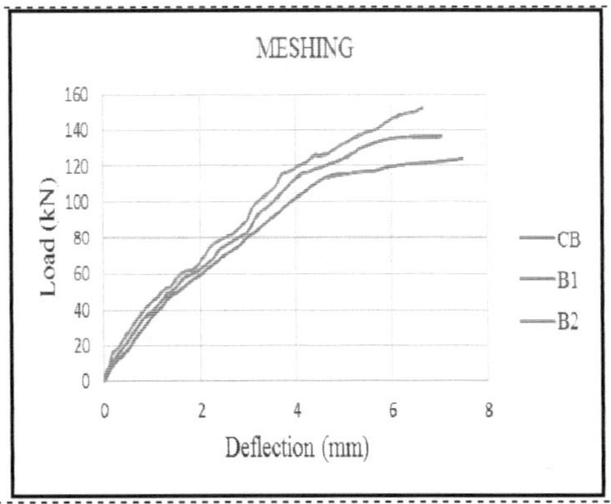

Figure 2. Load vs deflection response:
(c) CB, B1-B2,

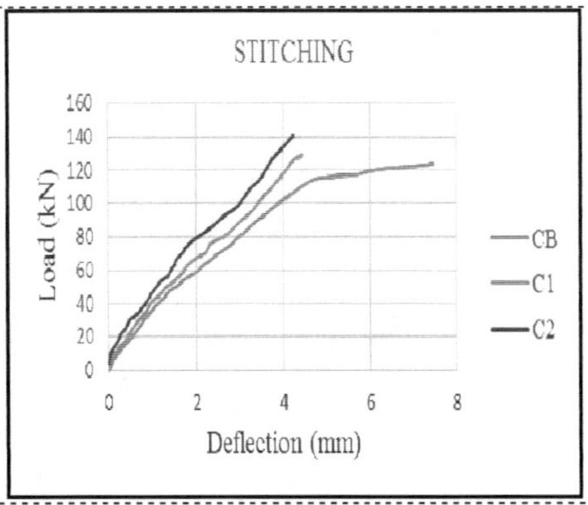

Figure 2. Load vs deflection response:
(d) CB, C1-C2

3.4 Adhesive epoxy bonding

The thickness overlay is selected based on literature [1, 10, 11 and 12]. Table 3 shows the detail of a retrofitted beams. Different methods are adopted in literature to improve strong bonding between normal concrete and Ultra - high performance concrete [13, 14]. In this study, a simple method to bond RPC strips with old concrete is explained. The partially loaded beams are cleaned from dust and angle grinder are used to roughen the surface before bonding the retrofit strips. To retrofit the partial loaded beams, three types of technique are adopted, they are grooving, meshing and stitching. The contact area into the faces of the beams are to be increased for grooving method. In such a way that a diagonal grooves 10–15 mm deep at a spacing of 200 mm is cut as shown in fig.3 (a). The meshing method is implemented by wrapping a G.I mesh in U- shape as shown in fig.3 (b) [15]. For stitching technique, a holes are made by drilling on either sides of the crack and a stitching dogs that is a 5 mm rod is bent in a U- shape as shown in fig. 3 (c) is inserted into that holes and extra spacing of holes are filled by epoxy. For good adhesive bonding, the two mixture of the epoxy is mixed thoroughly and applied to the bottom side of the partially loaded beams and at the sides of the beam maintaining a constant thickness of 3 mm. Uniform pressure is applied on the retrofitted specimens using L clamp during hardening of adhesive 24 hr to ensure good bonding.

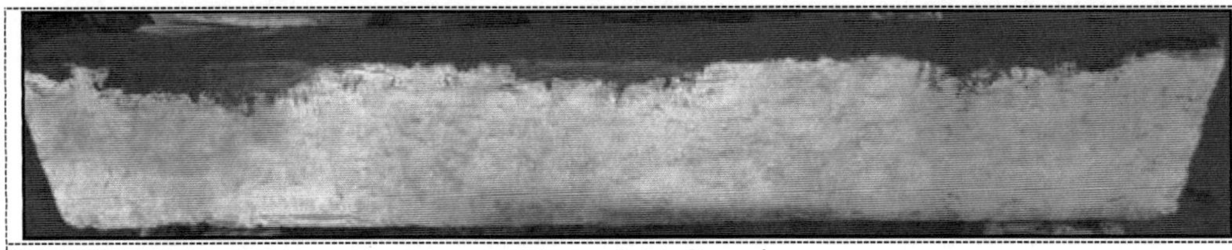

Figure 3 (a) Grooving

Figure 3 (b) Meshing

Figure 3 (c) Stitching

3.5 Performance of RPC overlay

Strengthening Criterion: Considering the strengthening criteria, the ultimate load carrying capacity of the retrofitted beams should be increased. From Table 5, it is clear that in grooving technique, the beam retrofitted with 20 mm thick overlay is able to regain the earlier attained original strength whereas 10 mm overlaid beam fails to get original strength when compared with control beam. The meshing with 20 mm overlay shows a higher load carrying capacity of 152 kN than all other retrofitted beams. The ultimate load carrying capacity of A2, B1, B2, C1 and C2 compared to control beam of series increased up to 8.6%, 10.2%, 22.9%, 4.1%, and 13.5% respectively. A peak load for control beam and all retrofitted beams are shown in Fig. 4. With the increase in overlay thickness in strengthening techniques, we observed the lower deflections with increase in ultimate load which shows the increased flexural capacity of retrofitted beams.

Table 5. Percentage increase in ultimate load of retrofitted beams

DESCRIPTION	OVERLAY THICKNESS (mm)	ULTIMATE LOAD (kN)	% INCREASE IN ULTIMATE LOAD
CONTROL BEAM	–	123.68	–
GROOVING (A)	10 mm	118.7	–
	20mm	134.3	8.6 %
MESHING (B)	10 mm	136.2	10.2 %
	20 mm	152	22.9 %
STITCHING (C)	10 mm	128.7	4.1 %
	20 mm	140.3	13.5 %

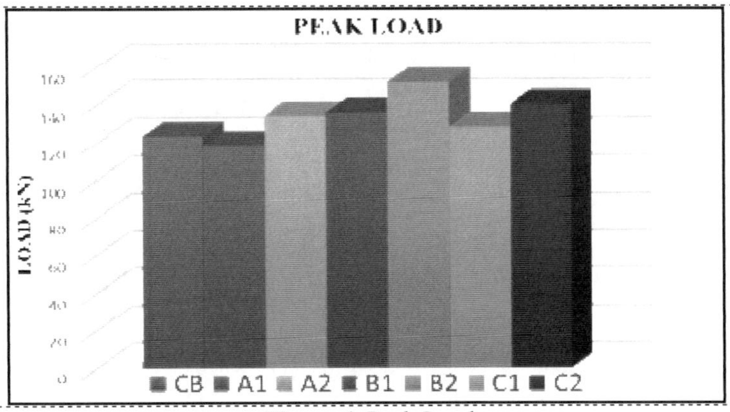

Figure 4. Peak Load

Figure 5. (a) Crack patterns showing control beam

Figure 5. (b) Crack patterns showing retrofitted beamA1

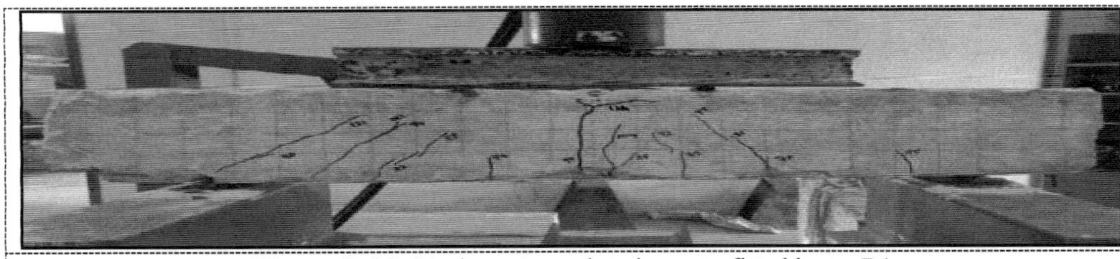

Figure 5. (c) Crack patterns showing retrofitted beam B1

Figure 5. (d) Crack patterns showing retrofitted beamC1

4. Conclusion

From the experimental results obtained it is clear that the Reactive Powder Concrete can be used as a retrofitting material and Nitobond EP epoxy adhesive is used for bonding new concrete with old and no debonding is observed during the experimental investigation. The results obtained are:

- The retrofitted beams were able to regain its original strength after strengthening in tension and acted monolith under bending.
- In Grooving method, the ultimate load for beam with 10 mm thick overlay is equal to the ultimate load of control beam but for beam retrofitted with 20 mm overlay thickness in tension side shows 8.6% increase in ultimate load.
- In meshing method increase in load is 10.2% and 22.9% for beams retrofitted with RPC thickness of overlay 10 mm and 20 mm respectively when compared with control specimen.
- In stitching method increase in load is 4.1% and 13.5% for beams retrofitted with RPC thickness of overlay 10 mm and 20 mm respectively when compared with control specimen.
- Beams retrofitted with meshing method with 20 mm RPC overlay shows high load carrying capacity than other retrofitted beams with a slight spalling of concrete.
- In consideration of cost, the grooving method and stitching method will be more economical when compared with meshing method.
- All the three retrofitting methods can be used for strengthening beams by overlay of RPC. This method can be successfully implemented to damaged structural member under flexure, since no debonding is observed.

Acknowledgement

The authors would like to thank to the authorities of Sastra University, Thanjavur for providing lab facilities to perform the present project work in the school of civil engineering.

References

[1] Jandaghi Alaee F 2003 Retrofitting of Reinforced Concrete Beams with CARDIFRC. *J. Compos. Constr.* 174-86.
[2] ACI 440.2R-08 2008 Guide for the Design and Construction of Externally Bonded FRP Systems for Strengthening Concrete Structures
[3] Mukhopadhyaya P 2001 Interface shear stress: a new design criterion for plate debonding. *J. Compos. Constr.* **5** 35–43.
[4] Prem P R 2016 Acoustic emission and flexural behaviour of RC beams strengthened with UHPC overlay. *Constr. Build. Mater.* **123** 481–92.
[5] Buyukozturk O 1998 Failure behaviour of precracked concrete beams retrofitted with FRP. *J. Compos. Constr.* 138-144.
[6] Richard P 1995 Composition of reactive powder concretes. *Cem. Concr. Res.* **25** 1501-11.
[7] Mostofinejad D 2016 Determination of optimized mix design and curing conditions of reactive powder concrete. *Constr. Build. Mater.* **123** 754-67.

[8] Yazici H 2009 Mechanical properties of reactive powder concrete containing mineral admixtures under different curing regimes. *Constr. Build. Mater.* **23** 1223–31.
[9] Singh S B 2015 Role of water/cement ratio on strength development of cement mortar. *J. Build. Eng.* **4** 94–100.
[10] Bruhwiler E. 2013 Rehabilitation and strengthening of concrete structures using ultra-high performance fibre reinforced concrete. *Struct. Eng. Int.* **23** 450-7.
[11] Farhat F 2007 High performance fibre reinforced cementitious composite cardifrc–performance and application to retrofitting. *J. Eng. Fract. Mech.* **74**(1) 151-67.
[12] Prem P R 2015 Flexural behaviour of damaged RC beams strengthened with ultra- high performance concrete. *J. Adv. Struct. Eng.* 2057–2069.
[13] Carbonell Munoz M A 2013 Bond performance between ultra-high performance concrete and normal-strength concrete *J. Mater. Civ. Eng.* **26**(8).
[14] Tayeh B A 2012 Mechanical and permeability properties of the interface between normal concrete substrate and ultra-high performance fiber concrete overlay. *Constr. Build. Mater.* **36** 538-48.
[15] Ezz-Eldeen H A 2015 An Experimental Study on Strengthening and Retrofitting of Damaged Reinforced Concrete Beams using Steel Wire Mesh and Steel Angles. *Int. J. Eng. Res. Technol.* **4**(5) 164-73.

Strengthening of beam - column joint with steel fibre reinforced concrete under seismic loading

K Prasanna*, R Ramasubramani, K S Anandh, Saisabarish and Vijay Krupaker Maddu

Department of Civil Engineering, SRM University, Kattankulathur, 603 203. India.

*Email: prasanna.env@gmail.com ram20jan90@gmail.com

Abstract. Earthquakes not only kill the human beings but the structures too. Considering this above fact, there is a necessity to take care of the beam column joint region in a structure. Under seismic excitation, the bar segment joint district is subjected to even and the vertical shear force extents are normally commonly greater than those in the adjoining bits of the structural segments. In the event that the joint is not precisely nitty gritty, the shaft segment joint may end up plainly frail. To keep this, an adequate control ought to be given in the joint area. The seismic examination and outline technique ought to be utilised as a part of the plan of the building structures and their segments ought to be recommended in the segment. The building structures incorporate finish sidelong and vertical drive opposing frameworks fit for giving satisfactory quality. Stiffness and the vitality dissemination ability to withstand the outline ground movements inside the recommended furthest reaches of twisting and quality request. The plan ground movements are accepted to happen along any even headings of a building structure. This work focuses on behaviour of M20 concrete in beam-column joint subjected to seismic loading, by using steel fibre (1.5%). The specimens detailed as per IS: 456-2000 and IS: 13920-1993 were casted and tested under cyclic and reverse cyclic loading. The parameters analysed were ductility, energy dissipation, load v/s displacement curve, beam-column reinforcement strain and crack pattern. From the experimental investigation, it is found that the fibre reinforced beam-column joint with fibre (1.50% of steel fibre) performs better ductility, load carrying capacity, energy dissipation and strength by 10% more than conventional reinforced beam-column joint.
Keywords: Fibre reinforced concrete, Seismic loading, Beam-column joint.

1. Introduction

1.1 General

Seismic tremors show a risk to open security and welfare in a huge segment all over. We can't stop tremors however we can shield ourselves from them, as "quakes don't slaughter individuals, yet the structures do". The conduct of fortified solid minute opposing casing structures in late seismic tremors everywhere throughout the globally highlighted the outcomes of meagre execution of bar segment joints. Shaft section joints in a strengthened solid minute opposing casing are urgent zones for an exchange of burdens viable between the interfacing components (i.e. pillars and segments) in the structure. In the investigation of reinforced concrete frames, the joints are for the most part accepted as inflexible [1].

1.2 Reinforced cement concrete

Consistently flexibility of concrete is 8% to 15% of its compressive quality. This inadequacy has been overseen over various decades by using a plan of fortifying bars (rebars) to make invigorated strong so that strong basically contradicts compressive tensions and rebars restrict flexible and shear stresses. The longitudinal rebar in a column restricts flexural (versatile uneasiness) however the stirrups, wrapped around the longitudinal bar, contradict shear stresses. In a portion, vertical bars restrict weight and catching burdens while ties contradict shear and offer detainment to vertical bars. With the use of steel strengthened, strongly makes for a conventional composite material with expansive use in the improvement business. Be that as it may, steel bars in the fortified strong give resistance against weight just locally. Softens up fortified strong people grow transparently until encountering a rebar. The necessity for multidirectional and immovably isolated support for bond rises. It's here the point the use of fibre fortress comes into for course of action [1,2]

1.3 Fibre Reinforced Concrete (FRC)

Fibre reinforced concrete is an active mix that consists of shortened, unattached fibres that are reliably spread and heedlessly orchestrated. The mix measured as a rate of the total volume of the composite (cement and fibre) named Vf. Vf commonly goes from 0.1 to 3% point of view extent (l/d) is learned by parcelling fibre length (l) by its expansiveness (d). Fibres with a unrounded cross range use comparable width for the figuring of point ratio [3].The essential part of the strands in solidified cement is to alter the breaking system. By modifying the breaking system, the full-scale splitting ends up plainly miniaturised scale splitting. The breaks are littler in width, in this manner lessening the penetrability of cement and a definitive splitting strain of the solid is upgraded. The filaments are equipped for conveying a heap over the crack. The rate of steel fiber utilised is around 1.5% is taken consistent for every one of the examples, the steel fiber length ranges 30mm, and viewpoint proportion runs around 60.The ratio of polypropylene fiber is changed from 0.2% and 0.4%. The length is around 20 mm, and the perspective portion is around 2500.

1.4 Beam-column joint

The joint is described as the piece of the section inside the significance of the column frames. In a contradicting edge, three sorts of joints can be identified viz. inside joint, outside joint and T joint showed up in Figure 1. When four bars layout into the vertical appearances of a segment, the joint is called as an inside joint. When one column traces into a vertical face of the segment and two unique shafts diagram from contrary headings into the joint that is an outside joint. Exactly when a column each edges into two adjacent vertical faces of a fragment, then the joint is called as a T joint

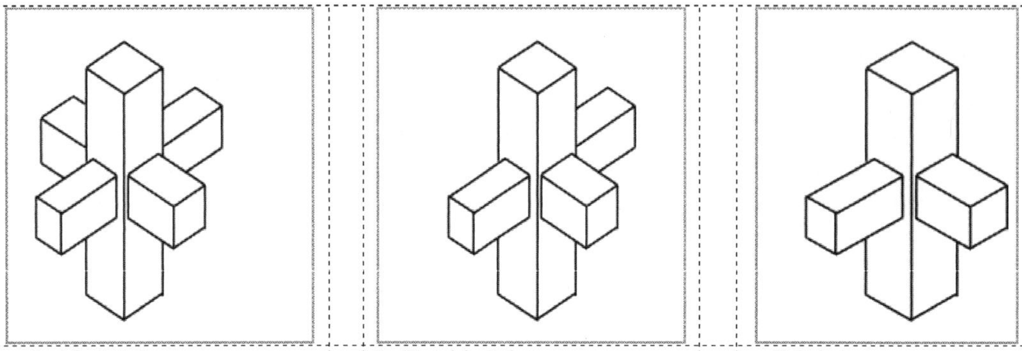

a. Inside Joint. **b.** Outside Joint. **c.** T Joint.

Figure 1. Types of joints in frames

The functional requirement of a joint, which is the zone of convergence of beams and columns, is to empower the associating individuals to create and maintain their definitive limit. The request on this limited size component is constantly serious, particularly under seismic stacking. The joints ought to have sufficient quality and firmness to oppose the inside strengths invited by the confining individuals.

2. Details of specimen

The test specimen was reduced to one fourth scale to suit the loading arrangement and test facilities. Four numbers of 8 mm diameter bars were used for main bars in column and 3 nos. of 8 mm diameter were used for top reinforcement and 2 nos. of 8 mm diameter were used for bottom reinforcement in beams. The dimension of beam was 110×90 mm and that of column was 90×90 mm[4].

2.1 Fibre

The steel fibres were used is shown in Figure 2. The properties of fibres are given in Table1.

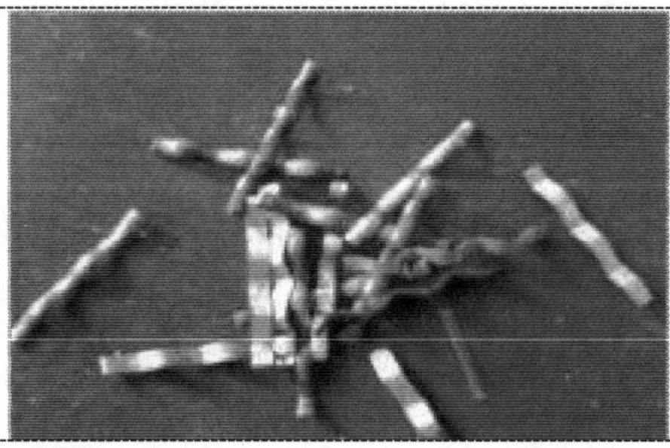

Figure 2. Steel Fibres.

Table 1. Properties of fibres

Types of mix	l/d ratio	Corresponding strength	Corresponding toughness
	0	1.00	1.00
Plain concrete with	25	1.50	2.00
randomly	50	1.60	8.00
dispersed fibres	75	1.70	10.50
	100	1.50	8.50

2.2 Concrete

The amount of aggregate added to the concrete is reduced while adding fibres. The conventional strength of concrete and concrete with steel fibre at 28 days was determined by compression test on cubes of size 6 inches×6 inches × 6 inches. The Table 2 shows the mix ration and Table 3 shows the specimen details for cubes of various proportions of the fibre with concrete. The type of test specimen and percentage of the fibre comparing to mixes[11,12].

Table 2. Mix ratio

Concrete grade	M20 grade
Normal concrete mix design	C : FA : CA : W/C 1 : 1.93 : 3.85 : 0.55
Fibre concrete mix design	C : FA : CA : W/C : SP 1 : 1.93 : 3.85 : 0.55 : 0.015

Table 3. Specimens details without seismic detailing

Specimens ID	Description of test specimens	% of Fibre
O1 and O2	Normal concrete	----
S1 and S2	Normal concrete	---
F11 andF12	Fibre Reinforced concrete	1.50% steel fibre

Table 4 shows the compressive strength for cubes 0f various proportions of fibre with concrete. We can conclude that concrete with steel fibre (1.5%) has more compressive strength comparing to other mixes.

Table 4. Compressive strength of concrete

Specimen	Type	Compressive strength of cube N/mm^2
O1,O2	Normal concrete	28.02
F11,F12	N.C + SF (1.5%)	36.22

2.3 Steel

The main reinforcement used for the specimen was tor steel of diameter 8 mm. shear reinforcement was mild steel of diameter 3.3 mm. Specimens of 300 mm length were cut out and tested in computerized. The reinforcement details and reinforcement cage of the ordinary joint are consider. The capacity of reinforcements is shown in Table 5.

Table 5. Capacity of reinforcements

Description	**Yield load (kN)**	**Ultimate load (kN)**	**Ultimate stress (N/mm^2)**	**Yield stress (N/mm^2)**
8 mm bar	21.00	26.9	353.09	417.73
3.3 mm bar	1.42	2.9	341.36	166.00

3. Reinforcement details

The rate of steel fibre in the bar and segment are 2.539 and 2.482 separately. The fundamental fortification given in the bar was 8 mm width bars, 3 Nos at the top and 2 Nos at the base. The stirrups are 3.3 mm width bars at 20 mm c/c for a separation of 2d, i.e. 220 mm from the substance of the segment and at 40 mm centre to centre for the rest of length of bar. The fortification subtle elements and support pen of the test example with steel fibre appear in Figure 4.4. The rate of steel in pillar and section are 2.539 and 2.482 separately. The fortification confines were put in the moulds, and cover amongst enclosure and shape gave 10 mm. The stirrups are 3.3 mm distance across bars at 40 mm c/c separating and 20 mm c/c middle segment of bar and section. The pressure or pressure in any bar at any section is created on either side of the part by a suitable improvement length. The first bars of the segment must be proceed up either inside or outside the support of floor shaft which outlines because of this section. It is essential that the width of section more than the width of the bar. The support is set and situated entirely the necessity in the auxiliary member [5].

4. Instrumental arrangement

All the specimens were tested in a reaction frame of 50 tonnes capacity. The Figure 3 describes the instrumental arrangement. A hydraulic jack were adopted for apply the load at the free end of the beam in both upward and downward direction individually [6].

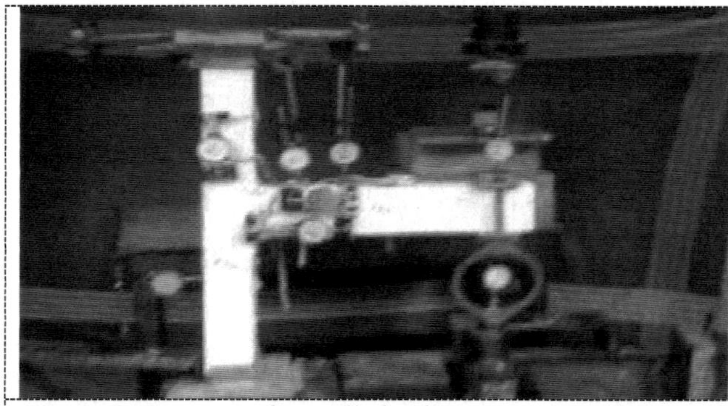

Figure 3. Test setup for cyclic load.

To record the heap correctly a proving was utilised. The heap is connected cyclic and turns around cyclic and measured for each 5mm diversion. The avoidance of the pillar at the purpose of stacking amid test was measured, such as 5 mm, 10mm and 15 mm individually. Inversion stacking is noted for each lessening in redirection. Every specimen was tried in a response edge of 50 tons limit.

4.1 Loading history

The exterior beam column joint specimen was regulated to cyclic loading and reverse cyclic loading simulating earth quake loads. The displacement sequence consists of 5 mm, -5mm, 10 mm, -10 mm, 15mm, -15mm, 45mm and -45mm. Each displacement level is indicated in dial gauge and corresponding loads are noted from proving ring. In the first cycle, beam was loaded gradually up to 5 mm deflection and then unloaded. and beam was loaded gradually up to -5 mm deflection and then unloaded the second cycle, beam was loaded gradually up to 10 mm and then unloaded and beam was loaded gradually up to -10 mm and then unloaded similarly each cycle of load is applied. The displacement history plotted as shown in Figure 4.

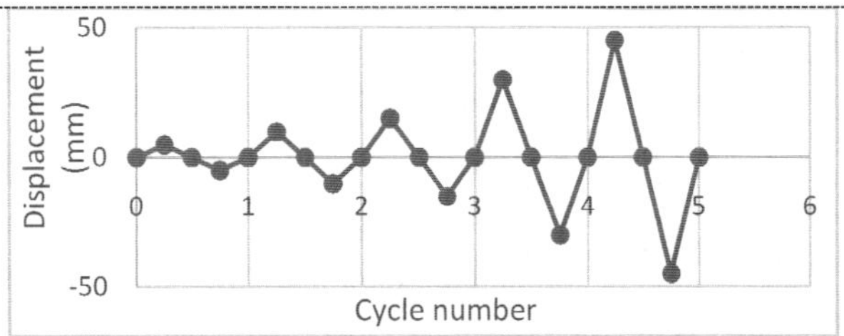

Figure 4. Displacement history.

5.Results and discussion

5.1 Load vs displacement cycle

The load Vs displacement cycle is drawn for every specimen, the load is noted for 5mm displacement, for each cycle and reverse cyclic of loading. The reversal of loading isnoted from the proving ring.For each samples were tested every displacement of 5 mm load in each cycle are noted the hystersis loop for all the specimen.The variation of load Vs displacement cycle for various beam column joints with various combination at beam and displacement.The area for each hystersis loop was calculated.The parameter of load Vs displacement cycle shown in Figure 5, Figure 6 and Figure 7 for O1,O2, S1,S2 and F11, F12 respectively.

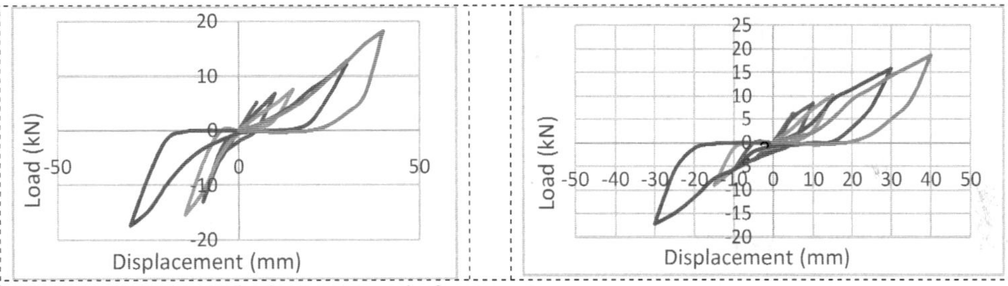

Figure 5. Load Vs Displacement cycle for O1,O2.

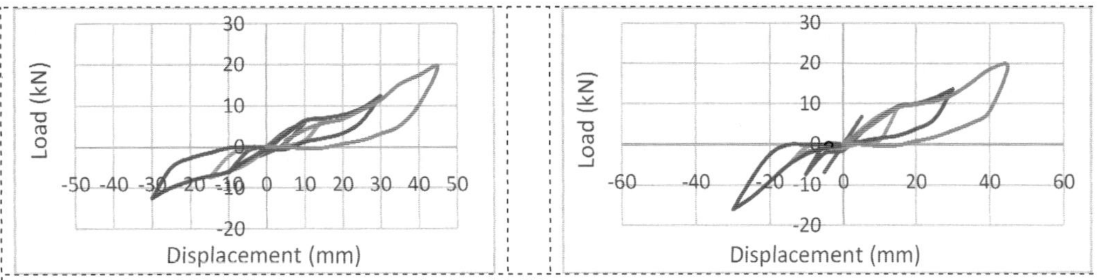

Figure 6. Load Vs Displacement cycle for S1,S2.

Figure 7. Load Vs Displacement cycle for F11,F12.

5.2 Energy dissipation

The the energy dissipated at each cycle for various specimens is shown in Figure 8. The energy dissipation capacity is calculated using the enclosed area of the load deformation curve during each cycle of laoding (i.e. area of each hystersis loop).

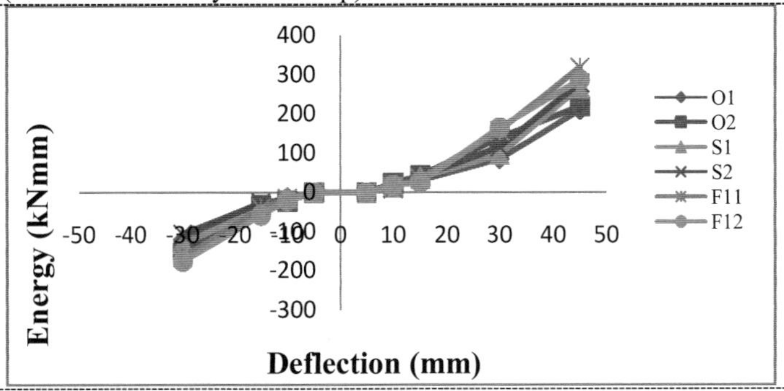

Figure 8. Energy dissipation Vs Displacement.

5.3 Beam and column reinforcement strain

To compute the strain, demec strain gauge was arranged in the beam lower reinforcement and in the column top outer face reinforcement. The gauge length 100 mm was maintained for all the specimens to fix the strain gauge.[7].

5.3.1 Beam bottom reinforcement strain. The displacement , cycle of various specimen are listed below for the specimen O2, S2 and F12 the yield strain occurs at 5 mm (1^{st}cycle) and ultimate strain at -30 mm (4th cycle). The stress condition may be the strength of concrete in the tension zone is greater than the strength of steel reinforcement the tensile resistance of concrete is lost due to formation of initial crack the steel may not be able to take over the tension which was earlier resisted by the concrete a failure will acquire suddenly with the concrete in the tension zone cracking and the steel fractured almost simultaneously. The Figure 9 shows the beam bottom reinforcement strain, cycle of various specimen are listed below for the specimen O2, S2 and F12

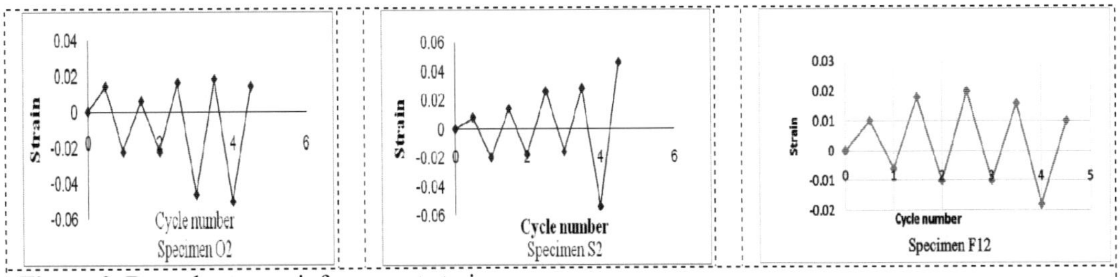

Figure 9. Beam bottom reinforcement strain.

5.3.2 Column Reinforcement Strain. Demec strain gage was utilised to gauge the strain in column top external support. For the specimen O2, the yield strain happens at 10mm (second cycle), and ultimate strain endures 30mm (fourth cycle). In the example F12, yield strain occurs at 5mm (first cycle), last encounter - 30mm (fourth cycle). The specimen S2 yield strain happens at - 5mm (first cycle), extreme resist 30mm (fourth cycle). In any column had a bigger cross sectional territory that needed for load support, the base rate of steel be founded on the zone of cement required to oppose the immediate anxiety and not upon the certain region the base number of bars given in a segment be four in rectangular column[8]. The longitudinal reinforcement by transverse connections to provide a limitation against out words clasping of each of the longitudinal bars the closures of the transverse connections the legitimately secured. The Figure 10 demonstrates the Column beat reinforcement strain cycle of different example is recorded underneath for the specimen O2, S2 and F12.

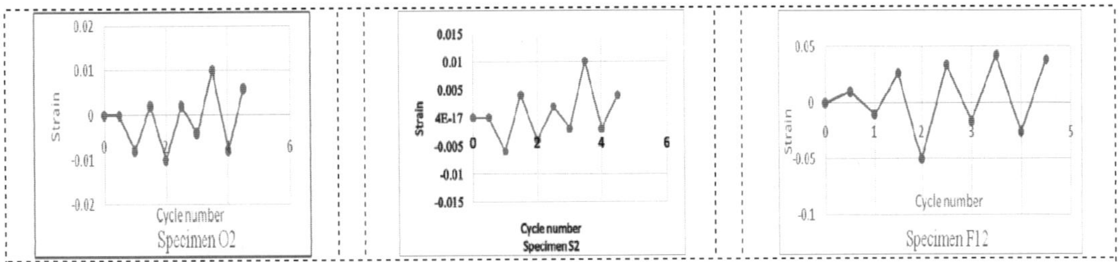

Figure 10. Column top reinforcement strain Vs Cycle number.

5.4 Load vs displacement curve

The Figure 11 shows Load v/s Displacement at each cycle for various specimens and the load v/s displacement curve during each cycle of loading.

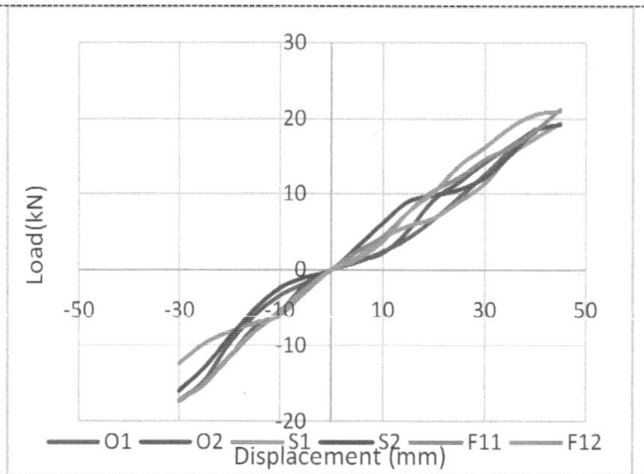

Figure 11. Load v/s Displacement curve.

5.5 Ductility

Ductility is an important characteristric of any structural element. It is described as the capacity of a structural element to undergo deformation beyond yield without loosing much of the load carrying capacity. Any type of brittle failure should be avoided, as it does not show warning before failure. If the structure posses sufficient ductile behaviour, it will be able to experience large deformation near ultimate loads. The amount of this inelastic deformation is proportional to the amount of ductility of the member.Ductility has generally been measured by a ratio known as ductility factor. It is usually expressed as a ratio of deflection (Δ) at failure to the corresponding yield, as shown below.

Displacement ductility factor $\mu_{\Delta} = \Delta_u / \Delta_y$ (1)

Increase in deformations after reaching ultimate load condition was not considered during experimentation. Hence the deformations at ultimate load have only been considered.The ductility for the various proportions were calculated and presented . It is observed that the ductility property for the beam column joint with fibre mix is more comparing to other specimens[9,10].

5.6 Crack pattern

In the beam-column joints, compression and tension developed in joint region during cyclic loading and the bond between concrete and reinforcement were reduced consequently. The first crack occurred near the beam-column joint and with further increase in loading, the cracks propogated and initial cracks started widening. The crack pattern of the specimens will be discussed as below. Figure 12 shows the crack pattern for ordinary joint. In this the first crack occurred vertically at 7mm deflection in second cyclic loading. When load is applied at the bottom of the beam. The second crack occurred at -10 mm deflection at third reverse cycle of loading. The third crack occured diagonally at beam column joint at 25 mm deflection at fourth cyclic of loading. The major crack occurred at joint region at 40mm deflection at fifth cyclic of loading.The Figure 13 shows the crack pattern for the seismic detailed joint. In this joint, while applying second cycle loading, the first crack occurred vertically at 5mm deflection in second cyclic loading. The second crack occurred at -15 mm deflection at third reverse cycle of loading. The third crack occured diagonaly at beam column joint occour in 30 mm deflection at fourth cyclic of loading. The crack pattern for the seismic detailed joint of major crack occurred at joint region in 45 mm deflection at fifth cyclic of loading. In this joint the crack width is small compare to ordinary joint, so it behaves better than ordinary joint.

By comparing the Figure 12 to Figure 14, we can observe that the width of the crack is reduced and the ductility is increased. From the Figure 14 the second, third, fourth cracks were only hair cracks. In the fibre reinforced specimens closely spaced finer cracks were developed and width of such cracks was smaller than those developed in conventional reinforced concrete joint. It was observed that the use of fibre reinforced concrete in the joint core could increase the joint stiffness and minimise damage to the concrete.By comparing the Figure 12 to Figure 14 , we can observe that In ordinary specimens wide cracks were developed at the joint and the crack width was more concentrated at the joint. But in fibre reinforced specimens exhibts finer cracks were developed and width of such cracks was smaller than those developed in conventional reinforced concrete joint.

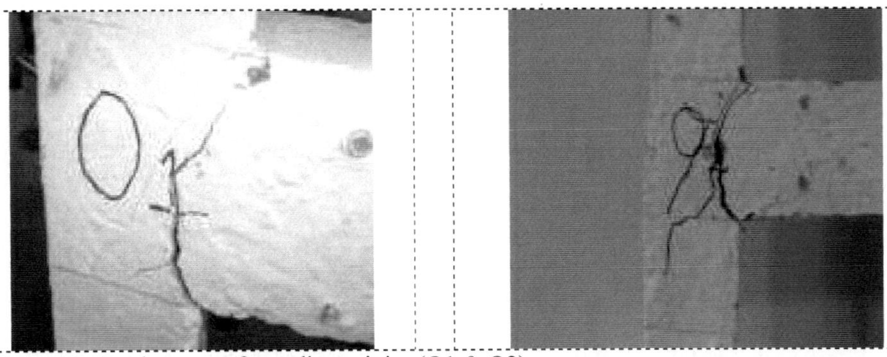

Figure 12. Crack pattern for ordinary joint (O1 & O2).

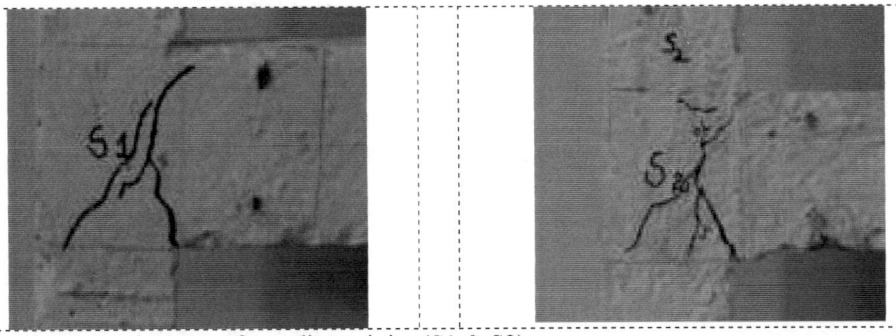

Figure 13. Crack pattern for ordinary joint (S1 & S2).

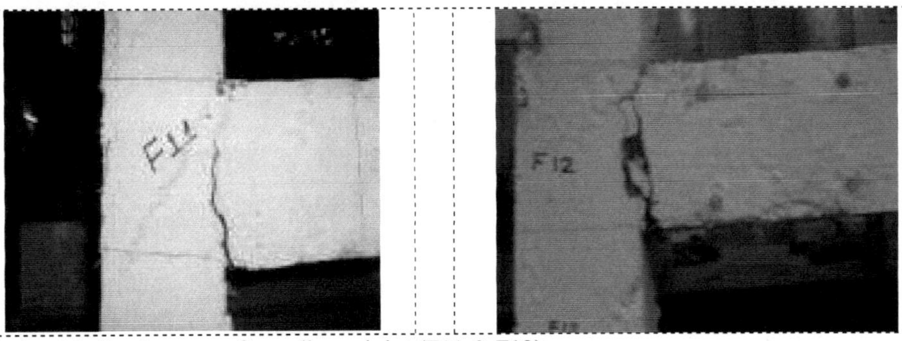

Figure 14. Crack pattern for ordinary joint (F11 & F12).

6. Conclusion

From the experimental investigation, it is found that the fibre reinforced beam-column joint with fibre (1.50% of steel fibre) performs better ductility, load carrying capacity, energy dissipation and strength by 10% more than conventional reinforced beam-column joint.

References

[1] Bayasi Z 1989 Development and Mechanical Characterization of Carbon Fibre Reinforced Cement Composites & Mechanical Properties and Structural Applications of Steel Fibre Reinforced Concrete. *Dissertation for the Degree of Ph.D.*, Michigan State University.
[2] Craig R, Mahadev S, Patel C C, Viteri M and Kertesz C 1984 Behaviour of Joints Using Reinforced Fibrous Concrete. Fibre Reinforced Concrete International Symposium SP-**81**, *Am. Concr. Inst.* 125-67.
[3] Durrani A J and Wight J K 1984 Behaviour of interior beam-to-column connections under earthquake-type loading. *ACI J. Proc.* **82**(3) 343-49.
[4] Gefken P R and Ramey M R 1989 Increased Joint Hoop Spacing in Type 2 Seismic Joints Using Fibre Reinforced Concrete. *ACI Struct. J.* 168-72.
[5] Henager C H 1977 Steel Fibrous, Ductile Concrete Joint for Seismic Resistant Structures. Reinforced Concrete Structures in Seismic Zones. *Am. Concr. Inst.* **53** 14371-86.
[6] Johnston C 1994 Fibre Reinforced Concrete: Significance of Tests and Properties of Concrete and Concrete-Making Material. *ASTM STP* **169C** 547-561.
[7] Prathibha S and Meher Prasad A 2004 Seismic Vulnerability of existing RC Building In India, *Proc. 13th World Conf. Earthquake Eng.* Vancouver, B.C. Canada 1207.
[8] Sudhir K Jain 1998 Learning from earthquakes. *The Indian Concr. J.* **72**(11) 294.
[9] Taylor H P J 1974 The behaviour of in situ beam column joints. *Techn Rep. Cem. Concr. Assoc.* **42** 492.
[10] Vollum R L 1997Design and analysis of reinforced concrete beam column joints. *The Struct. Eng.* **77** 23-4.
[11] Prasanna K and Anandh K S 2017 Analysis on Strength of Concrete by Partial Replacement of Riversand with Sawdust and Robosand *ARPN J. Eng. Appl. Sci.* **12** 2423-7.
[12] Prasanna K Anandh K S and Ravishankar S 2017 An Experimental Study on Strengthening of Concrete Mixed with Ground Granulated Blast Furnace Slag (GGBS) *ARPN J. Eng. Applied Sci.* **12** 2439-44.

Bamboo as a substitute for steel in reinforced concrete wall panels

P R Himasree*, N Ganesan and P V Indira

National Institute of Technology Calicut. India.

*Email:prhimasree@gmail.com

Abstract. The paper presents a review of the works done by various researchers on different types of reinforced concrete wall panels. Full scale bamboo reinforced concrete wall panels of three different aspect ratios of 1, 1.204 and 1.515 subjected to one way in-plane loading are considered in this study. Also an attempt is made to compare the ultimate loads estimated using the available equations with the experimental values of bamboo reinforced concrete wall panels. The investigation indicates that steel reinforcement could be replaced by bamboo in concrete wall panels.
Key words: *Bamboo Reinforced Concrete, In-plane loading, Wall panels.*

1. Introduction

The increasing demand for high speed, superior quality and cost effective construction led to the development of precast concrete structural elements. Of all the precast concrete structural elements, wall panels are the one that gained more importance in the field of construction to satisfy the housing needs of the ever increasing population.

Wall panels are structural elements subjected to in-plane action having negligible thickness compared to their length and breadth. In order to have sufficient strength to take in-plane loads, they may be reinforced with mild steel, steel fibers or steel fabric mesh. They are economical, not only from the structural design point, but also from the view point of overall construction.

The study presents the review of the works done by researchers on wall panels to determine the effect of various parameters like type and grade of concrete, reinforcing materials and its percentages, slenderness ratio (SR) and aspect ratio (AR). It also includes the comparison of the experimental ultimate loads with that of the loads computed by using the equations proposed by various researchers. It may be noted that the use of bamboo, an environmentally sustainable natural material that could replace the highly energy intensive material like steel in the construction of wall panels.

2. Experimental Programme

The experimental programme involved the casting and testing of three full scale Bamboo Reinforced Concrete Wall Panels (BRCWP) under one way in-plane loading. Details of the specimens are given in the Table 1

2.1 Materials and Mix Proportion

Portland Pozzolana Cement (PPC) conforming to IS: 1489 (Part 1):1991, M-Sand conforming to grading zone II of IS: 383-1970 (Reaffirmed 2002) and 12mm coarse aggregate conforming to Table 2 of IS: 383-1970 (Reaffirmed 2002) and portable water were used to obtain concrete of M20 grade. M20 mix was designed as per Indian Standard Concrete Mix Proportion Guidelines of IS: 10262-2009.

Table 1. Details of wall panels.

Panel designation	Specimen size (Lxhxt) (mm)	Slenderness Ratio, SR (h/t)	Aspect Ratio, AR (h/L)
BRCWPI	1320x2000x80	25	1.515
BRCWPII	1660x2000x80	25	1.204
BRCWPIII	2000x2000x80	25	1

2.2 Reinforcement

Varnished and sand blasted splints of BambusaBambos of 20mm width were used as reinforcement in wall panels. The properties of bamboo splints used as reinforcement in this study include an average ultimate tensile strength of 120MPa, modulus of elasticity of $6.73x10^4$MPa and an average compressive strength of culm of 40MPa. Spacing of splints was provided as per the National Building Code of India-Part 6. Splints were kept at equal spacing in both the horizontal and vertical directions. Figure 1 shows the arrangement of bamboo splints used as reinforcement in concrete wall panels.

Figure 1. Typical arrangement of reinforcement

2.3 Casting of Specimens

The machine mixed concrete was used for the casting of wall panels and the bamboo reinforcement cage was kept at the mid thickness of the mould. Concrete was compacted using needle vibrator and after 24 hours of casting, the panels were cured by covering it with wet gunny bags for 28 days.

2.4 Test Set Up for the Testing of Wall Panels

The panels were tested under pinned end conditions at both ends with uniformly distributed load applied at an eccentricity of t/6 to reflect the possible eccentric load in practice. All specimens were tested in the vertical position in a loading frame of 100 tons capacity installed in the Structural Engineering Lab. The wall panels were lifted using the Electric Overhead Travelling (E.O.T.) crane of 10 tons capacity. Figure 2 shows the schematic diagram of the test set up. The top and bottom hinged support conditions were simulated by placing a 16mm diameter polished rod in between four 6mm diameter rollers welded to the bearing plates. A stiffened I-beam was used to apply the load at the top and another stiffened I-beam was used below the bearing plates to act as support. Figure 3 shows the details of top hinged edge. A plumb bob was used to ensure the verticality of the wall panels. Loading was done gradually in stages up to failure and the experimental ultimate loads were recorded. Figure 4 shows the experimental test setup.

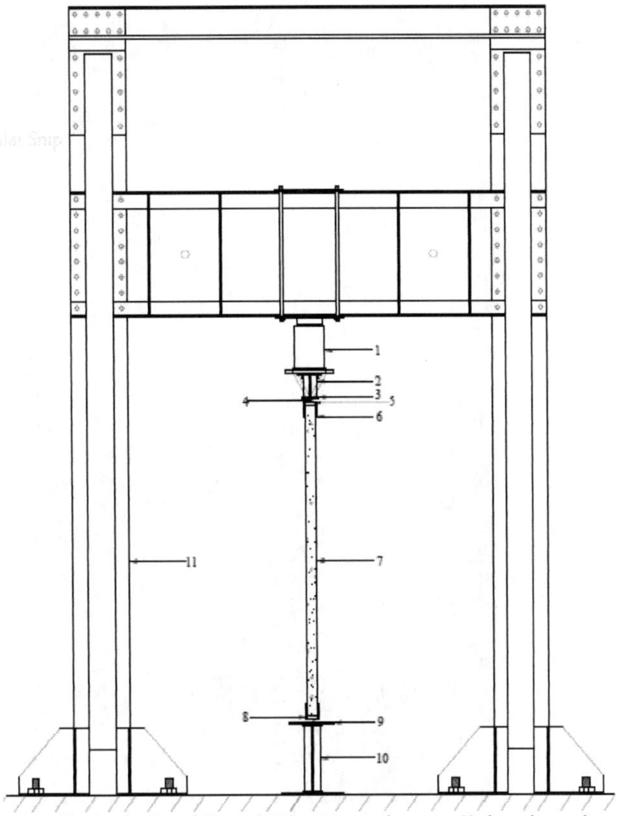

1) Hydraulic jack 2) stiffened I beam at the top 3) bearing plate 4) 16mm diameter roller 5) 4 number of 6mm diameter guide bars for roller 6)guide plate 7)specimen 8) bearing plate 9) 10mm thick steel plate 10) stiffened I beam at the bottom 11) Loading frame.

Figure 2. Schematic diagram of test set up

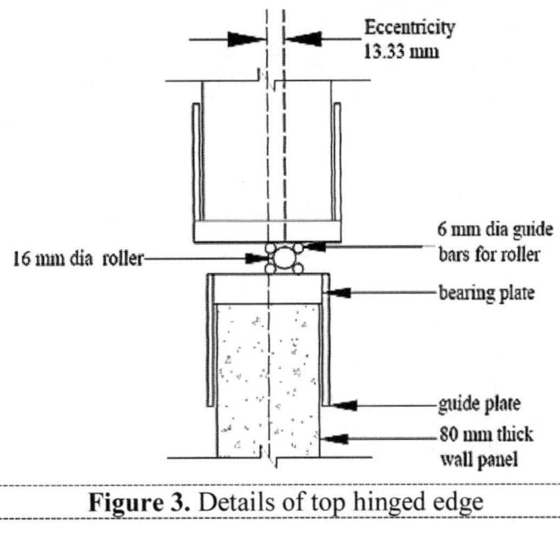

Figure 3. Details of top hinged edge

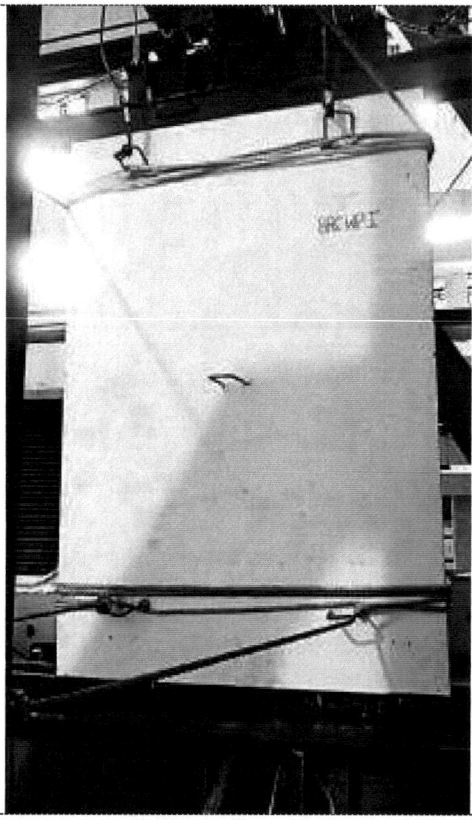

Figure 4. Test set up

3. Comparison of Earlier Studies

The studies conducted by various researchers on wall panels are summerised in the form of tables. The experimental study on bamboo reinforced concrete wall panels are compared with the studies done by various researchers on Reinforced Concrete (RC)wall panels.

3.1 Earlier studies

Researchers used various types of concrete like Normal Strength Concrete (NSC), High Strength Concrete (HSC), Steel Fiber Reinforced Concrete (SFRC), Self Compacting Concrete (SCC), Ultra High Performance Concrete (UHPC), Steel Fiber Reinforced Self Compacting Concrete (SFRSCC), Geo Polymer Concrete (GPC) and Steel Fiber Light Weight Concrete (SFLWC) as matrix for wall panel specimens with either steel bars or mesh or both as reinforcement in concrete. Table2 presents the type of concrete and the percentage of reinforcement used in the wall panel studies by various investigators. It may also be noticed that as the percentage of vertical steel increases, strength of the wall panel increases whereas, the percentage of horizontal steel has little effect on the strength of wall panel under one way in-plane loading. But as the percentage of horizontal steel increases, the strength of wall panels increases for the wall panels under two way in-plane action due to the contribution of horizontal reinforcement in resisting the two bending action of wall panels. The review reveals that the strength of the wall panel does not increase linearly with an increase in the strength of concrete. Also a study on full scale ribbed RC wall panels by Zielinski et al. found that the rigidity of wall panels improved by the presence of perimeter ribs.

Table 2. Type of concrete and percentage of reinforcement used in various investigations.

Researcher	Type of concrete	Percentage of reinforcement	
		Horizontal	Vertical
Swartz et al.(1974)	NSC	0.2 to 1	0.2 to 1
Oberlender and Everard (1977)	NSC	0.47	0.33
Pillai and Parthasarathy (1977)	NSC	0.14 to1.5	0.15
Zielinski et al. (1982)	NSC	-	-
Saheb and Desayi (1989)	NSC	0.19 to 0.50	0.19 to 0.50
Saheb and Desayi (1990)	NSC	0.19 to 0.50	0.16 to 0.85
Fragomeni et al. (1994)	NSC	0.208	0.15
Sanjayan and Maheswaran (1999)	HSC	0.85 to 1.69	0.85 to 1.69
Doh and Fragomeni (2004)	NSC and HSC	0.25	0.15
Ganesan et al. (2009)	NSC	0.74	0.88
Ganesan et al. (2010)	NSC and SFRC	0.74	0.88
Ganesan et al. (2010)	SCC	0.74	0.88
Ruby et al. (2011)	UHPC	0.25	0.15
Ganesan et al. (2012)	SFRSCC and SFRC	0.74	0.88
Ganesan et al. (2013)	GPC	0.74	0.88
Ganesan et al. (2013)	UHPC	0.25 to 0.65	0.15
Ganesan et al. (2014)	UHPC	0.25	0.15
Mamat et al. (2015)	SFLWC	0.5	

Details of SR and AR considered in the study of wall panels by various researchers are presented in Table 3. Except the studies done by Swartz et al. (1974), Saheb and Desayi (1990), Sanjayan and Maheswaran (1999), Doh and Fragomeni (2004) and Ganesan et al. (2013 and 2014) on the performance of wall panels under two way in-plane action all other studies done by various researchers focuses on the performance of wall panels under one way in-plane loading. The studies indicate that panels with two-way in plane action are more rigid and strong than the wall panels supported at the top and bottom edges alone, since their four edges are being supported.

The review of various investigations concluded that as SR increases, strength of wall panels under one way in-plane loading decreases whereas it increases for wall panels under two way in-plane loading. Limited studies are reported on the effect of AR on the strength of wall panels. It was found that as AR increases, strength of wall panels under one way in-plane loading decreases whereas it increases for the wall panels tested under two way in-plane action. Also it was noticed that the studies on wall panels done by researchers other than Zielinski et al. used scaled down model of wall panels.

Many investigators have proposed formulae to predict the ultimate load (Pu) of wall panels either by modifying the existing wall design formulae or by developing a new formula based on certain assumptions in their study. The formula proposed by various authors is listed in Table 4. Formulae developed by Oberlender and Everard (1977), Pillai and Parthasarathy (1977) and Kripanarayanan (1977), Saheb and Desayi (1989, 1990), IS 456 (2000) and ACI 318 (2008) was limited to NSC where as Saheb and Desayi (1990) had developed separate formula for NSC and HSC wall panels. Ganesan et a0l. (2010, 2012 and 2013) proposed formulae to predict the strength of SCC, SFRSCC and GPC wall panels. All these studies have used steel as reinforcement in concrete matrices.

Table 3. Details of the parameters considered in various investigations.

Researcher	Thickness (t) (mm)	SR (h/t)	AR (h/L)
Swartz et al. (1974)	19,25.4,31.8	38.33 to 76.67	2
Oberlender and Everard (1977)	76.2	8 to 28	1 to 3.5
Pillai and Parthasarathy (1977)	40,48,60,80	5 to 30	0.571 to 3
Zielinski et al. (1982)	38	72.23	2.25
Saheb and Desayi (1989)	50	9 to 27	0.67 to 2
Saheb and Desayi (1990)	50	9 to 27	0.67 to 2
Fragomeni et al. (1994)	35,40,50	8.4 to 12	1.4 to 2
Sanjayan and Maheswaran (1999)	50	8 and 30	1
Doh and Fragomeni (2004)	40	25 to 40	1 to 1.6
Ganesan et al. (2009)	40	12 to 30	0.75 to 1.875
Ganesan et al. (2010)	40	12 to 30	0.75 to 1.875
Ganesan et al. (2010)	40	12 to 30	0.75 to 1.875
Ruby et al. (2011)	25,35,45,55	13 to 30	1
Ganesan et al. (2012)	40	12 to 30	0.75 to 1.875
Ganesan et al. (2013)	40	12 to 21	1.07 to 1.875
Ganesan et al. (2013)	35	21.42	1
Ganesan et al. (2014)	35	14 to 28	1
Mamat et al. (2015)	40	15	0.75 to 1.875

Formulae predicted by Oberlender and Everard (1977), Pillai and Parthasarathy (1977) and Kripanarayanan (1977) to determine the strength of wall panels considered the effect of SR on the strength of wall panels but the contribution of reinforcement was not considered in the study. Even though formulae predicted by Zielinski et al. (1982) and Saheb and Desayi (1989) included the contribution of SR and reinforcement, the contribution of AR was not considered. Saheb and Desayi (1990) predicted a formula considering the contribution of SR, AR and reinforcement on the strength of wall panels but that formula was applicable only for the wall panels with aspect ratio less than 2.

Table 4. Equations proposed by various researchers.

Researcher	Equation
Oberlender and Everard (1977)	$P_u=0.6\Phi Af_c'[1-(h/30t)^2]$
Pillai and Parthasarathy (1977)	$P_u=0.57\Phi Af_c'[1-(h/50t)^2]$
Kripanarayanan (1977)	$P_u=0.55\Phi Af_c'[1-(kh/32t)^2]$
Zielinski et al. (1982)	$P_u=0.55\Phi[Af_c'+(f_y- f_c')A_{sc}][1-(h/40t)^2]$
Saheb and Desayi (1989)	$P_u=0.55\Phi[Af_c'+(f_y- f_c')A_{sc}][1-(h/32t)^2]$
Saheb and Desayi (1990)	$P_u=0.67\Phi[Af_c'+(f_y-f_c')A_{sc}][1-(L/120t)^2][1+0.12h/L)]$
Fragomeni and Mendis (1996)	$P_u=0.60\Phi f_c'(t-1.2e-2\ e_a)L$
	$P_u=30\Phi f_c'(t-1.2e-2\ e_a)(1+(f_c'-50)/80)L$
BS8110 (1997)	$P_u=0.3f_c'(t-2e)L$
IS 456 (2000)	$P_u=0.3f_{ck}(t-1.2e-2\ e_a)L$
Doh and Fragomeni (2004)	$P_u=2f_c'^{0.7}(t-1.2e-2\ e_a)L$
ACI 318 (2008)	$P_u=0.55\Phi Af_c'[1-(kh/32t)^2]$
Ganesan et al. (2010)	$P_u=0.57[Af_c'+(f_y-f_c')A_{sc}][1+(h/36t)-(h/29t)^2][1(h/9L)]$
Ganesan et al. (2012)	$P_u=0.56[Af_c'+(f_y-f_c')A_{sc}][1+(h/29t)-(h/26t)^2][1-(h/11L)]$
Ganesan et al. (2013)	$P_u=0.585[Af_c'+(f_y-f_c')A_{sc}][1+(h/40t)-(h/30t)^2][1(h/18L)]$

where, A= cross sectional area of wall panel, A_{sc}= area of compression reinforcement, f_c'= cylinder compressive strength, f_{ck}= characteristic compressive strength, f_y= yield strength of steel, h= height of wall panel, h/L= aspect ratio, h/t= slenderness ratio, k= effective length factor, L= length of wall panel, t= thickness of wall panel, Φ= strength reduction factor.

Fragomeni et al. (1994) proposed formulae for determining the strength of NSC and HSC wall panels but it didn't consider the effect of SR, AR and reinforcement. BS8110 (1997), Doh and Fragomeni (2004) and IS 456 (2000) developed formulas to predict the strength of wall panels considering the effect of the effect of SR, and the eccentricity at which load is applied to the wall panels, but the formulae didn't consider the contribution of AR and reinforcement on its strength. Formula suggested for predicting the ultimate strength of wall panels by ACI 318 (2008) included the strength reduction factor (Φ) and effective length factor (k) along with the term h/t for considering the effect of SR, but the formula doesn't consider the effect of AR and the contribution of reinforcement. Formula proposed by Mac Gregor and Wight (2009) does not consider the effect of AR, SR, strength reduction factor (Φ) and percentage of reinforcement hence it may overestimate the strength of wall panels. Ganesan et al. (2010, 2012 and 2013) proposed formulae to predict the strength of SCC, SFRSCC and GPC wall panels considering the contribution of SR, AR and reinforcement.

4. Test Results

The experimental values obtained for Bamboo Reinforced Concrete Wall Panel and the ultimate loads obtained by using the equations proposed by various researchers for Reinforced Concrete wall panels are given in Table 5.

Table 5. Ultimate loads obtained by using the equations proposed by various researchers

Researcher	Ultimate load (kN)		
	BRCWPI	BRCWPII	BRCWPIII
Oberlender and Everard(1977)	276.89	335.12	366.64
Pillai and Parthasarathy(1977)	645.66	781.45	854.94
Kripanarayanan(1977)	323.67	391.74	428.58
Zielinski et al(1982)	736.11	899.40	1006.63
Saheb and Desayi(1989)	587.51	739.14	842.89
Fragomeni(1994)	233.06	282.07	308.60
BS 8110(1997)	323.64	391.70	428.54
IS 456(2000)	647.12	783.21	856.87
Doh and Fragoemni(2004)	523.72	641.19	722.09
ACI 318(2008)	346.79	419.72	459.19
Experimental ultimate load	699.02	796.11	854.36

5. Conclusions

The ultimate load estimated by using the equations of Fragomeni, IS 456 and Pillai and Parthasarathy were comparable with the experimental values. The equation proposed by Oberlender and Everard, Kripanarayanan, BS8110, Doh and Fragomeni and ACI 318 underestimates the ultimate load. From the investigation it may be noted that BRC wall panels can carry loads as high as 850kN, which indicate that BRC wall panels can effectively replace steel reinforced wall panels leading to sustainability.

Acknowledgements
The authors acknowledge the Director, NIT Calicut, India for providing the facilities to carry out the work and the encouragement in completing this work.

References

[1] Ruby A, Beena P R and Mathew M E 2011 Experimental investigation on behaviour of ultra high performance concrete wall panels under one-way in-plane loading.*Int. J. Earth Sci. Eng.***4**(5) 907-15.

[2] ACI 318 2008 Building code requirements for structural concrete *Am.Concr.Inst.*.

[3] BS8110 1997 Part 1 Structural used of concrete: code of practice for design and construction*British Standard Inst.*

[4] Oberlender G D and Everard G D 1977 Investigation of Reinforced Concrete Walls *ACI J.***74**(28) 256-63.

[5] IS 10262 2009 Recommended guidelines for concrete mix design *Bureau of Indian Standards* New Delhi India.

[6] IS 383 1970 (Reaffirmed 2002) Specification for coarse and fine aggregate from natural sources for concrete *Bureau of Indian Standards* New Delhi India.

[7] IS 456 2000 Indian Standard Code of Practice for Plain and Reinforced Concrete Structures *Bureau of Indian Standards* New Delhi India.

[8] SanjayanJ G and MaheswaranV 1999 Load Capacity of Slender High-Strength Concrete Walls with Side Supports.*ACI Struct. J.***96**(63) 571-7.

[9] MacGregor J G and Wight J K 2009 Reinforced Concrete – Mechanics and Design *Pearson Prentice Hall* New Jersey.

[10] Doh J H and FragomeniS 2004 Evaluation of Experimental Work on Concrete Walls in One and Two-Way Action.*Australian J. Struct. Eng.***6**(1) 1-15.

[11] KripanarayananK M1977 Interesting Aspects of the Empirical Wall Design Equation.*ACI J.***74**(20) 204-7.

[12] Ganesan N, Ruby A, Beena P R and Anil R 2013 Influence of horizontal reinforcement on ultra high performance concrete wall panels under two way in plane action.*Int. J. Sci. Eng. Res.***4**(5) 149-52.

[13] Ganesan N, Indira P V and Anjana S 2013 Prediction of ultimate strength of reinforced geopolymer concrete wall panels in one-way action.*Constr. Build. Mater.***48** 91-7.

[14] Ganesan N, Indira P V and Prasad S R 2009 Ultimate strength of reinforced concrete wall panels.*Int. J. Earth Sci. Eng.***2**(4) 340-50.

[15] Ganesan N, Indira P V and Prasad S R 2010 Steel fiber reinforced concrete load bearing wall panels in one-way action.*ICI-ACECON*IIT Madras, Chennai, India 579-87.

[16] Ganesan N, Indira P V and Prasad S R 2012 strength and behavior of SFRSCC and SFRC wall panels under one-way in-plane action.*RILEM, HPFRCC***6**, 279-86.

[17] Ganesan N, Indira P V and Prasad S R 2010 Strength and behaviour of reinforced scc wall panels in one- way action.*Struct. Eng. Mech.***36**(1) 1-18.

[18] Ganesan N, Abraham R, Beena P R and Anisha T 2014 Behaviour of ultra high performance concrete wall panels under two-way in-plane loading.*Int. J. Struct. Eng.***5**(3).

[19] Mamat R, Hamzah S H, Ishak S S and SanyS J M 2015 Behaviour of steel fibre lightweight wall panel under one way in-plane action.*Int. J.Innov. Res. Sci.***4**(4) 1801-8.

[20] Saheb S M and DesayiP 1990 Ultimate strength of R.C. wall panels in two-way in-plane action.*J. Struct. Eng.***116**(5) 1384-402.

[21] FragomeniS and MendisP A 1996 Improved axial load formulae for normal and high strength reinforced concrete walls.*Australian Civil/Structural Eng. Trans. Inst Eng.***38**(5) 204-7.

[22] Fragomeni S, Mendis P A and Grayson W R 1994 Review of reinforced concrete wall design formulas.*ACI Struct. J.***91**(50) 521-8.

[23] Pillai S U and Parthasarathy C V 1977 Ultimate strength and design of concrete walls.*Build. Environ.***12** 25-9.

[24] Zielinski Z A,TroitskyM S and ChristodouloH 1982 Full-scale bearing strength investigation of thin wall-ribbed reinforced concrete panels.*ACI J.***79**(32) 313-21.

Numerical investigation into thermal load responses of steel railway bridge

K Saravana Raja Mohan[1*], J R Sreemathy[1] and U Saravanan[2]

[1]School of Civil Engineering, SASTRA University, Thanjavur, Tamil Nadu, India
[2]Department of Structural Engineering, Indian Institute of Technology Madras, Chennai, Tamil Nadu, India

*Email: srm@civil.sastra.edu

Abstract. Bridge design requires consideration of the effects produced by temperature variations and the resultant thermal gradients in the structure. Temperature fluctuation leads to expansion and contraction of bridges and these movements are taken care by providing expansion joints and bearings. Free movements of a member can be restrained by imposing certain boundary condition but at the same time considerable allowances should be made for the stresses resulting from this restrained condition since the additional deformations and stresses produced may affect the ultimate and serviceability limit states of the structure. If the reaction force generated by the restraints is very large, then its omission can lead to unsafe design. The principal objective of this research is to study the effects of temperature variation on stresses and deflection in a steel railway bridge. A numerical model, based on finite element analysis is presented for evaluating the thermal performance of the bridge. The selected bridge is analyzed and the temperature field distribution and the corresponding thermal stresses and strains are calculated using the finite element software ABAQUS. A thorough understanding of the thermal load responses of a structure will result in safer and dependable design practices.
Keywords: Steel railway bridge; Temperature variations; ABAQUS; Thermal stresses and strains

1. Introduction

Uncertainties with regards to both the magnitude and consequences of thermal stresses and strains in bridges are of a major trepidation to design engineers. Both the short-term transient daily temperature changes and the prolonged seasonal changes cause thermal stresses and strains in the bridge. For this very reason, thermal loads must be considered during bridge design and for the structural condition evaluation during the lifetime of the structure. The primary focus of this study is to enhance the comprehension of the thermal behavior of steel bridges subjected to temperature variations.

[1] proposed a method to calculate thermal stresses and deflections in a statically determinate beam based on rigorous analyses by assuming constant longitudinal and transverse temperature, uniform temperature through the steel girder and linearly varying temperature through the cross section. [4] measured two-dimensional steady-state temperature distributions of a steel simple span bridge. To achieve a steady-state thermal condition, the top and bottom surfaces of the bridge were exposed to known thermal boundary conditions. Temperature and strain variations in the mid-span section were recorded. The finite element method was used to simulate the bridge conditions to verify

its validity. [10] proposed a vertical temperature distribution for a steel girder bridge based on a synthesis of several theoretical and experimental studies on prototype bridges. [11] conducted two-dimensional heat conduction analyses by assuming constant longitudinal temperature to predict temperature distribution within steel bridges. They also reduced the two-dimensional models to single dimensional models by further assuming constant transverse temperature. Based on finite element method, [14] developed a numerical model for the prediction of the temperature distribution of steel bridges with steel decks. The calculated temperatures were compared with the measurements taken from scaled models and good agreement was achieved.

The empirical formulae and boundary conditions of heat transfer in bridges are first outlined and then solved using a general finite element software, ABAQUS to determine the temperature distribution and the consequent thermal stresses and strains. Following that, the effects of thermal load on the representative bridge model are discussed in detail. Thereby, a preliminary investigation on the temperature variation across the section of the selected representative bridge is conducted and the resulting thermal stresses and movements are discussed in detail. Though the track-bridge interaction is an important parameter defining the behavior of the entire structure, a detailed rail-deck interaction and dynamic response of the bridge is outside the scope of this work.

2. Methodology

2.1 Bridge description and field measurements

The numerical model is based on bridge number 81 on the Chennai-Mumbai line near Nagari in Maharashtra. It is a steel I-girder railway bridge spanning 13.31m in the longitudinal direction and 2.16m in the transverse direction. The total depth of the bridge is 1.212m. It is a simply supported structure resting on bearings placed on supports with the total bearing area being $1080mm^2$(332mm*325mm) on all the four edges of the bottom flanges. The values of Rayleigh damping coefficients alpha and beta are 6.13368 and 0.000407588 respectively. The assembly temperature is 30°C.

A variety of instruments were used in this study to determine some of the input variables. Thermocouples were used to measure temperature while a pyranometer was used to measure the total solar radiation on a flat surface. A wax trace box was used to measure the movements at the bearings to gain a measure of the actual boundary conditions at the support in comparison to the idealized support conditions. The temperature values measured using the thermocouples and boundary conditions determined using wax trace box were used as input variables to the numerical model.

2.2 Heat transfer mechanisms

Heat is transferred by conduction within a solid and by radiation and convection with the surrounding environment. The heat conduction can be modelled by applying the principle of Fourier's law and the heat exchange is formulated by boundary conditions.

In 1822, Fourier stated that the rate of heat transfer is proportional to the temperature gradient in a solid and established the well-known Fourier partial differential equation, which is

$$\frac{\partial}{\partial x}\left(k_x \frac{\partial T}{\partial x}\right) + \frac{\partial}{\partial y}\left(k_y \frac{\partial T}{\partial y}\right) + \frac{\partial}{\partial z}\left(k_z \frac{\partial T}{\partial z}\right) + Q = c\rho \frac{\partial T}{\partial t} \quad (1)$$

where k_x, k_y and k_z are thermal conductivity values corresponding to x, y and z cartesian axes, T is the temperature at position (x, y, z) at time t and c is the coefficient of specific heat of medium.

The material is a continuum, isotropic and homogeneous. After the hydration of cement in concrete bridge and the action of welding in steel bridge, the rate of heat generation Q can be set to zero.Considering heat exchange, the boundary condition is expressed as follows:

$$k\left(\frac{\partial T}{\partial x}n_x + \frac{\partial T}{\partial y}n_y\right)+q=0 \quad (2)$$

where n_x and n_y are direction cosines of the unit outward normal vector to the boundary surface and q is rate of heat exchanged between the boundary and environment per unit area.

2.3 Temperature components

According to the different effects, the thermal load can be classified as effective temperature, vertical temperature difference and horizontal temperature difference, defined below by [16].The effective temperature, which accounts for expanding and contracting of bridge components in the longitudinal direction, is the weighted mean value of temperature distributed along the cross section. The vertical temperature difference, which results in supplementary internal axial forces and bending moments in the vertical plane when the section ends are restrained, refers to the difference of temperatures between the top surface and other levels in the cross section. The horizontal temperature difference, which induces secondary internal axial forces and bending moments in horizontal plane if the deformation is constrained, represents the difference of temperatures between two positions on the same level in the cross section. According to the definition, the effective temperature T_e, vertical temperature difference T_v and horizontal temperature difference T_h can be expressed as:

$$T_e = \frac{1}{A}\iiint_A T(x,y)zdxdy \quad (3)$$

$$T_v = \frac{H}{I_x}\iint_A T(x,y)ydxdy \quad (4)$$

$$T_h = \frac{W}{I_y}\iint_A T(x,y)xdxdy \quad (5)$$

2.4 Thermal strain and deformation

Thermal strains are basically of two types, i) expansion or contraction of length due to an increase or decrease in overall average temperature respectively or ii) bending of members due to the presence of a temperature gradient throughout the depth.The geometric deformations of a structure are a function of temperature. Overall increase or decrease in temperature will lead to expansion or contraction respectively. The thermal strain Ɛt, that develops as a structure is subjected to change in temperature ΔT, can be estimated by,

$$\varepsilon t=\alpha.\Delta T \quad (6)$$

where α is the coefficient of thermal expansion.

Thermal strain is usually designated as positive when it represents expansion and negative when it represents contraction. Following the definition of engineering strain, the change in length ΔL is given by,

$$\Delta L = \varepsilon t.L = \alpha.\Delta T.L \quad (7)$$

where L is the length of the structure in the direction considered. This is the maximum change in length possible when the material is able to expand and contract freely, in the case of which no thermal stresses are produced.

A member whose ends are not restrained against rotation bend in the presence of a thermal gradient present within the member. This phenomenon is known is thermal bowing illustrated in figure1.For simply supported structures, when the bottom portion is warmer than the top, the stresses developed will be additive to those stresses caused by live and dead loads and vice versa.

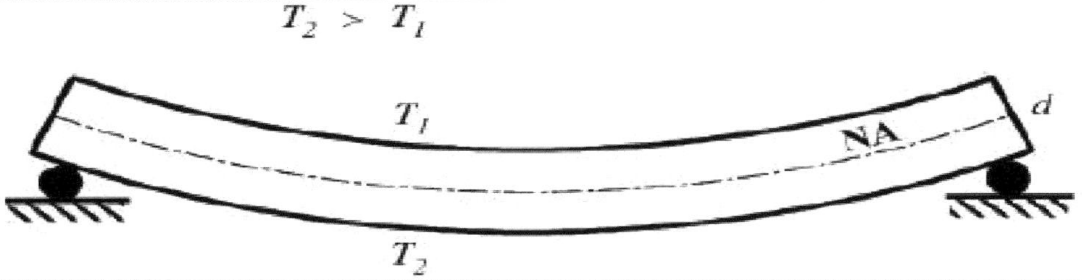

Figure 1. henomenon of thermal bowing

2.5 Thermal finite element analysis using ABAQUS

The partial differential equations discussed in the heat transfer section are solved using the general finite element software ABAQUS.ABAQUS has significant capabilities that are used to solve multi-physics problems. The entire bridge along with bearing was modelled and a fully coupled temperature-displacement analysis was conducted. The individual member components of the bridge namely the top and bottom flanges, web, stiffeners and the bracings are modelled as homogeneous deformable planar shell elements and then assembled to form the 3D bridge model. The shell element is considered rather than a solid element for better modelling of the thermal bowing phenomenon and to neglect the effect of temperature gradient along the thickness of the individual members owing to the negligible thickness of shell elements. The material properties of steel corresponding to both thermal analysis and the consequent stress analysis are given as input. The material properties of steel are tabulated below in Table 1.

Table 1. Material properties of steel

Material Property	Steel
Density	7850 Kg/m^3
Young's Modulus	200000 N/mm^2
Poisson's Ratio	0.3
Coefficient of Thermal Expansion	12e-6 mm/mm.K
Thermal Conductivity	43 W/m.K

A fully coupled steady state temperature-displacement analysis is executed to establish the temperature distribution field throughout the bridge and the consequent thermal stresses and strains in the structure. A good choice of time step is of vital importance, since too large of a time step may miss the peak point of interest while too small of a step leads to poor economy in computer time. The time step depends on the type of the governing partial differential equation and the features of the input. A reasonable time step was chosen for this process. Simply supported boundary conditions

(U1=U2=U3=0 at one end and U1=U2=0 at other end, where U1 is the displacement along transverse direction, U2 is the displacement in the vertical direction and U3 is the displacement along the longitudinal direction) were assigned at the bearing areas and three different types of analysis were conducted, them being: a temperature difference of 0.2^{0}C across the depth of the bridge for temperature values 1) 27^{0}C, 2) 30^{0}C and 3) 40°C. Element sizes have a significant effect on the accuracy of the results. To determine whether the element size is sufficiently fine, the number of elements is incrementally increased and comparisons are made between consequent analyses. Results obtained from the model with a certain number of elements can be compared to those obtained from the model with increased number of elements. If no significant difference is observed between them, then the mesh can be deemed adequately fine. Usually the mesh convergence process involves the comparison of strain energy in the whole body with respect to the number of elements corresponding to the element size. Detailed analyses evaluating mesh sensitivity showed that an element size of 28 mm is reasonable enough for the thermal analysis. Aspect ratio is another meshing factor that can influence the accuracy of the results. Therefore, to avoid excessive distortion of the elements, the aspect ratio of elements must be limited. A 4-node thermally coupled, doubly curved, general-purpose shell element with finite membrane strains named S4T is used for this process. The meshed bridge model is presented below in figure 2.

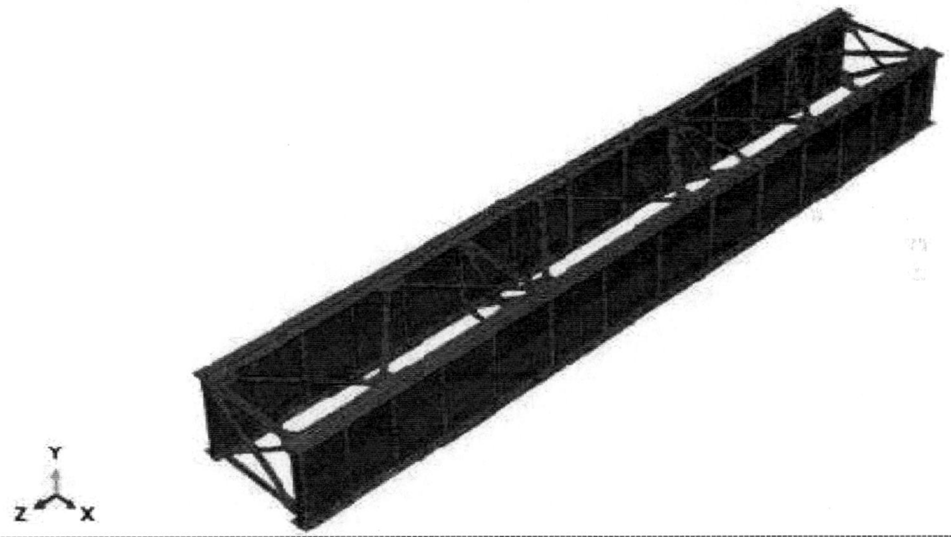

Figure 2. Meshed model

3. Results and discussion

The following results are obtained after conducting a coupled temperature-displacement analysis on the bridge model. The various stresses and strains along the local axes are plotted below. The temperature field distribution, which is the output from the heat transfer analysis and the thermal stresses and strains, which are the significant output from the general static stress analysis are shown below. These results are then studied in detail to draw conclusions about the impact of thermal loads on the structural behaviour of the structure.

3.1 Nodal temperature-NT11

Figure 3 to 5 shows that a linear vertical temperature gradient exists along the depth of the steel bridge. This is due to the high thermal conductivity value of steel, since steel is a very good conductor of heat. In the case of concrete which has a lower thermal conductivity when compared to steel, a non-linear vertical gradient will be prevalent. The minimum and maximum temperatures recorded using the thermocouples during the field measurements and which were input as a variable field in the numerical thermal analysis, which are 27°C and 40°C respectively, is well within the specified range of -5°C to 70°C as specified by [9] and -17°C to 48°C as per [2]. According to [6], the uniform temperature component depends on the minimum and maximum temperature that the bridge is expected to experience while the vertical temperature differences are considered by using an equivalent linear temperature difference components, ΔT_{heat} and ΔT_{cool}. When the bridge is heated up, the top surface is warmer than the bottom and ΔT_{heat} is taken as 15°C. When the bridge is cooled down, the bottom surface is warmer than the top and $\Delta T_{cool}$is taken as 18°C.

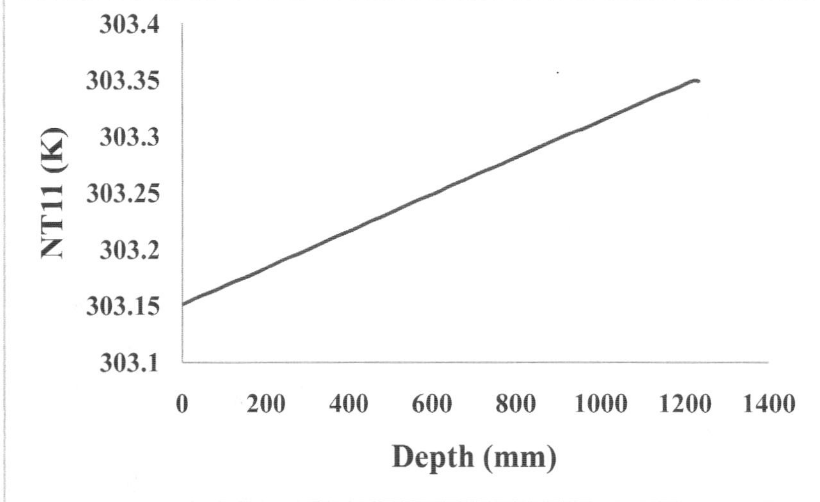

Figure 3. Linear vertical temperature gradient for case 1

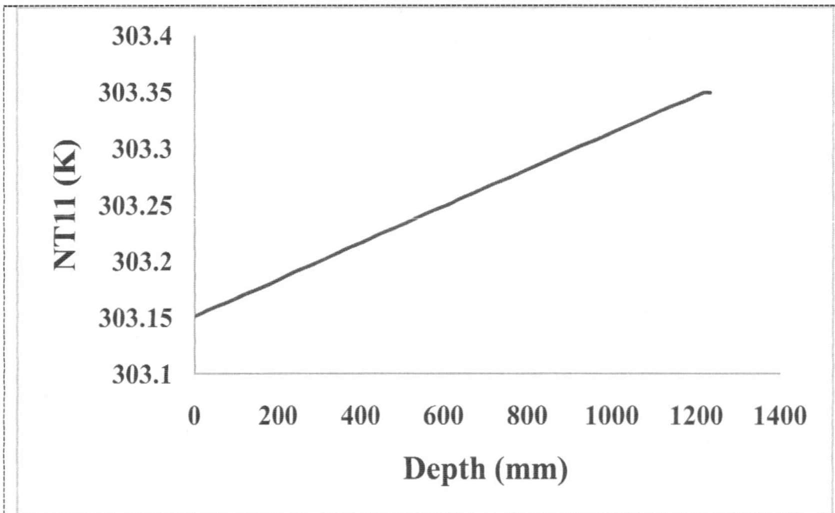

Figure 4. Linear vertical temperature gradient for case 2

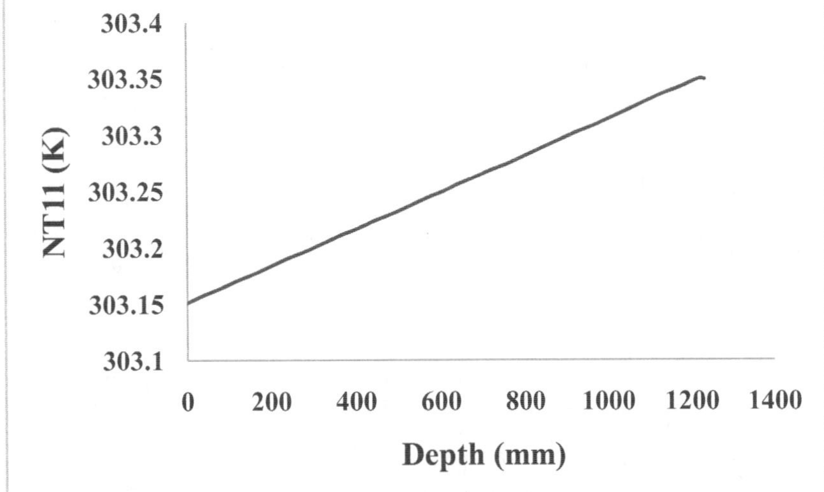

Figure 5. Linear vertical temperature gradient for case 3

3.2Normal stresses-S11&S22

From figures 6 to 8 it can be seen that the stresses vary from compression in the lower flange to tension in the upper flange, while the stresses along the web is approximately equal to zero. The tensile stresses are higher when compared to the compressive stresses since the temperature at the top flange is higher when compared to that of the bottom flange.

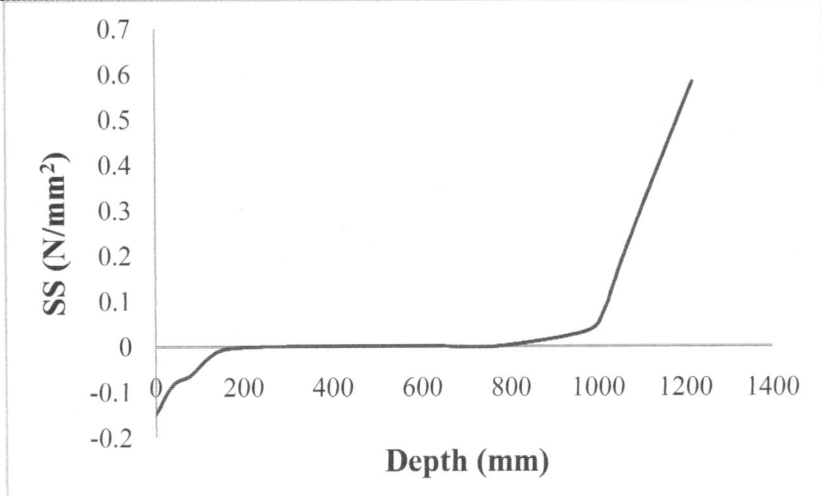

Figure 6. Variation of normal stress component S11 across the depth for case 1

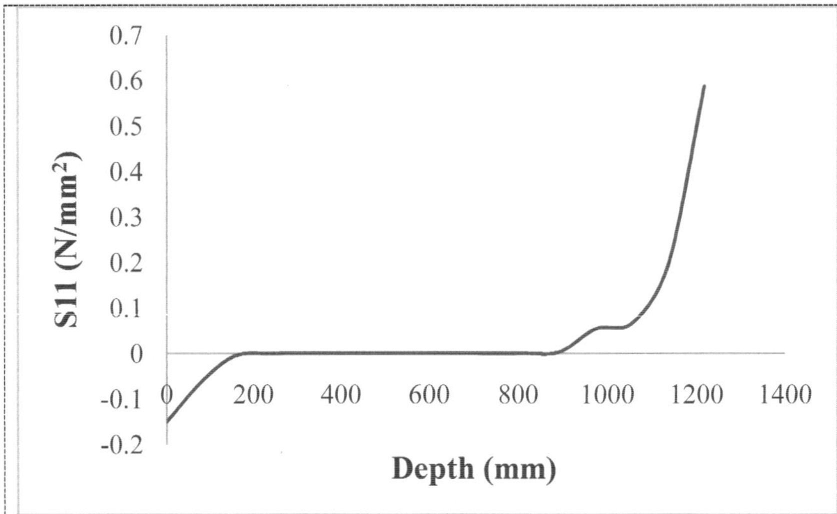

Figure 7. Variation of normal stress component S11 across the depth for case 2

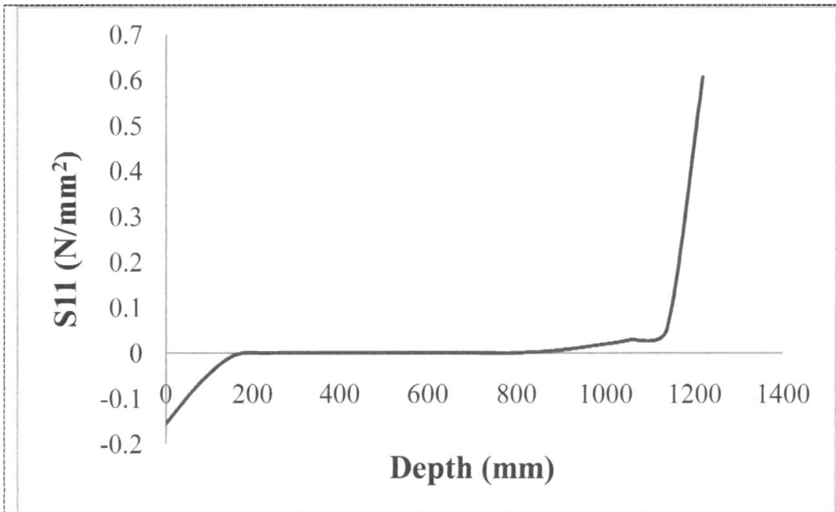

Figure 8. Variation of normal stress component S11 across the depth for case 3

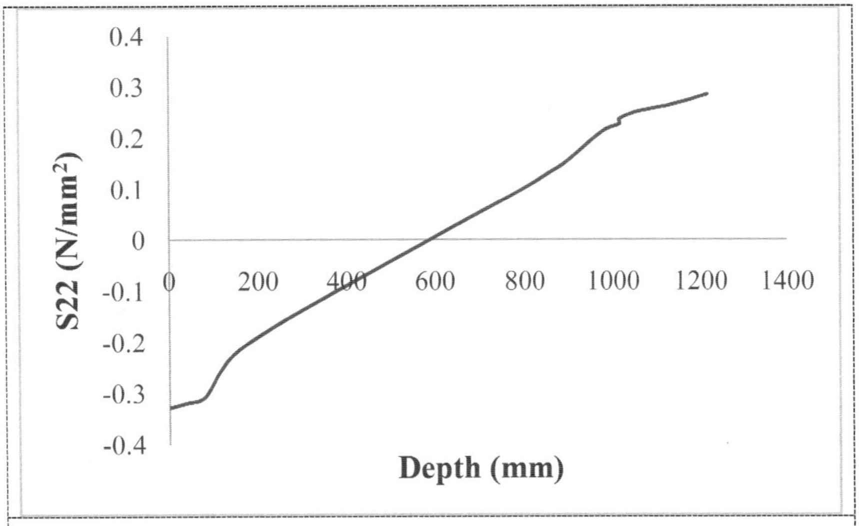

Figure 9. Variation of normal stress component S22 across the depth for case 1

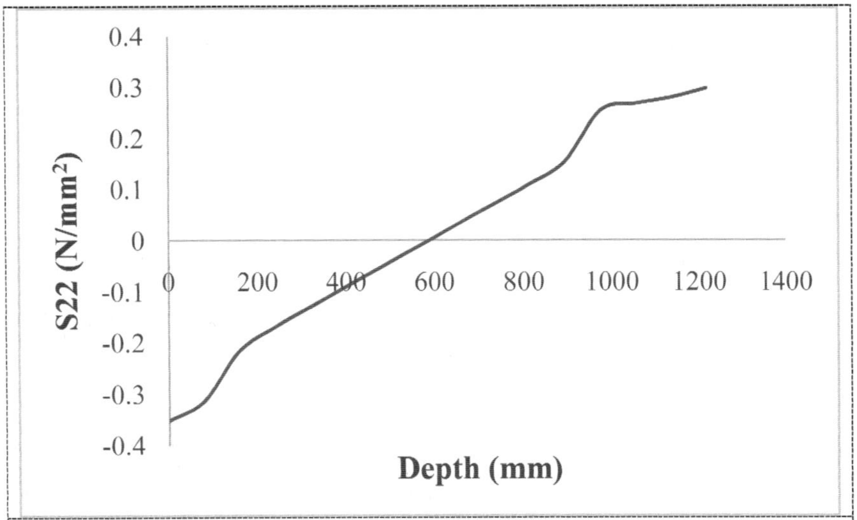

Figure 10. Variation of normal stress component S22 across the depth for case 2

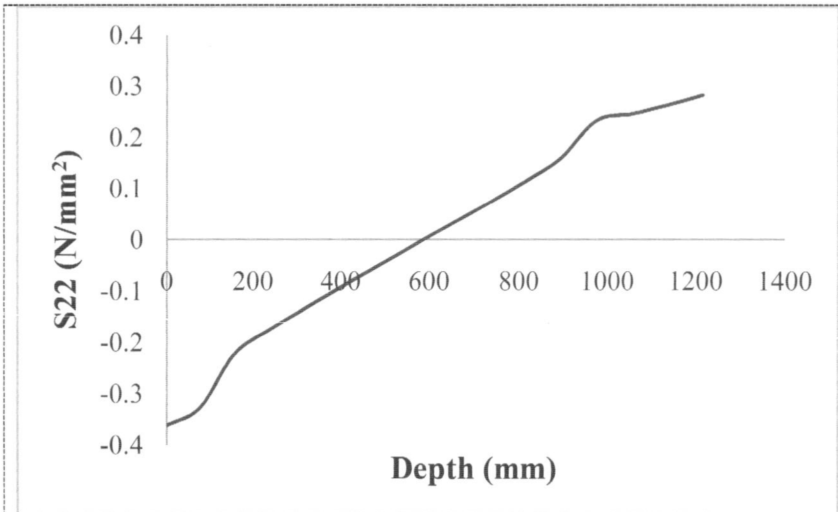

Figure 11. Variation of normal stress component S22 across the depth for case 3

A linear stress variation along the depth of the bridge is seen in figure 9 to 11 ranging from compression in lower flange to tension in upper flange. This also proves the phenomenon of thermal bowing indicating that the section arches up when the top surface is warmer than the bottom one leading to compressive stresses at the bottom and tensile stresses at the top.According to [9], design of steel railway structures usually is based on a working stress level that is some fraction of the minimum yield strength of the material. This value is usually taken as 0.55, allowing a safety factor of 1.82 against yielding of steel. Hence the working stress for structural steel with a yield strength of 250 N/mm^2 is 137.5 N/mm^2. The maximum stress obtained is 0.6 N/mm^2 experienced by the upper flange for a temperature of 313.35 K. Hence in this case thermal stresses accounts for 0.4% of the entire working stress and does not significantly affect the strength criterion of the structure. But when the field temperature is higher than the maximum value considered then there may be higher thermal stresses which may severely affect the ultimate limit states of the structure.

3.3 Shear stress-S12

Figures 12 to 14 confirms that there are shear forces at the supports or restraints giving rise to shear stresses across the bearing area. There are comparatively higher stresses at the right end which is the pinned support where the displacement along all the three mutually perpendicular directions is arrested giving rise to greater stresses when compared to the left end which is the roller support which allows displacement along the longitudinal axis.

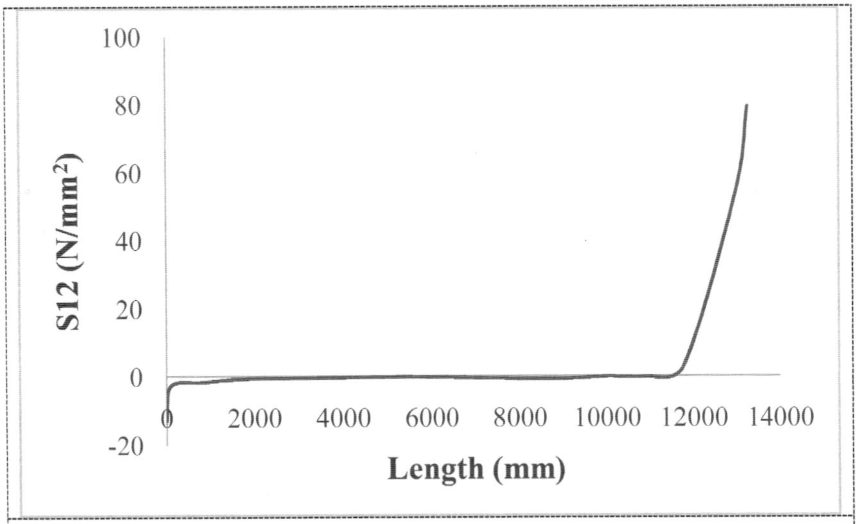

Figure 12. Variation of shear stress along the length for case 1

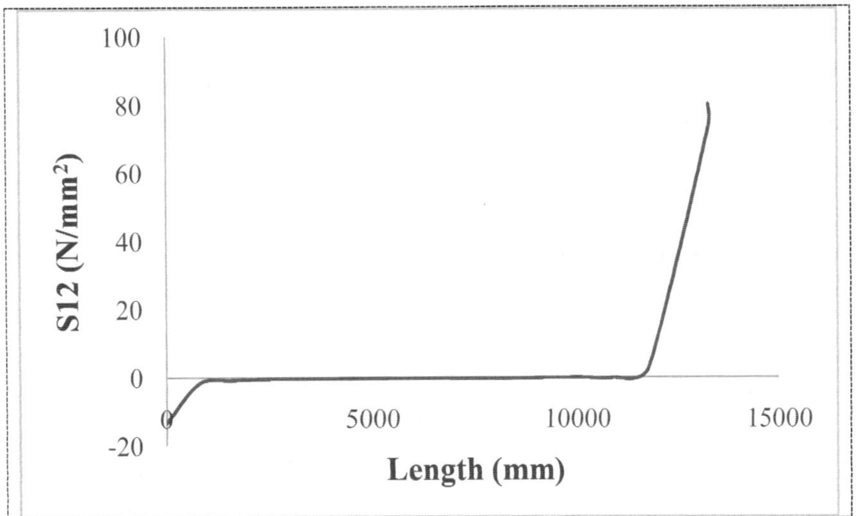

Figure 13. Variation of shear stress along the length for case 2

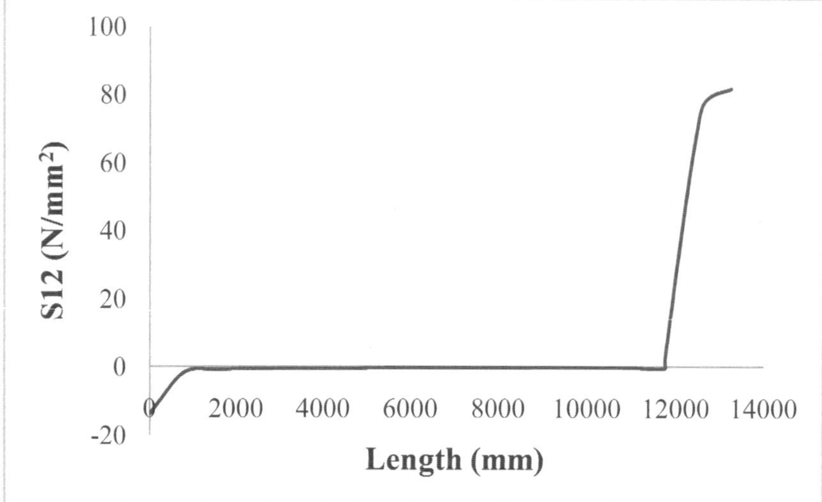

Figure 14. Variation of shear stress along the length for case 3

3.4 Displacement along vertical direction-U2

Figures 15 to 17 shows the vertical deflection along the length of the bridge. The deflection limit as specified by [9] should not be greater than 1/600 (deflection/length of the girder). The maximum deflection obtained in our case is 2.4569/13319=0.00018 which is less than the specified limit of 0.001. Hence the serviceability limit state of the bridge is not greatly affected. But in case of higher temperature, the deflection may exceed the permissible limit and this may influence the bridge-rail interaction which may cause vertical displacement of the rail leading to track irregularity.

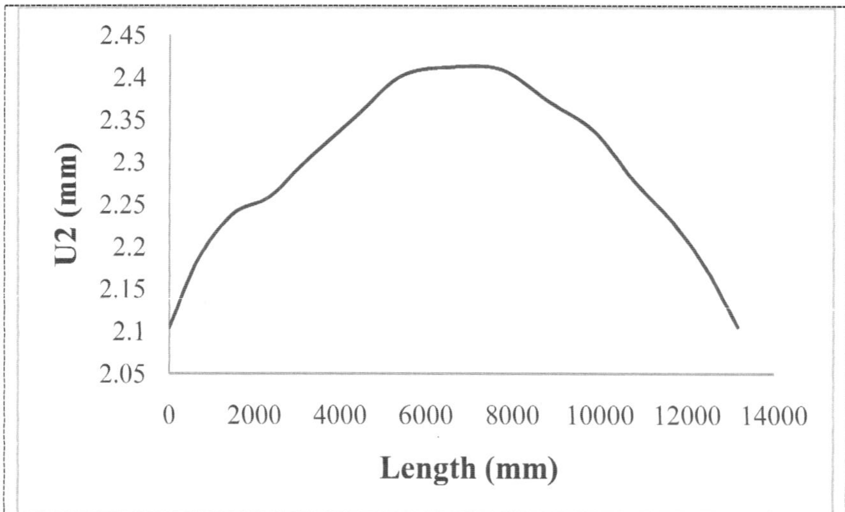

Figure 15. Deflection curves for case 1

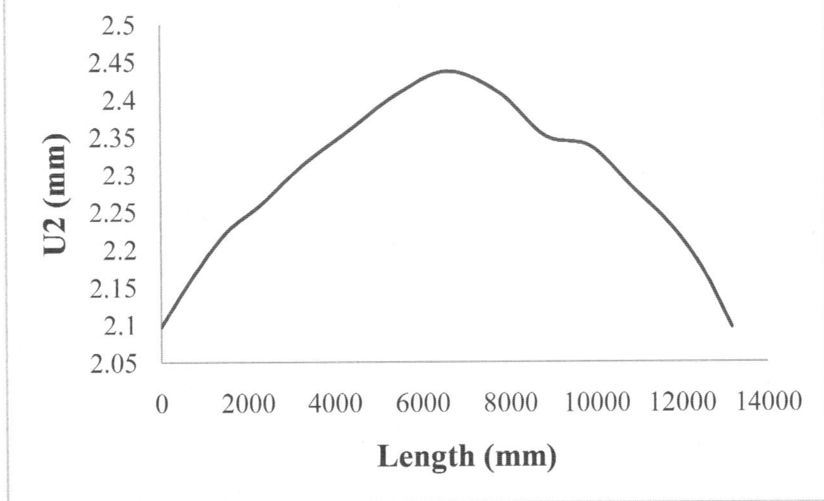

Figure 16. Deflection curves for case 2

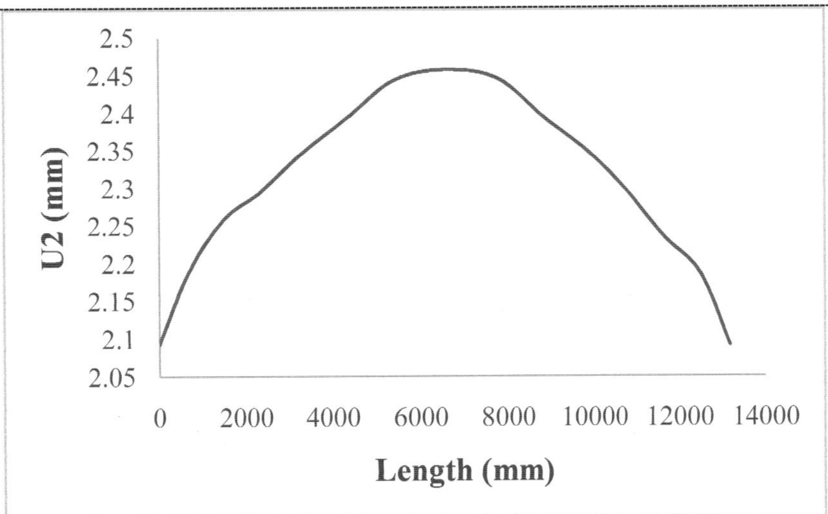

Figure 17. Deflection curves for case 3

3.5 Displacement along longitudinal direction-U3

It can be clearly noted from figures 18 to 20. That the maximum longitudinal displacement occurs at the left side roller support while there is zero displacement at the right side pinnedsupport. This movement is usually accommodated by expansion joints and bearings. According to [9], where provision for expansion and contraction, due to change in temperature and stress, is necessary, it shall be provided to the extent of not less than 25mm (1in) for every 30m (100ft) of span. The expansion bearings shall allow free movement in a longitudinal direction and at the same time prevent any transverse motion which corroborates our support conditions. [2] specifies a limit of 1 to 1¼ inches of movement for each 100 feet of span length. The maximum longitudinal displacement obtained is 49.327mm. Since this displacement is greater than the specified minimum displacement of 25mm, expansion joints should be provided at the ends of the girder.

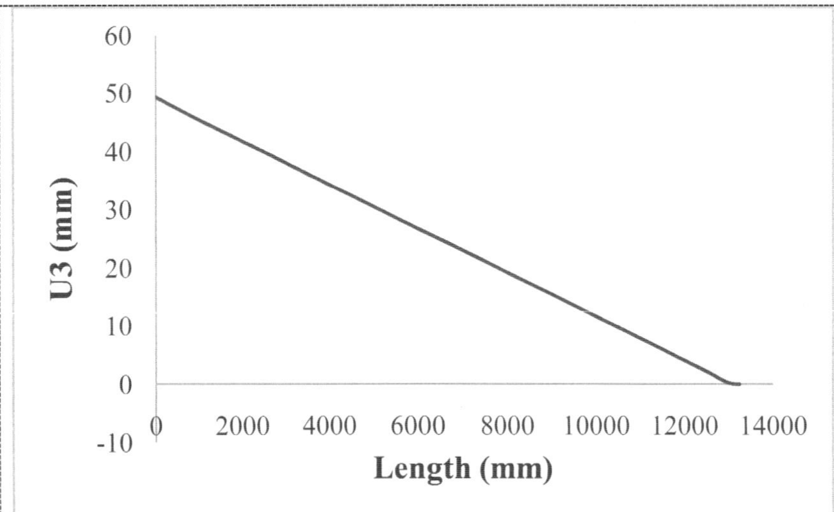

Figure 18. Longitudinal displacement for case 1

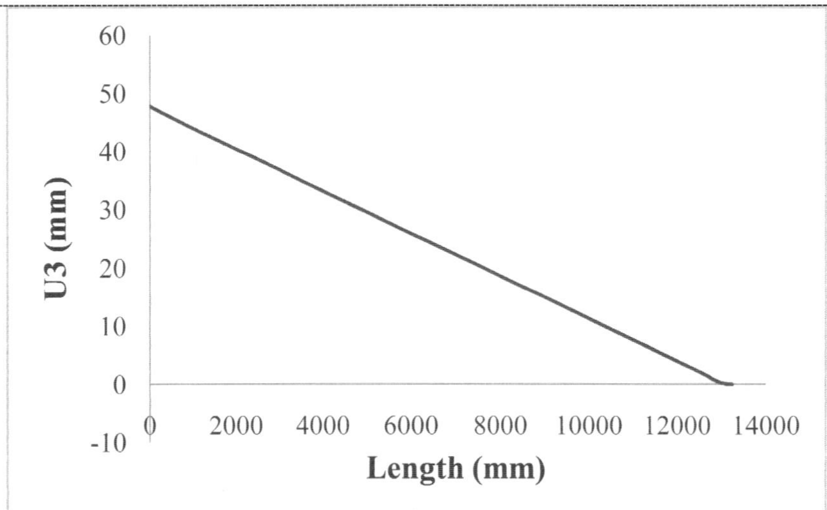

Figure 19. Longitudinal displacement for case 2

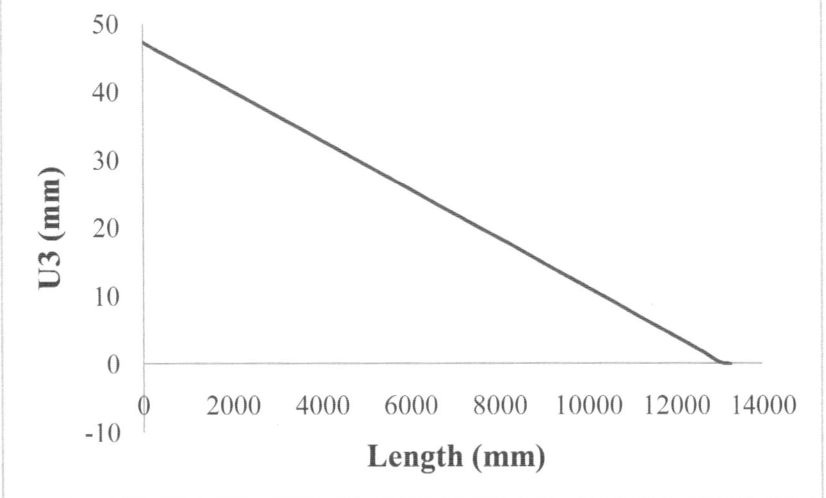

Figure 20. Longitudinal displacement for case 3

4. Conclusions

The main aim of this study is to demonstrate the effects of thermal gradient present within a steel railway bridge on its structural behaviour and its impact on the ultimate and serviceability limit states. Due to continuous climatic fluctuations in the surrounding environment, it is inevitable that structures are constantly subjected to varying temperatures resulting in temperature gradients within them. The thermal loads depend on various factors like amount of solar radiation, precipitation type and amount, wind effects, geographic location of the structure, orientation, material properties, geometry and so on. Thermal loads in structures result in thermal strains and in case these strains are restricted, stresses develop within the member. Hence it is of utmost importance that the thermal loads must be considered in the design in order to produce safe and reliable structures. This study includes field measurements followed by numerical analysis. A fully coupled temperature-displacement analysis is performed on the selected bridge model using the finite element software ABAQUS. After analysing the structural response of the steel bridge subjected to linear vertical temperature gradient it was concluded that, if the minimum and maximum temperature the bridge is subjected to during its lifetime remains within the permissible range as specified by the codal provisions, then the thermal stresses and strains also tend to be within the tolerable range without significantly affecting the performance of the structure. In other cases where there are extreme temperature fluctuations diligent care should be taken while designing the structure since increase in temperature leads to loss of strength and stiffness in members while varying temperatures causes varying thermal loads inducing fluctuating stresses which causes fatigue failure in members and connections. Continuous large expansion and contraction due to increase or decrease in temperatures respectively in expansion devices may lead to its early failure and repeated repair and replacement thereby increasing its maintenance cost. Numerical analysis provides a perfect approach to predict the thermal loads and the resulting thermal movements and stresses. However, these approaches are input parameters dependent and hence the predicted results and values of the various stresses and strains may deviate from the actual case since some of the input or predefined field variables are dependent on environmental factors and are subjected to changes from time to time.

Acknowledgements
The authors acknowledge the Vice Chancellor, SASTRA University, Thanjavur, India and the management of Indian Institute of Technology Madras for providing the facilities to carry out the work and the encouragement in completing this work.

References

[1] Arikan Y 2006 Field Monitoring of Thermal Effects in Steel Box Girder Bridges. *Master's Thesis.*
[2] AASHTO 2007 LRFD Bridge Design Specifications *American Association of Highway and Transport Officials* 4th Edition.
[3] Abaqus 2010 Analysis User's Manual for version 6.13 Dassault Systems Inc.
[4] Berwanger C and Symko Y 1975 Thermal Stresses in Steel-Concrete Composite Bridges. *J. Civ. Eng.* **2**(1) 66-84.
[5] Elbadry M M and Ghali A 1983 Nonlinear Temperature Distribution and its Effects on Bridges *IABSE Periodica* **66/83** 169-91.
[6] Eurocode-1 2003 Actions on Structures Part 1-5: General Actions-Thermal Actions ENV 1991 1-5:2003 *Comite European de Normalization* (CEN)
[7] Grisham G P 2005 Field Measurements of Bearing Displacements in Steel Girder Bridges. *Master's Thesis.*
[8] Hunt B and Cooke N 1975 Thermal Calculations for Bridge Design. *J. Struct. Division ASCE* **101** 1763-81.
[9] IRS 2003 Steel Bridge Code Indian Railway Standard Code of Practice for the Design of Steel or Wrought Iron Bridges carrying Rail Road or Pedestrian Traffic.
[10] Kennedy J B and Soliman M H 1987 Temperature Distribution in Composite Bridges. *J. Struct. Eng. ASCE* **113**(3) 475-82.
[11] Moorty S and Roeder C W 1992 Temperature-Dependant Bridge Movement. *J. Struct. Eng. ASCE* **118**(4) 1090-105.
[12] Reynolds J and Emanuel J H 1974 Thermal Stresses and Movements in Bridges. *J. Struct. Division ASCE* **100**(ST1) 63-79.
[13] Roeder C 2003 Proposed Design Method for Thermal Bridge Movements. *J. Bridge Eng.* **8**(1) 12-9.
[14] Tong M, Tham L G, Au F T K and Lee P K K 2001 Numerical Modelling for Temperature Distribution in Steel Bridges. *Comput. Struct.* **79**(6) 583-593.
[15] Tong M, Tham L G and Au F T K 2002 Extreme Thermal Loading on Steel Bridges in Tropical Region. *J. Bridge Eng.* **79**(6) 583-93.
[16] Zhou H F Principle Components of Temperatures *Engineering Structures.*
[17] Zuk W 1965 Thermal Behaviour of Composite Bridges – Insulated and Uninsulated *National Research Council* Vol**76** 231-253 Highway Research Record.

Flexural behavior of RCC beams with externally bonded FRP

Arun Vignesh S A Sumathi and K Saravana Raja Mohan
School of Civil Engineering, SASTRA University, Thanjavur –613 401, India

*Email: sumathi@civil.sastra.edu

Abstract: The increasing use of carbon and glass fibre reinforced polymer (FRP) sheets for strengthening existing reinforced concrete beams has generated considerable interest in understanding the behavior of the FRP sheets when subjected to bending. The study on flexure includes various parameters like percentage of increase in strength of the member due to the externally bonded Fiber reinforced polymer, examining the crack patterns, reasons of debonding of the fibre from the structure, scaling, convenience of using the fibres, cost effectiveness etc. The present work aims to study experimentally about the reasons behind the failure due to flexure of an EB-FRP concrete beam by studying the various parameters. Deflection control may become as important as flexural strength for the design of FRP-reinforced concrete structures. A numerical model is created using FEM software and the results are compared with that of the experiment.

Key words: FRP laminates Flexural study, External bonding, Numerical Analysis, Load carrying capacity

1. Introduction

Fibre reinforced composites have been widely used to strengthen reinforced concrete (RC) members. Mostly because they have a high strength-to-weight ratio, require relatively limited time to cure, and have mechanical properties that can be engineered to meet the desired structural performance. Fibre-reinforced polymer (FRP) composites are made up of continuous fibres and a thermosetting organic resin, are currently the most common type of composite system used for structural strengthening applications. Fibre reinforced cementitious matrix (FRCM) composite was another type of composite that was recently developed which contains continuous fibres with a cementitious (inorganic) matrix. The substantial increase in energy absorption capacity is the most significant improvement imparted by adding fibres to a concrete.

The present work aims to study experimentally about the reasons behind the failure due to flexure of an EB-FRP concrete beam by studying the various parameters stated above. Deflection control may become as important as flexural strength for the design of FRP-reinforced concrete structures.

The strengthening effect of EB-FRP using composite element involving glass fibre was studied [1] Use of glass fibre laminates in composite sections in external bonding is well established. Flexural response was better recorded using the epoxy adhesive. Use of carbon – glass hybrid was studied the stiffness contribution of the composite in bridge decks in rehabilitation [2]. The uniaxial tests under compression was performed based on deflection control condition, stress-strain behaviour for fiber concrete was studied [3]. The numerical model developed was referred the research done on the bending tests on FRP in slabs [4] Parameters varied here were thickness and the effect of changing the thickness is studied under the four point loading test [5]. An increase in the width of the FRP will produce an increase in the load-carrying capacity. Interfacial crack propagation and strain distribution during shear debonding is influenced by the width of the FRP laminate in comparison to that of the beam [6]. Variation of mechanical properties in terms of composite carbon, composite glass sheets and their hybrid combinations is by their hybrid combinations. Different failure modes that can occur is also studied [7]. Studies show that Strength increases provided by SRP bonded with cementitious grout were smaller than those obtained using epoxy. [8]. . Although use of cementitious matrix is in practice the transfer of stresses was found to be more efficient by use of epoxy strengthening adhesive [9]

2. Materials and methods

2.1 Materials

Ordinary Portland Cement (OPC) of grade 53 confirming to IS 12269 was utilized in the study. The specific gravity of the OPC sample was found to be 3.12 with an initial setting time of 40 minutes and a standard consistence of 28 % with its chemical composition is given in Table 1. FRP plates are non-capillary and non-hygroscopic. Therefore, they provide good moisture resistance. CFRP has a very high temperature resistance and is virtually inert.

Table1. Mechanical properties of FRP.

Property	CFRP	GFRP
Yield stress (σ_y)	200MPa	125MPa
Elastic modulus	1.5GPa	26GPa
Poisson's Ratio	0.28	0.28
Density	1.5 g/cm^3	1.8g/cm^3

The grading of the aggregate was categorized using sieve analysis test, and the results are detailed in Table 2 and accordingly the aggregate was categorized as graded under Zone III.

Table 2. Properties of aggregates.

Properties	**Fine aggregate**	**Coarse aggregate**
Specific gravity	2.58	2.72
Fineness modulus	2.56	6.97
Bulk density, kg/m^3	1697	1488
Water absorption, %	1.37	0.77

Aggregate passed through 16 mm sieve and retained on 12.5 mm sieve, were used as in the concrete mixture. The specific gravity of the coarse aggregate was 2.7. Wire basket method based on ASTM C 127 was used to determine the specific gravity. The steel bars of 12 and 8 mm high yield strength deformed bars were used as reinforcement in the cube specimens.

2.2 Methodology

High strength concrete of the grade M50 was designed in order to obtain a characteristic compressive strength of 50 MPa. The design mix ratio arrived was 1:0.8:2.13(Cement: FA: CA). The addition of high range water reducer reduced the w/c ratio to 0.32 and the workability was based on slump cone test according to ASTM C14. Beam specimens were cast and water cured for 28 days to study the flexural behavior.

Table 3. Properties of FRP laminates.

Property	Unit	CFRP	GFRP
Density	g/cm^3	1.5	1.8
Width	mm	100	100
Thickness	mm	2,1	2,1
Length	mm	1000	1000
Ultimate Tensile Strength	MPa	550	530
Poisson's ratio	-	0.28	0.28

The most widely applied basic FRP strengthening technique, involves the manual application of hand lay-up or prefabricated systems by means of cold cured adhesive bonding. The specimens were categorized as follows

CC	Normal RCC beam as control specimen
GFRP	Beam reinforced with GFRP plate of 2mm thickness
CFRP	Beam reinforced with CFRP plate of 2mm thickness
C-G1	Beam reinforced with CFRP and GFRP plate
C-G2	Beam reinforced with CFRP and GFRP plate

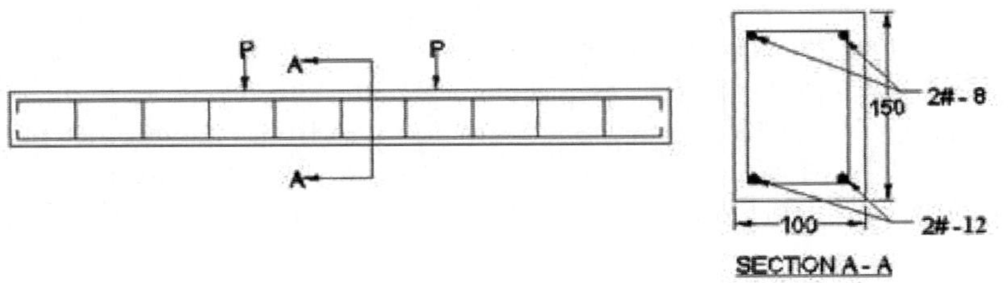

Figure 1. Reinforcement design

Figure 2. FRP laminates on the tension side of the beam

3. Results and discussion

3.1 Flexural Behaviour Test

The specimens cast were in the dimensions of 100 mm x 150 mm x 1200 mm beam prototypes. Curing was done for a total of 28 days. The test was carried out using 1000kN capacity flexural strength testing machine. The test setup includes two point loading using a single point loading system by which the loads are transferred equally to the two points using a spreader beam and two rollers. Deflections were measured in the beam by placing strain gauges at the bottom of the beam. Strains occurring at the points were also measured under using a LVDT strain gauge that was placed at particular intervals. The gauge length between the force points is 333.33 mm in the top and 100 mm from either corner of the beam at the supports. All the specimens were capped for uniform loading prior testing. The control of load over the test was 10 kN/min. All the data like displacement load and strain were recorded using Automatic data acquisition system which in turn connected to the computer.

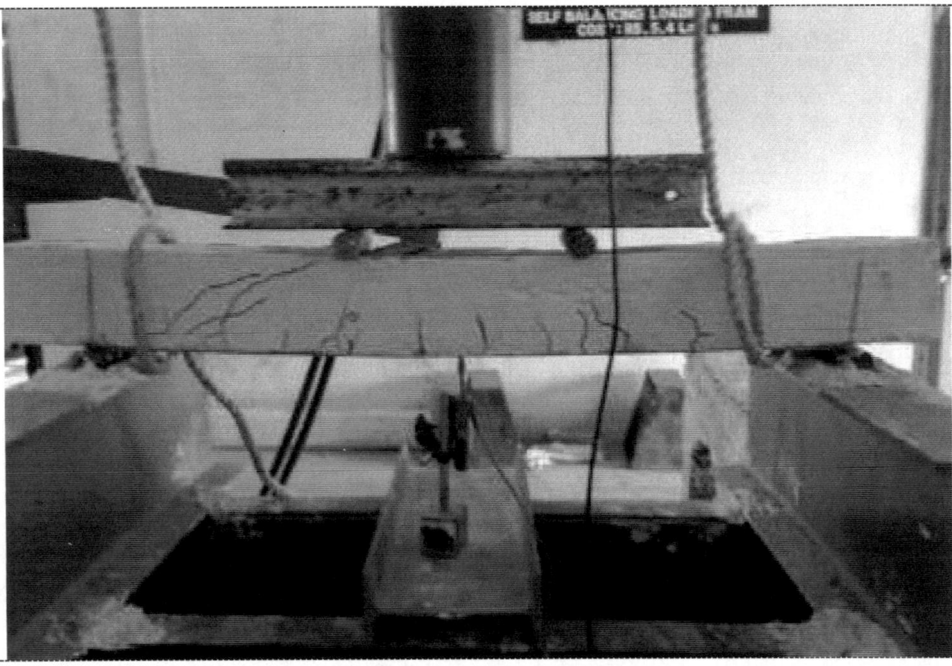

Figure 3. Flexural test setup

3.2 First crack load

The first crack load and moments for beams cast with different EB-FRP beams as well as for the control beam are given in Table 4

Table 4. Load values of the beams.

	First Crack Load (kN)	Ultimate load (kN)
Control	29.75	65
GFRP	34.2	70.12
CFRP	33.75	68.02
C-G1	45.7	74
C-G2	30.6	64.1

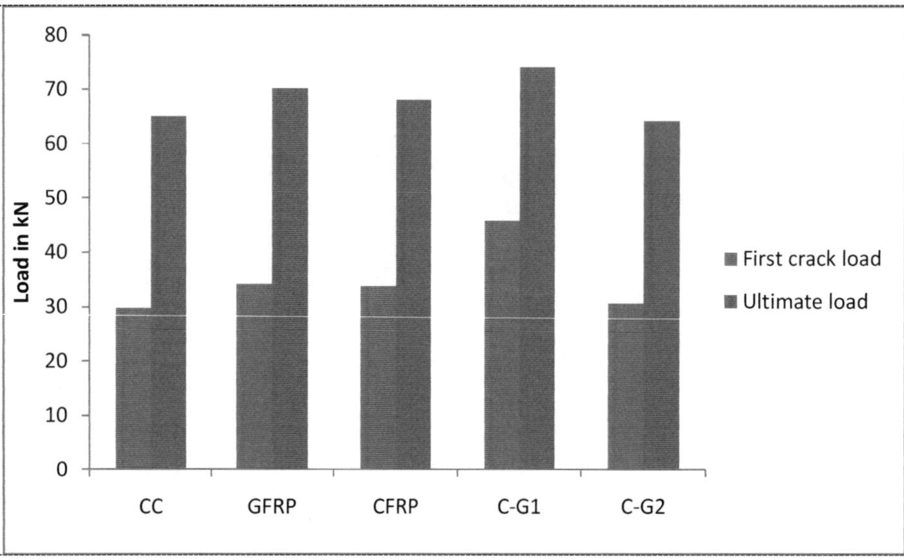

Figure 4. First crack and Ultimate load for beams

The load deflection curves for beams are shown in figure 5. It could be noted that with each application of load, the deflection value changes or in this case increases from the control beam. This implies that the addition of FRP makes the concrete beam more ductile. However, it could also be noted that the deflection for the CC and C-G2 beams are the same.

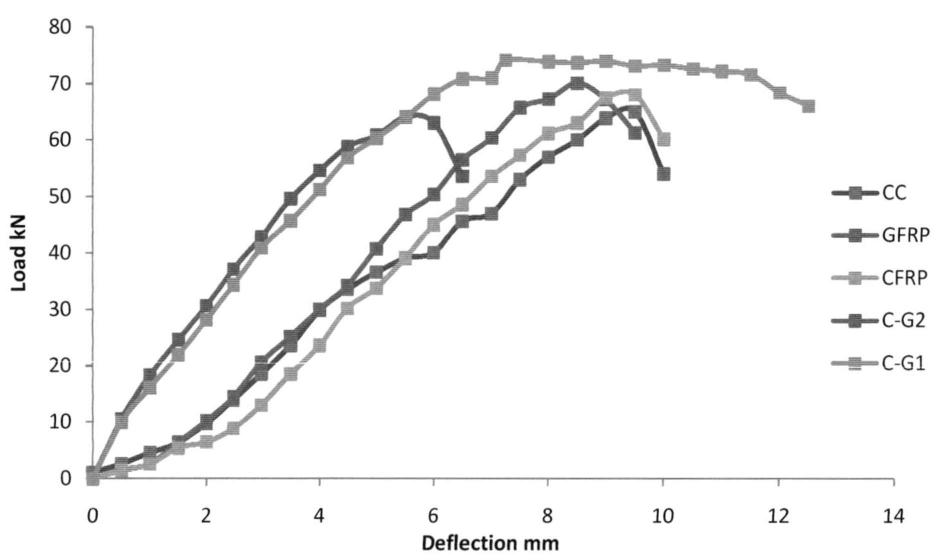

Figure 5. Load – deflection plot for the beams

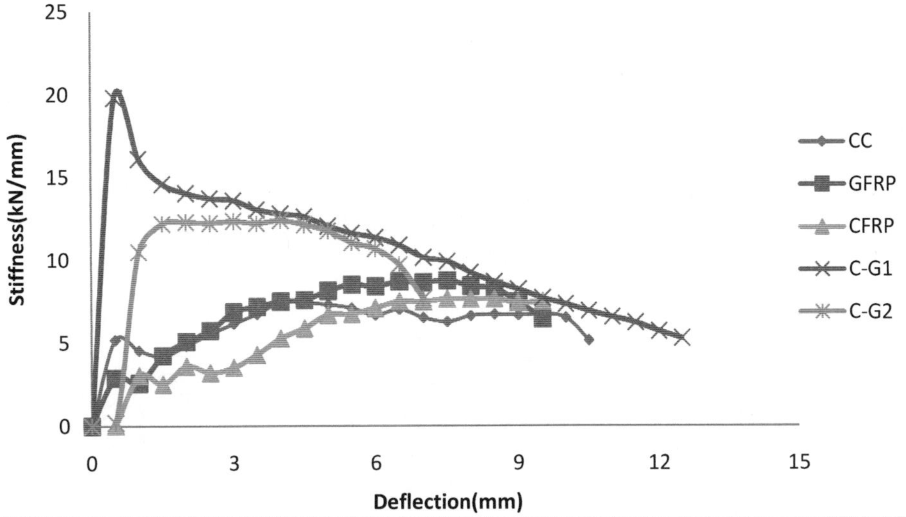

Figure 6. Stiffness – deflection plot for the beams

3.3 Energy absorption capacity

The energy absorption capacity of all the beams was calculated from the area under the load-deflection curve. The values are presented in the Figure below:

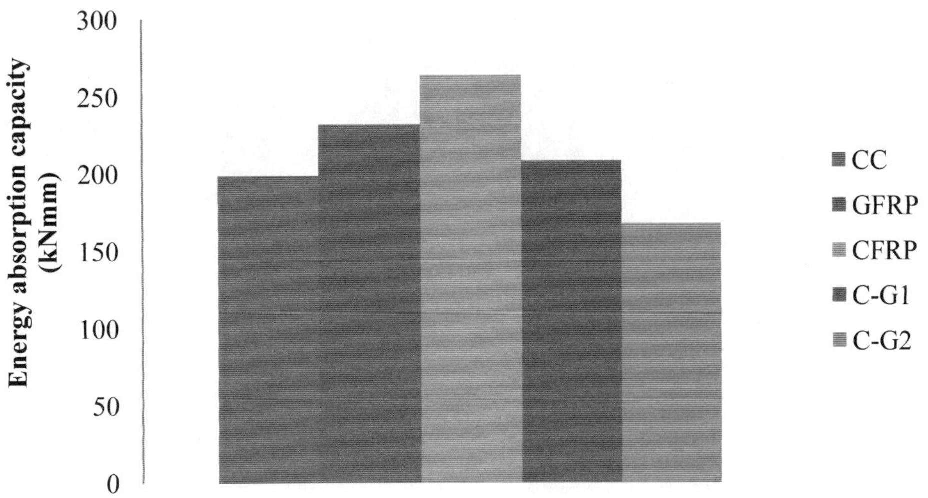

Figure 7. Energy absorption capacity of the beams

It can be observed that the energy absorption capacity increases significantly from the control beam. However, the value for the beam with CFRP content registers the highest value. The beam C-G2 registered a low value of energy absorption, this was due to the debonding occurred between the concrete and FRP. Also due to the brittle nature of the lesser thickness of carbon in the hybrid FRP laminate.

3.4 Load carrying capacity

The theoretical value for load carrying capacity has been arrived at by ACI method. From calculating the depth of the neutral axis, it could be identified that the section was under reinforced. Hence, the theoretical value was arrived at and presented in the Table below:

Table 5 Theoretical and experimental load carrying capacity values of the beams.

	Theoretical load in kN	Experimental load in kN
Control	42.34	65.0
GFRP	42.34	70.12
CFRP	42.34	68.02
C-G1	42.34	74.3
C-G 2	42.34	64.1

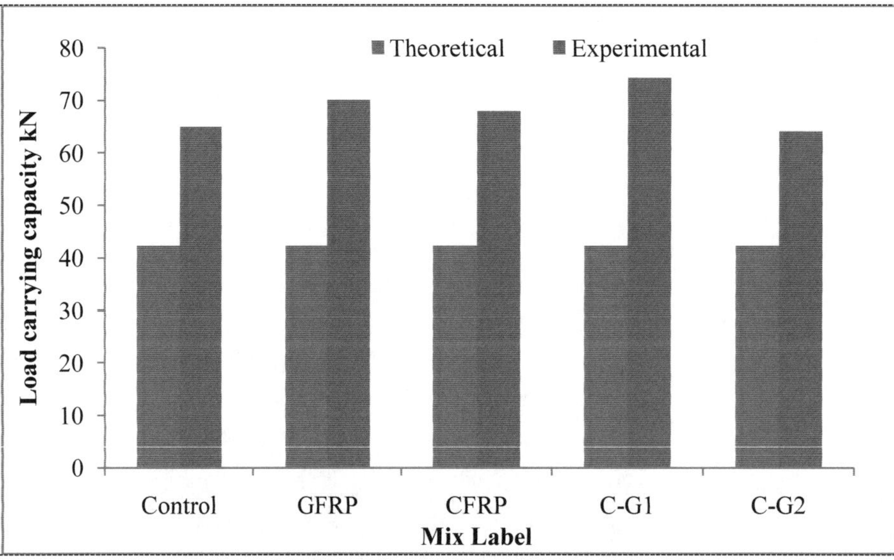

Figure 8. Load carrying capacity of the beams

It could be inferred from the above figure that the addition of FRP laminates increases the load carrying capacity of the beam. The addition of CFRP laminates was observed a lesser load value because CFRP was more brittle compared to GFRP. Similarly C-G 2 beam failed due to debonding of the FRP element.

IOP Conf. Series: Earth and Environmental Science **80** (2017) 012043 doi:10.1088/1755-1315/80/1/012043

4. Numerical analysis:

4.1 Proposed method of analysis:

For the current research, a beam of length 1200mm, width 100mm and depth 150mm is considered. The top longitudinal reinforcement consists of two hangar bars of 8mm diameter and the bottom longitudinal reinforcement consists of two bars of 12mm diameter with spacing 150 mm. The stirrups were made up of 6mm bars at the spacing of 150mm c/c.

In ANSYS 15, the beams are created as elements. The elements involved in creating the model are Solid 65, Link 180 and Solid 185. The concrete beam was modeled as the as three dimensional element, Solid 65. Solid 185 was used for creating the model for the CFRP and GFRP. Link 180 is a linear line element to be used as the element for the steel reinforcements. In this element, shear deformation effects are included. The elements are given in Table

4.2 Properties of the Elements:

Table 6. Elements and their details.

Material	Element	No of nodes
Concrete	SOLID 65	8
Steel	LINK 180	2
CFRP,GFRP	SOLID 185	8

The material properties of concrete given as input for it ANSYS model are given in Table 7.

Table 7. Property inputs for concrete.

Young's modulus(kN/m^2)	2.23*10^4
Poisson's ratio	0.15
Open shear transfer co-efficient	0.23
Closed shear transfer co-efficient	0.9
Uniaxial cracking stress	2.5
Uniaxial crushing stress	-1

The material properties of steel given as input for it ANSYS model are given in Table 8.

IOP Conf. Series: Earth and Environmental Science **80** (2017) 012043 doi:10.1088/1755-1315/80/1/012043

Table 8. Property input for steel bars.

Young's modulus of steel (kN/m^2)	2*10^5
Poisson's ratio	0.3
Yield stress (MPa)	415

The material properties of CFRP & GFRP given as input for it ANSYS model are given in Table 9.

Table 9. Property input for GFRP AND CFRP.

PROPERTY		GFRP	CFRP
Elastic modulus(kN/m^2)	E_X	2.1*10^5	2.3*10^5
Poisson's ratio	PRXY	0.28	0.28
Shear modulus kN/m^2	G_{XY}	1520	11790

The support, boundary conditions and loading points are shown in the image below

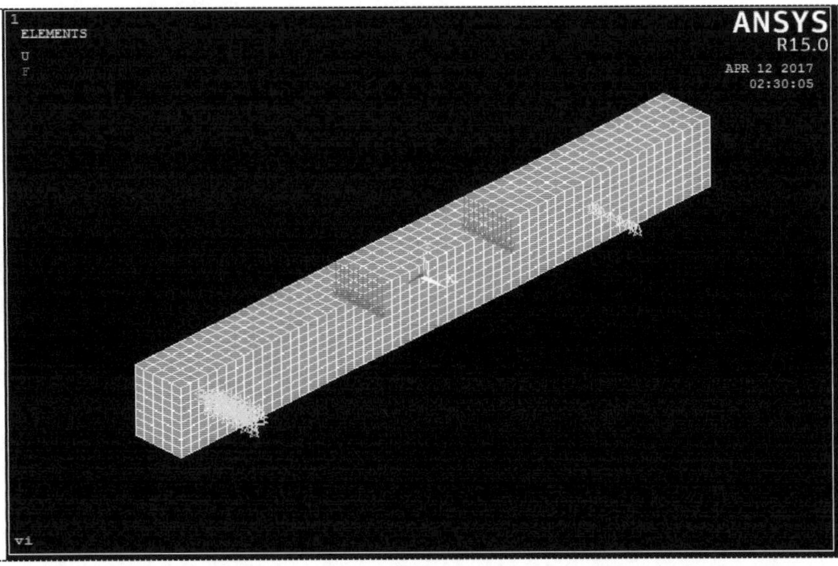

Figure 9. Boundary conditions of the beam

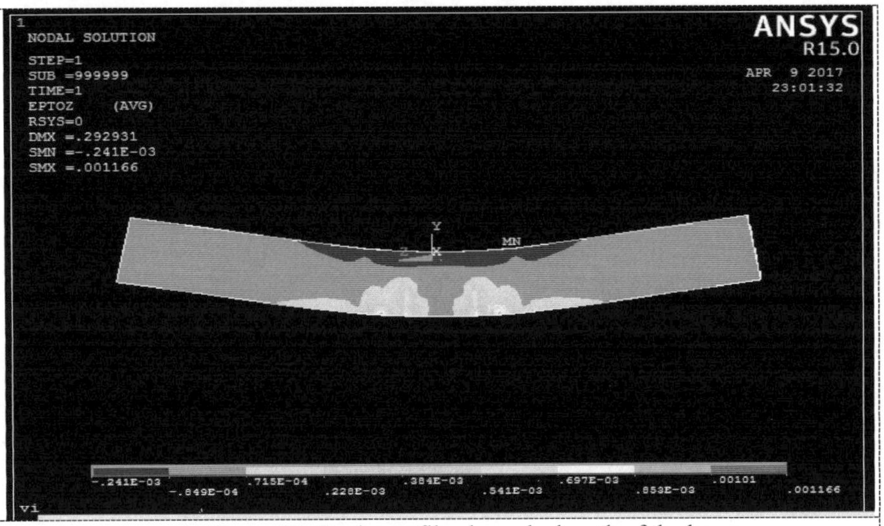

Figure 10. Strain profile along the length of the beam.

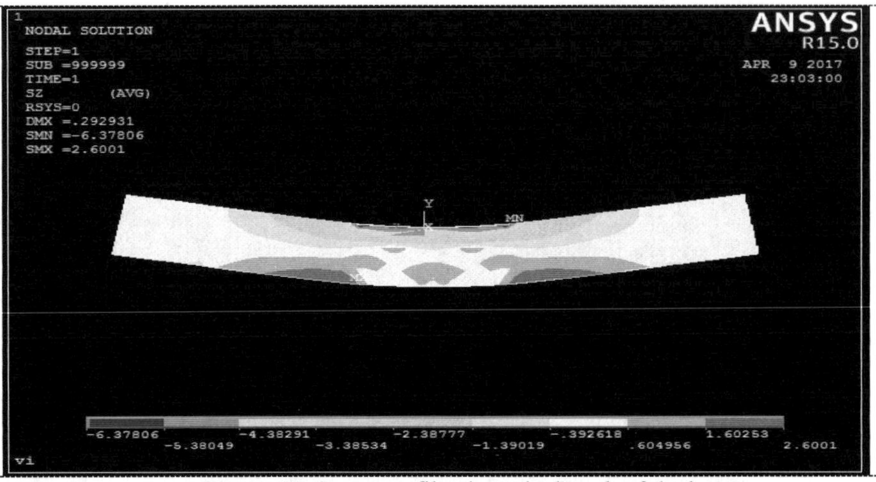

Figure 11. Stress profile along the length of the beam.

Table 10. Deflection at ultimate load obtained from the analysis.

Beam	Deflection(mm)
CC	9.44
GFRP	10.5
CFRP	10.76
C-G1	10.012
C-G2	7.55

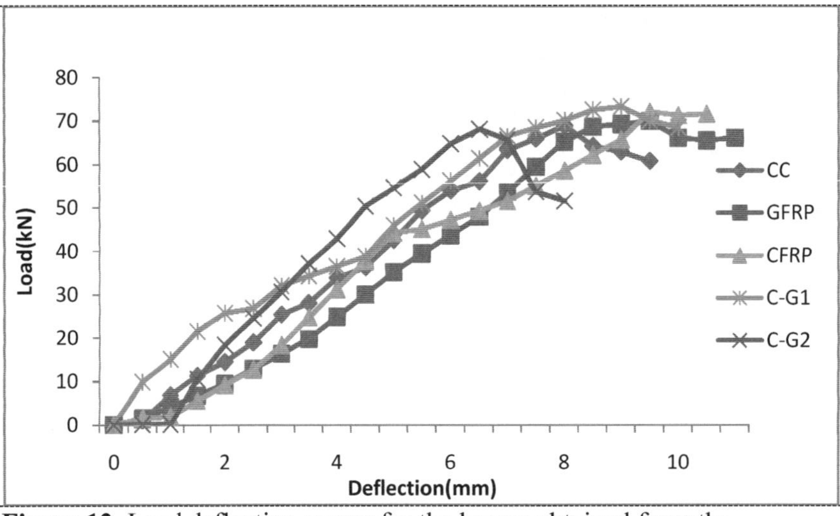

Figure 12. Load deflection curves for the beams obtained from the numerical analysis

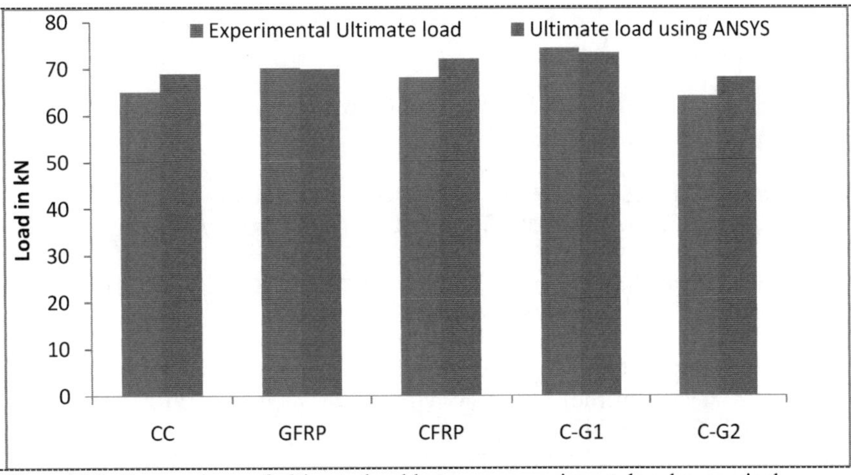

Figure 13. Comparison of ultimate load between experimental and numerical analysis

5. Conclusion

External bonding of FRP laminates affects the resistance in terms of durability characteristics. From the results the beam that was fitted with C-G1 beam and GFRP beam has higher load deflection behavior in the experimental analysis and in the numerical analysis. Two different failure modes were observed depending upon the type of FRP laminate used. Rupture and debonding of plates were observed. The ultimate load of the specimen with FRP laminates has also increased in the case of all the beams when compared to the control beam. However, comparing the trends observed in first crack load, the ultimate load, the resilience and the energy absorption capacity, the corresponding values increased till the C-G1 and were very close to the values of the control beam in case of C-G2 specimen. CFRP being more brittle compared to the GFRP tends to rupture first in case of the composite laminates .Hence laminate composite of CFRP- 1mm and GFRP – 2mm is considered to have a optimum load carrying capacity and energy absorption capacity.

Acknowledgements

The authors extend their sincere thanks to Vice Chancellor of SASTRA University for providing us laboratory facilities in school of civil engineering to successfully complete this work

References

[1] Tamim A A Abed F H Rahmani A A Effects of harsh environmental exposures on the bond capacity between concrete and GFRP reinforcing bars, 2014 *Adv. Concr. Constr.*, **2(1),** 1-11.

[2] Chakrabortty Khennane A Failure mechanisms of hybrid FRP-concrete beams with external filament-wound wrapping 2014, *Adv. Concr. Constr*, **2(1),** 57-75

[3] Hong, K. N Cho C G Lee S H Park Y Flexural Behavior of RC Members Using Externally Bonded Aluminum-Glass Fiber Composite Beams. 2014, *Polymers*, **6**, 667-685.

[4] Balsamo A Nardone F Iovinella I Ceroni F Pecce M Flexural strengthening of concrete beams with EB-FRP, SRP and SRCM:Experimental investigation 2013 *Compos.Part B,* **46**, 91–101.

[5] Barros J A O Figueiras J A Flexure Behavior Of SFRC: Testing And Modelling 1999, *J. Mater. Civ. Eng*, **11(4),** 331-339

[6] Sneed L H Verre S, Carloni C Ombres L Flexural behavior of RC beams strengthened with steel-FRCM composite 2016, *Eng. Struct.* **127**, 686–699.

[7] Subramaniam K V Carloni C Nobile L Width effect in the interface fracture during shear bonding of FRP sheets from concrete, 2007, *Eng. Fract. Mech.***74**, 578–594.

[8] Hawileh R A Obeidah A A Abdalla J A, Tamimi A A Temperature effect on the mechanical properties of carbon, glass and carbon–glass FRP laminates, 2015 *Const. Build. Mater.* **75**, 342–348.

[9] Prota A Tan K Y Nanni Pecce M Manfredi G Performance of Shallow Reinforced Concrete Beams with Externally Bonded Steel-Reinforced Polymer, 2016, *ACI Struct. J.* **103**, 163 – 170.

IOP Conf. Series: Earth and Environmental Science **80** (2017) 012044 doi:10.1088/1755-1315/80/1/012044

Ranking the Project Management Success Factors for Construction Project in South India

K Aneesha and M K Haridharan*

School of Civil Engineering, SASTRA University, Thanjavur – 613 401. India.

*Email: haridharan@civil.sastra.edu.

Abstract. In Today's construction industry, to achieve a greater advantage over the firms, success of each project and efficiency is required. Effective Project Management overcomes these types of challenges. This study identifies the success factors which are important for project management in construction project success. From the literature review, 26 factors were found to be critical. Project managers, construction managers, civil engineers, contractors and site engineers were the respondents. After analyzing the data in SPSS software, the dominant factors from the regression analysis are top management support, competent project team, abilities to solve problems, realistic cost and time estimates, information/communication, competency of the project manager are the 6 factors out of 12 in 26 factors. Effective communication between stakeholders got highest priority and client involvement, good leadership, clarity of project goals got second priority. Informal communication gives better results compared to formal communications like written formats. To remove communication barrier with the stakeholders, informal communication like speaking face-to-face with the language this fits for the stakeholders.

1. Introduction

Construction industry of India is the indicator of the development. It is the second largest industry in India after agriculture. Quick development in the construction sector can boosts up other industrial sectors like cement, paints, fitting and fixtures, chemicals and steel industries. Some of the major challenges in the industry are skilled and reliable workers, safety of workers as well as industry and limited time, resource and space constraints. To overcome these and for reducing cost, time and quality challenges in construction, effective project management should be required.Role of project management in construction is more now a day to meet specific constraints [1]. The empirical study about current practice in project management and found critical success factors and explained about the present level of project management in Indian construction industry [2-3]. Private sector using project management better compared to public sector. Defined the project management as use of knowledge, tools, skills and techniques to the project to meet specific project requirements and explained about the application of project management in construction industry [4].

In reality and in scientific view, project management practices are different [5]. Choosing of any work which is advanced in the project, project manager should have a lead role to choose that work[6]. [7-9] Found the difference between project success criteria and project success factors by using definitions and categorized criteria's and success factors. Explained about the key success factors of project management and how to overcome problems of project management in practice using these factors were explained in [10]. [11] Described about the relationship between project performance and critical success factors and identified the critical success factors of project success. [12] Explained about the critical success factors of project and project management. In addition to this, Project Excellence Model which is related about linking the project critical success factors with project success criteria was explained.

[13] Enlightened that why projects continued to failing if factors and criteria for success are known. Author proposed a Square Route which eliminates the Type I and Type II errors in project management success criteria. Using critical factors, construction companies change their policies for their project success and these critical success factors differs with respect to project manager, team, organization etc. [14-15]. The variables for project success, combination of four independent subsystems gives project success [16]. Those are strategic, structural, technical and managerial. [17] Explained about the criteria and factors of project management success. What to be achieves describes about the criteria whereas how to achieve describes about factors in project management. [18] Discussed about the best practices in project management. Technical and people oriented themes came into picture for best practices in project management. [19] Explained about the success evaluation factors in construction project management in small and medium Portuguese companies. Factors which are important for project success is complete the project within budget, on schedule, complete according to requirements, customer acceptance. It was shown that superior technology and performance is the major difference between larger and smaller companies. Smaller companies should compete harder in the market and to satisfy customer with innovative ideas. [20] Explained about the complexity of project management in developing countries. Research found that project manager should have ability to understand the complexity of project and their effects of project success. [21] Discussed about the critical success factors needed for effective construction management in Malaysian industry.

2. Problem Statement

Construction industry in all world faces problems of quality, budget and time overrun. Recent technology development in the construction industry reduces the time but failed to expect the perfect estimated time because of onsite and unexpected challenges. Effective project management overcome all these challenges by using knowledge, skills, experience etc. from the conception to completion of project. Finding the critical success factors are important for effective project management. Critical success factors help to find out the areas that should be performed well. These factors act as indicators to measure the organization performance in project management. These factors are measured, monitored and tracked for performance. Improvements in the critical success factors will give better results in the project management success in construction projects.

3. Objective of the study

- Understand the importance of project management in the construction industry.
- To examine the success criteria of project management and find out the critical success factors of project management success in construction project success.
- To rank the critical success factors of project management success.
- To give recommendations for improvements in the practices of critical success factors.

4. Methodology

The methodology is the general research strategy that outlines the way in which research is to be undertaken and identifies the methods to be used in it. Procedure and methods in this research are:

4.1. Study population

The research study consists of respondents in the construction companies specifically in South India.

4.2. Sampling Technique

To select respondents, random sampling method was used. In order to ensure reliability of the data, mostly respondents chosen are the project managers, construction managers.

4.3. Data collection method

To collect the primary data from the respondents, a structured questionnaire was implemented. The primary data collected from the construction professionals in the South Indian construction companies.

4.4. Questionnaire Design

A questionnaire was developed by finding out the common critical success factors from the literature review. 25 factors were found important from literature review and those were used in the questionnaire. Construction industry employees (project manager, construction manager, site engineer, contractor and civil engineer) have more knowledge on project management are the targeted group of respondents. In primary data, two sections which deals about the respondents' demographic profile and critical success factors of project management. This primary data was collected from the respondents through the mail and face-to-face interviews. Out of 150, 116 respondents gave their opinion. Five-point Likert scale was used to take the opinion of the respondents on factors in the questionnaire. This scale ranges from 1=strongly disagree to 5=strongly agree. Refer Appendix for detailed questionnaire.

4.5. Data analysis tool

Analysis of the data was done by using Statistical Package for the Social Sciences (SPSS). For finding out the internal consistency of the data Reliability analysis was used. Frequency Analysis was done for the demographic profile data to know the frequency of occurrence of a variable. Regression Analysis was used to find the significant variables in the factors. Critical success factors are ranked according to the highest mean of the given variables.

5. Results and Discussions

To analyse the data, first step is to check the reliability. It refers the degree to which it is error-free. It gives the consistency of a measuring test and tells that confidence of a researcher placed on the questionnaire to give same numeric value when the measurement is repeated on the same question. Cronbach's Alpha is used to find the reliability. Normally Cronbach's Alpha reliability coefficient ranges from 0 to1. Lower limit is not there to the coefficient. Closer the coefficient value to 1, greater is the internal consistency of the data. In below table 1.Cronbach's Alpha coefficient is 0.815.Found Cronbach's Alpha coefficient >0.8 is a reasonable goal[22]. So, all items are approved for the subsequent data analysis.

Table 1. Reliability Analysis

Cronbach's Alpha	N of Items
0.815	26

To transfer the raw data into grouped data, frequency distribution is used. Frequency distribution tells about the frequency of values or ranges of values for a variable.Here variable related to the respondents demographic profile. From the below table 2, it was shown that highest range of age of the respondents in this survey are 26-35 with percentage 41.4 followed by 18-25 with percentage

28.4. In this survey, most of the respondents are male with percentage 75.9 followed by female with percentage 24.1. 61.2% of the respondents completed graduate and 13.3% of respondents completed post graduate. Mostly 36.2% of the respondents are having 2-6 years of experience in their field. 20.7% of respondents having less than 1 year experience. Most of the survey was taken in urban construction companies. 62.9% from urban, 27.6% from semi-urban and 9.5% from rural. 40.5% of projects having budget of 26-50 and 25% of projects having budget of 51-75.

Table 2. Demographic profile of the respondents

S.No	Variables	Description	Majority wise first	Majority wise second	Majority wise third	Majority wise fourth
1	Age	Category	26-35	18-25	36-45	Above 55
		Percentage value	41.4	28.4	21.6	6
2	Gender	Category	Male	Female		
		Percentage value	75.9	24.1		
3	Education Qualification	Category	Graduate	Post Graduate	Diploma	
		Percentage value	61.2	26.7	12.1	
4	Numberof projects handled	Category	Less than 5	6-10	7-11	12-16
		Percentage value	43.1	26.7	22.4	7.8
5	Experience in years	Category	2-6	Less than 1	Above 16	7-11
		Percentage value	36.2	20.7	19.0	17.2
6	Location	Category	Urban	Semi Urban	Rural	
		Percentage value	62.9	27.6	9.5	
7	Average budget of the project	Category	26-50	51-75	Less than 25	76-100
		Percentage value	40.5	25	13.8	11.2

Relationship between dependent and independent variable comes by using regression analysis. In below table 3 shows the significant value of the independent variables with respect to dependent variable. This significant value tells about how much relationship exist with dependent variable. In this table, T comes from coefficient (B) divided by the standard error (Std. Error). B is called unstandardized coefficient and Beta is called standardized coefficient. Unstandardized coefficient tells about amount which dependent variable changes by changing one unit of independent variable keeping other independent variable constant. Standardized coefficient calculated using standard deviation.

Table 3. Regression Analysis

Coefficients

S.NO	Model	Unstandardized Coefficients		Standard Coefficients	T	Sig.
		B	Std. Error	Beta		
	(Constant)	.024	0.705		0.034	0.973
1	Top management support	-0.238	0.082	-0.235	-2.892	0.005
2	Monitor performance and feedback	-0.052	0.086	-0.059	-0.611	0.543
3	Competent project team	0.219	0.091	0.191	2.416	0.018
4	Problem solving abilities	0.309	0.106	0.221	2.925	0.004
5	Realistic cost and time estimate	0.211	0.091	0.238	2.336	0.022
6	Management of risk	0.050	0.106	0.041	0.469	0.640
7	Information/Communication	0.274	0.082	0.258	3.328	0.001
8	Detail plan in design and construction	0.002	0.099	0.021	0.225	0.822
9	Participation/involvement of the client	-0.051	0.093	-0.049	-0.542	0.589
10	Understanding of the project	0.028	0.084	0.027	0.329	0.743
11	Project Manager competency	0.389	0.098	0.343	3.977	0.000
12	Clarity of project goals	0.069	0.135	0.049	0.513	0.610
13	Effective communication between stakeholders	0.075	0.091	0.065	0.825	0.412
14	Schedule performance	0.263	0.100	0.229	2.642	0.010
15	Sufficient financial budget	-0.070	0.095	-0.064	-0.735	0.464
16	Technical capability of the project manager	-.0185	0.082	-0.178	-2.241	0.028
17	Commitment of all project participants	0.282	0.094	0.248	3.009	0.003
18	Well-laid out specifications	0.061	0.089	0.059	0.692	0.491
19	Good Leadership	-0.239	0.106	-0.188	-2.247	0.027
20	Clear and realistic goals	-0.133	0.100	-0.118	-1.335	0.185
21	Past experience of project management technology and tools	-0.269	0.095	-0.243	-2.837	0.006
22	Complete the project according to	0.056	0.088	0.053	.635	.527

	requirements					
23	Use of superior and appropriate technology	-0.167	0.080	0.053	-2.095	.039

Dependent Variable: Overall employee acceptance for usage of project management in project success.

$$Y=.024 + (.235)x_1 + (.059)x_2 + (.191)x_3 + (.221)x_4 + (.238)x_5 + (.041)x_6 + (.258)x_6 + (.021)x_7 + (-.049)x_8 + (.027)x_9 + (.343)x_{10} + (.049)x_{11} + (.065)x_{12} + (.0229)x_{13} + (-0.064)x_{14} + (.178)x_{15} + (.248)x_{16} + (.059)x_{17} + (.188)x_{18} + (.118)x_{19} + (.243)x_{20} + (.053)x_{21} + (.053)x_{22} + (-.053)x_{23} \quad (1)$$

Y is the overall usage of project management in project success. Above equation states that impact of project management variables on overall employee acceptance for usage of project management in project success. The result of t test disclose about the calculated significance of the regression coefficient (-0.235, 0.191, 0.221, 0.238, 0.258, 0.343, 0.229,-0.178, 0.248,-0.188,-0.243, 0.053) are having 0 and 5 percent respectively. From the above coefficient table it is concluded that variables of project management success in construction project success are top management support, competent project team, problem solving abilities, realistic cost and time estimates, communication, project manager competency, schedule performance, project manager technical capabilities, commitment of all project participants, good leadership, past experience of project management technology and tools, use of superior and appropriate technology. The top management support, project team, information/communication have direct impact on project success[23]. The top management involvement shows major differences of project management in the construction industry and software sectors [24-25].Time and quality is important but not considered as a critical success criteria for project success.

Factor analysis is used to group the factors which explain same information. It is useful to summarize the data into a small number of factors by considering a large number of factors. Below Table 4 reduced the 26 factors into 9 factors.

Eigen value which is shown in the above table 4 explains about measure of variance of the observed variables a factor explains. An Eigen value >=1 having more variance than a single observed variable. In this table, total 9 factors got Eigen value >=1.So, 26 factors are reduced to 9 factors.

1. Planning, estimation and monitoring
2. Time and cost management,
3. Client participation with definite objectives
4. Capability of project manager towards particular project
5. Competent team work with satisfaction
6. Communication towards reaching objectives
7. Management support and participants commitment
8. Updated technology and
9. Deliver directions with specification are the reduced number of factors from the 26 factors.

Ranking the factors is important to know which factors influence the project management success most. Below table 5 shows the ranking of factors using mean.

Table 4. Factor Analysis

Factor	Item	Rotated Factor Loadings	Eigen Value	% of Variation	Factor Name
I	Monitor performance and feedback	0.805	5.007	19.26	Planning, estimation and monitoring
	Realistic cost and time estimate	0.737			
	Detailed plans in design and construction	0.515			
II	Schedule Performance	0.812	2.148	8.263	Time and cost management
	Adequate budget	0.623			
III	Client involvement	0.682	1.895	7.288	Client participation with definite objectives
	Clarity of project goals	0.76			
IV	Project understanding	0.639	1.566	6.024	Capability of project manager towards particular project
	Project manager technical capabilities	0.774			
V	Competent project team	0.725	1.431	5.504	Competent team work with satisfaction
	Overall Employee satisfaction	0.735			
VI	Communication	0.633	1.307	5.025	Communication towards reaching objectives
	Clear and realistic objectives	0.71			
VII	Top management support	0.736	1.257	4.836	Management support and participants commitment
	project participants commitment towards goals	0.747			
VIII	Use of superior and appropriate technology	0.87	1.162	4.471	Updated Technology
IX	Well laid out specifications	0.581	1.019	3.919	Deliver directions with specification
	Good Leadership	0.804			

Table 5. Ranking Factors using mean

S.No		Mean	Std. Deviation	Rank
1.	Top management support	3.4310	1.23868	17
2.	Monitor performance and feedback	3.0862	1.40539	25
3.	Competent project team	3.5948	1.09527	7
4.	Problem solving abilities	3.4914	0.89923	16
5.	Realistic cost and time estimate	3.2069	1.41124	24
6.	Management of risk	3.2845	1.03677	23
7.	Communication	3.5000	1.18322	15
8.	Detailed plans in design and construction	3.5862	1.18739	9
9.	Systematic control over project execution	3.4310	1.12848	17
10.	Client involvement	3.6724	1.20705	2
11.	Project understanding	3.3276	1.23553	22
12.	Project manager competence	3.6466	1.10536	5
13.	Clarity of project goals	3.6724	0.88245	2
14.	Effective communication between stakeholders	3.7328	1.09034	1
15.	Schedule performance	3.5345	1.09092	12
16.	Adequate budget	3.5862	1.15019	9
17.	Project manager technical capabilities	3.5172	1.21213	14
18.	Project manager commitment towards goals	3.5690	1.12075	11
19.	Project participants commitment towards goals	3.6121	1.10155	6
20.	Well-laid out specifications	3.3966	1.20057	20
21.	Good Leadership	3.6724	0.98489	2
22.	Clear and realistic objectives	3.5259	1.11481	13
23.	Project management methods and tools	3.5948	1.13428	7
24.	Complete the project according to requirements	3.3707	1.18344	21
25.	Use of superior and appropriate technology	3.4224	1.37145	19

In the above table 5 shows highest mean having highest ranking. Here the top 5 success factors of project management success in construction project success are effective communication between stakeholders, good leadership, and clarity of project goals, client involvement and competency of project manager. [26] research found that communication among stakeholders, competency and commitment are more important in construction project management. The success factors having schedule performace and budgeted performance have 7th and 8th factor in project management for project success[27]. In this survey, schedule performance and adequate budget have 12th and 9th factor because of increasing of other important factors in this study. Monitoring of the project, effective coordination, effective decision making , feedback capabilities, plan and schedule, team work and control mechanisms are the critical success factors of project management[28]. The study about critical success factors of project management in sustainable housing development. Project manager authority, communication/information, effective planning, monitoring and perfomance and feedback are the success factors for sustainable housing development[29. [16] Explained about the factors of project management and project success. According to project management, cost, time, quality and client satisfaction are the success factors. [14] Adequate fund for execution have highest ranking in the success factors followed by professional and technical skills, relationship with stakeholders, companies financial strength, client decision making process. In this survey, adequate budget have 9th factor for project success. The success factors of sustainable housing in Nigeria. In his research [30]competent project team, support from the top management, adequate fund and resources, realistic cost and time estimates have 1, 5, 6, and 10. In this survey these factors have ranking of 7, 17, 9, and 24. [31]. The user involvement, good planning and estimation, good leadership and team members technical skills have top 4 success factors for project success. Top 4 critical factors in project management are project mission, top management support, project plan and schedule, client consultation. Classified critical factors into strategic and tactical[32]. Project mission, top management support and project scheduling have highest ranking in strategic factors. Client consultation, personnel selection and training have highest ranking in tactical factors.

6. Conclusion

The significant factors found by using regression analysis are support from top management support, competent project team, abilities to solve problems, realistic cost and time estimates, information/communication, competency of the project manager, schedule performance, technical capability of the project manager, commitment of all project managers, good leadership, past experience of project management tools and technology and use of superior and appropriate technology. Using factor analysis, 25 factors were reduced to 9 factors. These are

1. Planning, estimation and monitoring,
2. Time and cost management
3. Client participation with definite objectives
4. Capability of project manager towards project
5. Team work
6. Communication towards reaching objectives
7. Management support and participants commitment
8. Updated Technology
9. Deliver directions with specification.

From ranking of the factors using mean, effective communication between stakeholders got highest ranking. Clarity of project goals, good leadership and client involvement has second priority. Project manager competence, project participants' commitments towards goals have third and fourth priority.

7. Recommendations

- Bringing talented pool of employees within organization give better results. They have the ability to solve complex situations within less time. Good HRM should be developed. This play a major role in the construction industry.
- In small scale industries also project management training should be required. Most of the small industries face cost and time overrun because of inexperience in the initial stage.

- Proper training in the conception stage in the project management gives fruitful results in the upcoming future.
- Planning, scheduling and estimation are the main area to be focused by the project manager.
- The communication between stakeholders is most important in project management success. To improve this communication between stakeholders, maintain the long term relationship with stakeholders.

References

[1] Munns A K and Bjeirmi B F 1996 The role of project management in achieving project success. *Int. J. Proj. Manag.* **14** 81-7.
[2] White D and Fortune J 2002 Current practice in project management- an empirical study *Int. J. Proj. Manag.* **20**(1) 1-11.
[3] Auti A and Skitmore M 2008 Construction project management in India. *Int. J. Constr. Manag.* **8**(2) 65-77.
[4] MatosS and Lopes E 2013 Prince2 or PMBOK–a question of choice. *Procedia Technol.* **9** 787-94.
[5] Prakash M and Udayashankar D Project Management Practices and Risk Perception in Construction Companeis. *IOSR J. Mech. Civ. Eng. (IOSR-JMCE)* 37-42.
[6] Serbanoiu I, Verdes M, Serbanoiu A A, Serbanoiu B V and Munteanu M 2016 Actual Trends in Construction Project Management in Romania. *Adv. Eng. Forum* **21** 587-95.
[7] Muller R and Turner R 2010 Leadership Competency profiles of successful managers *Int. J. Proj. Manag.* **28** 437-48.
[8] Ika L A 2009 Project success as a topic in project management journals. *Proj. Manag. J.* **40**(4) 7-19.
[9] Basu R 2014 Managing quality in projects: An empirical study. *Int. J. Proj. Manag.* **32**(1) 178-87.
[10] Clarke A 1999 A practical use of key success factors to improve the effectiveness of project management. *Int. J. Proj. Manag.* **1 7**139-45.
[11] Alia Z, Zawawi E M A, Yusof K and Aris N M 2014 Determining critical success factors of project management practice: A conceptual framework. *Procedia-Social and Behavioral Sci.* **153** 61-9.
[12] Shahin A and Jamshidian M 2006 Critical success factors in project management: A comprehensive review. *Proc. 1st Int. Proj. Manag. Conf.* 1-14.
[13] Atkinson R 1999 Project management: cost, time and quality, two best guesses and a phenomenon, it's time to accept other success criteria. *Int. J. Proj. Manag.* **17** 337-42.
[14] Mohmood S and Sajid A D 2012 Exploring the Critical Success Factors of construction companies of developing countries. *Res. J. Soc. Sci. Manag.* **1** 8-16.
[15] Naoum S, Fong D, Walker G (2004). Critical Success Factors in project management. *Globalisation Constr.* 817-27.
[16] Abdulrahaman B A and Abdallah F A N 2014 Critical Success Factors in project management: critical review of secondary data. *Int. J. Sci. Eng. Res.* **5**(6) 325-31.
[17] Els M, der Merwe M F V and Hauptfleisch 2012 Critical success criteria and success factors in project management: A quest to enhance generic professional practices. 1-15.
[18] Loo R 2002 Working towards best practices in project management: a Canadian case study. *Int. J. Proj. Manag.* 93-8.
[19] Ribeiro P, Paiva A, Varajao J and Dominguez C 2013 Success eavaluation factors in construction project management- some evidence from medium and large Portuguese companies. *KSCE J. Civ. Eng.* **17**(4) 603-9.
[20] Yanwen W 2012 The study of complex project management in developing countries 2012 *Proc. Int. Conf. Solid State Devices and Mater. Sci.* 1547-52.
[21] Norizam A and M A Malek 2013 Developing Critical Success Factors(CSF) for effective

construction management in Malaysia Industry *Canadian Center Sci. Educ.* **9** 212-8.
[22] Gliem J A and Gliem R R 2003 Calculating, interpreting and reporting Cronbach's Alpha Reliability Coefficient for Likert-Type Scales. *Proc. Conf. Adult, Continuing, and Continuity Educ.* 82-8.
[23] Gemuenden H G and Lechler T 2009 Success factors of project management: The Critical Few-an Empirical investigation. *Proc. Portland Int. Conf. Manage. Technol.* 375-7.
[24] Varajao J, Dominguez C, Ribeiro P and Paiva A 2014 Critical success aspects in project management: similarities and dissimilarities between the construction and the software industry *Tehnicki vjesnik* **21**(3) 583-9.
[25] Shokri-Ghasabeh M and Kavoousi-Chabok K 2009 Generic project success and project management success criteria and factors: Literature review and survey. *World Sci. Eng. Academy S.* 456-68.
[26] Micheal J, Deepak T J and Tong I S Y 2014 Ranking the Factors that Influence the Construction Project Management Success: Malaysian Perspective. *Civ. Environ. Res.* **6**(1) 80-8.
[27] Gar KK 2014 Critical Success Factors of Project Management for Dam Constrution Projects in Myanmar Doctoral *Dissertation. BRAC University.*
[28] Omran A, Abdulbagei M A and Gebril A O 2012 An Evaluation of the Critical Scuccess Factors for Construction Projects in Libya. *Int. J. Econ. Behav* **2**(1)17-25.
[29] Ihuah, Paulinus W, Iyenemi I K, and David Eaton 2014 A review of Critical Project Management Success Factors (CPMSF) for sustainable social housing in Nigeria. *Int. J. Sust. Built Environ.* **3**(1) 62-71.
[30] Attarzadeh I and Siew H O 2008 Project management practices: the criteria for success or failure. *Communications of the IBIMA* **1**(28) 234-41.
[31] Pinto JK and Slevin D P1987 Critical factors in successful project implementation. *IEEE T. Eng. Manage.* **1** 22-7.
[32] Schultz R L, Slevin D P Sand Pinto J K 1987 Strategy and tactics in a process model of project implementation. *Interfaces* **17**(3) 34-46.

IOP Conf. Series: Earth and Environmental Science **80** (2017) 012045 doi:10.1088/1755-1315/80/1/012045

Examining the Challenging Hindrances facing in the Construction Projects: South India's Perspective

K Subramanyam and M K Haridharan*

School of Civil Enigneering, SASTRA University, Thanjavur – 613 401. India.

*Email: haridharan@civil.sastra.edu

Abstract. Developing countries like India require a huge infrastructure to facilitate needs of the people. Construction industry provides several opportunities to the individuals. Construction manager work is to supervise and organize the construction activities in construction projects. Now a day construction manager facing challenges. This paper aimed to study the challenges facing by the construction manager in the perception of construction professionals. 39 variables were taken from the literature review which found to be severe impact on construction managers' performance. Construction manager, project manager and site engineers are the respondents for this survey. Using SPSS, regression analysis was done and recognized significant challenges. These challenges were classified into 5 domains. In management challenges, resource availability and allocation, risks and uncertainties existing in the project onsite, top management support and cost constraints are the most significant variables. In skills requirement of a construction manager challenges, technical skills required to learn and adapt new technology in the project, decision making and planning according to the situation in site are the most significant variables. In performance challenges, implementation of tasks according to the plan is the important variable whereas in onsite challenges, manage project risks, develop project policies and procedures are the most important.

1. Introduction

Construction industry involves several departments within the organization along with peripheral departments align with that organization to perform works. To execute construction management related works in construction, a key entity is required called as Construction manager. He is the person responsible for initiation, execution, monitor the work periodically to check quality and deliver the project by meeting client constraints. In the process of executing these works construction manager faces more challenges. [1&2] Studied about the construction manager challenges in Indian construction industry whereas [3] identified construction management challenges in rural projects which were related to project manager and construction manager work. To achieve project quality, cost, time and scope requirements, construction manager has to manage the operations proactively [2]. [4] Stated that the project manager difficulty with challenging authority due to matrix structure, cross-functional and motivation which were comes under human resource challenges. In order to overcome the stated HR challenges the suggestions were reported.

The construction project managers suffered from different types of stresses. This stress had shown a negative influence on performance (task, interpersonal and organizational) of the construction project managers[5]. These parameters implied that it was important to implement preventive stress coping strategies. Till now, there are no perfect duties and responsibilities of construction manager related to job. [6] Concluded that there was some unanimity and disagreements among the parties to a construction project relative to CM duties. Disagreements were mainly between designers and contractors. But still, management related to site work come under construction manager duties. [7] explained that construction manager as a leader should aware of needs and characteristics of project. Leadership skills and technical competency must be required for the construction manager to handle the difficult situations. [8] Explained about the technical competencies of construction manager in macro and mesa level. As a construction manager, he should think about the safety in construction work place. [9] Stated about construction manager responsibilities related to safety. [10] Explained about the safety perception by 6 groups of job holders in construction industry. [11] Described in the public and private sector where project manager performance was affected.

[12] Explained about the evaluation of construction contractor performance beyond national boundaries. Japanese contractors have capability to make fewer defects compared to UK and USA. They can achieve specific goal in shorter time by increasing the human resource in the site. [13] Explained about the factors affecting delays in construction projects. Lack of commitment, poor site management, poor site coordination, inefficient planning, less clarity in project scope, improper communication, substandard contract are the factors that produces delays in construction projects. [14] Discussed about the communication patterns at construction manager level in Israel and explained about the effective and ineffective projects. Communication between design team and construction manager were found to be essential for attaining project objectives. [15] Disclosed the causes of contractor cost overrun in Ethiopian construction projects. As per the contractor, inefficiency in proper planning, variation in the material cost, poor productivity, inflationary pressure and project financing are the reasons for cost overrun. [16] Unveiled the causes of project delay in the construction industry in Afghanistan. Security, corruption, lack of technical capabilities of contractor's technical staff, delayed payments by clients; poor site management and poor supervision by contractor are the critical factors which cause delays in construction projects. [17] Explained about the causes of delays in construction industry. To reduce these delays adequate planning and coordination, monitoring of the projects by the experienced and qualified professionals should be required. [18] Discussed about the cost overrun in Malaysian public and private construction projects. Cost overrun can lead conflicts and litigation or in extreme project may be abandoned.

2. Problem Statement

Construction industry in India primarily focused on infrastructure, housing, and irrigation projects to facilitate the requirements and needs of the people. Due to complex activities involved in present construction projects, huge human resources are required to execute the concerned activities. To reduce the workloads of project manager in construction projects, firms created a designation called construction manager. Construction manager is responsible to look over the execution woks in the project. Construction firms in south India seriously neglected the challenges facing by the construction managers and providing necessary solutions to those challenges. Because of more duties and responsibilities, construction manager faces challenges. It is important to solve the problems facing by the construction manager in construction projects for better and optimal output. The main purpose of this research study is to understand and evaluate major challenges facing by the construction manager and suggest the appropriate recommendations to handle the challenges.

3. Objective of the study

The primary objective of this study is to scrutinize the challenging facing by the construction manager in the construction projects of South India. The objectives include

- To understand the demographic profile of the construction professionals working in South Indian Construction Projects.
- To investigate and find out the necessary construction practices and problem areas in construction projects of South India.

- To evaluate the challenges facing by the construction manager and deliver suggestive approaches to overcome the construction manager challenges

4. Methodology
This methodology finds the methods and approach of the research. Steps in this research are:

4.1. Study population
Respondents in this research study are construction professionals generally in South India.

4.2. Sampling Technique
Random sampling technique was used to select the respondents. Project managers and construction managers are the main respondents for this survey for good fit of reliability of the data.

4.3. Data collection method
Structured questionnaire was used to collect the primary data. Construction professionals in South India were the respondents for collection of primary data.

4.4. Questionnaire Design
This research study is a survey approach. The approach comprises of both literature search and structured questionnaires which was considered to be most suitable instrument to reach the population. The study was conducted with construction firms that are situated in Hyderabad, Chennai, Vijayawada, Tirupati, Thanjavur, Trichy and Bangalore which can be called as cluster of south India. This research study adopted a qualitative as well as quantitative search to find out the impact of various challenges facing by the construction manager in construction projects. The challenging variables extracted from the above literature. The extracted variables are cross examined with the construction professionals and specialized educational laureates. After thorough scrutiny, 39 challenging variables were listed in the questionnaire for final survey. The questionnaire comprises of two parts. Part-A explains about the demographic profile of the respondents whereas part-B explains about the challenging variables facing by the construction manager. The respondents for the study were construction manager, project manager and site engineers. A total of 107 questionnaires duly completed, returned and found suitable for analysis. The challenging variables questionnaire consists questions related to 5 domains namely Management challenges, Skill and Knowledge requirements, Performance challenges, Onsite challenges and External challenges. Respondents were asked to rate on 5 point Likert scale where 1 = strongly disagree, 2 = disagree, 3 = neither disagree nor agree, 4 = agree and 5 = strongly agree.

4.5. Data analysis tool
SPSS [21] software was used to analyze the data. Quantitative data analysis will be espoused to identify patterns and relationships for the data by statistical methods. The data was subjected to Reliability Analysis to check the internal consistency of the variables. The data was analysed using Descriptive Statistics, Regression, one way-ANOVA (ANalysis Of Variance), and Correlations with the help of SPSS(Statistical Package for Social Sciences) version 14.0. Data was analyzed using tables and percentages for Means and Relative Important Index (RII) in Excel sheets. Relative Importance Index (RII) was calculated for each challenging variable. The variables were ranked based on the Relative Importance Index (RII) value which ranges from 0 to 1.
Relative Importance Index (RII) = Sum of weights $(w_1+w_2+w_3+w_4+w_5)/(P*N)$(1)
Where w is the weighting given to each variable by the respondents, ranging from 1 to 5
P is the highest weight (i.e. 5) in the study; and
N is the total number of samples.

IOP Conf. Series: Earth and Environmental Science **80** (2017) 012045 doi:10.1088/1755-1315/80/1/012045

5. Results and Discussions

Analysis of the data should mainly focus on the reliability. If data is reliable, then further analysis should be prefer. Because reliability refers to the internal consistency of a research study or measuring test. Below Table 1 Indicates the Reliability Statistics. It shows the Cronbach's Alpha as 0.891 which is greater than the threshold value 0.7. Cronbach's Alpha is used as an estimate of reliability. For good fit of reliability Cronbach's Alpha should be greater than 0.7. So, these 40 items can be considered as good and fit for the further data analysis.

Table 1. Reliability Analysis

Cronbach's Alpha	N of Items
0.891	40

Using frequency distribution, raw data is transformed to grouped data. Grouped data is used to understand the range of values for a particular variable. Demographic profile of the respondents are chosen as the variable for this survey. Through this frequency of a particular variable chosen by the respondents can be known. Below table 2 shows that the respondents having age group of 26-35 years responded with a percentage of 35.5 and followed by 36-45 years with 27.1%. The percentage of Male and Female respondents surveyed was 83.2 and 16.8. Majority of the respondents were graduates with 52.3% followed by post graduates with 23.2%. It is coped with the views on educational programs[22 to25]. Out of 100, 36.4% of the respondents handled less than 5 projects and 30.8% handled 6-11 projects. Respondents having experience of 2-6 years were 13.6 percentage and 12-16 years of experience were 19.6 percentage. Most of the respondents worked in urban location with a percentage of 72 followed by semi urban with 20.6 percent. Majority respondents' deals projects having an average project budget of 26 – 50 crores with 55.1% followed by project budget less than 25 crores with 23.4 percent. Survey statistics show that commercial projects respondents are more with a percentage of 47.7 followed by industrial projects with 29.9 percent. Majority of the projects comes under medium category with 68.2% and followed by large projects with 17.8 percent.

Ranking the challenging factors of construction manager is used to know which factors are mainly challenging factors for construction manager. It is easy to take immediate actions or modifications for better results. Below Table 3 rankings are provided for the challenges based on relative importance index. From management challenges it is observed that resource availability and allocation, safety and cost constraints stands at 1, 2 and 3 places with relative importance index of 0.76, 0.74 and 0.72 which were previously ranked as 7, 4 and 3 respectively [2]. Decision making, technical capabilities of construction manager stand 1 and 2 places with 0.81 and 0.75 weightages. Variance is observed as 0.02 and 0.25 with [2]. Monitoring and Implementation of tasks got 1 and 2 rankings in performance challenges category. Develop project specific policies and procedures, Implementation of team meetings frequently, Commitment by the manpower to achieve the project goals, Material wastage, Clarity in drawings and specification and Equipment productivity stances in 1,2 and 3 rankings with RII of 0.75, 0.71 and 0.70. From onsite challenges, Codes and permits, Government regulations and Labor laws are ranked as 1 and 2 with 0.76 and 0.72 weightage factors from external challenges.

Table 2. Demographic profile of the respondents

S.No	Variables	Description	Majority wise first	Majority wise second	Majority wise third	Majority wise fourth
1	Age	Category	26-35	36-45	46-55	18-25
		Percentage value	35.5	27.1	21.5	12.1
2	Gender	Category	Male	Female		
		Percentage value	83.2	16.8		
3	Education Qualification	Category	Graduate	Post Graduate	Diploma	
		Percentage value	52.3	23.2	20.6	
4	Number of projects handled	Category	Less than 5	6-10	11-15	Above 15
		Percentage value	36.4	30.8	21.5	11.2
5	Experience in years	Category	2-6	12-16	Less than 1	Above 16
		Percentage value	34.6	19.6	16.8	15
6	Location	Category	Urban	Semi Urban	Rural	
		Percentage value	72	20.6	7.5	
7	Average budget of the project	Category	26 - 50	Less than 25	51-75	76-100
		Percentage value	55.1	23.4	18.7	2.8
8	Type of projects worked on	Category	Commercial	Industrial	Residenti al	Govt. Projects
		Percentage value	47.7	29.9	12.1	10.3
9	Size of the project	Category	Medium	Large	Small	
		Percentage value	68.2	17.8	14.0	

IOP Conf. Series: Earth and Environmental Science **80** (2017) 012045 doi:10.1088/1755-1315/80/1/012045

Table 3. Ranking using Relative Importance Index

	Mean	RII	Rank
Management Challenges			
Resource allocation	3.81	0.76	1
Time delay	3.05	0.61	7
Cost limitations and overrun	3.59	0.72	3
Quality proposed	2.90	0.58	8
Safety	3.72	0.74	2
Risk and uncertainties	3.50	0.70	4
Communication with top management	3.32	0.66	5
Top management support	3.29	0.66	5
Skill Requirement of a Construction Manager			
Technical skills	3.75	0.75	2
Planning	3.64	0.73	3
Communication with site engineers and workers	3.65	0.73	3
Coordination	3.65	0.73	3
Integration	3.51	0.70	5
Leadership	3.17	0.63	7
Problem-solving	3.26	0.65	6
Decision-making	3.55	0.71	4
Trustworthiness	4.03	0.81	1
Flexibility	3.00	0.60	8
Performance Challenges			
Planning on site	3.71	0.74	3
Implementation	3.74	0.75	2
Monitoring	3.83	0.77	1
Controlling	3.56	0.71	4
Onsite Strategy Challenges			
Develop Realistic estimates and forecasting	3.26	0.65	4
Managing project risks	3.30	0.66	6
Develop project specific policies and procedures	3.77	0.75	1
Assign roles and responsibilities to the manpower	3.46	0.69	3
Implementation of team meetings frequently	3.74	0.75	1
Commitment by the manpower to achieve the project goals	3.56	0.71	2
Dispute resolution in the project	3.39	0.68	5
Material wastage	3.55	0.71	2
Clarity in drawings and specification	3.49	0.70	3
Equipment productivity	3.49	0.70	3
Availability of technical staffs	3.47	0.69	4
Availability of skilled labor	3.44	0.69	4
External Challenges			

Political influence	2.51	0.50	4
Public involvement	3.42	0.68	3
Labor laws	3.61	0.72	2
Government regulations	3.62	0.72	2
Codes and permits	3.79	0.76	1

To know the relationship between dependent and independent variables in the data, regression analysis was used. This analysis tells about the most significant independent variables with respect to dependent variable. Here dependent variable is overall performance of handling the challenges.Below Table 4 shows the factors and their significant value. This significant values tells about how much relationship exist between factor and dependent variable. T calculated using coefficient (B) divided by standard error (Std. Error). Unstandardized coefficient B explains the amount which dependent variable changes by changing of one unit in independent variable. Standardized coefficient Beta compares the strength of each individual independent variable to the dependent variable.

Table 4. Regression Analysis

		Coefficients				
S.No.	Model	Unstandardized Coefficients		Standard Coefficients	T	Sig.
		B	Std. Error	Beta		
	(Constant)	1.273	.595		2.138	0.036
1.	Resources availability and allocation	-0.276	0.095	-0.306	-2.909	0.005
2.	Time Delay due to miscellaneous activities	0.050	0.088	0.052	0.565	0.574
3.	Cost constraints	0.301	0.092	0.353	3.272	0.002
4.	Meeting the proposed Quality specifications	-0.115	0.101	-0.116	-1.146	0.256
5.	Ensure the Engineers and Workers safety at Project site	0.013	0.083	0.015	0.161	0.873
6.	Risks and uncertainties existing in the project onsite	-0.196	0.082	-0.203	-2.396	0.019
7.	Communication with the Top Management	-0.026	0.077	-0.032	-0.329	0.743
8.	Top Management Support	0.236	0.072	0.295	3.299	0.002
9.	Technical skills required to learn and adapt New Technology in the project	0.439	0.089	0.501	4.941	0.000
10.	Plan the tasks and activities before executing in the site	-0.128	0.097	-0.126	-1.323	0.190
11.	Communication with site engineers and workers	-0.078	0.115	-0.083	-0.680	0.499
12.	Coordination of Manpower	-0.147	0.094	-0.150	-1.552	0.125
13.	Integration of several departments	0.059	0.087	0.058	0.679	0.500
14.	Leadership	-0.099	0.080	-0.117	-1.227	0.224
15.	Problem Solving Capabilities	0.067	0.067	0.091	1.008	0.317
16.	Decision making	0.245	0.080	0.276	3.063	0.003
17.	Trustworthiness	0.168	0.106	0.166	1.593	0.116
18.	Flexibility	0.008	0.059	0.011	0.134	0.894

19. Planning according to the situation in site	0.175	0.081	0.184	2.163	0.034
20. Implementation of tasks according to the plan	-0.254	0.087	-0.268	-2.917	0.005
21. Monitoring of activities	0.101	0.101	0.091	1.004	0.319
22. Controlling of Deviations	0.078	0.083	0.088	0.942	0.350
23. Develop realistic estimates and forecasting	-0.047	0.078	-0.059	-0.608	0.545
24. Managing project risks	-0.223	0.073	-0.272	-3.069	0.003
25. Develop project policies and procedures	-0.567	0.114	-0.544	-4.986	0.000
26. Assigning roles and responsibilities to the human resource	0.035	0.095	0.038	0.372	0.711
27. Implementation of Team meetings frequently	-0.038	0.085	-0.042	-0.451	0.653
28. Commitment by the manpower to achieve the project goals	-0.031	0.092	-0.034	-0.331	0.742
29. Dispute resolutions in the project	0.058	0.102	0.066	0.572	0.569
30. Wastage of Material	0.201	0.084	0.235	2.395	0.019
31. Clarity in Drawings and Specifications	-0.186	0.090	-0.225	-2.068	0.043
32. Equipment Productivity	-0.087	0.072	-0.099	-1.208	0.231
33. Availability of Technical Staff	-0.258	0.084	-0.276	-3.079	0.003
34. Availability of Skilled Labor	0.220	0.077	0.262	2.859	0.006
35. Political Influence	0.029	0.078	0.030	0.368	0.714
36. Public Involvement	-0.012	0.089	-0.014	-0.131	0.896
37. Labor Laws	0.275	0.107	0.304	2.575	0.012
38. Government Regulations	0.200	0.088	0.220	2.267	0.027
39. Codes and Permits	0.427	0.112	0.444	3.799	0.000

$$\hat{Y} = 1.273 + (-0.306)x_1 + (0.052)x_2 + (0.353)x_3 + (-0.116)x_4 + (0.15)x_5 +(-0.203)x_6 + (-0.032)x_7 + (0.295)x_8 + (0.501)x_9 + (-0.126)x_{10} +(-0.083)x_{11} + (-0.150)x_{12} + (0.058)x_{13} + (-0.117)x_{14} + (0.091)x_{15} + (0.276)x_{16} + (0.166)x_{17} + (0.011)x_{18} + (0.184)x_{19} + (-0.268)x_{20} + (0.091)x_{21} + (0.088)x_{22} + (-0.054)x_{23} + (-0.272)x_{24} + (-0.544)x_{25} +(0.038)x_{26} + (-0.042)x_{27} + (-0.034)x_{28} + (0.066)x_{29} + (0.235)x_{30} + (-0.225)x_{31} +(-0.099)x_{32} + (-0.276)x_{33} + (0.262)x_{34} + (0.030)x_{35} + (-0.014)x_{36} + (0.304)x_{37} + (0.220)x_{38}+ (0.444)x_{39} \quad (2)$$

where, $\hat{Y}$ is the estimated overall performance of handling the challenges. The above equation shows that the impact of challenging variables against the overall performance handling by the construction manager. The above equation states that, on an average, if the variable namely resources availability and allocation changes by 1 unit, there will be -0.306 unit increase in the overall construction managers' performance of handling the challenges. The result of the t test reveals that the calculated significance of the partial regression coefficients -0.306, 0.353, -0.203, 0.295, 0.501, 0.276, 0.184, -0.268, -0.272, -0.544, 0.235, -0.225, -0.276, 0.262, 0.304, 0.220 and 0.444 are valid at 1 and 5 percent level respectively.Resources availability and allocation, cost constraints, risks and uncertainties, top management support, technical skills, decision making, onsite planning, implementation of tasks according to the plan, managing project risks, develop project policies and procedures, material wastage, clarity in drawings and specifications, availability of technical staff, availability of skilled labour, labour laws, government regulations, codes and permits (0.005, 0.002, 0.019, 0.002, 0.000,

0.003, 0.034, 0.005, 0.003, 0.000, 0.019, 0.043, 0.003, 0.006, 0.012, 0.027, and 0.000) having the significance value less than 0.005 it has been concluded that these variables represent statistically dominant variables that evaluates the overall performance of handling the challenging by the construction manager.

6. Conclusion

From the findings and discussions it is concluded that the construction manager facing several challenges in construction projects of south India. The challenges are categorized in 5 domains. Resources availability and allocation, cost constraints, risks and uncertainties, top management support have more impact on construction managers' management approach. Technical skills and decision making skills are observed dominant skill requirements that required for a construction manager to handle the construction projects smooth and efficient. Onsite planning, implementation of tasks according to the plan is confessed as challenges for the construction manager in performing the activities. Manage project risks, develop project policies and procedures, material wastage, clarity in drawings and specifications, availability of technical staff and availability of skilled labour have higher impact on construction manager's performance in field. Labour laws, codes and permits and government regulations seem to be prominent from external challenges which pose difficulty for the construction manager to handle.

7. Recommendations

Resource allocation challenges and cost overrun challenges can be reduced by adapting resource management approaches and effective estimations. In order to overcome the risks and uncertainties in the construction projects, construction manager need to forecast the issues by inspecting progress of the project at regular intervals. Top management should have to increase its communication levels with construction manager to execute the works in a stipulated way. Construction manager need to learn the new technologies evolved in the construction practices to facilitate in making good decisions at right time. In order to reduce the technical staff scarcity in projects, job rotation among the engineers should be employed. On job training and development to the newly recruited engineers will give good results. Labour laws, Codes and permits and Government regulations should be reviewed periodically by the top management as well as construction manager and project manager. It reduces the ambiguity among the labours and engineers in the construction projects.

References

[1] Shirur S and Torgal S 2014 Enhancing safety and health management techniques in Indian construction Industry. *Int. J. Eng. Tech. Res.* ***2*(4)** 52-6.

[2] Okoye, P U, Ngwu C and Ugochukwu S C 2015 Evaluation of Management challenging facing Construction Practice in Nigeria. *Int. J. of Appl. Innov. Eng. Manage.* ***4***(1) 19-28.

[3] Tran, D Q, Hallowell M R, and Molenaar K R 2015 Construction Management Challenges and Best Practices for Rural Transit Projects. *Am. Soc. Civ. Eng. (ASCE)* ***31*(5)** 1-9.

[4] Minavand H, Farahmandian S and Minaei V 2013 HR Challenges of Project Managers. *IOSR Int. J. Bus. Manage* ***2***(5) 40-5.

[5] Senaratne S and Rasagopalasingam V 2016 The causes and effects of work stress in construction Project Managers: the case in Sri Lanka. *Int. J. of Constr. Manage.* **17**(1) 1-12.

[6] Arditi D and Ongkasuwan D 2009 Duties and Responsibilities of Construction Managers: Perceptions of parties involved in construction *J. Constr. Eng. Manage.* ***135***(12) 1370-4.

[7] Gharehbaghi K and McManus K 2003 The Construction Manager as a Leader. *Leadership Manage. Eng.* ***3***(1) 56-58.

[8] Yaman, S K, Abdullah A H, Mohammad H and Hassan F 2015 Technical compitency of Construction Manager in Malaysian Construction Industry. *Appl. Mech. Mater.* 1053-59.

[9] Smith G R and Roth R D 1991 Safety Programs and the Construction Manager. *J. Constr. Eng.*

IOP Conf. Series: Earth and Environmental Science **80** (2017) 012045 doi:10.1088/1755-1315/80/1/012045

Manage. ***117*(2)** 360-71.
[10] Chen W T, Lu C S, Liu S-S and Wang M S 2013 Measuring the perception of Safety among the Taiwan Construction Managers. *J. Civ. Eng. Manage.* ***19***(1) 37-48.
[11] Pheng L S and Chuan Q T 2006 Environmental factors and work performance of project managers. *Int. J. Proj. Manage* **24**(1) 24-37.
[12] Best R and Langston C 2006 Evaluation of construction contractor performance: a critical analysis of some recent research. *Constr. Manage. Econ.* ***24***(4) 439-45.
[13] Doloi H, Sawhney A, Iyer K C and Rentala S 2012 Analysing factors affecting delays in Indian construction projects. *Int. J. Proj. Manage.* **30**(4) 479-89.
[14] Shohet I M and Frydman S 2003 Communication Patterns in Construction at Construction Manager Level. *J. Constr. Eng. Manage.* ***129*(5)** 570-77.
[15] Zewdu Z T and Aregaw G T 2015 Causes of Contractor Cost Overrun in Construction Projects: The Case of Ethiopian Construction Sector. *Int. J. Bus. Econ. Res.* ***4***(4) 180-91.
[16] Odeh A M and Battaineh H T 2002 Causes of Construction Delays. *Int. J. Proj. Manage.* **20**(1) 67-73.
[17] Mohammed K A and Isah A D 2012 Causes of Delay in Nigeria Construction Industry *Interdiscipl. J. Contemporary Res. Bus.* **4**(2) 785-94.
[18] Shehu Z, Endut I R, Akintoye A and Holt G D 2014 Cost overrun in the Malaysian construction industry projects: A Deeper insight. *Int. J. Proj. Manage.* **32**(8) 1471-80.
[19] Larsen J K, Shen G Q, Lindhard S M and Brunoe T D 2016 Factors Affecting Schedule Delay, Cost Overrun, and Quality Level in Public Construction Projects. *J. Manage. Eng.* ***32***(1) 1-10.
[20] Aibinu A A, and Odeyinda H A 2006 Construction delays and their causative factors in Nigeria *J. Constr. Eng. Manage.* **132.7** 667-77.
[21] Pallant J and Manual S S 2001 A step by step guide to data analysis using SPSS.
[22] Ntuli B and Allopi D 2014 Impact of Inadequate Experience and Skill on the Construction Sector in KwaZulu-Natal, South Africa. *Eng. Technol. Appl. Sci. Res.* ***4***(1) 570-5.
[23] Gale A W and Fellows R F Challenge 1991 Challenge and innovation: The challenge to the Construction Industry. *J. Constr. Manage. Econ.* **8**(4) 431-6.
[24] laili Jabar I, Ismail F, and Mustafa A A 2013 Issues in managing construction phase of IBS projects. *Procedia-Social and Behavioral Sciences* **105** 81-9.
[25] Tran, D Q, Hallowell M R, and Molenaar K R 2015 Construction Management Challenges and Best Practices for Rural Transit Projects *Am. Soc. Civ. Eng. (ASCE)* ***31*(5)** 1-9.

IOP Conf. Series: Earth and Environmental Science **80** (2017) 012046 doi:10.1088/1755-1315/80/1/012046

Improving Thermal Performance of a Residential Building, Related to Its Orientations - A Case Study

S Akshaya*, S Harish, R Arthy, D Muthu and C Venkatasubramanian

School of Civil Engineering, SASTRA University, Thanjavur -613401, India.

*Email : akshayasiva@gmail.com

Abstract. Urban planners and stakeholders require knowledge about the effectiveness of city-scale climate adaptation measures in order to develop climate resilient cities and to push forward the political process for the implementation of climate adaptation strategies. This study examines the impact of modifications in orientation of buildings with respect to heat load. Heat load calculation is a mathematical process to determine the best capacity, application and style of HVAC system. The purpose is to ensure energy efficiency while also maximizing comfort inside the building. This study of load calculation is essential for a building because it helps to pick the best orientation and focuses to find an orientation that will reduce energy due to direct solar radiation. One of the factors affecting this assessment is the latitude of the location. The heat gain is effective through walls and fenestration. Improper management through ineffective orientation of the building's natural heat gain leads to excessive consumption of energy in the form of CL. The total heat gain for the above factors is calculated with the equations and assumptions as per ASHRAE code. After the calculation of heat load for different orientations, the best suited orientation of the building is found. By altering the building to suitable orientation, the dependence on electrical equipment can be minimized and thereby helps in energy conservation.

1. Introduction

Buildings consume more than half of the total energy generated in building. Energy efficiency is a high priority in many countries. Nowadays, energy conservation is of high importance. Efficient use of energy is important as the power generation using fossil fuels has adverse environmental effects. Conserving energy is the best way to ensure a secure and sustainable energy supply and to reduce global warming. Energy efficiency and the sustainability of buildings play a vital role in the construction industry. Energy efficiency measures reduce the amount of energy consumed by maintaining or improving the quality of service provided in the building. The importance of energy efficiency in building sector is especially significant in developing countries owing to rapid new construction with opportunities to employ efficient materials and best practices. [3]

2. Methodology

The methodology for carrying out the thermal auditing process in an exciting building in order to determine the thermal comfort temperature in that particular building is summarized below

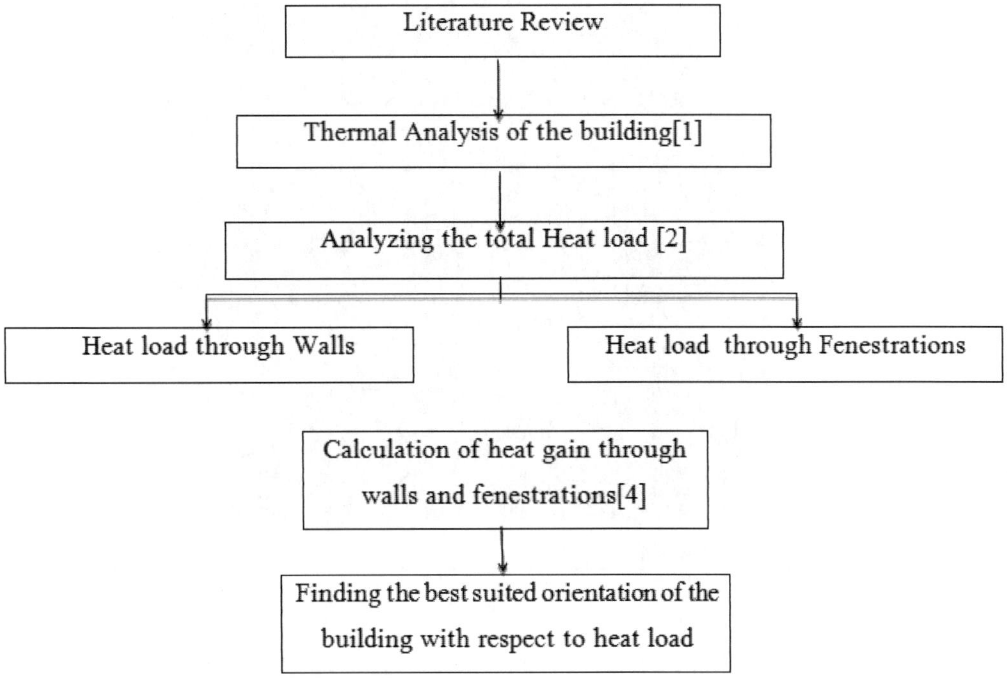

Figure 1. Flow chart representing methodology.

3. Method of calculations of Heat load

The CLTD/CLF/SCL method uses predetermined set of data to expedite and simplify the process of cooling/heating load approximation. The data is divided into many different sections based on many different variables. These variables include, building material of the envelope, thicknesses of the building materials, day of the year, time of day, orientation of the surface (e.g. wall or roof, 90 degrees or 180), and wall face orientation (cardinal directions, i.e. N, NW, S, SE, etc.), to name a few. In order to determine which set of CLTD/CLF/SCL data to look at, all the requisite variables must be defined.

The respective tables of data were generally developed by using the more complex transfer function method to determine the various cooling loads for different types of heating. The results gained by doing so are then normalized for each type of heat gain used for the tables, CLTD, CLF, and SF.

3.1 Case study

The following building with plinth area75'03" was chosen for calculating the total solar heat gain and show the variation of it with respect to the orientation of the building. It is located in Trichy district. It has an aspect ratio of 1.4 which is suitable for showing significant changes in the solar heat gain calculated.

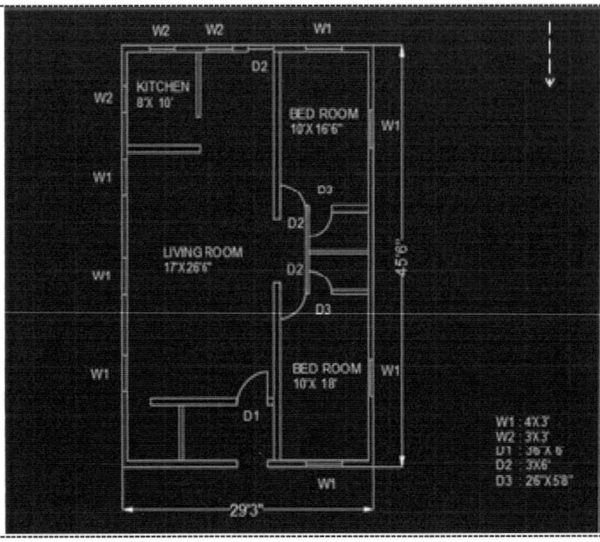

Figure 2. Building plan.

Table.1. Heat load through walls in different months about different orientations

Month	Heat load at 9am				Head load at 3pm			
	0°	90°	180°	270°	0°	90°	180°	270°
March	858.76	848.702	856.411	866.473	2090.007	1993.43	2115.643	2021.055
June	833.739	729.05	830.97	706.261	858.764	848.702	856.411	866.473
September	2090.00	1993.43	2115.64	2021.05	2064.98	1873.77	2090.20	1860.84
December	470.558	848.702	856.411	866.473	1701.801	1993.43	1727.72	2010.148

Note : All Heat Loads are in Watts, Area in m^2

Table.2. Heat load through fenestrations in different months about different orientations

Month	Heat load at 9am				Head load at 3pm			
	0°	90°	180°	270°	0°	90°	180°	270°
March	1803.46	1558.09	1301.14	1558.094	1515.03	1742.14	1913.70	1742.14
June	1575.54	1275.81	1126.75	1275.812	1286.13	1388.67	1642.32	1388.67
September	1734.85	1501.23	1258.19	1501.233	1465.76	1681.00	1844.06	1681.00
December	1676.21	1582.29	1285.35	1582.292	1537.27	1841.81	1847.48	1841.81

Note : All Heat Loads are in Watts, Area in m^2

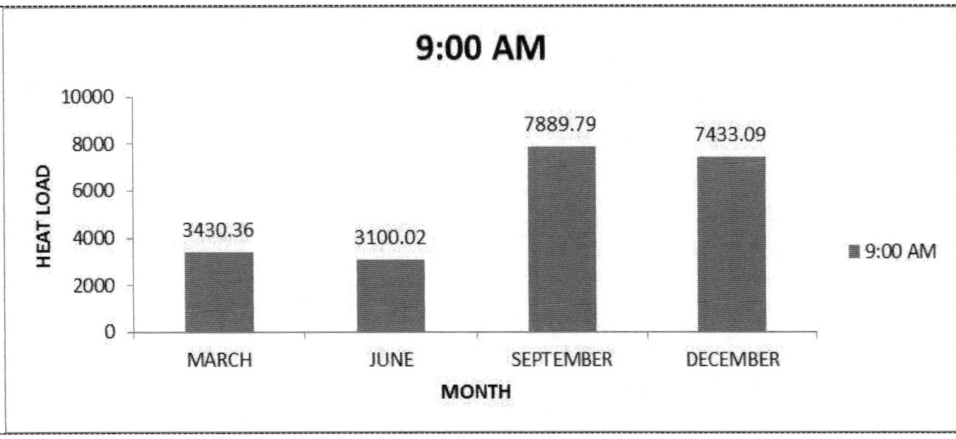

Figure 3. Total heat load during different months observed at 9 am.

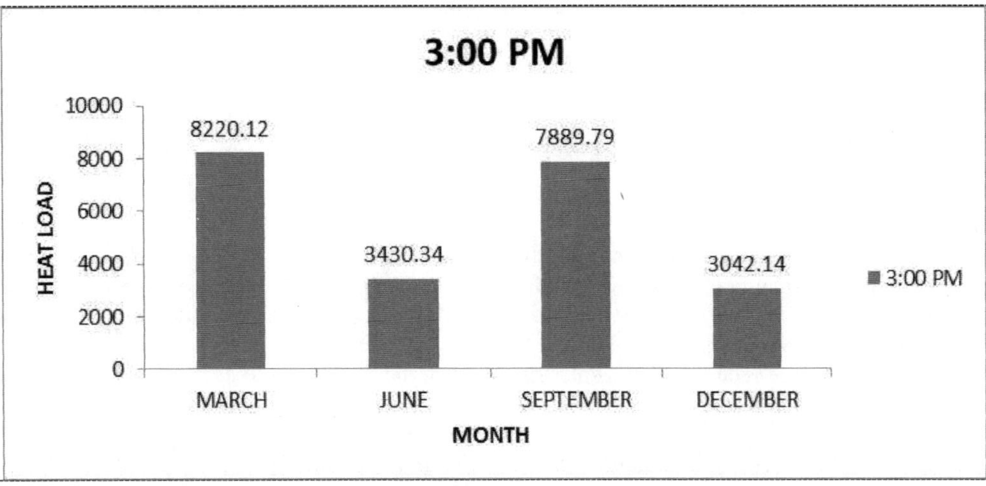

Figure 4. Total heat load during different months observed at 3 pm.

4. Results and Discussion

4.1 Solar Heat Gain

After obtaining the necessary heat gain results for the orientations of 0, 90,180,270 degrees for all equinoxes and solstices for the year 2016 and also for two different times in a day for the building as a whole, the average of the solar heat gain for each orientation is found. The table below shows the solar heat gain for different orientations and for different times during each day. Based on the above calculations, the graph is plotted between the Orientation and heat gain.

Table.3. Average heat load of various orientation (in watts)

Orientation	9am	3pm	Avg
0	9812.41	13751.51	11781.96
90	9153.795	14468.9	11811.35
180	7983.734	15296.79	11640.26
270	9212.204	14566.73	11889.47

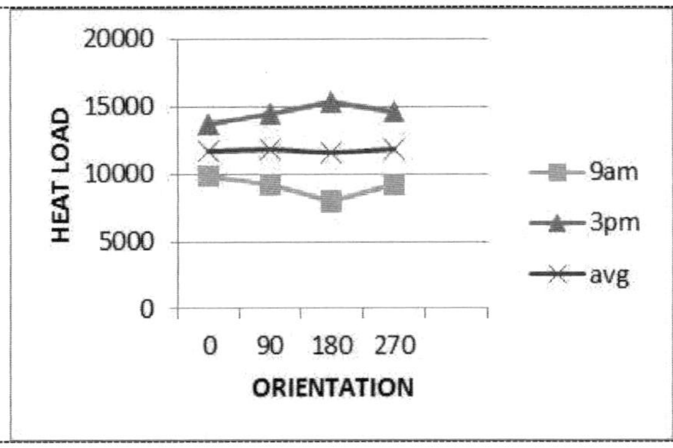

Figure 5. Max heat load with respect to orientation.

5. Recommendations

The admission of solar radiation into an interior space may cause problems such as indoor temperature, thermal and visual discomfort to the occupants, damage to sensitive objects and furnishings. Thus, it is of vital importance that the total heat load on the building must be controlled so as to increase the comfort conditions inside the building. The following recommendations can be taken into consideration to reduce the total solar heat gain on the particular building that we have chosen.

6. Orientation

From the obtained results, it be inferred that by orienting the building to 180deg, min heat load can be achieved on the building compared to the other orientations. The average for this particular orientation turns out to be 11640.26W which is the minimum of the orientations. In the present orientation of the building, the sunlight enters the bedroom in the afternoon and living room in the morning. Thus, if the building is constructed with the recommended orientation, the bedroom may receive sunlight during morning and reduce the heat load on air conditioners if used in bedroom.

7. Conclusion

This Case study provides an idea for implementing thermal auditing procedures for any type of building. In all respects, it involves in analyzing the thermal efficiency of the building. Equations used are well assessed before being adapted in the analysis of the building. This represents the detailed reference for performing an auditing process. Recommendations have been stated based on the thermal whereabouts of the building. This application acts as the initial step towards a larger research and also study in the existing building to save energy and to reduce the consumption of energy. This auditing procedure is procedure generic to all types of buildings. This thermal auditing provides recommendations about the provision of air conditioners in the building, it's capacity, with numbers, if the building is equipped with air conditioning system. If the building is not provided with air

IOP Conf. Series: Earth and Environmental Science **80** (2017) 012046 doi:10.1088/1755-1315/80/1/012046

conditioning system, then they can achieve the determined thermal comfort temperature using some passive cooling techniques. Therefore, this project presents a required thermal analysis that allows the detection of all the parameters which affect the thermal efficiency. Once all the characteristics of the building are known, the engineer or air conditioning technician would be provided with all the necessary details to model the insulation system adequately may be practically implemented in the residential, institutional and commercial buildings where the cooling loads are necessary. If all the above suggestions are applied in practical, the occupants can enjoy the thermal comfort for the entire year during their stay.

References

[1] El Hjoujii M E and Khaldoun A 2010 Thermal Auditing of Buildings: Essential Step towards Designing Energy Efficient Houses. American Society of Heating Refrigerating and Air Conditioning Engineers.(Section 2.Part D)

[2] Mohammed Tahsildoost and Zahra Sapadanim Zomorodian. 2015. Energy Retrofits Techniques: An Experimental Study of Two Typical School Buildings in Tehra. *Energ. Build.* **104** 65-72.

[3] http://www.math.nus.edu.sg/sun_and_architecture.pdf.

[4] http://www.accuweather.com/en/in/tiruchirappalli/190803/weather-forecast/190803.

IOP Conf. Series: Earth and Environmental Science **80** (2017) 012047 doi:10.1088/1755-1315/80/1/012047

Perception of Construction Participants in Construction delays: A case study in Tamilnadu, India.

V Rathinakumar*, T Vignesh and K Dhivagar

School of Civil Engineering, SASTRA University, Thanjavur – 613 401. India

*Email: rathinakumar@civil.sastra.edu

Abstract. Delays in the construction industry are universal fact, which affects the construction participants. The research work spotlights on determining the prevailing delays in the cities of Tamil Nadu, as perceived by the participants. After a few field level interactions, a questionnaire was framed and administered to the participants i.e., Consultants (50 Nos.), contractors (50 Nos.) and clients (150 Nos.) to understand their perception on the causes of delays. The factors for delay was categorized into 4 groups say Improper project planning, Design related issues, Finance related issues and Resource related issues. The responses were analysed using the SPSS software by applying ANOVA and Regression analysis. From the analysis, it was found that the personal financial problems of the client dominantly affect the entire construction progress and the subsequent design changes by the clients, Inadequate early project planning, labour related issues. Also, the delay groups were found to be the improper project planning and the Resource related issues.

1. Introduction

In India Construction Industry is the second largest growing, dominant, and ever green field next to agriculture, in which delay is the most enveloping factor in it. Delay may be pronounced as the let down to far-reaching the task in assured phase and projected cost. Delays are foreseeable and may take place at any stage of the construction progress and leads to the delay in the project delivery. Hence delay in a particular activity may also influence the initiation of another activity and it creates adversarial relationship, mistrust, litigations, mainly dissatisfaction, and finally leads to overall project delay. Delays in construction project are very pricey since that engages the financial investments of the clients and the resources of contractors and consultants.

1.1. Necessity of the Study

The intend of this research paper is to address the major factors influencing delay in the construction project and the relationship between the project delay and satisfaction of the respondents in the construction delivery based on the perception of the participants namely clients, consultant, and the contractor, which helps them to know the lively delays allied to the construction projects.

2. Causes and effects of delay

2.1. Causes of Delay

From the various literature studies, it is evident that a practical study is needed to know the actual scenario in the industry.

In [1] depicted the various successes and failure attributes for a project from the survey. The various success attributes are effectual monitoring, advice from the Project Manager and team, ethical attitude of the project participants, and PM's technical capability. The failure attributes include poor human resource management, negative attitude of the participants, inadequate early planning and conflicts between participants. Hence both are clearly pointed out in order to increase the success factors and decrease the failure attributes. And important result obtained is found to be the coordination among the project participants' which is found to be common irrespective of boundaries.

In [2] described the various dominant reasons for the causes for the project delay and suggests some practical strategies for the effective on time completion of the project. They are proper investigation of the ground conditions, good communication and proper management by the supervisors, contractors, and consultants, updated and resourceful man power at both technical and managerial level execution, effective and strong decision on the material specification, apt contractual system with efficient financial planning.

In [3] clearly portrayed the importance of the human resource planning and the various factors influencing the same form the interview from the construction industry. And they also present the important issues to be the uncomfortable environment, lack of protection and safety, unpredictable work with lower wages. Also suggests the importance of improving the project quality and completion of work on time.

In [4] suggested the various impacts of delay from the survey from the construction participants, which states slow decision by the owners, consultant's or architect's reluctance for change, poor labour productivity, poor site management and supervision, and reworks due to error in execution from the statistical regression model. From the research, they also mention that lack of commitment, and training to the site professionals, are to be the major reason for the delay.

In [5] declared the various factors that grounds for the construction delay and overriding among all is the finance related problem, also other important factors are found to be frequent design changes, slow delivery of materials, shortage of resources, failure to pay the completed works, price fluctuations, important of all is that the contractor's inability to complete in promised time.

In [6] declared that delays are organized into excusable and non-excusable delays from literature studies and categorizes delays to contractor related, client related, material related, and labour related with parameters that fall in it. Also, they assess the construction industry to know various delays that should be taken under consideration so that the non-excusable delay can be minimized. They conclude that selection of proper sub contractors, management of these parameters will reduce the indirect value, and harm the name and fame of organization.

In [7] stated some 73 causes of delays and categorizes those into 9 major groups as project, owner, contractor, consultant, design, material, equipment, labour, external group. And from the survey conducted from the construction industry the suggest keys to the major participants to avoid delay, i.e. to the owners, contractors, consultants, architect/design engineers. The common stricture found was the proper payment to the parties at correct time, frequent reviewing of activities to happen in correct time throughout.

In [8] have taken a case of the railway project that delayed for 135 months and identifies a new method of simulation to analyze and quantify the impact of delay factors. The critical path method and the delay analysis system is useful to know the duration of the project, critical activities that are done on time which are affected by the critical resources and to know the critical delay factor respectively. The results also reflect that the contractor rather than the consultant, and the owner are responsible for the construction delay. The sensitivity analysis also shows delays in material delivery, financial problem, man power shortage, slow site clearances are highly sensible.

In [9] clearly stressed that the cost and time over run are dominant in the construction industry and major reason for those are found to be delay in land acquisition, equipment erection, forest clearances, change in scope, cancellation of tender, law and order problem, price inflation. And also portrays the various roles of the consultants, contractors and clients for the effective completion of the project with the least delay, which signifies that there should also be coordination among the parties.

In [10] have taken two cases to know the various delay causes in the construction industry, effects of the delays, and various strategies to minimize the delays. The findings were that delay in the payment by the head office, poor site management, frequent staff changing, delay in supply of material, improper management by the engineers, and lack of man power. The effects of the delays were time over run, negative social impact, wastage of resources, and finally disputes. Hence depending on the various causes and the ill effects of the delay various commendations were proposed.

2.2. Effects of delay

The role of client, contractor, and consultant, infrastructure of the company, sudden change in environment and socio-political changes is very significant in the delay of the construction. The effect of these delays clearly reflects on the performance of construction project. Various common effects observed from the literature are the cost overrun, idling of resources, disputes, negative impact on the name and fame of the parties. In [5] stated various effects as the cost overrun due to inflations and fluctuation, extension of time, wastage of man power and resources, under-utilization of the equipment's, loss of confidence over one another, reduction in employment opportunities, arbitration, abandonment, etc. In [6] clearly declared that time overrun, cost overrun, dispute, arbitration, total abandonment, and litigation as the effects of delay. In [10] identified time overrun and cost overrun are the major recurring effect of the construction delay. From the survey conducted by [11] it is evident that the various effects of the delays are negotiations, cause of disputes, lawsuit, total desertion, abandonment. And also states in general that is the loss of wealth, time and capacity, resources.

3. Methodology

3.1. Population and sample

A comprehensive filed survey was conducted to collect the data from the respondents practicing in Trichy and Thanjavur. The selection of these two towns based on the category of participants in the survey in which a unique diversity was observed among them. Trichy and Thanjavur is located in delta region of Tamil Nadu with the latitude and longitude of 10.8050° N, 78.6856° E and 10.7825° N, 79.1313° E respectively. Tiruchirappalli is an Industrial town where we able observer lot of Industrial structures and the participants in this is region is capable to handle heavy and light structures. Thanjavur is an agricultural town in which Residential construction is prominent, the participants in Thanjavur have adequate knowledge in construction of both small, medium and high budget residential buildings. The primary data is obtained from the three main important construction participants namely the Client, the Contractor and the Consultant, and the information related to the causes of the delay were obtained from the literature. The sample size of consultants and contracts is fixed as 50 whereas 150 for the clients. The bigger sample size for clients is fixed based on to determine the margin of satisfaction when the construction was delayed. The first part contains the general information of the respondents and second part contains the general causes of the delays obtained from the study. A detailed demography was obtained from the participants including their age, education, experience in construction, budget of their construction. For any participant, experience plays a crucial role in construction, in this survey lot of contractors doing their practice with their immense of knowledge in their field which lead the path for their success. It is also observed that client is having more interest in their construction, apart from their working area. Good alignment was opinions were found between client and consultant which are taken as a good sign for our construction industry.

4. Analysis of questionnaire

4.1. *Causes of delays*

The opinions received from the participants of the construction are analyzed using the SPSS software, in which the Frequencies are obtained using the Descriptive Statistics tools, ANOVA, and Regression tools are used for obtaining the influencing causes on the Project Delay period and Satisfaction of the participants, and the percentage of dominant causes taken under the study respectively. Furthermore, the factors taken for the study is categorized into four groups namely, Improper Project Planning, Design Related issues, Finance Related issues and Resource Related issues, from which the percentage of influence of each category is determined using the Regression tool.

The test was carried on the perception of the construction participants on the roots of the delays in the project delivery. From the significant values obtained, various causes of the delays are mentioned in such a way that the causes with significant values less than 0.05 (5% significant level) are considered.

5. Discussions

5.1. *One-way ANOVA*

Based on the values obtained from the ANOVA on comparing the Demography and the Satisfaction of the Client the factors having significant values less than 0.05 were, Budget of the project, Number of floors, Purpose of the building (Own use, Rental, Both), Modes of loans taken (Bank, Jewel, Employer, Finance, Other Loans), Age of the client, Annual Income of the family has the influence. On matching up the Demography of the Client to the Project delay period, all the above mentioned causes on comparing with satisfaction do the same and also add the Number of children in the family as shown in Table 1.

Table 1. Significance values based on client's demography with respect to project delay and satisfaction

S. No.	Content	Client Delay	Client Satisfaction
1	Budget of the project	0.000	0.000
2	Location of the project	0.720	0.473
3	Number of floors	0.000	0.000
4	Purpose of the building	0.000	0.000
5	Modes of loans taken	0.000	0.003
6	Age of the client	0.000	0.008
7	Gender of the client	0.248	0.086
8	Education Qualification	0.918	0.301
9	Occupation of the client	0.060	0.251
10	Spouse Working condition	0.559	0.761
11	Annual Family income	0.000	0.000
12	Number of Children in family	0.005	0.143
13	Children Being Employed	0.402	0.835

Based on the values attained from the ANOVA on comparing the Demography of both Consultant and the Project Delay, Demography of Contractor and the Project Delay period, Demography of the Contractor and their Satisfaction the common and influencing parameter scrutinized was the Budget of the project as shown in Table 2. Budget of the project plays significant role in enhancing the features of the project thereby giving necessary clear cut dimensions towards its streamlined process and smooth progress of the projects under consideration. The values from the table inferring that, there exists a hold on the factors such as number of projects handled, Experience of the respondents, and also in certain cases location of the projects also plays significant role in contributing the desired value among the factors so taken for consideration.

Table 2. Significance values based on respondent's profile to that of project delay and satisfaction

	Consultant		Contractor	
Content	Delay	Satisfaction	Delay	Satisfaction
Budget of the Project	0.000	0.935	0.000	0.000
Location of the project	0.174	0.875	0.174	0.116
Age of the respondent	0.651	0.182	0.621	0.255
Experience	0.633	0.102	0.861	0.595
Number of projects	0.613	0.459	0.276	0.542
Qualification	0.779	0.375	0.709	0.674

5.2. Regression Analysis

On client's perception, considering the Project Delay period and all the causes taken for the study, it is found that 78.7% of the causes are superseding. The causes with the significant values less than 0.05 are Climatic factors, Inadequate planning by the construction participants, Government Bureaucracy, Traffic control and regulations on site, Reworks during construction, Subsequent design changes by the clients, Material Change in type and specification, Financial Constraints of the Client, Labour Availability, Equipment productivity, Demography of the labour are alleged to be overriding as shown in Table 3.

The main purpose of intending regression analysis is to shape the variables which are considered initially and thereby giving a best fit which contains ample construct among the initial variables. These constructs will have majority of impact over all other variables considered in this study. Since this research has major focus on the perspectives of client, it will have objective pattern of dependents which contributes significant relation towards delay in the construction projects. The various factors are shaped by matching it with other critical and viable factors which plays significant level in completing the projects without the delay factor. Those constraints are prioritized based on their relative importance over other factors adopted in the research tool.

On client's views, considering the Satisfaction of the client and all the causes taken for the study, it is found to that 54.9% of the causes are foremost. The causes with the significant values less than 0.05 are Climatic factors, Inadequate planning by the construction participants, Subsequent design changes by the clients, Financial constraints of Client are prevailing as listed in Table 3.

Table 3. Significance values on causes of delays to the participants Delay and Satisfaction

Content	Client		Consultant		Contractor	
	Delay	Satisfaction	Delay	Satisfaction	Delay	Satisfaction
REGRESSION (%)	78.7%	54.9%	55.4%	44.8%	65.8%	69.1%
ANOVA[b]	.000[a]	.000[a]	.048[a]	.266[a]	.003[a]	.001[a]
Climatic Factors	.000	.000	.204	.005	.164	.642
Lack of communication b/w parties	.727	.553	.601	.144	.541	.186
Inadequate planning by the Contractor Consultant & Client	.002	.015	.498	.192	.411	.332
Government Bureaucracy	.001	.479	.332	.068	.802	.881
Slow decision by the client	.321	.664	.683	.540	.007	.048
Traffic control and regulations	.000	.320	.754	.701	.509	.186
Reworks	.000	.244	.343	.065	.340	.153
Subsequent design changes by clients	.000	.011	.589	.044	.136	.036
Material change in type and specification during the construction	.021	.736	.527	.950	.497	.664
Financial Constraints of Client	.001	.047	.753	.106	.899	.208
Price Inflation	.210	.917	.802	.677	.365	.770
Labor Availability	.001	.509	.329	.321	.121	.022
Low Labor productivity	.665	.656	.144	.527	.019	.009
Material Availability	.267	.853	.424	.955	.770	.803
Slow delivery of materials	.147	.582	.582	.084	.668	.953
Site storage facilities	.262	.236	.889	.092	.265	.254
Equipment productivity	.041	.112	.528	.397	.407	.150
Demography of Labor	.021	.102	.486	.954	.083	.423
Poor Organization of the contractor or consultant	.103	.754	.979	.588	.084	.015

The individual groups based on the client's view are the Improper Project planning influences the project delay period and the satisfaction of the client by 63.5% and 45% respectively. The design related issues influence the delay period and the satisfaction by 40.9% and 32.6% respectively. The finance related issues influence the delay and the satisfaction by 17.9% and 10% respectively. Similarly, the resource related issues influence the same by 48% and 28.6% respectively. Hence the dominant factors less than 5% significant level under each category is tabulated in the Table 4 and Table 5.

Table 4. Significance values on categories with respect to Delay and Satisfaction of the participants

Content	Client		Consultant		Contractor	
	Delay	Satisfaction	Delay	Satisfaction	Delay	Satisfaction
	IMPROPER PROJECT PLANNING					
REGRESSION	63.5%	45%	44.9%	11.6%	42.9%	39.5%
ANOVA[b]	.000[a]	.000[a]	.000[a]	.474[a]	.000[a]	.001[a]
Climatic Factors	.000	.000	.023	.154	.029	.026
Lack of communication between the parties	.158	.436	.984	.042	.384	.228
Inadequate planning by the Contractor Consultant & Client	.009	.047	.736	.669	.311	.601
Government Bureaucracy	.217	.626	.023	.333	.493	.185
Slow decision by the client	.933	.975	.599	.695	.027	.047
Traffic control and regulations	.001	.013	.936	1.000	.316	.357
	DESIGN RELATED ISSUES					
REGRESSION	40.9%	32.6%	27.7%	6.1%	18.2%	26.9%
ANOVA[b]	.000[a]	.000[a]	.002[a]	.403[a]	.025[a]	.002[a]
Reworks	.000	.000	.745	.112	.373	.108
Subsequent Design changes by the client	.000	.000	.057	.141	.670	.734
Material change in type and specification during the construction	.000	.000	.039	.655	.021	.004

On consultant's perception, considering the Project Delay period and all the causes taken for the study, it is found that 55.4% of the causes are governing. On considering the Satisfaction of the consultant and all the causes taken for the study, it is found to that 44.8% of the causes are foremost. The causes with the significant values less than 0.05 are Climatic factors, Subsequent design changes by the clients are main as described in Table 3.

The individual groups based on the consultant's opinion are the Improper Project planning influences the project delay period and the satisfaction of the consultant by 44.9% and 11.6% respectively. The design related issues influence the delay period and the satisfaction by 27.7% and 6.1% respectively. The finance related issues influence the delay and the satisfaction by 21.7% and 2.7% respectively. Similarly, the resource related issues influence the same by 45.5% and 1.2% respectively. Hence the dominant factors less than 5% significant level under each category is tabulated in the Table 4 and Table 5.

Table 5. Significance values on categories with respect to Delay and Satisfaction of the participants

Finance Related Issues	Delay	Satisfaction	Delay	Satisfaction	Delay	Satisfaction
REGRESSION	17.9%	10.0%	21.7%	2.7%	27.2%	29.2%
ANOVA[b]	.000[a]	.000[a]	.003[a]	.528[a]	.001[a]	.000[a]
Financial Constraints of Client	.175	.061	.617	.288	.014	.008
Price Inflation	.000	.007	.002	.956	.913	.809
RESOURCE RELATED ISSUES						
REGRESSION	48.0%	28.6%	45.5%	1.2%	9.6%	11.6%
ANOVA[b]	.000[a]	.000[a]	.001[a]	.690[a]	.817[a]	.713[a]
Labor Availability	.000	.000	.291	.694	.534	.682
Low Labor productivity	.522	.355	.175	.790	.463	.659
Material Availability	.003	.022	.489	.252	.550	.557
Slow delivery of materials	.000	.003	.781	.442	.710	.635
Site storage facilities	.181	.831	.003	.551	.920	.523
Equipment productivity	.003	.086	.619	.151	.910	.532
Demography of Labor	.190	.014	.311	.903	.823	.646
Poor Organization of the contractor or consultant	.058	.778	.778	.799	.969	.968

While examining contractor's viewpoint on Project Delay period, all the causes taken for the study, it is found that 65.8% of the causes are superseding. Table 3 reveals that Slow decision by the client, and Low labour productivity are suspected to be intervening. On contractor's belief, considering the Satisfaction of the contractor and all the causes taken for the study, it is found to that 69.1% of the causes are prime. The causes with the significant values less than 0.05 are Slow decision making by the client, Subsequent design changes by the clients, Labour availability, Low labour productivity, and Poor organization of the consultant or the contractor are prevailing as illustrated in Table 3.

Based on the results from contractors for dominant groups, Improper Project planning influences the project delay period and the satisfaction of the contractor by 42.9% and 39.5% respectively. The design related issues influence the delay period and the satisfaction by 18.2% and 26.9% respectively. The finance related issues influence the delay and the satisfaction by 27.2% and 29.2% respectively. Similarly, the resource related issues influence the same by 9.6% and 11.6% respectively. Hence the dominant factors less than 5% significant level under each category is tabulated in the Table 4 and Table 5.

5.3. Correlation analysis

From Table 6, it is witnessed that there is an affinity between the satisfaction of the participants and the project delay period. And also, the results are significant based on the level of significance, the project delay period is significantly influencing the satisfaction of the client and the contractor at 1% percent level of significance and the same influences the consultant at 5% level of significance.

Table 6 Correlation between Project Delay Period and Satisfaction of the Participants

Participant	Level of Significance(%)	Significance Value	Delay Period	Satisfaction	Correlation Output
Client	1%	0.000	.784**	.784**	Significant
Consultant	5%	0.018	.332*	.332*	Significant
Contractor	1%	0.000	.892**	.892**	Significant

**. Correlation is significant at the 0.01 level (2-tailed).
*. Correlation is significant at the 0.05 level (2-tailed).

6. Conclusions

Clients stipulate the well-timed delivery of the project without cost swarming. The causes of delays are improper project planning and finance related issues, on which the consultants and the contractors mainly depend on. Also, the clients accept that they make frequent design changes during the construction process and there arises the delay.

On analyzing the views of the consultant, it is frozen that they concentrate on the satisfaction of their work as the percentage is lower when compared to the impact on the project delay period. Hence there is an affinity between the participants that they want their construction to be sound and durable irrespective of the delay. Another important problem noticed was the non-usage of the building codes as the normal values followed thorough out is taken. Importance of analyzing the soil conditions on every area of construction, its importance, also their effects should be known. Information collected from the contractor displays that causes of impact over both the satisfaction and the project delay are close enough, and the dominant cause are found to be frequent design changes by the clients, the availability of the labour and their productivity and they also accept that due to the improper organization by their groups and the consultant due to the other compulsion factors.

It is quite evident that the project delay period influences the satisfaction of the construction participants. Hence the consultant wants higher level of quality so that the satisfaction can be at the higher levels. It is not a limitation that only few parts of the districts were taken as sample, because these are the important and highly developing part of the state with huge commercial activities and agricultural activities with good water resources. The future study will focus on the detailed study about the each and every cause of the delay and their effects.

References

[1] Iyer K C and Jha K N 2005 Factors affecting cost performance: evidence for Indian Construction projects, *Int. J. Pro. Manage.* **23** 283-95.

[2] Tommy Y. Lo, Ivan W Fung and Karen C Tung 2006 Construction Delays in Hong Kong Civil Engineering Projects, *J. Constr. Eng. Manage.,* **132** 636-49.

[3] Norhidayah Md Dom Kasim N & Shamsudin A 2012 Factors influencing Human Resources Planning (HRP) for local workforce supply in Malaysian Construction Industry. *KICEM J. Constr. Eng. Pro. Manage.* **2** 5-13.

[4] Doloi H, Sawhney A, Iyer K C and Rentala S 2012 Analysing factors affecting delays in Indian Construction Projects. *Int. J. Pro. Manage.* **30** 479-89.

[5] Akinsiku O E and Akintunde Akinsulire. 2012 *KICEM J. Constr. Eng. Pro. Manage.* **2** 25-31.

[6] Enas Fatji Taher and Pandey R K 2013 Study of Delay in Project Planning and Design Stage of Civil Engineering Projects. *Int. J. Eng. Adv. Technol.* **2** 456-61.

[7] Pai S K and Bharath J R 2013Analysis of Critical Causes of Delays in Indian Infrastructure Projects. *Int. J. Innov. Res. Dev.* **2** 251-63.

[8] Megha D and Rajiv B 2013 A methodology for Ranking Causes of Delay for Residential Constrtion Projects in Indian Context. *Int. J. Emer. Technol. Adv. Eng.* **3** 396-404.

IOP Conf. Series: Earth and Environmental Science **80** (2017) 012047 doi:10.1088/1755-1315/80/1/012047

[9] Ashwathy R and Thomas C 2013 Development of a Delay Analysis System for a Railway construction project. *Int. J. Innov. Res. Sci. Eng. Technol.* **2** 531-41.

[10] Salunkhe A and Patil R 2014 Effect of Construction Delays on Project Time overrun: Indian Scenario. *Int. J. Res. Eng. Tech.* **3** 543-7.

[11] Indhu B and Ajai P 2014 Study of Delay Management in a Construction Project- A Case Study. *Int. J. Emer. Technol. Adv. Eng.* **4** 108-13.

Cost benefit analysis on different configurations of berthing structure

G Rajesh[1*], R Saravanan[2], Vijaya Ravichandran[2], SParameswaraPandian[3]and EvangelinRamani Sujatha[1]

[1]School of Civil Engineering, SASTRA University, Thanjavur,India
[2]National Institute of Ocean Technology, Pallikaranai, Chennai, India.
[3]BMT Consultants India,Chennai,India

*Email:grajesh1912@yahoo.co.in

Abstract. Port and harbors are essential for handling of the imports/exports of good transported through shipping. This paper discusses the different configuration of berthing structure, their design with respect to the site conditions and suitability. The analysis includes detailed load calculations conforming to the various codal provisions and design of the structure. The configuration of berthing structure considered are analyzed and designed using STAAD Pro for different combination of loads as per IS 4651.Bill of Quantities are prepared and final cost of construction is calculated. Factors affecting the construction and maintenance such as land availability, soil conditions, hydrodynamics of the site, dredging requirements, design ship size etc. are considered to finalize the configuration of the berthing structure. Result of the study shows that Diaphragm wall type of berthing structure is economic for Ennore port.

1.Introduction

Ennore port is the twelfth major port in India. Ennore port is protected by two breakwaters, such as3,080 m long northern breakwater and 1,070 m long Southern breakwater. It consist of six berths for handling coal, export of automobiles, Marine Liquid terminal (MLT), Iron ore etc. Due to increase in the traffic of vessels, import/export of cargo and coal, the numbers of berth are to be increased. Figure1 shows the layout of Ennore port.

A berthing structure is constructed to help with berthing and mooring of ships. It enables easy transfer/handling of cargo and easy movement of people from land to ships and vise-versa.Quays are defined as one or more berth, continuously bordering on and in contact with land or dock area. A structure that projects from the land out into water is called jetty.

Economic design of any structure requires a good preliminary analysis of the environs. Selection of configuration at planning stage plays an important role in reducing the cost and time consumed. However it is a challenge to provide the suitable configuration of structure in the early stages. [1] Discusses about the Analysis of forces acting on jetty using STAAD pro. [2] Created an expert system to structurally design the various component of berthing structure. [3]Discusses about the various loads acting on the berthing structure and manual design of RCC members. [4] Illustrates the various parameters affecting the planning of a berthing structure. [5]Discuss about the configuration of berthing structure with diaphragm wall. [6]Gives details about Young's modulus of dense sand.

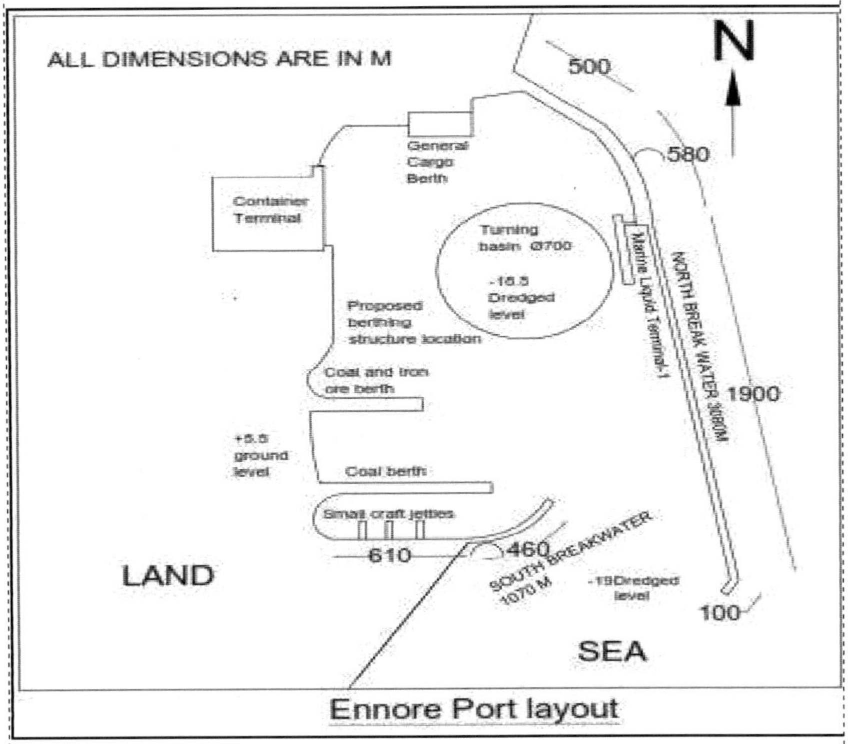

Figure 1. Ennore port layout.

Literature review shows that comparison between the different types of berthing structure and its economic impact based on site specific features of the port were not studied. This paper discusses about the suitable configuration of a berthing structure for a particular location and Dead Weight Tonnage (DWT) of ship 1.2 lakh tonnes. Three types of configurations of berths are considered namely Piled jetty, Piled jetty with approach, Diaphragm wall berthing structure and the most economical configuration is finalized as the best design for particular location, depending on the various factors affecting the design of structure. Analysis and design of all the structures is done using STAAD Pro and analytical calculations.

2. Methodology

Various data and information such as location of Ennore port with its various components, bathymetry report, soil report, Tide, wave and wind data, are collected. The maximum size of ship and their characteristic such as draft and keel clearance are found out.

[7] is a report given by environment clearance for the construction of coal berths and gives information regarding the site specific condition of Ennore port.The dredging level is found out to be -17m and top level of deck is found out as +7 m above Chart datum.

The different configurations considered are given below.

- Piled jetty of size 39m x 350m at a distance 98 m from the shore where there is a dredge level of -17 m and reclamation of land between shore and jetty could be used as container terminal.

- Piled jetty of size 39m x 350m with approach of width 21m and length 98m.
- A 350 m long Diaphragm wall jetty with piles for the crane beam including tie rod and dead man wall is considered for the design.

Analysis and design of all three types of berthing structures are carried out using STAAD Pro. The soil for the diaphragm wall is modeled as springs with appropriate spring stiffness value according to Vesic formula.

Bill of quantities is prepared with the rates for each part of construction, based on the current market rates. The approximate cost for all the type of berthing structure is found. Technical and cost comparison between the berthing structures is discussed and final results are obtained.Figure 2 shows the location of the proposed berthing structure.

3. Structural Analysis

The magnitudes of different loads acting on the structure are worked out. Concrete grade of M30 and Fe500 steel are considered. Table 1 gives the various loads acting on the structure and its values. Figure 3 shows the side view of the piled jetty. Seven rows of piles were used at different spacings that were decide after a number of trails.The total length of pile upto the depth of fixity is 29 m. Figure 3 also shows the sectional view of the approach,4 rows of piles at a spacing of 7 m c/c are used. Figure 4 shows the side view of diaphragm wall. Depth of diaphragm wall is 34 m. The soil is modelled as springs with dead man wall at a distance of 40 m connected by a post tenstioned tie rod.

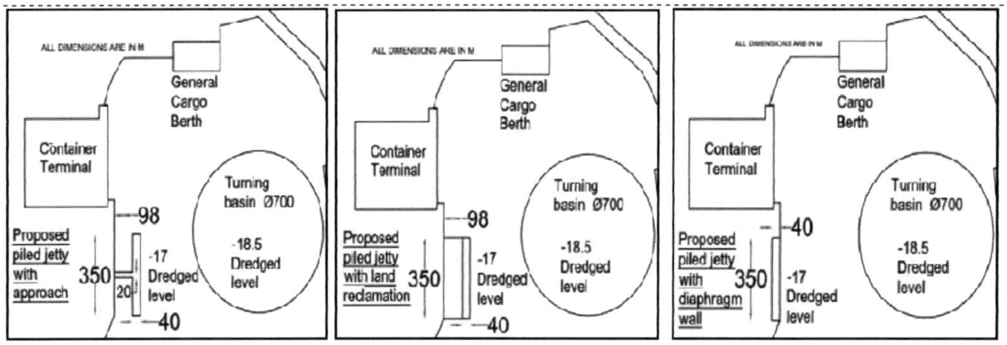

Figure 2. Proposed berthing structure layouts.

Mooring forces were found using [8] which gives the information about the calculation of current and wind forces acting on the ship. [9] informs about the various forces acting the berthing structure and the formulas for calculating the forces.

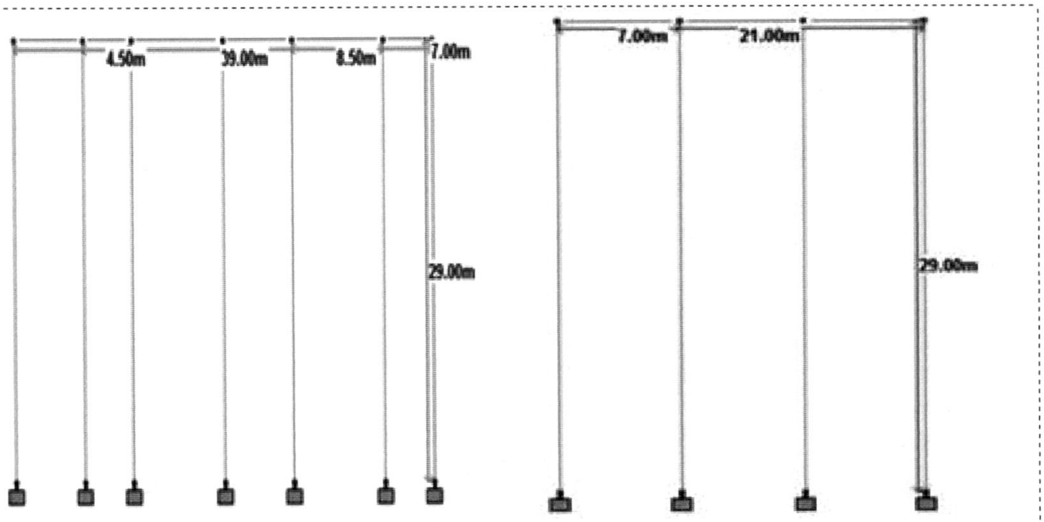

Figure 3. Piled jetty sectional view and approach sectional view.

3.1. Loads combination

General design considerations of berthing structure are given in [10].The load combination according to IS 4651(part4):1989 are given below.

Limit state of collapse

1.5DL+1.5LL+1EP+1HY+1.5BL,1.5DL+1.5LL+1EP+1HY+1.5ML,1.2DL+1.2LL+1EP+1.2HY,1.2DL+1.2LL+1EP+1HY+1.5WL, 1.2DL+1.2LL+1EP+1HY+1.5EL.

Limit state of serviceability

1DL+1LL+1EP+1HY+1SF, 1DL+1LL+1EP+1HY+1BM

DL- Dead load, LL-Vertical Live load, EP- Earth pressure, HY- Hydrostatic and hydrodynamic forces, BL- Berthing force, ML- Mooring forces, SF-Secondary stresses, WL-Wind load, EL- Seismic forces. Other load combination included the extreme and normal wind loads, wave loads and crane loads. Due to different direction of wind, earthquake, wave forces and various possibilities of berthing, mooring and crane loads about138 combinations of loads were considered for analysis using STAAD Pro.

Table 1. Loads acting on structure

S. No.	Load Type	Pattern of Load / Formula	Load	Remarks
		Berthing structure type I and II- Piled jetty		
1	Dead load	Self-weight	25 kN/m^3	
2	Live load	Uniformly distributed load	50 kN/m^2	IS 4651 part 3
3	Crane load	Post Panama crane	12000 kN	Maximum extreme load per corner on sea side
		Post Panama crane	11200 kN	Maximum operating load per corner on sea side

4	Berthing load	$E= (W_D V^2/2g) \times C_m \times C_e \times C_g$ Using DCN1600H fender	2494 kN	As per IS 4651 and shore protection manual
5	Mooring load	Force on bollard	1500 kN	As per BS6349 wind and current forces on ship
6	Wave forces on piles	$F_{DM}= 0.5 \times C_D \times \rho \times D \times H^2 \times K_{DM}$ $F_{IM}=0.5 \times C_M \times \rho \times D^2 \times H \times K_{IM}$	2.67 kN	Extreme wave load
		$F_{DM}= 0.5 \times C_D \times \rho \times D \times H^2 \times K_{DM}$ $F_{IM}=0.5 \times C_M \times \rho \times D^2 \times H \times K_{IM}$	2.024 kN	Operating wave load
7	Wave force on 7m wall	$P_m = 101 \times H_b \times w \times d \times (D+d)/(L_D D)$ and $P_s = wH_b/2$	2616.12 kN/m	Operating wave load
		$P_m = 101 \times H_b \times w \times d \times (D+d)/(L_D D)$ and $P_s = wH_b/2$	3270 kN/m	Extreme wave load
8	Wind load	$V_Z = V_b \times K_1 \times K_2 \times K$	0.4822 kN/m^2	As per IS 875 Extreme load
		$V_Z = V_b \times K_1 \times K_2 \times K_3$	0.3086 kN/m^2	As per IS 875 Operating load
		Berthing structure type III- Diaphragm wall		
9	Earth pressure	$Ep= K_a \gamma h$	67.74kN/m^2	According to layers of dense sand
10	Differential water pressure	Good drainage condition	5.5kN/m^2	According to change in tide levels

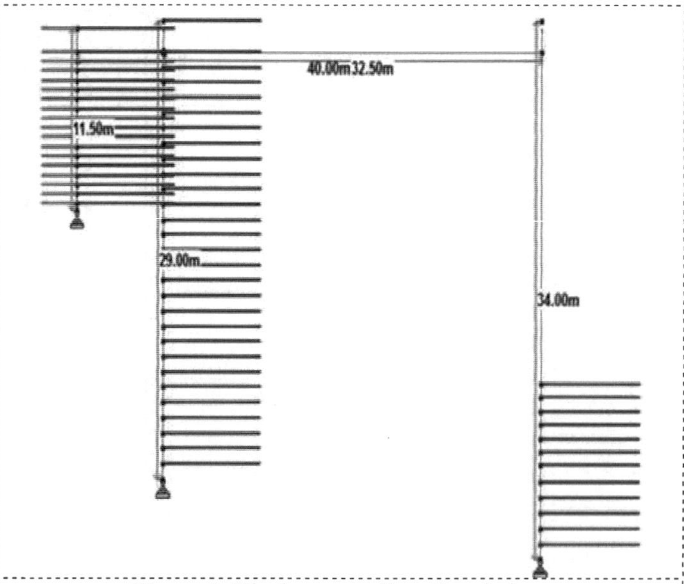

Figure 4. Diaphragm wall sectional view

3.2. Depth of fixity

The depth to the point of fixity is given in Figure 5. "e" is the effective eccentricity of the point of load application obtained either by converting the moment to an equivalent horizontal load or by actual position of the horizontal load application. The Indian standard for construction of piles [11] is used to calculate the depth of fixity. [12] Gives information about the forces acting on the jetty and also about depth of fixity. Depth of fixity is found to be 5.89 m in dense sand as per the soil condition given in [7].

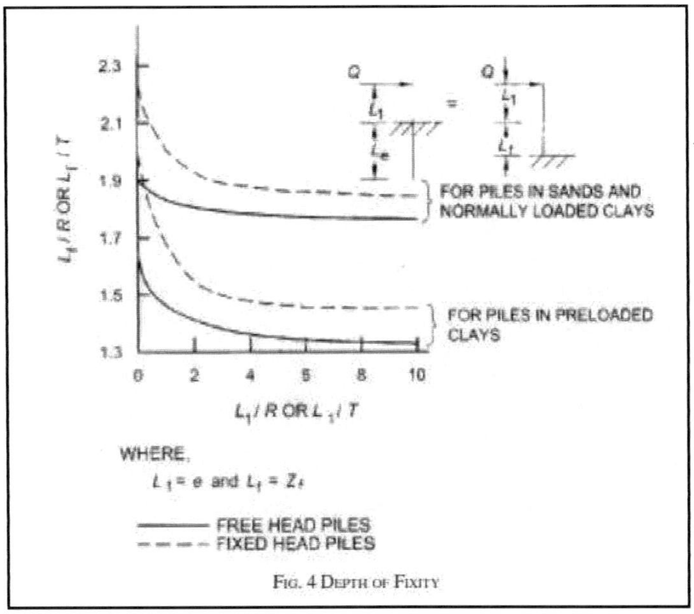

Figure 5. Depth of fixity (IS2911:2010).

4. Design of RCC Sections

The design of pile, crane beams, beams are designed for the piled jetty structures. The diaphragm wall with its tie rod and dead man wall were analysed and percentage of reinforcement is calculated.

The finalized geometry of the structure is given the Table 2.The general rectangular and circular shape of sections was used. The geometry of diaphragm wall is considered from [5]. The geometry was finalized after a series of trails. The maximum percentage of reinforcement for each section is given in Table 2.

Table 2. Geometry of members

S. No.	Member	Size of member	Reinforcement %
	Berthing structure type I and II- Piled jetty		
1	Pile(crane beam)	1.3m diameter. Depth of foundation 34m	3%
2	Pile	1.2m diameter. Depth of foundation 34m	2.25%

3	Crane beam	1.3m x 1.4m	0.659%
4	Beam	0.9m x 1m	0.888%
5	Slab	Depth of 0.5m	0.5%
	Berthing structure type III- Diaphragm wall		
6	Diaphragm wall	1.1m thick wall with 4 m panel. Expansion gap every 60 m	Vertical =0.7% Horizontal=0.15%
7	Tie rod	Post tensioned 75 mm diameter rod	
8	Dead man wall	0.6m thick wall 1m in length for every 7 m	0.15%

The rendered viewof the piled jetty and piled jetty with approach is shown in Figure 6.

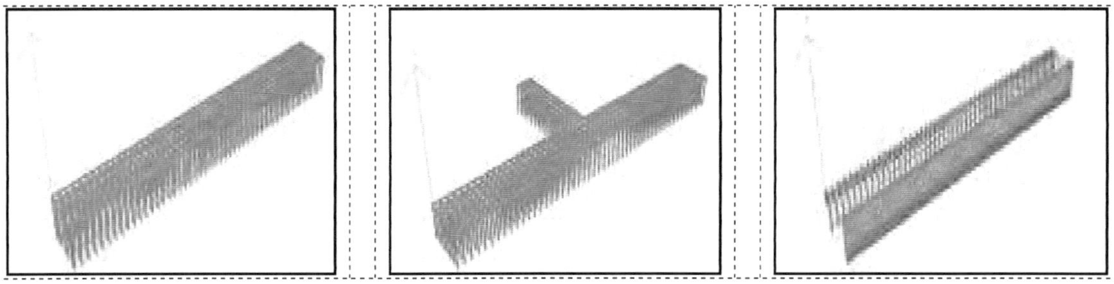

Figure 6. Rendered view of piled jetty, piled jetty with approach, diaphragm wall berthing structure.

5. Comparison between the Berthing Structures

Bill of quantities for the three type of berthing structure is prepared and the total cost of construction is estimated according to present market rate. Some of the main Technical difference are land reclamation, dredging. Cost is calculated for each type of berthing structure and approximate overall cost is given in Table 3.

Table 3. Cost of berthing structures.

S. No.	Type of structure	Cost in Rs. (Crores)
1	Piled jetty with reclamation	83.12
2	Piled jetty with approach	86.74
3	Diaphragm wall berthing structure	71.71

6. Conclusion

Detailed analysis and design of the three types of configuration of berthing structure are done using STAAD Pro. Approximate cost of structures is calculated. It is seen that the cost of Diaphragm wall berthing structure lower than that of piled structures but the amount of land available on the land side and the soil condition will be major criteria in deciding the use of diaphragm wall type of berthing structure. Soil stabilization must be done in case of loose sand and more number of piles must be designed if it is clayey soil which will increase the construction cost.

Environmental impact assessment must be carried out using numerical modeling to decide feasibility of each type of structure. Reclaimed region in piled jetty with reclamation can be used as container terminal. Piled jetty with approach is very useful for future development as additional jetty can be constructed parallel to approach and more number of ships can be accommodated.

The factors discussed can be used to create an expert system to decide the configuration of berthing system in the planning stage of the project.

References

[1] Himesh B Chopra and Patel P G 2015 Application of forces acting on jetty structure. *Int. J. Sci. Technol. Eng.* **11** 83-91.

[2] RangaRao A V and Sundaravadivelu S 1999 A knowledge based expert system for design of berthing structures. *Ocean Eng.* **26**653-57.

[3] Santosh Kumar B and Ashok Kumar S 2016 Analysis and Design of Dock Berth Structure. *Int. J. Sci. Adv. Res. Technol.* **2**(3)8-16.

[4] VivekT andDurga Prasad R 2016Analysis and Design of Marine Berthing Structure. *Int. J. Eng. Res. Appl.* **12**46-56.

[5] AlielahiH, Nouri H andEbrahimian B 2009 Performance based design of ShahidRajaeediaphragm wall:Pseudo-static and dynamic perspective. *Proc.Int.Conf. Performance-Based Des. Earthquake Geotechnical Eng.* (London: Taylor and francis group) 491-98.

[6] Morten Anker and Jørgensen Arne Hansen 2016Geotechnical Engineering for a new container terminal in Lomé, Togo, West Africa. *Proc. 17th Nordic geotechnical meeting challenges in Nordic Geotechnic*(Reykjavik, Iceland : Icelandic geotechnical society) 879-88.

[7] Asian Consulting Engineers Private Limited,New Delhi 2014 *Environmental Impact Assessment for Proposed Additional Coal Berths CB 3 (9 MTPA) and CB 4 (9 MTPA), Ennore Port, Tamil NaduFinal EIA Report*(Tamil Nadu,India).

[8] BS 6349 (Part 4) 1994 Maritime structures —Part 4: Code of practice for design offendering and mooring systems.

[9] IS 4651 (Part 3) 1974 Code of Practice for Planning and Designof Ports and Harbours, Part III: Loading.*Bureau of Indian Standards.*

[10] IS 4651 (Part 4) 1989 Code of practice for planning and designof ports and harbours: General design consideration.*Bureau of Indian Standards.*

[11] IS 2911 (Part 1,3) 2010 Design and construction of pile foundations — code of practice, part 1: concrete piles,Section 3: Driven Precast Concrete Piles.*Bureau of Indian Standards.*

[12] YogeshChaudhari, DebaratiDatta and Vasusmitha Joshi 2016 Estimation of Marine loads on Jetty Structure.*J. Civ. Eng. Environ. Technol.* **5**440-6.

A computer based approach for Material, Manpower and Equipment managementin the Construction Projects

Jaladanki Sasidhar*, D Muthu, C Venkatasubramanian and K Ramakrishnan

School of Civil Engineering, SASTRA University, Thanjavur-613401, India

*Email: jaladankisasidhar@gmail.com

Abstract. The success of any construction project will depend on efficient management of resources in a perfect manner to complete the project with a reasonable budget and time and the quality cannot be compromised. The efficient and timely procurement of material, deployment of adequate labor at correct time and mobilization of machinery lacking in time, all of them causes delay, lack of quality and finally affect the project cost. It is known factor that Project cost can be controlled by taking corrective actions on mobilization of resources at a right time. This research focuses on integration of management systems with the computer to generate the model which uses OOM data structure which decides to include automatic commodity code generation, automatic takeoff execution, intelligent purchase order generation, and components of design and schedule integration to overcome the problems of stock out. To overcome the problem in equipment management system inventory management module is suggested and the data set of equipment registration number, equipment number, description, date of purchase, manufacturer, equipment price, market value, life of equipment, production data of the equipment which includes equipment number, date, name of the job, hourly rate, insurance, depreciation cost of the equipment, taxes, storage cost , interest, oil, grease, and fuel consumption, etc. is analyzed and the decision support systems to overcome the problem arising out improper management is generated. The problem on labor is managed using scheduling, Strategic management of human resources. From the generated support systems tool, the resources are mobilized at a right time and help the project manager to finish project in time and thereby save the abnormal project cost and also provides the percentage that can be improved and also research focuses on determining the percentage of delays that are caused by lack of management of materials, manpower and machinery in different types of projects and how the percentage various from project to project.

1. Introduction

Construction management is the management of construction techniques for completing projects within estimated budget, time and according to plans and specifications without compromising the quality and safety[5].The improvement in the microcomputers, computer hardware and software,because of their low costs have made use of computers in various fields of construction management like project management, cost reducing,scheduling, bid analysis, accounting and financing,management of materials, equipment and manapower.

Materials management is a system that is essential in controlling and planning the quantity &quality of the material, fair price and the right quantity as required and also it includes purchasing, material and shippingcontrol from suppliers which is required for material cost control[4]. Three essential phases that leads to advantageous materials management are materials purchasing, materials usage, and storage.

When we consider any project equipment is a major cost for construction firms. The equipment management system andcontractor's equipment policy will have a great influence on the profit of the organization, mainly for contractors because of more amount of investments in equipment. The components of a construction equipment policy mainly consists of cost of the equipment, equipment financing, timely replacement of equipment's components or equipment, equipment standardization, records of the equipment, safety , inventory, and maintenance management,. A suitable approach in equipment management system is FLEET, which can supplement the contractor's equipment policy and which results in productive and effective utilization of equipment and more profitability. Fleet consists of four units: (1) Inventory management unit; (2) maintenance management unit; (3) cost, time, and production records unit; and (4) report generator unit[5].

The dominating factor is, the influence of labor and their cost on total project costs. For the past decades labor has been a inexpensive commodity which is a primary solution to quality and schedule problems by increasing the size of the work force. Strategic planning is a process in which the objectivesof an organization are identified and developed and appropriate actions are taken to meet the identified objectives. In this process resources are allocated for carryingout actions. The process will begin with the identifying organization's mission, objectives and formulationg the strategieswhich will result in the attainment of identified objectives and for implementing these strategies action plans with short-term goals and targets should be taken. Tier II approach is future oriented approach, which is designed for improving workers' productivity, skills and creating a situation in which value of the workers is increased[15]. The other approach is Tier I which is designed for efficiently and effectively managing the existing workforce, regardless of its skill levels. The implementation of this approach is on organization by proper management of workers in the field and communication. Both approaches will result in development of human capital which is a part of an overall human resource management. Supervisor's, superintendents and foremen should have strong management abilities for effective coordination of the activities. Higher-level hierarchy should focus on planning, scheduling and procurement for maximization for the success of the project. The Tier I approach is used for improving workforce retention, short-interval planning, and for improving the overall productivity and increasing the success rate for both project and organization.

2. Background and Literature Review

Management of materials is the process of coordinating and integrating of materials related functions such as takeoff, purchasing, evaluation of vendors, shipping,expediting of materials, materials receiving, stocking and distribution. These material related functions can be achieved efficiently byplanning of the early, training and communications[4]. The recent trend in materials management is developing computer-based approach which provides information needed for preventing materials shortages, excess of materials, labor delays, and cash flowproblems. The cost for developing and executing thecomputer based approaches has been justified qualitativelyby the results provided by them and the lack of control in their absence and also the methodologies they adopt for material management systems also has a significant on the project it includes Analysis of site and management, Inventory controlling, purchasing procedures, Procurement and Tracking Analysis on costs. Nearly material cost 35% to 40% of the project and if proper care should be taken for the management of materials by use certain management systems. The equipment management system will have influenceon the profit of the organization, and also for contractors with large amount of assets in equipment. FLEET is an Equipment management system which consists of four units: (1) Inventory unit; (2)maintenance management unit; (3) production unit and (4) report generating unit[5]. Inventory unit consists of the data for the equipment which includesnumber of the equipment, manufacturer, description, equipment price, market value, purchase date, and life of the equipment. Maintenance management unit will have a series of files for recording and scheduling preventive maintenance for various components of equipment,overhauls, and field/shop repairs. production unit module consists of time, cost, and production data for the equipment which includes number of the equipment, date, hourly rate, job name, insurance, interest, taxes, depreciationcostofthe equipment,fuel, oil, and greaseconsumption, storage cost, repair service, operating labor&hours, overhead, downtime, and production[5]. These data are lumped together and mapped in a file called master file. The data recorded in the other three units is accessed through the report generating unit which will produce suitable reports necessary for the purpose of effective management and to keep daily activitiesup-to-date. Development and use of these systems will increase the productivity and profit for the contractors and also for the organization.

The Tier I approach is used for improving the productivity of the construction labor by effective management and proper supervision . The Tier I strategy is used for discussing the stratigies and the methods needed to implement this approach. By this approach productivity can be increased, reduced turnover, and absenteeism of the workforce can be reduced. Tier II approach is a future oriented which is designed for improving workers' skills and productivity, which increases value of the workers. The goal of this increased value is to increase the wages and retention of the labor in the industry. It will utilize labor with some management skills and key responsibilities, work high performance, and there will be a increase in utilization of information technology. The other approach, Tier I, is designed for efficiently managing the existing manpower regardless of its skill level. These approaches depends on organization, communication, and proper utilization of manpower in the field. Both approaches will result in development of human capital which is a part of an overall human resource management. The basic characteristics of Tier I and Tier II includes highly skilled workforce, training for the workers, management of planning, training available to workers.

3. Methodology

3.1. Data collection

This research is based on a survey designed by gathering all necessary information in an effective manner. The survey presents the percentage of time or cost that can be saved by using computer based approaches for the management of three M'S in different types of construction projects based on the purpose. Based on the purpose the construction sites are divided into residential, commercial, industrial projects. The survey have been done on fifteen sites and the data has been collected for finding the percentage of reduction in the cost and time duration in any project due to lack of management systems for the management for the material, machinery, manpower. The reasons for the delays in project are also determined from the surveys and are tabulated for this purpose six case studies have been done in the sites.

3.2. Data characteristics

The data was collected primarily from the engineers who has experience of not less than five years in the construction and also from the contractors who are experience of residential, commercial, industrial buildings.

4. Statistical Analysis

After completing the data collection process, the data set was inspected for drawingresults &conclusions regarding the impact of the material management systems for the management of materials, machinery, and manpower and how much they have impact on the project or how it affects the project in terms of time and cost.

5. Limitations

In this paper the circumstances due to unexpected situations are not considered like floods, earthquakes, strikes from the labor and also the effect of material management systems on the heavy buildings and different sizes of project are not considered during research. All these will be considered in further research and approach will be made in order to minimize these effects on the project.

6. Case study

6.1. Case Study one

This case study is done on residential building which consists of G+2 stories, in this case study -1 they were not using any management systems due to which there was a delay in project of total one hundred and eight days out of which lack of proper management of materials has resulted in delay of seventeen days and due to improper management of labor has resulted in delay of eighty days and delay due to improper management of machinery resulted in a delay of five days. Remaining delays of six days are due to unavoidable circumstances i.e., strikes, floods rains, etc.

Table 1. The estimated duration and actual duration of case study one.

CASE STUDY	ESTIMATED DURATION	ACTUAL DURATION
1	295	403

6.2. Case Study two

This case study is done on residential building which consists of G+2 stories, in this case study -1 they were not using any management systems due to which there was a delay in project of total forty five days out of which lack of proper management of materials has resulted in delay of eight days and due to improper management of labor has resulted in delay of thirty one days and delay due to improper management of machinery resulted in a delay of four days. Remaining delays of two days are due to unavoidable circumstances i.e., strikes, floods rains, etc.

Table 2. The estimated duration and actual duration of case study two.

CASE STUDY	ESTIMATED DURATION	ACTUAL DURATION
2	497	542

IOP Conf. Series: Earth and Environmental Science **80** (2017) 012049 doi:10.1088/1755-1315/80/1/012049

Figure 1a.casestudies done for residential buildings one

Figure 1b. casestudies done for residential buildings two

6.3. Case study three

This case studies are of commercial projects. Case study three is a hospital which consists of basement and G+2, in this project they were not using any material management systems due to which there was a delay of one hundred and fifty three days out of which lack of proper management of materials has resulted in delay of forty seven days and due to improper management of labor has resulted in delay of seventy days and delay due to improper management of machinery resulted in a delay of thirty two days. Remaining delays of four days are due to unavoidable circumstances i.e., strikes, floods rains, etc.

Table 3. The estimated duration and actual duration of case study three.

CASE STUDY	ESTIMATED DURATION	ACTUAL DURATION
3	942	1095

Figure 2a. Layout for Commercial buildings one

Figure 2b. Layout for Commercial buildings two

6.4. Case study four

This case studies are of commercial projects. Case study four is a hospital which consists of basement and G+2, in this project they were not using any material management systems due to which there was a delay of forty two days out of which lack of proper management of materials has resulted in delay of twelve days and due to improper management of labor has resulted in delay of twenty one days and delay due to improper management of machinery resulted in a delay of eight days. Remaining delay of one day are due to unavoidable circumstances i.e., strikes, floods rains, etc.

Table 4. The estimated duration and actual duration of case study four.

CASE STUDY	ESTIMATED DURATION	ACTUAL DURATION
4	447	489

6.5. Case study five

This case studies were done on industrial buildings in which both are not using any management systems due to which there was a delay of one hundred and sixty six days out of which lack of proper management of materials has resulted in delay of fifty four days and due to improper management of labor has resulted in delay of sixty three days and delay due to improper management of machinery resulted in a delay of forty seven days. Remaining delays of two days are due to unavoidable circumstances i.e., strikes, floods rains, etc.

Table 5. The estimated duration and actual duration of case study five.

CASE STUDY	ESTIMATED DURATION	ACTUAL DURATION
5	621	787

6.6. Case study six

This case studies were done on industrial buildings in which both are not using any management systems due to which there was a delay of one hundred and eighty nine days out of which lack of proper management of materials has resulted in delay of sixty one days and due to improper management of labor has resulted in delay of eighty two days and delay due to improper management of machinery resulted in a delay of forty two days. Remaining delays of four days are due to unavoidable circumstances i.e., strikes, floods rains, etc.

Table 6. The estimated duration and actual duration of case study six

CASE STUDY	ESTIMATED DURATION	ACTUAL DURATION
6	674	863

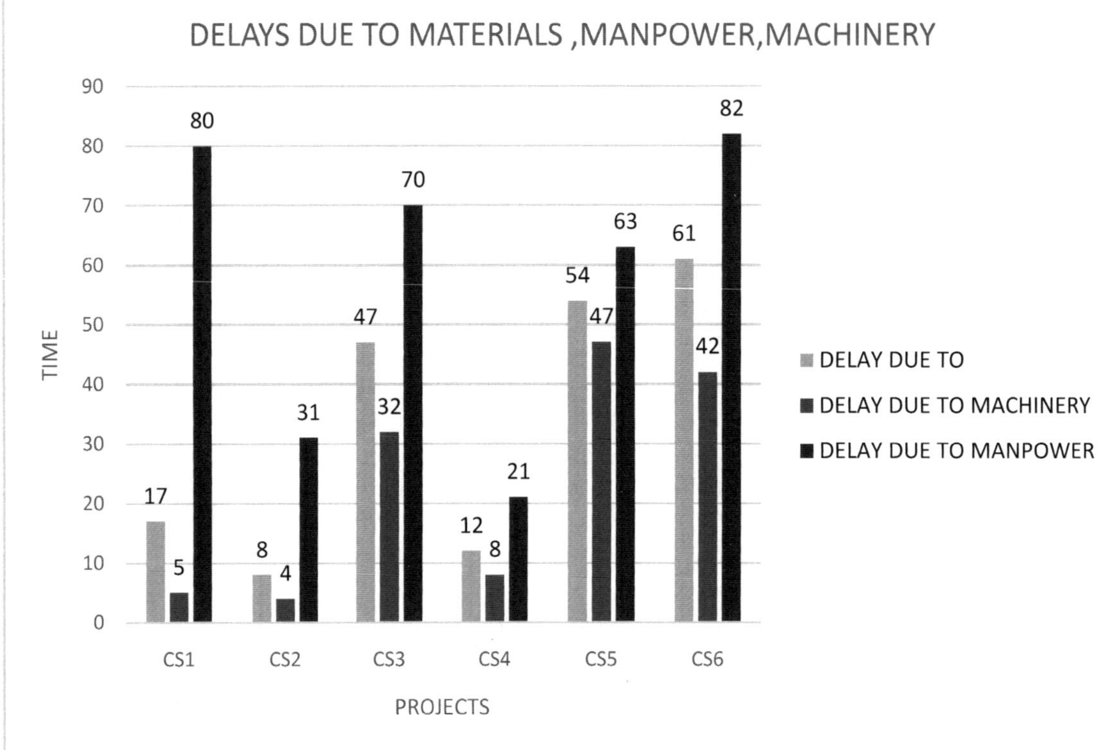

Figure 3. The Bargraph between time and delays due to materials, machinery and manpower from the case studies.

From the case studies the graph has been plotted and it can be seen that the percentage of delays various from project to project and the percentage of delay due to machinery increases from residential to industrial projects.

7. Results and conclusions

From the case studies and also from the surveys taken from the engineers and contractors it has been concluded that many of the engineers and contractors are not using any management systems for the management of materials, manpower, machinery and also it has been concluded that the percentage of delays due to lack of management systems varies according to the type of the project.
So by using the management systems the delays in any project can be reduced in turn the cost gets reduced and also for the purpose the computer based management system has been designed and given to the engineers and contractors.

In Residential buildings the delays due to lack of management systems is ninety two percentage of total delays. Table 7 gives the percentage of delays that occur due to delay in Material, Machinery, and Manpower in residential buildings.

Table 7. Gives the percentage of delays that occur due to delay in Material, Machinery, and Manpower in residential buildings.

S. No.	Description	Percentage
1.	Delay due to materials	17
2.	Delay due to manpower	70
3.	Delay due to machinery	6

Commercial buildings the delay in the project due to lack of management systems is ninety six percentage of total delays. Table 7.2 gives the percentage of delays that occur due to delay in Material, Machinery, and Manpower in Commercial buildings.

Table 8. Gives the percentage of delays that occur due to delay in Material, Machinery, and Manpower in Commercial buildings.

S. No.	Description	Percentage
1.	Delay due to materials	30
2.	Delay due to manpower	47
3.	Delay due to machinery	19

Industrial buildings the delay in the project due to lack of management systems is ninety five percentage of total delays. Table 7.3 gives the percentage of delays that occur due to delay in Material, Machinery, and Manpower in Industrial buildings.

Table 9. Gives the percentage of delays that occur due to delay in Material, Machinery, and Manpower in Industrial buildings.

S. No.	Description	Percentage
1.	Delay due to materials	34
2.	Delay due to manpower	37
3.	Delay due to machinery	27

It can be concluded from the above results that delays in any project are mainly due to manpower and materials but the percentages of delays changes from project to project. In case of equipment the percentage of delays increases from residential to industrial buildings. The management systems are developed for the management of Materials, Manpower and Machinery which are presented in Figures 3 to 5.

Figure 3a. The modules which were developed for the management of Materials, Manpower, and Machinery.

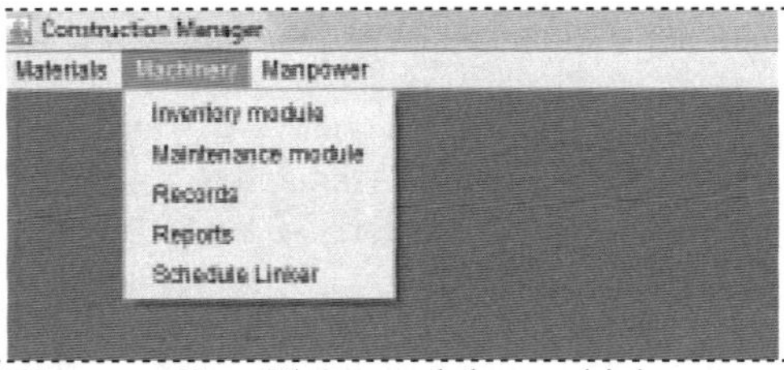

Figure 3b. The modules which were developed for the management of Materials, Manpower, and Machinery.

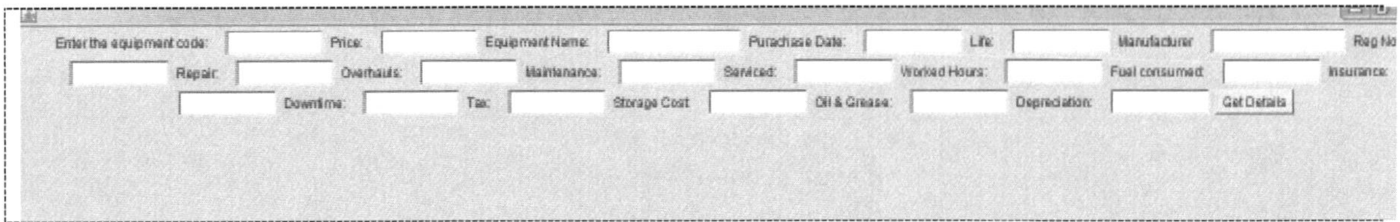

Figure 4. The modules which were developed for the management of Materials, Manpower, and Machinery.

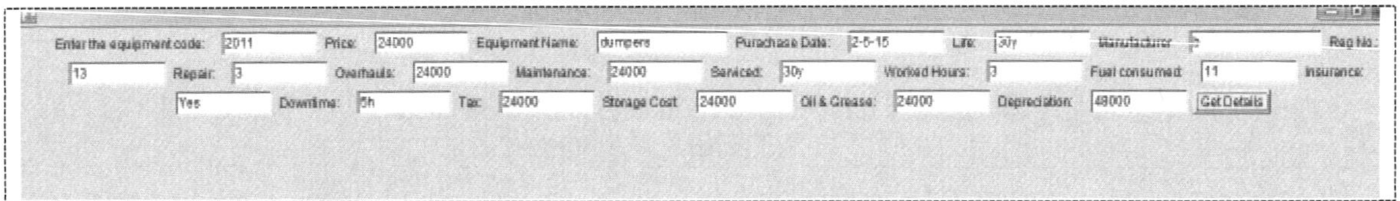

Figure 5. The modules which were developed for the management of Materials, Manpower, and Machinery.

References

[1] Caldas C H, Menches C L, Reyes P M, Navarro L and Vargas D M 2015 Materials Management Practices in the Construction Industry. *J. Constr. Eng. Manage.* ASCE **20**(3) 04014039.

[2] Lansford B C and Stukhart G 1986 Attributes of Material Management Systems. *J. Constr. Eng. Manage.* ASCE **112**(1)

[3] Hisham Said and Khaled El-Rayes 2011 Optimizing Material Procurement and Storage on Construction Sites. *J. Constr. Eng. Manage.* ASCE **137**(6) 421-31.

[4] Madhavi T P, Mathew S V and Sasidharan R 2013 Material Management in Construction a case study. *Int. J. Res. Eng. Technol.* 400-3.

[5] Tavakoli A, Masehi J J and Collyard C S 1990 FLEET: Equipment Management System. *J. Constr. Eng. Manage.* ASCE **35**(2) 211-20.

[6] Monnot J M and Williams R C 2011 Construction Equipment Telematics. *J. Constr. Eng. Manage.* ASCE **137**(10) 793-6.

[7] Kannan G 2011 Field Studies in Construction Equipment Economics and Productivity. *J. Constr. Eng. Manage.* ASCE **137**(10) 823-8.

[8] Thomas H R, Riley D R and Messner J I 2005 Fundamental principles of site material management. *J. Constr. Eng. Manage.* ASCE **131**(7) 808-15.

[9] Thomas H R, Riley D R and Sinha S K 2006 Fundamental principles for avoiding congested work areas—a case study. *Pract. Periodical Struct. Des. Constr.* **4**(197) 197-205.

[10] Maloney W F 1997 Strategic Planning for Human Resource Management in Construction. *J. Constr. Eng. Manage.* ASCE **13**(3).

[11] Borcherding J D, Glover R W, Haas C T and Tucker R L 2001 Metric-based implementation of the Tier II Work Force Strategy. Rep. 20 University of Texas at Austin Austin Tex

[12] Vereen S C, Rasdorf W and Hummer J E 2016 Development and Comparative Analysis of Construction Industry Labor Productivity Metrics. *J. Constr. Eng. Manage.* ASCE **142**(7) 04016020

[13] Michael V C 2009 Three keys for making telematics work The Equipment Executive Construction Equipment Jan. 1, 2009, 47, http://www.constructionequipment.com/three-keys-making-telematics-work?page=show

[14] Thomas H R 2015 Benchmarking Construction Labor Productivity source–Practice. *Periodical on Struct. Des. Constr.* **20**(4) 04014048

[15] Brandenburg S G, Haas C T and Byrom K 2006 Strategic Management of Human Resources in Construction. *J. Constr. Eng. Manage.* ASCE **22**(2) 89-96.

Implementation and implication of total quality management on client- contractor relationship in residential projects

Swetha Murali* **and V Ponmalar**

College of Engineering, Guindy, Anna University, Chennai, Tamil Nadu, India-600025

*Email: swetha.murali06@gmail.com

Abstract. To make innovation and continuous improvement as a norm, some traditional practices must become unlearnt. Change for growth and competitiveness are required for sustainability for any profitable business such as the construction industry. The leading companies are willing to implement Total Quality Management (TQM) principles, to realise potential advantages and improve growth and efficiency. Ironically, researches recollected quality as the most significant provider for competitive advantage in industrial leadership. The two objectives of this paper are 1) Identify TQM effectiveness in residential projects and 2) Identify the client satisfaction/dissatisfaction areas using Analytical Hierarchy Process (AHP) and suggest effective mitigate measures. Using statistical survey techniques like set of questionnaire survey, it is observed that total quality management was applied in some leading successful organization to an extent. The main attributes for quality achievement can be defined as teamwork and better communication with single agreed goal between client and contractor. Onsite safety is a paramount attribute in the identifying quality within the residential projects. It was noticed that the process based quality methods such as onsite safe working condition; safe management system and modern engineering process safety controls etc. as interlinked functions. Training and effective communication with all stakeholders on quality management principles is essential for effective quality work. Late Only through effective TQM principles companies can avoid some contract litigations with an increased client satisfaction Index.

Keywords: Quality, Management, Analytical, Hierarchy, Process

1. Introduction

Customer satisfaction is only achieved if the project on progress has some minimal defects and the contract supplied for the projects is economic in price with offerings from other parties. The intention of TQM is to achieve customer satisfaction through continuous improvement, client and contractor involvement, team work, training and education. It is often regarded as a culture encouraging a total commitment to client gratification through innovation and amendment in all aspects of the business. The customer in an 'perfect' environment is often considered as the final recipient of the services from the project, but it is also considered to be every being or department or stakeholders in the organization. The construction trade often differs from the production industries in such a way that the introduction and application of TQM in the former industry is way more challenging. The uniqueness of the construction industry is that it has its processes done only once while the manufacturing process is branded by stable condition processes. The industry often tends to muddle TQM with Quality Assurance (QA) and Quality Control (QC), considering the acquiescence of standards such as ISO 9001 and 9002 are the only application of quality on construction projects. Bahri, et al., 2012 emphasizes on the objective that TQM is to ensure continuous progress in the organization's people, systems, environment and processes so as to attain better-quality customer service and improved profits through efficiency and efficacy in the whole organization. Flint et al. (2011), exhibited that although rising attention is paid to customer satisfaction, devotion, value by managers, and researchers, their interrelationship still demands an answer. The Total Quality culture differs from an organization to another. Irrespective of the differences, it intends to accomplish common purposes such as: implementation of lean management, cost management, improvement of client satisfaction and increased market value. As it is observed, the objectives are dynamic in nature and requires frequent upgrading in the processes. There are wide ranges of TQM tools which can be used for diverse application in the field of construction industry. There is no best tool for every application. A knowledgeable practitioner is aware of each of the tool and uses an appropriate one for the activity needed.

2. Research Methodology

The data for the project is collected via questionnaire survey which targets the clients and contractors of companies in and around Chennai. The scheme of the project includes: 1. Identify the level of effectiveness for implementing practices of TQM in construction industry and 2. Area or phases with which the clients are dissatisfied. The effectiveness of parameters relating to best definition of quality and total quality management workability in the organization were identified and analysed. Client dissatisfaction index was developed using Analytical Hierarchy Process (AHP) on the basis of four targets to be achieved in the project cycle. The four targets are cost, time, quality and attitude towards satisfaction of work.

3. Analysis of Questionnaire

This study used descriptive statistical method to analyse important characteristics and summarise survey results. Reliability is a prerequisite for measuring validity. For this analysis, the Cronbach's Alpha is 0.761 inferring acceptability and dependability of the data for this model. About 55 respondents were surveyed for the implementation of TQM in residential projects and about 30 respondents were reviewed for the development of client satisfaction index.

4. Demographic Profile

The respondents were typically between in the age of 18-30 with experience level of 1-10 years and are university graduates. This helps in understanding that the respondents are in the preliminary stages in the organization; Knowledge and training of TQM in the organization is questionable with regard to the effective implementation of the managerial practices in the company.

5. Descriptive Statistics

As in Table 1, in pertaining to the definition of Quality, Partnership between the supplier and organizer (Mean= 3.09), teamwork (Mean=3.07) and frequent maintenance (Mean=2.81) satisfy the definition of quality. When intended for cost optimization, clients tend to go for products from bargained suppliers. Some projects with frequent maintenance assure that the resources do not go fault during the operational time and decrease the turnout during the delay. It is usually perceived that individuals made to a team achieve higher common goals than working alone. Team approach should not be restrained inside organization but should loop traders and external customers to the ring. Similarly, all the parameters affecting the effectiveness of TQM in organization include no necessary quality program, application of knowledge and training in organizations to an extent influencing the cost divergence, decreased customer satisfaction and delay in completion of projects. In Table 2, Potential improvement for quality include onsite safety (Mean=2.75). Certifying materials (Mean=2.20) and Testing procedure in site (Mean=2.18) can help in understanding the quality grade of the materials.

Table 1. Definition of Quality

Attributes	Mean	Standard Deviation
Expensive	2.18	0.520
Customer Satisfaction	2.00	0.518
Appearance	1.92	0.702
Increased Profit	2.41	0.745
Value for Money	2.41	0.980
Team Work	3.07	1.071
Partnership b/w supplier and organizer	3.09	0.882
Frequent Maintenance	2.81	1.020

Table 2. Potential Improvement in Quality

Attributes	Mean	Standard Deviation
Onsite Supervision	2.01	0.720
Redesign	1.45	0.539
Testing Procedure in Site	2.18	0.556
Certification of Material	2.20	0.689
Close out of Projects	1.86	0.832
Onsite Safety	2.75	0.551
Personal Management of Employees	1.88	0.466
Onsite Supervision	2.01	0.720
Personal Management of Employees	1.88	0.466

6. Graphical Representation

In Figure 1, the difference between functional quality and process based quality is elucidated. The implementation or management of quality does not meet the expected goal because an organization perceive quality through a process of chains converge on inputs, functions and association among processes as they are progressed and stretched through the organization. This makes the particular department focus on their internal targets alone and exclude the external targets. This discourages the transmit of information, impart knowledge and weaken team work culture. It is impossible to institute function based management due to strong chain of command.

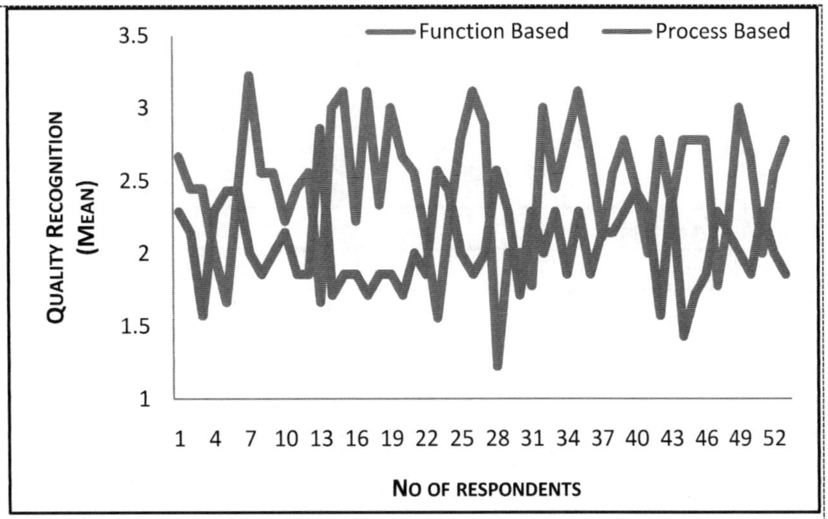

Figure 1. Quality Recognition.

7. Analysis Using Analytical Hierarchy Process (AHP)

As in Figure 2, indices suggest that time pressure (44.45%) is main factor which affects quality target of the project. The clients get disinterested when there is shortage of staff members. If there is less supervision or poor workmanship, there is inadequate job completion and affects the target of the project. With the other targets, such as cost, time and satisfaction towards work, the clients emphasise the importance of proper assessment during the design and planning stage so cost efficacy can be managed. Seldom contractual and variation claims (8.62%) affects the target when there is a monetary hit. Inefficient Coordination (38.64%), Poor project planning (22.48%) and delays in resolving problems in site (18.48%) are the highest drawbacks which hit time target. Poor communication (24.75%), less customer care (13.11%), empathy towards clients (15.95%) are some areas where the attitude of clients seems dissatisfied. Figure 3 to 5 shows the percentage of variations of each parameters of individual targets.

ATTRIBUTE- QUALITY TARGET									
AHP TABLE FOR FACTORS AFFECTING QUALITY TARGET									
Parameters	Poor Workmanship	Time Pressure	Wrong Specs	Wrong Choice of Method	Absence of Quality Assurance	Product	5th Root	Weight	Percentage
Poor Workmanship due to inadequate supervision	1	0.5	3.00	2	4	12	1.64	0.25	25.08
Time Pressure: Crashed Job Execution	2	1	5	3	7	210	2.91	0.44	44.45
Wrong Specifications	0.33	0.2	1	3	5	1	1	0.15	15.26
Wrong Choice of Construction Method/equipment	0.50	0.33	0.33	1	3	0.17	0.70	0.11	10.66
Absence of Quality assurance Process	0.25	0.14	0.2	0.33	1	0.00	0.30	0.05	4.56
Sum of Columns	4.08	2.18	9.53	9.33	20		6.56		100.00
Sum of column x row weight	**1.02**	**0.97**	**1.45**	**1.00**	**0.91**		**5.35**		
						CI	0.09		
						RI	1.12		
						CR	**0.08**	<0.1, Hence Consistent.	

Figure 2. AHP for Quality Target.

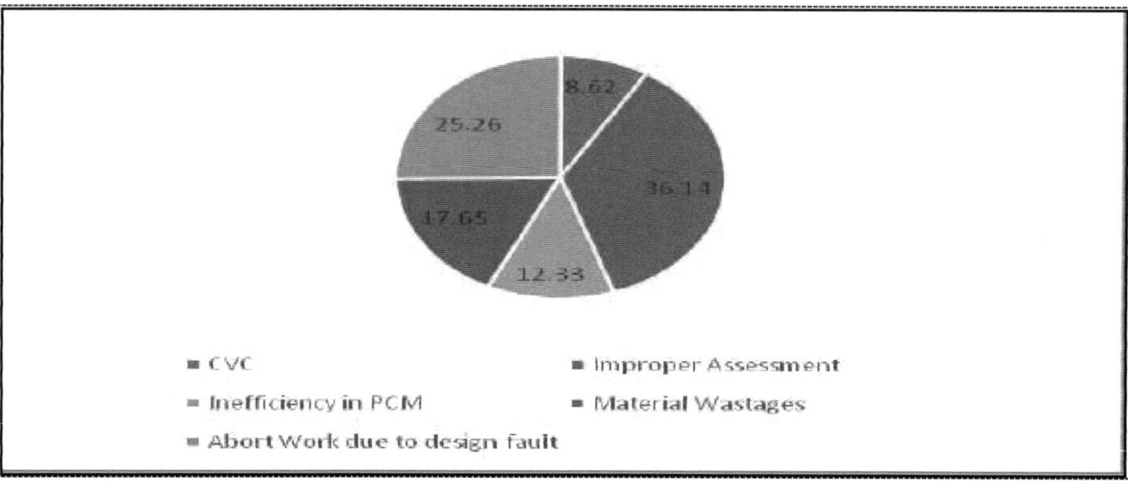

Figure 3. Factors affecting Cost Target.

IOP Conf. Series: Earth and Environmental Science **80** (2017) 012050 doi:10.1088/1755-1315/80/1/012050

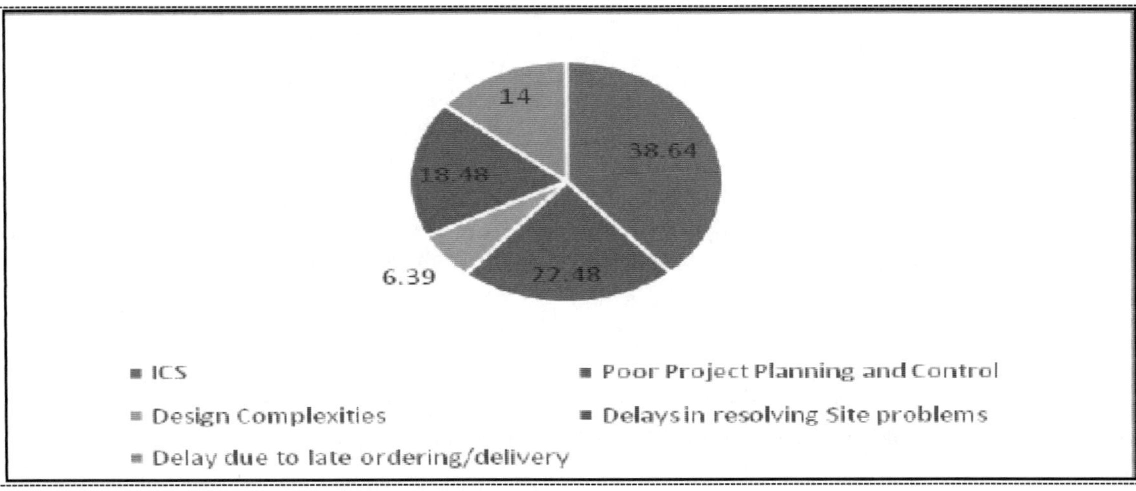

Figure 4. Factors affecting Time Target.

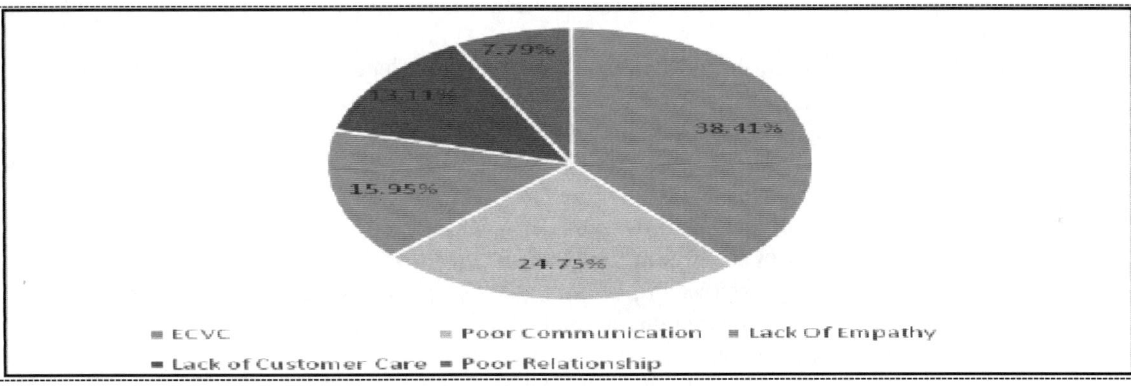

Figure 5. Factors affecting Attitude towards Satisfaction.

8. Results and Discussion

Figure 6 suggests that amongst the major factors contributing to dissatisfaction of clients, Quality is regarded as the most important attribute in satisfying the clients. If there is compromise in quality, it is the highest disappointment to the clients since a quality work gives the utmost satisfaction.

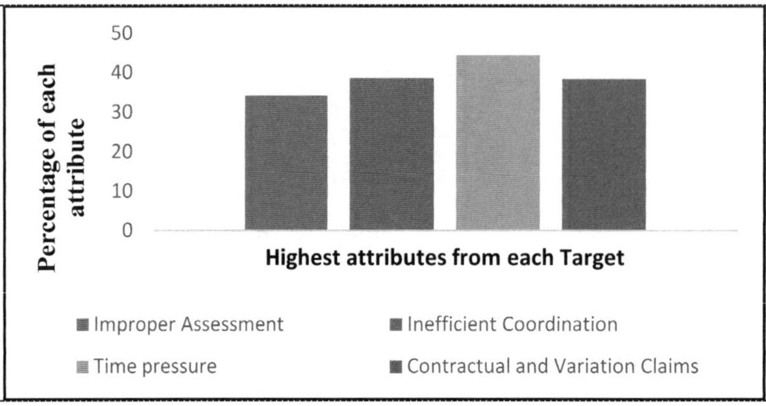

Figure 6. Major attribute Affecting Individual Target.

9. Conclusions

From the analysis, it can be concluded that:

- From the research work, it was noticed few important points were noticed by the contractor are listed below:
  - The highlighted attributes for quality definition include partnership between organizer and supplier, teamwork and frequent maintenance. These attributes boost the quality performances and helps in maintaining a satisfied client relationship.
  - Most of the respondents agree that TQM is applied in organization to some extent even when they are knowledgeable and well trained.
  - For the improvement in quality factors, Onsite Safety is an important attribute so as to curb accidents which can be developed by improvising the onsite supervisions.
  - Functional quality parameters like expensiveness, appearance, etc are not enough to run the organization. Process based quality methods such as testing and certification of materials are preferred so that the organization hierarchical interlinks are maintained.
- For the implications of Client-Contractor relationship:
  - In order to restrain late changes in activities and material wastages, proper planning and execution of work should be done with an experienced guidance and proper lean principles must be adopted for the material handling.
  - Since, time is a major concern for project completion, improper planning and scheduling can lead to overtime. Due to this factor, solutions like crashing of activities are rendered which infer to a substandard quality progress or increase the labour demand and increase in resource levelling is demanded.
  - Planning of material delivery should have a standard cycle including the reorder level. Occasionally, if the past experiences are not recorded or studied properly, poor choice of method or usage of incorrect equipment for activities and in case of small residential projects if the quality audits are minimal, there is a chance of lacking in quality wise construction.
  - If there is poor communication between the client and contractors there are chances of high risk in misleading information, high risk in contract breaches and the variation claims may exceed 14%. This will lead to poor relationship amongst the two parties. The contractors should make the client's satisfaction as the top priority and empathize their feelings and fulfill their needs and requirements.

References

[1] Arshida M M and Agil S O 2012 Critical Success Factors for Total Quality Management Implementation within the Libyan Iron and Steel Company *Tun Abdul Razak University, Graduate School of Business MLB-254.*

[2] Bahri S, Hamzah D and Yusuf M R 2012. Implementation of Total Quality Management and Its Effect on Organizational Performance of Manufacturing Industries through Organizational Culture in South Sulawesi, Indonesia. *J. Bus. Manage.* **5**(1) 10-24.

[3] Baidoun S 2003 An Empirical Study of Critical Factors of TQM in Palestinian Organizations. *Logistics Information Manage.* **16**(2) 156-71.

[4] Flint D J, Blocker C P and Boutin P J 2011. Customer value anticipation, customer satisfaction and loyalty: An empirical examination. *Ind. Mark. Manage.* **40**(2) 219-30.

[5] James Harrington H, Frank Voehl and Hal Wiggin 2012 Applying TQM to the construction industry. *The Total Qual. Manage. J.* **24**(4) 352-62.

[6] Jun M and Cai S 2010 Examining the relationships between internal service quality and its dimensions, and internal customer satisfaction. *Total Qual. Manage.* **21**(2) 205-23.

[7] Nazhad M B, Mosavi J S and Kordabadi S S.2012 Organizational Culture and Quality Management (ISO/9001) Case Study: Tehran Universities Employees. *J. Basic and Appl. Sci. Res.* **2**(12)12590-9.

[8] Lenka U, Suar D and Mohapatra P K 2010 Customer satisfaction in Indian commercial banks through total quality management approach. *Total Qual. Manage.* **21**(12) 1315-41.

[9] Wang S J and Yu Chu P 2000 Critical Factors Affecting the Implementation Decisions and Processes of ISO Quality Management Systems in Taiwanese Public Sectors. *Institute of Public Affairs Management, National Sun, Yat-sen University.*

[10] Zakuan N, Muniandy S, Saman N Z and Md Arif M S 2012 Critical Success Factors of Total Quality Management Implementation In Higher Education Institution: A Review. *Int. J. Academic Res. Bus. Soc. Sci.* **2**(12).

Biodegradation of paper waste using Eisenia foetida by vermi-composting Technology

Mahalakshmi Mathivanan*, G Aravind Vishnu Saravanan, Aravindh Baji and J Manoj kumar

School of Civil Engineering, SASTRA University, Thanjavur – 613401, India.

*Email: mahalakshmi@civil.sastra.edu

Abstract. The paper wastes are being a big concern over past decades. The process of reuse of the paper wastes is employed by 'eisenia foetida' in Vermiculture. The paper waste in SASTRA is collected around 50kg and organic wastes like vegetable wastes and cow dung wastes are also collected. In the adjacent area of Nirman Vihar, SASTRA, the experimental setup is done in a Geo-synthetic polymer bag. The area is divided into three segments and in each segment appropriate amount of paper waste and organic waste were added along with 25 numbers of earthworms. The setup is watered daily and monitored periodically and it is kindled for proper aeration. The soil samples were collected on 20 days, 45 days and 60 days from the day the earthworms were added. After 60 days of the experiment, the paper wastes, compost and earthworms are separated. The quantity of the wastes was compared to the initial amount and the composts are collected. The elemental analysis of the soil used as Vermi-bed is analyzed for improvement of soil nutrients. The vermiwashed water of the setup is analyzed for total protein. The number of earthworm is also compared to initial quantity. Out of all, the loss percentage of the organic waste and paper waste shows the degradation of the paper wastes.

1. Introduction

Aristotle has said before 2350 years, "Earthworms are intestines of the earth". Only in the twentieth century has the truth in this statement been verified and found correct[1]. Vermiculture is principally the science of breeding and raising earthworms. It defines the stimulating potential for waste reduction, fertilizer production, as well as an assortment of possible uses for the future. The population of the earth is increasing and with this increase comes an increased need for management of wastes especially paper wastes[2]. The process of waste paper recycling involves mixing used paper with water and chemicals to break it down. The share of ink in a wastepaper stock is up to about 2% of the total weight. The fundamental specifications of the experiment are re-cycle of paper, food wastes, environmental wastes, vermin-compost, enormous power of reproduction, rapid rate of multiplication, production of bio-mass[3]. From the environmental perspective, paper wasting is important as systematic collection and recycling of waste paper can significantly reduce the generation of municipal solid wastes. It is estimated to be around tons of waste paper recycling can lead to saving of mammoth part of raw materials and energy. The tiger Worm (eisenia foetida) is best suited for vermiculture researches and vermin-treatment for solid and liquid organic wastes. The earthworms are bisexual and so rapidly multiply[4]. Many studies found out that, earthworms double themselves within 60 days with respect to the environment they live in. The castings let out by earthworms are accountable for the change in soil nutrition[5]. So, they can be considered a nutritive plant food since it enriches the basic nutrition like nitrogen, potassium, phosphorus and proteins too. The production of worm bio-mass can be considered pro-biotic feed for fishery, poultry[6]. The main objective of the proposed work is to study on the Bio-conversion of paper wastes through the process of Vermin-compost, employing earthworm *eisenia foetida* with organic wastes like cattle dung and vegetable wastes to stabilize the nutrient requirements and hasten the decompose of paper.

2. Materials and Methods

2.1. Materials Used

The following materials were utilized for this study

(i) **Geosynthetic bag:** A geo-synthetic polymer bag can accommodate a space of 4'*3'1' to evenly distribute for each pit of size 4'*1'*1'.

(ii) **Soil:** Concerning for finding optimum conditions for agricultural purpose, Sandy soil (evacuated to a depth of 500cm to maintain chemical characteristics of the soil) is collected in SASTRA University, Thanjavur, Tamil Nadu. In the allotted space, soils are filled for a height of 15cm to act as Vermi-bed for the earthworms.

(iii) **Earthworms:** 'eisenia foetida' – Of 25 numbers weighing 25g are collected in Vallam, Tamil Nadu. Each earthworm individually varied between 0.05g to 1.5g.

(iv) **Wastes:**

- Paper waste: 50kg of paper wastes are collected from SASTRA UNIVERSITY BOYS HOSTEL and cut down into pieces with the help of cutting machines in School Of Mechanical Engineering in SASTRA. 13.5kg is filled in each pit.
- Cow dung wastes(organic): 21kg of cow dung is collected in dairy farm, SASTRA University.
- Vegetables wastes(organic): 21kg of Vegetable waste is collected from Paramanadha Vihar and Kamadhenu hostel, SASTRA University.

(v) **Shelter:** A sheet of plastic mesh is used to cover the experimental setup in order to prevent from predators and pests.

2.2. Methodology Adopted

Table 1. Test Programme

In this pit, first layer is vermi bed of height 15cm and 25 earthworms are spread over the bed[7]. Then, Paper wastes of 13.5kg are spread over and cow dung wastes of 13.5kg are also added. The setup should not be compressed in order to maintain good aeration inside the setup.

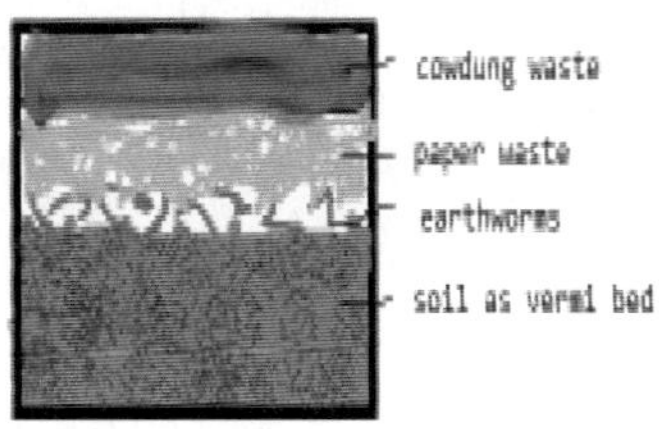

In this pit, first layer is vermi bed of height 15cm and 25 earthworms are spread over the bed. Then, Paper wastes of 13.5kg are spread over and Vegetable wastes of 13.5kg are also added. The setup should not be compressed in order to maintain good aeration inside the setup.

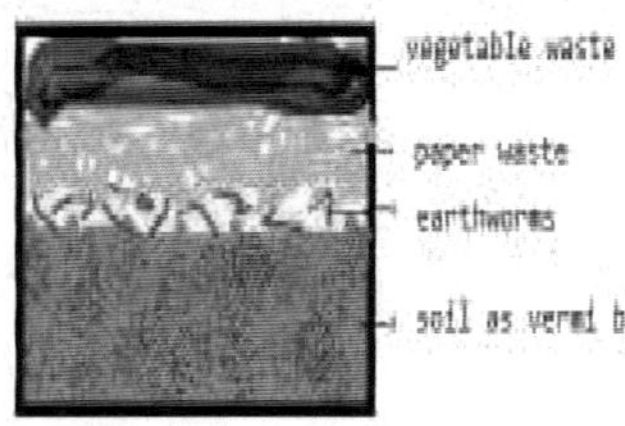

In this pit, first layer is vermi bed of height 15cm and 25 earthworms are spread over the bed. Then, Paper wastes of 13.5kg are spread over and Vegetable wastes of 6.75kg and cow dung wastes of 6.75kg are mixed and spread above the paper wastes. The setup should not be compressed in order to maintain good aeration inside the setup.

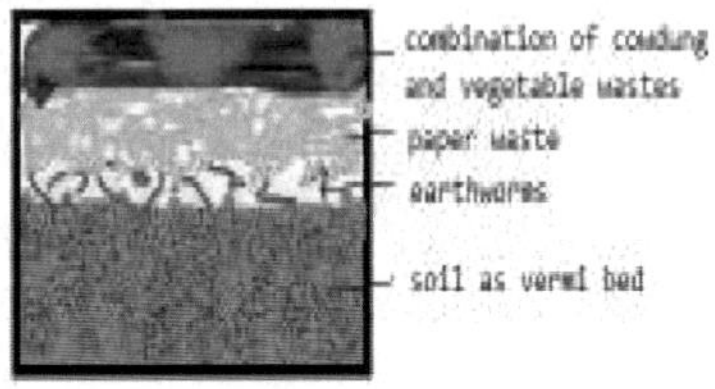

Soil characteristics and nutrients contents

The soil characteristics such as Calcium Carbonate, electrical conductivity, pH and nutrient contents such as nitrogen, phosphorus, potassium, iron, manganese, zinc and copper are found for 0th day, 20th day, 45th day and 60th day in Rice Research Institute, Aadudhurai, Tamil Nadu.

Bio-chemistry auto analyzer for total protein

The total protein from the vermi-washing fluid of the earthworms is found from auto analyzer.

Growth comparisons

The physical properties of the earthworms are noted down on 0th day and 60th day to measure the intensity changes in the number and weight in each pit.

Weight loss on organic substrate

The difference in the weight of the organic substrates from initial and final stages shows the weight loss considering the fact that earthworm takes on in building its body[8].

3. *Results and Discussion*

Table 2. Soil characters and nutrient contents

Contents	0th day	PIT-1	PIT-2	PIT-3
TEXTURE	Sandy clay	Sandy clay	Sandy clay	Sandy clay
$CaCO_3$	No	No	No	No
EC	0.35	0.18	0.26	0.21
pH	6.7	8.7	8.6	8.6
Sodic land	-	500	500	500
N	32.2	60.2	77	40.6
P	117	35	35	45
K	325	90	95	85
Fe	3.89	0.87	0.87	1.6
Mn	2.0	2.64	2.88	1.4
Zn	0.8	0.72	0.89	0.3
Cu	1.21	1.54	1.35	0.33

3.1. Impact of pH on soil:

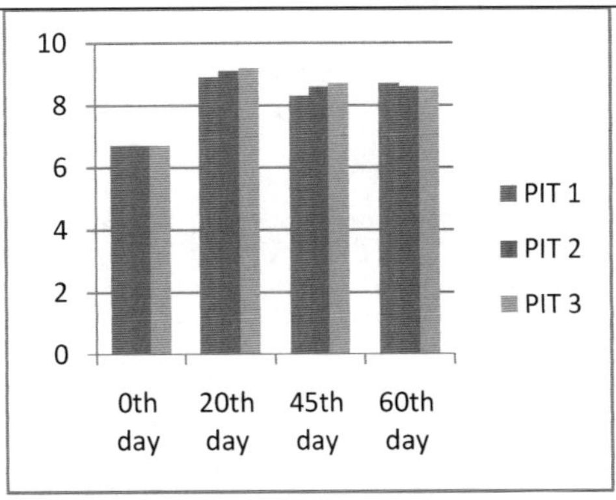

Figure 1. Impact of pH on soil

- The accessibility of the micronutrients zinc (Zn), iron (Fe), manganese (Mn), copper (Cu) and boron (B) be likely to reduce as soil pH increases[9].
- Studies have shown that pH declination is occurring in the recent years rapidly in direct seeded land and continuously cropped land[10]. Seepage of alkaline salts can increase the pH above the optimum range. So, nowadays the pH of the soil with an optimum range could be too acid or alkaline.

3.2. Biochemistry auto-analyzer results for total protein:
Vermiwash, a liquid bio-fertilizer that is collected through the column of activated earthworm[11]. The vermiwashed fluid is given for Bio-chemistry Auto analyzer to find the total protein contents after 60th day is shown in the Table 3.

Table 3. Total protein in vermin washwater

Vermiwash	Sample 1	Sample 2	Sample 3
Total Protein	0.52	0.55	0.56

The reason for increased total protein content in vermin wash is due to the presence of photolytic enzymes, which are secreted in earthworms gut.

3.3. Growth comparisons of earthworms:
The earthworms coming out of the cocoon will be of size 0.02cm and within 5 to 6 weeks earthworms mature to the size of 18cm approximately[12]. Earthworms start reproduction by the week of second. At the end of 60th day, composts are separated and earthworms are harvested. Total number of earthworms in each pit is calculated and mentioned in the Table 4.

Table 4. Number of earthworms in each pit

Earthworms	PIT 1	PIT 2	PIT 3
Total number	136	214	432
Total weight (g)	108.8	175.5	\341.3
Initial weight(g)	22.2	21.1	24.7

At the end of 60th day, 25 numbers of earthworms in each pit has risen to 136, 214 and 432 respectively. And the difference between the initial weight and final weight of earthworms shows the tremendous increase in population[13].

3.4. Impact of vermicomposting on weight loss of organic substrate

Paper wastes and organic wastes were separated from the set up and the loss percentage is calculated from the final weight attained on the 60th day.

Table 5. Percentage of weight loss

Type of waste	Initial Weight of Substrate (kg)			Final weight (kg)			Loss %		
	Pit	Pit 2	Pit 3	Pit 1	Pit 2	Pit 3	Pit 1	Pit 2	Pit 3
Paper Waste	13.5	13.5	13.5	11.6	10.2	8.6	14.1	24.4	36.3
Organic waste	13.5	13.5	13.5	3.16	1.45	5.8	76.6	89.2	42.9

In the comparison, paper wastes are decreased of 36.3% in PIT 3. Because, the variety of foods is more in PIT 3 comparing to PIT 1 and PIT 2. The bacteria in the decaying process of vegetables will be multiplied in numerous amounts and hence, the other wastes are also easily decayed and appeared to be abundant food for earthworms[14]. So, the number of earthworms has also increased in more numbers. In PIT 3, the organic wastes loss percentage seems to be less comparing to the other three, because the lost quantity are converted into compost.

In the PIT 2, the loss percentage of organic wastes is very high, because, vegetables are always comprised of 80% to 95% of water content. And, the loss percentage of paper wastes is also in reasonable quantity. In PIT 3, the loss percentage of organic wastes is medium and paper wastes less comparing to the other two PITS[15].

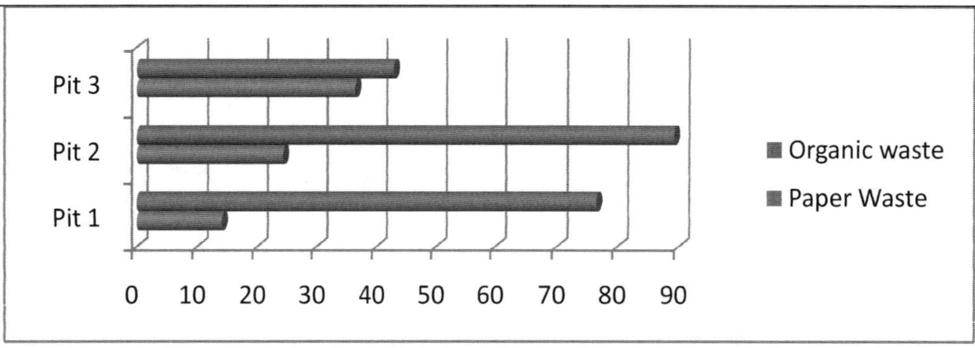

Figure 2. Comparison of loss percentage of substrates

4. Conclusion

Taking worms out of earth and placing them in the vermi beds is a human responsibility. With their own unique needs, they live under the earth. So it is important to provide healthy habitat to do their work. Providing the right conditions for the worms will thrive and ensure us excellent results. Vermiculture and vermicomposting is a substantial way of reducing paper and organic wastes. The process is capable of producing high quality fertilizers and also maintaining the balance of the ecological environment.

In this experiment, from the results of organic substrate loss percentage, pH change, growth comparisons, biochemistry analysis and soil characteristics, we can infer that

- The second combination, vegetable waste and paper waste is best suitable for improving paper degradation and also helps in soil enriching.
- The third combination, can be exclusively used for harvesting earthworms as it shows enormous growth in its size and number, the notable amount of compost reaped also shows the rate of decomposition of paper and other waste.
- Though the yield and soil nutrients are comparatively less in the combination of cow dung and paper wastes, the same can be used when no other options are available.

Acknowledgements
The authors thank SASTRA University for providing infrastructure and research facility to carry out the research work.

References

[1] Sinaga T P T, Nugroho A, Lee Y W and Suh Y 2011 GIS Mapping of Tsunami Vulnerability: Case Study of the Jembrana Regency in Bali, Indonesia. *KSCE J. Civ. Eng.* **15** 537–43.

[2] ARRPET 2011 Vermicomposting as an Eco-tool in Sustainable Solid Waste Management. The Asian Regional Research Program on Environmental Technology *Asian Institute of Technology, Anna University.*

[3] Bhawalkar U1995 Vermiculture eco-technology. Publication of Bhawalkar *Earthworm Research Institute (BERI) Pune India.*

[4] Roy Hartenstein and Michael S Bisesi 1989 Use of earthworm biotechnology for the management of effluents from intensively housed livestock *Outlook Agriculture* **18** 72-6.

[5] Fraser-Quick G 2002 Vermiculture—a sustainable total waste management solution *What's New in Waste Management* **4(6)** 13–6.

[6] Nair Jaya, Bharambe, Gokul Ramkrishna, Patil, Swapnil Suresh, Bapat and Priyadarshan Satishchandra 2008 Vermiculturerevolution: a low-cost & sustainable technology for management of municipal & industrial organic wastes (solid & liquid) by earthworms *Invited Paper by NOVA Science Publishers NY USA* 159–27.

[7] Taylor M, Clarke WP and Greenfield PF 2003 The Treatment of Domestic Wastewater Using Small-Scale Vermicompost Filter Beds *Ecological Eng.* **21** 197-03.

[8] Singh JS, Gopal B and Dash MC 1978 Role of earthworms in the decomposer system Glimpses of ecology *India International Scientific Publication, New Delhi* 399–06.

[9] Turgut T Onay and Frederick G Pohland 1998 In Situ Nitrogen Management in Controlled Bioreactor Landfills *Water Resour.* ***32*** 1383-92.

[10] Bhawalkar US 1993 Turning Garbage into Gold. An Introduction to Vermiculture *Biotechnology Bhawalkar Earthworm Res. Inst. Pune.*

[11] Bintoro G and Clive A Edward 2002 The effect of multiple application of different organic waste on the growth fecundity and survival of *Eiseniafoetida* (Savigny) (Lumbricidae) *Pedobiol.* **31** 199-09.

[12] Ghatnekar SD, Mahavash FK and Ghatnegar GS 1998 Management of solid waste through vermiculture biotechnology *Ecotechnology Pollut. control Envir. Manage.* 58-67.

[13] Garg P, Gupta A and Satya S 2006 Vermicomposting of different types of waste using *Eiseniafoetida:* A comparative study. *Biores. Technol.* **97** 391- 95.

[14] Hughes RJ, Nair Jaya, Kuruvilla Mathew and Geon Ho 2007 Toxicity of domestic wastewater pH to key earthworm species within an innovative decentralised vermifiltration system. *Water Sci. Technol.* **55** 211-18.

[15] Rajiv K Sinha, Sunil Herat, Sunita Agarwal, Ravi Asadi and Emilio Carretero. Vermiculture Technology for Environmental Management: Study of Action of Earthworms Elsinia fetida, Eudrilus euginae and Perionyx excavatus on Biodegradation of Some Community Wastes in India and Australia. *The Environmentalist U.K* **22** 261-68.

IOP Conf. Series: Earth and Environmental Science **80** (2017) 012052 doi:10.1088/1755-1315/80/1/012052

Comprehensive study on the detrimental effects of fossil fuel exploration and pipe laying in deltaic region

L Sandeep Khanna*, K Ramnath, J Monica, D Muthu and C Venkatasubramanian

School of Civil Engineering, SASTRA University, Thanjavur-613 401. India.

*E-mail: sandeep.star96@gmail.com

Abstract. Thanjavur is the **"Granary of South India"**. As the prosperous capital of Chola kingdom it was praised as "*Chola Naadu Sorudaithu*" (the land that had abundant food). Now, due to Cauvery water shortage issues, the farmers had to be content with single crop a year. Adding to the woes are urbanization and development programs which lack foresight or long term plans that exploit natural resources without a well-articulated thought process. Presently the net sown area in the deltaic region is about 11.87 lac hectares. In the guise of national interests, there is a pursuit of these regions by agencies- public sector undertakings with vested interests. The oil exploration in Cauvery basin (Narimanam block) by Public Sector Undertakings, estimated lignite reserves of 36000 million tonnes and gas reserves and 104.7 billion cubic metres (CBM Coal Gas Methane), which has placed the deltaic region in the corporate radar. Environmentalists and legislators have also turned a blind eye towards the detrimental aftermaths upon the execution of crude product explorations on our cultivable lands.

1. Introduction

The Geological Survey of India has identified fossil fuel reserves in the deltaic region. The government has hence commissioned agencies to extract these resources, impacting green fields and cultivable areas, food production and livelihood. Few of the factors that makes Tamil Nadu the focus region is the confirmed availability of large reserves of fossil fuels, cost benefit, high feasibility as the exploration is on shore compared to the difficult mountainous terrains (Assam) or offshore explorations (Mumbai) are high and favourable success rates. This project is vigorously pursued by the government due to national interests, soaring energy needs, demand for self-reliance, economic development and the need to utilize the indigenous resources.

2. Methodology

The source of information for this study were obtained from the existing data by government agencies Indian Meteorology Department and Directorate General of Hydrocarbons and also from direct site visits and laboratory tests. To study the effects of pipe laying, the undergoing project at Thirumakottai has been chosen was studied. Soil samples from the site were collected for this purpose. To study the effects of fossil fuel explorations, the abandoned drilling site at Thiruvanduthurai was studied. The residents of that village were interviewed. The groundwater samples were collected and laboratory

IOP Conf. Series: Earth and Environmental Science **80** (2017) 012052 doi:10.1088/1755-1315/80/1/012052

tests were conducted and studied for the change in basic parameters.

3. Overview of E&P companies in India

The Petroleum ministry of India has assigned two agencies for these project- ONGC, the public-sector undertaking formed in 1956 whose core business is oil gas exploration and extraction and GAIL, a public-sector undertaking formed in 1984 whose core business is gas transmission and marketing.

It started its exploration activities in Cauvery basin in 1965 to investigate the presence oil and gas reserves. The first well was discovered in 1977 at Karaikal after which 50 unsuccessful wells were drilled. In 1985 oil discoveries were made in Kovilkalapal and Narimanam. Presently there are 200 oil and gas producing wells which provides 39 LCMD gas.

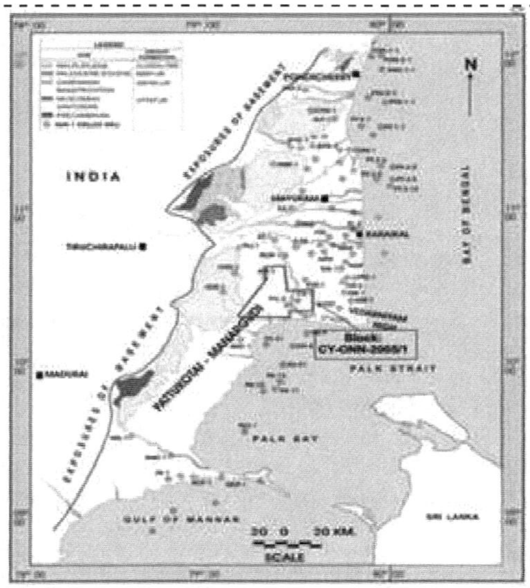

Figure 1. Geological map of Cauvery basin

4. Case study of pipelining site

Place : Thirumakottai

Route : Thirumakottai TNEB Power Plant to Nallur GAIL Gas plant

Object of this scheme : Providing gas for electricity generation.

Mode : Laying a gas pipeline near to the existing pipeline laid in 2002

Pipe Dimensions : Clear length 40 ft, dia-1 foot (0.3048 m), length provided for welding- 0.5 foot, Total length of the pipe 41ft (12.49 m) length

Weld specifications : Fillet weld, after welding the 1 foot width provided is covered with a epoxy resin and is then laid.

Pit dimensions : Width: 5ft, Depth: 7ft pipe laid at 5ft from the top.

Details:

- A compensation of 10 to 40 % of the land value as a one-time payment
- For the further digging or inspection only crop compensation is provided
- 13m width of land is acquired from the owner and only about few feet of land is actually used
- The digging and the closing of the pit will make the lower layer of the soil to lifted up to ground level which is saline which affects the fertility
- The improper compaction while closing the pit will lead to settling of land due to inundation or irrigation
- Trees cannot be grown near the pipeline as the roots spread out and affects the pipeline

Figure 2a. Pipes before laying

Figure 2b. Laid pipes at Thirumakottai

5. Case study for exploration drilling

Place : Thiruvanduthurai

Location : **Latitude :**10.650305□ N, **Longitude:**79.545655□ E

Testimony: "To access the drilling site roads were laid through the agricultural land. The road itself caused extensive damage. Now the site is not under use. In Sendhamangalam also there was a similar site which is also not feasible. The hard rock a stratum under the ground has been shattered like glass for the exploration and the groundwater levels has gone down. Before this project groundwater level was at 30ft now the level has gone down below 80 ft and has become saline. More than these abandoned sites, the flaring in the sites are more dangerous and reduce rainfall. In Narimanam, Vellakudi and few other places there are active flaring sites. When they drill, the vibrations of the ground can be felt for few kilometers. Earlier per acre the farmers used to use only half a bag of Urea and one bag of Triphosphate, but now even after using advanced fertilisers they cannot have a good yield".

IOP Conf. Series: Earth and Environmental Science **80** (2017) 012052 doi:10.1088/1755-1315/80/1/012052

6. Detrimental effects

6.1. Change in rainfall pattern

The graph below is the plot between annual rainfall for the past 50 years[1].The rainfall pattern between Thiruvarur and Thoothukudi is compared as both are in geographical proximity but Thiruvarur has considerable petrochemical exploration infrastructure whereas Thoothukudi has less or negligible petrochemical operations. The exploration and extraction process involves burning of residual gases from the well, known as flaring. The flaring operations emit considerable amounts of CO_2 and other greenhouse gases which has changed the climatic pattern, resulting in reduction in annual rainfall gradually. The exploration in kaveri region was started in 1959 and has eventually increased to a current production of 176 MMCD. Due to flaring, rainfall in Thiruvarur district which was relatively higher than Thoothukudi district has fallen over the past decades. The difference in precipitation between Thoothukudi district and Thiruvarur district was plotted against number of years averaged for five years since production. The plot showed that the precipitation in Thiruvarur has significantly reduced compared to Thoothukudi. Thoothukudi was chosen because it had similar climatic condition but with no exploration activities nearby. The reduction in difference shows that earlier Thiruvarur had higher precipitation when compared to Thoothukudi and has been reducing eventually.

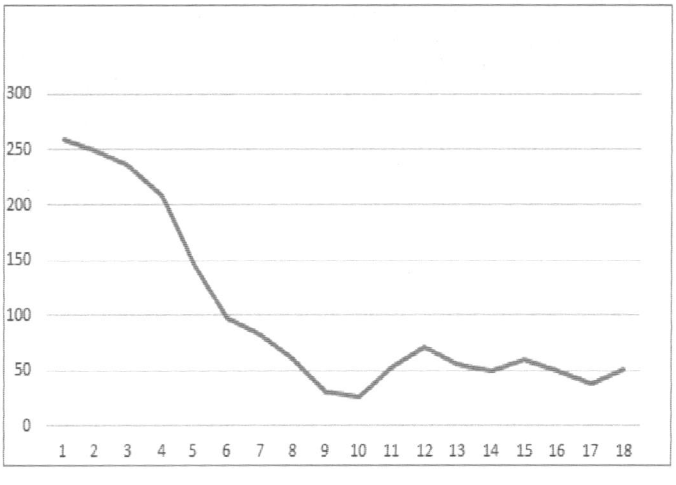

Figure 3. Fall of annual precipitation in Thiruvarur over the years

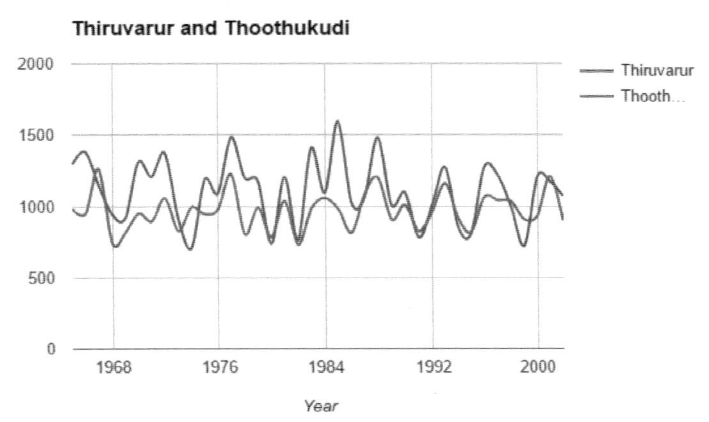

Figure 4. Difference of annual precipitation between Thoothukudi and Thiruvarur in mm plotted against years since beginning of operation.

6.2. Soil fertility change due to oil exploration

When a drilling operation is not feasible the land is returned to the farmers. But the drilling mud are not disposed properly and is buried in the site itself. The laboratory test of drilling mud sample contained 156mg/Kg of cadmium in it. Cadmium is a micronutrient which is harmful for both plants and humans when it exceeds permissible limits. 0.8mg/Kg is the safe limit of Cadmium in soil. The plants can't tolerate excess Cadmium in the soil and the yield becomes very poor[2].This was visibly observed at the site of study.

6.2.1 Test results from Thirumakottai

Table 1: NPK content at field

pH	**Salt**	**$CaCO_3$**	**K**	**P**	**N**	**SOIL**
9.3	0.16	No	55	40	72.4	**U1**
8.4	0.15	No	100	30	78.8	**U2**
9.4	0.32	Yes	170	5	40.4	**D1**
9.4	0.19	Yes	110	5	40.4	**D2**

(Tested at Rice Research Institute, Aaduthurai)

LEGEND:

U1 Soil specimen at ground level

U2 soil specimen at ground level

D1 soil specimen at a depth of 7m

D2 soil specimen at a depth of 7m

NPK in units of kilo/acre.

Dissolved salts expressed as electrical conductivity, in units of dSm^{-1}

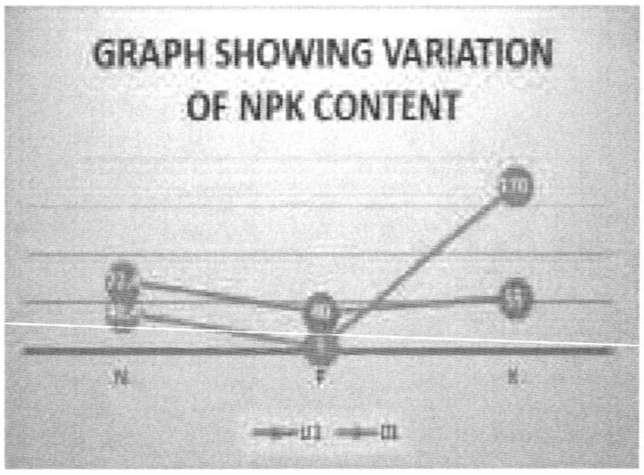

Figure 5a. Variation of NPK content

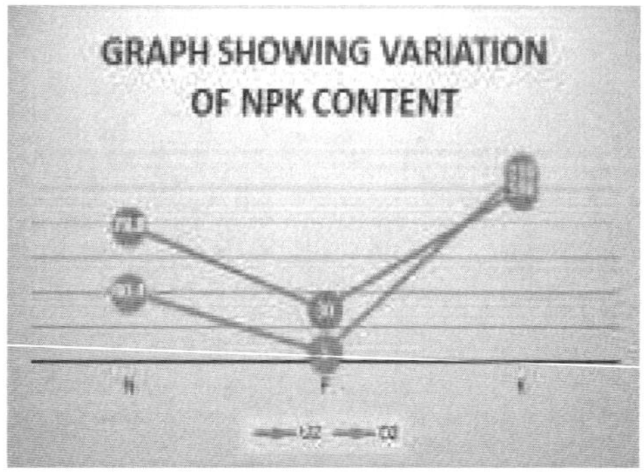

Figure 5b. Variation of NPK content

6.2.2 Inference from NPK test.

The NPK test result show that the Nitrogen and Phosphorus content of the soil is higher at the ground level than in the excavated part. Due to the excavation process the soil is overturned and the soil in the depth reaches the surface and the fertile soil gets buried. Nitrogen and Phosphorous content of soil is due to the presence of organic matter whereas the Potassium content is the intrinsic characteristic of the soil. So the Nitrogen and Phosphorous content of the topsoil is higher than the disturbed soil. This needs to be compensated with additional manure which further adds to the woes of the farmers

6.3. Improper land use

The success rates of exploratory drills is about 1/3rd. Out of 600 wells drilled, only 200 were successful and of these only 74 were active. Abandoned wells are not fully or properly restored to resume agriculture activities. Improper restoration is done without gravel red soil and pebbles. After drilling, the utility area is only 10% with 90% left unrecovered for social or environmental activities. 1 well requires 5 acres of land. Nearly 15 lac acres of cultivable land is lost in 25 years. Colossal loss of livelihood results from crop production loss per season.

6.4. Water quality deterioration

Drinking water scarcity is another social problem. Depletion of ground water table even at a depth of 1500 feet is reported, as a result of exploration. Contamination due to chemicals released while drilling (Cadmium, Mercury, Polycyclic aromatic hydrocarbon, Lead etc.) degrades groundwater quality as drilling involves use of various chemical admixtures and also the drilling wastes contains many hydrocarbons and radioactive metals[3].

6.5. Pollution and health hazards

6.5.1. Flaring effects. A waste gas flare affects the plant growth in the surrounding region. In a study conducted in Nigeria on Yam plantation the plant growth were suppressed with the proximity to the flaring site and the plant size increased with distance from the site. It is also observed that the nutritional value of the plants was altered due to factors like emission and change in temperature. The sugar and the protein content of the plant showed a significant change with proximity to the flaring site[4]. Flaring and other exploration activity has a negative impact on the net yield of the crops, i.e the produce per unit area of cultivable land [5] (Canice Sonny, 2014).

IOP Conf. Series: Earth and Environmental Science **80** (2017) 012052 doi:10.1088/1755-1315/80/1/012052

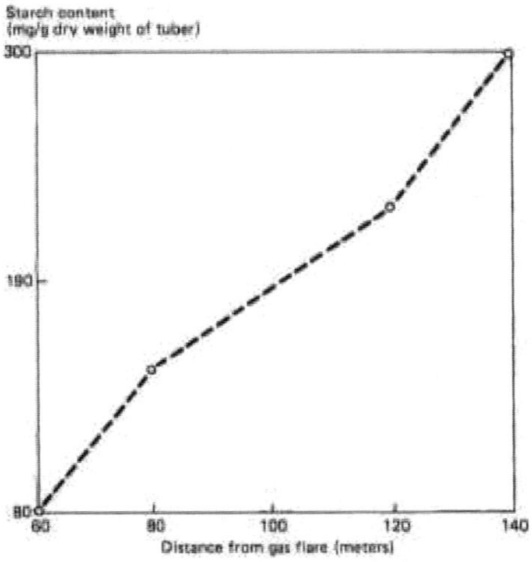

Figure 6a. Variation of starch content in plants.

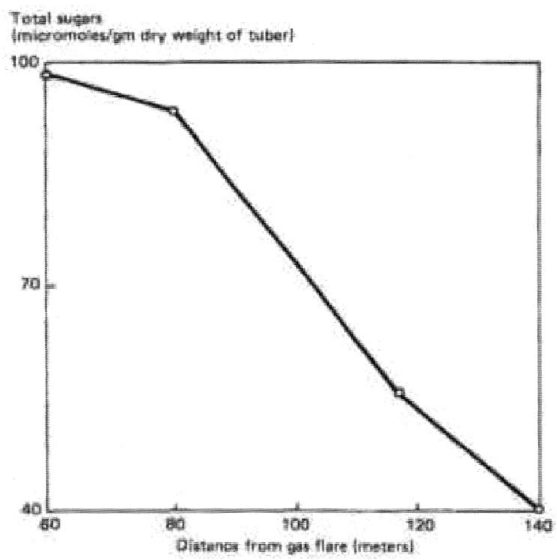

Figure 6b. Variation of total sugar in plants.

Water and air pollution take a toll on people's health. Compressor station discharges organic chemicals into air which includes formaldehyde, benzene, nitrogen oxide and butane/ propane which are toxic in nature. Health impacts of compressor station include hearing impairment, visual impairment, respiratory system impacts, headaches, decreased motor skills, irregular pulse, epidermal problems, dizziness and allergic reactions. Acute and chronic diseases include lungs, kidney and liver ailments.

6.5.2. Cancer.
Experimental results shows that the contact of crude oil with the skin of mice causes skin tumours[6]. Experimental studies shows that Volatile Organic Compounds (VOC) increases the chance of leukaemia and Polyaromatic Hydrocarbons (PAH) having carcinogenic effects on the labours working in petroleum industries. Benzene which is a VOC, is responsible for leukaemia[7].

Studies conducted in China show a high chance of leukaemia among oil field workers and studies across the world shows varied results, such as one study among petroleum and natural gas workers shows excess risk of testicular cancer in men[8].

Studies conducted in San Carlos, a village in north-eastern Ecuador reveals that water contamination due to oil extraction wells around the village causes cancer in various parts of body which has been increasing gradually[9].

6.6. Induced seismicity
Apart from crude oil, there are proven reserves of Coal-Bed Methane in the Cauvery Delta Region. The extraction of Coal-Bed Methane involves injecting water along with propellants and chemical admixtures at high pressure into the coal bed below the rock layers. The high-pressure fluid fractures the rock formation and forces the Coal Bed Methane from the ground. The pumped water deteriorates the soil pressure and the stress in the rock layers. The used waste water is again pumped into the deep sedimentary layers as it is very expensive to treat and dispose the contaminated water. The artificial change of rock stresses resulting from CBM is similar to the rock stresses induced from large

reservoirs.

The seismicity induced by reservoirs have been studied for a really long time and concluded to cause an increase in seismicity. Deep well injections have been directly linked with the increase in seismicity in Youngstown, Ohio [10] (Won-Young Kim, 2013). From the above seismic map of peninsular India it is clearly evident that there are four minor faults and three major faults in the Cauvery Delta Region. The risks of earthquake and related seismicity in the region are low as the region comes under zone- II of seismic risk (IS 1893, 2002). In the event of CBM extraction the stress variation in the faults may lead to induced seismicity in the region.

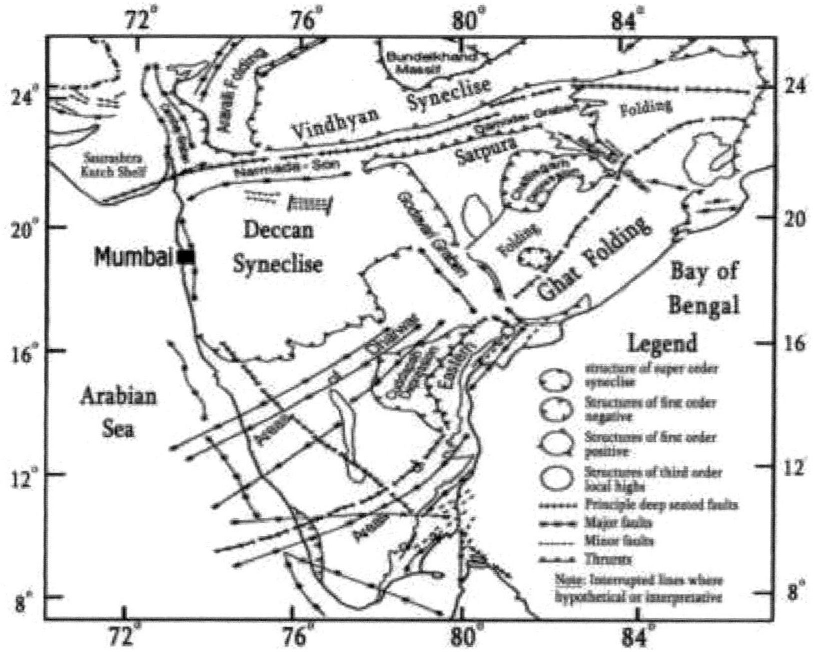

Figure 7. Major evolutionary units and associated tectonic features in the peninsular shield of India

7. Suggestions

- The pipeline can be laid on sides of the roads, causing only minimum disturbance to cultivable lands.
- The excavated soil should be properly refilled and compacted to retain the fertility of the soil such that the crop yield potential can be retained.
- Drilling mud should be properly disposed by reusing methodology.
- All excavation should be carried out during day time only to prevent or to decrease the impact of any accidents, either manually or mechanically.
- Refilling and consolidation of earth should be carried out during non-cropping season only.
- The farmers must be educated about the seriousness and safeness of the project.

- Safety and precautionary measures should be taught to the people of nearby community.
- Booklet, films, lectures, poster and discussions should be given through continuous education program.

8. Conclusion and Recommendations

It is acclaimed that these projects upon successful completion would supply clean and cheap fuel for industries in a faster and a safer manner than distribution by rail/ road which requires extensive infrastructure and additional receiving facilities. Certainly, the secondary sector stands to benefit. However, the negative aspects outweigh the advantages claimed by proponents. Pipeline failures, accidents, spills and explosions can lead to property damage, loss of life and also impact health and habitat. continued excavation leads to further loss of fertility. Large amount of soil disturbance paves way to erosion and sedimentation, particularly in steeper regions and around water bodies. Rights of way on either side of pipeline for maintenance (cleared habitat) and fragmentation (of large areas into smaller ones) results in loss of habitat and adversely affects species movement.

Apart from health, safety and environmental concerns, there are valid apprehensions about the livelihood of present and future generations. It is a well-known fact that primary sector provides social security for the Indian majority. No compensation can ever make up for the loss of source of income or heal the pain of holding a condemned property by farmers. Primary sector is our lifeline, essential for our sustenance and self-reliance. Ironically the world's largest solar power plant at Kamuthi (678 MW capacity) is in the same state. Given the abundant solar potential and wind potential, other forms of clean energy can be explored rather than investing in the Petroleum industry.

References

[1] Records of rainfall data http://www.indiawaterportal.org/met_data/

[2] Piotrowska M and Dudka S 1994 Estimation of maximum permissible levels of cadmium in a light soil by using cereal plants. *Water, Air and Soil Pollution* **73**(1) 179-88.

[3] Asia I O, Jegede S I, Jegede D A, Ize-Iyamu O K and Akpasubi E 2007 The effects of petroleum exploration and production operations on the heavy metals contents of soil and groundwater in the Niger Delta *Int. J. Physic. Sci.* **2**(10) 271-75.

[4] Lawanson A O, Lmevbore A M A and Fanimokun V O, The Effects of Waste-Gas Flares on the Surrounding Cassava Plantations in the Niger Delta Regions of Nigeria. *6th Symposium Proc.* **41**(3) 239-45. http://www.istrc.org/images/Documents/Symposiums/Sixth/6th_symposium_proceedings_0041_section_3_239.pdf

[5] Canice S N and Sonny A N 2014 Resource Productivity under Yam Based crop Mixture in crude and non-crude oil Producing communities of Imo State, Nigeria. *Agricultura tropica et subtropica* **47**(1) 20-8.

[6] Wilson J S and Holland L M 1988 Periodic response difference in mouse epidermis chronically exposed to crude-oils or BaP: males vs females. *Toxicology* **50**(1) 83–94.

[7] Austin H, Delzell E and Cole P 1988 Benzene and leukaemia: a review of the literature and a risk assessment *Am. J. Epidemiol.* **127**(3) 419–39.

[8] Mills P K, Newell G R and Johnson D E 1984 Testicular cancer associated with employment in agriculture and oil and natural gas extraction. *Lancet* **28**(1) 207–10.

[9] San S, Armstrong B, Córdoba J and Stephens C 2001 Exposures and cancer incidence near oil fields in the Amazon basin of Ecuador. *Occup. Environ. Med.* **58**(8) 517-22.

[10] Kim W Y 2013 Induced seismicity associated with fluid injection into a deep well in Youngstown, Ohio. *J. Geophy. Res.Sol. Ea.* **118**(7) 3506–18.

IOP Conf. Series: Earth and Environmental Science **80** (2017) 012053 doi:10.1088/1755-1315/80/1/012053

Analysis of noise pollution level in a University campus in South India

D Thattai*, J S Sudarsan, R Sathyanathan and Visalatchi Ramasamy

Department of Civil Engineering, SRM University, Kattankulathur – 603203. India.

*Email: deepthathattai.v@ktr.srmuniv.ac.in

Abstract. Noise comprises those sounds occurring around us that are not part of the environment under consideration. Noise is also a type of pollution and impacts on our health and wellness. The prevalence of noise is increasing in magnitude and severity because of growing population and urbanization. Noise pollution leads to many chronic and socially significant impacts. This study analyzes the level of noise at different points in SRM University. As the University encompasses a hospital also, it is more important to identify the sources of high noise levels and control them. As per Indian standards the desirable noise pollution for educational institutions and hospitals in daytime is 50 dbA. Noise levels were measured with a sound level meter at 19 points within the campus at three different timings (8–10 am, 12–2 pm, and 3–5 pm) over two cycles of measurements. The preliminary results show higher noise levels during morning and evening. Noise during Cycle 2 (latter half of semester) was 20% more compared to that of Cycle 1 (beginning of semester).

1. Introduction

The World Health Organization (WHO) aims to make all peoples attain the highest possible level of health. Health is defined in the WHO Constitution as "[a] state of complete physical, mental and social well-being and not merely the absence of disease or infirmity". By this broad definition that includes well-being, noise impacts can be clubbed as 'health' issues [1]. The word "noise" is derived from the Latin word "nausea," which means seasickness or a sensation of discomfort [2]. Noise comprises those sounds occurring around us that are not part of the environment under consideration. It is also a type of pollution and impacts on our health and wellness and the ability to do productive work. Sources of noise pollution include industries, traffic and vehicles, construction and domestic appliances. The effects of noise are both direct and indirect; they affect the health and make our living environment miserable [3–5].

Various measures have been taken and standards evolved by the U.S. Environmental Protection Agency and Indian standards IS: 4954-1968, limiting the noise levels for different types of areas. In order to legally enforce control on noise levels, the Government of India has notified the Noise Pollution (Regulation and Control) Rules 2000, under the Environment (Protection) Act, 1986. Under these rules, ambient air quality standards with respect to noise have been specified for different types of areas. For residential areas, the day time noise should not exceed 55 dB(A) L_{eq} as per these rules.

Universities, colleges, schools, libraries, national laboratories and hospitals come under Silent Zone. It may be mentioned that these are buildings that require on average quiet conditions but are also at the same time noise sources at certain times, for e.g. student movement in universities. In the first instance, efforts should be made to find a site that is situated in a comparatively quiet environment. At least the site should be away from the noisy roads, rail tracks, aerodromes etc.Where there is no choice, due attention shall need to be paid to the layout of the rooms and facilities. Adequate open space should be left between the building and the roads/noise source. The rooms shall be arranged to be farther away from the noise source, while other unimportant rooms like gymnasium may be laid out closer to it and in a manner to provide a noise shield to the class-rooms and operation theatres.

In this study, the noise level measurements were carried during January–February 2015 and March–April 2015 to assess the noise generated in SRM University campus.

2. Study area

The study area is the University campus covering over 42 blocks. The campus consists of the Faculty of Engineering and Technology, Faculty of Medicine and Health Sciences, Faculty of Science and Humanities and Faculty of Management. It also houses the Medical College Hospital, Research Centre and Hostels. The hospital can accommodate 1200 in-patients. The University has a central library with vast resources and an auditorium that can accommodate over 4,000 people. The study area also covered a few points at a radius of 5 km around the campus, including the railway station (Figure 1).

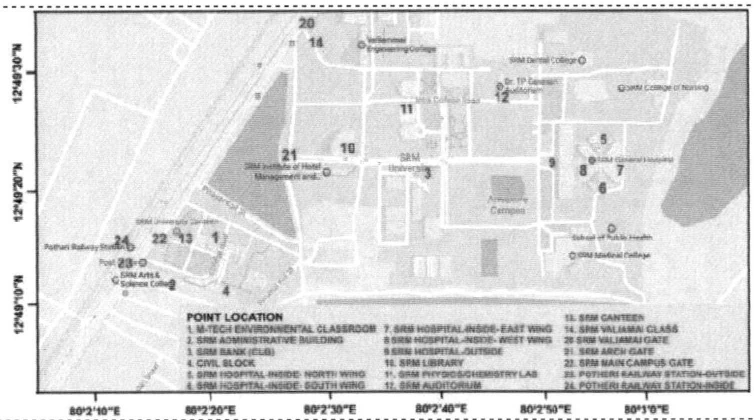

Figure 1. Sampling locations

3. Methodology

A total of 24 sampling collection points was taken in two cycles of readings viz. January–February and March–April, 2015. Three timings were chosen for the study: 8–10 am, 12–2 pm and 3–5 pm, to capture the high noise events of student entry and exit and lunch hour. The sound level meter was used to measure noise level within 24 minutes in a location for taking 1 reading, when noise level was consistent for 5 seconds. Reading was instantaneously tabulated. Five such sample readings were taken in each location.The sound level meter was used in slow response setting, which is ideal for educational institutions, and the instrument was used to record readings in dbA mode.

The values obtained from sound level meter were converted to Leq values. The equivalent continuous noise level (Leq) is the "sound pressure level of a steady sound that has, over a given period, the same energy as a fluctuating sound" under study. It is an average and is measured in dB(A) scale. It can be calculated using the formula [6]

$$L_{eq} = 10\log_{10}\left[\frac{1}{T_M}\int_Q^{T_M}\left(\frac{P(t)}{P_0}\right)^2 dt\right] \quad (1)$$

Where Leq is the equivalent continuous linear weighted sound pressure level at 20μPa, determined over a measured time interval T_M (s), P(t) is the instantaneous sound pressure of the sound signal, P_o is the reference sound pressure of 20μPa.

Adding Leq values requires taking an anti-log of each value. The addition can be performed asshown [7]:

$$\text{Total}L_{eq} = 10\ \log\left(\frac{10^{\frac{Leq1}{10}}+10^{\frac{Leq2}{10}}+10^{\frac{Leq3}{10}}+\cdots+10^{\frac{Leqn}{10}}}{n}\right) \quad (2)$$

4. Results and discussion

This study analysed the noise levels in 24 stations. The equivalent noise level (L_{eq} for 2 hours) across various stations during three different time periods across the two cycles are shown in Figures 2 and 3.

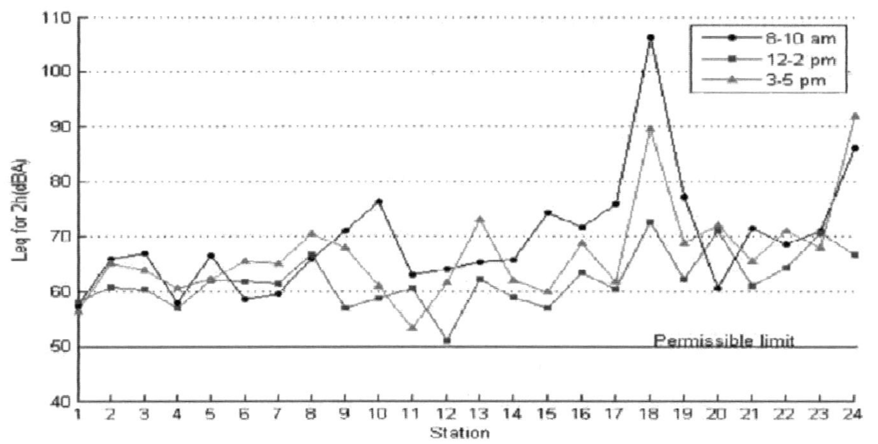

Figure 2. L_{eq} vs Station (March–April).

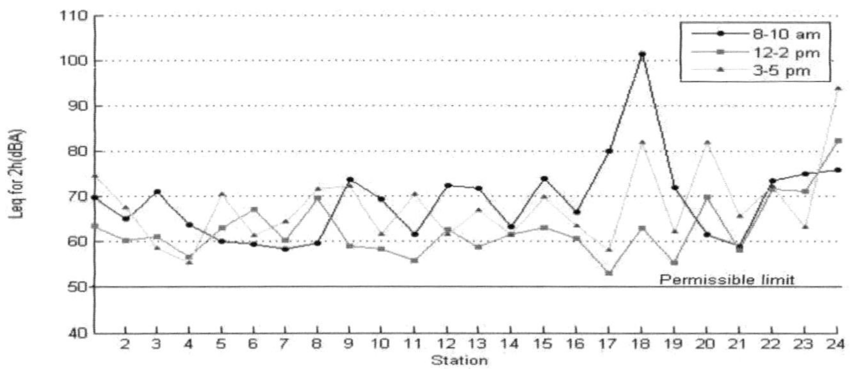

Figure 3. L_{eq} vs Station (Jan–Feb).

High noise levels were recorded during morning and evening, corresponding to student entry and exit times. Though the peaks are seen in the sampling locations outside the campus the permissible limit is not maintained in any of the stations. The peak value reaches around 106 dBA. Road traffic and presence of railway station is the reason behind the peak values. Yet, the permissible limit for train noise level is given by IS:4954-1968 as

Freight trains (900 to 1200 m long) running at 25 to 40 km/h – 90 dB at 6 m, 77 dB at 150 m
Freight trains (900 to 1200 m long) – 96 dB at 7.5 m running at 65 km/h
Engine Noise Level – Under all operating conditions, 97–105 dB at 15 m.
Whistle Noise Level – 110 dB at 15 m.

The problem here is the location of the campus right across both the railway track and NH45, leading to doubling of noise. The advantages of being close to the station and bus stop are offset by the intrusion of noise. The minimum value of 52dBA is noticed in Station 12, which is the auditorium. Even this is above the permissible limit. Lunch hour seems to be a quiet time; probably because the noise shifts to the hostels and canteens (which were not measured). Comparing the two cycles, there is not much difference between starting of the semester and end of the semester, as seen in figures 4–6.

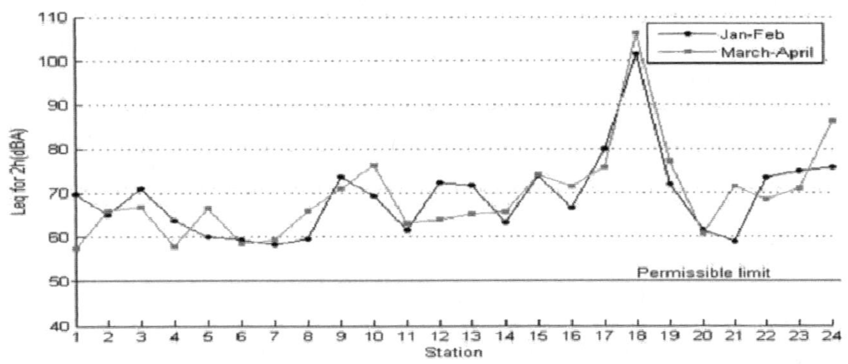

Figure 4. Comparison of L_{eq} (8–10am) in two different cycles.

The noise levels are doubled in just a few locations. The maximum values are also noticed at Potheri, SRM Valliammai Gate and Guduvancherry Railway Station and West inside of Hospitals Zone. In the duration 8–10 am and 3–5 pm (Figures 4 and 6), it is observed that maximum values were recorded in railway stations in Potheri and Guduvancherry due to students (~3000) arriving to and departing from the campus, respectively. Similar trends were observed by others [5, 8–11]. The noise levels in the stations are only slightly higher than that in the campus in the afternoon, reflecting the lesser number of students in the stations.

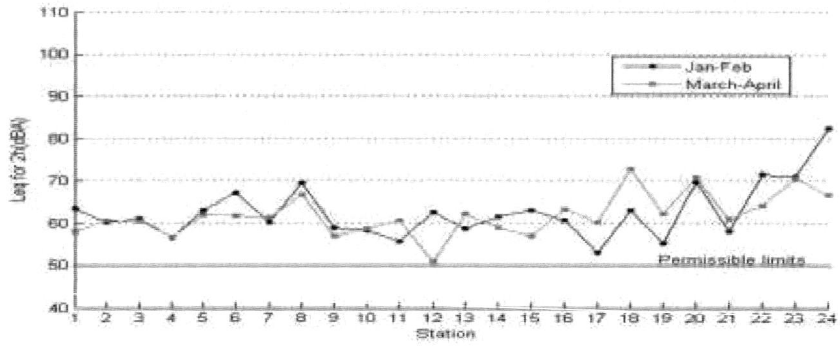

Figure 5. Comparison of L_{eq}(12–2pm) in two different cycles.

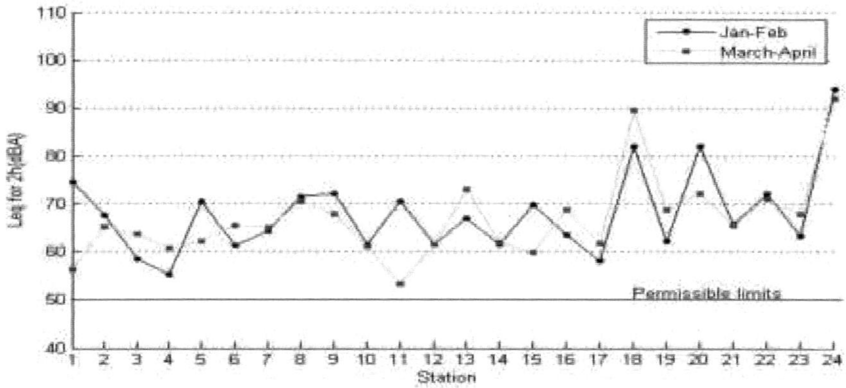

Figure 6. Comparison of L_{eq}(2–3 pm) in two different cycles.

The L_{eq} for 8 h corresponding to the work day is also consistently above the permissible limit (Figure 7). There is also not much variation between the two cycles.

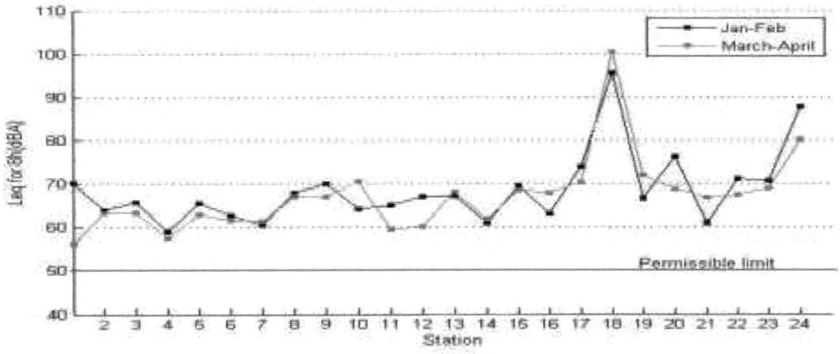

Figure 7. Leq for 8h vs Station.

The advantages of a University campus being near the railway station and/or highway are offset by the level of increase in noise caused by the same. When the University was established, it was a quiet zone with only a few trains plying through the day and the surrounding areas were empty. After 30 years, however, the University itself has grown by leaps and bounds. Residential buildings have mushroomed in Potheri. The stretch of NH45 on which the University lies has also exponentially grown, thereby leading to heavy and constant traffic. The population is increasing every year in the campus. All these are contributing to high noise levels such that the noise levels are higher than the permissible limits always during the day. This has to be brought down if the ill effects of noise pollution are to be prevented. The presence of a hospital within the campus also mandates low noise levels. Some measures to reduce the noise pollution include banning student vehicles within the college premises; planting more trees and sound barriers wherever possible; installing noise-absorbing fixtures in the classrooms; and sensitizing the campus community to the effects of noise pollution.

5. Conclusions

The ill effects of noise pollution are well recorded. The noise levels within and near a University campus were measured and analysed in this study. The advantages of a University campus being near the railway station and/or highway are offset by the level of increase in noise caused by the same, as evidenced in this study. Measures should be taken in such locations to reduce the noise levels so that classrooms will not be subjected to external noise.

References

[1] DeKoning H W 1987 Setting Environmental Standards, *Guidelines for Decision Making World Health Organization* Geneva, Switzerland 76–89.
[2] https://www.techopedia.com/definition/2025/noise
[3] Tiwari T N and Kudesia V P 1994 Noise Pollution and Its Control Pragati Prakashan Meerut, India.
[4] Huma B 2012 Noise Pollution and Human Health in Pakistan. MPhil. Thesis GC University, Pakistan
[5] Alam J B, Alam M J B, Rahman M M, Dikshit A K and Khan S K 2006 Study on traffic noise level of Sylhet by multiple regression analysis associated with health hazards. *Iran J. Environ. Health. Sci. Eng.* **3**(2) 71–8.
[6] Bernard P 2014 L_{eq}, SEL What? Why? When? Application Notes Brüel and Kjær BO0051–14.
[7] https://www.osha.gov/dts/osta/otm/new_noise/appendixb.pdf.
[8] Sagar T V and Rao G N 2006 Noise pollution levels in Visakhapatnam city (India). *J. Environ.*

Sci. Eng. **48**(2) 139–42.
[9] http://www.southwoodresources.com.au/southwood/pdf/planning/APPMSEMF.PDF.
[10] Goswami S, Kumar Nayak S, Surjendu and Kumar D 2011 A study on traffic noise of two campuses of University, Balasore, India. *J. Env. Biology* **32**(1) 105–9.
[11] Lad R J, Patil V N and Raut P D 2011 Study of Noise Pollution during Deepawali Festival in Kolhapur City of Maharashtra, India. *Indian Streams Res. J.* 1 VII 82.

IOP Conf. Series: Earth and Environmental Science **80** (2017) 012054 doi:10.1088/1755-1315/80/1/012054

Natural disaster management in India with focus on floods and cyclones

Deeptha V Thattai*, **RSathyanathan, R Dinesh and L Harshit Kumar**

Department of Civil Engineering, SRM University, Chennai - 603203. India.

*Email: deepthathattai.v@ktr.srmuniv.ac.in

Abstract. Disasters are of two major kinds, natural and manmade, and affect the community. Natural disasters are caused by natural earth processes like floods, droughts, cyclones, tsunamis, earthquakes and epidemics. Manmade disasters occur due to chemical spills, accidents, terrorism activities etc. India is prone to almost all the major natural disasters. The high population density combined with poor preparedness, planning and management, and rescue and relief measures inevitably lead to huge losses of lives and property every year in the country. This paper analyses the disaster management policy of India and its implementation using two recent case studies – one where a relative degree of success has been achieved (cyclones) and the other where we are still struggling to have even a basic preparedness system in place (floods).

1. Introduction

The Disaster Management Act, 2005 defines disaster as "a catastrophe, mishap, calamity or grave occurrence in any area, arising from natural or manmade causes, or by accident or negligence which results in substantial loss of life or human suffering or damage to, and destruction of, property, or damage to, or degradation of, environment, and is of such a nature or magnitude as to be beyond the coping capacity of the community of the affected area".[1] In this paper, two case studies – cyclones and floods – are taken up for comparison of disaster management strategies adopted in the country and the areas that need improvement are highlighted.

2. Natural disasters

Disasters are of two major types – natural and manmade. As the names imply, natural disasters are caused by the earth's natural processes that occur on a regular basis whereas manmade disasters are due to human actions. From a larger perspective, no natural process can be a disaster by itself; it just occurs as a result of the causal effects. For example, the movement of plates gives rise to earthquakes and tsunamis; climate processes give rise to cyclones, floods and droughts. These have been occurring and will keep occurring through time. When such a resultant process interacts with the human populations and their belongings so as to cause a widespread loss of lives and property, we call that a disaster. If an earthquake or tsunami occurs in the middle of an ocean and we are not affected by it, we don't consider that a disaster. Hence, disaster is purely from an anthropogenic point of view, and thus all disasters are "manmade" to a certain extent, as it is we who decide to settle down in the path of a natural process. Apart from that, by our sheer mismanagement of the natural resources, what ought to be a normal event with little consequence sometimes ends up as a disaster, e.g. the floods that recur in our cities. Thus, disaster management, in addition to planning, preparedness and mitigation of disasters, should also include proper management of natural resources as a preventive measure including consideration of the consequences of the (mis)management of such resources.Figure 1 shows the natural disasters and their classification (adapted from ASDR 2013 [2]).

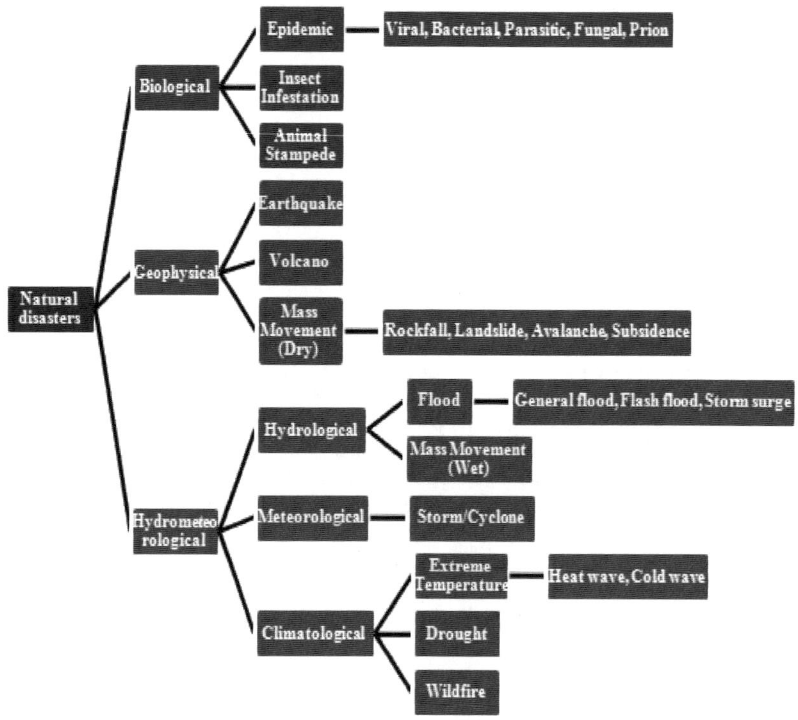

Figure 1. Natural disasters and the subgroups (adapted from ASDR 2013).

2.1. *Natural disasters in India*

As per India's National Policy on Disaster Management [3], the natural disasters that India is prone to are earthquakes, floods, droughts, cyclones, tsunamis, landslips and avalanches.Almost 59% of India's landmass is prone to earthquakes; over 12% of land is prone to floods; about 76% of the coastline is prone to cyclones and tsunamis; 68% of the cultivable area is drought-prone; and hilly areas are subjected to wet and dry landslides and avalanches. Around 2% of the GDP is lost due to disasters [4]. Figure 2 shows the relative position of India among the top ten disaster-prone countries in 2013(adapted from ASDR 2013 [2]). India ranks fifth in the number of incidences (12), but what we should focus is on the importance of the nature of the disaster. Except for countries more prone to earthquakes, most of the disasters are caused by hydrological and meteorological factors, i.e., floods and storms/cyclones.

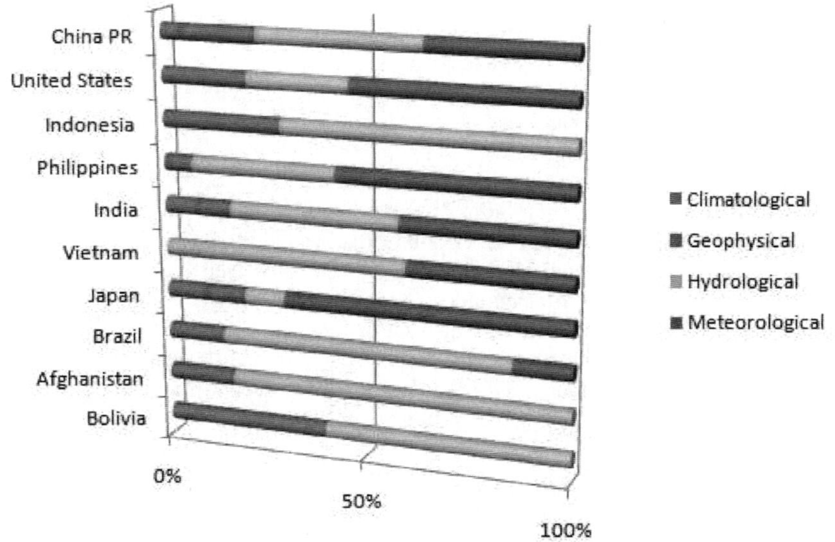

Figure 2. Percentage distribution of disasters among the top ten countries in 2013 (adapted from ASDR 2013).

Table 1 gives some details of major disasters in India in recent years. Again, we can see that in the last nine years, hydrometeorological factors have caused the worst damage.

Table 1. Details of recent major disasters in India.

YEAR	PLACE	TYPE
1993	Latur, Maharashtra	Earthquake
1999	Odisha	Cyclone
2001	Gujarat	Earthquake
2003	India (mainly Andhra Pradesh)	Heat wave
2004	Southeast coast	Tsunami
2005	Kashmir	Earthquake
2005	Maharashtra	Flood
2008	Bihar	Flood
2009	10 states	Drought
2010	Jammu &Kashmir (J&K)	Flood
2013	Uttarakhand	Flood
2013	Maharashtra	Drought
2014	J&K	Flood
Every year	Major cities	Floods

Earthquakes of course create huge losses when they occur with high magnitude but they are far fewer in occurrence. Floods, droughts and cyclones, however, occur year after year and cause insidious damage just by the sheer frequency and geographical spread. Climate change is enhancing the risk and unpredictability now. Hence it is very important to focus on implementing preventive and relief measures for these events on a war footing.

3. Disaster management in India

During the British era and post-Independence years, disaster management was largely limited to post-disaster relief works, food-for-work programmes etc., under a Central Relief Commissioner who headed the State Relief Commissioners. A disaster management cell was set up in the 1990s following the declaration of the International Decade for Natural Disaster Reduction by the United Nations. This was later moved to the Ministry of Home Affairs [4].

The National Disaster Management Authority (NDMA) is currently the nodal government agency dealing with disaster management in India. It was created by the Government of India by enacting the Disaster Management Act on 23 December 2005. It is headed by the Prime Minister and oversees the State Disaster Management Authorities (SDMAs) headed by the respective Chief Ministers. The National Institute of Disaster Management (NIDM) for capacity building and National Disaster Response Force (NDRF)forresponse were also set up subsequently. The primary aim of the NDMA is to lead and implement a holistic and integrated approach to disaster management.

Following the setting up of the NDMA, measures for prevention and mitigation were formulated for the major disasters. The following paragraphs focus on cyclones and floods.

3.1. Cyclones

The almost 8000 km-long Indian coastline is exposed to about 10% of the world's tropical cyclones. Of these, the majority originate over the Bay of Bengal and strike the east coast. The cyclones bring heavy rains and storm surges into the coastal regions. The major destructive force, however, is the wind that accompanies the cyclone. It causes huge damage. States affected by tropical cyclones in India are Gujarat, AP, Karnataka, Kerala, Maharashtra, Odisha and Tamil Nadu.

Mitigation Project was launched by the Ministry of Home Affairs (MHA) covering Odisha and AP to upgrade cyclone forecasting, to track cyclones and provide warning systems. The project also aimed to build capacity and construct cyclone shelters and embankments.The total cost of the project was Rs. 1496.71 crores with 80% assistance from the World Bank and the rest from the two states.

The India Meteorological Department (IMD) is the nodal agency for early warning of cyclones.

Floods

India experiences flood havoc every year due to natural and manmade causes. The total flood affected area is 456.40 lakh ha and an average of 72.75 lakh ha is affected annually [5]. Apart from flooding of river banks during monsoon rains and cloudbursts, the urban cities in the country are increasingly witnessing flooding events leading to huge losses. As per a report in the magazine Down to Earth, 10, 22, 35, 37 and 17+ cities in India bore the brunt of flooding in 2005, 2006, 2007, 2010 and 2014, respectively [6].

The National Flood Risk Mitigation Project wasstarted to mitigate risk, severity and effects of floods.

The Ministry of Water Resources launched the Flood Management Programme (FMP) at a total cost of Rs.8000 crores for the 11th Plan period (2007–12). It aimed to monetarily assist the state governments for undertaking flood management works in critical areas. As of 31 March, 2010, 117 works for 10 states were reportedly complete; 1.33 billion hectares restoredandprotected, andabout12.89 million people would be rendered safe, according to the report [4]. 42 new flood management schemes were included under the FMP from Assam, Bihar, Goa, Gujarat, J&K, Kerala, Manipur, Odisha, Pondicherry, Tamil Nadu, Uttarakhand, Uttar Pradesh and West Bengal for 2010–11.

The Central Water Commission (CWC) is the nodal agency for flood forecasting and warning. On average, 6000 forecasts are issued during monsoon every year.

4. Case studies

The pre- and post-disaster management strategies adopted for two annually recurrent disasters in India are explained below with case studies.

4.1. Cyclones

Due to more awareness and monitoring of cyclones in the last several years, the track and magnitude are known beforehand and evacuation and relief measures are taken up on a large scale whenever a cyclone is set for landfall along the Indian coast. The deaths due to cyclones have decreased as a result. Cyclone Hudhud that made landfall near Visakhapatnam inAndhra Pradesh (AP) in October 2014 was a very severe cyclonic storm. It impacted AP, Odisha and Chattisgarh. Around 7 lakh people were evacuated prior to landfall and this minimized the death toll to around 100 [7]. The NDRF was massively deployed for rescue and relief operations. The cyclone caused extensive damage to Visakhapatnam city and the neighbouring districts and the total damage is estimated at Rs. 70,000 crore, the highest till date in India. After Cyclone Phailin hit Odisha in 2013, there were reports of looting of relief materials and absence of relief measures. Thus, even though human fatalities have been reduced because of advance information and the steps taken by the government and the people, the destruction of infrastructure in the aftermath of a cyclone has still not been reduced. This leads to the weaker sections of society suffering in the post-disaster scenario.

One of the shocks for India in 2016, especially Chennai, was the Vardah Cyclone [8]. A low pressure area created in Malay Peninsula and the Sumatra region moved towards southeast Bay of Bengal and created the tropical disturbance and then evolved into a cyclone with a maximum wind speed of 192 kmph and impacted AP and TN. Chennai was in the eye of the cyclone. Due to the warning given by the Cyclone Warning Division of India Meteorological Department, over 15 teams of NDRF had been already deployed to the coastal areas of Tamil Nadu and Andhra Pradesh. The cyclone uprooted more than 12,000 trees and disrupted telecommunication in TN and made a landfall in Chennai. Around 20,000 people were evacuated and this minimized the death toll to around 16.The total damage was estimated at Rs. 6,749crores to the state [9]. The people who were affected by Vardah were also helped by the Indian Armed Forces. The losses were huge but people helped each other to bring up the situation to a normal state.

4.2. Floods

In the first week of September 2014, the Srinagar weather station in J&K recorded more than 500 mm of rainfall [6] leading to disastrous floods, which killed more than 215 people and displaced tens of thousands more in the state. The state emergency forces were ineffective and people in responsible positions were prominent by their absence.Around 2.5 lakh people were rescued by 30,000 troops of armed forces, the NDRF and the locals. The Central Water Commission (CWC) had no flood forecasts for the state.Infrastructure losses are pegged at Rs.6000 crores. Deforestation in the catchment areas of rivers – especially Jhelum, Chenab and Indus – and streams is a big problem in the Kashmir valley. The loss of treeshas caused more erosion from catchment areas and raised the riverbeds; consequently the water carrying capacity of the streams and rivers is much reduced leading to breaching of river banks and inundation.Illegal and unplanned construction in the floodplains has altered the natural flow as the existing contour drainage is cutoff. Unplanned road construction in the Himalayas destabilises the upstream slope and chokes the river bed with debris.More importantly, failure to preserve wetlands and lakes has played a major role in ecological degradation as their sponge action is cut off.Wularlake, which was 20,200 ha has been reduced to 2,400 ha and the Dal lake in Srinagar has been shrunk to almost half its size.Almost half of Srinagar's lakes, ponds and wetlands have been encroached over the years and untreated sewage and sullage are being discharged into it. This has led to eutrophication of water bodies leading to further reduction in storage capacity. Excessive usage of chemical fertilisers in the fruit orchards throughout Kashmir has compounded this problem. Relief measures have been implemented slowly and there is now a fear that the problems will only increase because of the harsh winter. The imminent danger to Srinagar was highlighted in 2010 itself [10] but no action was taken.

The Nov-Dec 2015 floods of Chennai is described as the worst in a century [11]. According to Skymet data, Chennai recorded a whopping rainfall of about 1218.6mm in November 2015.The ensuing flood affected the Coromandel coast of TN and the Union Territory of Puducherry, with the city of Chennai and surroundings particularly hard hit. Urban development that cares little for natural processeshas led to many wetlands being built over. In addition, old, underdeveloped civic infrastructure and drainage systems have led to increased flooding in major cities. Earlier, numerous lakes and tanks were built by the rulers, which ensured that excess water from one lake entered the next lake and so on, thus ensuring a good distribution of the excess water and minimum wastage of precious rainwater. As per a CSE report, there were over 600 lakes in Chennai the 1980s, but but by 2008 only a few were still in good condition. More than 500 people lost their lives and over 1.8 million people were evacuated with the help of NDRF and Indian Armed Forces during the Chennai floods. The total damage was estimated in the range of Rs. 200 billion to over Rs. 1 trillion [12]. The insurance losses due to the floods were the maximum encountered so far in Chennai.

5. Lapses

The Comptroller and Auditor General's report [5] on the audit of disaster management found several shortcomings. As can also be seen from the case studies above, the management of disasters has to be thoroughly strengthened from all angles. The following are the major lapses:

SDMAs not fully functional; state disaster response funds mismanaged in many states; state disaster response forces raised only in seven states.

Interference among agencies leading to duplication of same works.

No actionable plan for flood management in the Ministry of Water Resources.

Emergency action plans for only 25% of dams; inflow data for only 0.006% of barrages and reservoirs as of Sep 2011.

No data till date for many rivers; no information on water bodies; no proper monitoring mechanisms in place.

FMP not properly implemented; more than 50% of flood prone areas not protected.

No cyclone-related programs initiated by Ministry of Earth Sciences even though it was identified as the nodal ministry.

Management and mitigation plans not prepared for cyclones.

6. Suggestions for improvement

While it is true that whatever preparation we undertake will not be enough in a catastrophe, we cannot take that as an excuse and not have basic protection measures in place. The majority of India's natural disasters are annual occurrences – floods, droughts and cyclones – for which, even after so many years after Independence, we donot haveenough planning, protection and relief and rehabilitation. In Chennai, for example, there was excess water in 2015, which not only incapacitated the city but also emptied into the sea. Whereas in 2016, even Cyclone Vardah was not enough to make up for the deficit in the northeast monsoon and the city is staring at one of the worst droughts even before summer has started. The agencies that deal with disaster are found wanting in implementing their responsibilities; there is no coordination among various sectors and agencies; there is no proper control structure; there is lack of communication and information at critical moments. All these and more need to be addressed and steps taken so that when any part of the country encounters a disaster, the suffering of the people is minimised through a strong infrastructure and timely relief and rehabilitation. While there are calls for the Railway Minister to step down owning responsibility for a single accident, no one is held responsible in the aftermath of a disaster. What can we do to change this?

Technically capable people with a set term should be allowed to head NDMA and implement its policies.

The hazards database of the country should be updated as soon as possible with inputs from district and state levels and from field experts and civil society organisations (CSOs).

A chain of command during disasters should exist with powers of implementation.

The agencies at the district and state levels should be trained for effective response measures.

The projects undertaken should be finished within deadlines and with strict quality.

A nationwide and statewide database of CSOs and individuals experienced with handling disasters should be maintained and their help sought during crises.

Private sector should be roped in for implementing pre- and post-disaster measures.

Reputed CSOs must be involved from the planning stage itself and provided with necessary wherewithal for their involvement during and after a disaster.

Specific to cyclones and floods:

Data on river discharges should be collected for all major rivers and should be freely available. Similar data should be provided for all hazards so that more research will be conducted across the country.

Where long term data is not available, discharges for all rivers should be empirically estimated and probable flood levels in highly populated areas should be identified.

The existing flood zone maps should be modified to include areas of repeated urban flooding.

Urbanisation should be planned so that existing water bodies are not destroyed. If possible, restoration of damaged wetlands and water bodies should be taken up on a war footing.

Desilting of ponds, tanks and other water bodies should be done periodically with the support of CSOs, local people and other stakeholders to maintain their maximum storage capacity.

In urban areas, choked storm water drains should be cleaned before monsoon starts so that inundation can be avoided.

Afforestation in the catchment areas should be promoted to prevent siltation of water bodies due to excessive soil erosion.

Drainage in urban areas should be improved drastically without affecting the existing natural drainage system.

Stocks of relief materials should be made available in adequate supply and in working condition; looting should be prevented.

Efforts should be undertaken to minimise infrastructure damage due to cyclones and floods.

The effects of climate change on the country's hydrometeorology and coastal processes need to be researched and measures taken accordingly.

7. Conclusions

With the second highest population, increasing urbanisation and added high risk for extreme events, India is in a vulnerable situation of facing disasters year after year. The unplanned and substandard infrastructure combined with poor implementation of planning, relief and rehabilitation measures increases the human, physical and economic losses sustained during and after a disaster. It is high time that we wake up to reality and enforce strict measures to reduce losses on all fronts.Accountability at all levels is the need of the hour. While the economically stronger sections of society will have more resilience in bouncing back to near normalcy from a disaster as a community, it is the economically weaker and vulnerable (elderly people, women, children, and the differently abled) sections that are hit the most during and even many years after a disaster occurs. Care should be taken to see that the weaker sections are also able to recover fast from disasters.

References

[1] UNISDR Reporthttp://www.unisdr.org/we/inform/terminologyAccessed 02 Nov 2014.

[2] Guha-Sapir D, Hoyois Ph and Below R 2014*Annual Disaster Statistical Review 2013: The Numbers and Trends*Brussels: CRED

[3] National Policy on Disaster Management NDMA 2009

[4] MHA *Disaster Management in India* Ministry of Home Affairs Report 2011

[5] CAG 2013 http://www.downtoearth.org.in/themes/DTE/images/cag-report_20130625.pdf

[6] http://www.downtoearth.org.in/themes/DTE/dte-infographics/20141015_coverstory.htm

[7] Wikipedia http://en.wikipedia.org/wiki/2014_India%E2%80%93Pakistan_floods

[8] Wikipedia https://en.wikipedia.org/wiki/Cyclone_Vardah Accessed 6 Feb 2017
[9] http://indianexpress.com/article/india/damage-caused-by-cyclone-vardah-less-than-tamil-nadu-estimate-mha-4451023/
[10] https://in.news.yahoo.com/warnings-ignored--kashmir-flood-was-predicted-in-2010-032531732.htmlAccessed 10 Nov 2014
[11] Wikipedia https://en.wikipedia.org/wiki/2015_South_Indian_floods Accessed 5 Feb 2017
[12] http://indianexpress.com/article/india/india-news-india/chennai-floods-rains-jayalalithaa-imd-reasons-rescue-news-updates/ Accessed 5 Feb 2017

IOP Conf. Series: Earth and Environmental Science **80** (2017) 012055 doi:10.1088/1755-1315/80/1/012055

Production of Biogas from wastes Blended with CowDung for Electricity generation-A Case study

D Muthu*, C Venkatasubramanian, K Ramakrishnan, Jaladanki Sasidhar

School of Civil Engineering, SASTRA University, Thanjavur-613401, India

*Email: dmuthu@civil.sastra.edu

Abstract. The country's production of solid waste generation is piling up year after year and the generation of Bio-Gas finds a fruitful solution to overcome this problem. This technology can contribute to energy conservation if the economic viability and social acceptance of this technology are favorable. Our campus has a number of hostel buildings which generates large quantum of kitchen waste and sewage per day. This research will have process ofcarrying out survey, characterization of kitchen waste from several kitchens & Canteens and knowing the potential for biogas production. The waste generated from kitchen and sewage from the hostels is given as feedstock to produce 600 m3 of biogas per day with cow dung as byproduct. The methane gas generated from Biogas is purified and this is used for power generation. Two biogas engine generators of 30 kVA and 50 kVA were installed. This power is used for backup power for girl's hostel lighting load. From this study it is concluded that the generation of Biogas production and its usage for power production is the best option to handle these large quantum of sewage, kitchen waste generated from various buildings and also treated effluent from biogas plant and the biomass generated is a wealth for doing agriculture for any community ultimately it protects the environment.

1. Introduction

In further generations there will be demand for excessive energy because of fast depletion of fossil fuel deposits, fuel quality, price which is to be paid for basic materials plus their transportation cost and all the above have degradation of environment which is caused by the use of conventional energy sources. The solution lies by the use of non-conventional energy (or) renewable resources.

Renewable sources of energy can be defined as the sources of energy which are capable of being produced continuously. The renewable sources are inexhaustible and clean as they are primary sources from sun, biomass and wind, etc. Hence appropriate new technologies and policies are to be propagated, promoted, and adopted for the purpose of optimizing the use of available energy resources. Some of the sources area) Tidal energy b)Wind energy c) Bio mass d)Solar energy e) Geo thermal energy.

Biogas (or) Biomass is defined as the gas produced by the process of breakdown of organic matter in the absence of oxygen. Organic waste includesanimal material, dead plant, kitchen waste and animal feces which can be converted into a fuel called biogas. Biogas generally originates from biogenic material which is type of biofuel.

In the process of production of biogas anaerobic digestion takes place with anaerobic bacteria (or) fermentation of biodegradable materials in the absence of oxygen. It primarily consists of carbon dioxide methane and small amounts of hydrogen sulphide, siloxanes,and moistures. The gases methane,carbon monoxideand hydrogen can be oxidized with oxygen. This energy which is released allows biogas to be used as a fuel, purpose of heating and also can be used in a gas engine for the purpose of converting the energyinto electricity and heat.

The major problem is the water scarcity so recycling the waste water and effective treatment and utilization of the treated water will help in various areas like agriculture and household purposes. As the treated effluent contains potassium and phosphate reduces the quantity of manure and it increases the yield of the crops.

2. Research Objectives and Scope

The primary objective of this paper is to find the amount of electricity produced, its usage and the subsequent amount of cost reduction, thereby finding the rate of electricity produced per unit load. Global warming-increasing C02 emissions are the main cause of increased sea level and subsequent global warming. Rising prices of fossil fuels-including crude oil prices thus impacting the increase in rate of petrol and diesel and by using biofuel it will have positive impact on our economy and low cost of operation. The study on each crops-regarding the type of irrigation, amount of manure required, life period etc., study on soil-type of soil, fertility index of the soil. Checking the parameters of treated effluent water and comparing it with FAO irrigation water quality standards. Suggesting the suitable type of crop for the soil when treated effluent water is supplied, to get the maximum yield.

3. Research Methodology

This research applies a case-study approach with an in-depth statistical analyses of data obtained from a project. The research methodology includes

1) Properties of biogas.
2) Digestion tank.
3) Generation of electricity.
4) Composition of effluent treated water characteristics and its impacts if used directly.
5) Study on soil sample. Methods adopted two soil samples are collected from the field
    a) Soil sample with the application of TEW once a day
    b) Existing dry soil sample
6) Suggestions
7) Results and discussions
8) Conclusions, advantages and disadvantages of biogas and treated effluent.

4. Properties of Biogas

The production of biogas is done by the anaerobic digestion biodegradable materials like manure, green waste,sewage, plant material,municipal waste and crops. Biogas mainly consists ofcarbon dioxide methane and small amounts of hydrogen sulphide, siloxanes,and moistures.

The general composition of Biogas is
Methane gas- 55-70%
Carbon dioxide gas- 30-45%
Hydrogen Sulphide- 2-8%

The gases methane, carbon monoxide and hydrogen can be oxidized with oxygen. This energy which is released allows biogas to be used as a fuel, purpose of heating and also can be used in a gas engine for the purpose of converting the energy into electricity and heat. Biogas can be used as natural gas by compressing and the generated gas is used to power motor vehicles. Biogas is one of the renewable energy source, so it can be used for renewable energy subsidies in some parts of the world. Cleaned biogascan be upgraded to natural gas when it is changed to bio methane. Biogas is normally be produced as landfill gas (LFG) or digested gas. A biogas plant is the often named as anaerobic digester because it treats farm wastes. Anaerobic digesters are used for producing biogas.

4.1 Digestion tank

In digestion tank a series of processes takes place in which microorganisms are used for break down of biodegradable material in the absence of oxygen. It is used for managing wastes in both industrial and domestic purposes to release energy. In industries to produce food and drink products, and also home fermentation mainly anaerobic digestion is used. Anaerobic digestion produces silage.

The process of digestion begins with bacterial hydrolysis of the influent materials to break down insoluble organic polymers like carbohydrates, and it will make them available for other bacteria. Then Acidogenic bacteria converts the amino acids and sugars intohydrogen, ammonia, carbon dioxide and organic acids. Then Acetogenic bacteria converts resulting organic acids into acetic acid, and also additional amounts of ammonia, hydrogen, carbon dioxide. In the last stage methanogens convert these available products to carbon dioxideand methane. The methanogenicarchaea populations plays an important role in anaerobic wastewater treatments.

Anaerobic digestion is the process widely used as a source of renewable energy process which produces a biogas consisting of methane, carbon dioxide and traces of other gases. This gases availablecan be used directly to replace fossil fuels.

Anaerobic digestion consists of four key biological and chemical processes:

1. hydrolysis
2. acidogenesis
3. acetogenesis
4. methanogenesis

Hydrolysis:

It is the process of breaking the chains and dissolving the small molecules into a solution. Therefore, the first step in anaerobic digestion is the hydrolysis of these high molecular Weight polymeric components. In hydrolysis the complex organic molecules are broken down into amino acids, simple sugars, and fatty acids.

Acidogenesis:

It is a biological process which results in further breakdown of the remaining components bybacteria called acidogenic. Here, VFAs are created, along with carbon dioxide, ammonia, and hydrogen sulphide, and also some other by products. The acidogenesisprocess is similar to the way milk sours.

Acetogenesis:

In anaerobic digestion the third stage is acetogenesis. In this process simple molecules created in the acidogenesis phase are further digested by acetogens for producing large amount of acetic acid, hydrogenandcarbon dioxide.

IOP Conf. Series: Earth and Environmental Science **80** (2017) 012055 doi:10.1088/1755-1315/80/1/012055

Methanogenesis:
In anaerobic digestion the terminal stage is the biological process of methanogenesis. In this process methanogens uses intermediate products from the previous stages and convert them into water, methane and carbon dioxide. In this process majority of the biogas is emitted.
For this process PH plays an important role it should be between PH 6.5 and PH 8.
A simplified chemical equation for the overall processes of anaerobic digestion is as follows:

$$C_6 H_{12}O_6 = 3CO_2 + 3CH_4$$

The standardization for the volume of dry biogas is done via a simple mathematic algorithm. The volume of water vapor will be the function of biogas temperature correction of measured gas volume for both water vapor content and thermal expansion.

Biogas contains siloxanes formed from anaerobic decomposition of materials which are commonly found in detergents and soaps. During the process of decomposition of materials containing siloxanes silicon is released which combines with free oxygen and also some other elements in the combustion gas. In this process deposits are formed containing mostly silica, silicates and small amounts of calcium, zinc, phosphorous.

As mentioned earlier, biogas consists of 55-75% methane gas, 30-45% carbon dioxide gas, 2- 8% water vapour and traces of hydrogen. Comparing this with natural gas, which contains 75to 90% methane gas. The energy content of the energy source mainly depends on the content of methane gas. High methane content would be more desirable. A presence of carbon dioxide and water vapour content is unavoidable, but content of sulphur must be minimized particularly when used in engines. The average calorific value of biogas is about 22-24.5 MJ/m3, so that 1 m3 of biogas corresponds to 0.55-0.65diesel fuel which is about 6 kWh.

5. Generation of electricity

In most of the cases in combustion engines biogas can be used as a fuel, in which mechanical energy is converted into electrical energy using electric generator to produce electricity. Appropriate electric generators are available in all countries with different outputs and sizes. This technology is well known and maintenance for this type of engines is simple. In most case 3-phase electric motors can also be used for generation of electricity.

Far more challenging is the first stage of the generator set that is the combustion engine using the biogas as fuel. In all types of combustion engines biogas can be used as fuel. Combustion engine includesdiesel engines,gas engines (Otto motor) and Sterling motors etc. Gas turbines are mostly used as biogas engines, especially in the US because they are very small and can have strict exhaust emissions requirements. Small biogas turbines with outputs of 35-75 KW are available in the market. But this can be used for small-scale applications in developing countries because of Speeds and the high operating temperature they are expensive. Further manufacturing of gas turbines is challenging and maintenance require specific skillsdue to spinning at very high speeds and the high operating temperatures. Sterling motors have the advantage of being tolerant of fuel composition and quality because they are External combustion engines and they are low efficient. Due to the low efficiency used for small applications.
Internal combustion motors have become the standard technology either as gas or diesel motors in most commercially run biogas power plants.

6. Appropriate combustion engines

Diesel engines operate with biogas only in dual fuel mode. For the process of facilitating the ignition of the biogas, a small amount of ignition gas should be used often diesel fuel is injected together with the biogas. Modern pilot injection gas engines ("Zundstrahlmotoren") need about 2-3% additional ignition oil. Almost diesel engine can be converted into a pilot injection gas engine. The advantage is that motors running in dual fuel mode can use gas of low heating value, in such cases they consume a considerable amount of diesel. The engine sizes of around 200kW, pilot injection engines seem will have more advantage over gas motors due to their slighter higher efficiency (4-5% higher) and lower investment costs.

Gas motors with spark ignition (Otto system) can be operated only with biogas. In practice, a little amount of petrol (gasoline) is often be used to start the engine. Normally this technology will be used for very small generator sets (0.5-10 kW) as well as for large power plants. Especially in Germany, these engines are more advantageous as these do not require additional fossil fuels.
For use in gas or diesel engines, the gas must have certain requirements:

- The methane content plays an important role and it should be as high as possible as this is the main combustible part of the gas;
- The carbon dioxide and water vapour content should be kept low as possible, mainly because they causes low calorific value for the gas;
- The sulphur content mainly in form of Hydrogen sulphide, must be low, as it is converted to corrosion causing acids by condensation and combustion. The water vapor content can be reduced in the process by condensation in the gas storage or on the way to the engine. The reduction of the hydrogen sulphidecontent in the biogas can be done by various technical methods. These can be classified as biological, chemical and physical and divided into internal and external methods. In the last two decades experimentation has been carried,however, as complete elimination is unnecessary for use in robust engines.The following simple methods are generally established themselves as standard:
- To produce gas of homogeneous quality an optimized steady fermentation process with continuous availability of appropriate feedstock or influent is important.
- The injection of a small amount of oxygen (air) into the storage fermenter headspace leads to oxidation of Hydrogen sulphide by microorganisms and hence the elimination sulphur from the gaseous phase can be done. This is the most commonly used method for desulphurization. It is cheap and can eliminate up to 90-95% of the sulphur content in the biogas. However, the challenging is the right proportioning of air.
- Another option is filtration that is external chemical treatment. The active material may be:
  - Activated carbon: Certain companies provide activated carbon filters as a component in their gensets.
  - Iron-hydroxide: In this process filter can be regenerated by adding oxygen and it is reversible Adsorption material is iron rich soils, waste material from aluminumorsteel production;

7. Composition of effluent treated water characteristics and its impacts if used directly

The composition of sewage water various depending upon the source, collection type and treatment provided. Large proportions of these sewage waters is organic in nature and it contains essential plant nutrients. The sewage water contains more than 90% water. The solid portion contains25-40% inert materials, 35-50% organics, 10-15% bio-resistant organics and 5-8% miscellaneous substances on oven dry weight basis. The chemical composition of sewage waters is tested and it has found many in pH,suspended solids, electrical conductivity, organic C, CO, Mg++, Ca++ and other essential and toxic elements in the sewage water. The pH of the sewage water ranged from 7.2 to 8.3 and it was within normal range and irrigation standards. These waters doesn't have any significant change in the soil pH due to high buffering capacity of the soils. The EC of sewage waters collected from various villages varied from 0.1 to 0.4 ds ml and their continuous use in the agricultural fields may have a significant increase in salinity of the soils and ultimately restricting the plant growth. The BOD values vary from 50 to 135mg/l. BOD value above 100mg/1 is actually not suitable for irrigation, thus we have to reduce it below 100mg/1 by sedimentation and filtration process. Apart from these properties, the sewage waters contains amount of N, P and K. The amounts of N, P and K in sewage waters ranges from 9 to 105, 4.3 to 54 and 20 to 2600 mg/L respectively. Moreover due to several factors the composition of sewage water is not constant and changes within the year. The extent of the contaminants will be less during the rainy seasons.so on consolidation the waste water has the following constituents

- Nutrients (Nitrogen, Phosphorus, Potassium)
- Organic matter
- Inorganic matter (dissolved minerals)
- Pathogens
- Toxic chemicals

The wastewater constituents, parameters, and possible impacts are given explained below Plant food nutrients like N, P, K etc. creates following impact
-Excess N: It has a potential to cause nitrogen injury, delayed growing season,excessive vegetative growth and maturity, and potential to cause economic loss to farmer
-Excessive amounts of N and P will cause excessive growth of undesirable species (eutrophication).
-Nitrogen leaching will cause groundwater pollution with adverse effect to health and environmental impacts.
-Lack of nitrogen will causeYellowing of younger leaves (attains a state of cholrotic) poor plant growth.
-Lack of phosphorus will cause change in colour of leaves, stunted growth,small sized fruit, purple veins in the leaves of the plant and reduced blooms of fruits.
Biodegradable organics parameters like BOD, COD causes depletion of dissolved oxygen in surface water - development of septic conditions unsuitable for the workers to work in brings very bad odour.
Hydrogen ion pH concentration causes - possible adverse impact on plant growth due to acidity or alkalinity impact sometimes beneficial on soil flora and fauna.

8. Impacts

8.1. Agriculture

Some of the potential impacts of wastewater use in agriculture in discussed below. If the waste water is directly used for agriculture it will have some pathogenic microorganisms such as bacteria, viruses and parasites, which are responsible to cause diseases. Human parasites like protozoa and helminthes eggs will have special importance in this regard as it is more difficult to remove if not treated properly and will cause a large number of infectious gastrointestinal diseases. However, in the process of evaluating health impacts it must be remembered that the actual risk that make people fall ill must be quantified and not the whether the pathogens are present in water.

8.2. Crops

Wastewater treated and untreated is extensively used in agriculture because it consistsof nutrients and provides the moisture necessary for crop growth. By the use of wastewater irrigation crops gives more yields which reduces the use for chemical fertilizers providing net cost savings to farmers. If the total

nitrogen delivered to the crop through wastewater irrigation is more than the recommended nitrogen dose for optimal yields, it may but delay ripening and maturity,stimulate vegetative growth and in extreme circumstances, cause yield losses. Scientists have made attempts to quantify the effects of treated and untreated Wastewater on a number of qualities and parameters.
Wastewater can be used for producing better quality crops with higher yield.
Following benefits can be obtained by Proper agronomic and water management practices:
1) Value of fertilizer saved.
2) Higher yields
3) Additional water for irrigation, and
Alternatively, yields may negatively be affectedif plant food nutrients delivered through wastewater irrigation may result in nutrient over supply.

8.3. Groundwater Resources
The quality of groundwater resources are affected by the wastewater application by the excess salts and nutrients which are found in wastewater leaching below the roots of plants. The impact on the groundwater depends on various factors which includes water table depth, groundwater quality, drainage of soil and quality of wastewater irrigation. The quality of groundwater can be determined by the impact and magnitude from leaching of nitrates. The brackish groundwater indicates the leaching of nitrates would be of little concern. The wastewater also effect on potable water supplies such as tube wells and also the evaluation is done based on the groundwater pollution. The above are to be considered before using the treated effluent water.

9. RESULTS

- TEW improves the soil fertility by reducing the salinity content of the soil
- It increases the potassium and phosphate content of the soil thus reduces the quantity of manure used.
- Nitrogen content of the soil though slightly reduced is still within optimum limits.
- Farmers spend nearly 5 lakh for installation and maintenance of bore wells which seems to be really difficult for some peasants.
- So the use of TEW minimizes their expenses to some extent.

The major disadvantage is the bad odour of the treated waste water, which could be reduced by the following methods:

- Tanks must be rinsed properly and it must be cleaned from any stagnant solids at the bottom so as to prevent the growth of bacteria.
- Sulphate in the water, or a carryover of sulphur from oils, can act as a source for bacteria growth producing a sulphideodour.
- When high total suspended solids occur in the treated waste water, bacteria grow and give off the hydrogen gas odour.
- This odour creates uncomfortable environment for the workers to work in.

Odour can be controlled by one of the following:
1) Neutralization of the offending odour.
2) Masking of the odour.
3) Switching the oxygen source that contributes to the growth of bacteria.

10. Recommendations

AQUA PUCKS:

- Aqua pucks are used to reduce odours caused by anaerobic conditions.
- These contain oxygenating technology with bio-degradable bacteria to eliminate noxious odours.
- Aqua pucks also breakdown fats and oils. The action generated from aqua pucks purges water from odours and gases.

11. Further Conclusions

By using biogas we have economic, agronomic, environmental advantages:

Economic advantages include:

- Income can be increased.
- Increase in the cost of fossil fuels which can be reduced by use biogas
- Manure cost can be reduced by use of effluent outlet as a manure.

Agronomic advantages include:

- Liquid manure can be transformed into manure and which can be used as a fertilizer.
- Organic waste processing for competitive prices.

Environmental advantages include:

- Sustainable management of organic waste.
- Reduction of pollution due to nitrogen stripping.
- Biogas is a renewable energy which replaces fossil fuels.

Waste disposal, improper sewerage and drainage are the major environmental concern because of high levels of water pollution in an urbanizing India. The sewage contains essential nutrients and possesses properties which are easily utilized for the purpose of irrigating the field crops. This project highlights the present and future water supply demand.

The large amount of wastewater collected in the effluent treatment plant will be diverted into nearby villages for their agricultural purposes after the treatment is completed. Remedial measures need to be taken for improving the wastewater reuse pattern.

References

[1] Ghimire P C 2013 SNV supported domestic biogas programmes in Asia and Africa. *Renewable Energ.* **49** 90-4.
[2] Deublein D and Steinhauser A 2010 Biogas from waste and renewable resources. 2nd Edition. *Wiley.*
[3] Abatzoglou N and Boivin S 2008 A Review of biogas purification processes. *Biofuels, Bioproducts and Biorefining* **3**(1) 42-71.doi:10.1002/bbb.117.ISSN1932-104X
[4] Huertas J I Giraldo N and Izquierdo S 2011 Removal of H_2S and $C0_2$from Biogas by amine absorption. *Mass Transfer in Chem. Eng. Process.* **7** 133-50.
[5] Overview of greenhouse gases, Methane emissions. Climate change, United States environmental protection agency, 11 December 2015
[6] Biomass Energy: Manure for fuel 2009 State energy conservation office. State of Texas.
[7] Hanjra M A 2012 Wastewater irrigation and environmental health: Implications for water governance and public policy. *Int. J. Hyg. Environ. Health* **215**(3) 255-69.

IOP Conf. Series: Earth and Environmental Science **80** (2017) 012056 doi:10.1088/1755-1315/80/1/012056

Influence of Random Inclusion of Coconut Fibres on the Short term Strength of Highly Compressible Clay

Evangelin Ramani Sujatha*, S SaiSree, C Prabalini and Z Aysha Farsana

School of Civil Engineering, SASTRA University, Thanjavur 613401. India

*Email: r.evangelin@gmail.com

Abstract. The choice of natural fibres for soil stabilization provides an economic, safe and eco-friendly alternative to improve the properties of soil. They are an important step forward toward sustainable development. An attempt was made to study the influence of the random addition of untreated coconut fibres on the short term strength of soil, its stress-strain behavior, compaction characteristics and index properties. The soil selected for the study is a highly compressible clay sample with a liquid limit of 52.5 % and plasticity index of 38 %. The soil has no organic content. The study reveals that the compaction curves tend to shift to the right side, indicating more plastic behavior with the addition of fibres. The addition of fibres also re-orient the soil structure to a more dispersed fashion. A significant increase in the unconfined compressive strength is also observed. An increase of nearly 51 % in the unconfined compressive strength is observed at 0.75 % coir inclusion. The stress-strain behavior of the soil shows a shift toward more plastic behavior. The mode of failure of the soil specimen is by cracking and with fibre inclusion, length of the failure cracks is restrained as the fibre tends to hold the cracks together, resulting in shorter cracks, with significant bulging of the specimen at failure.

1. Introduction

Reinforcing soil with tensile elements is a widely used successful ground improvement technique. Traditional soil reinforcements include rods, strips, sheets, membranes, etc. The use of natural fibres like jute, bamboo, coir, etc. is also widely prevalent in countries like Sri Lanka, Philippines and India, where these fibres are abundantly available. The use of these fibres as reinforcement does not only provide an economic option but is also a major step forward in the direction of sustainable development [1]. It does not cause environmental degradation and also provides opportunities of employment in the local community. Studies show that the strength of fiber reinforced soil increases with increase in aspect ratio, fiber content, fiber modulus; and soil fiber surface friction. The reason for the increase in compressive strength of specimens reinforced with different proportions of fibers as compared with compressive strength of specimens of unreinforced soil is attributed to better adhesion of soil matrix with fibers [1-4].

Of all the natural fibres available, coir is abundantly available and has a high tearing strength[2]. Rao and Balan [5] report that coir contains 54 % cellulose and 40 % lignin. The high lignin content makes it a useful reinforcement in geotechnical applications like enhancing stability of embankment and erosion control. It is very effective in improving the short-term stability and deformation behaviour [2, 5] and has greater applicability in temporary structures. Phenol and bitumen coating can improve the durability of these fibre [2].

Case studies on the use of coir to reinforce bunds [6] demonstrate that it is cost effective than stone pitching, strengthen pavements [7] and other ground improvement techniques. Strength behaviour of coir reinforced was studied by authors like Babu and Vasudevan [2], Chauhan et al. [8], and Arif and Singh [4]but a comprehensive study on its compressibility, post peak behaviour and failure pattern has not been reported in detail. This paper reports the experimental results of coir reinforced soil, elaborating of the effects of coir on the compressibility, strength and failure. A number of tests on consistency limits, standard compaction tests and unconfined compression tests were performed with different percentages of coir (0.25, 0.5, 0.75, 1.0 and 1.25) to study the effect of coir reinforcement on improving the geotechnical properties of the soil.

2. Materials

2.1.Soil

The soil used for the study was extracted from Angarai village in Thiruchirapalli District, India. Sample was retrieved at a depth of 2 m below the ground level from trenches made for this purpose. The soil is dark brown in colour with a characteristic pungent odour. The soil has no organic content. The specific gravity of the soil is 2.00. The geotechnical properties of the soil are listed in Table 1. The toughness index of the soil (Table 1) indicates that the soil does not have adequate resistance to deformation.

Table 1. Geotechnical Properties of the Soil

DESCRIPTION	VALUE
Specific gravity,Gs	2.00
Effective size,D_{10} (mm)	0.0012
Uniformity co-efficient,C_u	11.67
Co-efficient of curvature,C_c	0.134
Liquid limit,w_L	43
Plastic limit,w_p	26.83
Flow Index,I_f	50.8
Toughness Index,I_t	0.32
Maximum Dry density,γ_d (kN/m^3)	19.12
Optimum Moisture Content	22%
Compression Index,C_c	0.295

The particle size distribution curve of the soil is shown in figure 1. The soil falls under inorganic clay (CI) according to Indian Standard Soil Classification systems and low compressible clay according to USCS. The unconfined compressive strength of the soil indicates that it is a stiff clay.

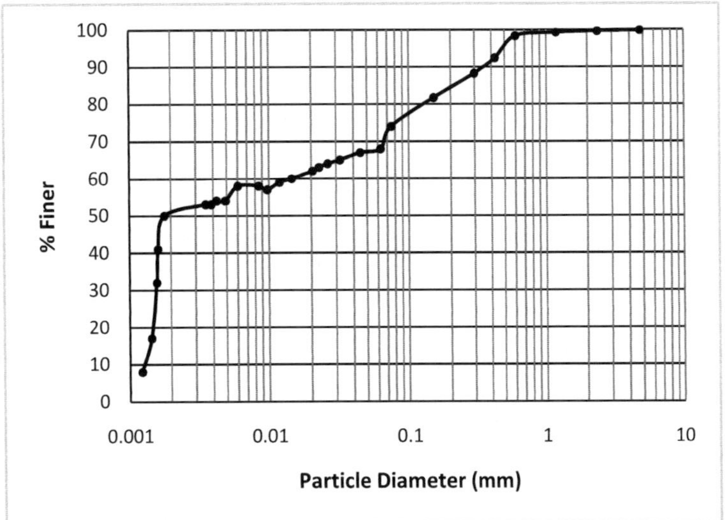

Figure 1. Gradation Curve of the Soil Sample

2.2Coir

Coconut coir was obtained from Thiruchirapalli, Tamilnadu.It was cleaned and cut to length of 15 mm. The coir fibres are elastic with high initial strength and durability. The average diameter of the coir fibres used is 0.01 mm. The percentage of water absorption of the coir fibres was 54 %.

3. Sample preparation and testing

Soil from the borrow area was air-dried, pulverized, sieved and mixed thoroughly with coir at the required water content. Percentage of coir used for the study are 0.25 %, 0.5 %, 0.75 % and 1 %. The soil-coir fibre mix was kept in a container for the mix to reach equilibrium moisture content. This soil was used for determining the consistency limit, optimum moisture content, maximum dry density and unconfined compressive strength. The tests on unreinforced and reinforced soil was carried out in accordance to IS: 2720-2006 [9]. The unconfined compressive strength was tested at optimum moisture content.

4. Results and Discussions

Results of the experimental investigations carried out to study the effects of coir reinforcement on compaction characteristics, unconfined compressive strength and stress-strain behaviour is presented.

4.1Compaction Characteristics

Light compaction tests were carried out to on unreinforced and reinforced soil (at selected fibre percentages) to determine the maximum dry density (MDD) and optimum water content (OMC). On fibre inclusion, compaction curves tend to shift right at all percentages of fibres tested. Random inclusion of coir fibres cause an increase in OMC and decrease in MDD. Fibres have a tendency to absorb moisture and this can be attributed to the increase in OMC on random fibre inclusion [10-12]. Fibres have less unit weight. Replacing fibres in the soil matrix because a decrease in weight occupied in the standard volume, causing a decrease in the maximum dry density [12-13]. The change in OMCand MDD is not very appreciable (Fig. 2) for all percentages of fibre studies.

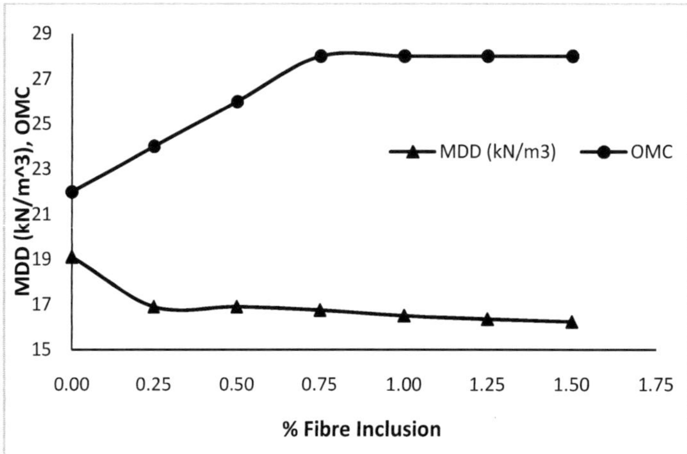

Figure 2. Effect of Coconut Coir Inclusion on the OMC and MDD

4.2Stress Strain Behaviour

Figure 3 depicts the stress-strain behaviour of unreinforced and coconut coir reinforced soil at various percentages of fibre inclusion. It can be observed that fibre reinforcement increases the plastic nature of the soil. Inspite of the increase in shear strength of the reinforced soil, the stress-strain behaviour remains the same. A decrease in the rate of loss shear strength post failure is observed (Fig. 3). This behaviour can be noted for all percentages of fibre inclusion. Figure 3 also shows that the decrease in the rate of loss of shear strength increases with fibre content (i.e) higher the fibre content, more plastic the fibre-soil matrix is. At lower strain rates, the behaviour of the unreinforced and reinforced soil is markedly similar. The study also shows that fibre reinforced soil takes longer to fail. Energy absorption capacity, which reveals the toughness of the reinforced soil matrix, also increase with the increase in fibre content (Fig. 3). The strain at failure also increases with the increase in fibre content. Again, this is an indicator of increase in plastic behaviour of the fibre reinforced soil.

4.3Unconfined Compressive Strength

The unconfined compressive strength of the soil reinforced with coconut coir shows a marked change. The unconfined compressive strength is defined as the maximum load per unit area or the load per unit area at 10 % axial strain, whichever occurs first during the performance of a test [13]. Random inclusion of coconut coir results in the increase of the unconfined compressive strength of the soil upto 0.75 %. At 0.75 % of fibre inclusion, there is an increase of nearly 60 % in the unconfined compressive strength. The presence of fibres in the soil matrix increase the interaction between the soil and the fibre, thereby increasing the strength. But, this is true only upto a certain percentage of fibre inclusion (in this case, 0.75 %). Beyond this fibre content, fibre clumping takes place and this results in the decrease of shear strength. The improvement of the unconfined compressive strength of the soil is depicted as a shear strength improvement index (SII) in Figure 4. It can be observed that fibre inclusion is effective at percentages of 0.25, 0.50 and 0.75 and 0.75 % is the optimum.

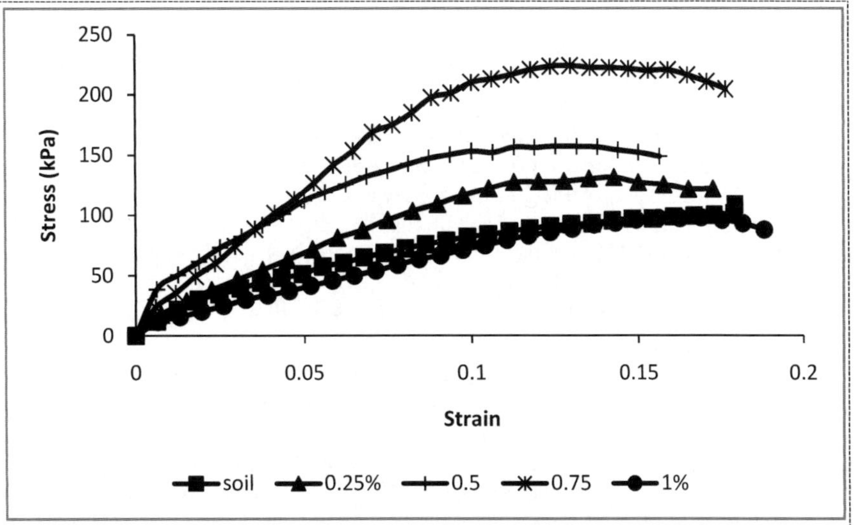

Figure 3. Stress-Strain Behaviour of Coconut Fibre Reinforced Soil

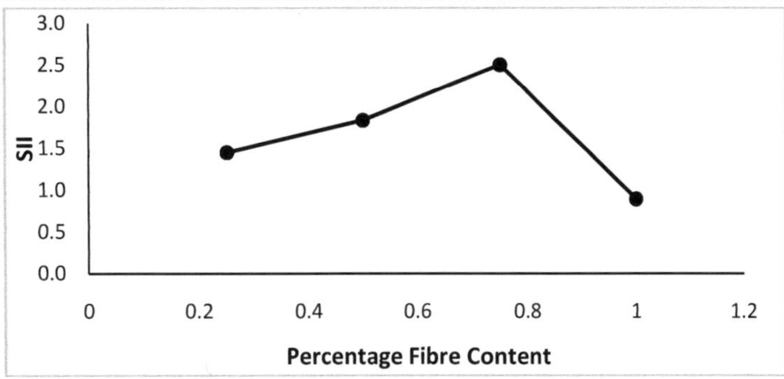

Figure 4. Shear Strength Improvement Factor

5. Conclusion

Results of the experimental investigation illustrate that random inclusion of coconut coir improves the unconfined compressive strength and stiffness of the soil. The optimum dosage of coconut coir inclusion is 0.75 %. Random inclusion of fibres improve the energy absorption capacity, inhibit formation of long and wide cracks and considerably improve the geotechnical performance of the soil. Though coir is biodegradable, it has high durability and is resistant to deterioration. Hence, coir reinforcement can be suitably used for improving the strength of temporary structures.

References

[1] Sayyed Mahdi Hejaz, Mohammad Sheikhzadeh, Sayyed Mahdi Abtahi and Ali Zadhoush 2012 A simple review of soil reinforcement by using natural and synthetic fibers. *Constr. Build. Mater.* **30** 100–16.

[2] SivakumarBabu G L and Vasudevan A K 2008 Strength and Stiffness Response of Coir fiber Reinforced Tropical Soil, *J. Mat. Civil Eng.* ASCE 571-8

[3] Millogo Y, Morel J C, Aubert J E and Ghavami K 2014 Experimental analysis of pressed adobe blocks reinforced with Hibiscus cannabinus fibers. *Constr. Build. Mater.* **52** 71–78.

[4] Singh S and Arif S 2014 Inclusion of coconut coir fiber in soil mixed with coal ash. *Int. J. Res. Eng. Technol.* **3**(11) 209-13

[5] Venkatappa Rao G and Balan K 2000 Coir Geotextiles, Emerging Trends. *Kerala state coir corporation, Alappuzha*, pp. 179-88.

[6] Lekha K R 2004 Field instrumentation and monitoring of soil erosion in coir geotextile stabilized slopes – A case study. *Geotext. Geomembr.* **22** 399-413

[7] Mittal Shelly and Singh R R 2014 Improvement of local subgrade soil for road construction by the use of coconut coir fiber. *Int. J. Res. Eng. Technol.* **3**(5) 707-11.

[8] Mahipal Singh Chauhan and Satyendra Mittal 2011 California Bearing Ratio of randomly oriented fiber reinforced clayey subgrade for rural roads. *Proc. Indian Geotechnical Conf.*

[9] IS: 2720-2006 Compendium of Soil Testing. *Bureau of Indian Standards*. New Delhi.

[10] Kumar A, Walia B S and Mohan J 2006 Compressive strength of fibre reinforced highly compressible clay. *Constr. Build. Mater.* **20**, 1063-8.

[11] Aqeel Al Adili, Rafig Azzam, Giovanni Spagnoli and Joerg Schrader 2012 Strength of soil reinforced with fiber materials (papyrus). *Soil Mech. Found. Eng.*, **48**(6) 241-7.

[12] Suchit Kumar Patel and Baleshwar Singh 2014 Unconfined Compressive Strength Behaviour of Fibre-Reinforced Lateritic Soil. *J. Civil Eng. Environ. Technol.* **1**(4): 93–8.

[13] Mona Malekzadeh and Huriye Bilsel 2012 Effect of Polypropylene Fiber on Mechanical Behaviour of Expansive Soils. *EJGE* **17**(A) 55-63.

IOP Conf. Series: Earth and Environmental Science **80** (2017) 012057 doi:10.1088/1755-1315/80/1/012057

Hazardous E-waste and its impact on soil structure

K Dharini*, J Bernadette Cynthia, B Kamalambikai, J P Arul Sudar Celestina and D Muthu

School of Civil Engineering, SASTRA University, Thanjavur -613401. India.

*E mail:dharinikanagaraj@gmail.com

Abstract. E-waste disposal has been a significant issue over the past few decades with the development of technology and the plethora of electronic products produced. The inclusive term E-Waste encapsulates various forms of electrical and electronical equipment which provides no value to the current owners and it is one among the fastest growing waste streams. E-Waste is a complex, non-biodegradable waste which is generally dumped in mountain like heaps. These wastes are said to have a large quantities of lead, cadmium, arsenic etc.it is mandatory to dispose such scrupulously since they have the ability to affect the soil and water parameters. Solid waste management is a blooming field which strives to reduce the accumulation of used electronic gadgets. Rainwater gets infiltrated through the e-waste landfill and it leaches through the soil which in turn reaches the groundwater directly thereby affecting the water intended for drinking and domestic purposes. This study focuses on the consequences of toxic waste by comparing the difference in properties of the soil structure prior to and after the e-waste landfill at various concentrations.

1. Introduction

Every year 2.7 million tonnes of e-waste are produced in India and disposal becomes a huge problem. India becomes the fifth largest producer of e-waste. Generally treating and processing e-waste is an expensive task and space becomes a huge challenge. The problem is convoluted by the toxicity of the waste that is being handled by workers and dumped unsafely in municipal yards. That informal sector has been handling such waste are out of capital and there is not much support from the government.
Lead and Mercury takes a long time to decay and the people who gets exposed to it will have long term irreversible health effects. Most of the e-waste is dumped in the landfill and metals like lead serve to harm the nature of soil. Due to the scarcity of lands there are limited chances of getting fresh dump yards hence the available yards are overflowing with such wastes. The chemicals produced are non-biodegradable and they persist in the environment for prolonged time periods thus increasing exposure risks.

Burning of e-waste on an open landfill for obtaining gold and other precious metals produce fine particulate matter from the smoke produced and cause cardio-vascular and pulmonary ailments for children in that specific area.The wind pattern of a particular area carry toxic particles and they enter the soil-crop-food pathway affecting both humans and animals as they enter the food chain.The mother boards have abnormal level of mercury and improper disposal may cause skin and respiratory diseases.Drinking water contaminated with lead affects the central and nervous system and causes poor brain growth, dwarfism, hearing disability and impaired formation and function of blood cells.The

IOP Conf. Series: Earth and Environmental Science **80** (2017) 012057 doi:10.1088/1755-1315/80/1/012057

exhaust produced as a result of combustion of mother circuit board causes lung cancer when inhaled. It was also responsible for dermatological diseases. [1] [4]

2. Experimental Investigation

Procedure adopted for soil: In order to identify the change in parameters for the variations in the concentration of lead, three pits of similar soil composition were dug with equal dimensions of 2 x 2 x 1.5 feet. Solder was melted and shredded into pieces for mixing with the soil.

- Pit -1 was mixed with 50g of un-shredded lead and 100g of PCB and the pit was closed and compacted.
- Pit -2 was mixed with 10g of shredded lead and was closed and compacted.
- Pit 3 was mixed with150g of shredded lead and was closed and compacted.

An original soil sample i.e the sample without any mixture of lead was excavated and sent for testing its parameters. The parameters of the soil mixed with lead to that of the original soil is later analysed. Pits were wateredwith four litres of water, twice a day, continuously for a month to analyse the change in the soil parameters.Mustard seeds were sprinkled above all the three pits to check if they could grow on the ground containing toxic substances. Acknowledging the concept of ground water leachate, three pits parallel to the original pits with a distance of 1m were dug out to procure a soil sample at a depth of 2 feet. These samples were again sent for testing its parameters.

Table 1. Observation table for soil parameters

Parameters	Original Sample(kg/acre)	Pit 1 (50g of lead+ 100g of PCB) (kg/acre)	Pit 2 (100g of lead) (kg/acre)	Pit 3 (150 g of lead) (kg/acre)
Calcium carbonate	Medium	High	High	High
Phosphrous	255	155	100	155
Nitrogen	72.4	66	50	35.6
Potassium	245	60	115	75
pH	8.1	9.3	9.3	9.3
Electrical conductivity	0.21	1.2	1.4	0.3

Procedure adopted for water: To find the change in parameters of the water sample containing toxic substances such as lead, mercury, tin etc. a series of tests were conducted. A setup was made where a separate sample mixed with lead was placed in a bucket which was provided with multiple holes of 1cm diameter at the bottom. The soil was watered regularly for a month and the water was let to percolate through the soil and collected down below through the perforated holes. Thus the water obtained was sent for testing to compare the change in parameters to that of the parameters of the water which doesn't contain lead. Tests such as turbidity, chlorides test, total dissolve solids, pH were conducted on the samples obtained.

Table 2. Observation table for water parameters

Description	Normal water	Lead water
pH	8.3	10.2
TDS(ppm)	300	1400
Hardness(ppm)	140	260
Chlorides(ppm)	187	195
Conductivity(μS/cm)	1.2	4.7

3. Results and discussion

The results obtained from the water test are presented in Table 2. It can be understood that the parameters such as pH, TDS, hardness, chlorides and conductivity were stable in normal water and these parameters were observed to have undesirable values in lead water. If the pH value is less than 7 it indicates acidity and if it is greater than 7 indicates base. The presence of toxic metal has lead to the increase in alkalinity of the lead water when compared with normal water. High amount of dissolved solids reduces the quality of water thereby affecting people with health disorders. The desirable value of total dissolved solids in normal water is 500ppm whereas the amount of TDS in lead water is 1400ppn that makes it unsafe.Water hardness may cause serious effects in industries handling water especially boilers, cooling tower etc. The desirable value of hardness in normal water should be within 300ppm and the hardness of water containing lead is found to be 260 ppm. Excessive concentration of chloride ions makes the water unpleasant to drink. The desirable value of chloride is 250ppm and we can clearly see the increase in chloride content from normal water to lead water. The desirable values of electrical conductivity in drinking water is 0.5 – 3 μS/cm. "Mineral taste" in water is a common problem due to the presence of higher dissolved solids and also causes problems with the boilers and other equipment. The conductivity value in lead water is beyond this value and hence it is undesirable.

The results obtained from the soil test conducted by National Rice Research Institute, Aaduduarai are presented in table 1. Presence of both low and high Calcium contents may cause problems.Calcium is generally required for proper functioning of plant cell walls but it may harm the plant if the Calcium content increases which leads to increase in pH.According to the test results there is high amount of calcium carbonate in the test samples affected by E-waste which has lead to the depletion of nutrients essential for plant growth. When plants undergo the process of photosynthesis to drive its metabolism,Phosphorous helps in harnessing that energy. The deficiency of its nutrient can lead to impaired growth and have abnormal dark green colour. We can clearly see the depletion of phosphorus across the table in the soil subjected to landfill. Nitrogen is very vital since it forms a major component of chlorophyll, the compound by which the plants uses sunlight as energy to produce photosynthesis.. It is also the building blocks of proteins. In our case the presence of harmful elements has led to the depletion of nitrogen which is essential for plant growth. Potassium entrants nitrogen as the nutrient absorbed by plants. High levels of potassium indicate a healthy atmosphere for the wellbeing of plants. We can see the depletion of potassium in test sample subjected to landfill. The availability of plant nutrients is completely affected by soil PH. Soil acidity or alkalinity directly affects the plants growth. , they are unavailable to plants. It is for the nutrients to disperse through the soil under alkaline conditions. Although the optimum range of pH in soil is 5.5 to 7 some plants will grow in a more acid or in alkaline level. pH is not an indication of fertility but it does affect the availability of fertilizer nutrients. The pH levels of the landfill soil are being increased. The electrical conductivity of water is the measure of salinity. Excessively high salinity can affect plants in many ways. The EC values of lead affected soil are increased.

4. Conclusion

From the observed test results it can be inferred that the soil samples subjected to harmful landfills are infertile and are made unfit for vegetation permanently. The lead content depletes the soil of its natural nutrients and makes it sterile. Flora and fauna of the area near the landfill gets affected since the landfill releases particulate matter casing air pollution. Electronics that are thrown in landfills, produce toxins that may leach into groundwater and affect local resources. The water which was made to react with lead is unsuitable for drinking due to the abnormal pH, turbidity, TDS, conductivity, chloride levels. E-waste leaches the groundwater and has its effect on the ground water table. It thereby affects the water intended for domestic and drinking purposes as even a small quantity of lead in drinking water can cause potential health hazards to children as well as adults. Improperly disposed e-waste has been responsible for water pollution, air pollution, soil pollution and a threat to humans and animals.Because of the alarming rates at which e-waste is being illegally and dangerously dealt with, the natural environment is being affected. Therefore E-waste has to properly processed, recycled and disposed for the safety of the environment.

Acknowledgement:

The authors of this paper are extremely thankful to the authorities of National Rice Research Institute, Aadudurai for conducting the soil tests which helped us to conduct this study.

References

[1] Upton A C, Kneip T and Toniolo P 1989 Public health aspects of toxic chemical disposaltes. *Annu. Rev. Public Health* **10** 1–25.
[2] Environmental hazards in city likely if electronic waste is not tackled quickly, The Hindu, May 12, 2005.
[3] Aditya Environmental Services Pvt. Ltd. (AESPL), Inventory of Hazardous Wastes in Maharashtra, sponsored by Maharashtra Pollution Control Board (MPCB).
[4] Wath S B, Dutt P S and Chakrabarti T 2011. E-Waste scenario in India, its management and implications. *Environ. Monit. Assess.* **172** 249–62.

Laboratory study on subgrade soil stabilization using RBI grade 81

J Bernadette Cynthia[1*], B Kamalambikai[1], R Prasanna Kumar[2] and K Dharini[1]

[1] School of Civil Engineering, SASTRA University, Thanjavur -613401, India
[2] Department of Civil Engineering, Malla Reddy Engineering College, Maisammaguda, Dhulapally, Secunderabad-500 100 Telangana, India

*E mail: bernadettecynthia1996@gmail.com

Abstract. The present study investigates the effect of reinforcing the sub grade soils with RBI 81 material. A soil nearby was collected and preliminary tests were conducted to classify the soil and it was found from the results that the sample collected was a poorly graded clay. Subsequently Tests such as Proctor Compaction, CBR, and UCC were conducted to study the various engineering properties of the identified soil. In addition to the above tests were also conducted on the soil by reinforcing with varying percentages of RBI 81. From the analysis of test results it was found that this material (RBI 81) will significantly improve the CBR value of the soil.

1. Introduction

Pavement construction is increasing day by day due to increase in traffic volumes. Rapid urbanization and industrialization resulted in exponential growth of vehicular population that created huge demand for more number of highways in India. AS a consequence of there is lot of pressure on the construction materials such as road aggregates, gravel etc. Many times the sub grade soils on which the highways rest are found to be weak, requiring very high thickness of pavements. In order to reduce the thickness of pavements founded on weak Sub grade soils, they are stabilized using various types of materials such as lime, cement, bitumen, fibers etc. Over the years researchers have found the application of several new materials for reinforcing with soil to improve their strength characteristics. RBI grade 81 is one such material which eliminates the need to replace soil material and suitable for almost all types of soils. In addition, it also reduces the cost profile of the entire road construction through its savings in raw materials, labour and equipment costs. There is also less amount of maintenance involved in all weather conditions and an increased life expectancy makes RBI 81 a practically feasible material for road construction. It is an eco friendly and economical pavement material and an inorganic soil stabilizer. It eliminates the need for excessive aggregate material and has a wide range of benefits as mentioned in the abstract. Experimental studies were conducted on plastic clayey soils by using Cement Kiln Dust and RBI 81. The authors came to the conclusion that that there was considerable increase in CBR and UCC values when respect to different curing periods. Stabilization done on another clayey soil using pond ash and RBI grade 81 also gave good improvement in the quality of a flexible pavement. Micro level studies such as XRD, SEM and EDS were carried out on expansive clayey soils in addition to the tests for engineering properties and observed considerable improvement in strength in relation to amount of stabilizer and also age of curing [1-3].

2. Experimental Investigations

Locally available soil was collected and various index properties of the soil were found out by conducting basic tests on soil to assist in classifying and identifying the soil. Specific gravity of the soil was determined . Dry sieve analysis was carried out in the lab under standard conditions to study the particle size distribution. The Coefficient of Uniformity C_u and coefficient of curvature C_c were determined from the plot. Consistency limits of soil were found out as per IS guidelines. Maximum dry density was with 2% of RBI 81 grade stabilizer as per IS standards and they were presented in Table – 1. Subsequently CBR test and unconfined compression test were conducted for soil alone and soil mix with same percentages of RBI grade 81 stabilizer. CBR and UCC test results were given in Table -2. [5-9]

Table 1. Index Properties of Soil

Property	Test Value
Liquid Limit	56 %
Plastic limit	18.20 %
Plasticity index	37.80
Flow Index	49.82
Toughness Index	0.80
Coefficient of Uniformity	5.70
Coefficient of Curvature	1.01
Specific Gravity	2.75

Table 2. Properties of aggregates used in the mixes.

Soil type	Optimum moisture content (%)	Maximum dry density(g/cc)
Control Soil	12	1.96
Soil + 2% of RBI 81	16	1.88
Soil + 2% of RBI 81	22	1.71

3. Results and Discussion

The results obtained from the index properties soil are presented in Table -1. It can be understood from the uniformity coefficient and curvature coefficient that the soil is well graded with more silt and little sand. Its specific gravity was also found to be 2.7, indicating that it is silty sand. From Table – 2, it was found that the optimum moisture content of soil reinforced with RBI 81 was increasing than that of control soil. This may be because of the fibrous nature of the stabilizer and also dry density is decreasing with increase in moisture content. However the main goal is to find the improvement in CBR vale. Hence further investigation was carried out. From Table -3 it can be understood that UCC value of RBI reinforced soil increases marginally only. But the CBR values increased considerably indicating that the material can be used in sub grade soils to improve the strength characteristics of pavement layers and also to reduce the thickness of pavements.

Table 3. UCC and CBR test results Control soil and RBI 81 reinforced soil

Soil type	C_u (kN/m^2)	Q_u (kN/m^2)	Initial tangent modulus (kN/m^2)	Initial secant modulus (kN/m^2)
Control soil	121	242	6400	4200
Soil + 2% of RBI 81	123	245	11428	5714
Soil + 6% of RBI 81	126	253	45751	3636

4. Conclusion

From the analysis of results of tests conducted on control soil (soil alone) and soil reinforced with varying proportions of RBI 81, it was found that the CBR value of RBI 81 reinforced soil was increased from 10.56% to 19.02 %, indicating a significant increase in the penetration resistance of the soil. This will help in considerable reduction in the thickness of pavements laid over this type of soil resulting in great economy in the construction of pavements. The results also proved that the RBI 81 reinforced soil showed considerable improvement in unconfined compression strength, however after further increase in the dosage of RBI 81, it resulted in decrease in UCC value. Therefore it can be concluded that RBI 81 can be very effectively used to reinforce the weak sub grade soils.

Acknowledgement

The authors are extremely thankful to the authorities of Alchemist technology Limited for supplying the RBI grade 81 stabilizer which helped us to conduct this laboratory study.

References

[1] Lekha B M and Ravi Shankar A U 2014 Laboratory performance of RBI 81 stabilized soil for pavements *Int. J. Civil Eng. Res.* **5**(2) 2105-10.
[2] Patil B Mand and Patil K A 2013 Improvement in properties of subgrade soil by using Moorum and RBI 81. *Int. J. Sci. Eng. Res.* **7**(12) 966-971
[3] Gunturi M, Ravichandran P T, Divya Krishnan K, Annadurai R and Rajkumar K P R 2015 Micro level Analysis of RBI 81 Stabilized Expansive Soil. *Int. J. ChemTech Res.* **7**(2) 666-72.
[4] IS 2720 Part 3 1980 Determination of specific gravity *Bureau of Indian Standards*
[5] IS 2720 Part 4 1985 Grain size analysis *Bureau of Indian Standards*
[6] IS 2720 Part 5 1985 Determination of liquid and plastic limit of soil *Bureau of Indian Standards*
[7] IS 2720 Part 7 1980 Determination of water content- dry density relation using light Compaction *Bureau of Indian Standards*
[8] IS 2720 Part 16 1987 Laboratory determination of CBR *Bureau of Indian Standards*
[9] IS 2720 Part 10 1991 Determination of unconfined compressive strength of soil *Bureau of Indian Standards.*

Improving pedestrian facilities in congested urban areas: a case study of Chennai city

B Subramanyam[1*] and R Prasanna Kumar[2]

[1]School of Civil Engineering, SASTRA University, Thanjavur – 613 401. India.
[2]Professor, Department of Civil Engineering, Malla Reddy Engineering College (Autonomous), Telangana, India

*Email: subramanyam@civil.sastra.edu and subramanyamjy@gmail.com

Abstract. Traffic congestion and lack of public pedestrian space are some problems faced by most urban metropolises. Conventionally walking has been a mode of transportation in Indian cities. The percentage of pedestrians may vary from 16 to 57 depending upon the city. Encounters between vehicular traffic and pedestrian traffic are at its rise currently. Rapid industrialization and urbanization in India has resulted in neglecting of pedestrian facilities. Consequently pedestrian are at greater risk for their safety more especially in the commercial zones of large cities. A change in perspective spotlight will create a sense of awareness that the pedestrian traffic is also vital as the vehicular traffic. Soothing the traffic would moderately cut the driving expediency but the pedestrians will get a much safer and peaceful route to their terminuses. Safety and comfort are the two pans of a balance while considering the pedestrian traffic. Considering these aspects, this study deals a study in improving pedestrian facilities by analysing the existing skeleton of the selected locations. The adequacy of facility is checked based on IRC latest guidelines and counteractive measures are postulated.
Keywords: pedestrian, vehicular traffic, Safety and comfort

1. Introduction

The major characteristics of Indian cities are very high densities and a variety of land use. Pedestrians and vehicles are the two main inhabitants of the urban roads. Earlier, there was much focus on the development of urban roads, focusing only on solving the issues faced by the vehicular traffic. The result of this is that nowadays vehicular traffic occupies a larger number of roads in cities, leading to a number of conflicts arising for pedestrians and other non-vehicular users. The most basic step to creating a safer city is by providing proper amenities to pedestrians. This research paper focuses on identifying the problems faced by the pedestrians in congested urban cities like Chennai. A management technique for the pedestrians is the skywalk. Therefore, the safety of pedestrians has become a very integral part of any transport planning. Many urban areas focus on development at a very small scale because of which results in congestion and leads to pedestrian-vehicular conflicts especially at Signalized Intersections. Two behavioural sides of pedestrians are of interest to planners, i.e. how pedestrians move relative to the others, their choice of route, and their decision whether to cross the road or not etc. and how pedestrian travel times are affected by the signal control measures.

2. Literature Review

The quality of service provided to the pedestrians is evaluated in terms of average delay to the pedestrians and also in terms of the comfort experienced by them. These are defined in terms of Level of Service (LoS) and is designated with letters A , B, C etc indicating poor quality of service in the descending order.

[1] Map the conditions at the site and comparing it with standards. Pedestrian's proneness to accidents is the most considered factor in determining the dearth of facilities. Various road users include commuter, shopper, Disabled person and child. Differing priorities such as safety versus directness, based on different requirements are considered. Factors such as reliability, safety, greenery environment are the major contributors to a sustained skeleton. This is objectively measured but to an extensive scale. The existing features or the road furniture's related to the pedestrians such as pedestrian facilities, Hazards to pedestrians were noted. The analysis provided remedial measure that ensures safety of the pedestrian. The template of the data collection, analysis of the existing condition, was generalized as a model for any situation that could predict the pedestrian proneness to accidents. [2] States that delay is the key feature for designing or improving the existing facilities. Various intersection requirements such as bus stop, signalized systems and regular bus service routes are necessary to find the delay. Data regarding opinion of the pedestrians, Road inventory data and data related to speed of vehicle and the pedestrian are the foremost components to be considered. This can be removed [3]carried out pedestrian traffic flow studies in two medium sized cities to assess the comfort and safety levels of pedestrians of these cities in Haryana. Various facilities are rated based on the analysis and the results are given in the form of percentages. Satisfaction and safety are given much importance. Lack of awareness and less importance to the pedestrians are the concerns that are to be taken care of.

[4] Propose a micro-simulation based approach for preparing footpath design guidelines integrating street vendors in these facilities. The micro-simulation model is calibrated and validated using data collected on selected locations in Bangalore city. The model was then used for two purposes: (a) to demonstrated strategies to improve LOS at the study locations in Bangalore with re-organized vendor spaces, (b) to develop general footpath design guidelines for different footpath widths and pedestrian flows. Offsets are provided in the footpaths to allocate separate threshold spaces to make the flow free from any obstructions. HCM guidelines are followed to match the data and suitable results are provided.

3. Data collection

The study location for the present work chosen was Chennai. The locations which had conflict between both vehicles and pedestrian were identified. Eleven locations were chosen on the basic of conflict between vehicular traffic and pedestrian traffic. From the observation it was found that 8am to 10am was the peak period for the pedestrians in the morning and 6pm to 8pm was the peak period in the evening. Pedestrian volume count was made for every 15minutes interval during the peak identified peak period. Geometric details of the study locations such as Footpath width, kerb height etc were collected and presented in table 1. Other details such as type of footpath surface, obstruction, electrical poles, location of trees, encroachment, subway, Foot over Bridge etc. were also observed. The volume count includes the pedestrian movement in both the directions of the pathway. The pedestrian traffic volume data were collected at morning and evening peak hours. The pedestrian traffic volume data were presented in table 2 and table 3.

IOP Conf. Series: Earth and Environmental Science **80** (2017) 012059 doi:10.1088/1755-1315/80/1/012059

Table 1. Geometric Details of Identified study locations

S. No	Locations	Width(m)	Footpath Surface
1	Olympia tech park	2.5	Raised
2	Race course guindy	1	Raised
3	Purasiwalkam high road 1	1.2-1.8	Raised
4	Purasiwalkam high road 2	1.2-1.8	Raised
5	Central	NIL	Not raised
6	Kodambakkam (meenakshi college)	NIL	Not raised
7	Koyambedu bus stand	2.5-4	Raised
8	T.nagar	0.6-1.2	Raised
9	Mgr statue	NIL	Not raised
10	Beach railway station	NIL	Not raised
11	Light house	1.3	Raised

Table 2. Pedestrian Traffic volumes during morning peak hour

Location	8.00-8-15	8.16-8.30	8.31-8.45	8.46-9.00	9.01-9.15	9.16-930	9.31-9.45	9.46-10.00
Olympia tech park	298	350	368	384	414	438	446	430
Race course guindy	570	584	620	660	740	712	680	672
Purasiwalkam high road 1	502	512	560	552	590	620	642	658
Purasiwalkam high road 2	456	486	520	470	440	428	460	530
Central	3100	2988	2882	2852	2845	2951	2892	2974
Kodambakkam	424	484	521	541	512	486	476	497
Koyambedu bus stand	542	524	502	492	482	508	524	574
T.Nagar	0	298	265	304	385	348	396	428
MGR statue	-	-	-	-	-	-	-	-
Beach railway station	152	162	172	165	210	260	284	292

Table3. Pedestrian Traffic volumes during evening peak hour

Locations	6.00-6.15	6.16-6.30	6.31-6.45	6.46-7.00	7.01-7.15	7.16-7.30	7.31-7.45	7.46-8.00
Olympia tech park	450	480	442	468	520	490	512	532
Race course guindy	624	676	686	692	668	720	742	786
Purasiwalkam high road 1	674	682	725	714	734	762	742	692
Purasiwalkam high road 2	570	524	598	542	620	624	618	590
Central	3452	3354	3426	3374	3452	3512	3596	3698
Kodambakkam	514	520	489	496	467	472	452	483
Koyambedu bus stand	542	596	572	569	620	604	629	672
T.nagar	560	512	579	620	614	630	642	615
Mgr statue	120	145	198	212	254	241	225	214
Beach railway station	220	254	264	256	248	273	289	278
Light house	189	186	168	196	215	226	210	242

4. Data analysis

4.1 Peak hour factor

This parameter helps us to understand the variation of traffic volume with in the peak hour. To account for these variations in the peak hour traffic engineers usually calculate peak hour factor to identify the critical volume.

PHF is calculated using the following equation.

$$\text{PHF}= \frac{\text{Hourly volume}}{\text{Peak rate of flow within the hour}} \quad (1)$$

If 15-minute periods are used, the PHF is computed as:

$$\text{PHF}=V/(4\ X\ V15)$$

Where,
V = peak-hour volume (vph)
V_{15} = volume during the peak 15 minutes of flow (ped/15 minutes)

Greater variation in traffic volume within the peak hour is indicated by lower values of PHF and vice versa. The values of the Peak Hour Factor obtained in all the study locations ranges between 0.9 to 1.0. It signifies that there is not much variation in the flow rate. Subsequently to understand the fluctuation in pedestrian traffic volume within the peak period, Peak Hour Factors (PHF) for all the study locations were determined and presented in table 4.Speed is an important variable in assessing the delay to the Pedestrians which is essential in arriving at appropriate Level of Service (LOS), based on the quality of service offered by the facility to the users. Hence speed data was collected from all the study locations and presented in table 5. The LOS calculations were presented in table 6.

Table 4. Peak hour factors at different study locations

Location	Peak Hour Volume (M)	Peak Hour Volume I	Peak 15 mins Volume (M)	Peak 15 mins Volume I	Peak Hour Factor (M)	Peak Hour Factor I
Olympia tech park	1728	2054	446	532	0.96	0.96
Race course guindy	2804	2916	740	786	0.94	0.927
Purasiwalkam high road 1	2510	2953	658	762	0.953	0.96
Purasiwalkam high road 2	1932	2452	486	624	0.99	0.98
Central	11822	14258	3100	3698	0.95	0.963
Kodambakkam	2060	2376	541	496	0.95	1.19
Koyambedu bus stand	2088	2525	574	672	0.91	0.93
T.Nagar	1557	2501	428	642	0.9	0.97
MGR statue	-	934	-	254	-	0.919
Beach railway station	1046	1088	292	289	0.89	0.941

Table 5. Observed Pedestrian speeds at study locations

Location	Men (m/s)	Women (m/s)	Avg speed (m/s)	Width (m)	Los
Olympia tech park	298	350	368	384	414
Race course guindy	570	584	620	660	740
Purasiwalkam high road 1	502	512	560	552	590
Purasiwalkam high road 2	456	486	520	470	440
Central	3100	2988	2882	2852	2845
Kodambakkam	424	484	521	541	512
Koyambedu bus stand	542	524	502	492	482
T.Nagar	0	298	265	304	385
MGR statue	-	-	-	-	-
Beach railway station	152	162	172	165	210

Table 6. Pedestrian delay at study locations

Location	Length (m)	Avg. speed (min)	Actual speed (min)	Delay (min)
Race course guindy	560	6.43	8.14	1.31
Purasiwalkam high road 1	435	5.09	6.11	1.02
Purasiwalkam high road 2	267	3.12	3.45	0.33
Kodambakkam	552	6.51	7.43	0.52
T.nagar	421	5.50	7.36	1.46
Beach railway station	620	7.43	9.32	1.49

5. Conclusion

From the analysis of data collected at various study locations and based on the Level of Service calculations suggestions for improvement of pedestrian facilities were made.

- Instalment of rumble strips
- construction of pedestrian islands
- installment of pedestrians sensors in crosswalks
- installment of more perspicuous traffic signs and signals
- removal of trees and utility poles

Scope of future work: more advanced data collection techniques to collect robust data and soft tools need to be utilized in the analysis for understanding the intricacies of pedestrian characteristics and improve them to meet the demands of the present cities.

References

[1] Dinesh Kumar V and Satheesh 2014 Pedestrian Safety in Urban Situation. *Indian Highways*, 27-31.

[2] Dipika Gupta and Patel 2014 Pedestrian Simulation in Congested Urban Area. *Int. J. Eng. Tech. Res. (IJETR)* **2**(3) 111-5.

[3] Chitranshu Saluja and Bhavana Arora 2014 Quality and Safety Assessment of Pedestrian Sidewalk in Two Mid-Sized Cities of Haryana. *Int. J. Innov. Res. Sci. Eng. Technol.* **3**(8) 15518-22.

[4] Ashish Verma and Shirin Mary Antony 2014 Integration of Street Vendors in Footpath Design Guidelines for an Indian City. *J. Intell. Transportation and Urban Plan.* **2**(3) 129-48.

Effect of addition of GGBS and lime in soil stabilisation for stabilising local village roads in Thanjavur region

R Saravanan*, T Udhayakumar, S Dinesh, C Venkatasubramanian and D Muthu

School of Civil Engineering, SASTRA University, Thanjavur – 613 401. India.

*Email: rsaravanan39@gmail.com

Abstract. Construction of pavements uses various filling materials and due to the cost factor, the local soil is used for pavement construction. The strength of the soil is improved by stabilisation. This stabilisation increases the load bearing capacities of soil for heavy wheeled vehicle traffic. GGBS, silica fume, rice husk are the basic waste materials used as a waste material, which improves the quality of soil and reduces the cost of pavements. In this study, a detailed investigation is made on the Ground Granulated Blast-furnace Slag (GGBS), activated by lime, in the stabilisation of low bearing capacity sand and clay soils collected from Thanjavur district (Budalur, Sengipatti, Vallam and Palliahgraharam villages). The tests are carried out as per Indian Standards. The test procedures separated into two phases, namely Stage-I and Stage-II. In Stage-I the soil tests include soil type, particle size distribution, soil index properties, standard proctor tests, shear tests and CBR test. In Stage-II the soil tests include shear tests and CBR test for the suitable required proportions of GGBS along with lime in the collected soil samples. The test results from stage-I and stage-II are compared and from the study, it is inferred that the application of GGBS is a useful material for soil stabilisation.

1. Introduction

In India, major problems associated with the construction of village road are the preparation of subgrade. In many of the cases, the newly constructed road deteriorates rapidly due to the poor subgrade. If the in-situ subgrade soil doesn't have enough strength properties, that local soil is needed to be removed and is replaced by the better and healthier soil from the nearby available areas. The transportation of subgrade soil to those local areas is mostly uneconomic and increase the cost of the project exponentially. Soil stabilisation is the best way to make the construction more feasible and cost effective. For efficient and economic stabilisation, we need to utilise the Stabilising techniques for existing soil by adding some chemicals or some waste materials-fly ash, GGBS, iron slag, natural fibres, etc., In this project, the soil is Stabilised by adding GGBS, with lime activators to enhance the soil properties and strength aspects. The process of adding GGBS and lime will induce the calcium silicate hydrate bond in the soil, this would help in the formation of cementitious substance in soil.

GGBS or GGBFS (Ground Granulated Blast Furnace Slag) is the waste product drawn by the rapid cooling of molten iron slag. It is obtained from steel manufacturing process. In order to make use of GGBS, an attempt has been made for adopting it in soil stabilisation. From the previous studies, it is clear that the GGBS induces the strength of cement and is extensively used as cement additives. This can be implemented in the soil for stabilisation, but the results of addition of GGBS alone in soil will not giving sufficient satisfactory results. GGBS is mixed with lime (CaO) in the form of lime powder, lime is used as the activator in this process to initiate the bonding of soil with GGBS. Due to the inclusion of lime and GGBS, there is a significant decrement in linear expansion of lime stabilised clays.

Lime is also one of the old construction material in practice from the ancient time. Lime stabilisation is one of the common methods of stabilising soil, due to its resourceful and economic usage. The finding from the previous studies shows that, on addition of lime is added to clay soils induces flocculation. This induces variations in the engineering and index properties of soil. The studies reported in the literature indicates the addition of lime increased the optimum water content, swelling potential, liquid limit, plasticity index, shrinkage limit and maximum dry density of the soil. The findings from studies shows the optimum usage of lime in soil will be 1% by its dry weight, from this finding the 1% of lime is used as an activator in this study. The work was intended to study the suitability of GGBS over different types of soil and to improve the soil strength by increasing its CBR value, shear resisting properties. To make efficient utilization the waste material from steel industries (GGBS) and finding out the optimal percentage of addition of GGBS to the soil for obtaining better results. To reduce the impact on the environment by excavating soil and hazardous chemical addition to soil and make pavement construction as cost effective and increase the life of local village roads in Thanjavur region.

2. Materials and Methodology

2.1 Materials

2.1.1 Soil. For the purpose of deep investigation, 4 representative samples collected from various places of Thanjavur district. Soil 1 is collected from Tirumalaisamudram village, Soil 2 from Vallam, Soil 3 from Palliahgraharam villages and Soil 4 from Muniyampatti village.

2.1.2 Lime. The lime is procured and used in the form of quicklime (CaO), in the form of a white powder and is obtained from limestone quarries site from Ariyalur zone. This lime has the specific gravity of 1.57.

2.1.3 GGBS. The GGBS is obtained as a fine white powder. The GGBS is obtained from Indian Steel Company in Salem. The GGBS used in this study is having a specific gravity of 2.68 and GGBS is used as basic additives for stabilisation.

2.2 Material Collection

A sequence of tests was carried out on soil specimens without adding any additives to collected soil and then followed by tests for soil samples with additives, in which GGBS is added at different proportions and lime in a constant percentage. The tests have been executed in two stages which includes Initial and preliminary soil test as stage-I and in stage-II soil is mixed with 2%, 4% and 6% of GGBS and 1% of lime. Soil samples are prepared by mixing 1% of lime by its dry unit weight and their respective proportions with GGBS by its dry unit weight. The suitability of the soil is accessed principally by the CBR (California Bearing Ratio) test value. In this study, the soil is classified and identified according to IS:1498 part 1-1970. Soil samples are prepared according to IS:2720 part 1-1983. Specific gravity test was done using IS:2720 part 3-1980. Subsequently, the liquid and plastic limit test and shrinkage limit tests were done in accordance with IS:2720 part 5 and 6 respectively. The optimum moisture content and maximum dry density of the soil samples using light compaction were determined using to IS:2720 part 7-1980 test methods. Using optimum moisture content the Unconfined Compressive Strength (UCC) of soil samples was determined according to IS:2720 part 10-1991. To conclude the study, the CBR value is found out for the result comparison, using the lab testing in accordance with IS:2720 part 16-1992.

2.2.1 Stage-I. For initial and preliminary soil test without any additives, following tests were performed in this stage.

- Specific gravity, sieve analysis, test for index properties for soil classification.

- Standard proctor test for accessing of maximum dry density and optimum moisture content.
- UCC/ direct shear and CBR test for determination strength properties.

2.2.2 Stage-II. In this stage, soil samples mixed with 2%, 4% and 6% of GGBS and 1% of lime by its dry unit weight. The soil with additives are tested for

- Standard proctor test for accessing of maximum dry density and optimum moisture for strength property tests.
- UCC/ direct shear and CBR test for determination strength properties.
- Based on the results obtained the strength characteristics of soil is analysed.

3. Test Results

The initial test results for all 4 samples are listed in Table 1. Test results for Soil 1, 2, 3 and 4 with 2, 4 and 6% of GGBS, with a constant 1% of lime are listed in Table 2-5.

Table 1. Soil properties without additives

Property	Soil 1	Soil 2	Soil 3	Soil 4
Soil Color	Red	Yellow	Black	Brown
Specific Gravity (no unit)	2.27	2.22	2.35	2.41
% of Gravel	1.37	7.07	0.25	0.53
% of Sand	48.60	76.96	42.61	21.70
% of Fines	52.60	14.50	56.75	75.14
Optimum Moisture Content (%)	13.60	12.20	14.20	17.80
Maximum Dry Density (g/cc)	2.46	2.45	2.35	1.80
Soil Type According to USCS	CH	SW	CL	CH

Table 2. Test Results for Soil 1

% of GGBS	0%	2%	4%	6%
Liquid Limit (%)	34.18	48.24	49.26	47.27
Plastic Limit (%)	19.50	25.69	23.35	25.43
Plasticity Index (%)	14.67	22.51	25.91	21.79
Unconfined Compressive Strength (KN/m^2)	128.00	210.00	275.00	270.00
CBR Value (%)	7.69	11.19	11.82	11.68

Table 3. Test Results for Soil 2

% of GGBS	0%	2%	4%	6%
Angle of Internal Friction (°)	33.07	36.75	41.00	38.26
Cohesion (KN/m^2)	26.33	23.47	4.44	2.66
Shear Strength (KN/m^2)	96.46	210.76	221.78	181.81
CBR Value (%)	10.58	16.42	24.82	23.36

Table 4. Test Results for Soil 3

% of GGBS	0%	2%	4%	6%
Liquid Limit (%)	34.95	28.80	38.01	29.54
Plastic Limit (%)	23.88	12.66	20.06	13.56
Plasticity Index (%)	11.07	16.14	17.95	15.98
Unconfined Compressive Strength (KN/m^2)	205.00	261.00	365.00	345.00
CBR Value (%)	6.81	10.22	14.11	9.85

Table 5. Test Results for Soil 4

% of GGBS	0%	2%	4%	6%
Liquid Limit (%)	57.84	66.42	70.31	65.50
Plastic Limit (%)	39.56	29.31	40.21	40.27
Plasticity Index (%)	18.29	37.11	30.10	25.24
Unconfined Compressive Strength (KN/m^2)	260.00	265.00	300.00	180.00
CBR Value (%)	8.91	14.26	17.52	14.10

4. Results and discussion

The above results signify that there is a significant amount of increase in CBR value up to 4% addition of GGBS and 4% is obtained as the optimum result, beyond 4% there is a reduction in CBR value. The variation of CBR for all samples are shown in the Figures below. The Figures 1, 2, 3 and 4 shows the penetration versus load Figure for soil 1, 2, 3 and 4 respectively.

The shear resting property of the soil is also increased with respect to the addition of GGBS and lime. The UCC strength also shows an increase in strength in clay soil samples. Up to 4% addition of GGBS, there is an increases the strength in soil 1, 3 and 4. The unconfined compressive strength of soil 1, 3 and 4 are shown in Figures. The Figure 5, 6 and 7 shows the strain versus stress Figure for soil 1, 3 and 4 respectively.

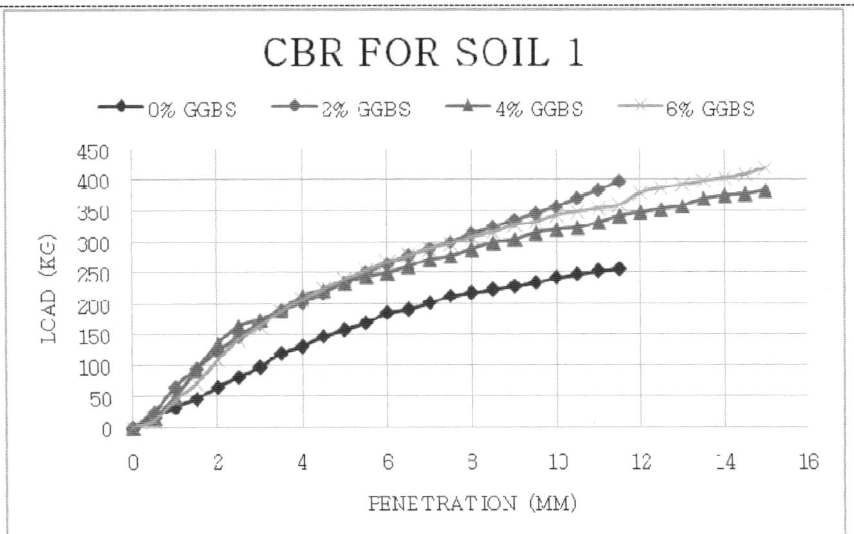

Figure 1. CBR results for Soil 1.

IOP Conf. Series: Earth and Environmental Science **80** (2017) 012060 doi:10.1088/1755-1315/80/1/012060

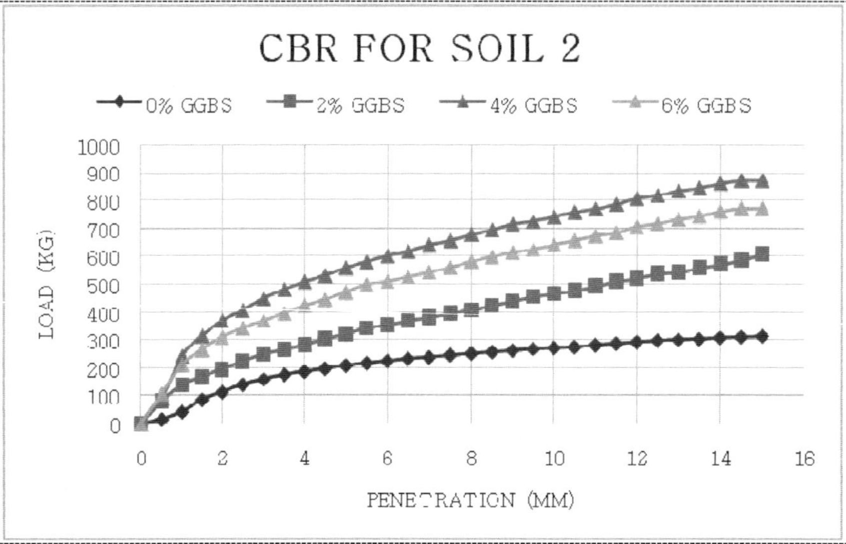

Figure 2. CBR results for Soil 2.

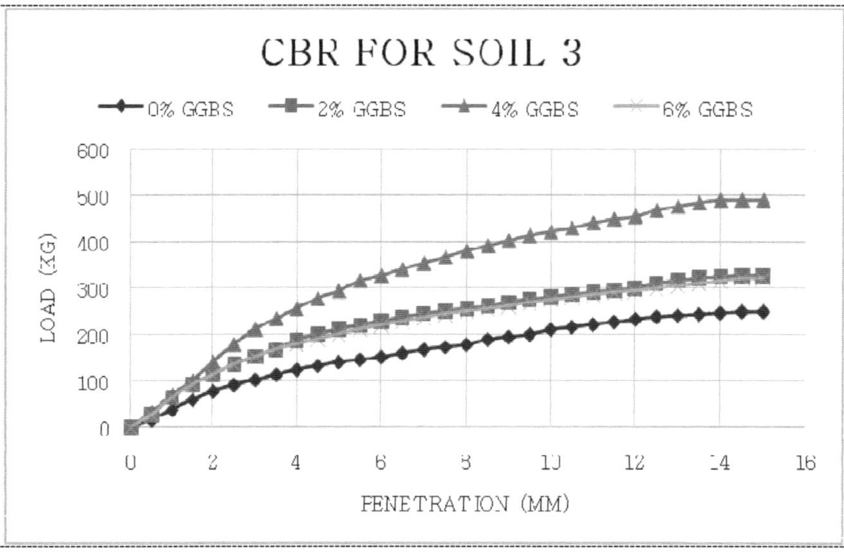

Figure 3. CBR results for Soil 3.

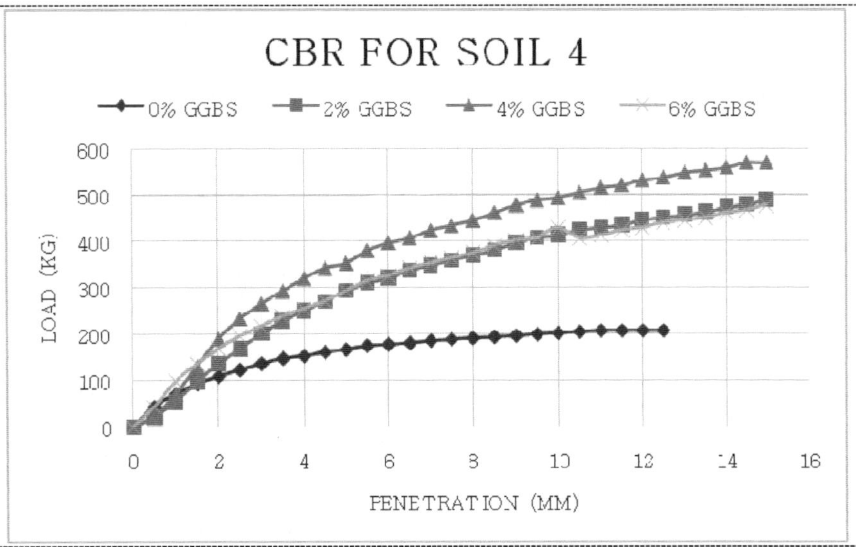

Figure 4. CBR results for Soil 4.

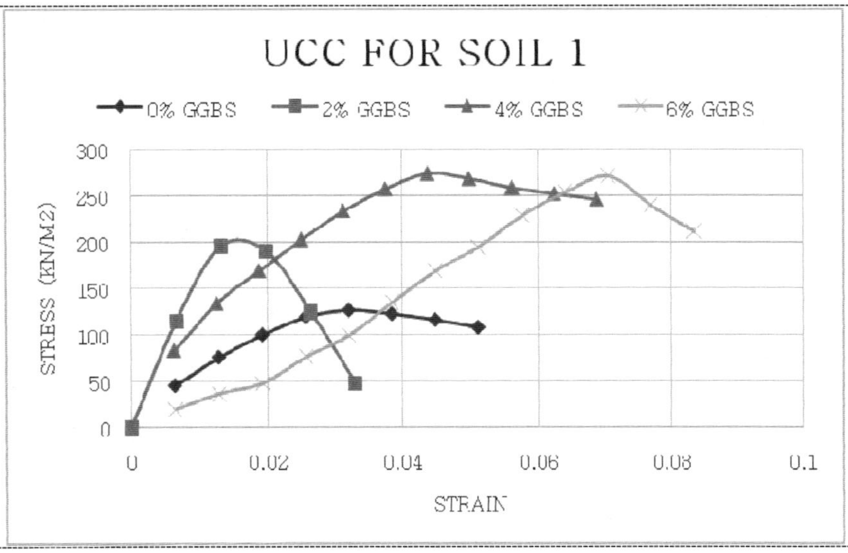

Figure 5. UCC results for Soil 1.

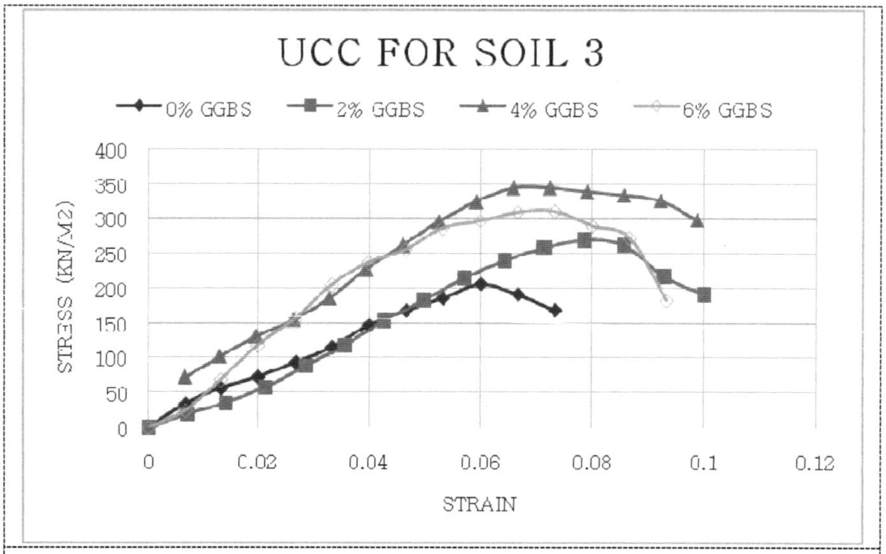

Figure 6. UCC results for Soil 3.

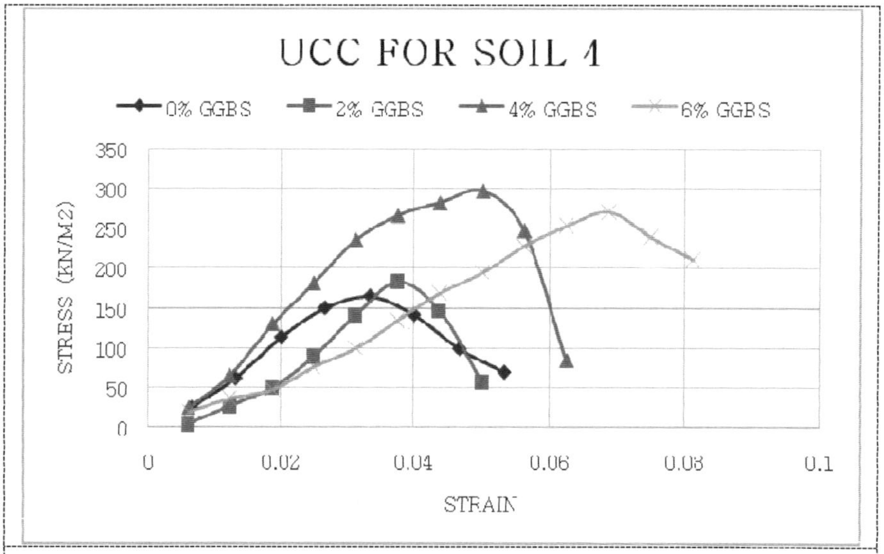

Figure 7. UCC results for Soil 4.

5. Conclusions

Based on the above analysis it has been concluded with,

1. Effective stabilisation is obtained with an addition of 4% of GGBS by dry weight of soil.
2. CBR value increases constantly with an addition of up to 4% of GGBS and 1% of lime to the soil.
3. The addition of GGBS beyond 6% will have no significant effect on soil than 4% addition, beyond that 6 % there is a slight reduction in strength on soil has been visualised.

4. The addition of a small quantity of lime of 1%, increases the strength and alters the index property of soil and reduces the plastic property of soil. The addition of GGBS and lime shows a better result in sandy soil than clay soil.
5. The unconfined compressive strength of clay soils, with a mix of maximum 4% GGBS and 1% lime shows a reasonable increase in shear strength of soil. The shear strength of sandy soil is increased with addition of up to 4% of GGBS and 1% of lime.
6. Plastic index of the soil is reduced with the increase in GGBS addition, after 4% there is notable decrement has been found out.
7. This addition of waste material for soil subgrade, considerable cost reduction is achieved in formation of roads. Therefore, subgrade for roads in Thanjavur region can be stabilized using GGBS and lime to achieve improved strength.

References

[1] Celik E and Nalbantoglu Z 2013 Effects of ground granulated blastfurnace slag (GGBS) on the swelling properties of lime-Stabilised sulfate-bearing soils. *Eng. Geol.* **163** 20-5
[2] Dash S K and Hussain M 2012 Lime stabilisation of soils: Reappraisal. *J. Mater. Civ. Eng.* **24**(6) 707–14.
[3] Eren S and Filiz M 2009 Comparing the conventional soil stabilisation methods to the consolid system used as an alternative admixture matter in Isparta Daridere material. *Constr. Build. Mater.* **23**(7) 2473–80.
[4] Obuzor G N, Kinuthia J M and Robinson R B 2011 Utilisation of lime activated GGBS to reduce the deleterious effect of flooding on stabilised road structural materials: A laboratory simulation. *Eng. Geol.* **122** 334–8.
[5] Obuzor G N, Kinuthia J M and Robinson R B 2012 Soil stabilisation with lime-activated-GGBS—A mitigation to flooding effects on road structural layers/embankments constructed on floodplains. *Eng. Geol.* **151** 112–9.
[6] Poh H Y 2006 Soil stabilisation using basic oxygen steel slag fines. *J. Mater. Civ. Eng.* **18**(2).
[7] Sharma A K and Sivapullaiah P V 2016 Ground granulated blast furnace slag amended fly ash as an expansive soil Stabiliser. *Soils and Foundations* **56**(2) 205–12.
[8] Wild S, Kinuthia J M, Jones G I and Higgins D D 1998 Effects of partial substitution of lime with ground granulated blast furnace slag (GGBS) on the strength properties of lime-stabilised sulphate-bearing clay soils. *Eng. Geol.***51** 37–53.
[9] Yi Y, Liska M and Al-Tabbaa A 2014 Properties of Two model soils Stabilised with different blends and contents of GGBS, MgO, lime, and PC. *J. Mater. Civ. Eng.* **26**(2) 267–74.
[10] Yu B W, Du Y J, Jin, F and Liu C Y 2016 Multiscale study of sodium Sulfate soaking durability of low plastic clay Stabilised by Reactive Magnesia-Activated Ground Granulated blast-furnace slag. *J. Mater. Civ. Eng.* **28** 04016016.

Deflection Control in Rigid Pavements

Nulu Varunkrishna* and R Jayasankar

Department of Civil Engineering, National Institute of Technology Tiruchirappalli, India,

*Email: nulu.varunkrishna@gmail.com

Abstract. The need for modern transportation systems together with the high demand for perpetual pavements under the drastically increasing applied loads has led to a great deal of research on concrete as a pavement material worldwide. This research indeed instigated many modifications in concrete aiming for improving the concrete properties. Pavement Quality Concrete requires higher flexural strength and fewer deflections in hardened state. Fiber reinforcement and latex modification are two reliable approaches serving the required purposes. The concrete made with these two modifications is called Polymer-modified Fiber-reinforced concrete. The present study deals with the usage of polypropylene as fiber and SBR (Styrene Butadiene Rubber) Latex as polymer. M30 grade concrete was modified by replacing cement with two different percentages of fiber (0.5%, 1.0% of weight of cement) and with three different percentages of SBR latex (10%, 15% & 20% of weight of cement).

1. Introduction

1.1. *General*

Transportation is one of the fastest growing and highly demanded fields since the start of the now-a-day modern era. Roadway, the most preferred and agile among all the modes of transport has the highest need for sophisticated improvements and modifications. So, as a part of these techniques, developing pavements with high service life and little or no maintenance requirement has been in highest demand which are otherwise called perpetual pavements. These are been in use to cope up with the drastically increasing modernization as well as to serve longer and better [1]. Rigid pavements are a better option when compared to flexible pavements for the design of perpetual pavements as the former has higher service life [1,2]. Besides, rigid pavements in spite of their high initial cost, are proven to be more economic than flexible type pavements when due considerations are made to life cycle cost and all types of maintenance costs [1-3]. Since almost all rigid pavements are laid with cement concrete as the material, our focus needs to be shifted to concrete. Further, there is a high potential for research on concrete because concrete, originally has some undesirable properties like excessive brittleness, shrinkage cracking etc., In order to overcome these flaws and make concrete, a more useful material several modifications are being proposed. Polymer modification and fiber reinforcement are two of the most effective modifications done to concrete as they concentrate on serious problems of concrete (especially when it is used as a pavement material) like low tensile strength and high shrinkage [5].

1.2. Fiber Reinforced Concrete (FRC)
Reinforcing concrete with fiber is found to be finer reinforcement when compared to all other ways of reinforcing. Fiber reinforcing in concrete, found to be improving many of the desirable properties of concrete like tensile strength, impact resistance, fracture strength, toughness, load carrying capacity, fatigue performance, etc., [4-10] and also to reduce undesirable properties like plastic shrinkage and permeability. However, these advantages can be observed in the concrete mix only when the fiber percentages are well designed in the mix. Besides, enough mixing has to be done while casting as it is difficult for the fibers to be spread uniform and enough dispersion only can help the fiber to improve the properties of concrete in desirable way.

1.3. Polymer Modified Concrete (PMC)
Synthetic polymers have been in use for many decades to modify concrete mixes. The mostly used polymers are thestyrene–butadiene copolymers (SBR), styrene–acrylic copolymers (SA), acrylic polymers (PA), vinyl acetate copolymers, and vinyl acetate homopolymer [11]. Modified systems have always been formulated considering thewell-known association and combination of cement, fine aggregate, coarse aggregate,water and latex combined with a polymer/cement ratio(generally ranging from 5 to 20%), depending on the weight of cementitious materials in the mix [11-18]. The most widely proven and accepted fact about these modified systems is that because of the interpenetration of cement hydrates and polymer, which in turn is a result of many physicochemical reactions within the compounds, there is a formation of a strong network structure after the curing process. The cement paste and the aggregates showed improved values of cohesion at the interface due to the above mentioned reason and this is one of the main reasons the mix shows increased compactness in its hardened state. As an outcome of all these processes, these modified systems have been showing improved tensile and flexural strengths, adhesive and durability properties while showing a considerable reduction in permeability.

1.4. Polymer Modified Fiber Reinforced Concrete (PM-FRC)
For Pavement Quality Concrete (PQC), improving the strength properties is not merely sufficient. Instead, the performance characteristics are to be focused more on PQC to make concrete a better material for pavements. One such attempt is controlling the deflections in rigid pavements which cannot be achieved by modifying the concrete using only polymer or by simply reinforcing it. Polymer modification and fiber reinforcement individually are found to be giving satisfactory results in improving the flexural strength and plastic shrinkage properties of concrete, but have no considerable effect in controlling the deflections in PQC. These two techniques, when combined and used together in the same mix was proved to be better than using them individually as the latter shows advantages of both the modifications resulting in a superior mix[19,20].

2. Materials
2.1. Cement
In the present study, Ordinary Portland Cement (OPC) of 43 grade was used. The cement was tested as per IS 4031-1988 [21].

2.2. Fine Aggregate
Locally available river sand was used as a fine aggregate. The sand was screened at the site to remove deleterious materials and it was made sure that sand contained no or negligible water content before it was used for the mixing process.

2.3. Coarse Aggregate
Coarse aggregate is the strongest and least porous constituent of concrete. Moisture movement is a considerably serious problem in concrete mixes causing drying shrinkage along with other dimensional changes. These ill effects due to the moisture movement can be effectively controlled by

use of coarse aggregate in the mix. When properly graded, the use of coarse aggregate is even found to check down the permeability of concrete. Two sizes of crushed, angular shaped coarse aggregate: 20mm and 10mm were used in the study.

2.4. SBR Latex

Latex type polymer is selected as a modifier in this study which has styrene butadiene emulsion as its main component along with some special admixtures in it. Latexes have been showing outstanding bonding properties in mortar and concrete mixes.SBR Latexes have a typical composition of flexible butadiene monomers and rigid styrene monomers which shows many desirable features in the overall mix. The physical properties of SBR Latex are given in the below table.

Table 1. Properties of S.B.R latex

Property	Value
Specific Gravity	1.02
pH value	9-11
Solid Content	48%

2.5. Polypropylene Fiber

Fiber reinforcing in concrete has been a reliable approach to improve the tensile and flexural properties of the mixes even better than the traditional bar type reinforcement. This is mainly because of the ability of the fibers to reach out and spread throughout the mix. This is a sort of disadvantage in case ofpolypropylene fibers. Because their hydrophobic nature, polypropylene fibers need to be mixed for long time so that the fiber disperses well throughout the mix. This mixing time should not even be too long because longer mixing times may result shredding of the fibers. The physical properties of the polypropylene fibers are shown in the following table.

Table 2. Properties of Polypropylene fiber

Property	Value
Specific Gravity	0.9
Young's Modulus	5500-7000MPa
Tensile Strength	360 MPa
Melting Point	160°c
Fiber Length	12mm

3. Present Study

This study deals with seven different mixes out of which six are Polymer modified Fiber reinforced concrete and one is for nominal M30 grade concrete. The six PM-FRC mixes dealt with 3 different percentages of SBR latex (10, 15 & 20%) and also with 2 different percentages of polypropylene fiber (0.5% & 1%) A total of nine prism specimens were casted for each mix and were tested for flexural strength as well as maximum deflection value. The samples were named as per the latex and fiberpercentage in their respective mixes where Lx-PPy means the sample contains x% of SBR Latex and y% of Polypropylene fiber by weight of cement in the mix.

(Example: L15-PP1 means the mix contains 15% SBR Latex and 1% Polypropylene fiber)

4. Mix Proportioning

The mix proportions, in the present study were calculated based on the design guidelines given by IS 10262:2009[22]

Table 3. Mix Proportions

Specimen ID	Cement (kg/m3)	Fine aggregate (kg/m3)	Coarse aggregate 20 mm (kg/m3)	Coarse aggregate 10 mm (kg/m3)	Water (kg/m3)	S.B.R Latex (Kg/m3)	P.P Fiber (Kg/m3)
L0-PP0	442.8	628.37	655.36	436.9	191.6	0	0
L10-PP0.5	428.2	628.37	655.36	436.9	191.6	14.38	0.635
L15-PP0.5	421.22	628.37	655.36	436.9	191.6	21.57	0.635
L20-PP0.5	414.03	628.37	655.36	436.9	191.6	28.76	0.635
L10-PP1	428.42	628.37	655.36	436.9	191.6	14.38	1.27
L15-PP1	421.22	628.37	655.36	436.9	191.6	21.57	1.27
L20-PP1	414.03	628.37	655.36	436.9	191.6	28.76	1.27

5. Results and Discussions

5.1. Flexural Strength

Prism specimens of dimensions 100 mm x 100 mm x 500 mm were cast and tested according to IS:516 (1959) [23]. The specimens were taken out of the curing tank well before the testing process and dried because water has shown high reluctance in escaping from the concrete matrix as a result of the use of polymer and fiber in it. All the mixes were tested at 7, 14 & 28days in order to monitor the rate of strength gain in the specimens. The results of the flexural strength tests are given in Table 4.

Table 4. Flexural Strength of various mixes

MIX	Flexural Strength (MPa)		
	7-Days	14-Days	28-Days
L0-PP0	3.42	4.85	5.45
L10-PP0.5	2.87	4.99	6.26
L15-PP0.5	3.46	5.95	7.48
L20-PP0.5	2.83	4.26	5.38
L10-PP1.0	3.31	5.6	6.66
L15-PP1.0	3.75	6.41	7.75
L20-PP1.0	2.87	4.72	5.72

The results show that there was an increase in flexural strength for the modified concrete irrespective of the polymer and fiber percentages. The comparisons of the flexural strength results are shown below:

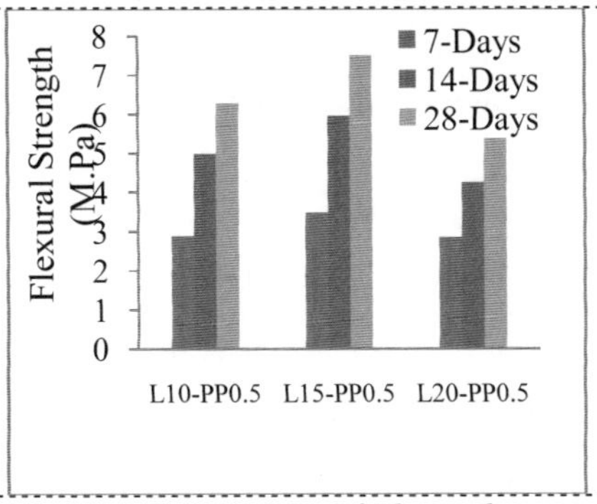

Figure:1 *Comparison of Flexural Strength for mixes with various latex contents (0.5% fiber)*

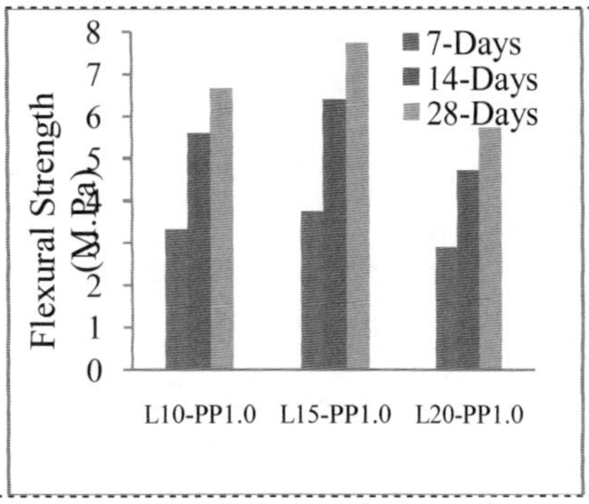

Figure 2. *Comparison of Flexural Strength for mixes with various latex contents (1% fiber).*

5.2. Deflection Studies

The maximum value of deflections obtained for each mix at 7, 14 and 28 days are tabulated in the Table 5 below:

Table 5. Deflections of Various mixes

MIX - ID	7-Days	14-Days	28-Days
L10-PP0.5	2.41	1.86	1.57
L15-PP0.5	2.24	1.76	1.44
L20-PP0.5	2.95	2.43	1.85
L10-PP1.0	2.39	1.95	1.68
L15-PP1.0	2.05	1.42	1.05
L20-PP1.0	3	2.27	1.76
L0-PP0	2.51	2.21	1.95

Mixes with polymer and fiber percentage had shown lesser deflection values when compared to normal mix except for the mix with 20% Latex and 1% fiber. The deflections reduced with increase in latex content up to 15% and then reached a higher value. This shows that 20% latex content is too high for the mixes. Out of all the mixes used, 15% latex and 1% fiber showed the least deflections and the mixes with 15% latex showed lesser deflections than the other specimens of its kind for both the fiber contents. This difference is much larger for the fiber content of 1% which is recommended as the optimum combination of polymer and fiber.

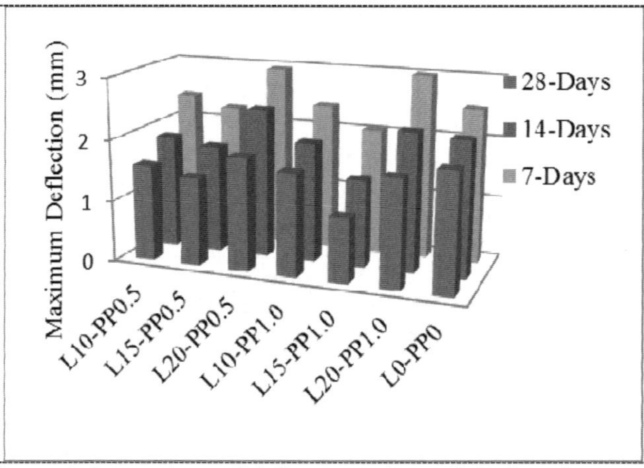

Figure 3. Comparison of deflection values of all the mixes at 7, 14 & 28days

6. Conclusions

The following conclusions were drawn out of the present study:

- Concrete, when modified with latex and fiber had shown improved properties of flexural strength and lesser deflections.
- The increase in flexural strength when compared to the conventional concrete varied from a minimum of 5.6% to a maximum of nearly 45%.
- By increasing the fiber content from 0.5-1%, the flexural strength and resistance to deformation have increased slightly.

- Latex content of 15% is found to be giving the best results in terms of deflection and flexural strength. This is because of the sufficiency of latex at 15% and increase of latex beyond that point, resulted in disturbance of the latex-fiber structure.

References

[1] Chandan B, Atasi D and Thirumalasetty P 2013 Perpetual Pavement - A Boon for the Indian Roads. *Proc. 2nd Conf. Transportation Res. Group of India (2nd CTRG)* Procedia – Soc. Behavioral Sci. **104** 139-48.
[2] Wimsatt A J, Albitres C M C, Krugler P E, Scullion T, Freeman T J and Valdovinos M B 2009 Considerations for Rigid Vs. Flexible Pavement Designs when Allowed As Alternate Bids *Technical Report* 0-6085-1.
[3] Milind M V and Kadam K N 2016 A Comparative Study on Rigid and Flexible Pavement: A Review. *IOSR J. Mech. Civ. Eng.* **13**(3) 84-8.
[4] Choi S Y, Park J S and Jung W T 2011 A Study on the Shrinkage control of Fiber Reinforced Concrete Pavement. *Proc. 12th East Asia-Pacific Conf. Struct. Eng. Constr.* Procedia Eng. **14** 2815-22.
[5] Saman K, Roohul A K Amadur R K Mishabul I and Saman N 2015 Mechanical properties of Polypropylene fiber reinforced concrete for M25 & M30 mixes: A comparitive study. *Int. J. Sci. Eng. Appl. Sci. (IJSEAS)* **1**(6) 327-40.
[6] Anthony N E and Abimbola O I 2014 Optimal Polypropylene Fiber content for Improved Compressive and Flexural Strength of Concrete. *IOSR J. Mech. Civ. Eng.* **11**(3)129-35.
[7] Milind M V 2015 Performance of Polypropylene Fiber Reinforced Concrete. *IOSR J. Mech. Civ. Eng.* **12**(1) 28-36.
[8] Alhozaimy A M, Soroushian P and Mirza F 1996 Mechanical properties of Polypropylene Fiber reinforced concrete and the effect of pozzolanic materials. *Cem. Concr. Compos.* **18**(2) 85-92.
[9] Ahmed S, Bukhari I A, Siddiqui J I and Qureshi S A 2006 A study on properties of polypropylene fiber reinforced concrete . *Proc. Int. 31st Conf. Our world in Concr. Struct.* 1-11.
[10] Shende A M and Pande A M 2011 Comparitive Study on Fiber Reinforced cum Control Concrete under Flexure and Deflection. *Int. J. Appl. Eng. Res.* **1**(4) 942-50.
[11] Won J P, Kim J H, Park C G and Kim H Y 2009 Shrinkage cracking of Styrene Butadiene Polymeric Emulsion-Modified Concrete using Rapid Hardening Cement. *J. Appl. Polym. Sci.* **112** 2229–34.
[12] Bhikshma V, JagannadhaRao K and Balaji B 2010 An Experimental Study on behaviour of Polymer Cement Concrete. *Asian J. Civ. Eng. (Building and Housing)***11** 563-73.
[13] Huang B, Wu H, Shu X and Brudette E G 2010 Laboratory evaluation of permeability and strength of polymer-modified pervious concrete. *Constr. Build. Mater.* **24**(5) 818-23.
[14] Al-Nu'man B S and Al-Hadithi A I2009 Flexural Behaviour of Polymer Modified Reinforced Concrete Beams. *J. Eng. Dev.* **13**(1) 89-110.
[15] Benali Y and Ghomari F 2016 Latex influence on the mechanical behavior and durability of cementitious materials. *J. Adhesion Sci. Technol.* **31**(3) 219-41.
[16] Diab A M, Elyamany H E and Ali A H 2012 Experimental investigation of the effect of latex solid/water ratio on latex modified co-matrix mechanical properties. *Alexandria Eng. J.* **52**(1) 83-98.
[17] Kapil S and Joshi Y 2014 Performance analysis of Styrene Butadiene Rubber Latex on Cement Conrete Mixes. *Int. J. Eng. Res. Appl.* **4**(3) 838-44.
[18] Shuyi Y and Yong G 2012 Effect of Styrene Butadiene Rubber Latex on Mortar and Concrete properties. *Adv. Eng. Forum* **5** 283-88.
[19] Hanfeng XU 2003 Properties of Fiber Reinforced Concrete with Latex Modification. *M.S Thesis*, Department of Civil Engineering The University of British Columbia.
[20] Harish G R, Sadath A K Z and Chandramouli S V 2016 Experimental Study on Compressive and Flexural strengths on SBR Latex modified Polypropylene Fiber reinforced concrete. *Int. J.*

Eng. Res. Technol. **5**(4) 632-6.
[21] Bureau of Indian Standards (BIS) 1988 Code of Practice- Methods of Physical Tests for Hydraulic Cement New Delhi BIS Standard No IS 4031.
[22] Bureau of Indian Standards (BIS) 2009 Code of Practice – Concrete Mix Proportioning – Guidelines, New Delhi BIS Standard No IS 10262.
[23] Bureau of Indian Standards (BIS) 1959 Code of Practice – Methods of test for strength of concrete New Delhi BIS Standard No IS 516.
[24] Bureau of Indian Standards (BIS) 2000 Code of Practice- Plain and Reinforced Concrete New Delhi BIS Standard No IS 456

Evaluation and Recalibration of Empirical Constant for Estimation of Reference Crop Evapotranspiration against the Modified Penman Method

K Sasireka*, C Jagan Mohan Reddy, C Charan Reddy and K Ramakrishnan

School of Civil Engineering, SASTRA University, Thanjavur – 613 401, India

*Email: sasireka@civil.sastra.edu

Abstract. The major demand in our country is irrigation demand. Looking to the low irrigation potential and small water resources, it is felt necessary to see that water must be used economically and efficiently. This may be achieved by using latest methods of determination of water requirements for crops and applying the proper water management practices. Evapotranspiration (ET) is a basic for calculation of water requirement for crops. The various popular empirical equations for reference crop evapotranspiration (ET_r) belong to three categories namely, Temperature, Radiation based methods and Combined methods. The above methods are site specific; hence it is necessary to recalibrate the coefficients for applying them in India. In the present paper, the standard combined method namely FAO modified Penman method was used to recalibrate the constants in temperature based (TB) methods and it can also be used to determine the ET_r for the selected station. Four TB evapotranspiration models namely Blaney-Criddle, Romanenko, Kharrufa, and, Thronthwaite methods are recalibrated and the constant in each method are redefined for the data from Lekkur station, Cuddalore district in India. The result shows that, large error existed when ET_r has been calculated with original constants. Hence regression equations were developed to minimise these variations in magnitude. It was found that out of four methods the Blaney-Criddle method suits better for the selected region.

Keywords: Reference crop evapotranspiration (ET_r), FAO modified Penman method, temperature-based methods

1. Introduction

Evapotranspiration (ET) encompasses the total evaporation from plant body and surrounding areas, either potential or actual evapotranspiration may be of concern. Potential evapotranspiration or reference crop evapotranspiration (ET_r) is the rate that would occur under some specified standard condition of vegetation with an unrestricted supply of water in the soil and without advection or heat storage effects. Even though ETr is considered as vital parameter for the terrestrial hydrologic system, measurement of ETr is difficult, due to its spatial and temporal variation. ETr is a major determinant variable in water cycles as well as it plays a major role for the redistribution of water on the earth's surface. ET is the most significant component in the areas where the irrigation is a major component of agriculture due to low precipitation [1]. ET of other crops can be obtained by multiplying ETr with crop coefficients, which vary with the crop and its growth stage. Even though standard Penman model gives accurate result of ET value [2], it needs more numbers of data and extensively experienced person to use that equation correctly [3]. Therefore, in order to select suitable method to use it in India, it is mandatory to analyse and evaluate the various forms of existing temperature-based (TB) models with the standard Penman model. During the past several decades, many literature showed that the evapotranspiration has being extensively studied in hydrology and water resources management [4-10].

2. Study Area

The meteorological data collected from Lekkur station has been used to calculate ET in the present study. This station is located in Thittakudi which is taluk headquarters of Cuddalore district in Tamil Nadu, India at a latitude of 11.416° N and longitude of 79.116°E. Several meteorological variables namely air temperature, wind speed, relative humidity, sunshine hours and wind speed for the period of 1975 to 2003 have been used for estimation of ET and development of regional equation. The monthly average main climatic parameters are shown in Table 1.

Table 1. Monthly average climatic parameters for Lekkur station

Month	Humidity in %	Maximum Temperature in °C	Minimum Temperature in °C	Sunshine in hrs/ day	Wind Velocity in km/h
Jun	52.99	34.89	26.47	6.85	8.51
Jul	55.71	33.70	25.47	5.56	7.49
Aug	59.00	33.13	24.75	6.06	7.06
Sep	65.70	32.76	24.05	6.63	5.19
Oct	72.51	30.94	23.37	5.82	3.73
Nov	75.56	28.74	22.22	5.55	4.43
Dec	71.38	27.13	20.83	6.48	5.68
Jan	65.74	28.05	20.70	8.14	5.21
Feb	60.85	30.19	21.33	9.34	5.11
Mar	56.91	33.02	22.98	9.67	5.47
Apr	56.54	35.17	26.23	9.30	6.55
May	54.04	35.14	26.74	8.59	7.43

3. Reference Crop Evapotranspiration using Temperature – Based methods

Two important methods namely temperature-based and radiation-based methods along with the FAO modified Penman method were studied by Jothiprakash *et al* [11], [12]. For calculating irrigation demand ETr is to be determined. In this paper following four TB models along with FAO modified Penman methods are used to calculate ETr.

3.1 Blaney – Criddle Method

Blaney-Criddle method is considered as the one of the simplest method because ET_r is calculated using measured temperature data alone. However this is not a very accurate method; it gives a rough estimate only. This method is inaccurate under "extreme" climatic conditions. This method gives underestimated ET_r value in windy, dry, sunny areas, and overestimated value in calm, humid, clouded areas. The mathematical expression of the Blaney-Criddle method for the estimation of ET has the form [13].

$$ET_o = a + b\,(p\,(0.46\,T + 8.13) \qquad (1)$$

where,

ET_o = Reference crop evapotranspiration in mm/day
a & b = Calibration factors
p = Mean daily percentage of total annual daytime hours.
T = Mean daily air temperature in °C
a = $0.0043\ RH_{min} - (n/N) - 1.41$ (1.1)
b = $((0.82 - (0.0041RH_{mean}) + (1.07\ n/N) + (0.066U) - (0.006RH_{mean}\ n/N) - (0.0006RH_{mean}\ U)$ (1.2)
n/N = Mean ratio of actual to possible sunshine hours.
RH_{min} = Minimum daily relative humidity
U = Wind speed at a height of 2 m from ground surface.

3.2 Thornthwaite Method

This method was developed for the east-central U.S. It is simpler than Penman's equation because this method requires less climatic data. Bautista et.al [14] concluded that this method worked very well during the rainy months. The equations is as follows

$$PET = 1.6L_d\left(\frac{10T}{I}\right)^a \quad (2)$$

where,
PET - Potential evapotranspiration in cm/month - to get a per day amount, we assume there are 30 days per month, so PET in mm/day = PET*10 (mm/cm)/30 days/month)
L_d - Daytime hours in units of 12, for example if there are 10 daytime hours, then L_d=10/12
I - The annual heat index that is computed from the monthly heat indices
T - Average temperature for the day or month in °C

$$I = \sum_{j=1}^{12} i_j \quad (2.1)$$

where i_j is computed as

$$i_j = \left(\frac{T_j}{5}\right)^{1.514} \quad (2.2)$$

'a' is computed as either:

$$a = 67.5\times10^{-8}I^3 - 77.1\times10^{-6}I^2 + 0.0179I + 0.492 \quad (2.3)$$

3.3. Romanenko method

An evapotranspiration equation was derived by Romanenko [15] based on the relationship using mean temperature and relative humidity is as follows.

$$E = 0.0018(25+T)^2(100-rh) \quad (3)$$

where
T= Mean air temperature in °C and
Rh = Mean monthly relative humidity in %.
The units of E are mm/month and based on a 31 day month - divide by 31 for mm/day.

3.4. Kharrufa method

Kharrufa [16] derived an equation through relationship of ET/p and T in the form of

$$ET = 0.34pT_a^{1.3} \quad (4)$$

where,
ET = Kharrufa Potential evapotranspiration (in mm/month) and T_a and p have the same definitions earlier.

3.5 FAO modified Penman method

This method include both climatological effect of aerodynamic and temperature variations. The equation used to estimate ETr is given by:

$$ET_r = C_p\,[WR_n + (1\text{-}W).\,f(u).\,(e_s - e_a)] \quad (5.1)$$

where,
ET_r = Reference crop evapotranspiration in mm/day
W = temperature and altitude dependent weighting factor
R_n = Net radiation in equivalent evaporation in mm/day
$R_n = R_{ns} - R_{nl}$ (5.2)
R_{ns} = Net short wave radiation mm/day = $(1\text{-}\propto)R_s$ (5.3)
$\propto$ = reflection factor (0.25) (albedo)
R_s = Solar radiation in mm/day = $(a+b*n/N)R_a$ (5.4)
n =Actual sunshine hours
N = Maximum possible sunshine hours
a & b = constants depending on location
(if no data is available the values of a and b can be assumed to be equal to 0.25 and 0.5 respectively) [2]
R_a = Extra terrestrial radiation in mm/day
R_{nl} = Net long wave radiation mm/day
= $f(T).f(e_a).f(n/N)$ mm/day (5.5)
f (T) = Effect of temperature on long wave radiation (R_{nl})
f (e_a) = Effect of actual vapour pressure on long wave radiation (R_{nl})
f (n/N) = Effect of ratio between actual and maximum bright sunshine hours on long wave radiation (R_{nl})
f(u) = Wind related function = 0.27 [1 + U/100] (5.6)
U = Wind speed in km / day at 2 m height
e_s = Saturated vapour pressure at mean air temperature in mbar
e_a = Actual vapour pressure of the air in mbar = $e_s\,(RH_{mean}/100)$ (5.7)
C_p = Adjustment factor for ratio U_{day}/U_{night}, RH_{max} and R_s

If the wind speed data are available at height other than 2 meter, they need to be converted to wind speed corresponding to 2 meter height using $1/7^{th}$ power law as recommended by F.A.O. [13].

$$U_{2m} = U_x\,[2/x]^{1/7} \quad (5.8)$$

where U_{2m} is the wind speed at 2 m height from ground surface and U_x is the wind speed measured at 'x' m height from ground surface. In the present study the U_{day}/U_{night}, ratio is assumed as 1.5 [2]. The 'a' and 'b' coefficient in the radiation formula are assumed to be 0.25 and 0.51 respectively. In the FAO modified Penman formula, values of 'W' temperature and altitude dependent weighting factor, net long wave radiation in mm/day (R_{nl}), extra terrestrial radiation in mm/day (R_a), the mean daily duration of maximum possible sunshine hours (N) and the value adjustment factor C_p were taken from the FAO 24 [13].

4. Results and Discussion

The mean monthly values of ET_r were estimated using the various methods discussed above with original constant values. Table 2 shows estimated ETr values with original constant using above discussed method as well as the percentage error for all method with reference to FAO modified Penman method. From the Table 2 it can be seen that the average annual error between FAO modified Penman method and other methods varies from –27.1% to 19.6%. The overestimated and underestimated values of ETr were represented by positive error and negative error respectively with reference to the standard method. With the original constant values almost all the methods has overestimated the ET_r values except Thronthwaite method. The highest error was noticed when Thronthwaite method was used and Blaney – Criddle method yielded least error.

Table 2. Mean monthly estimated ET_r in 'mm' using selected methods with original constant values

Month	$ET_{r\ Pen}$	$ET_{r\ Bla}$	Error (%)	$ET_{r\ Thron}$	Error (%)	$ET_{r\ Rom}$	Error (%)	$ET_{r\ Kha}$	Error (%)
Jun	203.4	298.1	46.6	162.4	-20.1	271.6	33.6	207.7	2.1
Jul	181.2	187.8	3.6	113.9	-37.2	247.1	36.4	162.0	-10.6
Aug	178.6	179.7	0.6	111.9	-37.4	223.3	25.0	171.4	-4.0
Sep	158.7	122.4	-22.9	112.9	-28.9	183.3	15.5	183.4	15.6
Oct	132.2	51.6	-61.0	78.0	-41.0	139.1	5.3	150.7	14.0
Nov	115.3	48.9	-57.6	54.6	-52.6	115.5	0.1	131.9	14.4
Dec	121.1	94.7	-21.7	47.6	-60.7	126.9	4.9	141.9	17.2
Jan	138.0	143.1	3.7	66.3	-52.0	155.4	12.5	183.5	32.9
Feb	164.7	199.2	20.9	101.1	-38.6	188.9	14.6	228.0	38.4
Mar	200.5	256.5	27.9	153.7	-23.4	226.6	13.0	262.5	30.9
Apr	207.1	313.3	51.3	229.8	10.9	253.5	22.4	285.4	37.8
May	210.9	342.2	62.3	235.2	11.5	274.9	30.4	271.3	28.7
Annual	2011.7	2237.5	11.2	1467.2	-27.1	2406.1	19.6	2379.8	18.3

$\mathbf{ET_{r\_Pen}}$ =FAO modified Penman method, $ET_{r\ Bla}$ = Blaney-criddle, $\mathbf{ET_{r\_Thron}}$ =Thronthwaite, $\mathbf{ET_{r\_Kha}}$ = Kharrufa method, $\mathbf{ET_{r\_Rom}}$ = Romanenko method,

4.1. Correlation of monthly evapotranspiration estimated with original constants

By using following linear regression equation, the monthly ET_r values estimated using above mentioned methods were correlated with FAO modified Penman method.

$$Y= AX + B \tag{6}$$

Where, Y - ET_r estimated using FAO modified Penman method and X - ET_r estimated from above mentioned four TB methods. As far as the R^2 values are concerned, the Romanenko method has the highest R^2 value of 0.95 with FAO modified Penman method which is given in Figure 1. The Blaney – Criddle method gave least percentage error of mean monthly ETr estimation, has resulted with a R^2 of 0.898.

Figure 1 shows the scattered plot of monthly ETr values obtained by using above TB methods against FAO modified Penman method. Figure 1 shows a poor correlation within the TB methods except Romanenko method.

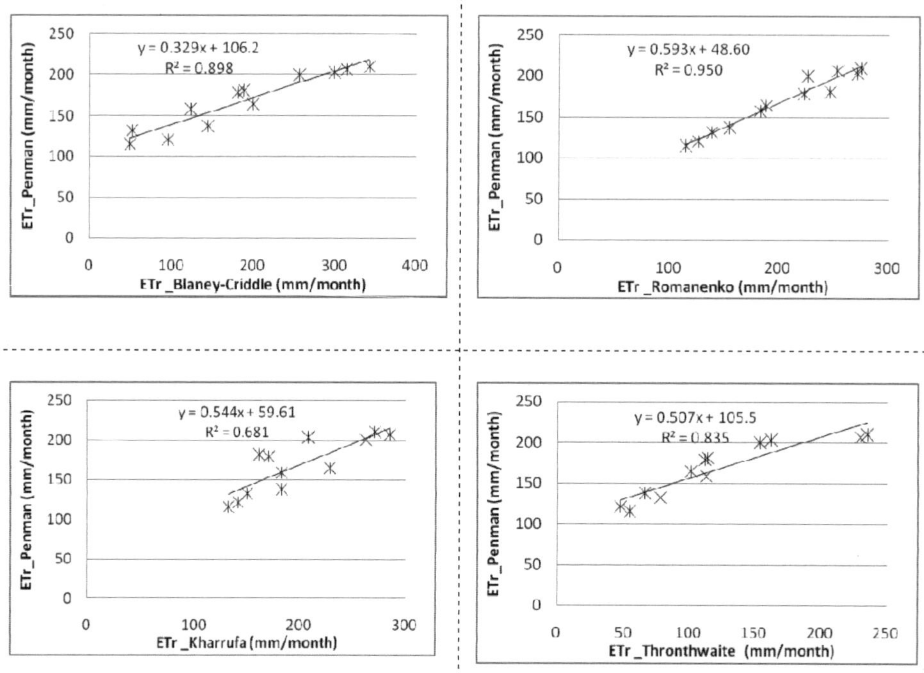

Figure 1. Monthly ETr estimated using various TB methods against FAO modified Penman method with original constants.

4.2. Modification to the constants used in temperature-based ET_r equations

The empirical equations, as used in this study, may be consistent in the areas and over the intervals for which they have been evolved, but massive inaccuracy can be predicted when they are extrapolated to different climatic regions without recalibrating the constants involved in the formulae [17]. Hence it is vital to recalibrate the constants involved in each equation. Therefore, constants involved in each equations were recalibrated against FAO modified Penman method using "Linear Regression equation" [18] to improve the performance of each method. The constant values of 0.46, 0.0018, 0.34, 16 used in Blaney – Criddle, Romanenko, Kharrufa, and Thronthwaite method respectively are recalibrated. The mean monthly ET_r estimated using various TB methods with the recalibrated constant values are shown in Table 3. Table 3 shows that the average percentage of error varies form – 0.09% to -0.06%. As far as recalibrated constant values are used, all the methods are gives under estimated ET_r values. But the results shows that there is a significant improvement with recalibrated constant value.

The R^2 value shows that there is no much variation with that of R^2 estimated using original constants. But there is a big trade in intercept and slope. The calculated ET_r with recalibrated constant from the above discussed method are depicted as a scatter plot against FAO modified Penman method is shown in Figure 2. Figure 2 illustrates that there is no significant changes in the scattering pattern of ET_r after calibration of constant and it shows only a change in magnitude of ET_r with that of the FAO modified Penman method.

The original constant and recalibrated constant values against FAO modified Penman method is shown in Table 4. From the Table 4 it is evident that there is great variation in the recalibrated constant of Blaney – Criddle and Thronthwaite method compared to original constant values. The annual difference of ET_r values among the above referred TB methods against FAO modified Penman method before and after recalibration are shown in Table 5. Table 5 shows that the level of difference of annual ET_r is large for all the methods except Blaney-Criddle against FAO modified Penman method with original constant value and it can be seen from the same table that the difference of annual ET_r is significantly reduced for all the methods with help of recalibrated constant values.

Table 3. Mean monthly estimated ET_r in 'mm' using selected methods with recalibrated constant values

Month	$ET_{r\,Pen}$	$ET_{r\,Bla}$	Error (%)	$ET_{r\,Thron}$	Error (%)	$ET_{r\,Rom}$	Error (%)	$ET_{r\,Kha}$	Error (%)
Jun	203.4	204.3	0.44	187.8	-7.64	209.7	3.10	172.6	-15.12
Jul	181.2	168.0	-7.29	163.2	-9.92	195.1	7.69	147.7	-18.47
Aug	178.6	165.3	-7.44	162.2	-9.17	181.0	1.36	152.9	-14.40
Sep	158.7	146.5	-7.73	162.7	2.51	157.3	-0.89	159.4	0.42
Oct	132.2	123.2	-6.79	145.1	9.77	131.1	-0.81	141.6	7.14
Nov	115.3	122.3	6.05	133.2	15.51	117.1	1.52	131.4	13.90
Dec	121.1	137.4	13.48	129.6	7.09	123.9	2.33	136.8	13.01
Jan	138.0	153.3	11.06	139.1	0.79	140.7	1.95	159.4	15.51
Feb	164.7	171.7	4.25	156.8	-4.84	160.6	-2.51	183.6	11.47
Mar	200.5	190.6	-4.95	183.4	-8.52	183.0	-8.73	202.4	0.96
Apr	207.1	209.3	1.03	222.0	7.17	199.0	-3.95	214.9	3.74
May	210.9	218.8	3.75	224.7	6.57	211.6	0.35	207.2	-1.74
Annual	**2011.7**	**2010.5**	**-0.06**	**2009.9**	**-0.09**	**2010.0**	**-0.08**	**2009.9**	**-0.09**

Table 4. Comparison of parameter values before and after calibration in temperature-based methods

Method	Original constant	Recalibrated constant
Blaney-Criddle	0.46	-0.104
Thornthwaite	16	25.689
Romanenko	0.0018	0.002
Kharrufa	0.34	0.294

Table 5. The annual difference in ET_r (mm/year) value between temperature-based method against FAO modified Penman method with original and Recalibrated constants

Method	Difference with original constants	Recalibrated constants difference
Blaney-Criddle	-225.81	1.15
Thornthwaite	544.45	1.79
Romanenko	-394.46	1.64
Kharrufa	-368.10	1.76

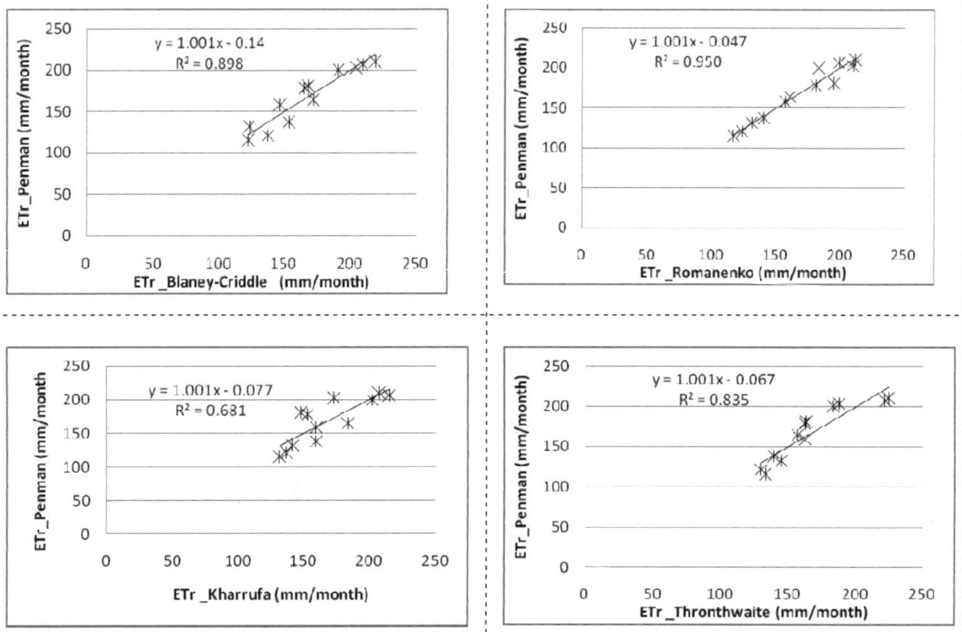

Figure 2. Monthly ET_r estimated using various TB methods against FAO modified Penman method with recalibrated constants.

5. Conclusion

The reference crop evapotranspiration depends upon the meteorological variables. Hence different methods have been developed and applied depending upon the available data. In the present study four temperature-based methods developed in other countries has been evaluated and compared with FAO modified Penman method. After evaluation, the constants in each method were recalibrated, so that it suits better to the present study area Thittakudi, Cuddalore district. Based on the comparison of estimated ET_r with original and recalibrated constants for temperature based methods, it was observed that there were large errors developed when original constants of 0.46, 16, 0.0018 and 0.34 of Blaney-Criddle, Thronthwaite, Romanenko and Kharrufa respectively used. In order to minimize the error, the constants were recalibrated and obtained resepectively as -0.104, 25.68, 0.0015, 0.293. Thus based on the recalibrated constants and R^2 values it can be concluded that Romanenko as well as Blaney-Criddle method may be used for ET_r estimation in the study area.

References

[1] Alexandris S, Stricevic R and Petcovic S 2008 Comparative analysis of reference evapotranspiration from the surface of rainfed grass in central Serbia, calculated by six empirica methods against the Penman-Monteith formula. *Eur. Water* **21**(22) 17-28.

[2] Mohan S and Rama Prasad 1987 Study for Evapotranspiration Models. Research Report No WRI Department of Civil Engineering. *Indian Inst. Sci.* Bangalore.

[3] Jothiprakash V, Ramachandran M R and Shamuganathan P 2002 Artificial Neural Network Model for Estimation of REF-ET. *J. Inst. Eng.* (India **18** 17-20.

[4] Wigmosta M S, Vai, L W and Lettenmaier D P 1994 A distributed hydrology-vegetation model for complex terrain. *Water Resour.Res.* **30**(6) 1665-79.

[5] Romanowicz R A 1997 MATLAB implementation of TOPMODEL. *Hydrol. Process.* **11**(9) 1115-29.

[6] Seibert J, Bishop K H and Nyberg L 1997 A test of TOPMODEL's ability to predict spatially distributed groundwater levels. *Hydrol. Process.* **11**(9) 1131-44.

[7] Senarath S U S, Ogden F L, Downer C W and Sharif H O 2000 On the calibration and verification of two-dimensional, distributed, Hortonian, continuous watershed models. *Water Resour.Res.* **36**(6) 1495-510.

[8] Alexandris S, Stricevic R and Petkovic S 2008 Comparative analysis of reference evapotranspiration from the surface of rainfed grass in central Serbia, calculated by six empirical methods against the Penman-Monteith formula. *Eur. Water* **21/22** 17-28

[9] Ahmed H, and Liu J 2013 Evaluating Reference Crop Evapotranspiration (*ETo*) in the Centre of Guanzhong Basin—Case of Xingping&Wugong, Shaanxi, China. *Eng.* **5**(5) 459-68.

[10] Abhinaya Subedi, José L, and Chávez 2015 Crop Evapotranspiration (ET) Estimation Models: A Review and Discussion of the Applicability and Limitations of ET Methods. *J. Agric. Sci.* **7**(6) 50-68.

[11] Jothiprakash V, Devaman, M G, and Sasireka K 2007 Inter-comparison of Temperature based Reference Crop Evapotranspiration Methods. *J. Inst. Eng.* (India) **88** 19-26.

[12] Jothiprakash V, Devamane M G and Sasireka K 2008 Inter-comparison of Radiation based Reference Crop Evapotranspiration Methods. *J. Inst. Eng.* (India) **89** 22-7.

[13] Doorenbos J and Pruitt W O 1997 Crop Water Requirements. Irrigation and Drainage Paper 24, Food and Agriculture Organization of the United Nations. Rome. Italy: 144.

[14] Bautista F, Bautista D and Delgado-Carranza C 2009 Calibration of the equations of Hargreaves and Thornthwaite to estimate the potential evapotranspiration in semi-arid and subhumid tropical climates for regional applications. *Atmosphera* **22**(4) 331-48.

[15] Romanenko V A 1961Computation of the autumn soil moisture using a universal relationship for a large area. *Proc. Ukrainian Hydrometeorological Res. Inst.* (Kiev) 3.

[16] Kharrufa N S 1985 Simplified equation for Evapotranspirationin Arid Regions. *Beitrage Zur Hydrologie Sonderheft.***5.1.** 39

[17] Hounam C E 1971 Problems of evapotranspiration assessment in the water balance, *Report No. 3,* on WMO/IHD Projects. WMO, Geneva.

[18] Xu C -Y and Singh V P 2001 Evaluation and generalization of temperature- based methods for calculating evaporation. *Hydrol. Process.* **15** 305-319.

IOP Conf. Series: Earth and Environmental Science **80** (2017) 012063 doi:10.1088/1755-1315/80/1/012063

Delineation of Groundwater Potential Zone in Sengipatti for Thanjavur District using Analytical Hierarchy Process

G Siva, N Nasir and R Selvakumar*

School of Civil Engineering, SASTRA University, Thanjavur – 613 401. India.

*Email:selvakumar@civil.sastra.edu

Abstract. Purpose of the ground water is very important to in our world. The condition of the ground water level and occurrence is differing from geological nature of earth. The present study, delineate the ground water potential zone in the hard rock terrain of Sengipatti and surrounding area using Remote sensing, Geographical Information System (GIS) and Multi Criteria Decision Making (MCDM) technique. The Analytical Hierarchy Process (AHP) technique is use to determined the weights of the different themes. Thus thematic layers drainage, geomorphology, land use/land cover, slope, lithology are prepare using Geological Survey of India Toposheets and IRS-IC satellite image. The weights are applied to the thematic layers after linear combination of the all layers. Finally delineate the possible ground water potential zone in the study area. Thus prospective zone are classify the three categories high, moderate, low. It has been total study area of the sengipatti region 32.28% of the area is high ground water potential zone and 34.1% area was moderate ground water potential zone and 33.63% area is low ground water potential zone.

1. Introduction

Water, one of the panchaboothas is vital for sustaining lives. Earlier water was abundant and was used judiciously. But as time progressed the increased population and its associated agricultural and industrial expansion impose demand on limited resources[1]. Thus the increased usage leads to overexploitation of various sources of water and thereby creates a condition of water scarcity. Due to scarcity and pollution of surface water resource, people moved on to exploit the ground water resource [2]. Scientist and technocrats adopted many methods to identify the hidden resource. That too in crystalline terrain, targeting the resource is still enigmatic because apart from primary porosity, various parameters play a vital role in controlling accumulation and movement of groundwater. Amongst the surface electrical resistivity has yielded better results in targeting the groundwater resource[3]. This method is a bit tedious because of its extensive field work. To eliminate the field work, indirect methods were developed. Amongst satellite image based interpretation proven an effective tool. In recent years, owing to its multispectral and multi-temporal characteristics, high resolution satellite images are gaining attention in exploring the groundwater resources especially its potential in precisely identifying various ground features are serving as an indirect indicators in targeting the ground water [4-6]. Likewise the geographic information system (GIS) has proven as an effective tool in delineating groundwater prospect zones by integrating and analysing thematic layers [7-9].

The present study discuss on the identification of groundwater potential zones in and around Sengipatti area, Thanjavur district (Figure.1) using the advanced technology of remote sensing and Analytical Hierarchy Process (AHP). The occurrence, rate of infiltration and storage potential of groundwater is governed by multiple parameters like lithology, geomorphology, slope, drainage density and so on. Hence each terrain parameters have to be systematically assessed based on their importance over the occurrence of groundwater. Further in the study area, no previous studies were conducted to find out the groundwater potential areas and also the relationship between groundwater potential and surface and sub-surface units. For the same, using remote sensing and collateral data, various thematic layers were prepared on both the surface and subsurface parameters. Owing to its complexity, a Multi Criteria Decision Analysis (MCDA) method is applied to target the groundwater. In general, a knowledge based weights were assigned to individual thematic layers and ranks to individual features in each theme [10]. Then by multiplying them, scores of each feature were derivedand based on their derived scores, the groundwater potentiality zones was demarcated. In the above basic method weights and ranks to factors governing the problem are simply assigned based on the expert's knowledge and further the connections between the factors are not considered. But in reality they are highly interdependent with each other. Recently there have been advancements in MCDA leading to various methods[11]. Hence a method called Analytic Hierarchy Process (AHP) was developed. In this method a hierarchy of all the factors is considered. AHP is suitable for taking complex decisions which involve the comparison of parameters that are difficult to quantify[11].AHP is a combination of Mathematics and Psychology. The parameters which we have used for the identification of ground water prospects zone are drainage, geomorphology, land use/land cover, Slope and lithology (Description of rock units).Thenusing ANP, the parameters mentioned above are given weightage in accordance with the other parameters as far their importance in influencing the groundwater is concerned [10]. Later weights were derived in terms of numerical numbers indicating their influence of groundwater over the other features. Using ArcGIS 10.2 software, derived values are assigned to the above themes and finally integrated and classified into different classes of groundwater potential zones.

2. Study area

Sengipatti is a mid-sized village located in the district of Thanjavur in the state of Tamil Nadu in India. The geographical position of the Sengipattiis 10° 43' north Latitude and 78° 57' east Longitude.Sengipatti has a total population of 4,271 peoples. There are about 1,029 houses in Sengipatti village.It is located in midway between Thanjavur and Tiruchirapalli districts along the National Highway. Agriculture is the prime activity however in recent times has more small scale industries. Hence the need the water is heaping day by day.

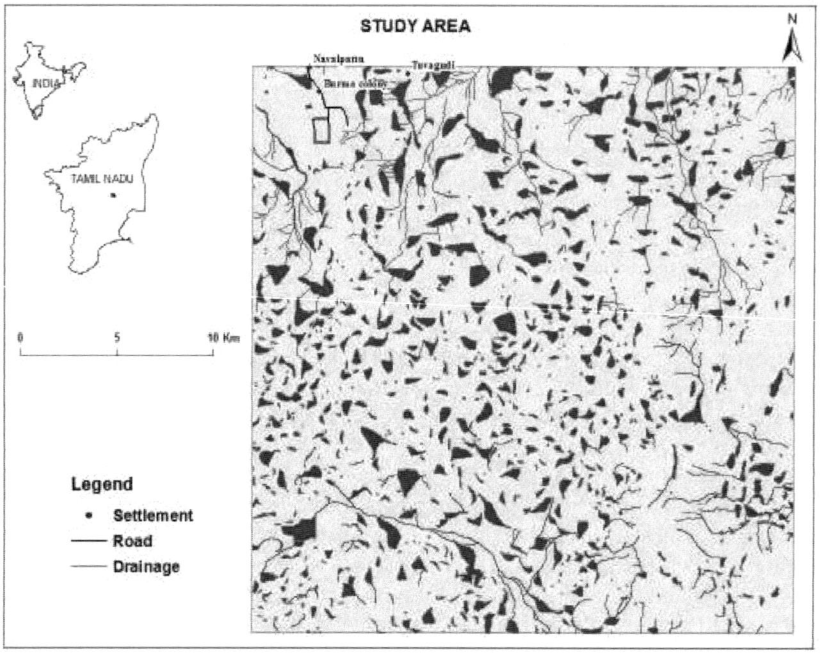

Figure 1. . Study Area.

3. Methodology

In the present study Geocoded satellite data of 20th June 2000 (IRS 1C LISS-III) and Survey of India Toposheet 58J/14 were used. For analysis, Analytic Hierarchy Process (AHP) method is adopted. The parameters which we have used for the identification of ground water prospects zone are drainage, geomorphology, land use/land cover, slope and lithology (description of rock units). The weightage is provided to various parameters by manual process and it is qualitative. It is converted into a quantitative number by creating a table indicating the inter dependency between various parameters. To check whether the assigned weightage is appropriate or not, a ratio known as consistency ratio (C.R) is calculated. This C.R value will be less than 0.1 if the assumed weights are appropriate. Each parameter is further divided into criteria and the individual sub criteria weights are obtained and they are multiplied with the main parameters weightage. Finally, the derived weights are assigned to all the parameters in each thematic layer and subsequently converted into raster datasets. Then using raster addition tool, they were integrated and cumulative factor scores of each pixel was calculated. Consequently, the derived scores were categorized into three and thereby the final groundwater prospect map showing (i) High potential (ii) Moderate potential and (iii) Low potential zone was prepared.

3.1.Surface Parameters

3.1.1 Geomorphology

Geomorphology is the term which mainly deals with the physical shape of the earth and the processes associated with it. There are many factors that play a role in geomorphology. Some among them includes Weathering and erosion, Action of wind, Gravity action etc. (Water Resources- Garald G Parker) Geomorphological characteristics of an area affect its response to a considerable extent. Integrating geomorphological background with hydrological characteristics of an area provides a simple way to understand the groundwater behavior. The parameter geomorphology is further categorized into 6 sub criterions and they are Shallow flood plain, Pediplain moderate, Pediplain shallow, Upland, Pediment Inselberg complex and Residual / Structural hill. The perusal of the geomorphological map shows that the Pediplaincovers majority of the area occupying the central and the south west portion of the study area. While the Shallow flood plains in the south central portion and the uplands in the central and south east parts Sengipatti Village. Structural Hills were observed in the extreme south west portion of the locality (Figure 2).

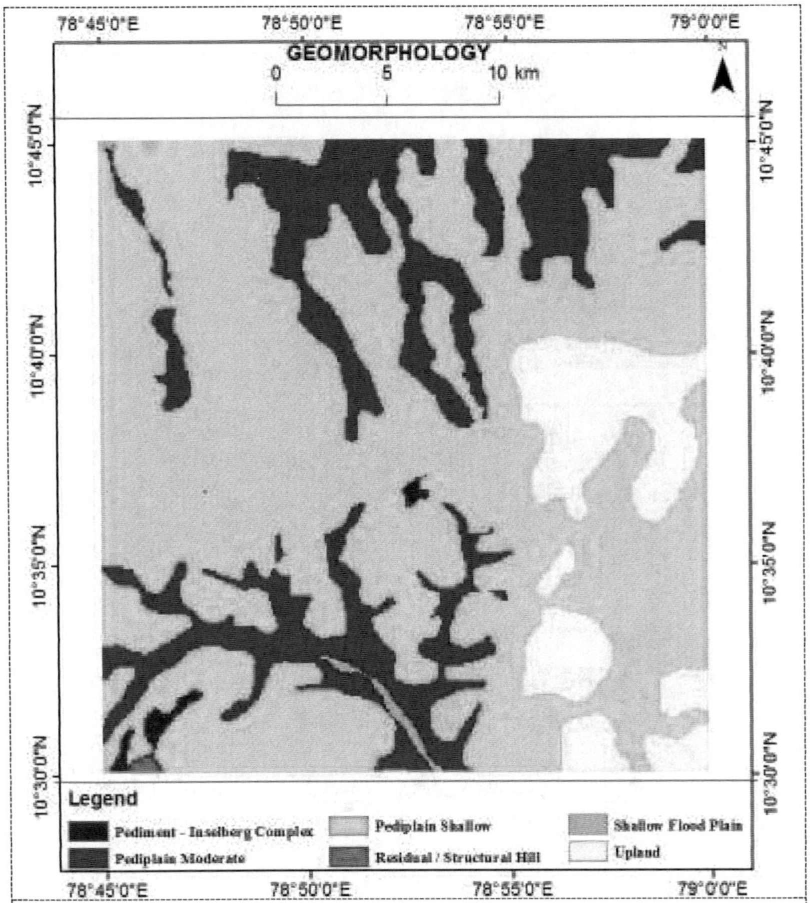

Figure 2. Map showing Geomorphology of the study area

3.1.2 Slope

Slope of an area refers to the natural gradient observed in that locality, which is one of the important parameters as far ground water prospect is concerned. If the slope is too steep then the water runs off quickly and does not infiltrate the ground easily. If the slope is too low then the water stagnates and the condition is suitable for penetration easily and hence ground water improves. We have four sub categories under the parameter slope. They are 0-1%, 1-3%, 3-5% and above 5%. In analyzing the map of the study area taken, the following observations are obtained. Slope of 0-1% is visualized in the central northern part, extreme south eastern part and single central part of the Sengipatti village. The slope percentage 1-3 % is observed in most of the places of the village and it covers the central part and other extremities of the locality. Slightly higher slope percentage of 3-5% is spotted in the central spot, extreme south west and south eastern part of the village. Higher slope percentage exists only in few spots of the village like extreme edge of the map (5-10 %) and other one in western edge with a value of 10-15 % (Figure 3).

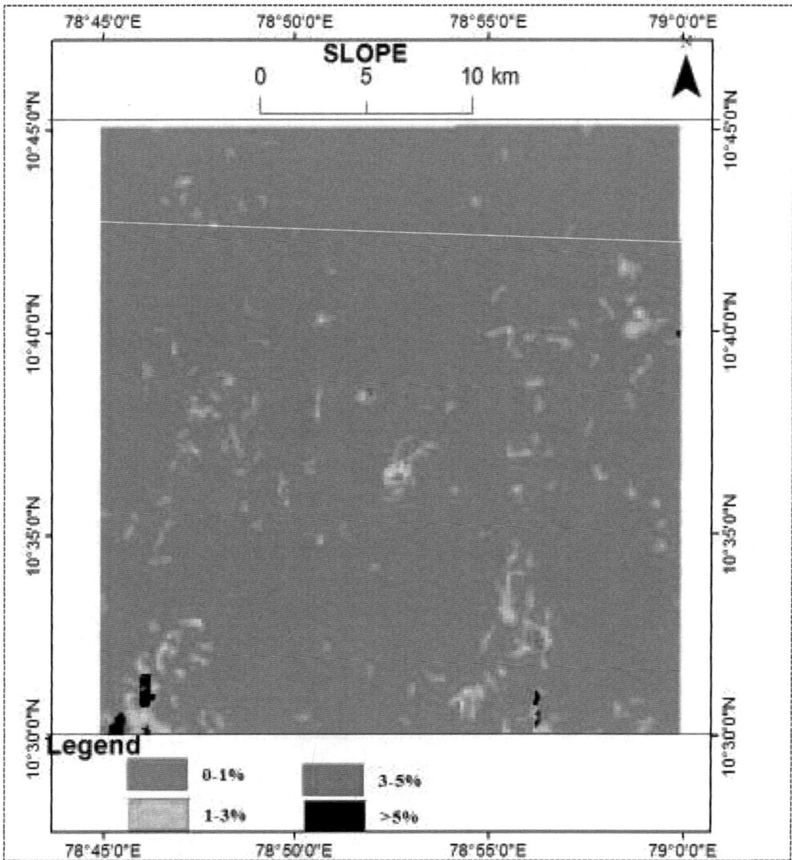

Figure 3. Map showing Slope of the study area

3.1.3 Land Use and Land Cover

Land use/ land cover describes the surface of the earth cover. The land cover is the natural layer which includes features like water bodies, forest and so on while land use is the modified natural environment comprising parameters like crop lands, fallow lands, plantations, built-up lands and so on. These parameters either directly or indirectly influence the groundwater. Agriculture influences ground water by the demand for crop needs. Due to wasteland there is depletion of groundwater in many areas. In forest areas a rise in water table is expected. A rapid deterioration of groundwater might occur if the land is built-up with answered sanitation or poultry farms. The study area consists of agricultural lands, wastelands, built-up and forest lands. The study area is almost covered by agricultural land in all directions. Waste land is found as scattered in the northwest, southeast and central parts of the study while, few built-up lands are witnessed in the northern and southern regions. The forest regions are predominantly found in the eastern and western parts and seldom present in the middle and eastern parts of the study (Figure.4).

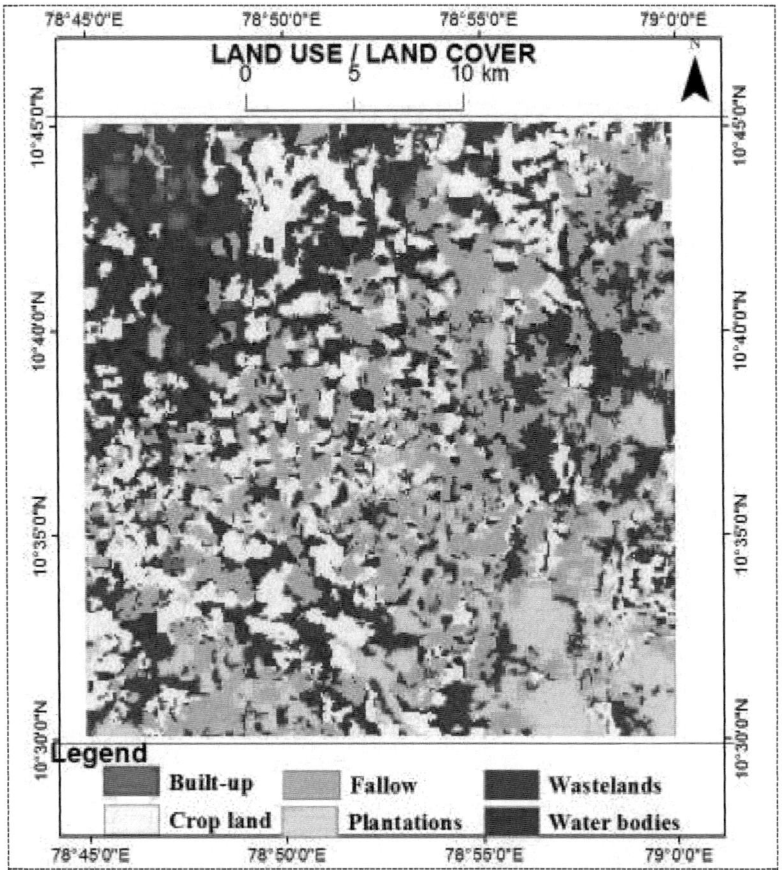

Figure 4. Map showing Land use and Land Cover of the study area.

3.1.4 Drainage

Drainage is the movement of water in terms of and drainage system is the pattern formed by the drainage in an area which includes streams, rivers and lakes thus forming an important part in ground water analysis. Drainage and its pattern depend on number of factors like slope, lithology, soil, land use/land cover and so on. Larger areas of drainage tanks are found in the northern half of the study area.In the study area small sized tanks are found scattered over the entire area. The south eastern part has scarcely distributed tanks. The south western half of the study area has large as well as small sized tanks (Figure.5)

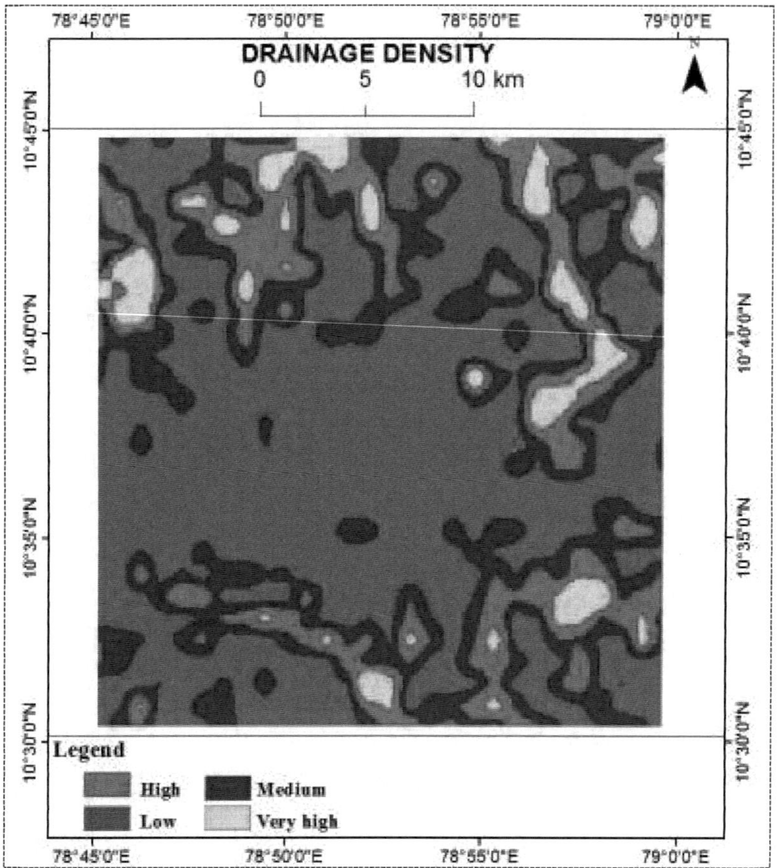

Figure 5. Map showing Drainage Pattern of the study area

3.1.5 Lithology

The lithology describes the physical characteristics of a rock type. There are fivelithological units in the study area viz: clay, Granite/ Quartzite, Laterite, Quartzo-felspathic Gneiss, Hornblende biotite Gneiss. Around 70 percent of the study area is covered with hornblende-biotite Gneiss. Quartzo-felspathic and laterite covers 5 percent each. Quartzite and granite is distributed sporadically over the study area(Figure.6**)**.

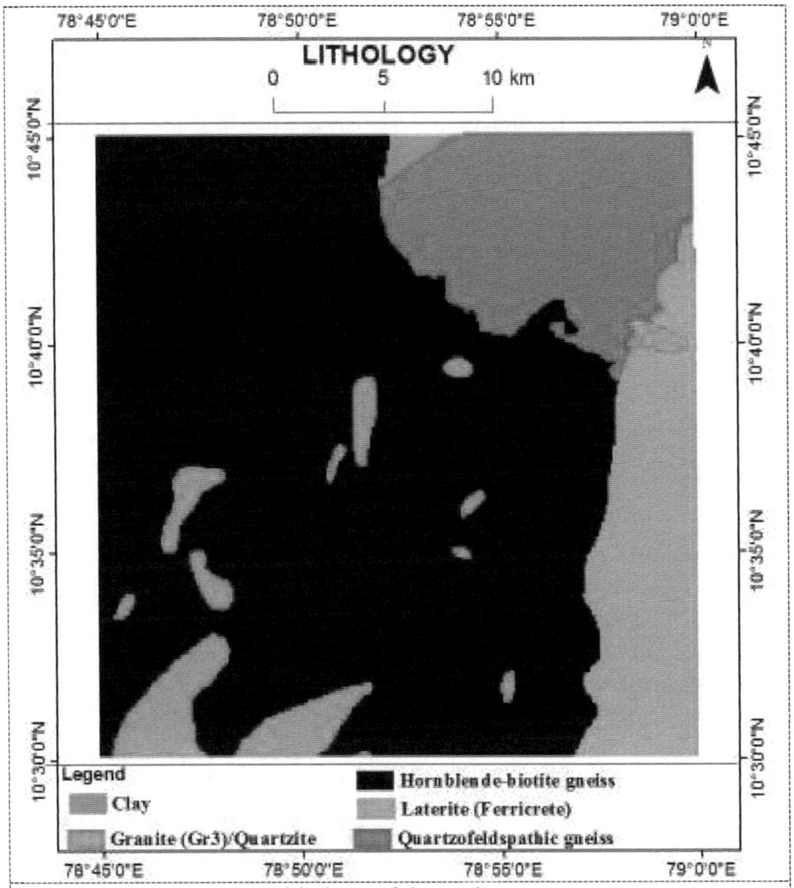

Figure 6. Map showing Lithology of the study area

4. Analytic Hierarchy Process

Analytic hierarchy process is a structured method used for analyzing and organizing complex decisions using the principle of psychology and mathematics. It was developed by Thomas L.Saaty in 1970's. Rather than giving the correct solution to the issue at hand it comes out with the best solution based on the clients need. Hierarchy process deals with step wise operations to give out a solution. It works on the principle of Multi Criteria Decision Making (MCDM) and is useful in analyzing the real world problems which involves multiple criteria. In this process, information is fragmented into packets of alternatives and criteria. This information is further synthesized to derive the relative ranking of alternatives. A final weight obtained gives the best outcome.

4.1 Assigning Weights

The process of assuming weights are done qualitatively as already stated. The weights are given manually and it is verified with the help of certain terms. The terms include Final Weightage for each parameter, Consistency Index, Random Consistency Index and Consistency ratio.

The above criteria are assumed with weights in accord to our study area. The Slope Gradient was observed to be almost constant throughout our study area thus resulting with the weightage of 1. Lithology is nothing but the rock units prevailing in the study area and it seemed to be varying in parts of the study area. Drainage and Land use almost go on the same line, resulting with equal weightage to them. Geomorphology is given with maximum weightage. The above results are tabulated by obtaining the final weightage from each of the parameter with respect to their sub criteria (Table 1)

Table 1. Criteria Weights

	Slope	Lithology	Drainage	Land use	Geomor phology	product	5th root	Weights
Slope	1.000	0.200	0.143	0.143	0.111	0.0004	0.214	0.0275
Lithology	5.000	1.000	0.330	0.200	0.143	0.0471	0.542	0.0697
Drainage	7.000	3.000	1.000	0.330	0.200	1.386	1.067	0.1370
Land use	7.000	5.000	3.000	1.000	0.330	34.65	2.032	0.2608
Geomorphology	9.000	7.000	5.000	3.000	1.000	945	3.936	0.5051

4.2 Slope Weights

0-1% slope is a very low gradient unit and hence the water will not flow under the action of gravity thereby increasing the chances of water penetrating into the underlying layers. Thus when the slope increases the probability that ground water exists gradually decreases. Higher Slopes permits easy flow of water and reduces the water existing in that region. The weightage above are also allocated on the same principle (Table 2).

Table 2. Weights derived for slope sub criteria.

	>5%	3-5%	1-3%	0-1%	Product	4th Root	Weight
>5%	1.000	0.333	0.200	0.143	0.009	0.312	0.0550
3-5%	3.000	1.000	0.333	0.200	0.2	0.668	0.1177
1-3%	5.000	3.000	1.000	0.333	5	1.495	0.2633
0-1%	7.000	5.000	3.000	1.000	105	3.201	0.5638

4.3 Lithology Weights

Clay layers absorb the water within its particles just like hygroscopic sponge and hence the level of groundwater below is fairly reduced. Granite Is a part of the bed rock above the groundwater level. Unlike limestone and chalk that easily get weathered or dissolved, the former withstands a large amount of water not paving its way for penetration. Laterite is formed by intensive weathering, having porous and permeable layers. They enable the function of acquifers also and hence increase the ground water level (Table 3)

Table 3. Weights derived for Lithology sub criteria

	Clay	Granite/ Quartzite	Laterite	QF gneiss	HblBioti te gneiss	Product	5th root	Weight
Clay	1.000	1.000	0.333	0.200	0.143	0.009	0.394	0.0537
Granite /Quartzite	1.000	1.000	0.333	0.200	0.143	0.009	0.394	0.0537
Laterite	3.000	3.000	1.000	0.333	0.200	0.6	0.902	0.1230
QF gneiss	5.000	5.000	3.000	1.000	0.333	25	1.903	0.2594
Hbl.Biotite gneiss	7.000	7.000	5.000	3.000	1.000	735	3.743	0.5100

4.4 Drainage Density Weights

Drainage density showsthe closeness in spacing of steams within a given square grid area. The runoff will be more if density is high while infiltration will be more if density is low. For the same, grid map with 250 x 250m grid size is overlaid over the drainage map and therefrom total length of the drainage in each grid was calculated. Subsequently, contours were generated using SURFER Software and categorized into four as very high, high, moderate and low density areas.Accordingly weights were assigned with Very High density (1485 – 3845 m/sq.km) are given 1, High density (775-1485 m/sq.km) is given 3, Medium density (275- 775m/sq.km) is given 5, and Low density (0-275 m/sq.km) is given 7 (Table 4).

Table 4. Weights derived for Drainage Density sub criteria

	Very High	High	Medium	Low	Product	4th Root	Weight
Very High	1.000	0.333	0.200	0.143	0.009	0.312	0.0550
High	3.000	1.000	0.333	0.200	0.2	0.668	0.1177
Medium	5.000	3.000	1.000	0.333	5	1.495	0.2633
Low	7.000	5.000	3.000	1.000	105	3.201	0.5638

4.5. Land Use / Land Cover Weights

Built up lands are those in which structures are raised and commercial activities take place, so in these lands there is a less possibility that ground water exists leading to the least weightage. Waste lands are almost the same but only with a difference that it is not usable for both agriculture and constructional purposes. Plantation is provided with comparatively higher weightage as there exists few plants which comes out of moisture prevailing in the ground. Fallow lands are used for agricultural purposes in certain period of time so weightage given is the same as plantation and its 3. Crop Lands are those which are completely utilized for agriculture and without water agriculture doesn't exists, so its given with a weightage of 7. Finally Water Bodies are sure of having water in the underlying depths and thus given with the highest weightage of 9 (Table 5).

Table 5. Weights derived for Land use / Land Cover sub criteria

	Built up land	Waste land	Planta Tion	Fallow land	Crop land	Water bodies	Product	6th Root	Weight
Built up land	1.00	1.00	0.333	0.333	0.143	0.111	0.001	0.347	0.0361
Waste land	1.00	1.00	0.333	0.333	0.143	0.111	0.001	0.347	0.0361
Plantation	3.00	3.00	1.000	1.000	0.200	0.143	0.257	0.797	0.0828
Fallow land	3.00	3.00	1.000	1.000	0.333	0.200	0.6	0.918	0.0953
Crop land	7.00	7.00	5.000	3.000	1.000	0.200	147	2.297	0.2386
Water bodies	9.00	9.00	7.000	5.000	5.000	1.000	14175	4.919	0.5109

4.6 Geomorphology Weights

Shallow Flood plain can be defined as the area where the water exists at shallow depths above the ground level. So in these areas Ground water will always be in charged state. Pediplain surfaces are categorized into two sub divisions as moderate and shallow. Uplands are the areas of topographically unstable terrain, with high peaks and valleys. In simple terms, elevated mountainous plateau. As these are elevated, due to the inclination the water flows down from these areas into low lying areas. Pediment Inselbergs are isolated hill that stands above well-developed plains and its height varies according to different terrains. This category is the one in which ground water is in deficit as it is an isolated huge rock. Hills are the structure that extends above the surrounding terrain. Some amount of water underlies beneath the hills and mountains, but they are not easily accessible and can't be directly used for potable purposes. Accordingly the weightage is provided considering all the above stated reasons (Table 6).

Table 6. Weights derived for Geomorphology sub criteria

	Residual/ Structural Hill	Pediment Inselberg complex	Up land	Pediplain shallow	Pediplain moderate	Shallw flood plain	Product	6th Root	Weight

Residual/ Structural hill	1.00	1.00	0.5	0.333	0.143	0.111	0.002	0.371	0.0389
Pediment Inselberg complex	1.00	1.00	0.0	0.333	0.200	0.143	0.004	0.410	0.0429
Upland	2.00	2.00	1.0	0.500	0.200	0.143	0.057	0.620	0.0650
Pediplain shallow	3.00	3.00	2.0	1.000	0.333	0.143	0.857	0.974	0.1021
Pediplain moderate	7.00	5.00	5.0	3.000	1.000	0.200	105	2.172	0.2276
Shallow flood plain	9.00	7.00	7.0	7.000	5.000	1.000	15435	4.989	0.5230

5. Results and Discussion

In the present study, using satellite imagegeomorphological and Landuse/Landcover parameters were precisely mapped. Geomorphologically, the area is as represented by alluvial plain, pediplain, and Upland. While the land use/land cover is represented by crop land, land with scrub, land without scrub, water bodies and buildup lands.On caring out the above study it can be concluded that remote sensing, GIS and AHP, when used together is a very effective technique for mapping. Using satellitesand conventional data various thematic maps were generated on base, geology, geomorphology and Land use/Land cover. All the above various thematic maps have been digitized, GIS data base have been creating using Arc GIS 10.2 software. The GIS tools provides a variety of scenario's to be generated and studied before actual finalizing of plans for implementation. The ranks obtained for all the main parameters and their sub-criteria were given to the respective thematic layers and converted into raster format using ArcGIS software. Using raster calculator, thematic layers were added. Finally the derived groundwater potential map evaluated by the weighted linear combination of these weights is shown in Figure.7.

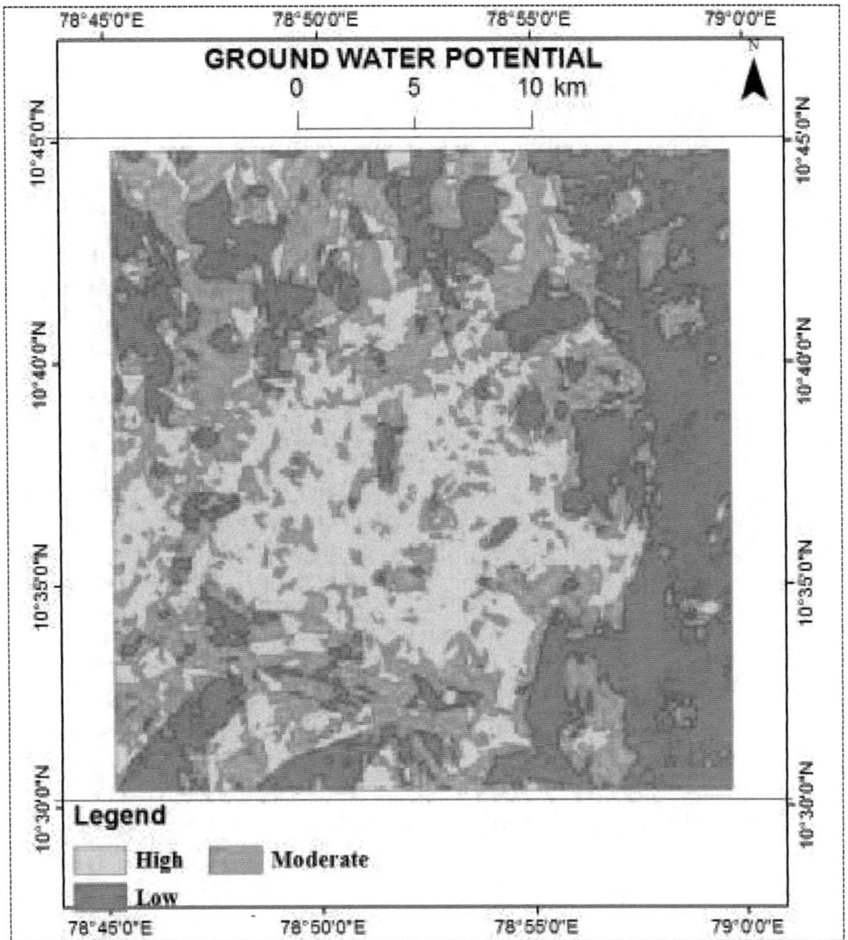

Figure 7. Map showing Groundwater Prospects of the study area.

Results indicated that only32.28% of the area is high ground water potential zone and 34.1% area was moderate ground water potential zone and 33.63% area is low ground water potential zone. Thus the above study demonstrates the potential remote sensing and GIS along with AHP technique in identifying the groundwater potential zone especially in hard rock terrain.

6. Conclusion

The present study details the potential of remote Sensing, GIS and AHP for precisely demarcating the groundwater potential zone, Using remote sensing and collateral data, thematic mapping was done and consequently converted into vector database using GIS. Weights were derived for each parameter in each thematic layers using AHP technique and finally by overlay analysis, the groundwater potential zone was demarcated. In Sengipatti area, three groundwater potential zones have been delineated. The results conclude that the Sengipatti are in Thanjavur district is a water potential zone. The perusal of Groundwater potential map indicates that most of the area falls under moderate to high potential category. Hence it is suggested that if the resources are properly utilized, the growth of industrial, commercial and agricultural sectors can be achieved.

References:

[1] Siebert S Burke J FauresJ MFrenken K. HoogeveenJ Doll P and Portmann F T 2010 Groundwater use for irrigation – a global inventory. *Int. J. Hydrol. Earth Syst. Sci.* **14**, 1863-80.

[2] Bharti N and Katyal D 2011 Water quality indices used for surface water vulnerability assessment. *Int. J. Environ. Sci.* **2**(1) 154-73.

[3] Vijay Kumar V, Ramadass G and Jagadish S V 2015 Delineation of Groundwater Potential Zones through Electrical Resistivity Parameters in hard rock terrain , Osmania University Campus, Hyderabad, Telangana State, India. *IOSR J. Appl. Geol. Geophysics* **3**(6) 1-10.

[4] Bahuguna I M NayakS Tamilarsan V and Moses J 2003 Groundwater prospective zones in Basaltic terrain using remote sensing. *J. Indian Soc. Remote Sensing* **31**(2) 107-18

[5] JagadeeswaraRao P, Harikrishna P and Suryaprakasa Rao B 2004 An integrated study on groundwater resources of Pedda Gedda watershed. *J. Indian Soc. Remote Sensing* **32**(3) 307-11.

[6] Dhakate R, Singh V S, Negi B C, Chandra S and Ananda Rao V 2008 Geomorphological and geophysical approach for locating favorable groundwater zones in granitic terrain, Andhra Pradesh, India. *J. Environ. Manage.* **88** 1373-83.

[7] Ramasamy S M, Nagappan N and Selvakumar R 2005 Fracture Pattern Modelling and Ground water Hydrology in Hard Rock Aquifer System, Central Tamil Nadu, India. SM. Ramasamy (ed.) *Remote Sensing in Water Resources*. Rawat Publ.121-36.

[8] Rokade V M Kundal P and Joshi A K 2007 Groundwater potential modeling through remote sensing and GIS: A case study from Rajura Taluka, Chandrapur District, Maharashtra, *Jour. Geol. Soc. India* **69**(5) 943-8.

[9] Thushan Chandrasiri Ekneligoda and Herbert Henkel 2010 Interactive spatial analysis of lineaments. *Comput. Geosci.* **36** 1081-90.

[10] Amir Hossein Hatefi Ardakani and Mohamad Reza Ekhtesasi 2016Groundwater potentiality through Analytic Hierarchy Process (AHP) using remote sensing and Geographic Information System (GIS). *J. Geope.* **6**(1) 75-88.

[11] Saaty T L 1980 The Analytic Hierarchy Process, McGraw-Hill, New York, NY.

Water supply pipe dimensioning using hydraulic power dissipation

J R Sreemathy, G Rashmi and C R Suribabu*

School of Civil Engineering, SASTRA University, Thanjavur – 613 401, India

*Email: suribabu@civil.sastra.edu

Abstract. Proper sizing of the pipe component of water distribution networks play an important role in the overall design of the any water supply system. Several approaches have been applied for the design of networks from an economical point of view. Traditional optimization techniques and population based stochastic algorithms are widely used to optimize the networks. But the use of these approaches is mostly found to be limited to the research level due to difficulties in understanding by the practicing engineers, design engineers and consulting firms. More over due to non-availability of commercial software related to the optimal design of water distribution system,it forces the practicing engineers to adopt either trial and error or experience-based design. This paper presents a simple approach based on power dissipation in each pipeline as a parameter to design the network economically, but not to the level of global minimum cost.

1. Introduction

Water distribution system is a hydraulic infrastructure that consists of various components such as pipes, tanks, reservoirs, pumps, and valves etc.[24]. Consumer satisfaction in terms of quantity and quality is an important responsibility of a service provider. Effective water supply is of paramount importance in designing a new water distribution network or in expanding the existing one. It is also essential to investigate and establish a reliable network ensuring adequate pressure head at consumer nodes. Determination of flow and pressure in network pipes has been of great value and interest for those involved with design, construction and maintenance of public water distribution systems. Analysis and design of pipe networkare considered a relatively complex problem, particularly if the network consists of wide ranges of pipes as it frequently occurs in water distribution systems of large metropolitan areas. In the absence of significant fluid acceleration, the behaviour of a network can be visualized as sequence of steady state conditions, which forms a simple practice but is a vital component for assessing the adequacy of a network [15]. Use of optimization techniques in deciding the size of pipes in the network from a set of commercially available diameters have been attempted for more than three decades due to its computational and engineering complexity [1-5,8,11,12,18-21]. Software and programs developed based on demand driven analysis are widely coupled with optimization algorithm to explore pipe diameters that satisfy the hydraulic-head requirements at least cost. [13] and [10] developed a rule based search approach for economical pipe sizing. [7] developed an iterative model in which a network is designed by setting all the pipes to minimum acceptable pipe diameters followed by aniterative process progressing in such a way that the additional cost is as low as possible while the diameter of pipe is increased. [19] presented a new approach in which flow velocity in a pipe is considered as implicit information for pipe sizing. As there is a growing interest in developing a methodology based on heuristics search instead of experimenting with various new populations based algorithms, the present work aims to dimension the pipes using hydraulic power dissipation of energy in the pipeline. The head loss in the pipe is not onlyattributed to the roughness of the pipe, but also due to water demand, pipe length, gradient and diameter. The network configuration initially set by either smaller size or larger size is progressively improved to reach economical configuration using power dissipated in the pipeline to overcome the friction loss. The proposed methodology is illustrated using some well-known bench mark networks.

2. Hydraulic simulation

This study presents the use of EPANET software in the design of the water distribution networks which performs extended period simulation of hydraulic and water quality behaviour within pressurized pipe networks [3]. A network usually consists of pipes, nodes (pipe junctions), pumps, valves and storage tanks or reservoirs. EPANET tracks the flow of water in each pipe, the pressure at each node and the height of water in each tank throughout the network [14]. This software is used to calculate the power dissipation in each link after updating from initial network configuration. The results in each simulation run are used to verify the pressures at all junctions and the flow at all pipes.

3. Method of approaches

This paper presents two methods to dimension the pipes in the network. Both the approaches consider hydraulic power dissipation as the main parameter for sizing the pipe from its initial configuration. In the decreasing diameter method, maximum size is assigned to all pipes of the network as initial configuration, whereas in the case of increasing diameter approach, minimum size is assigned to compose the initial configuration for optimization.

3.1 Decreasing diameter method

1. The pipe system is plotted using EPANET software [13].
2. The corresponding length of the pipes, Hazen William's roughness coefficients, the base demand for various nodes and head available at the tankare given as input.
3. Initially the diameter of all pipes is set to the maximum available commercial size.
4. A hydraulic analysis is then performed and the pressure head at various nodes and the flow and head loss at various links are noted.
5. The power loss ($\rho g Q h_f$) due to friction is calculated and tabulated.
6. According to the decreasing diameter approach, the link with minimum power loss is found out and its diameter is decreased subsequently to a lower diameter.
7. Hydraulic analysis is performed again and the pressure head, flow and friction losses are tabulated.
8. Again, the link with minimum power loss is found and its diameter is reduced.
9. Steps 5 and 6 are repeated till the pressure head reaches a value close to the minimum required pressure head.
10. After completing the above steps, the size of each pipe is reduced to next size and violation of pressure is verified, if no violation of pressure is found, then the new diameter is assigned and this process is repeated for all the pipes in the network for further reduction of cost.

3.2 Increasing diameter method

1. The pipe network is plotted using EPANET software.
2. The corresponding length of the pipes, Hazen William's roughness coefficients, the base demand for various nodes and head available at the tankare given as input and the flow is measured.
3. Initially the diameter of all the pipes is set to a minimum available commercial size.
4. Hydraulic analysis is performed and the pressure head at various nodes and the flow and friction loss at various links are noted.
5. Power loss due to friction is calculated and tabulated.

6. According to the increasing diameter approach, the link with maximum power loss is found and its diameter is increased subsequently to a higher diameter.
7. Steps 5 and step 6 are repeated till the pressure head in all the nodes reaches minimum service pressure.
8. After completing the above steps, the size of each pipe is reduced to next size and violation of pressure is verified, if no violation of pressure is found, then the new diameter is assigned and this process is repeated for all the pipes in the network for further reduction of cost.

4. Benchmark networks

4.1 Hanoi network

The Hanoi water distribution network is taken from the literature for illustration of the recommended method. It is arranged in 3 loops with thirty-twonodes and thirty-four pipes. It has only one fixed head source at an elevation of 100 m and no pumping provisions. The required minimum head at all the nodes is set as 30 m. Commercially obtainable pipe diameters of 12, 16, 20, 24, 30 and 40 inches are taken. Figure 1, shows the layout of network. Table 1a, provides the complete data for the design of the network. The Hazen William roughness factor of each pipe is considered as 130. Table 1b, provides the cost details for the network.The results of proposed methods are presented in Table 1c, and for comparison purpose, two results available in the literature are also taken.

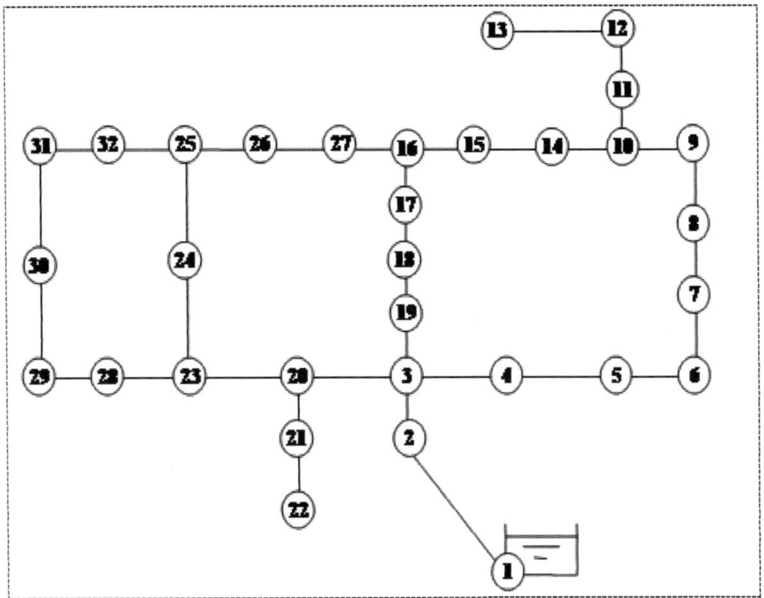

Figure 1. Layout of Hanoi network

IOP Conf. Series: Earth and Environmental Science **80** (2017) 012064 doi:10.1088/1755-1315/80/1/012064

Table 1a. Pipe length and nodal demand details

Node no.	Demand (cu.m/h)	Link index	Arc	Length (m)
1	-19940	1	(1,2)	100
2	890	2	(2,3)	1350
3	850	3	(3,4)	900
4	130	4	(4,5)	1150
5	725	5	(5,6)	1450
6	1005	6	(6,7)	450
7	1350	7	(7,8)	850
8	550	8	(8,9)	850
9	525	9	(9,10)	800
10	525	10	(10,11)	950
11	500	11	(11,12)	1200
12	560	12	(12,13)	3500
13	940	13	(10,14)	800
14	615	14	(14,15)	500
15	280	15	(15,16)	550
16	310	16	(16,17)	2730
17	865	17	(17,18)	1750
18	1345	18	(18,19)	800
19	60	19	(19,3)	400
20	1275	20	(3,20)	2200
21	930	21	(20,21)	1500
22	485	22	(21,22)	500
23	1045	23	(20,23)	2650
24	820	24	(23,24)	1230
25	170	25	(24,25)	1300
26	900	26	(25,26)	850
27	370	27	(26,27)	300
28	290	28	(27,16)	750
29	360	29	(23,28)	1500
30	360	30	(28,29)	2000
31	105	31	(29,30)	1600
32	805	32	(30,31)	150
		33	(31,32)	860
		34	(32,25)	950

Table 1b Cost data for Hanoi network

Diameter (mm)	Cost per m length ($/m)
304.8 (12 in)	45.73
406.4 (16 in)	70.43
508 (20 in)	98.38
609.6 (24 in)	129.333
762 (30 in)	180.8
1016 (40 in)	278.3

Table 1c. Optimal results for Hanoi network

Node/ Pipe no.	Increasing diameter approach		Decreasing diameter approach		Differential evolution algorithm (suribabu, 2010)		Heuristic based approach (suribabu, 2012)	
	Diameter (mm)	Pressure (m)	Diameter (mm)	Pressure (m)	Diameter (mm)	Pressure (m)	Diameter (mm)	Pressure (m)
1	1016	100	1016	100	1016	100	1016	100
2	1016	97.14	1016	97.14	1016	97.14	1016	97.14
3	1016	61.67	1016	61.67	1016	61.67	1016	61.67
4	1016	57.94	1016	57.94	1016	56.92	1016	57.59
5	1016	53.34	1016	53.34	1016	51.02	1016	52.54
6	762	48.61	762	48.61	1016	44.81	1016	47.31
7	1016	44.32	762	44.33	1016	43.35	1016	46.11
8	762	43.18	1016	39.71	1016	41.61	1016	44.78
9	762	39.72	762	38.86	1016	40.23	1016	43.74
10	762	37.36	762	36.50	762	39.20	762	43.02
11	609.6	35.8	609.6	34.94	609.6	37.64	762	41.46
12	762	32.37	762	31.51	609.6	34.21	508	40.3
13	304.8	30.95	304.8	30.09	508	30.01	304.8	30.08
14	304.8	35.48	304.8	34.64	406.4	35.52	304.8	31.93
15	406.4	39.04	304.8	38.22	304.8	33.72	304.8	32.00
16	609.6	41.63	609.6	48.78	304.8	31.30	508	34.73
17	762	52.98	1016	57.38	406.4	33.41	609.6	43.42
18	1016	57.99	1016	58.40	609.6	49.93	609.6	50.78
19	762	59.18	762	59.46	508	55.09	762	60.05
20	1016	52.46	1016	51.43	1016	50.61	1016	51.21
21	508	43.11	508	42.48	508	41.26	508	41.86
22	304.8	37.95	304.8	37.32	304.8	36.10	304.8	36.70
23	1016	47.89	1016	46.76	1016	44.52	1016	45.63
24	762	44.75	762	42.95	762	38.93	762	40.70
25	609.6	39.96	762	40.82	762	35.34	762	37.66
26	304.8	33.21	406.4	36.81	508	31.70	406.4	30.55
27	304.8	36.25	304.8	37.54	304.8	30.76	304.8	30.54
28	406.4	38.48	304.8	37.83	304.8	38.94	304.8	40.10
29	406.4	33.1	406.4	32.83	406.4	30.13	406.4	39.49
30	406.4	31.45	406.4	31.60	304.8	30.42	304.8	31.81
31	304.8	31.81	304.8	32.02	304.8	30.70	304.8	32.98
32	304.8	36.05	304.8	36.78	406.4	33.18	304.8	35.50
33	304.8		304.8		406.4		406.4	
34	508		508		609.6		609.6	
Cost	$ 6.477 million		$ 6.717 million		$6.0801 million		$6.2323 million	

4.2 GoYang network

GoYang network consists of thirty pipes, twenty-two demand nodes.A single fixed pump of 4.52 kW is connected to thereservoir with a constant head of 71 m. The Hazen-Williams roughness coefficient for each new pipe is taken as 100. The minimum pressure head for each node is set as 15 m. Figure 2, shows the layout of network. Table 2a, provides the complete data for the design of the network. Table 2b, provides the cost details for the network. The results of proposed methods are presented in Table 2c, and for comparison purpose, two results available in the literature are also taken.

Table 2a. Pipe length and nodal demand details

Node no.	Demand (L/s)	Link index	Length (m)
1	0	1	165
2	1.77	2	124
3	0.82	3	118
4	0.68	4	81
5	0.87	5	134
6	0.78	6	135
7	0.73	7	202
8	0.56	8	135
9	0.49	9	170
10	0.35	10	113
11	0.49	11	335
12	0.43	12	115
13	0.43	13	345
14	0.73	14	114
15	5.16	15	103
16	1.25	16	261
17	0.92	17	72
18	0.64	18	373
19	1.37	19	98
20	1.44	20	110
21	0.37	21	98
22	9.25	22	246
		23	174
		24	102
		25	92
		26	100
		27	130
		28	90
		29	185
		30	90

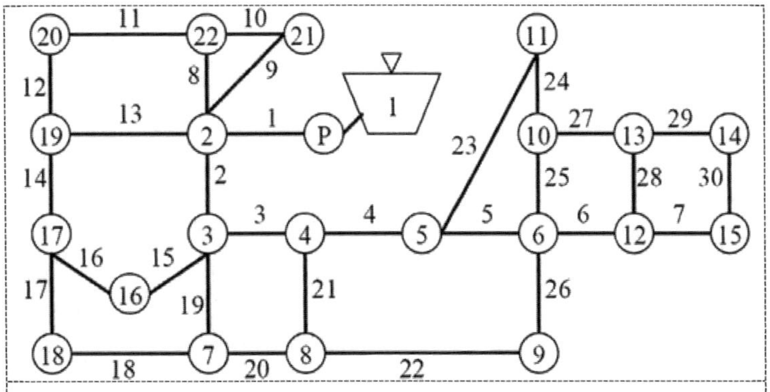

Figure 2. Layout of GoYang network

Table 2b. Cost data for GoYang network

Diameter (mm)	Cost ($/mm)
80	37.89
100	38.933
125	40.563
150	42.554
200	47.624
250	54.125
300	62.109
350	71.524

4.3 New York Tunnel network

This network includes twenty-one pipes arranged in two loops, nineteen demand nodes and one reservoir with a fixed head of 300 ft. (1 ft.=0.3048 m). In order to meet the projected increased demands, all the existing pipes are considered for replication. The Hazen-Williams roughness coefficient for both new and current pipes is 100. The minimum pressure headto be maintained at all nodes is fixed at 255 ft. except for node 16 and 17, where the minimum pressure head that has to be maintained is fixed at 260 ft. and 272.8 ft. respectively.Figure 3, shows the layout of network. Table 3a provides the complete data for the design of the network. Table 3b provides the cost details for the network. The results of proposed methods are presented in Table 3c and for comparison purpose; two results available in the literature are also taken.

IOP Conf. Series: Earth and Environmental Science **80** (2017) 012064 doi:10.1088/1755-1315/80/1/012064

Table 2c. Optimal results for GoYang network

Node / Pipe no.	Increasing diameter approach		Decreasing diameter approach		Heuristic based algorithm (Menon,2016)		Harmony search algorithm (Geem,2006)	
	Diameter (mm)	Pressure (m)	Diameter (mm)	Pressure (m)	Diameter (mm)	Pressure (m)	Diameter (mm)	Pressure (m)
1	250	15.62	250	15.62	200	15.62	150	15.61
2	150	29.79	150	29.79	150	28.93	150	24.91
3	100	31.26	100	31.26	100	30.4	125	26.32
4	80	27.78	80	27.78	100	26.89	150	24.11
5	80	23.6	80	23.6	80	24.54	100	22.78
6	80	21.32	80	21.32	80	21.9	100	20.67
7	80	29.76	80	29.76	100	28.92	80	25.34
8	100	28.05	100	28.05	80	27.26	80	24.41
9	80	21.42	80	21.42	80	21.64	80	20.01
10	80	15.91	80	15.91	80	16.55	80	15.43
11	80	15.65	80	15.65	80	16.39	80	15.06
12	80	17.93	80	17.93	80	18.54	80	18.16
13	80	17.23	80	17.23	80	17.84	80	17.38
14	80	15.11	80	15.11	80	15.72	80	15.72
15	80	15.25	80	15.25	80	15.86	80	15.42
16	80	30.69	80	30.69	80	29.83	80	25.88
17	80	28.92	80	28.92	80	28.07	80	24.29
18	80	28.56	80	28.56	80	27.72	80	23.99
19	80	29.54	80	29.54	80	28.69	80	24.89
20	80	29.13	80	29.13	80	28.29	80	24.43
21	80	21.72	80	21.72	80	20.87	80	16.89
22	80	22.02	80	22.02	80	21.16	80	17.21
23	80		80		80		80	
24	80		80		80		80	
25	80		80		80		80	
26	80		80		80		80	
27	80		80		80		80	
28	80		80		80		80	
29	80		80		80		80	
30	80		80		80		80	
Cost	$ 178,193.9		$ 178,193.9		$ 177,275.6		$ 176,994.6	

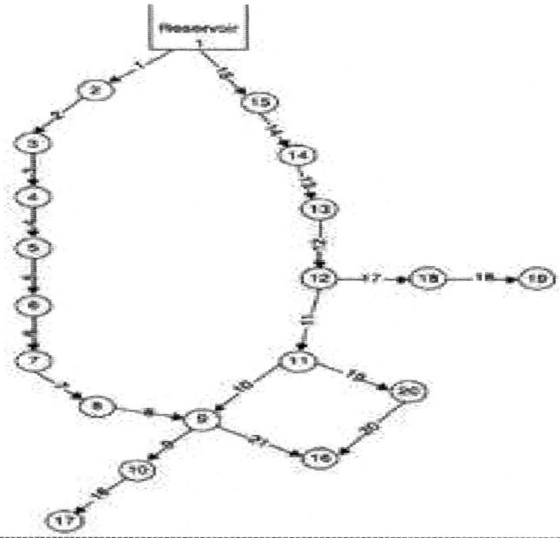

Figure 3. Layout of NYT network

Table 3a. Pipe length and nodal demand details

Node no.	Demand (CFS)	Link index	Length (ft.)
2	92.4	1	11600
3	92.4	2	19800
4	88.2	3	7300
5	88.2	4	8300
6	88.2	5	8600
7	88.2	6	19100
8	88.2	7	9600
9	170	8	12500
10	1	9	9600
11	170	10	11200
12	117.1	11	14500
13	117.1	12	12200
14	92.4	13	24100
15	92.4	14	21100
16	170	15	15500
17	57.5	16	26400
18	117.1	17	31200
19	117.1	18	24000
20	170	19	14400
		20	38400
		21	26400

Table 3b. Cost data for NYT network

Diameter (in.)	Cost ($/ft.)
36	93.59
48	133.70
60	176.32
72	221.05
84	261.61
96	315.80
108	365.46
120	416.46
132	468.71
144	522.11
156	576.59
168	632.09
180	688.54
192	745.91
204	804.14

Table 3c. Optimal results for NYT network

Node/ Pipe no.	Increasing diameter approach		Decreasing diameter approach		Differential evolution algorithm (Suribabu, 2010)		Ant colony optimization algorithm (Maier et al.,2003)	
	Diameter (mm)	Pressure (m)	Diameter (mm)	Pressure (m)	Diameter (mm)	Pressure (m)	Diameter (mm)	Pressure (m)
1	0	300	0	300.00	0	300.00	0	300.00
2	0	294.42	0	295.96	0	294.21	0	294.21
3	0	286.69	0	290.61	0	286.15	0	286.15
4	0	284.44	0	289.13	0	283.79	0	283.79
5	0	282.46	0	287.92	0	281.70	0	281.70
6	0	280.94	0	287.08	0	280.07	0	280.07
7	0	278.57	0	285.96	144	277.51	144	277.51
8	48	275.1	0	284.71	0	276.67	0	276.67
9	84	272.86	0	284.20	0	273.78	0	273.78
10	0	272.84	0	284.17	0	273.74	0	273.74
11	0	272.96	0	284.50	0	273.87	0	273.87
12	0	274.32	0	286.64	0	275.14	0	275.14
13	0	277.4	168	290.72	0	278.10	0	278.10
14	0	285.12	204	294.86	0	285.56	0	285.56
15	0	293.13	204	297.65	0	293.33	0	293.33
16	204	260.82	0	261.09	96	260.08	96	260.08
17	96	272.8	120	276.91	96	273.68	96	273.68
18	96	260.36	60	280.39	84	261.18	84	261.18
19	84	256.58	96	254.00	72	255.05	72	255.05
20	0	264.05	0	277.49	0	260.73	0	260.73
21	72		36		72		72	
Cost	$ 52.59 million		$ 68.90 million		$ 38.64 million		$ 38.64 million	

5. Evaluating the performance of pipe networks using resilience indices

The process of optimization usually involves bringing down the cost by reducing the pipe sizes or by completely eliminating some pipes which tends to render the system with inadequate capacity to respond to pipe failures or user demands which exceed projected values. Hence it is of utmost importance to include reliability considerations in all stages of planning, design and implementation of water distribution systems. A water distribution system is efficient and reliable enough only when it supplies water to all the nodes and meets the demand and pressure under satisfactory conditions. However, sometimes there is a possibility of a pipe failure because of which the entire flow changes resulting in high energy losses. This problem can be overcome by assigning more power than that required at each node and this loss in energy is met up by the surplus energy available in the system. So, assessment of pipe networks in this manner enables us to understand the concept of resilience which clearly indicates how quickly and easily a pipe network revives itself from failure conditions. The formulae proposed to find Resilience Index (RI) [20] and Modified Resilience Index (MRI) [6] are as follows:

$$RI = \frac{\sum_{j=1}^{N} q_i\left(h_{avl,j} - h_{min,j}\right)}{\left(\sum_{i=1}^{R} Q_r h_{res,i} + \sum_{b=1}^{B} \frac{P_b}{v}\right) - \sum_{j=1}^{N} q_j h_{min,j}} \tag{1}$$

$$MRI = \frac{\sum_{j=1}^{N} q_i\left(h_{avl,j} - h_{min,j}\right)}{\sum_{j=1}^{N} q_j h_{min,j}} \tag{2}$$

where q_i is the demand at node i, $h_{avl,j}$ is the available pressure head at node j, $h_{min,j}$ is the minimum pressure head at node j, Q_r is the flow from reservoir i, $h_{res,i}$ is the sum of reservoir elevation and its water level of reservoir i, P_b is the capacity of pump b and v is the specific weight of the liquid.

Table 4. Performance measure values for Hanoi network

Analysis	Increasing Diameter	Decreasing Diameter	Differential evolution Suribabu (2010)	Heuristic-based approach Suribabu (2012)
Resilience Index(RI)	0.235	0.231	0.192	0.216
Modified Index (MRI)	0.549	0.539	0.447	0.504

Table 5. Performance measure values for GoYang network

Analysis	Increasing Diameter	Decreasing Diameter	Heuristic based algorithm Menon (2016)	Harmony search algorithm Geem (2006)
Resilience Index (RI)	0.489	0.489	0.465	0.432
Modified Index (MRI)	0.530	0.530	0.471	0.463

Table 6. Performance measure values for NYT network

Analysis	Increasing Diameter	Decreasing Diameter	Differential evolution Suribabu (2010)	Ant Colony Optimization algorithm Maier et al. (2003)
Resilience Index (RI)	0.421	0.617	0.42	0.42
Modified Index (MRI)	0.072	0.106	0.07	0.07

6. Results and discussion

To increase the reliability and resilience of a network it is advisable to maintain a higher pressure than what is required (i.e. the minimum pressure that has to be maintained at every node). But, maintaining a higher pressure at a node invariably leads to increasing the diameter of the pipe which consequently increases the cost and the design ceases to be economical. Hence, it is necessary to strike a balance between cost and reliability. Higher resilience values usually lead to higher costs but then the networkis hydraulically efficient when compared to a network with lower resilience index. In the case of Hanoi network, the highest pressure maintained is 97.14m for a pipe of diameter 1016mm at node 2 in all the cases. It is seen that the pressure head decreases at every successive node when water flows through larger number of pipes owing to friction loss in every pipe. The friction loss is greater in pipes of smaller diameter. The lowest pressure is seen at node 13 in all the cases. While it would have been safer to assign a larger diameter to pipe 10, 11, 12, 21 and22 which connects node 13 and 22 respectively to enable maintenance of higher pressure since there is only one pipe which supplies water to that point and any leakage or breakage of pipe en-route would affect that consumer point severely, we have resorted to lower diameter according to the various procedures followed.It is observed that pipes which are in quick succession with the reservoir namely pipes 1,2,3, 19 and 20 are of larger diameters when compared to pipes which are relatively far from the reservoir namely pipes 26 to 34 and consequently higher pressure is maintained at nodes connected by pipes 1,2,3, 19 and 20and lower pressure is maintained at nodes connected by pipes 26 to 34.Since the direction of flow changes according to the shortest route possible to reach a node, consecutive nodes are supplied water from two opposing directions which makes the pipe in between them unnecessary and hence lower diameter may be assigned to them for a more economic cost. The difference in cost between the first two solutions found by increasing and decreasing diameter approaches and the solutions obtained by differential evolution method and heuristic based approach results is due to drastic variation in diameter of pipes 16 to 19. The increasing diameter approach provides least cost solution compared with decreasing diameter approach.In the case of GoYang network, it is seen that both the approaches provide the same result which invariably leads to higher diameters for pipes in quick succession to the reservoir namely pipes 1, 2 and 3 and the least possible diameter 80 mm for other pipes which results in sufficient pressure being maintained at all the nodes higher than the required pressure head of 15m.Focussing on the New-York City Tunnel network, the primary objective is to revamp it by taking into account population growth and the consequent increase in water demand. If the current network is no longer able to meet increased demand, it leads to pressure violations at consumer nodes. Hence, the network must be improved by replicating some of the pipes, i.e., putting new pipes in parallel with the existing ones, at a minimum cost. This process consists of the following steps: 1) Determine the pipes that need to be duplicated and 2) For the selected pipes fix a diameter within the available diameter set. It was found that age and increased demands left the existing gravity flow tunnels in New-York City inefficient in meeting the pressure requirements at nodes namely 16, 17, 18, 19 and 20. The network consist of 21 pipes that may be duplicated with one of the 15 available commercial diameter pipe sizes or can be deemed to exist on their own without the need to be duplicated. All the proposed techniques indicate that pipes 17, 18, 19 and 21 require duplication, with the addition of pipes 8, 9 and

16(increasing diameter approach) and pipes 13, 14 and 15(decreasing diameter approach) in our viewpoint.The power dissipation based approaches follow shortest path route to assign diameter to the pipe. It is evident from the study that shortest path alone does not make a network to have a configuration that will lead to an economical design. Since commercial diameter pipe sizes are used to configure the network, assigning higher size than that required for certain links are inevitable in such an optimization method which ultimately deviates from shortest path configuration in order to achieve economical design. While assessing the designed networks from performance point of view, the network having higher cost provides better performance measures than least cost design.

7. Conclusion

The present study proposes a simple method based on power dissipation in the pipe to dimension the water distribution network without the use of an optimization technique. The hydraulic energy available from the source is dissipated while water travels along the pipe to overcome major and minor losses. Sizing the pipe based on power consumed in each pipe uses the shortest path from the source to demand nodes. Though it is the correct way to size the pipe as any commodity needs to be transported along shortest path and it will invariably lead to economical results, in the pipeline dimensioning this idea will not hold good for all the cases as commercial pipe sizes are used, which may be of a greater size than that actually required from the economical point of view.Hence this aspect of the proposed sizing method always leads to higher cost than network design by optimization techniques. But, this method has advantages of its simplicity and also does not use any operation research techniques.Prominent observation of the proposed approaches is that the increasing diameter approach usually provided more economic results (lower cost) than decreasing diameter approach in most of the cases. After carefully studying the reliability characteristics of the various water distribution networks, it was noted thatnetworks with low resilience values were relatively cheaper and exhibited satisfactory hydraulic operational performance. If there should be an occurrence of multiple sources in a water distribution network, the total surplus power available at the demand nodes is not precisely gauged by the resilience index. In such a case, utilization of the modified resilience index is favourable since it considers only the surplus powerat the demand nodes and not the surplus power available for internal dissipation. Hence, it is consequently beneficial to use the modified resilience index as a pointer of the capacity of the network to deal with increased demands or pipe failures. The concerns detailed in this research need to be further advanced and expanded by researchers and engineers who are committed to produce both economic and efficient networks.

References

[1] Alperovits E and Shamir U 1977 Design of optimal water distribution systems. *Wat. Res. Resour.* **13**(6) 885-900.

[2] Cunha M and Sousa J 1999 Water distribution network design optimization: simulated annealing approach. *J. Water Resour. Plan. Manage.* **125** (4) 215-21.

[3] Darshan Mehta, Krunal Lakhani, Divy Patel and Govind Patel Study of water distribution network using Epanet. Int. J. Adv. Res. Eng. Sci. Manage. 1-9.

[4] Eusuff M M and Lansey K E 2003 Optimization of water distribution network design using the shuffled frog leaping algorithm. *J. Wat. Resour. Plan. Manage.* **129**(3) 210-25.

[5] Fujiwara O and Khang D B 1990 A two-phase decomposition method for optimal design of looped water distribution networks. *Water Resour. Res.* **26**(4) 539-49.

[6] Gomes H P, Bezerra S T M, Carvalho P S O and Salvino M M 2009 Optimal dimensioning model of water distribution systems. Water SA **35**(4) 421-31.

[7] Jayaram N and Srinivasan K 2008 Performance-based optimal design and rehabilitation of water distribution networks using life cycle costing. *Water Resour. Res.* **44** 1417.

[8] Kadu M S, Gupta R and Bhave P R 2008 Optimal design of water networks using a modified genetic algorithm with reduction in search space. *J. Water Resour. Plan. Manage.* **134** (2) 147-160.

[9] Kashfina Kapadia Menon and Narulkar S M 2016 Application of heuristic based algorithm in water distribution network design. *J. Water Resour. Pollut. Stud.* **1**(3) 1-14.

[10] Keedwell E and Soon-ThiamKhu 2006 Novel cellular automata approach to optimal water distribution network design. *J. Comput.Civ. Eng.* **20**(1) 49-56.

[11] Maier H R, Simpson A R, Zecchin A C, Foong W K, Phang K Y, Seah H Y and Tan C L 2003 Ant colony optimization for design of water distribution systems. *J. Water Res.* **129**(3).

[12] Mohan S and Jinesh Babu K S 2010 Optimal water distribution network design with Honey-Bee mating optimization. *J. Water Res. Plan. Manage.* 2**4**(1) 117-26.

[13] Monbaliu J, Jo J, Fraisse C W and Vadas R G 1990 Computer aided design of pipe networks. *Proc. Int. Symp. Water Res. Syst. Appl.* S.P. Simonovic, I.C.Goulter and B.J. LenceFriesen Printers, Winnipeg, Canada.

[14]Rossman L A 2000 EPANET 2-User Manual, National Risk Management Research Laboratory, Office of Research and Development, U.S. Environmental Protection Agency, Cincinnati, Ohio, USA.

[15] Saminu A, Abubakar, Nasiru and Sagir L 2013 Design of NDA water distribution network using EPANET. *Int. J. Emerging Sci. Eng.* **1**(9) 5-9.

[16] Savic D A and Walters G A 1997 Genetic algorithms for least cost design of water distribution networks. *J. Water Res. Plan. Manage.* **123**(2) 67-77.

[17] Simpson A R, Dandy G C and Murphy L J 1994 Genetic algorithm compared to other techniques for pipe optimization. *J. Water Res. Plan. Manage.* **120** (4) 423-43.

[18] Suribabu C R 2010 Differential evolution algorithm for optimal design of water distribution networks. *J. Hydro informatics* **12**(1) 66-82.

[19] Suribabu C R 2012 Heuristic-based pipe dimensioning model for water distribution networks. *J. pipeline Syst. Eng. Practice* **3**(4) 115-24.

[20] Suribabu C R and Neelakantan T R 2006a Design of water distribution networks using particle swarm optimization. *Urban Water* **3**(2) 111-20.

[21] Suribabu C R and Neelakantan T R 2006 Particle swarm optimization compared to other heuristic search techniques for pipe sizing. *J. Envir. Informatics In.Soc. Envir. Inform. Sci.* Canada **8**(1) 1-9.

[22] Todini E 2000 Looped water distribution networks design using a resilience index based heuristic approach. *Urban Water* **2**(2) 115-22.

[23] Vairavamoorthy K and Ali M 2005 Pipe index vector: a method to improve genetic-algorithm-based pipe optimization. *J. Hydraul. Eng.* **131** (12) 1117-25.

[24] Vipinkumar G Yadav, Darshan Mehta and Sahita I Waikhom 2015 To Assess the Prevailing Water Distribution Network using EPANET. *Int. Res. J. Eng. Technol.* **2**(8) 777-81.

IOP Conf. Series: Earth and Environmental Science **80** (2017) 012065 doi:10.1088/1755-1315/80/1/012065

Assessing Performance of Multipurpose Reservoir System Using Two-Point Linear Hedging Rule

K Sasireka* and T R Neelakantan

School of Civil Engineering, SASTRA University, Thanjavur – 613 401, India

*Email: sasireka@civil.sastra.edu

Abstract. Reservoir operation is the one of the important filed of water resource management. Innovative techniques in water resource management are focussed at optimizing the available water and in decreasing the environmental impact of water utilization on the natural environment. In the operation of multi reservoir system, efficient regulation of the release to satisfy the demand for various purpose like domestic, irrigation and hydropower can lead to increase the benefit from the reservoir as well as significantly reduces the damage due to floods. Hedging rule is one of the emerging techniques in reservoir operation, which reduce the severity of drought by accepting number of smaller shortages. The key objective of this paper is to maximize the minimum power production and improve the reliability of water supply for municipal and irrigation purpose by using hedging rule. In this paper, Type II two-point linear hedging rule is attempted to improve the operation of Bargi reservoir in the Narmada basin in India. The results obtained from simulation of hedging rule is compared with results from Standard Operating Policy, the result shows that the application of hedging rule significantly improved the reliability of water supply and reliability of irrigation release and firm power production.
Keywords: Type II two-point linear hedging rule, reservoir operation, Standard Operating Policy.

1. Introduction

Reservoir operation studies aim for reliable supply of water for various uses like municipal requirements, irrigation requirements, hydropower generation requirements, flood control, and recreation storage requirements in the reservoir. A planning model can help to make decisions like fixing the reservoir size, maximum demand, and changing the cropping pattern. Depending upon the situations, several operating policies are used. In planning model, Standard Operating Policy (SOP) is considered as the one of the simplest policy, which plans to release a amount of water equal to the water demand, if possible [1]. The reservoir operation using hedging rules has been established as one of the important advances in the field of reservoir operation during the past three decades. SOP increase the single period severe deficit but hedging rule minimize the impact of deficits by distributing the deficit over a longer duration. The literature concering reservoir operation extensively demonstrated and detailed literature reviews were provided by [2-5]. In the case of hydropower reservoir, the water demand is not constant, but the power demand is constant. Hence by considering nonlinearity of head (H) and flowrate (Q) in power equation ($P \propto QH$), new form of SOPp is introduced by [6]. Sasireka et al [7] demonstrated two point linear hedging rule and discrete phased hedging rule for hydropower reservoir operation with a case study. In this paper, Type II two point linear hedging rule is applied for Bargi reservoir for hydropower generation.

1.1 Type II Two-point Linear hedging rule

Shiau [8] in his paper depicted about reserve-storage based two-point hedging rule, which is represented as 'type II two-point hedging rule' (Figure. 1). When the available water is less than $P1_x$, there is no reservoir release as per this rule and the available water is stored as reserve for the next period. Ramakrishnan [9] studied a multipurpose reservoir system in which a similar rule was adopted for irrigation supply. During severe drought water will be preserved for future drinking water supply, a higher priority water use, and will not be released for irrigation requirement in the current period. This hedging rule may suit such irrigation water releases that are controlled by drinking water reserves. In Figure 1, ($P1_x$, $P1_y$) referred as Starting Available Storage, SAS and ($P2_x$, $P2_y$) referred as Ending Available Storage, EAS. The x-axis is representing available water and the y-axis is representing the sum of release towards the demand and spill. D_t is the demand in time-period t and K is the capacity of the reservoir. The "Available Storage (AS)" is the sum of the initial storage in the reservoir and the expected inflow during the period considered minus evaporation losses.

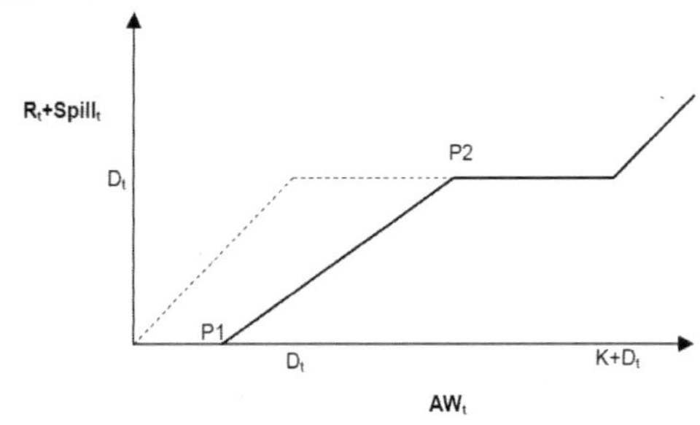

Figure 1. Type II two-point linear hedging rule.

The Type II two-point hedging rule can be mathematically represented as follows.

$$P1_x = aD_t;\ P2_x = bD_t;\ P1_y = 0;\ P_{2y} = D_t;\ 0 \le a \le 1;\ 1 \le b \le \frac{K + D_t}{D_t} \tag{1a}$$

$$\begin{aligned} R_t &= 0, \quad if\ AS_t \le P1_x \\ R_t &= P1_y + \frac{D_t - P1_y}{P2_x - P1_x} \times (AS_t - P1_x), \quad if\ P1_x \le AS_t \le P2_x \\ R_t &= D_t, \quad if\ AS_t \ge P2_x \end{aligned} \tag{1b}$$

$$Spill_t = \begin{cases} AS_t - K - D_t, & if\ AS_t \ge (K + D_t) \\ 0, & otherwise \end{cases} \tag{1c}$$

When the available Storage (AS_t) falls below $P1_x$, the available water has been kept as storage to meet the future demand. To adopt this policy, the number of parameters for which the optimal values are to be identified per period is two, namely the points $P1_x$ and $P2_x$..

1.2 Modified Type II Two-Point Linear Hedging Rule For Hydropower Reservoir Operation

In hydropower reservoir operation water release is not a fixed quantity, hence when the head available is more, a smaller quantity of release is sufficient compared to a release when the head is low. In water supply reservoir operation water demand is constant and it is considered as a control parameter for hedging rule. However in case of hydropower reservoir operation the power demand is constant but water demand is continuously varying due to change in head, hence, two quantities like release at maximum capacity of reservoir and minimum release to produce firm power were worked and used as a control parameter In Model-1, the hedging control parameter C_P is estimated such that if all the C_P quantity of water available in the reservoir is released that should produce the firm power. In Model-2, the hedging control parameter C_P is estimated such that it is quantity of water to be released to generate the firm power when the head available is assumed equal to the maximum possible head. Thus, the release as per the modified Type II two-point hedging rules can be written as follows.

$$P1_x = aC_P;\; P2_x = bC_P;\; P_{2y} = Power\max;\; 0 \le a \le 1;\; 1 \le b \le \frac{K + C_P}{C_P} \tag{2a}$$

$$R_t = 0, \quad if\ AS_t \le P1_x$$

$$R_t = P1_x + \frac{C_P - P1_x}{P2_x - P1_x} \times \left(AS_t - P1_x\right), \quad if\ P1_x \le AS_t \le P2_x$$

$$R_t = C_P, \quad if\ AS_t \ge P2_x \tag{2b}$$

2. Study Area

Bargi reservoir in India constructed across the river Narmada, Madhyapradesh is selected as a study area for demonstrating simulation of modified Type II two-point linear hedging rule. The latitude and longitude of the dam is 22°56.5'N and 79 55.5'E respectively. This reservoir is a multipurpose reservoir, it serves water supply for domestic and industrial purposes, irrigation and the generation of hydropower. Presently, one left bank canal system which consists of two units of capacity of 45 MW each are operational.

Table 1. Monthly Evaporation rate, Monthly Inflow and Monthly target demands for the Bargi Reservoir (Source: NIH, 1995-96) [10]

Month	Mean Inflow (10^6 m^3)	Mean Evaporation rate (cm/month)	Irrigation Demand (10^6 m^3)	Hydropower demand (M kWh)
Jun.	146.7	19.8	55.9	20
July	1568.8	11.4	192.5	40
Aug.	3038.1	10.0	69.3	60
Sep.	1500.7	10.8	240.0	60
Oct.	433.1	7.8	435.9	60
Nov.	120.5	6.0	191.0	30
Dec.	81.2	7.8	274.7	30
Jan.	56.9	6.8	322.7	20
Feb.	41.1	6.9	200.4	20
Mar.	25.0	12.6	79.2	20
Apr.	12.5	17.7	53.4	20
May.	7.3	23.9	76.9	20

The monthly water demand from the reservoir through the left bank canal for domestic water supply is 4.5 $x10^6$ m^3. The annual water demand from the reservoir for irrigation is 2160.1 $x10^6$ m^3 and annual firm energy demand of the reservoir is 363 MkWh. Table 1 shows the mean inflow, evaporation rate and demand in the reservoir. Reservoir capacity and storage levels are given in Table 2. In this study, 475 numbers of monthly data from June 1951 to December 1990 was used for demonstration of hedging rule.

Table 2. Storage capacity and reservoir level of Bargi reservoir

Storage capacity (10^6 m^3)			Reservoir level (m)		
Gross	Live	Dead	Maximum	Full	Dead storage
3920	3180	704	425.7	422.76	403.55

3. Methodology

Simulation is a mathematical technique which is used in this study, to compare the performance of reservoir using hedging rule with standard operating policy. Bargi reservoir is a multipurpose reservoir hence water release is made as per the order of priority as follows: municipal water supply, irrigation and hydropower production. Water supply for municipal and irrigation purpose is done based on SOP, shown in figure 2.

The continuity equation used for simulation model is given below

$$S_{mon+1} = S_{mon} + I_{mon} - Evp_{mon} - R_{mon} - Spill_{mon}, \forall mon \tag{3}$$

where, R_{mon} is the release made for municipal and irrigation requirement and power generation, I_{mon} is the reservoir inflow and Evp_{mon} is monthly evaporation volume, which is calculated from the water spread area (WSA) multiplied by the rate of evaporation. The WSA equation is given below

$$WSA = (-8\times10^{-6}\times S^2) + (0.113\times S - 39.854) \tag{4}$$

where S= (Beginning storage + End storage)/2.

Hydropower release is made as per the Standard Operating Policy for power (SOP_P) [3]. In figure 3, S represents the release corresponding to the firm power, R_c represents release corresponding to maximum capacity of the reservoir. From P_1 to P_2 the curve is falling curve due to increasing of head, hence to calculate the varying flow rate between these points a equation is arrived with help of dam data using curve fit.

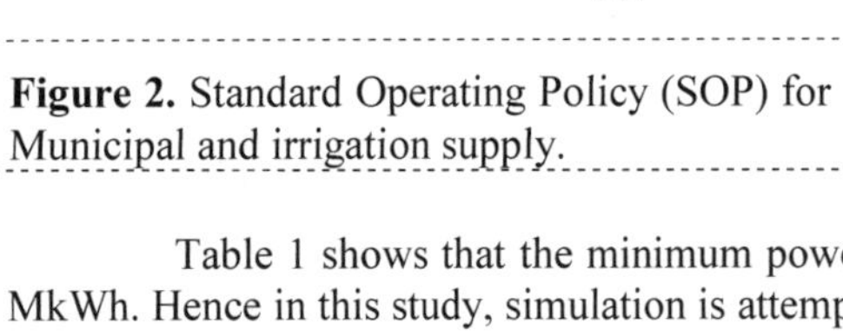

Figure 2. Standard Operating Policy (SOP) for Municipal and irrigation supply.

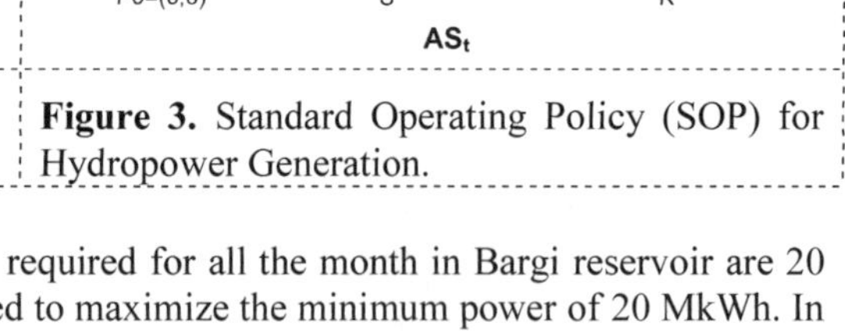

Figure 3. Standard Operating Policy (SOP) for Hydropower Generation.

Table 1 shows that the minimum power required for all the month in Bargi reservoir are 20 MkWh. Hence in this study, simulation is attempted to maximize the minimum power of 20 MkWh. In simulation of type II two-point hedging rule, the control parameter (C_p) corresponding to minimum release, in Model-I is 406.53 x10^6 m^3 and for release corresponding to maximum capacity in Model-II is 255 x10^6 m^3 which is obtained from the power equation $P=\eta\gamma QH$ (where P- Power in MW, η- Efficiency of the turbine (60%), γ - unit weight of water Q –Flow rate in 10^6 m^3 and H – Operating head). By varying the coefficients ('a' and 'b') in both the model, the best performance of the reservoir is identified. Both the Models the coefficient 'a' varies from 0.1 to 1. In Model-I 'b' varies from 1 to 10.65, and Model-II it varies from 1 to 16.39.

4. Results and Discussion

The performance of the reservoir has been identified by using different performance indices namely, average power production, average spill, reliability of release for various purposes and number of times zero power produced. The reliability of the system [11] given as follows.

$$Reliability \quad of \quad the \quad system = \frac{Numberot \quad tmet\arg et \quad releaseachieved \quad (success)}{Total \quad No \quad of \quad Operating \quad Period} \tag{5}$$

Table 3 shows the comparison of simulation of SOP results with best results obtained from type II hedging rule by both the model. The coefficient (a,b) at which the best results are obtained are also given in table 3. For average power production and considering spill Model-I perform better compared to Model-II and SOP. Maximum average irrigation release is better in both the Model compared to simulation of SOP. For remaining all other performance Model-II showed better results compared to Model-I and SOP.

Table 3. comparison of Type II two-point hedging rule with simulation model

Performance Indices	Model-I			Model-II			Simulation using SOP & SOP_p
	Values	Coefficients		Values	Coefficients		
		A	b		a	b	
Maximum average power production (MW)	25.51	0.1	1	22.03	0.1	1	22.94
Minimum average Spill (10^6m^3)	119.9	0.1	1	160.88	0.1	1	147.66
Maximum average irrigation release (10^6m^3)	183.04	0.8 – 1	10.65	183.04	0.1-1	13-16.39	173.38
Maximum Reliability of Firm power (%)	12.84	1	10.65	15.16	0.5- 1	16.39	10.95
Maximum Reliability of target irrigation release(%)	100	0.9 -1	10.65	100	0.1-1	16.39	90.53
Maximum Reliability of municipal water supply (%)	100	0.5- 1	10.65	100	0.1-1	16.39	97.05
Minimum No. of times zero power produced	8	0.2 , 0.3	10.65	0	0.1-0.6, 0.1-1	15, 16.39	45.00

IOP Conf. Series: Earth and Environmental Science **80** (2017) 012065 doi:10.1088/1755-1315/80/1/012065

There is a trade-off between average power production and reliability of water supply for municipal, irrigation and hydropower production. The average power production decreases with increasing of coefficient but it is reverse in the case of reliability water supply for the above mentioned purposes. The most important aim of the hedging rule is to reduce the severity of drought that is continuous drought by accepting number of smaller droughts. By applying hedging rule it is observed the there is no zero power production in the Model-II and only 8 numbers of zero power production in Model-I and it is more about 45 numbers of zero power production when the reservoir is operated with SOP.

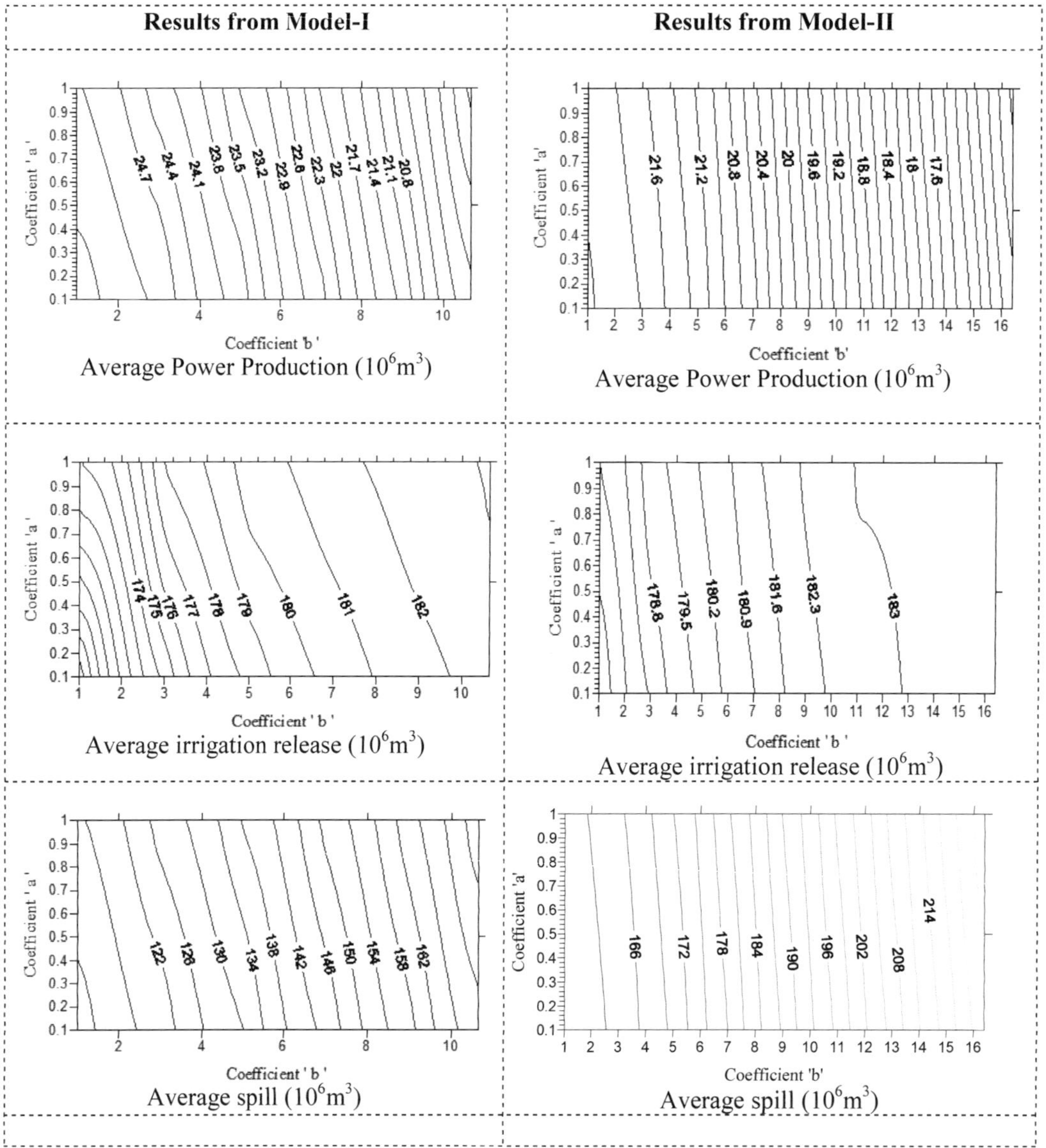

Average Power Production (10^6m^3) — Average Power Production (10^6m^3)

Average irrigation release (10^6m^3) — Average irrigation release (10^6m^3)

Average spill (10^6m^3) — Average spill (10^6m^3)

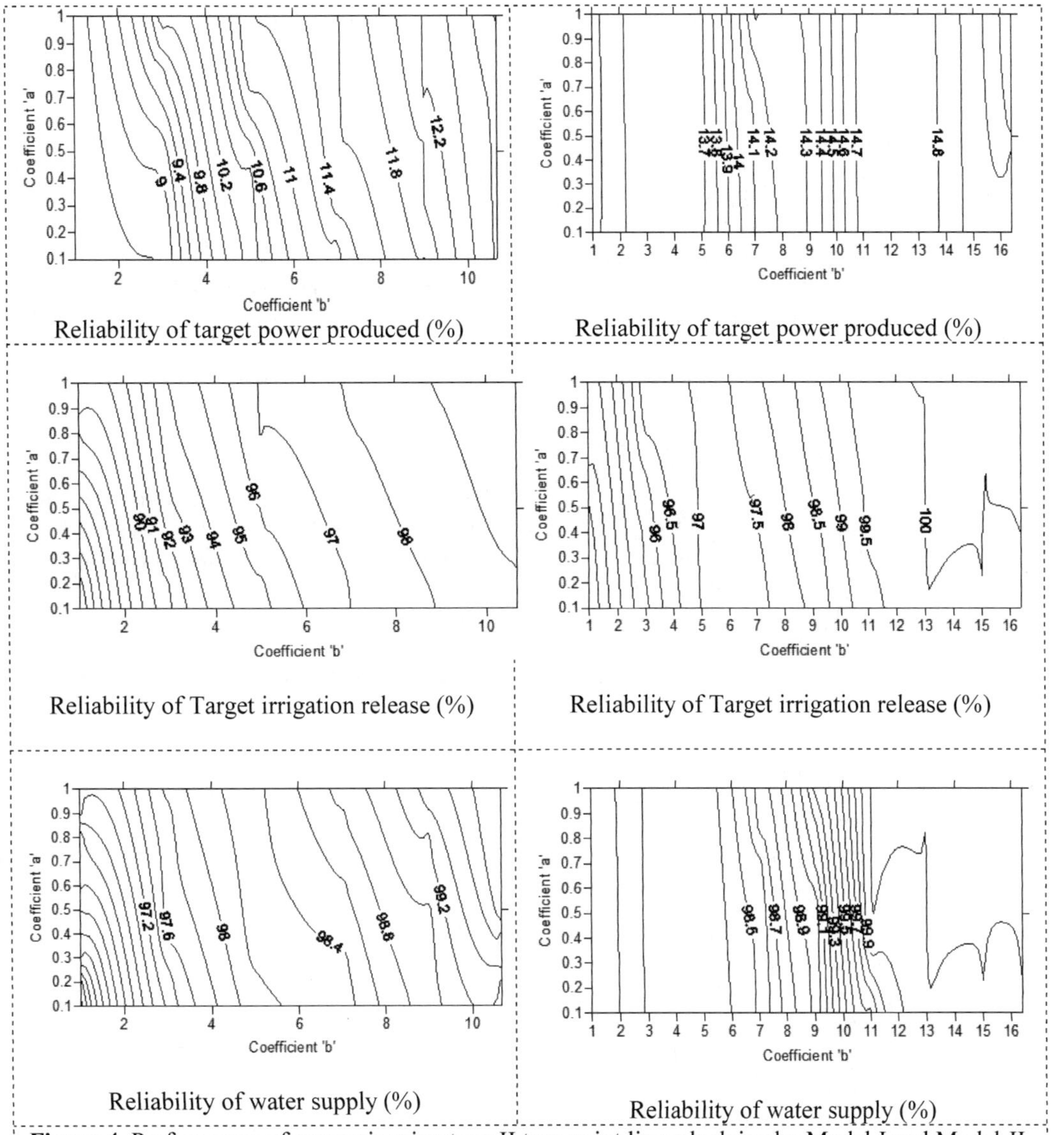

Figure 4. Performance of reservoir using type II two-point linear hedging by Model-I and Model-II.

The performance of the reservoir operation using type II two-point linear hedging rule by varying the values of coefficient(a=0.1 to1, b=1 to10.65) for Model-I and (a= 0.1 to 1, b= 1 to 16.39) for Model-II are shown as a contour plot in Figure 4. Figure 4 is helpful for choosing the alternative reservoir operation rule to get maximum benefit from the reservoir.

5. Conclusion

Application of hedging rule for hydropower operation is considered as a new research area in the field of water resource management. Hence in this paper one of the simplest 'type II two-point hedging rule' is used for hydropower release of a multipurpose reservoir as case study to identify the performance of the reservoir. The results show that the water supply for municipal and irrigation is satisfied with 100% reliability using hedging rule but it is 97.05% and 90.53% respectively using SOP.

The overall performance shows that the type II two-point hedging rule perform well compared to Standard Operating Policy.

References

[1] Stedinger J R 1984 Performance of LDR models for preliminary design and reservoir operation. *Water Resour. Res* .**20**(2) 215-24.

[2] Yeh W W G 1985 Reservoir management and operations models: A state of-the-art review. *Water Resour. Res.* **21**(12) 1797-818.

[3] Simonovic S P 1993 Flood control management by integrating GIS with expert systems: Winnipeg City case study. In HydroGIS93: Application of Geographic Information Systems in Hydrology and Water Resources, edited by K. Kovar and H.P. Nachtnebel, IAHS Publication no. 211, 61-72.

[4] Wurbs R 1993 Reservoir-system simulation and optimization models. *J. Water Resour. Plan. Manag.* **119**(4) 455-72.

[5] Labadie J W 2004 Optimal operation of multireservoir systems: State-of-the-art review. *J. Water. Resour. Plan. Manag.* **130**(2) 93-111.

[6] Neelakantan T R and Sasireka K 2013 Hydropower Reservoir Operation using Standard Operating and Standard Hedging Policies. *Int. J. Eng. Technol.* **5**(2) 1191- 6.

[7] Sasireka K and Neelakantan T R 2016 Application of Hedging Rules for Operation of Hydro-Power Reservoir. *Jordan J. Civ. Eng.* **10**(3) 305-12.

[8] Shiau J T 2011 Analytical optimal hedging with explicit incorporation of reservoir release and carryover storage targets. *Water Resour. Res.* **47** W01515.

[9] Ramkrishnan K, Suribabu C R and Neelakantan T R 2010 Crop Calendar Adjustment Study for Sathanur Irrigation System in India Using Genetic Algorithm. *Water Resour. Manage.* **24**(14) 3835-51.

[10] Jain S K, Goel M K, Senthil Kumar A R and Agrawal P K 1995-96. Multiobjective optimization of operation of a dam. National Institute of Hydrology, Roorkee, TR (BR) 143.

[11] Srinivasan K, Neelakantan T R, Narayan P S and Nagarajukumar C 1999 Mixed-integer programming model for reservoir performance optimization. *J. Water Resour. Plan. Manage.* **125**(5) 298-301.

Rainfall Analyses of Coonoor Hill Station of Nilgiris District for Landslide Studies

Evangelin Ramani Sujatha* and C R Suribabu

School of Civil Engineering, SASTRA University, Thanjavur 613401. India.

*Email: r.evangelin@gmail.com

Abstract. The most common triggering factor of landslides in a hill terrain is rainfall. Assessment of the extreme and antecedent rainfall events and its quantum is imperative to evaluate the temporal occurrence of landslides. It also plays a vital role in the choice of the preventive measures to be adopted. This study focuses on an in-depth rainfall analysis of Coonoor hill station. The analysis includes the study of monthly, seasonal and annual rainfall patterns for a period of 80 years, between 1935 and 2013. Further, one day maximum, 5 day and more antecedent rainfall and its amount is calculated for the years between 2007-2012, 2014 and 2015.The result of the study indicates an increase in the normal rainfall based on the mean of 30 years of data (for the recent decades) and erratic pattern of rainfall during pre-monsoon, post-monsoon south-west monsoon periods. A detailed analysis of daily rainfall for the selected period indicates that extreme highest daily rainfall of more than 300 mm above occurred after consecutive rainfall trigged massive landslides comparing highest rainfall amount around 100 to 180 mm rainfall events.

1. Introduction

For an efficient drainage design, watershed management and to propose landslides prevention measures in the hilly region, it is vital to know the extreme precipitant events, antecedent rainfall and its period of occurrence. Accurate estimation of debris flow, potential zone of soil scour and landslide susceptibility along hill roads are crucial aspects for hill area development and to protect the ecosystem from environmental degradation. Due to non-availability of short duration rainfall at most of places around the world, it is common practice to utilize the amount extreme daily rainfall events for modelling the landslide events in the hilly terrain. Rainfall during strong northeast monsoon season in Coonoor-Mettupalayam region of Nilgris district of Tamilnadu often triggers many landslide. Hence, short duration rainfall is essential to predict and understand the landslide and debris flow occurrences in the mountainous region. Intense rainfalls of short durations usually occur within longer-duration storms rather than as isolated events. The relationship between rainfall and landslides has attracted the interest of numerous researchers because rainfall is the most common landslide-triggering factor in many regions in the world [1]. Literature shows that the most commonly investigated rainfall parameters for landslide initiation are cumulative rainfall, antecedent rainfall, rainfall intensity and rainfall duration. Chowdhury and Flent Je [2] emphasized that landslides triggered by rainfall needs complete understanding of the link between rainfall and landslide, frequency of landslides with respect to frequency of heavy rainfall event, monitoring of subsurface shear movements and pore pressure at the existing landslide points and studying the relationship between rainfall magnitude and landslides.

Jaiswal [3] described that landslides in Nilgiris as debris slides or debris flow triggered by high intensity or prolonged rainfall.

2. Study Area

Coonoor taluk of Nilgiris district, the second largest hill station in Nilgiris is a preferred tourist destination. It is also a municipality. It is also one of the regions most affected by landslides in Tamilnadu. It is located at an altitude of 1850 m above MSL and has a population of about 45,494 people according to 2011 census. The climate falls under the sub-tropical highland type according to Köppen classification system. The average elevation is 1850 m above MSL. Charnockite is the most prevalent bedrock material, covered with laterite which forms an irregular soil horizon. The rocks are of metamorphic origin. The economy of the region heavily relies on the tea plantations and tourism related activities.

3. Normal Rainfall for Coonoor Station

Figure 1 presents the historical rainfall for Coonoor station. It is clear from the plot that annual rainfall exceeds almost 1000 mm. Mean of 30 years of annual rainfall denotes the normal rainfall of a station. Recent 30 years of rainfall will be used for computation of normal rainfall of the station. From available monthly historical rainfall data, the normal rainfall for recent 30 years (1985-2014) and previous two stages of 30 years are calculated and presented in the Table 1.

Table 1. Normal Rainfall for Coonoor station

Year of analysis	No. of years	Annual Normal rainfall (mm) with (standard deviation)	Pre-monsoon normal rainfall (mm) with (standard deviation)	South-West Monsoon rainfall (mm) with (standard deviation)	North East Monsoon rainfall (mm) with (standard deviation)
1935 – 1964	30	1545 (307)	484 (191)	325 (77)	735 (236)
1955 – 1984	30	1678 (469)	446 (225)	356 (101)	876 (393)
1985 – 2014	30	1773 (325)	416 (181)	369 (99)	988 (319)
1935 - 2014	80	1665 (390)	443 (199)	351 (94)	870 (350)

This analysis indicates how normal rainfall hada change over 80 years. The recent normal rainfall is 1773 mm. This value is higher than the normal rainfall worked out for different 30 years of span. Further, the average of 80 years of rainfall account as 1665 mm. It is again lesser than recent 30 years average. This depicts that the rainfall in the Coonoor station has been increasing over the years. To confirm the increasing trend of annual rainfall, 5- year and 10-year moving average plot are prepared and present in figures 2 and 3. The trend clearly shows the increasing amount of rainfall over each five years. Normal rainfall for pre-monsoon, south-west and north east monsoon periods is evaluated and for selected 30 years of interval and also for 80 years of data. Normal rainfall for pre-monsoon period is found to be lesser for recent 30 year period than other considered periods. But, there is an increased value of rainfall ascertained for south-west and north-east monsoon periods. This increase in rainfall during recent years has resulted due to highest annual rainfall amount reported for 7 years out of 30 years which exceeded 2000 mm and it is greater than 12 % the normal rainfall of this thirty years average. The standard deviation except south west monsoon period is having higher value irrespective of period of analysis. This also indicates that rainfall pattern is highly erratic during pre-monsoon and north-east monsoon periods. The number of rainy days per year varies from 65 to 100 days. The plot (Fig.4)shows the average monthly rainfall over the year. The average plot indicates that minimum rainfall of around 60 mm occurs in January and June and maximum rainfall of about 325 mm is expected during October and November months. The monthly rainfall variability (ratio between standard deviation and mean rainfall) is worked out based on 80 years of data and same is presented in

Fig.5.High variability in monthly rainfall is noticed during January, February, March, May and December months. The remaining months of the year poses variability around 50 %.

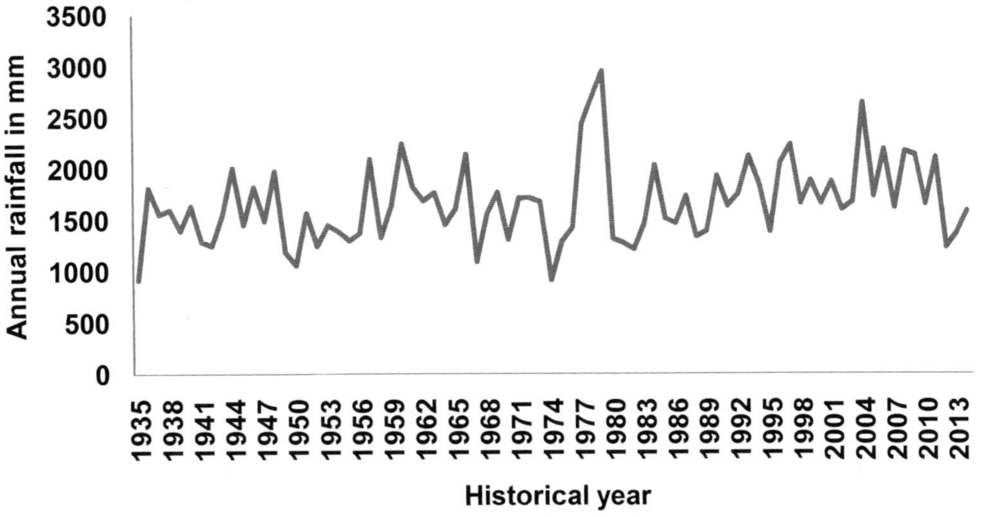

Figure 1. Annual historical rainfall of Coonoor hill station

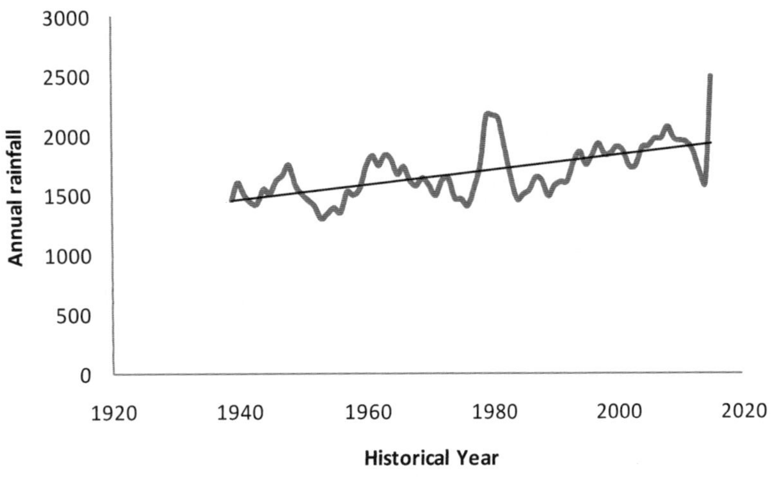

Figure 2. 5-year Moving average annual rainfall plot

IOP Conf. Series: Earth and Environmental Science **80** (2017) 012066 doi:10.1088/1755-1315/80/1/012066

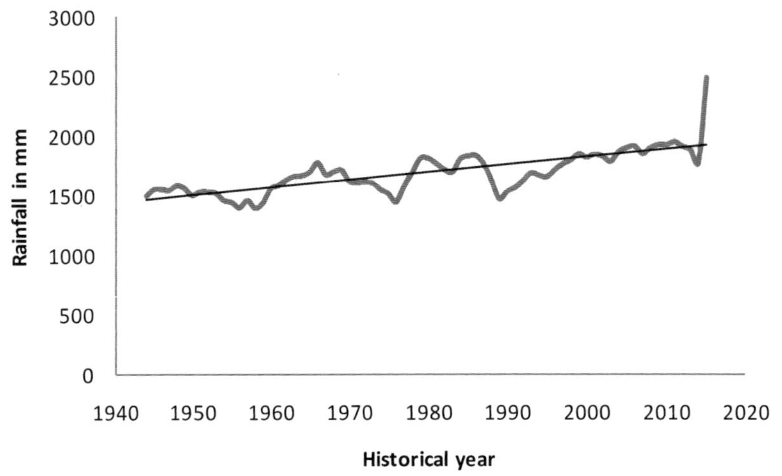

Figure 3. 10-year Moving average annual rainfall plot

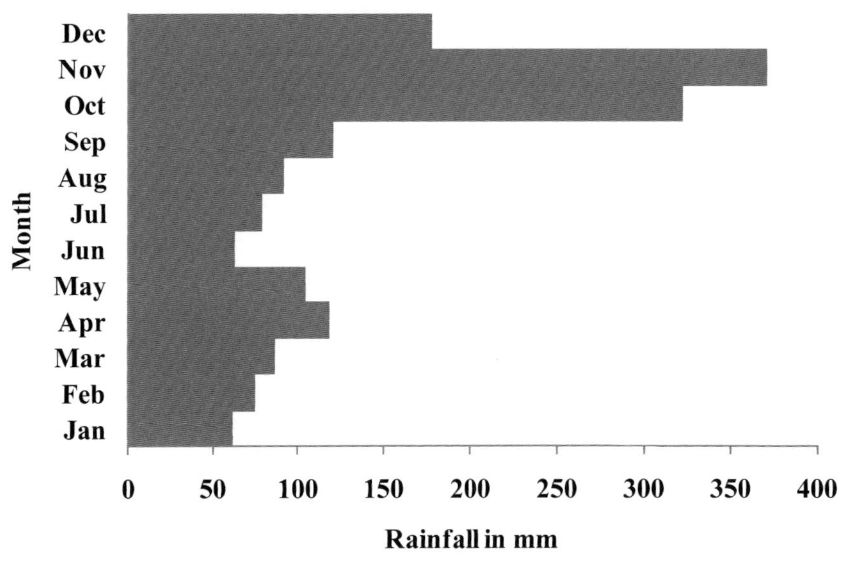

Figure 4. Average Monthly rainfall in mm for Coonoor Station-80 years

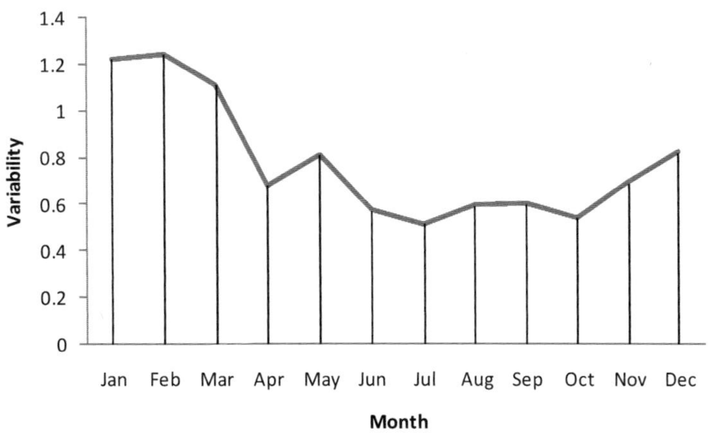

Figure 5. Monthly rainfall variability of Coonoor Station

4. Year Wise Rainfall Analysis

4.1 Year 2007

Year 2007 received annual rainfall of 1521 mm (Table 2) in which October and December month rainfall exceeded 300 mm. Though these two months received more than 300 mm, there is large difference in the number of rainy days. The rainy days in December is only 6, in which three days reported consecutive rainfall. But, October received 13 days rainfall consecutively. The moisture content of the soil could be uniform in case of October than December of that year. Further, 3 months are having 5 days antecedent rainfall whose cumulative amount are 20.8 100.8 and 308.6 mm respectively. The last two months data (August and September) are more significant in saturation of soil than in the July month in which 20.8 mm received. Rainfall in July is continuous but its total amount is less than notable 5 days antecedent rainfall as per soil conservation treatment concern.

Table 2. Year 2007 Rainfall statistics for Coonoor Station

Month	Total amount (mm)	No. of rainy days	No. of consecutive rainy days	Amount of rainfall (mm)	Maximum daily rainfall (mm)
January	23.6	1			23.6
February	129.8	5	3	126.6	83.2
March	0.0	-	-	-	-
April	66.8	6	2	50.2	40.8
May	73	5	5	40.6	21
June	87	7	4	26	49.6
July	70	8	5	20.8	11.4
August	132.8	13	8	100.8	29.8
September	117.6	9	3	56.6	27.8
October	318.2	13	13	308.6	62.6
November	110.	7	4	84	42.4
December	392.2	6	3	305	170
Total	1521	80	Year Maximum	308.6	170

4.2 Year 2008

Year 2008 received more rainfall during post monsoon months February and March. These two months account for most rainfall of the year followed by October. Six months received more than 5-

day antecedent rainfall, but not in sequential order. The annual rainfall of this year is 1910.5 mm (Table 3) which is widely spread throughout year. January, May and September months received low rainfall. The maximum daily rainfall of 113 mm occurred inMarch and two days received rainfall around 110 mm.

Table 3. Year 2008 Rainfall statistics for Coonoor Station

Month	Total amount in mm	No. of rainy days	No. of Consecutive rainy days	Amount of rainfall (mm)	Maximum daily rainfall (mm)
January	13.6	2			8.2
February	339.3	10	9	303.4	113
March	433.5	12	12	221.2	111.5
April	86.4	3	3	84.8	55
May	36.6	3			17
June	97.1	6	2	49.6	45
July	144.5	12	10	52.4	45.5
August	231.5	17	6	89	48.4
September	19	2			11.4
October	336	13	7	150.5	82.6
November	97	6	8	61.4	30.6
December	76	5	3	27.4	40.6
Total	1910.5	91	Year maximum	303.4	113

4.3 Year 2009

Table 4 gives the monthly rainfall amount and number of rainy days during the year 2009. Year 2009 received very heavy rainfall during November month that exceeded 1000 mm rainfall amount. The number of consecutive rainy days during that month is 9 days out of 15 days of rainfall. Five days antecedent rainy days are witnessed in 6 months consecutively starting from July to December. The amount of rainfall in each of these months exceeded minimum rainfall of 36 mm which is considered in estimating 5 days' antecedent soil moisture condition in the area to address the degree of saturation of the soil. It is clear that the area was completed continued in the deep wet for continuous period of 6 months and several landslides were report in this region during November. The maximum daily rainfall of 395 mm occurred in November. November 9, 10, and 11 received 141.2 mm, 363.0 mm and 395 mm rainfall respectively. High soil moisture with fully saturated soil mass and extremely heavy downpour triggered massive, major and mini landslides at several places of the hilly road sections. This year alone there have been nearly 111 landslides in Ooty, Coonoor and Kotagiri region claiming 80 lives, disrupting traffic along the State highway connecting Mettupalayam and Ooty through Coonoor and hampering the livelihood of this hill community. North-East monsoon rains in November 2009 (monthly Rainfall of 1181 mm and 9 days consecutive rainfall of 991.2 mm) caused a major landslide affecting the entire slope, disrupting traffic on the Mettupalayam– Coonoor –Ooty Highway (NH 67) and completed destroyed the tea plantations on the slope. This slope lies along the Mettupalayam– Coonoor-Ooty highway which is a busy traffic corridor and the damage to this stretch can severely hamper the traffic flow to and from Coonoor and Ooty along the 14th hairpin bend. The slope also houses a hospital and an old age home, in addition to private horticulture farms and tea plantations on the entire slope. Necessary prevention measures are required in order to safe guard this region in case similar or higher magnitude consecutive rainfall events occur in near future. This makes the slope an important location to be monitored and studied.

Table 4. Year 2009 Rainfall statistics for Coonoor Station

Month	Total amount in mm	No. of rainy days	No. of Consecutive rainy days	Amount of rainfall	Maximum daily rainfall (mm)
January	10.2	2	-		4.6
February	0.0	0	-		-
March	84.7	4	3	80.2	49.6
April	20	1	-		20
May	86.2	7	3	53.2	25
June	39.8	3	3	36.4	23.2
July	74	9	7	36.4	10
August	147.6	9	8	118	43.2
September	110.8	8	8	104	39
October	138.3	11	5	74.4	27
November	1181	15	9	991.2	395
December	190	10	5	130.2	48.6
Total	2081.8	79	Year Maximum	991.2	395

4.4 Year 2010

The maximum daily rainfall for this year is found to be 81 mm which is lesser than 140 mm. Generally, it is considered that 140 mm rainfall a day could trigger landslide in cut slopes. There are five events of five consecutive day rainfall, in which two events rainfall amount exceeded 140 mm. The annual rainfall is 1409.4 mm (Table 5) and November month contribution is found to be significant. In the November month itself, there are two long periods of rainy events. This year rainfall could be deficit with respected to normal rainfall of the region.

Table 5. Year 2010 Rainfall statistics for Coonoor Station

Month	Total amount in mm	No. of rainy days	No. of Consecutive rainy days	Amount of consecutive rainfall (mm)	Maximum daily rainfall (mm)
January	40.6	2			25
February	1.0	1			1
March	0.0	-			0
April	4.5	1			1
May	133.6	8	5	51.4	37.6
June	57.2	7	3	22.8	35
July	123.6	8	5	41.4	36.4
August	50	4			25
September	101	10	6	54	23
October	224	12	4	195	70
November	475	15	7	211.6	81
December	198.9	8	7	156.6	72
Total	1409.4	76	Year maximum	211.6	81

4.5 Year 2011

This year received very high rainfall comparing the previous 4 years. The rainfall in the months January, October and November received more than 450 mm (Table 6) which made extraordinary contribution to the total amount. There are six events of 5 days consecutive rainfall. The highest rainfall falls in the consecutive rainfall periods. It is to be noted that there was a continuous rainfall of 18 days starting from end of October to the beginning of November. The cumulative rainfall of this event is 649.1 mm. Such a long spell certainly could invite landslides if there are no landslide prevention measures. As the rainfall is continuous, the soil will be full saturated. Further, this rainfall event would have produced large surface runoff and base flow event after rainfall over. A maximum rainfall of 145 mm occurred two times in this year.

Table 6. Year 2011 Rainfall statistics for Coonoor Station

Month	Total amount in mm	No. of rainy days	No. of Consecutive rainy days	Amount of consecutive rainfall (mm)	Maximum daily rainfall (mm)
January	461.4	14	8, 5	179 &277.8	145
February	193.4	5	6	193.4	133
March	31	4	-	-	8.8
April	192.5	7	5	91	74
May		-	-	-	-
June	129.6	8	4	68.8	33.4
July	84	6	-	-	27.6
August	96.1	9	4	48	28
September	117.8	11	4	36.6	25
October	691.5	21	11	470.1	84.8
November	460.6	14	8	179	145
December	87.2	5	-	-	40.6
Total	2545.1	104	Year maximum	470.1	145

4.6 Year 2012

The rainfall during post and pre monsoon periods (Table 7) is found to be low comparing previous years of data. The data related to south west monsoon periods is not available in the analysis.

Table 7. Year 2012 Rainfall statistics for Coonoor Station

Month	Total amount in mm	No. Of rainy days	No. Consecutive rainy days	Amount of consecutive rainfall (mm)	Maximum daily rainfall (mm)
January	0.0	0			
February	13.4	2			8
March	81.4	3			51
April	83.2	6	5	76.8	62.2
May	49.4	5			17
June	NA				
July	NA				
August	NA				
September	NA				
October	776.5	14	7 & 7	286.4 & 436.7	180
November	134.6	6			51
December	NF				
Total			Year maximum	436.7	180

There are three events which had a five day consecutive rainfall, in which two events occurred in October with one day interval between them. Almost 15 day continuous rainfall with one non rainy at the middle of the period has brought 723.1 mm of rainfall. The highest downpour of 180 mm is recorded during this period. The October month of this year poses extremely wet weather conditions.

4.7 Year 2014

This year received total rainfall of 1581.6 mm (Table 8), in which North-east monsoon period recorded more rainfall than other seasons. Year maximum rainfall of 102 mm has recorded once. 3 events of 5-day consecutive rainfall have occurred. It is to be noted here that 2 events of 5-day consecutive rainfall are less than 160 mm. The six day continuous rainfall in the October month is recorded as 353 mm. But, this continuous rainfall did not bring much landslide. Heavy rainfall within continuous rainy days is recorded only 91 mm and its distribution over the day could be crucial in trigging landslide.

Table 8. Year 2014 Rainfall statistics for Coonoor Station

Month	Total amount in mm	No. Of rainy days	No. of Consecutive rainy days	Amount of consecutive rainfall (mm)	Maximum daily rainfall (mm)
January	0.00	0			
February	108	5	4	82.2	37.4
March	93	4	3	84.8	47.6
April	-				
May	190	7	5	153.2	97
June	16	2			10
July	42.8	6			7.6
August	154.5	7	5	77.8	39
September	64.1	5			21
October	460.8	11	6	353	91
November	236.6	11	4	202.6	84.2
December	215.8	7	3	179.2	102
Total	1581.6	65	Year maximum	353	102

4.8 Year 2015

The total rainfall received during year 2015 is 2488 mm (Table 9) which is 40 % higher than the normal rainfall. Number of rainy days is 99 and maximum one day rainfall is recorded as 175 mm in the March month. Though the total rainfall during November month is highest (676 mm) in the year, the maximum rainfall received during that month is 145 mm which is found to be lesser than the rainfall recorded during March month.

Table 9. Year 2015 Rainfall statistics for Coonoor Station

Month	Total amount in mm	No. of rainy days	No. of Consecutive rainy days	Amount of consecutive rainfall (mm)	Maximum daily rainfall (mm)
January	0.0				
February	4.5	1			3.1
March	489	3			175
April	167.3	6			121.2
May	301	15	9	247	95
June	144	11	5	52	22
July	90	7	3	34	25
August	51	4			17
September	207	12	2 x 4	59,102	45
October	236	12	3	53	47
November	676	18	5	357	145
December	122	10	6	91.5	24
Total	2487.8	99	Year Maximum	357	175

There are four incidence of 5-day consecutive rainfall in the year. November month's consecutive rainfall magnitude is found to be high. September received two incidents of 5-day consecutive rainfall events, but its value is small. There were 52 days of rainy days in four months. The total rainfall recorded during North-East monsoon season is 1034 mm. It is to be noted that several minor landslide incidents were reported during this season.

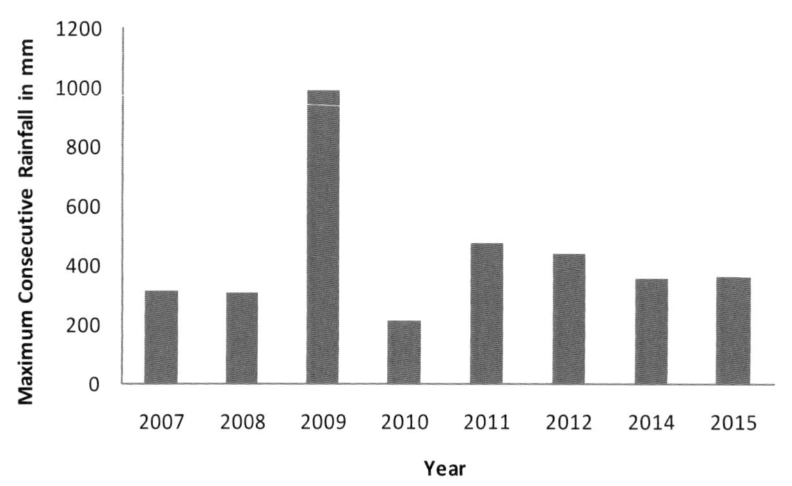

Figure 6. Consecutive maximum rainfall events during year 2007 to 2015

5. Discussion

Landslides induced by heavy rainfall often occur along Coonoor-Mettupalayam State highway during North-East Monsoon season. As this road one of shortest route connecting these two towns, the traffic flow between 4.00 am and 8.00 pm is always high. Any major obstructions along the road seriously affect the traffic flow movement. Despite the stable slope that consist of various soils such as colluvial and residual soil available along the road section, the consecutive rainfall coupled with intermittent heavy day rainfall during vigorous monsoon period accelerates mud flow along the slope that could uproot plants and tree due its high velocity of mud flow. Further, continuous rainfall fills the pore of the soil and mounting the pore pressure at critical zone in the slope that could suddenly trigger the movement of boulder and disturbing the stability of slope. The rainfall recorded during year 2009 in Coonoor station on November 9, 10, and 11 received 141.2 mm, 363.0 mm and 395 mm rainfall respectively. And another raingauge station located at RunneyMedu which is near to highway section received 170, 415 and 285 mm on November 8, 9 and 10, respectively. The heavy downpour noted in these two stations make fool-proof evidence to landslides events along the State Highways and Hospital complex located near vicinity of 14th Hairpin bend. Due to continuous soil moisture and extreme heavy downpour triggered several massive, major and mini landslides at several places along this road section during these days. While comparing maximum consecutive rainfall that occurred between 2007 and 2015(Fig.6) excluding year 2013 indicates that the year correspond to the rainfall exceeding 900 mm triggered large number of landslide events. High pore water pressures generate during prolonged and intense rainfall which could be the main reason to trigger massive landslide during these periods. Due to non-availability of pore water pressure data, it is very difficult to assess the threshold rainfall which can trigger landslides. It is clear from the present analysis that the slope instability occur only during very heavy down pour with continuous rainfall event. Hence, it is important to prioritize the critical slopes for preventive and remedial action, developing early warning systems using sensors based on pore pressure. Six days continuous rainfall during October, 2014 triggered minor landslides in the Coonoor region. The rainfall amount of 91, 48.4,81.2, 79.6, 32.6 and 20.2 are recorded between 19th October and 24th October, 2014. The landslide reported on 20th October, 2014 had rainfall of 48.2 mm with one day previous rainfall of 91 mm. This landslide happened with one day antecedent moisture. The antecedent moisture prevailed before occurrence of 91 mm rainfall due to two independent rainfalls occurred in the same week. It is to be noted that triggering of any landslide in the study area requires sufficient antecedent moisture and followed a heavy downpour. The type of landslides depends on the magnitude of heavy rainfall which would have been occurred within consecutive rainfall events. There were 38 events of 5-day and more consecutive day rainfall noticed. But all these consecutive rainfall events did not produce landslide events. Those events which recorded heavy rainfall within continuous rainfall alone produced landslides.

6. Conclusion

The daily rainfall data for a selected recent year and monthly rainfall data for 80 years used for analysis of rainfall data to evaluate rainfall which could be potential triggering factor for minor, medium and major landslide in and around Coonoor station of Nilgris district of Tamilnadu, India. Most of landslides events have occurred during continuous rainfall events and its cumulative rainfall amount exceeding 500 mm. Major landslides reported whenever daily rainfall exceeds 250 mm with several days of antecedent rainfall. Based on 5-yrs and 10-yrs moving average analysis of 80 years of annual rainfall, there is increasing trend of rainfall amount witnessed in the selected station. The contribution of North-East monsoon rainfall is found to predominant and most of landslides are triggered during this period. In most of the year, number of rainy days in the month of October and November is found to be greater than 10 days. August and September months are next to these months recorded as most rainy month of a year. The station normal annual rainfall based on recent 30 years is evaluated as 1773 mm and pre-monsoon, South-West monsoon and North-East Monsoon seasonal rainfall are obtained as 416, 369 and 988 mm respectively. Further, analysis concludes that landslides were happened only those events of continuous rainfall in which maximum rainfall noticed. It is to be noted here that rainfall threshold for landsliding are dynamic and it could depend on geo-environmental conditions of local terrain within close vicinity of landslide spot. Improving measuring of rainfall at closer spatial condition in conjunction with pore pressure measurement is needed at longer period.

Acknowledgment

This study was supported by DST-NRDMS (155/18-2015). The authors would like to acknowledge with thanks, the financial support rendered by NRDMS, DST for the research.

References

[1] Corominas J 2001 Landslides and Climate. *In: Bromhead EN (ed) Keynote Lectures delivered at the 8th Int. Symp. on Landslides, Cardiff*, June 2000
[2] Chowdhury Robin and Flentje Phil 2002 Uncertainties in rainfall induced landslide hazard, *Q. J. Eng. Geol. Hydr.* **35**, 61-70.
[3] Jaiswal Pankaj 2011 Landslide risk quantification along transportation corridors based on historical information, PhD dissertation, Faculty of Geo-information Science and Earth observation, University of Twente. Netherlands.

IOP Conf. Series: Earth and Environmental Science **80** (2017) 012067 doi:10.1088/1755-1315/80/1/012067

Statistical Analysis of 30 Years Rainfall Data: A Case Study

G Arvind, P Ashok Kumar, S GirishKarthi and C R Suribabu*

School of Civil Engineering, SASTRA University, Thanjavur – 613 401, India.

*Email: suribabu@civil.sastra.edu

Abstract. Rainfall is a prime input for various engineering design such as hydraulic structures, bridges and culverts, canals, storm water sewer and road drainage system. The detailed statistical analysis of each region is essential to estimate the relevant input value for design and analysis of engineering structures and also for crop planning. A rain gauge station located closely in Trichy district is selected for statistical analysis where agriculture is the prime occupation. The daily rainfall data for a period of 30 years is used to understand normal rainfall, deficit rainfall, Excess rainfall and Seasonal rainfall of the selected circle headquarters. Further various plotting position formulae available is used to evaluate return period of monthly, seasonally and annual rainfall. This analysis will provide useful information for water resources planner, farmers and urban engineers to assess the availability of water and create the storage accordingly. The mean, standard deviation and coefficient of variation of monthly and annual rainfall was calculated to check the rainfall variability. From the calculated results, the rainfall pattern is found to be erratic. The best fit probability distribution was identified based on the minimum deviation between actual and estimated values. The scientific results and the analysis paved the way to determine the proper onset and withdrawal of monsoon results which were used for land preparation and sowing.

1. Introduction

Water is vital for any life process and there can be no substitute for it. Water is also used for transportation, is a source of power and serves many other useful purposes for domestic consumption, agriculture and industry. The main important source of water in any area is rain and it has a dramatic effect on agriculture. Plants get their water supply from natural sources and through irrigation. The yield of crops particularly in rain-fed areas depends on the rainfall pattern, which makes it important to predict the probability of occurrence of rainfall from the past records of hydrological data using statistical analysis. Frequency or probability distribution helps to relate the magnitude of the extreme events like floods, droughts and severe storms with their number of occurrences such that their chance of occurrence with time can be predicted easily. By fitting a frequency distribution to the set of hydrological data, the probability of occurrences of random parameter can be calculated. To fit the distribution, the hydrological data is analyzed and the variability in the data is studied from the statistical parameters. Suchit Kumar Rai et al., studied the change, variability and rainfall probability for crop planning in few districts of Central India [1]. Nyatuame et al. [2] performed the statistical analysis and studied the variability in the distribution of rainfall. Rajendran et al. [3] carried out the frequency analysis of rainy days and studied the rainfall variation.

2. Study Area

The present study is carried out for Musiri (Tiruchirapalli district, Fig.1) town situated at a distance of 29 km from Tiruchirapalli city. The region has a latitude and longitude of 10.9549°N and 78.4439°E respectively. Agriculture is the main occupation in this town situated on the northern bank of Cavery river. The crops include paddy, sugarcane, banana and vegetables. Musiri receives rainfall from both the northeast and southwest on an average (30 years) of 245.49 mm and 352.62 mm respectively. The daily rainfall data is collected from the Indian Meteorological Department (IMD), Musiri station, for a period of 30 years (1984-2013). This data is used for the Yearly, Monthly and Seasonal Rainfall-Probability analysis. Figure 2 presents the historical annual rainfall for the station.

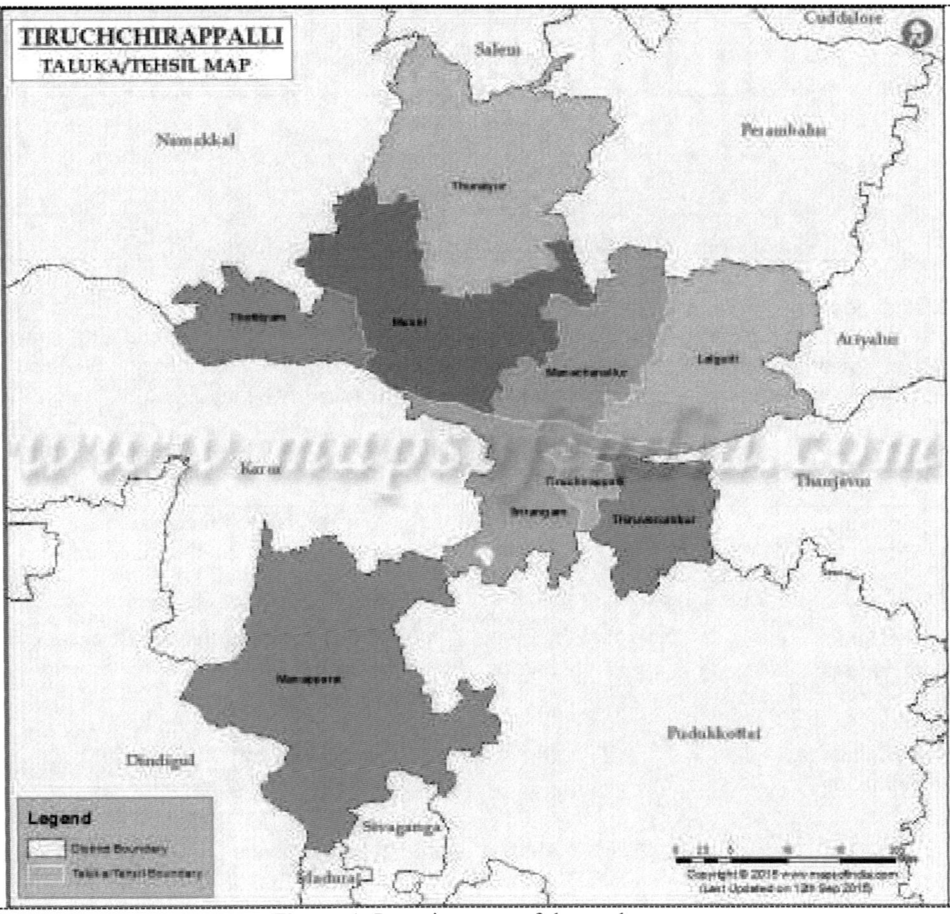

Figure 1. Location map of the study area

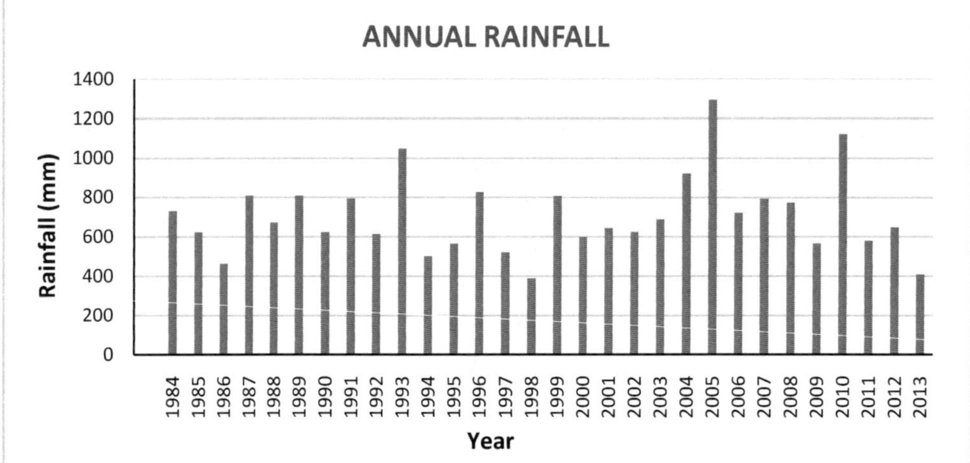

Figure 2. Annual Rainfall of Musiri town

3. Methodology

The methodology adopted in this study is Rainfall Statistics (Table 1), Probability analysis using plotting position and probabilistic methods (Table 2). From the Preliminary study and analysis, variation in results among the plotting position methods is found to be insignificant.

Table 1. Formulae for Statistical Parameters

Description	Symbol	Formula	Explanation
Arithmetic Mean	X_{avg}	$\sum X_i / n$	X is the rainfall magnitude in mm, i=1, 2, to n and n is the length of the sample.
Standard deviation	Σ	$[\sum (X_i - X_{avg})^2 / (n-1)]^{1/2}$	X is the rainfall magnitude in mm, i=1, 2, to n and n is the length of the sample.
Co-efficient of Variation	C_v	$100 \times (\sigma / X_{avg})$	X_{avg} is the Mean σ is the Standard deviation
Co-efficient of Skewness	C_s	$(1 / \sigma^3) \times [(N / (N^2 - 3N + 2)] \times \sum (X_i - X_{avg})^3$	σ is the Standard deviation N = Total no. of years X_{avg} is the Mean X is the rainfall magnitude in mm, i=1, 2 to n

3.1 Annual Rainfall Analysis

The annual rainfall data is analyzed and the variation in distribution over the area is studied with the statistical parameters. The best fit distribution method is found using various plotting position and probabilistic methods.

Table 2 Plotting position and Probabilistic methods

S. No.	Plotting position methods	Probabilistic methods
1	California = m/N	Normal Distribution
2	Hazen = (m-0.5)/N	Log-Normal Distribution
3	Weibull = m/(N+1)	Pearson Type-III Distribution
4	Beard = (m-0.31)/(N+0.38)	Log-Pearson Type-III Distribution
5	Chegodayev = (m-0.3)/(N+0.4)	Extreme Value Type-I Distribution
6	Blom = (m-3/8)(N+1/4)	
7	Tukey = (3m-1)/(3N+1)	
8	Gringorten = (m-0.44)/(N+0.12)	
9	Cunnane = (m-0.4)/(N+0.2)	
10	Adamowski = (m-1/4)/(N+1/2)	

where, m is rank of the data and N = length of the sample (no. of years).

3.2 Monthly Rainfall Analysis

From the Preliminary study and analysis, variation in results among the plotting position methods is found to be insignificant and hence, only Weibull method is adopted for the analysis among them. From the Probabilistic methods, Gumbel and Normal distribution methods are used.The rainfall data are arranged into a number of intervals with definite ranges. Mean and standard deviation were found out for the grouped data. Chi-square values are calculated for the above methods, with the obtained probabilities. The method that gives the least Chi-square value is found to best fit the distribution.
Weibull Distribution is a continuous probability distribution type where in rainfall amounts are assigned with a rank and the corresponding probabilities are found out using probability density function:

$$(X) = m/\,(n + 1) \quad (1)$$

Where,

m and n represents the rank and total number of data used in the analysis

Gumbel Distribution is used to model the distribution of the extremities of a number of samples of various distributions.

$$P(X) = \exp.\,((-\,(a + x)/c) - e - (-\,(a + x)\,/\,c))$$
$$a = 0.450055\,\sigma - Xavg \& c = 0.7797\,\sigma \quad (2)$$

Where,

P(X) is the probability density function for Gumbel method and Xavg represents the average rainfall in mm

Normal Distribution is a very common continuous probability distribution. Normal distributions represent real-valued random variables whose distributions are not known.

$$B = 0.5[1 + 0.196854\,|Z| + 0.115194\,|Z|^2 + 0.000344\,|Z|^3 + 0.015927\,|Z|^4$$
$$Z = (X - Xavg)\,/\,\sigma,\ F\,(Xi) = B for Z< 0\ \& F\,(Xi) = 1 - B for Z> 0 \quad (3)$$

The Probability Density function for the Normal Distribution method is as follows:

$$(X) = F\,(Xi + 1) - (X) \quad (4)$$

Where,

Xi is the rainfall at any instant i = 1,2,3 to n

Goodness of Fit is a test used to find out the best fit probability distribution. The best fit distribution varies for different time period. Chi-squared test is used in the determination of best fit distribution for weekly and seasonal rainfall in this study. Chi-Squared Test is used for continuously sampled data only and is used to determine if a sample comes from a population with a specific distribution.

$$Xc^2 = \sum(O-E)^2 / E \quad (5)$$

Where,

$\sum$ from i = 1 to k, O = Observed frequency,
E = Expected frequency,
i = Number of observations and k = the total number of data used

Chi-Square Formula adopted in the study is as follows:

$$Xc^2 = [(fs\,(Xi) - P\,(Xi))^2 / P\,(Xi)] \times \sum Ni$$
$$fs\,(Xi) = Ni/\sum Ni \quad (6)$$

Where,

Xc is the Chi – squared value,
P(X) is the probability density function

3.3 Seasonal Rainfall Analysis

In this analysis, the variation of distribution of rainfall is studied with the statistical parameters using formulae mentioned in Table 1.

3.4 Effective Rainfall

Water requirement for various crops is found and they are related with the effective rainfall, which is calculated from the rainfall data. Effective rainfall is the amount of rainfall effectively used by the crops. The effective rainfall in the study area is calculated using the formula:

$$R_e = 0.8*P-25, \text{ if } P \geq 75\text{mm} \quad (7)$$
$$R_e = 0.6*P-10, \text{ if } P < 75\text{mm} \quad (8)$$

where,

R_e is the Effective Rainfall (mm),
P is the Total Monthly Rainfall (mm)

4. Results and Discussions

4.1 Annual Rainfall Analysis

The rainfall data are ranked in descending order and various plotting position and probabilistic methods are applied to determine the return period. Rainfall magnitudes were calculated for different return periods using the rainfall-return period equation obtained from the graphs for all plotting position methods (Table 3).

Table 3. Maximum Annual Rainfall based on Plotting Position methods

Method/Return period	10	30	50	100	300	500
WEIBULL	1019.914	1275.352	1394.124	1555.288	1810.726	1929.499
CALIFORNIA	1027.544	1282.982	1401.754	1562.918	1818.356	1937.129
HAZEN	973.8311	1199.124	1303.879	1446.022	1671.315	1776.07
CHEGODAYEV	**994.2638**	**1233.036**	**1344.059**	**1494.708**	**1733.48**	**1844.503**
BLOM	987.0367	1221.063	1329.879	1477.533	1711.56	1820.376
GRINGORTEN	980.3763	1210.008	1316.781	1461.663	1691.295	1798.067
BEARD	993.3244	1231.482	1342.218	1492.479	1730.636	1841.373
TUKEY	991.092	1227.788	1337.845	1487.184	1723.88	1833.937
CUNNANE	984.5198	1216.887	1324.932	1471.54	1703.907	1811.952
ADAMOWSKI	998.8729	1240.656	1353.078	1505.626	1747.408	1859.831
AVERAGE MAX RAINFALL	**995.078**	**1233.838**	**1344.855**	**1495.496**	**1734.256**	**1845.274**

For Musiri, California method gives the maximum value for rainfall for different return periods and Hazen method is found to give the least value and is hence not acceptable for the analysis. It is seen that Chegodayev method gives a maximum rainfall which is approximately 99.9% to that of the average maximum rainfall unlike other methods of distribution and is hence the best fit distribution for annual rainfall data. It is also seen that whenever there is an increase in return period, the rainfall amount also increases and vice-versa. Hence, Rainfall and return period are proportional to each other.

Table 4. Maximum Annual Rainfall based on Probabilistic methods

Distribution/Return Period	50 years (x) in mm	100 years (x) in mm
NORMAL	1119.037	1173.951
GUMBEL	**1315.274**	**1441.859**
PEARSON TYPE-III	1222.606	1322.139
LOG-NORMAL	1193.561	1286.205
LOG-PEARSON TYPE-III	1225.471	1334.455

Considering the results of Plotting positions to be actual, Gumbel distribution (Extreme value type-I) gives a value that is closer to the actual value and is hence the best method of fit for the annual rainfall data (Table 4).

Table 5. Statistical Parameters for Annual Rainfall Analysis

S. No.	Description	Normal series	Log-transferred series
1	Mean Rainfall (mm)	704.35	2.83168
2	Standard deviation (mm)	201.88	0.119355
3	Co-efficient of Variation	0.2866	0.042150
4	Co-efficient of skewness	1.0582	0.181678

4.1.1 Statistical parameters. The above Table 5 shows that the Standard deviation value is considerable large which indicate there is larger variation in rainfall pattern. Skewness represents the distribution of data about the mean. It is equal to zero in the case of normal distribution. When the peak of the sample is towards the right of a plotted graph, it is said to be negatively skewed and when the peak of the sample is towards the left of a plotted graph, it is said to be positively skewed. From the above table, it is clear that Normal series and Log-transferred series data are positively skewed.

4.2 Monthly Rainfall Analysis

The rainfall data are arranged into a number of intervals with a range of 25 mm and the frequency of occurrence is found out initially, to convert the normal data into a grouped data. Mean and Standard Deviation were found out for the same to check the variation in rainfall. Chi-square values obtained are compared for all the methods and the Least Chi-square value in all the cases is given by Gumbel distribution.

Table 6. Chi-square values for Monthly rainfall analysis

Distribution	Chi-Square value
WEIBULL	2443.51
GUMBEL	**258.93**
NORMAL	367.4

Thus, from the above Table 6, it is inferred that the chi-square value is least for Gumbel distribution, showing that it best fits the monthly rainfall data.

Table 7. Statistical Parameters for Monthly Rainfall Analysis

Months	Mean (mm)	Standard Deviation (mm)	Coefficient of Variation	Coefficient of Skewness
January	8.877	23.637	2.663	2.823
February	6.563	18.887	2.877	3.350
March	13.047	27.319	2.094	2.698
April	25.250	35.078	1.389	2.239
May	52.493	49.681	0.946	2.100
June	23.327	38.466	1.649	2.399
July	37.213	41.139	1.105	0.852
August	66.147	48.218	0.729	0.250
September	118.820	74.452	0.626	0.896
October	145.003	81.348	0.561	0.630
November	156.077	108.673	0.696	0.856
December	51.537	64.968	1.260	2.694

4.2.1 Statistical parameters. From the above Table 7, it is observed that maximum average rainfall is received in the months of August, September, October and November during the monsoons, having the standard deviation values less than their corresponding mean values. Whereas, the standard deviation values for the remaining months are higher than their corresponding mean values showing larger variation in the distribution of rainfall over the months. From the above table it is clear that, Normal series data are positively skewed.

Table 8. Statistical Parameters for Seasonal Rainfall Analysis

S. No.	Statistical Parameters	SW Monsoon	NE Monsoon
1	Mean Rainfall (mm)	245.49	352.62
2	Standard deviation (mm)	116.12	166.70
3	Co-efficient of Variation	0.0157	0.0157
4	Co-efficient of skewness	0.406	0.896

4.3 Seasonal Rainfall Analysis

From the above Table 8, it is shown that Musiri receives the rainfall seasonally on an average of 245.49 mm and 352.62 mm during south-west (SW) and north-east (NE) monsoons respectively and it also shows that to how much extent the distribution of rainfall is varied over the study area during the respective monsoons. The monthly rainfall series data of both the monsoons are positively skewed.

4.4 Crop Planning

Average effective rainfall for 30 years were found and tabulated. From table 8, it is observed that the average effective rainfall is higher in the months of August, September, October and November. Whereas, the average effective rainfall in the months of January, February and March is zero. Based on the effective rainfall the crop planning is done which is shown in Table 9.

Table 9. Average effective rainfall for 30 years

Months	Average Rainfall (mm)
January	0
February	0
March	0
April	5.90
May	23.64
June	4.96
July	13.53
August	32.94
September	72.15
October	92.05
November	101.42
December	23.76

Musiri taluk has net cultivable area of 27,344 hectares which constitutes 13.4% of total agricultural area in Trichy district. The soil type in Trichy district is found to be Red sandy soil and black soil. Crops like Paddy, Sugarcane, Maize, Pulses (Black-gram and Soybean), Groundnut, Sesame(Gingelly) and Sunflower are very much suitable to be grown in Red Sandy soil and Black soil is suitable for growing Cotton. Cultivation of crop varies according to the climate and water requirement. From the collected data, it has been found that Paddy constitutes about 39% the total cultivable area and stands top in net production followed by Groundnut which is cultivated for about 2338 ha, Maize for about 1315 ha, Sugarcane for about 1002 ha, Cotton for 442 ha, Gingelly for merely 57 ha and other cash crops like Cholam, Cumbu etc for the remaining area. Based on the effective rainfall data, crop water requirement for various crops were identified and segregated into rain-fed and irrigated. The Table 8 above shows the best sowing and harvesting time for cultivation of various crops and water requirement through rain and irrigation.

Table 10. Crop planning for the Musiri Region

S. No.	Crops	TWR*	best suited month		Rain-Fed (mm)	Irrigated (mm)
			Sowing	Harvesting		
1	Maize	500mm for 100 days	Early August	November	299	201
2	Pulses:					
	1. Blackgram	280mm for 65 days	Mid August	October	197	83
	2. Soybean	320mm for 85 days	September	November	266	54
3	Groundnut	510mm for 105 days	September	December	290	220
4	Sesame (Gingelly)	150mm for 85 days	July	September	119	31
5	Cotton	600mm for 165 days	Mid July	Mid December	335	265
6	Sunflower	450mm for 110 days	August	November End	299	151

*TWR-Total Water Requirement

5. Conclusion

From the Rainfall Probability analysis on the Annual and Monthly rainfall for Musiri Region, it is evident that Gumbel Distribution (Extreme Value Type – I) is ascertained as the best fit distribution type considering its Least Chi-Square Value among all other methods of analysis. Chegodayev Distribution from Plotting Position methods is found to best fit the Annual rainfall data. The present statistical analysis provides clear picture on rainfall data and it is found that rainfall available in the region is insufficient to carry out wet crop. Conjunctive use of surface water, available rainfall and ground water is essential for better agricultural and irrigation management for this area, Thus, the analysis helps in understanding the rainfall pattern of Musiri region and also in efficient crop planning and water availability of the region.

Acknowledgements

The authors acknowledge the Vice Chancellor, SASTRA University, Thanjavur, India for providing the facilities to carry out the work and the encouragement in completing this work.

References

[1] Suchit Kumar Rai, Sunil Kumar, Arvind Kumar Rai, Satyapriya and Dana Ram Palsaniya 2014 Climate Change Variability and Rainfall Probability for Crop Planning in Few Districts of Central India. *Atmos. Climate Sci.* **4** 394-403.

[2] Nyatuame M, Owusu-Gyimah V and Ampiaw F 2014 Statistical Analysis of Rainfall Trend for Volta Region in Ghana. *Int. J. Atmos. Sci.* **67**(2) 1-11.

[3] Rajendran V, Venkatasubramani R and Vijayakumar G 2016 Rainfall variation and frequency analysis study in Dharmapuri district (India). *Indian J. Geo. Mar. Sci.* **45**(11) 1560-5.

Protection of Urban Water body Infrastructure – Policy Requirements

T R Neelakantan* and K Ramakrishnan

School of Civil Engineering, SASTRA University, Thanjavur – 613401, India.

*E mail: trneelakantan@yahoo.com

Abstract. Water body is an important infrastructure of urban landscape. Water bodies like tanks and ponds are constructed to harvest rainwater for local use. Such water bodies serve many environmental functions including flood and soil erosion control and are useful for irrigation, drinking water supply and groundwater recharge. A large number of water bodies recently have been lost due to anthropogenic activities and the remaining water bodies are under stress due to risk of degradation. There are many phases to solve or control the problem; starting from stopping the abuse, to restoration to monitoring and maintenance. In this situation, the existing urban and peri-urban water bodies are to be preserved and rehabilitated. In this study, policy requirements for the protection (preservation and rehabilitation) of water bodies are analyzed with special reference to Thanjavur city. Thanjavur city has many water bodies and moat around the Big-Temple and the palace, and stands as an evidence for water management in ancient days. These water bodies are to be protected and used properly for sustainable growth of the city. This paper envisages the following three: (a) need for evaluation of hydraulic and hydrologic properties of the water bodies for conserving rainwater and controlling flood water in the existing urban water bodies; (b) need for evaluation of potential of socio-environmental services by the water bodies, and (c) need for developing a relative importance index for protection of water bodies to prioritize the remedial actions.

1. Introduction

Water body is an important feature of urban landscape. While urban lakes are present all over the world, temple tanks are special to Indian context. A large number of water bodies recently have been lost due to anthropogenic activities and the remaining water bodies are under stress due to risk of degradation. Throughout the world, many water bodies are suffering from pollution and encroachment problems from industrial, urban and agricultural development. There are many phases to solve or control the problem; starting from stopping the abuse, to restoration to monitoring and maintenance.

Water bodies like tanks and ponds are constructed to harvest rainwater for local use. Such water bodies serve many environmental functions including flood and soil erosion control and are useful for irrigation, drinking water supply and groundwater recharge. These water bodies were maintained by local communities earlier. The shifting of maintenance responsibility from local community to the government agencies led to the menace of encroachment and abuse. The encroachment in urban and peri-urban areas is more prominent. The urban population growth demands more land for housing and other infrastructure. The scarce and costly land resources stress the water bodies ultimately. Even the government bodies find the water bodies as an easy source for extra land for development. In this situation, the existing urban and peri-urban water bodies are to be preserved and rehabilitated.

Thanjavur city (10°46N, 79°08E), shortlisted for 100 *Smart City Mission*, by Ministry of Urban Development, Government of India is one of the cities undergoing the process of losing the water bodies for urban growth. Thanjavur became a corporation in the beginning of 2014, has an area of about 37 sq km and has a population of around five lakh. Thanjavur city is one of the old cities in India. Thanjavur is famous for the more than 1000-year old Big-Temple (Lord Bhrahadeeswara Temple) made of stones. This city was the capital of the Chola dynasty for many years. Thanjavur city with many water bodies and moat around the Big-Temple and the palace stands as an evidence for water management in ancient days.

A moat (water body) around the temple and the palace area are still seen though they are not in proper shape for a considerable stretch. There are number of tanks around the temple and palace and these tanks were connected by pipes. The surplus of upper tank was made to flow to a lower tank through the pipes. The upper tank acted as settling tank and pure water from the top layer was sent to the lower tanks, which served as water source for the town dwellers. The moat received the surplus water and served as a water blockade. There are more than 20 water bodies including a handful of temple tanks in Thanjavur with a capacity of more than one million m^3. Apart from these temple tanks and lakes, a moat of about 4.5 km length and 40 m width, which is hydraulically connected to Cauvery river, is present in the main town area. These water bodies are to be protected and used properly for sustainable growth of the city. Keeping the status of the water bodies in Thanjavur in mind which is a common scenario in many of towns in southern states of India, the policy requirements for protection of urban water bodies is envisaged. The objective of this work is to identify (a) need for evaluation of hydraulic and hydrologic properties of the water bodies for conserving rainwater and controlling flood water in the existing urban water bodies; (b) need for evaluation of potential of socio-environmental services by the water bodies, and (c) need for developing a relative importance index for protection of water bodies to prioritize the remedial actions.

Very limited works have been reported in this field both in India and outside. In the Indian context, this is important as water scarcity and flooding both affect the cities and many tanks are encroached or being encroached and disappearing fast due to urbanization. In many developed countries, water quantity availability is not a major issue while the quality of water is a major issue. Hence, many international literatures are focusing on water quality issues.

2. Indian Scenario

Urban and peri-urban water bodies were studied by various researchers [9,13,17] in the recent years. Two tanks, one in Chennai city and another just out of Chennai city were studied [9] and the author tried to find the cause for the dry status of urban tanks. The author [9] reported that the human intervensions like desilting and concrete flooring of tank beds as major reasons. Considering both quantity and quality aspects, the author [9] suggested methods to re-establish the hydrological role of tanks.

In Bangalore city, Varthur Lake which is fed by sewage, was studied for self-remediation through physical, chemical and biological analyses including BOD removal and nutrient loads [13]. It was identified that the lake behaves like an anaerobic-aerobic lagoon.

To monitor and prioritize conservation works of six water bodies in and around Chennai city, a Water Bodies Protection Index was developed [17]. To formulate the index five factors namely (1) quality of water, (2) bio-diversity, (3) encroachment of water body, (4) involvement of local community in the management and preservation of water body, and (5) role of government agencies in protecting water body were used by the authors and they conducted a Delphi study to assign wightages to the above factors.

It is difficult to understand a long list of numerical values of various water quality and other parameters. An integrated single index value makes information more easily and quickly understandable. Such an index can facilitate comparison among various water bodies. Therefore, a single index value is better for transmitting information to general audiences or policy makers. In India, few researchers worked on developing a single index for assessing water quality. A water quality index was formulated and used [1], which considers the following six water quality parameters: dissolved oxygen, BOD, total coliform, turbidity, TDS and pH. The researchers [21] developed a water quality index to quantify overall water quality for human consumption and applied that for their study area in Tamilnadu. They used the following twelve parameters: total dissolved solids, HCO_3, Cl, SO_4, PO_4, NO_3, F, Ca, Mg, Na, K, and Si.

3. International status

Developing a *Water Poverty Index* by linking water availability and socioeconomic variables was discussed to assist the policy makers on equitable water allocation [18]. The researcher [18] discussed the issues on acceptability and relevance, and the problem of scale of index and different approaches in forming a meaningful index in assessing the links between water and poverty. This work was followed latter by many researchers [4,14,19]. Further, many researchers proposed use of different water quality indices to get a single numerical expression for the overall quality of a water [1,2,12,15,21].

A single indicator called *Watershed Sustainability Index* for aggregating hydrologic, environmental, life and policy parameters was developed [3] and applied to a basin in Brazil. Ten ponds in Vienna, Austria were studied to get information on the seasonal variations in the environment [16]. Multivariate statistical analysis proved that ponds were affected by salinity and by algal biomass and the authors identified eutrophication and inflow quality as the cause. A model to relate environmental conditions and occurrence of algal species was developed which shows variation in states from season to season.

Maintaining or restoring towards a natural water balance not only keeps-up the environment, it also improves the living standards of urban area [8]. The effect of urban expansion on surface water bodies in Wuhan, China was studied [6]. The researchers reported that the reduction, disappearance and pollution of surface water bodies results in undervaluation of water bodies and increases the threat for further damages and recommended an integrated and proactive land-use planning and management system. Integrated use of social, economical and environmental parameters for sustainable water resources management of West Java, Indonesia was attempted [10] and [11]. They developed a conceptual framework for developing a *Water Sustainability Index* and applied Delphi technique to finalize the components/indicators of the index.

Self-purification potential that reflects the capacity of an ecosystem to assimilate all inputs was discussed in a research [20]. The potential for self-purification depends on the restoration and/or protection of the functioning of the ecosystem. Such quality objectives of the self-purification potential would allow the system to not to cross the limits of the ecological "resilience domain" which were set by the maximum allowable stress on environment so that the system can recover fast.

Policies on water quality management for an urban lake were proposed and evaluated [7] giving due consideration for the different interests of stakeholders. Alternative policies on water quality management were analysed and ranked using fuzzy Borda count and evidential reasoning methods and compared.

The literature review indicates that every city has its own unique hydrological system, human interference and problems. Thus, it can be understood that it is very difficult to provide a generalized method that is suitable for every situation. However, a more comprehensive work will provide guidelines or approaches to many other studies.

4. Base line data requirement

Knowing that the problems associated with water bodies are location specific, the policy requirements also location specific. However, before making the policies, certain baseline data may need to be collected to understand the problem. To make policies for the protection of urban water bodies certain baseline data are required which may cover the following.

- locations and size of the water bodies, encroachment of water bodies, sediment deposits necessity of deepening, inflow sources and arrangements, outflow arrangements, possibilities of improving the inflow into the water bodies
- utilization of water from water body for drinking, bathing, and washing of cloths, suitability of water quality, disposal of solid wastes, sewage, and open defecation in water bodies, eutrophication, contaminations and source control measures required, water quality variation with respect to seasons, groundwater recharge potential, potential of water bodies as flood control reservoirs, reduction of heat island problem by the water bodies, menace due to mosquito breeding in water bodies

5. Tentative steps required for protection of urban water bodies

Tentative steps required for the protection of urban water bodies comprises of three major parts viz. (a) estimating hydraulic and hydrologic properties of the water bodies, (b) estimating potential of environmental services by the water bodies and (c) ranking the water bodies for remedial actions. The first part involves field observations and analytical estimations. The second part involves analyses of quality of water samples and field observations. The third part involves survey through questionnaires and analytical study.

5.1 Estimating hydraulic and hydrological properties of the water bodies

The following components have to be envisaged under this part.

- Inventory of the water bodies: Identifying the locations and size of the water bodies
- Encroachment of water bodies: Identifying dumping of construction or urban solid wastes into the tank; dividing the water body into two by dumping soil and use the dividing bund for road transport – which will make one part of the water body dry as there is no inlet; implementing government plans in the water bodies.
- If a water body is not maintained, silting or sediment deposits may reduce capacity. Hence, identifying the necessity of deepening of a water bodies becomes essential.
- An urban water body may be disconnected from inflow sources. Hence, the inflow sources and inlet arrangements are to be identified. Further, arrangements for overflow are also to be identified. Few water bodies may fall in a chain, that is, they may have connections like surplus or release from one can reach the next downstream water body. Such arrangements are to be identified. In Thanjavur, it is claimed that few water tanks had links through underground pipe systems, which are to be explored, and possibilities of revamping is to be analyzed.
- Some of the water bodies are hydraulically connected to irrigation canals. Status of the connecting channels is to be explored.
- The recent water fillings due to rainfall and channel flow are to be studied. The possibilities of improving the inflow into the water bodies through connections will be studied. Using the land-use patterns, rainfall data and SCS curve number technique, potential of inflow will be estimated.

5.2 Estimating potential of environmental services by the water bodies

The following components have to be envisaged under this part.

- It is common that people use water directly from water bodies for their domestic water needs like drinking, bathing, and washing of cloths. The direct use of water from the water bodies will be studied. Suitability of water quality for such uses will be checked. Some of the water bodies are abused by disposing solid wastes, sewage, and open defecation. Urban water bodies may receive higher nutrient loads and hence eutrophication may cause problem. If contaminations are observed, the sources of contamination will be identified and source control measures will be suggested. The water quality variation in the water bodies with respect to seasons of a year will be analyzed.
- One of the important services of the water bodies is groundwater recharge. The potential of water bodies for recharging the groundwater will be studied. Variation of groundwater table in the neighboring wells with respect to water availability in the water body will be observed. The vertical hydraulic conductivity of the soil layer below the water body will be analyzed through groundwater flow modeling.
- If the bed of a water body is too porous, water will quickly flow down to the aquifer below and the water body will become dry soon. Mound in the water-table will not stay long and hence, the impact of recharge will not be realized much. If the bed of water body is of non-conductive in nature (clay layer), ground water recharge will not take place. It may be good to have a moderate vertical permeability to make the water available either as surface water or as groundwater. If the bed is too pervious or impervious, the bed should be modified with external materials. The thickness of material to be modified will be analyzed using groundwater modeling and suggestions will be reported.
- The potential of water bodies as flood control devices will be analysed using model study.
- Heat island problem of urban areas can be minimized by the water bodies. The effect of water bodies on the micro-climate (temperature variations) will be studied and reported.
- Water bodies may serve as breeding place of mosquitoes, habitat for frogs and insects. The menace due to these to the surrounding dwellings will also be studied.

5.3 Ranking the water bodies for remedial actions

Many of the urban and peri-urban water bodies are disappearing due to urbanization. Government agencies generally do not have sufficient resources to monitor, control and rehabilitate all the water bodies. In this situation, it is very important to prioritize and concentrate on few critical water bodies. The few water bodies to be selected for this purpose are to be identified based on many parameters. The money spent on the water bodies should return maximum tangible and intangible benefits. Hence, to prioritize the rehabilitation and to save precious fresh water resources, developing a comprehensive ranking index is of great necessity today. Indicators can help for ranking the water bodies and priorities can be assigned in a scientific logical manner. A comprehensive method for ranking the water bodies based on (a) hydraulic and hydrologic properties of the water bodies, (b) potential of environmental services by the water bodies and (c) survey from experts and locals is to be evaluated.

6. Conclusion

Policy for the protection (preservation and rehabilitation) of water bodies is gaining more and more importance as the water bodies are being lost due to anthropogenic effects. Since the funding available for preservation and rehabilitation are limited, prioritization is also very important. Hence, in this paper, the points to be considered in the policy making, through three major divisions viz. (a) evaluation of hydraulic and hydrologic properties; (b) evaluation of potential of socio-environmental services by the water bodies, and (c) developing a relative importance index for prioritizing or ranking the water bodies are discussed in detail.

References

[1] Avvannavar S M and Shrihari S 2008 Evaluation of water quality index for drinking purposes

for river Netravathi, Mangalore, South India. *Environ. Monit. Assess.* **143**(1) 279-90.

[2] Babaei S F, Hassani A H, Torabian A, Karbassi A R and Hosseinzadeh Lotfi F 2011 Evolution of a new surface water quality index for Karoon catchment in Iran. *Water Sci. Technol.* **64**(12) 2483-91.

[3] Chaves H M L and Alipaz S 2007 An integrated indicator based on basin hydrology, environment, life, and policy: The watershed sustainability index. *Water Resour. Manage.* **21**(5) 883-95.

[4] Cho D I, Ogwang T and Opio C 2010 Simplifying the water poverty index. *Soc. Indic. Res.* **97**(2) 257-67.

[5] Cude C G and Oregon 2001 water quality index: A tool for evaluating water quality management effectiveness. *J. Am. Water. Resour. As.* **37**(1) 125-37.

[6] Du N, Ottens H and Sliuzas R 2010 Spatial impact of urban expansion on surface water bodies-A case study of Wuhan, China. *Landscape and Urban Planning* **94** (3-4) 175-85.

[7] Estalaki S M, Kerachian R and Nikoo M R 2016 Developing water quality management policies for the Chitgar urban lake: application of fuzzy social choice and evidential reasoning methods. *Environ. Earth Sci.* **75** 404-16.

[8] Fletcher T D, Andrieu H and Hamel P 2013 Understanding, management and modelling of urban hydrology and its consequences for receiving waters: A state of the art. *Adv. Water Res.* **51** 261-79.

[9] Ganesan M 2008 The temple tanks of Madras, India: Rehabilitation of an ancient technique for multipurpose water storage. *Indian J. Sci. Technol.* **1**(7) **29594** 1-8.

[10] Juwana I, Perera B J C and Muttil N 2010 A water sustainability index for West Java. Part 1: Developing the conceptual framework. *Water Sci. Technol.* **62**(7) 1629-40.

[11] Juwana I, Perera B J C and Muttil N 2010 A water sustainability index for West Java - Part 2: Refining the conceptual framework using Delphi technique *Water Sci. Technol.* **62**(7) 1641-52.

[12] Kannel P R, Lee S, Lee Y - S, Kanel S R and Khan S P 2007 Application of water quality indices and dissolved oxygen as indicators for river water classification and urban impact assessment. *Environ. Monit. Assess.* **132**(1) 93-110.

[13] Mahapatra D M , Chanakya H N and Ramachandra T V 2011 Assessment of treatment capabilities of Varthur Lake, Bangalore, India, *Int. J. Environ. Technol. Manage.* **14**(1-4) 84-102.

[14] Molle F and Mollinga P 2003 Water poverty indicators: Conceptual problems and policy issues. *Water Policy* **5** 529-44.

[15] Pesce S F and Wunderlin D A 2000 Use of water quality indices to verify the impact of Cordoba City (Argentina) on Suquia River. *Water Res.* **34**(11) 2915-26.

[16] Schagerl M, Bloch I, Angeler D G and Fesl C 2010 The use of urban clay-pit ponds for human recreation: Assessment of impacts on water quality and phytoplankton assemblages *Environ. Monit. Assess.* **165**(1-4) 283-93.

[17] Sudha M C, Ravichandran S and Sakthivadivel R 2013 Water Bodies Protection Index for assessing the sustainability status of lakes under the influence of urbanization: A case study of south Chennai, India. *Environ. Dev. Sustain.* **15**(5) 1157-71.

[18] Sullivan C 2002 Calculating a Water Poverty Index. *World Dev.* **30**(7) 1195-210.

[19] Sullivan C A and Meigh J 2007 Integration of the biophysical and social sciences using an indicator approach: Addressing water problems at different scales. *Water Resour. Manage.* **21**(1) 111-28.

[20] Tixier G, Lafont M, Grapentine L, Rochfort Q and Marsalek J 2011 Ecological risk assessment of urban stormwater ponds: Literature review and proposal of a new conceptual approach providing ecological quality goals and the associated bioassessment tools. *Ecol. Indicators.* **11**(6) 1497-506.

[21] Vasanthavigar M, Srinivasamoorthy K, Vijayaragavan K, Rajiv Ganthi R, Chidambaram S

Anandhan P, Manivannan R and Vasudevan S 2010 Application of water quality index for groundwater quality assessment: Thirumanimuttar sub-basin, Tamilnadu, India. *Environ. Monit. Assess.* **171**(1-4) 595-609.

Time Series Forecasting of Daily Reference Evapotranspiration by Neural Network Ensemble Learning for Irrigation System

N Manikumari[1*], A Murugappan[1] and G Vinodhini[2]

[1]Department of Civil Engineering, Faculty of Engineering and Technology, Annamalai University, Annamalainagar, India
[2]Department of Computer Science & Engineering, Faculty of Engineering and Technology, Annamalai University, Annamalainagar, India

*Email: aumani67@yahoo.in

Abstract. Time series forecasting has gained remarkable interest of researchers in the last few decades. Neural networks based time series forecasting have been employed in various application areas. Reference Evapotranspiration (ET_O) is one of the most important components of the hydrologic cycle and its precise assessment is vital in water balance and crop yield estimation, water resources system design and management. This work aimed at achieving accurate time series forecast of ET_O using a combination of neural network approaches. This work was carried out using data collected in the command area of VEERANAM Tank during the period 2004 –2014 in India. In this work, the Neural Network (NN) models were combined by ensemble learning in order to improve the accuracy for forecasting Daily ET_O (for the year 2015). Bagged Neural Network (Bagged-NN) and Boosted Neural Network (Boosted-NN) ensemble learning were employed. It has been proved that Bagged-NN and Boosted-NN ensemble models are better than individual NN models in terms of accuracy. Among the ensemble models, Boosted-NN reduces the forecasting errors compared to Bagged-NN and individual NNs. Regression co-efficient, Mean Absolute Deviation, Mean Absolute Percentage error and Root Mean Square Error also ascertain that Boosted-NN lead to improved ET_O forecasting performance.
Key words: *Time series; Reference Evapotranspiration; Neural networks.*

1. Introduction

Water is a valuable contribution of nature essential for life on earth. The increasing demand of population and climate change makes it necessary to manage water effectively [1, 3, 6, 10]. Evapotranspiration is one of the key elements in the hydrologic cycle. Evapotranspiration rate from a reference surface, not short of water, is known as Reference crop Evapotranspiration (ET_o). The Penman-Monteith (FAO-56) method is method universally accepted for computing ET_o. More accurate forecasting models of ET_o would enable more efficient water resources management. Time series forecasting of Daily ET_o based on data and past values is the focus of this work. It is very vital for irrigation scheduling, water-resources distribution planning and cropping-system design, agricultural water use, regulation and planning [2, 13, 17]. In recent years, attention has been paid on analyzing time series trends of ET_o. More recently a number of machine learning techniques were applied to time series forecasting and on a number of occasions they have showed considerable improvement compared to traditional regression models. While there is an extensive knowledge available in machine learning and pattern recognition, it has been rarely used in temporal aspects of time series prediction. This work presents an ensemble based forecasting model that is designed to handle time series in various scenarios and generate accurate predictions. The proposed ensemble model has been found to be more suitable for forecasting ET_o.

2. Related Work

Time series forecasting has lot of essential applications in modern water resources management, such as environment protection, water supply planning, water quality management, irrigation system, hydropower generation, and sustainable utilization of water resources etc. [21, 25]. Machine learning methods play a major role in water management modeling [10]. New machine learning techniques have been created in practice for increasingly complex computations in the modeling. The key feature of forecasting models are that these models only require data of the time series and the modeling of hydrologic time series can be directly computed [22, 4, 19]. Many researchers have investigated on autoregressive model to establish a time series forecasting to analyze and forecast ET_o for various regions and seasons [5, 14, 15, 16, 20]. Meanwhile, some researchers have proven that the forecasting accuracy can be improved by eliminating noise from time series data using appropriate data preprocessing techniques, such as principal component, moving average, singular spectrum analysis, wavelet analysis, Gamma test [7, 12, 23, 24, 18]. Few researchers concluded that the wavelet based time series algorithm can be used to model events such as droughts with reasonable accuracy [8, 9, 23, 24].

However, with the continuous rise of water demand due to economic development, population growth, irrigation needs and industrial uses, modern water resources management would require more efficient techniques for improving accuracy in forecasting. Ensemble learning is one such technique used to improve the performance of the machine learning algorithms in forecasting. This paper investigates time series forecasting of Daily ET_o using ensemble techniques. Boosting and bagging ensemble methods are investigated to predict ET_o with time series data. The experiments were designed in two phases to evaluate the performance of ensemble techniques. First, various diverse learning techniques such as Support Vector Machine (SVM), Artificial Neural Networks (ANN), M5 model, Decision Tree (DT), k-Nearest Neighbor (k-NN) were employed in forecasting ET_o. The performances of these models were evaluated to identify the best learning method. In the second phase, Bagged and Boosted NN ensemble models were developed using identified optimum predictor as base learners. The results obtained were evaluated in order to assess the forecasting performance.

3. Problem Design

The research design is shown in Fig. 1. The data used in this research work was collected for the period 2004 to 2014. Then collected data was prepared for forecasting by preprocessing. In data preparation, various climatic parameters used for empirical estimation using P-M method were selected. Based on the attributes used for estimation of ET_o by P-M method, the data collected were grouped into four data models for forecasting by ensemble methods. Then the results were evaluated and compared to identify the best ensemble approach for forecasting ET_o.

4. Data Collection and Preparation

The command area of Veeranam tank was taken for the study. Veeranam tank is one of the major irrigation tanks of Tamilnadu which forms part of Cauvery basin, the largest river basin of Tamilnadu. The daily meteorologic and lysimetric data observed at Indian Meteorological Observatory, Annamalainagar of Tamilnadu was taken for computation of Evapotranspiration for Lowland rice.

The FAO Penman-Monteith method is the standard method for computing ET_o, as recommended by FAO (FAO,1977)since it involves many climatological parameters representative of the location in addition to latitude and elevation. Penman-Monteith method considers the crop physiological factors also, in addition to the meteorological factors considered in the Modified Penman method. The detailed description of the data used is given in Table 1. The dataset contains 9 independent attributes and one calculated dependent attribute for a period of 10 years from 2004-2014.The statistical properties of attributes are listed in Table 2. Forecasting Daily ET_o was carried out for 365 days from Jan 2015 to Dec 2015. Four different data models were developed based on the attributes used for ETo prediction. The detailed description of the data used for development of various data models is given in Table 3.

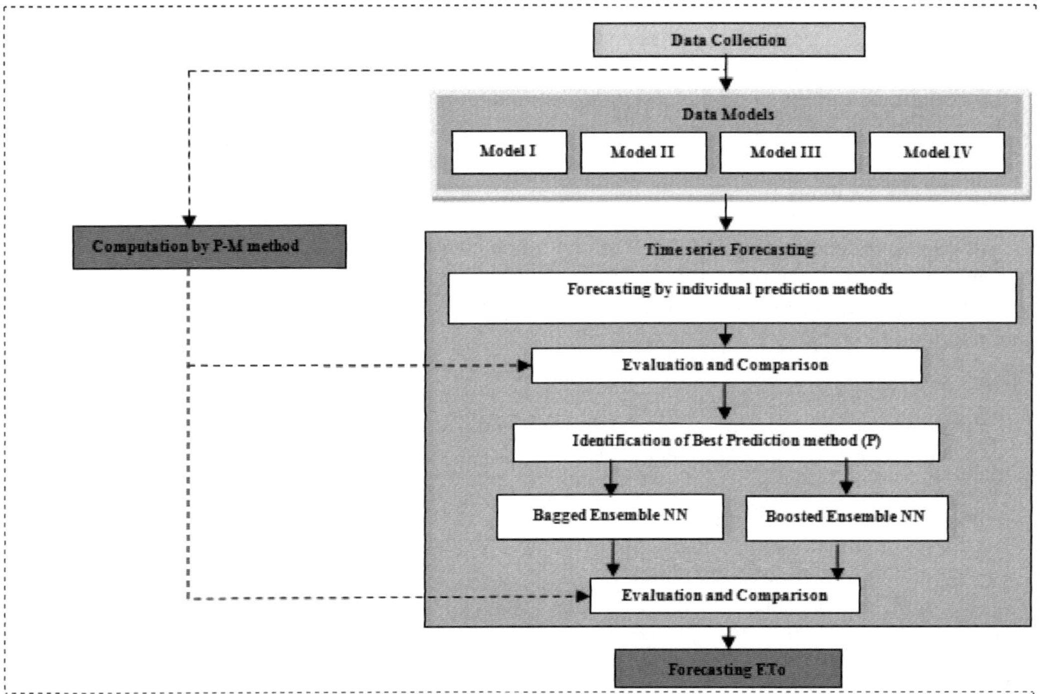

Figure 1. Problem Design.

Table1. Details of Dataset

No. of years of data used	10 years
Period	2004 – 2014
Total No. of Samples	4018 (Daily ET_o)
Total No. of Attributes	9 + 1 (predicted label)
Attribute List	DATE, RHMX ,RHMN, WS, TMAX, TMIN SSH, LAT, ELE, ET_o

Table 2. Statistical Description of Dataset

Attribute	Description	Type	Mean	Min ; Max Values
ETo	Reference Evapotranspiration	Label	03.725	[1.170 ; 5.485]
DATE	Day	Label	-	-
RHMX	Maximum Relative Humidity	Regular	86.985	[8.000; 100.000]
RHMN	Minimum Relative Humidity	Regular	58.094	[5.000; 100.00]
WS	Wind Speed	Regular	05.555	[0.000 ; 80.000]
TMAX	Maximum Temperature	Regular	32.847	[1.800 ; 50.700]
TMIN	Minimum Temperature	Regular	23.787	[1.500 ; 35.000]
SSH	Bright Sunshine hours	Regular	07.128	[0.000 ; 12.000]
LAT	Latitude	Regular	11.400	[0.0 , 11.4]
ELE	Elevation	Regular	05.790	[0.0, 5.79]

Table 3. Description of Data Models

	Data Model I	Data Model II	Data Model III	Data Model IV
No. of samples	4018	4018	4018	4018
Attributes used	3	5	6	8
Attributes list	TMAX, TMIN	TMAX, TMIN RHMN, RHMX	TMAX, TMIN SSH, RHMN RHMX, WS	TMAX, TMIN, SSH, RHMN, RHMX,WS, LAT, ELE

5. Forecasting Models

Models were developed using single classifier methods like SVM, ANN (MLP), M5 model trees, k-NN, Decision tree and Ensemble methods. An ensemble is a set of individual models taking a decision by averaging the results of individual models. Depending on the way the ensemble is designed, it can be classified into boosting, bagging or other algorithms. In Bagging a model is fit for all potential data points, on the original training set (Fig. 2). Original training set of size upto the size of the training set was replaced by generating Bootstrap samples. Some of the data points can appear more than once while others don't appear at all. The instability of the decision rule is effectively removed in bagging by averaging across re-samples. The purpose of boosting is to increase the strength of a weak learning algorithm.

Build the model for m=1 to M
a. Bootstrap sample D_m of size N with replacement from the original training set D.
b. Train a prediction model $G_m(x)$ to the bootstrap sample D_m.

Figure 2. Bagging Algorithm.

A weak learner is trained a number of times, using a reweighted version of the original training set in Boosting (Fig. 3). Boosting trains the first weak learner with equal weight on all the data points in the training set, then trains all other weak learners based on the updated weight. Heavier weight is given to the data point wrongly classified by the preview weak learner, and lighter weight is given to the correctly classified data point. By such a process, the next classifier attempts to fix the errors made by the previous learner.

1. Set the initial distribution on the training set.
2. Iterate until the stopping criterion is reached
a. Develop a model given from the data selected on the training set according to the current distribution.
b. Update the distribution on the training set.
c. Assess the stopping criterion.
3. Combine the models.

Figure 3. Boosting Algorithm.

Further validation criteria of forecasting models were evaluated and compared by standard metrics like Root Mean Squared Error (RMSE) selected as the performance criterion of ETo prediction. RMSE is defined as:

$$RMSE = \sqrt{\sum_{i=1}^{N}(ETi_{calculated} - ETi_{predicted})^2 \ /\mathrm{N}} \quad (1)$$

where, ET_icalculated represents the ET_o computed by FAO-56 PM method and ETi predicted represents the model predicted ETO. Mean Absolute Percentage Error (MAPE) is selected as the accuracy criterion of ETo prediction. MAPE is defined as:

$$MAPE = \frac{1}{N}\sum_{i=1}^{N}\frac{ETi_{calculated} - ETi_{predicted}}{ETi_{predicted}} * 100 \quad (2)$$

Mean Absolute Deviation (MAD) of a set of data is the average distance between each data value and the mean. The coefficient of correlation (R) is selected as the degree of collinearity criterion of ET_o prediction. R is defined as:

$$R^2 = 1 - \frac{\sum_{i=1}^{N}(ETi_{calculated} - ETi_{predicted})^2}{\sum_{i=1}^{N}(ETi_{calculated} - meanET_{predicted})^2} \quad (3)$$

6. Performance of Forecasting Models

6.1. Individual Forecasting Models

All the forecasting methods were implemented using WEKA machine learning tool with default parameters available. Table 4 gives the RMSE and MAPE error values of all forecasting models on various data models. The results show that ANN model has least error measure in terms of RMSE and MAPE for the entire four data models. For Data model I, a minimum RMSE value of 0.184 was obtained for ANN model compared to other forecasting models. For data model II, a minimum RMSE value of 0.176 was obtained for ANN predictor compared to other forecasting models.

Table 4. Performance measure of individual Forecasting models

	Data Model - I		Data Model - II		Data Model - III		Data Model - IV	
	MAPE	**RMSE**	**MAPE**	**RMSE**	**MAPE**	**RMSE**	**MAPE**	**RMSE**
SVM	0.092	0.183	0.082	0.177	0.066	0.175	0.074	0.179
ANN	**0.091**	**0.184**	**0.078**	**0.176**	**0.053**	**0.167**	**0.069**	**0.169**
DT	0.181	0.322	0.142	0.267	0.136	0.244	0.14	0.250
M5	0.091	0.185	**0.076**	**0.175**	0.058	0.171	0.062	0.173
k-NN	0.186	0.325	0.135	0.233	0.121	0.212	0.19	0.227

The MAPE value of 0.05 obtained for ANN predictor of Data model III is the least among other forecasting models. The highest RMSE value was observed for decision tree method for all data models. Similar observation was noted for MAPE values also. Thus ANN performs better than other forecasting models in all data models. Thus, ANN model was chosen as the base learner for developing the ensemble methods proposed in the next phase of this work.

6.2. Ensemble Forecasting Model

Bagged Neural Network model (Bagged-NN) was constructed using 10 MLP learners. Each iteration was constructed by employing multiple learning methods with k value as 10, the learning models were evaluated using 10 fold cross validation. For Boosted-NN, % of weight mass to base training was set to be 100. The number of iterations was set as 10. MLP was used as the base learner for boosting. Table 5 shows the results obtained for Bagged and Boosted NN models in terms of various error metrics such as MAD, RMSE and MAPE. Among ensemble models, Boosted-NN performed better than Bagged-NN for all data models. A minimum RMSE value of **0.049**, MAD of **0.107** and MAPE of **7.228%** were observed for Boosted-NN of Data model III.

Among all data models, Data model III was identified as the best with minimum error values for both ensemble data models. The performance of Data model III was much better than Data models I and II, but it was observed that Data model IV did not perform better than Data model III. Thus, including all nine attributes was not an optimum attribute set for forecasting ET_0. At the same time, the six attributes grouped under Data model III play a significant role in obtaining ET_0. This indicates that including the latitude and elevation does not influence ET_0 prediction.

The simulation results of the various data models developed with Boosted-NN and Bagged-NN in comparison with those of the FAO-56 PM ETo are shown in Figures 4 to11 for all four data models. The predictions made by the Bagged-NN and Boosted-NN methods for Data model I are shown in Figures 4 & 5 respectively. It was clear that the estimates of the Boosted-NN models were less scattered than those of Bagged-NN models. Similar observation was noted for all the other data models such as Data model II, Data model III and Data model IV. Figures 6 to 11 show that data points in the plot of Boosted-NN are much closer towards the diagonal line. Thus Boosted-NN predicts more data points effectively than Bagged-NN for all data models. This is also proved by the fact that R value is higher (**0.991**) for Boosted-NN than Bagged-NN (**0.989**) as seen in Table 6.

Table 5. Performance of Ensemble Forecasting Models

Data	MAD		RMSE		MAPE	
	Boosted NN	**Bagged NN**	**Boosted NN**	**Bagged NN**	**Boosted NN**	**Bagged NN**
I	0.079	0.081	0.132	0.135	11.72	11.99
II	0.065	0.072	0.118	0.126	9.59	10.60
III	**0.049**	0.057	**0.107**	0.116	**7.228**	8.387
IV	0.054	0.063	0.112	0.191	7.921	8.856

Table 6. Regression Co-efficient of Ensemble forecasting model

Data Model	Boosted R	Bagged R
I	0.987	0.986
II	0.989	0.987
III	**0.991**	0.989
IV	0.989	0.987

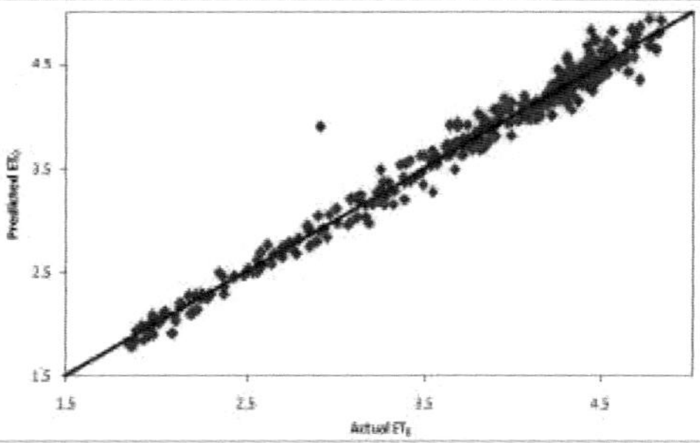

Figure 4. Regression plot for Bagged-ANN of Data Model I

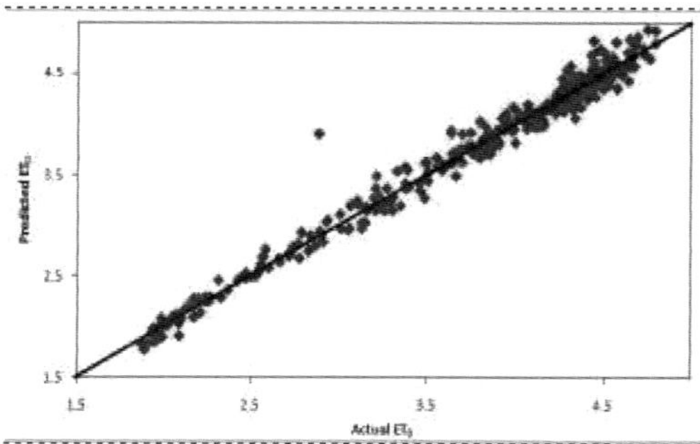

Figure 5. Regression plot for Boosted-ANN of Data Model I

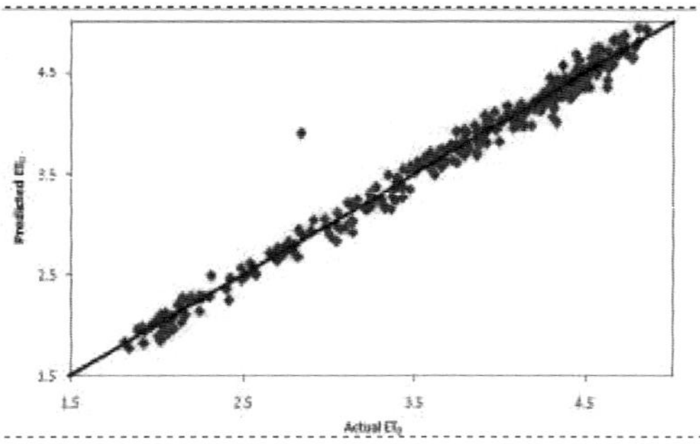

Figure 6. Regression plot for Bagged-ANN of Data Model II

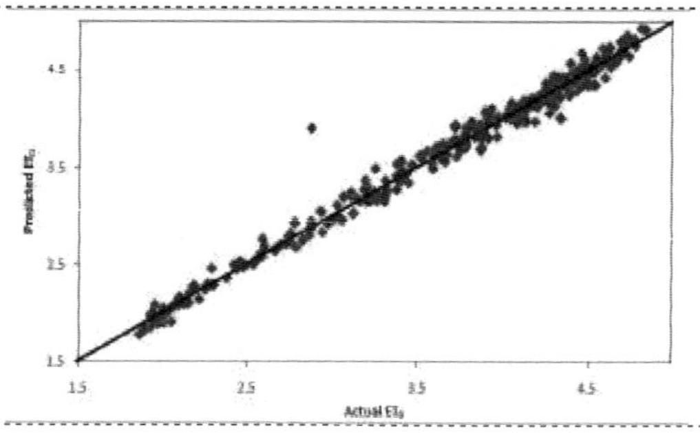

Figure 7. Regression plot for Boosted-ANN of Data Model II

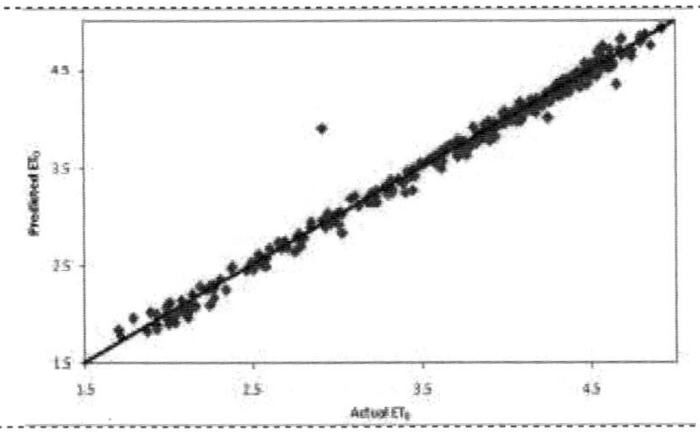

Figure 8. Regression plot for Bagged-ANN of Data Model III

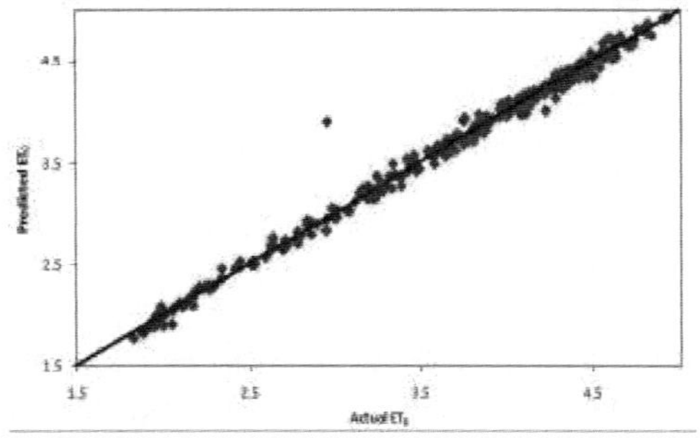

Figure 9. Regression plot for Boosted-ANN of Data Model III

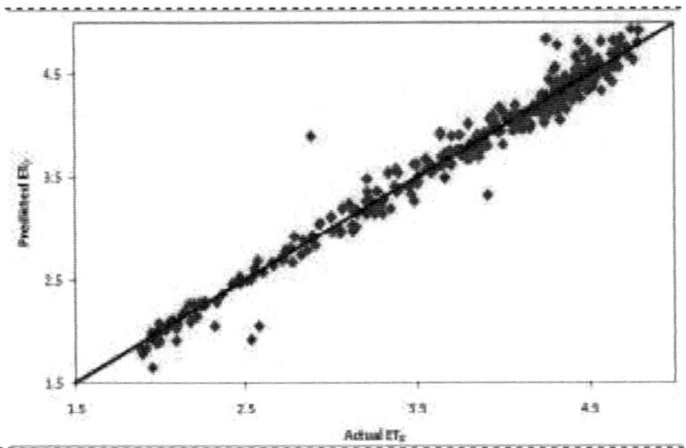

Figure 10. Regression plot for Bagged-ANN of Data Model IV

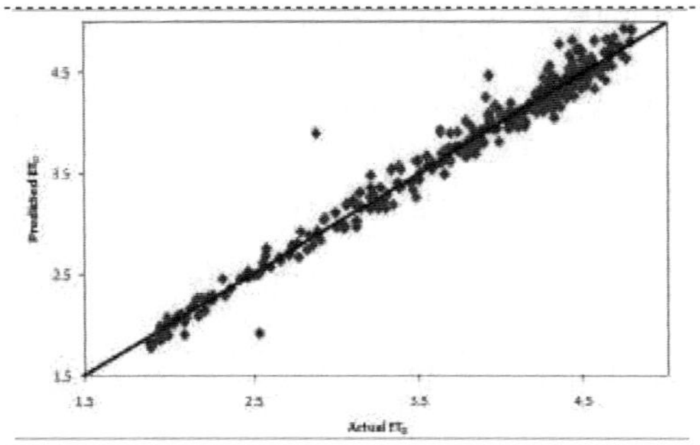

Figure 11. Regression plot for Boosted-ANN of Data Model IV

Fig. 12 illustrates the forecasted and estimated ET_o values of Data model III by Boosted-NN, which were identified as best by model evaluation.

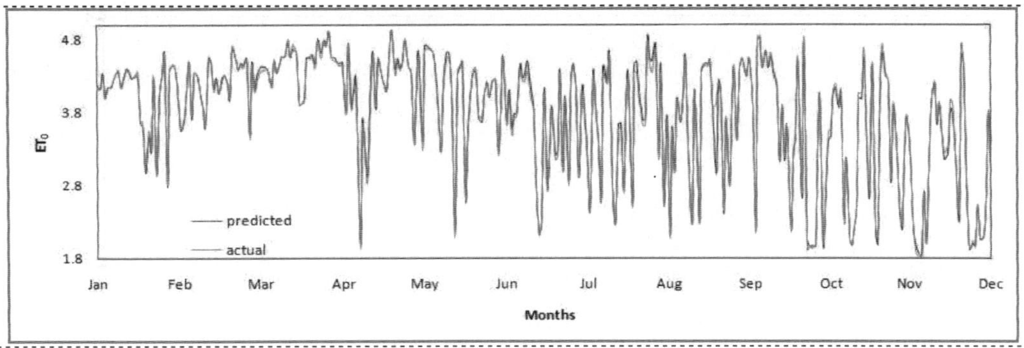

Figure 12. Forecasting plot of ET_o for Boosted-NN Data model III

The forecasting plot shows the Daily ET_o obtained for the year 2015 from the past time series data i.e. from 2004 to 2014. It is clear that the forecasted Daily ET_o estimates of the Boosted-NN Data model III were almost similar.

7. Conclusion

In practice, accurately computing Daily ET_o is difficult because it is characterized by many climatic parameters. Therefore, models based on machine learning provide a new alternative to ET_o forecasting problems. The potential of ensemble based modeling for forecasting Daily ET_oupto365 days in advance was investigated in this paper. This work promotes a new forecasting model for time series prediction which combines ensemble of NN models. Bagging and Boosting were adopted to the problem of learning time series data for predicting future values of ET_o.

Back Propagation Neural Network (BPNN) was used as base learner in ensemble forecasting models. Results obtained show that the Boosting-NN algorithm actually improves the performance of model in forecasting ET_o as compared to Bagged-NN and individual learners. Future work may include the alteration of the boosting algorithm for dealing with BPNN with parameter optimization.

References

[1] Ahmed K F, Wang G and Silander J 2013 Statistical Downscaling and Bias Correction of Climate Model Outputs for Climate Change Impact Assessment in the U.S. Northeast, Global Planet Change. **100** 320-32.

[2] Birsan M V, Molnar P, Burlamdo P and Pfaundler M 2005 Stream flow Trends in Switzerland. *J. Hydrol.* **314** 312-29.

[3] Burn D H. and Hesch N M 2007 Trends in Evaporation for the Canadian Prairies. *J. Hydrol.* **336** 61-73.

[4] Chen H L and Rao A R 2002 Testing Hydrologic Time Series for Stationarity. *J. Hydrologic Eng.* **7**(2) 129-36.

[5] Etuk E H and Natamba B 2015 Modelling Monthly Ugandan Shilling/Us Dollar Exchange Rates by Seasonal Boxjenkins Techniques. *Int. J. Life Sci. Eng.* **1**(4) 165–70.

[6] Hobbins M T, Ramírez J A and Brown T C 2004 Trends in an Evaporation and Actual Evapotranspiration across the Conterminous Us: Paradoxical or Complementary ? *Geophys. Res. Lett.* **31** L13503.

[7] Hu Q, Feng S, Guo H, Chen G and Jiang T 2007 Interactions of the Yangtze River Flow and Hydrologic Processes of the Poyang lake, China. *J. Hydrol.* **347**(1) 90-100.

[8] Karthikeyan L and Kumar D N 2013 Predictability of Nonstationary Time Series Using Wavelet and Emd Based Arma Models. *J. Hydrol.* **502** 103-19.

[9] Makkeasorn A, Chang N B and Zhou X 2008 Short-Term Streamflow Forecasting with Global Climate Change Implications–A Comparative Study between Genetic Programming and Neural Network Models. *J. Hydrol.* **352**(3) 336-54.

[10] Manikumari N and Vinodhini G 2016 Regression Models for Predicting Reference Evapotranspiration. *Int. J. Eng. Trends Technol.* **38**(3) 134-9.

[11] Manikumari N 2016 A Comparative Study on Decision Tree Methods in Predicting Daily Reference Crop Evapotranspiration. *Int. J. Sci. Adv. Res. Technol.* **2**(10) 274-8.

[12] Noori R, Karbassi A R, Moghaddamnia A, Han D, Zokaei-Ashtiani M H, Farokhnia A and Gousheh M G 2011 Assessment of Input variables Determination on the SVM Model Performance using PCA, Gamma Test and Forward Selection Techniques for Monthly StreamFlow Prediction. *J. Hydrol.* **401**(3) 177-89.

[13] Norrant C and Douguedroit A 2006 Monthly and Daily Precipitation Trends in the Mediterranean (1950–2000). *Theor. Appl. Climatol.* **83** 89-106.

[14] Parera A, Western A, Nawarathna B and George B 2013 Forecasting Daily Reference Evapotranspiration for Shepparton, Victoria, Australia using Numerical weather prediction outputs. *20th Int. Cong. Model. Simul.* Adelaide, Australia.

[15] Popale P G and Gorantiwar S D 2014 Stochastic Generation and Forecasting of Weekly rainfall for Rahuri Region. *Int. J. Innov. Res. Sci. Eng. Technol.* **3**(4) 185-96.

[16] Psilovikos A and Elhag M 2013 Forecasting of Remotely sensed Daily Evapotranspiration data over Nile Delta region, Egypt. *Water Res. Manag.* **27** 4115-30.

[17] Renard B, Lang M, Bois P, Dupeyrat A, Mestre O, Niel H, Gailhard J, Laurent C, Neppel L and Sauquet E 2006 Evolution of hydrometric extremes in France from Observed Data (Observed Changes in Hydrological Extremes in France) **6** 48-54.

[18] Shiri J and Kisi O 2010 Short-Term and Long-Term Stream flow Forecasting using a Wavelet and Neuro-Fuzzy Conjunction model. *J. Hydrol.* **394**(3) 486-93.

[19] Srikanthan R and McMahon T A 2001 Stochastic Generation of Annual, Monthly and Daily climate data: A Review. *Hydrol. Earth Syst. Sci. Discussions* **5**(4) 653-70.

[20] Valipour M 2012 Ability of Box-Jenkins Models to Estimate Reference Potential Evapotranspiration (A Case Study: Mehrabad Synoptic Station, Tehran, Iran). *IOSR J. Agr. Vet. Sci.* **1**(5) 1-11.

[21] Wang W C, Chau K W, Cheng C T and Qiu L 2009 A Comparison of Performance of several Artificial Intelligence methods for Forecasting Monthly Discharge Time Series. *J. Hydrol.* **374**(3) 294-306.

[22] Wei X, Liu S, Zhou G and Wang C 2005 Hydrological Processes in Major types of Chinese forests. *Hydrol. Process.* **19**(1) 63-75.

[23] Wu C L and Chau K W 2011 Rainfall–Runoff modeling using Artificial Neural Network Coupled with Singular Spectrum Analysis. *J. Hydrol.* **399**(3) 394-409.

[24] Wu P, Wood R, Ridley J and Lowe J 2010 Temporary Acceleration of the Hydrological cycle in response to a CO_2 Ramp down. *Geophys. Res. Lett.* **37**(12).

[25] Zhang H, Huang G H, Wang D and Zhang X 2011 Uncertainty Assessment of climate change impacts on the Hydrology of Small Prairie Wetlands. *J. Hydrol.* **396**(1) 94-103.

IOP Conf. Series: Earth and Environmental Science **80** (2017) 012070 doi:10.1088/1755-1315/80/1/012070

Technical and Sociological Investigation of Impacts in Using Lignite Mine Drainage for Irrigation – A Case Study

A Murugappan[1]*, A Manoharan[1], G Senthilkumar[1] and J Krishnamurthy[2]

[1]Department of Civil Engineering, Annamalai University, Annamalainagar, India
[2]Department of Sociology, Annamalai University, Annamalainagar, India

*Email: profam@gmail.com

Abstract. Irrigated farming depends on an ample supply of water compatible quality. Presently, a lot of irrigation projects have to depend on inferior quality and not so enviable sources of water supply. In order to prevent troubles during usage of such water supplies of poor quality, there must be meticulous preparation to ensure that the water available with such quality characteristics is put to best use. The effect of water quality upon soil and crops must be better understood in choosing fitting options to manage with impending water quality associated troubles that might decrease soil and crop productivity under existing circumstances of water use. Two tanks (small sized reservoirs) namely, Walajah Tank and Perumal Tank in Cuddalore District, used for irrigation, receive mine drainage water pumped out continuously from the open cast lignite mines of the NLC India Limited, Neyveli, Tamilnadu State. This water has been used by the farmers in the irrigated commands of both Walajah Tank and Perumal Tank for more than three decades. Recently, the beneficiaries had raised fears on the quality of mine drainage waters they had been using for raising crops in the commands of both the tanks. They opined that the coal dust laden mine water used for irrigation had affected the crop yields. This incited us to take up a study to (i) assess the status of quality of surface waters released from the two tanks for irrigation in the respective command areas and (ii) assess the likely impacts of quality of water on soil and on growth and productivity of crops cultivated in the command areas. Further to the technical evaluation of the impacts, a structured questionnaire survey was also conducted among the farmers and the common public in the study area. The findings of the survey confirmed with the outcome of the technical assessment in that the mine drainage had a poor impact in the cultivable command area of Walajah tank system while such impacts were less significant in most parts of the command area of Perumal tank system.
Key words: *Lignite mine drainage, irrigation, command area, impact assessment*

1. Introduction

Fast urbanization and industrial growth during the previous few decades have aggravated several severe apprehensions for the environment [1]. In the previous few decades, due to fast growth in industrial sector and increase in population, a tremendous stress has been imposed on the fresh water requirements [2, 3]. Water quality plays a significant part in advancing farming production and health standards of human beings. Quality of water is much dependent on the preferred utilization of water as diverse water uses warrant special standards of water quality requirements [4, 5].

The quality and quantity of surface and ground water resources in several parts of India get intimidated by mining activities [6, 7, 8, 9]. The mine water quality depends on a string of geological, hydrological and mining conditions, which differ considerably from one mine to another [10, 11]. There is a great variation in the concentration of contaminants present in the water discharged from a mine and in certain cases the quality characteristics of mine water may be such that it satisfies the drinking water standards [12]. However, on many occasions, the water discharged from a mine is not usable as such and it may contain beyond maximum permissible concentrations of heavy metals, toxic anions, and organic and biological contaminants [7].

The main proportion of our coal and lignite wealth goes into the production of power and has a massive impact on quantity and quality of surface and ground water resources. Water is used in coal mines for numerous purposes such as washing, spraying, and for preparation of coal. This may result in a disparity with water users in other sectors and requirements of environment. Mining activities result in dewatering of nearby groundwater aquifers, thereby reducing the groundwater table in the surrounding areas. Due to the reduction in water table, agricultural activities, in particular, and other activities that depend upon groundwater get badly affected. Water pumped out from mines, spent water discharged from coal handling plants, water discharged from dust extraction and dust suppression systems, water washed off from overburden dumps, workshops and effluents from domestic systems and washery are the major sources of water pollution due to mining activities. Further, run off resulting from discarded waste dumps and pits turns out acidic causing soil erosion and contamination of surface and sub-surface water systems. Several examples of such invasive impacts are seen in coalfields of Jharkhand, Orissa, West Bengal, Maharashtra, Uttar Pradesh, Madhya Pradesh, Maharashtra and Andhra Pradesh. Luckily many of these are not associated with pyrite and acid mine drainage is not a stern problem. However, a few mines have acid mine drainage problems and a significant example being the coal mines in Meghalaya.

Lignite is primarily utilized for power production in pit head power plants. The proportion of lignite actually used for power production has ranged from 80% to 86% since 2001-02 (during this period, more than 98% of lignite used in Tamil Nadu State was for power production). Mining of lignite in Neyveli is an essence of how mining impacts water resources. Here, mining for lignite situated above artesian aquifers has actually led to destruction of the aquifers above and beneath the lignite and has also incited mining of water in an exceptional scale. Numerous other lignite deposits in India, mostly in coastal Kutch also have artesian conditions and the constant expansion and deepening of lignite mines will ultimately lead to an immense devastation of the storage potentials, loss utilizable groundwater and destruction of the recharge systems.

At present, NLC (India) Limited (previously known as "Neyveli Lignite Corporation Limited) is India's biggest lignite mining organization, and an apex power production entity. Per year, about 24 million tons of high-grade lignite is produced from the three open cast mines operated by the company in Neyveli. Mine I (Northern in the image), which is the first mine of the NLC (India) Limited, extends over an area of about 17 square kilometers and has a reserve of nearly 300 million tons. Mine II which was first struck in early 1984 and got extended in the early 1990s, had a reserve of nearly 400 million tons. The total lignite reserves in the Neyveli mine field are valued at larger than two billion tons. As is evident, a vast area has been opened up totally transforming the surface water flow regimes. The area between the Gadillam and Vellar Rivers has fully altered. Thus, the first and the primary blow of the large scale opencast lignite mine is the devastation of the surface topography and thereby loosing the potential for surface storages and recharge. These changes are irreversible and the swift expansion envisions additional irreparable changes in the surface flow and recharge regimes. The impact of lignite mining in Neyveli on the water regime is particularly severe because of the unique hydrogeological conditions of the region. In the unconfined and semi-confined aquifers lying above the lignite deposit, groundwater is present. In the layers below the lignite deposit, groundwater is

IOP Conf. Series: Earth and Environmental Science **80** (2017) 012070 doi:10.1088/1755-1315/80/1/012070

present in artesian conditions. This means that in order to reach the lignite, the aquifers be totally destroyed as these layers are the `over burden' for the miner. Thus, second irreversible impact is that the shallow aquifers that would be replenished stop to survive and that renewable storage capacity of the ground water is lost. Presently, about 180000 to 205000 m^3/day of water is pumped out from NLC and another 160000-180000 m^3/day of water flows from agricultural wells, which account for a total withdrawal from aquifer of about 140 M m^3/yr. The paradox of lignite based power development, predominantly for the local people, is the community blessed not only to have surface sources but also groundwater sources which would be available at the source without use of external energy and simply have to dig pits to receive their water as the supply is at such low-pressures.

The mine waters of NLC India limited are continuously drained through Upper Paravanar to the surface water body, namely Walajah Tank. As the water drained out is in huge quantities, the surplus water gets continuously drained to the downstream surface water body, namely Perumal tank through middle Paravanar. The water in these two tanks is utilized mainly for irrigation in the respective command areas of the tanks. Even during failures of monsoon, the farmers in the irrigated commands of these two tank systems could benefit from using the mine drainage by raising two crops successfully each year. However, the farmers, particularly in the command area of the Walajah tank system, raised fears that the yields of crops have decreased in the recent years. This necessitated a pilot study with the following objectives:

i. to assess the status of quality of surface waters released from the Walajah Tank System as well as the Perumal Tank System for irrigation in the respective commands, and
ii. to assessg the status of soil properties in the command areas of the Walajah Tank System and Perumal Tank System and their impact on growth and yield of crops.

2. Study area

The Walajah Tank irrigation system and Perumal Tank irrigation system form part of the Sethiathope Project. These are situated on the left bank of the River Vellar, in the Cuddalore District, bordered on the east by coastal sand dunes. The climate of the area is monsoon type with a mean yearly rainfall of 1284 mm with nearly two-thirds of this yearly rainfall occurring mostly during the North-East monsoon period of October-December. As the study area is situated in the flood plain of the River Vellar, it suffers chronically from a shortage of water before September and from frequent floods during November-December either by the Vellar or by the Paravanar River. Paravanar River is a major drain crossing the area from West to North-East. Thanks to the continuous supply of water from the Neyveli mine drainage, the commands of both Walajah and Perumal Tank Systems could be irrigated throughout the year. Figure 1 shows the location map of study area.

Walajah Tank has a very limited storage capacity but a relatively huge command area. Due to human encroachment, the original storage capacity of 2.57 million cubic metres has reduced to 1.66 million cubic metres. Further, as a major portion of the tank has been infested by weed growth, the capacity of the tank has been further reduced. Under these conditions, Walajah Tank does not act as a tank and does not fulfill the function for which it is constructed. The tank receives water (i) through runoff generated from its own catchment during the rainy season, (ii) through inflow from Neyveli mines throughout the year and (iii) through surplus waters released from Veeranam tank through the Vellar Rajan Canal. Walajah tank has a total command area of 11,292 acres which is served by 12 channels benefiting a total of 22 villages.

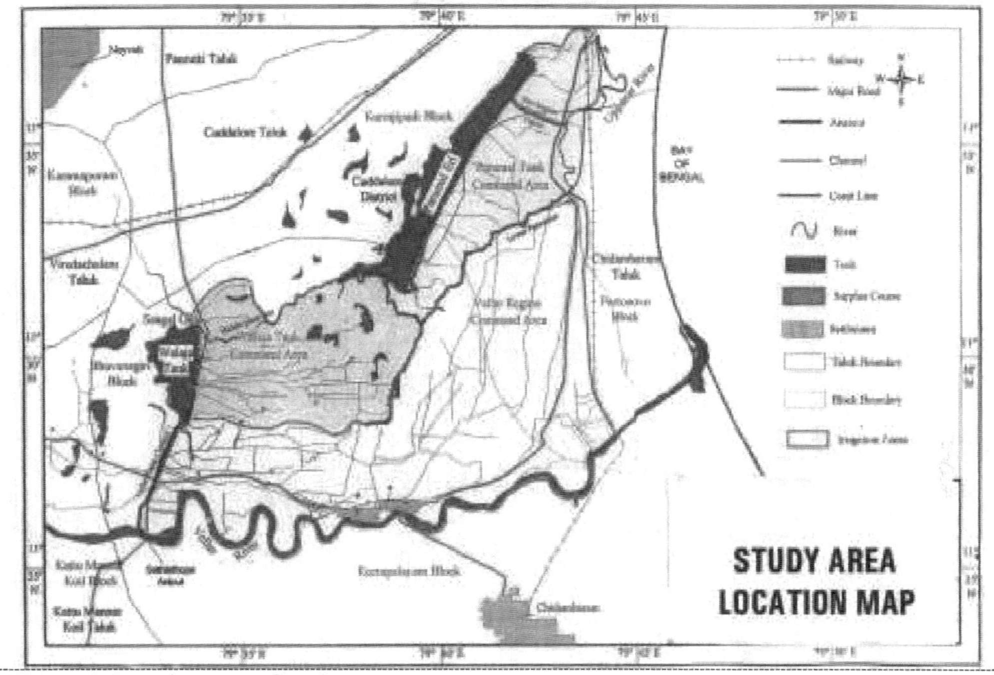

Figure 1. Location map of study area.

The Perumal tank has a storage capacity much larger than that of the Walajah tank. The combined catchment of Perumal tank is 552.9 sq. km and it has an original capacity of 16.25 million cubic metres. During the years, the tank has silted up in part and its storage capacity has decreased to 13.5 million cubic metres. The tank feeds water for irrigating a command area of 6,503 acres through 11 off-taking channels. A total of 20 villages are benefited by the tank. Two weirs located on the southern flank of the tank, near the point where the Paravanar River joins the tank, protect the tank against flooding. These two weirs discharge the surplus waters into the Lower Paravanar River. The tank receives runoff from its catchment and water surpluses from Walajah Tank through the Middle Paravanar. Water is released almost throughout the year through the surplus arrangement from Walajah tank to Perumal tank due to the continuous inflow to Walajah Tank from Neyveli mines.
Soils in the Sethiathope Command consists predominantly of fine grade alluvial soils, clays and clay loams traversed by zones of coarser soils of sandy loams.

Paddy is the principal crop grown in the commands of both tank systems. As the ryots of both tank systems are blessed with continuous availability of water in the form of mine drainage from Neyveli, they raise two crops in a year. The first crop of paddy is usually a short term variety grown in June/July to September/October. The second crop in the commands of Walajah Tank is a medium variety paddy crop grown during September/October to January/February, while the second crop in the commands of Perumal Tank is also a medium variety of paddy but grown during December-January to March/April. As the water availability is scarce during the South-West monsoon period, the first crop covers only about 33 percent of the command area. The second crop, which is grown during the North-East monsoon period, covers about 90% of the command area.

3. Water quality assessment of surface waters of Walajah tank and Perumal tank systems

A problem of salinity subsists, when a loss in yield is caused due to excess concentrations of salt built up in the crop root zone. The problem of salinity in irrigated areas occurs from a saline, high water table or from salts present in the water applied for irrigation. When the accumulation of salts in the crop root zone occurs to such a level that the crop is no longer able to extract water from the salty soil solution, the crops are in water stressed condition for a longer time period thereby effecting yield reductions in the crop. The growth rate of crop is greatly reduced due to marked reduction in uptake of water. The salts which cause the problem of salinity are soluble in water and are transported at ease by water. During irrigation, when more quantity of irrigation water gets infiltrated into the soil than that being used by the crop, it would be helpful in removing a portion of salts accumulated in the crop root zone during previous irrigations by leaching them below the root zone. Leaching is the key factor in mitigating a water quality-related salinity problem. The quantum of leaching required depends upon the quality of irrigation water and tolerance of the crop grown to salinity.

A problem of infiltration in relation to quality of water occurs when the rate of infiltration of the water applied for irrigation or rainfall is significantly reduced and water remained on the soil surface for a pretty longer duration or infiltrated very slowly thereby hampering the supply of adequate water to the crop for obtaining acceptable yields. Soil factors such as structure, degree of compaction, organic matter content and chemical make-up can also significantly affect the water intake rate.
The two main water quality parameters that affect the normal infiltration rate are the salinity of water (total quantity of salts in the water) and its sodium content relative to the calcium and magnesium content. Water with high salinity will augment infiltration. Water with low salinity or water with a high sodium to calcium ratio will reduce infiltration. Both factors may function simultaneously.
The assessment of quality of waters of both Walajah and Perumal tank systems used for irrigation is done referring to the "Guidelines for interpretation of water quality for irrigation" given by FAO Irrigation and Drainage Paper 29 Rev. 1 [13]. The classification of irrigation water based on Electrical Conductivity (EC) and Sodium Absorption Ratio (SAR) proposed by USSL [14] has been used in the present study.

Rice is categorized as a moderately sensitive crop with regard to salt tolerance. Table 1 shows the tolerance and yield potential of rice crop as influenced Electrical Conductivity (EC) of water used for irrigation and that of soil in which crop is grown.

Table 1. Crop tolerance and Yield Potential of rice as influenced by Irrigation Water Salinity (ECw) and Soil Salinity (ECe)

Field Crop	**Yield Potential (in percent)**									
	100		90		75		50		0	
	ECw	ECe	ECw	ECe	ECw	ECe	ECw	ECe	ECw	ECe
Rice	2.0	3.0	2.6	3.8	3.4	5.1	4.8	7.2	7.6	11.0

The EC of water samples drawn from various reaches of different channels of Walajah Tank System showed a variation from a low of 1.016 dS/m to a high of 2.200 dS/m. For majority of the samples of water, the Electrical Conductivity was found to be in the narrow range of 1.500 to 1.800 dS/m.

The EC of water samples drawn from different reaches of various channels taking off from Perumal Tank System ranged between a low of 0.526 dS/m to a high of 1.531 dS/m The water samples drawn from six Channels were found to be within 0.750 dS/m. The EC of water in Kundiamallur and Manvaikkal channels lie in a close range of 1.466 dS/m to 1.531 dS/m. in the other channels mamely, Kallaiyankuppam Channel, sirupalaiyur Channel and Theerthanagiri Channel, the EC values lie in the range 0.740 dS/m to 1.527 dS/m. Comparatively speaking, the EC values of waters of Perumal Tank System were less than the waters of Walajah Tank System. Both tanks almost continuously receive the Neyveli Mine Drainage with Walajah Tank located on the upstream of Perumal Tank. As the storage capacity and catchment area of Perumal tank are much higher (about 6 times and 3 times respectively) than that of the Walajah Tank, whenever rainfall occurs in the catchment of Perumal tank it augments the local inflow to the tank in addition to the Neyveli drainage water. This rain water dilutes the concentration of total dissolved salts present in Neyveli Mine Drainage resulting in lesser EC of irrigation waters of Perumal Tank. Because of lesser storage capacity and lesser catchment area, the inflow from the local catchment of Walajah Tank during periods of rainfall does not add to the storage much and has less diluting effect on the waters of Neyveli Mine Drainage. Therefore, the EC of irrigation waters of Walajah Tank System are much higher.

The SAR values of water samples drawn from irrigation channels of Walajah Tank system fell in the range between 1.650 and 3.515. But, for most of the water samples, the SAR lied in a close range of 2.028 and 2.942.

For irrigation waters of Perumal Tank system, the SAR values were in the range 1.696 to 3.249. The water samples drawn from different reaches of six out of eleven channels had SAR values in the range of 3.249 to 2.307 (Theerthanagiri Channel) while in the other 5 channels, the SAR values were found to be relatively lesser in the range 1.696 to 2.576. Comparatively speaking, in general, the SAR values of irrigation waters of Perumal Tank System were found to be less than those of Walajah Tank System.

On the average, the SAR values of most of the water samples tested indicate SAR values less than 3.0 for both Walajah and Perumal Tank irrigation systems. Therefore, as per the rules for interpretation of water quality for irrigation referred previously, for SAR between 0 and 3 and EC of water greater than 0.7 dS/m, there is no limitation on utilization of water for irrigation.

As per the USSL Salinity diagram for irrigation water, in general, the irrigation water of Walajah Tank may be categorized as Class C3-S1 (that is, low sodium hazard and medium-high salinity hazard). The quality of water of Walajah Tank for irrigation is under moderate category. In case of Perumal Tank System, the irrigation waters of six channels fell under the Class C2-S1 (that is, low sodium hazard and moderate salinity hazard). The quality of waters of these channels for irrigation is under good category. The quality of irrigation waters of the other five channels of Perumal Tank fell under the category C3-S1 (moderate quality).

Hence, going by the above standards and considering also the crops grown in the commands of both Walajah and Perumal Tank irrigation systems, there is no problem of salinity and no problem of infiltration caused by SAR.

But, the farmers of Walajah Tank System, in particular, say there has been reduction in the yields of rice crops grown in the command area in the recent years. *Why?*

It was observed that the turbidity of irrigation waters flowing in many of the channels of Walajah Tank System was high. The high turbidity of waters could be attributed mainly to the colloidal suspended particles, mainly the coal dust present in the Neyveli mine drainage. This revealed that the Walajah Tank was not functioning as an effective storage unit and simply acted as a receiving unit before passing on the mine drainage from it directly to the off-taking irrigation channels. No effectual settling of drainage water took place in the tank. This was essentially due to the insufficient storage capacity of the Walajah Tank to permit successful settling of the suspended particles in waters received by it, particularly the mine drainage which is received continuously by the tank all throughout the year. Further, even the original storage capacity of the tank was not available due to sedimentation, weed growth and encroachments. Simply, the mine waters drained to Walajah Tank flowed in the form of a channel course nearer to the eastern bund of the tank before getting discharged through the various sluices. Altogether, there is no storage at all in the Walajah Tank. Hence, Walajah Tank does not serve the primary function expected of a tank, that is, storage.

It was found that the turbidity of waters released through the channels of Perumal Tank system for irrigation was very less. The analysis of water samples drawn from different reaches of different irrigation channels revealed that the suspended solids were very less and could be practically ignored. That is, the higher available storage capacity of Perumal Tank was successful in trapping the suspended particles present in water.

To identify the possible effects of these colloidal particles (including the micro particles due to coal wash present in mine drainage) on irrigated soils, it necessitated the evaluation of fundamental physical properties of field soil samples, namely, bulk density, particle size distribution and permeability.

4. Physical analysis of soil samples

Table 2 shows the optimum bulk density for maximum plant available water and field range bulk densities for a range of textures [15].

Comparing the data from Archer and Smith [15] in Table 2, it is obvious that for sandy soils with less plant available water capacity, an increase in bulk density can be helpful. However, on heavier soils, the optimum bulk density seems to be lesser than the bulk density of fields under cultivation and hence the problem is to keep the bulk density as low as feasible. Thus, a rise in bulk density in fine grained soils will lead to a fall in the amount of macro pores thereby decreasing plant available water at low suctions. At high suctions, a rise in bulk density boosts soil water retention (not plant available) and the net effect is a decrease in plant available water. Although a rise in bulk density might enhance water withholding properties of certain soils, high densities are objectionable in all soils as resistance to root penetration is increased leading to hindering growth and distribution and hence decreasing water use efficiency.

Table 2. Optimum Bulk Density[1] for maximum Plant Available Water and Field Range Bulk Densities for a Range of Textures

Soil	Optimum density (g/cm^3)	Field Density (under cultivation)	
		Range (g/cm^3)	Mean (g/cm^3)
Loamy sand	1.75	1.23 - 1.59	1.52
Sandy loam	1.50	1.05 - 1.72	1.34
Silt loam	1.40	No data	No data
Clay loam	1.20	0.94 - 1.57	1.30

[1]*Optimum Bulk Density is the bulk density where plant available water is maximum*

It was observed that the bulk densities of almost all soil samples collected from the Walajah Tank command were high and more than the recommended permissible range for effective uptake of nutrients by the crops grown. In Perumal Tank command area, the bulk densities of soil samples were relatively lesser.

Sieve analysis was also performed on undisturbed soil samples collected from the command in the head reaches of certain channels of Walajah Tank System, to determine the particle size distribution. it is found that the content of clay in field soil samples drawn from cultivated lands of Walajah Tank Command is high. Considering the particle size distribution and the recommended range of field density, it is most likely that the higher bulk densities of soils will have a negative impact on growth of crops grown.

The permeability of soil gets reduced with increase in percentage of clay-sized particles. Permeability of soil is one of the key factors in making the nutrients available to the root zone of the crops grown. Depending upon the soil type and groundwater levels, the inundated rice fields will have a continuous water percolation to the deeper layers. Agronomists opine that the percolation process is favorable for the plant growth, as the water movement will keep oxygen content within the soil at a reasonable soil. Normal percolation rates are 1-3 mm/day on soils with high clay content. It was observed that the permeability of all soil samples were very less. This might have hampered the growth of the crops grown in the study area.

5. Socio economic impact assessment

A structured interview schedule was prepared and translated into the local vernacular (Tamil) language to obtain information from the villagers of the command. The respondents were asked to provide information on the suitability of NLC mine drainage for irrigation, impact on soils, impact on crop yield, employment potential and health. All the 22 villages of the Walajah command and 21 villages of Perumal command were included in the study. A total of 430 households were chosen using a stratified random sampling technique in both the commands.

In Walajah tank command area, over 40 (42) percent of the respondents expressed that the NLC mine drainage was suitable for raising crops while about 60(58) percent of the respondents felt that it was not suitable for crops. For Perumal tank command, about two-thirds of the respondents viewed positively the suitability of NLC mine water for irrigation and only one-third (35%) of the respondents felt negatively.

It was also aimed to find out the respondents' views regarding the impact of NLC mine waters on the yields of crops. Under the Walajah tank command, a well over 80 (80 & 85) percent of both large and small farmers expressed their concern with regard to reduction in crop yields due to NLC mine waters. On enquiry they expressed that in general they lose 5 to10 bags of paddy per acre per season. On the other hand in Perumal tank command nearly three quarters (78% & 77%) of respondents each belonging to large and small farmers category highlighted that there was no significant change in the yield due to irrigation waters. Only about 20% of the respondents in each category revealed that there was decrease in yield.

In the Walajah Tank Command, majority of the respondents (62%) viewed that the quality of the NLC mine drainage water had affected the soil quality whereas in the Perumal Tank system only 40 percent of the respondents had opined that the water affected the soil fertility. However, a majority of the respondents (60%) belonging to Perumal command reveal that the water had not damaged the soil quality and they got yields normally.

The results of the impact assessment of NLC mine drainage on the commands of Walajah tank and Perumal tank, on the basis of the field study, reveal that in general the mine drainage water is suitable for irrigation. Though some respondents in the Walajah command revealed that there was some loss in yield due to the quality of irrigation waters, the respondents of Perumal tank system revealed that there was no such reduction in the yield due to irrigation waters. Also, the respondents of both the commands, Walajah tank command in particular, revealed that irrigation water had affected the soil quality and yield to some extent, the extent of damage caused was not alarming. The respondents suggest that in order to improve the impact of nlc mine waters, treating water before it was released for irrigation from the Walajah tank and strengthening the agricultural extension services of the state agricultural department in order to suggest appropriate farming strategies that would go a long way in sustaining their future. Since the NLC mine drainage happens to be their main source of irrigation, which provides them employment throughout the year, the respondents expressed their concern for proper and wise use of irrigation waters. They expressed that NLC mine water is the only lifeline for the people of their region.

6. Conclusions and Recommendations

- There is no problem of salinity and no problem of infiltration caused by SAR in the commands of both Walajah and Perumal Tank irrigation systems.
- The turbidity of irrigation waters flowing in many of the channels of Walajah Tank System was high mainly due to the colloidal suspended particles, mainly the coal dust from the Neyveli Mine Drainage. This is basically due to the inadequate storage capacity of the Walajah Tank to allow effective settling of the suspended particles in waters received by it. The turbidity of waters released through the channels of Perumal Tank system for irrigation was very less as the available storage capacity of Perumal Tank was adequate in effective trapping of the suspended particles present in water.
- The higher bulk density and reduced permeability of soil samples drawn from various irrigated fields of Walajah Tank indicated that there was a likely problem of infiltration. The physical nature of the soils was such that it was not permeable to the extent needed by the crops grown in the command area. As the permeability was less, it had reduced the uptake of water by the root tips and caused a water stress situation to the crops thereby affecting the yield of the crops.
- Removing the weeds, removing the encroachments and desilting of Walajah tank would help in resotaration of the original capacity of the Walajah tank. Further, the storage capacity of the tank can be enhanced by deepening the lowest bed level of the tank. These two measures combined together will make Walajah tank to perform efficiently in settling of colloidal coal dust particles in mine water received by it.
- For decreasing the bulk density and increasing the permeability of soil in agricultural lands in order to create a favourable physical environment for crop growth, more of organic like farmyard manure, crop residuals, press mud, organic waste, and other farm waste and green manures can be added adequately.
- The results of socio-economic impact study reveals that to some extent the water has damaged the quality of soil especially in Walajah Tank command and affected the crop yields. The respondents suggest appropriate treatment of water be made before it is released from Walajah Tank for irrigation.
- To summarize, restoring the original storage capacity and further deepening of Walajah Tank for effective settling of sediments in mine water, incorporation of various organic waste and farm waste, inclusion of green manure to improve both physical status and nutrient availability of soil to sustain the soil health and to realize better yield thereby improving the standard of living of the farming community.

References

[1] Reza R and Singh G 2010 Assessment of Heavy Metal Contamination and its Indexing Approach for River Water *Int. J. Environ. Sci. Tech.,* **7**(4) 785-792.

[2] Singh P, Tiwari A K and Singh P K 2014 Hydrochemical characteristic and quality assessment of groundwater of Ranchi township area, Jharkhand, India *Curr. World Environ.,* **9**(3) 804–13.

[3] Chandra S, Singh P K, Tiwari A K, Panigrahy B and Kumar A 2014 Evaluation of hydrogeological factor and their relationship with seasonal water table fluctuation in Dhanbad district, Jharkhand, India *ISH J. Hydraul. Eng.* **21**(2) 193–206.

[4] Tiwari A K, Singh A K, Singh A K and Singh M P 2015 Hydrogeochemical analysis and evaluation of surface water quality of Pratapgarh district, Uttar Pradesh, India *Appl. Water Sci,.* 1-15, doi:10.1007/s13201-015-0313-z

[5] Singh P K, Panigrahy B P and Tiwari A K 2015 Hydro chemical investigation and qualitative assessment of surface water resources in Jharia Coal field region, Dhanbad *J. Chem. Pharmaceu. Res*. **7**(2) 36-41.
[6] Choubey V D 1991 Hydrological and environmental impact of coal mining, Jharia coalfield, India *Environ. Geol.* **17** 185–94.
[7] Gupta D C 1999 Environmental aspects of selected trace elements associated with coal and natural waters of Pench valley coalfield of India and their impact on human health. *Int. J. Coal Geol.* **40** 133–49.
[8] Singh G 1994 Augmentation of underground pumped out water for potable purpose from coal mines of Jharia coalfield. *Proc. 5th Int. Mine Water Congress* Nottingham UK **2** 679–89.
[9] Tiwary R K 2001 Environmental impact of coal mining on water regime and its management *Water Air Soil Pollut.* **132** 185–1\99.
[10] Younger P L, Banwart S A and Hedin R S 2002 Mine water–hydrology, pollution, remediation *Kluwer Acad. Publ., Dordrecht, The Netherlands.*
[11] Younger P L and Wolkersdorfer C 2004 Mining impacts on the fresh water environments: technical and managerial guidelines for catchment scale management. *Mine Water Environ.* **23** S2–S80.
[12] Singh A K, Mahato M, Neogi B and Singh K K 2010 Quality assessment of mine water in the Raniganj coalfield area, India. *Mine Water Environ.* **29** 248–62.
[13] FAO 1985 Water quality for agriculture. *Irrigation and Drainage Paper* **29 Rev. 1**, Rome, Italy.
[14] USSL 1954 Diagnosis and Improvement of Saline and Alkaline Soils. *US Department of Agriculture Handbook*, No. **60** 160 p.
[15] Archer J R and Smith P D 1972 The Relation between Bulk Density, Available Water Capacity, and Air Capacity of Soils. *European J. Soil Sci.* **23**(4) 475–480.

Application of Geoelectrical Resistivity Technique for Groundwater Explorationin Lower Ponnaiyar Sub-Watershed, Tamilnadu, India

S Poongothai*and N Sridhar

Department of Civil Engineering, Annamalai University, Annamalainagar – 608002. India.

*Email: spoong86@yahoo.com

Abstract. GeoelectricalResistivity survey was carried out in Lower Ponnaiyar Sub-watershed, to intercept the sub-surface lithology and groundwater prospective zones. Twenty Vertical Electrical Sounding (VES) of the Schlumberger configuration were performed with an AB/2 separation 100 m. The acquired geoelectric data were interpreted with curve matching and computer iteration using Inverse Slope software. The results showed the presence of six geoelectric layers comprising Alluvium, Sand, Silt, Clay, Clayey sand and Hard Rock. The investigations also revealed the range of resistivityvalues from - 40 to 400 Ωm, while the depth of aquifer ranged between 6 and 100 m. The prospective for theoccurrence of groundwater in the Lower Ponnaiyar watershed was categorized as high, good, moderate and low by interpreting the sub-surface geophysical studies. The groundwater recharge structures like percolation pond, check dam have to be constructed in the moderate and low zones of groundwater potential so as to augment the groundwater resources.
Keywords: *Groundwater potential,Geoelectric resistivity survey, VES,Recharge structures*

1. Introduction

Geophysical methods can be used to measure the physical properties of the subsurface, specifically related to the position of water and its quality, and the position and properties of geological units. Essentially, the geophysical methods comprise of measurement and interpretation of signals from or induced physical phenomena generated as a result of the spatial changes in one or more physical properties of sub-terrain formation. The various physical properties, currently used in different geophysical techniques for groundwater exploration are electrical conductivity, magnetic susceptibility, density, gravity, elasticity, dielectric constant and radioactivity [9]. In this study,one of the geophysical methods, namely the electrical resistivity technique has been employed to delineate the subsurface lithology and assessment of groundwater potential of the watershed.

2. Study area

The study area lies in between 79° 15' 13" and 79° 48' 28" E longitudes and 11° 50' 18" and 11° 55' 18" N latitudes with a total area of extent of 598.162 km^2 (Figure.1). It includes five taluks namely Tirukoilur, Ulundurpettai, Villupuram, Panruti and Cuddalore with a general elevation of 81.25 m above MSL sloping from West to East. The river ponnaiyar is almost dry throughout the year. Water flows during the monsoon season only in Tamilnadu. However, this water flow raises the groundwater

IOP Conf. Series: Earth and Environmental Science **80** (2017) 012071 doi:10.1088/1755-1315/80/1/012071

table throughout the river basin and feeds numerous reservoirs /tanks. The maximum rainfall recorded is 1851 mm (1996-97), while the minimum is 548 mm (1989-90). The maximum temperature may rise up to 40 °C and decrease during winter season up to 27 °C. The winter water level ranges from 2.50 to 5.00 m and the summer water level ranges from 3 to 7 m. The Index map of the Lower Ponnaiyar Sub-Watershed is presented in Figure 1.

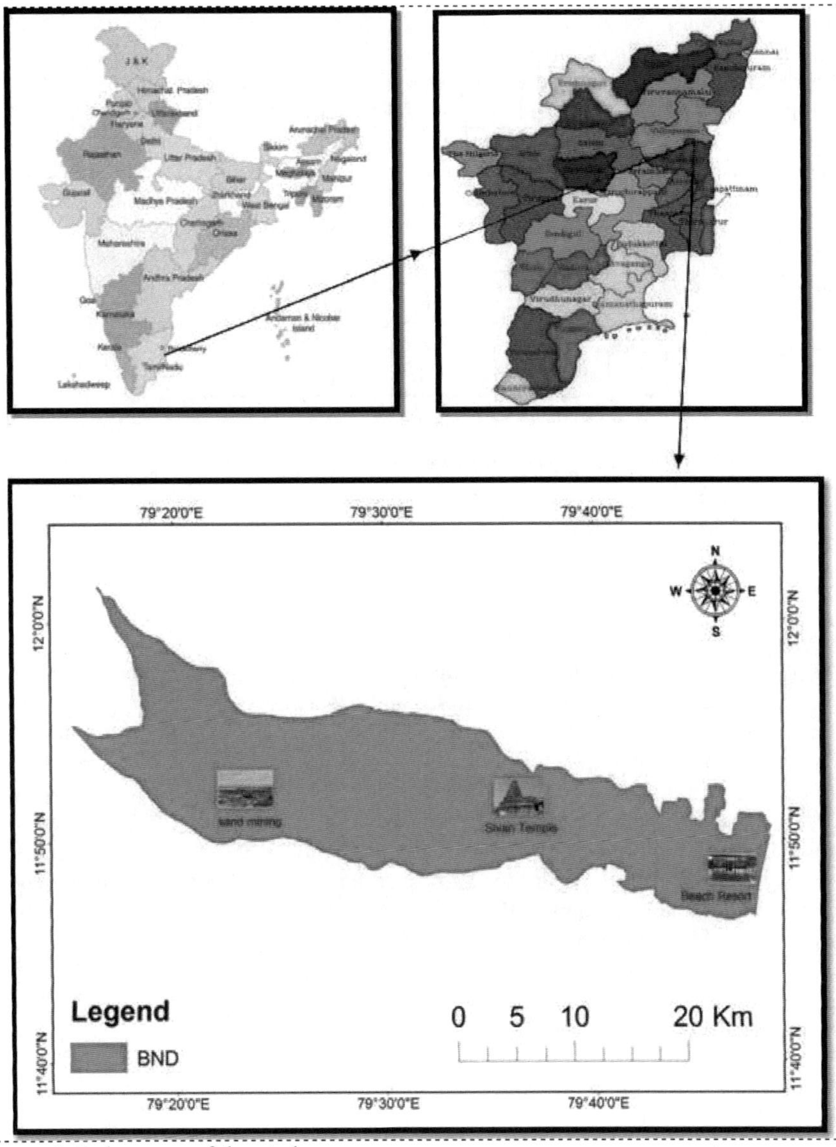

Figure 1. Index map of the study area.

3. Methodology

Totally, 20 VES data was done in the areas ofLower Ponnaiyarwatershed todelineate the subsurfacelithology and explore the groundwater potential. The VES data were processed using the software packages. In this study, VES were conducted using the Schlumberger array

3.1 Schlumberger array

The Schlumberger method has a greater penetration than the Wenner. The one-dimensional VES were carried out in this research with Schlumberger electrode Array. The 4 electrodes (A, M, N, and B) were located lengthwise a straight-line proportionally over centre point 'O' (Vide Figure 2). The separation between potential electrodes (MN) was kept small when compared to the current electrode separation AB. Using L = AB/2 and l = MN/2 the configuration factor for Schlumberger array is $(\pi (L^2 - l^2) / 2l)$*R and the apparent resistivity was calculated with the formula. In this method, a series of potential differences are acquired at successively larger electrode spacing while maintaining a fixed central reference point. The induced current passed through progressively deeper layers at greater electrode spacing. Apparent resistivity values calculated from measured potential differences which could be interpreted in terms of overburden thickness, water table depth and thicknesses of subsurface strata [7].A total of twenty VES were done in the study area. The traverse separation is 5 m. The twenty traverses are represented as VES 1-20. An average spread of 100 m (AB/2) is covered.

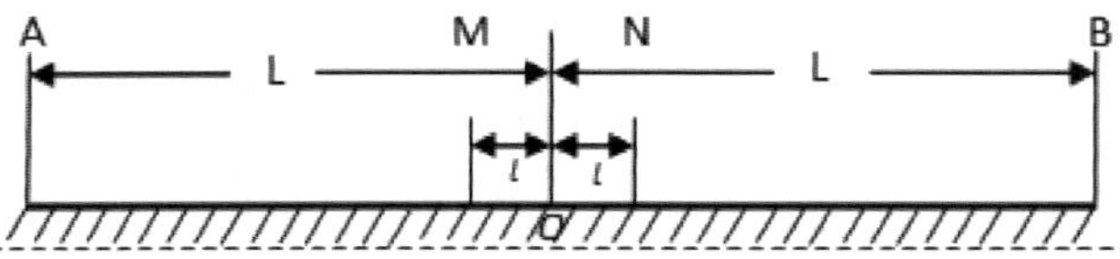

Schlumberger electrode array - $\rho_a = [\pi(L^2 - l^2)/2l]*R$

Figure 2. Layout of Schlumberger electrode Arrangement.

3.2 Electrical resistivity survey in the field

The resistivity survey was done in this research to study groundwater potential in Lower Ponnaiyar sub-basin, such as depth, thickness, resistivity and sediment at which water can be obtained. The geoelectrical methods used in the survey are VES and with the aim of determining the sub-surface lithology and groundwater potential. Twenty VES were conducted using the Schlumberger configuration and horizontal spread covering the entire area. All these twenty locations were within the boundary of the study area (Figure 3).The VES technique measures the vertical variations in ground apparent resistivities with respect to a fixed Centre of array. VES data obtained are qualitatively and quantitatively iterated using Inverse Slope software for its partial curve matching. The values of the apparent resistivity obtained are plotted against current electrode spacing using a log-log graph paper. The field data are first curve matched and the result subjected to computer iteration software to obtain the layer true resistivity and thickness which are derived from VES 1 to VES 20 diagrams. Selected curve diagram (VES1) is showed in Figure 4. Based on the geoelectrical characteristics, the lithology and groundwater potential are interpreted.

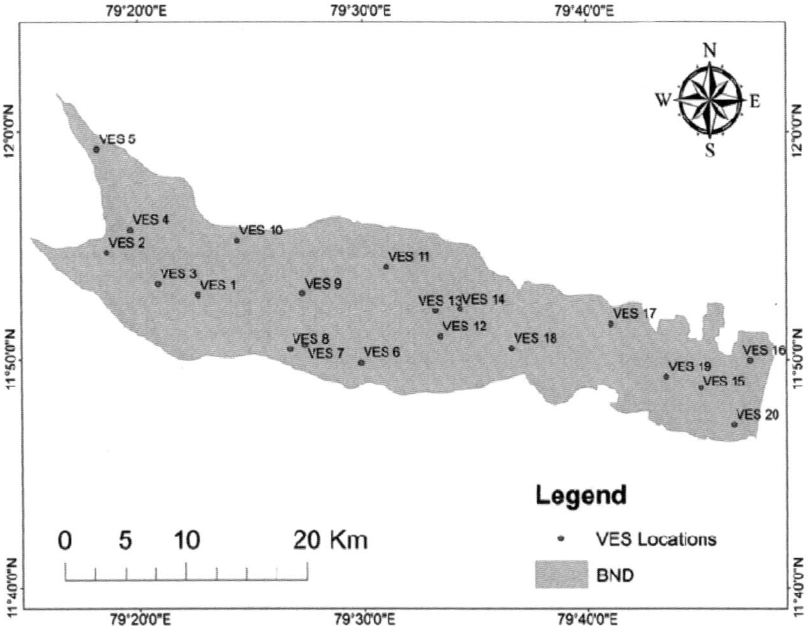

Figure 3. 1D-VES locations of the Study Area.

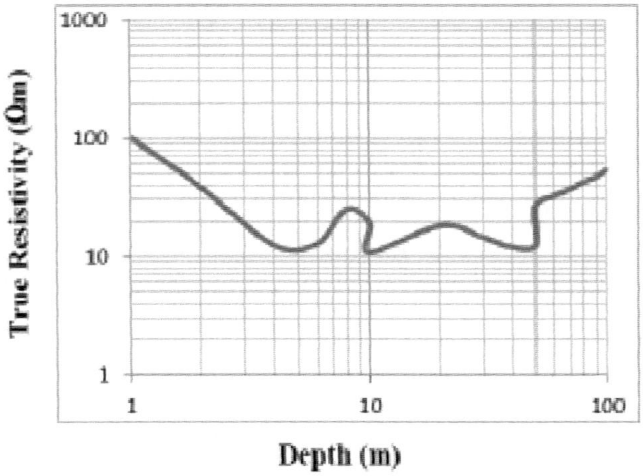

Figure 4. Curve diagram for VES1(Emappur).

4. Results and Discussion

The VES data were collected from 20 locations were interpreted to obtain layered resistivity parameters and to delineate potential zones in the deep aquifer of Lower Ponnaiyar sub-watershed.

IOP Conf. Series: Earth and Environmental Science **80** (2017) 012071 doi:10.1088/1755-1315/80/1/012071

4.1 Interpretation of VES curves

The Qualitative interpretation of VES data showed multi-layered resistivity curves for most of the locations based on the depth below the ground surface. Analysis of the layer curves (more than five curves) showed that potential fracture zones found in this resistivity profile. The apparent resistivity maps of each geoelectrical layers are prepared using Inverse slope software. The apparent resistivity and AB/2 values are plotted on thedouble-log sheet in Inverse slope software. Out of 20 curves, 8 are 5 layer type, 5 are 6 layer type and 7 are 7 layer type curves. The geo electrical sections and its interpreted lithology of all VES locations 1 is depicted in Figure 5.

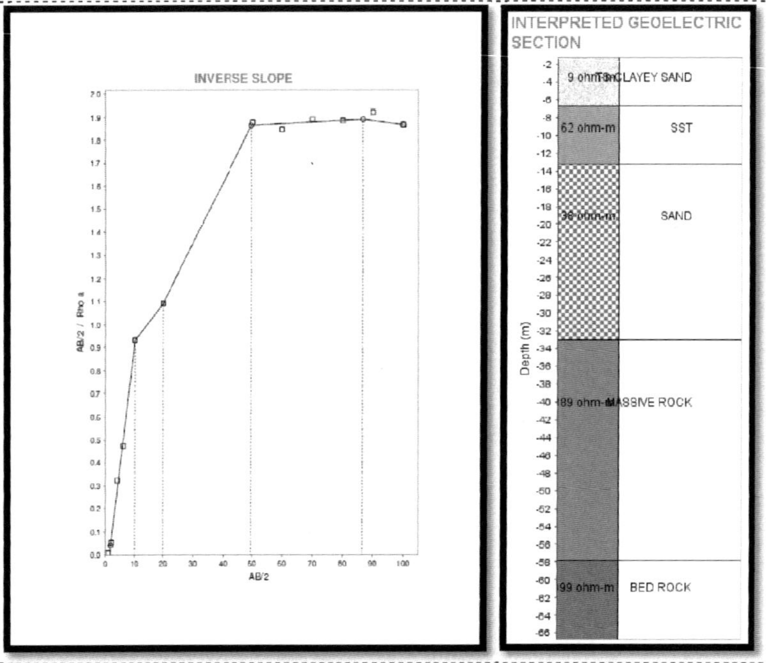

Figure 5. Interpreted Geoelectric Section of VES 1 -Emappur.

4.2 Delineation of sub surface lithology and groundwater potential

Totally, six geoelectrical layers are createdto recognize and establish the anomaly zones.

- The thickness of the1[st] layer is up to 6m from ground level and its apparent resistivity values ranged from -40 to 400 Ωm (vide Figure 6).This could be attributed, the presence of Clay, sand, silt and hard rock. Sands are exposed in VES-5 with high resistivity values in the study area, which yields good groundwater potential. The very low value less than 100 Ωm indicated the hard rock terrain of the study area. The sands are excellent aquifers which have moderate resistivity value of more than 200 Ωm and the clay has a low resistivity range, which has moderate and low groundwater potential. Some places, in the T.V.Nallurblock, have covered good and moderate groundwater potential areas such as Karumbur, Semangalam, Anatur and some villages in Annagramam block.
- The resistivity of the 2nd geoelectrical layer still get decreased from -10 to 180 Ωm, which is with a thickness of up to 10 m when compared to the first layer (Figure 7), which could be interpreted as clay, sand and hard rock.
- The resistivity range of the3[rd] layer (Figure 8) is further decreased from 0 to 170 Ωm, which is with a thickness of up to 20 m. It consists of soil types as clay and hard rock.

IOP Conf. Series: Earth and Environmental Science **80** (2017) 012071 doi:10.1088/1755-1315/80/1/012071

- The 4th layer has a low resistivity range between 0 and 300 Ωm consisting of clay, sand and hard rock with a thickness of up to 60 m (Figure 9). Most of the villages in Mugaiyur and T.V. Nallur blocks have moderate groundwater potential areas.
- The resistivity of the 5[th]layer reached up to 360 Ωm with the thickness 80 m, it could be attributed, the presence of sand, silt and rock.
- The resistivity increases in the Sixth geoelectrical layer between -40 and 400 Ωm with depth up to 100m compared to thefifth layer. Table 1 and 2 represented the consolidated interpreted subsurface lithology for various depth and the corresponding potential of the study area.
- GIS analysis also demarked for identifying areas of groundwater prospective to validate with this study which showed reliable results.

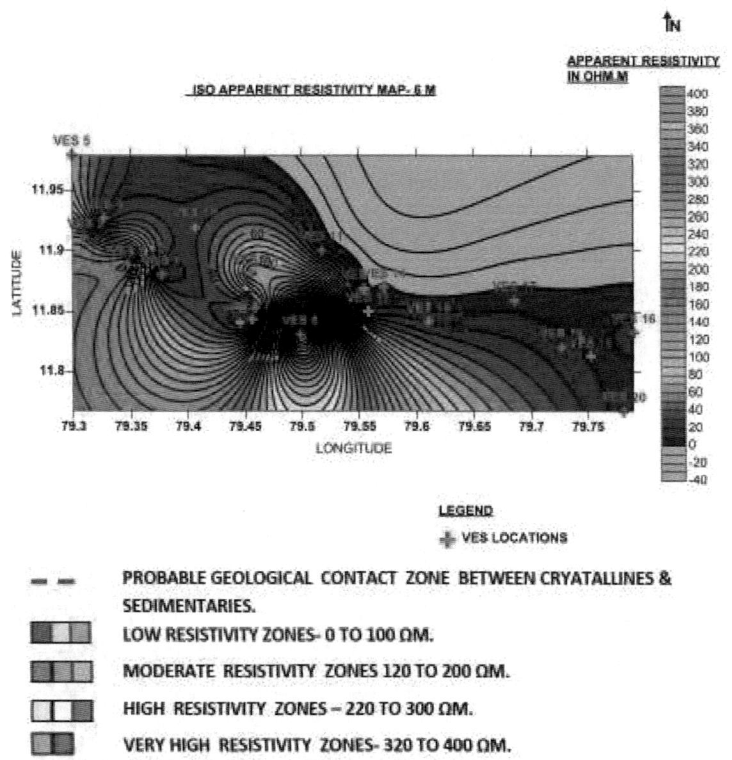

Figure 6. First layer map of the study area (6 m bgl).

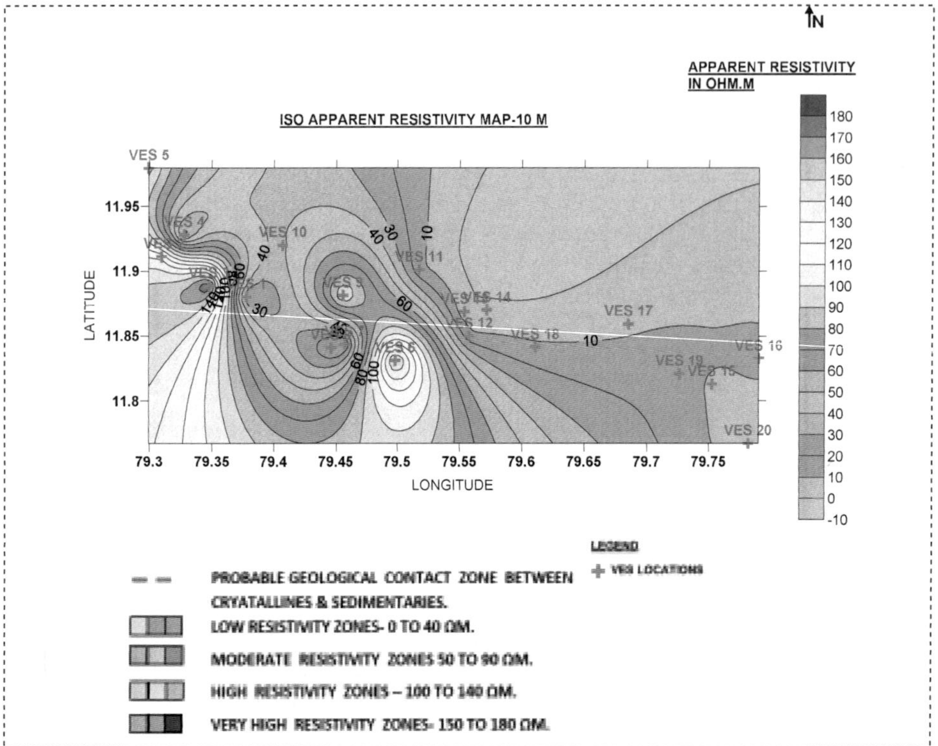

Figure 7. Second layer map of the study area (10 m bgl).

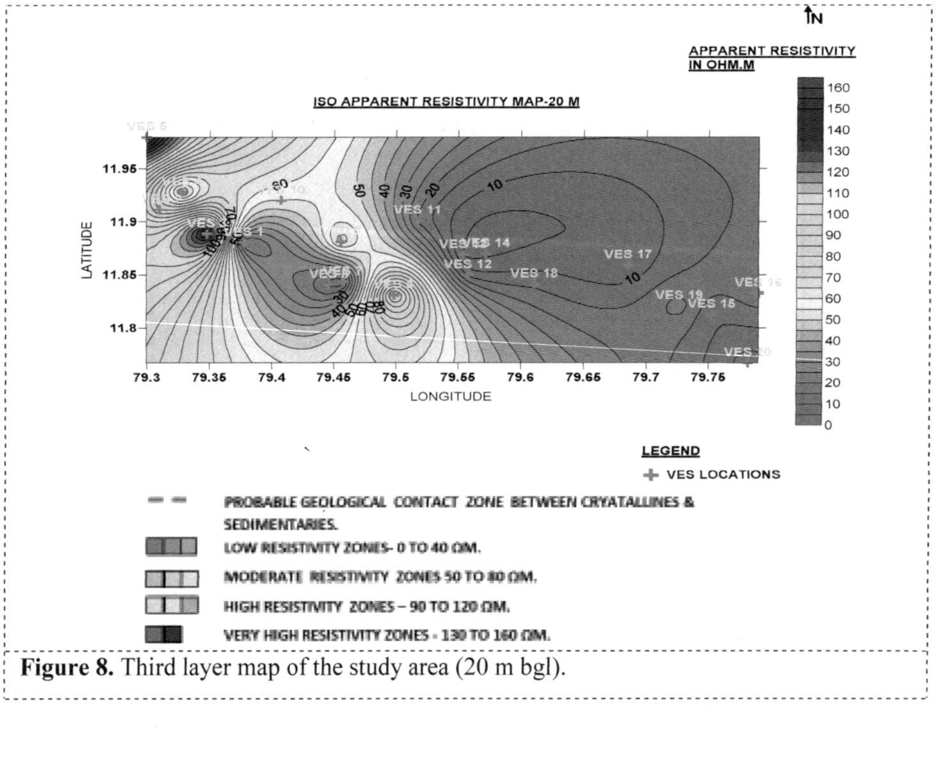

Figure 8. Third layer map of the study area (20 m bgl).

IOP Conf. Series: Earth and Environmental Science **80** (2017) 012071 doi:10.1088/1755-1315/80/1/012071

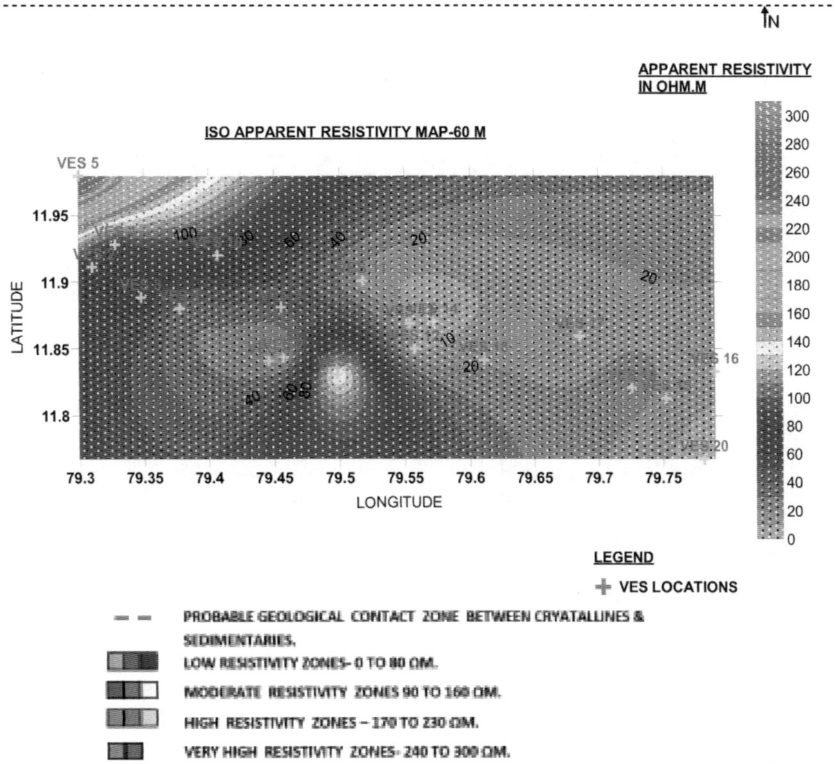

Figure 9. Fourth layer Map of the Study Area (60 m bgl).

Table 1. Interpreted Subsurface Lithology of the Lower Ponnaiyar Watershed for Various Depths and Resistivity Values

S. No.	Geoelectric Layers	True resistivity Range (Ωm)	Thickness of layers (m)	Interpreted subsurface lithology
1	I	-40 to 400	6	Clay, Sand, Silt & Hard rock
2	II	-10 to 180	10	Clay, Sand, Silt & Hard rock
3	III	0 to 160	20	Clay, Sand & Silt
4	IV	0 to 300	60	Clay, Sand and Gneiss
5	V	0 to 360	80	Clay, Sand and Gneiss
6	VI	-40 to 400	100	Clay, Sand, Silt & Hard rock

Table 2. Lithology and Groundwater prospective of the Lower Ponnaiyar Watershed

VES No.	No. of Layers	Lithology	Groundwater potential
1	5	Clay, Sand, Silt & Hard rock	High
2	5	Clay, Sand, Silt & Hard rock	High
3	7	Clay, Sand, Silt & Hard rock	High
4	5	Gneiss	Low
5	7	Gneiss	Low
6	6	Sand & Hard rock	Moderate
7	6	Clay, Sand, & Hard rock	Moderate
8	7	Clay, Sand, Silt & Hard rock	Moderate
9	7	Clay, Sand, Silt & Hard rock	Good
10	6	Gneiss	Low
11	5	Clay & Sand	Good
12	5	Clay & Sand	Good
13	6	Clay, Sand, Silt	Good
14	5	Clay, Sand & Silt	Good
15	5	Clay, Sand, Silt	Good
16	5	Clay, Sand, Silt	Good
17	7	Clay, Sand, Silt	High
18	7	Clay & Sand	High
19	6	Clay, Sand, Silt	High
20	7	Alluvium, Clay & Sand	Good

5. Conclusions

Based on the interpretation of geoelectrical data, VES indicated the presence of sand sediment (good water) with aresistivity of more than 200 Ωm. It also revealed six geo-electric layers consisting of the sub-surface layer (Alluvium, sand, silt, clay, clayey sand and bed rock). Schlumberger's resistivity vertical electrical surveys proved to achieve better results for groundwater prospects in hard rock terrains. The potential for theoccurrence of groundwater in the study area was classified as high (3%), good (50%), moderate (45%) and low (2 %) by interpreting the subsurface geophysical investigations. This study suggests favourable locations for artificial recharge structures to maintain the current status of groundwater.

References

[1] Ezhisaivallabi K and Poongothai S 2016 Geoelectric signatures of the coastal aquifer system of Adappa watershed. *Int. J. ChemTech Res.* **9** 607-13.

[2] Nejad H T 2009 Geoelectric Investigation of the Aquifer Characteristics and Groundwater Potential in Behbahan Azad University Farm, Khuzestan Province, Iran. *J. Appl. Sci.* **9** 3691-8.

[3] Ochuko Anomohanran 2013 Hydrogeophysical and Hydrogeological Investigations of Groundwater Resources in Delta Central, Nigeria. *J. Taibah Univ. Sci.* **9** 57–68.

[4] Ruby D, Chitra C, Vasantha A, Ramasubbulakshmi T and Manivel M 2010 Integrated analysis of Geophysical Data of Ponnaiyarriver basin using Arcview GIS Software. *Int. J. Geomatics and Geosciences* **1** 456-65.

[5] Sankaranarayana P V and Ramanujachari K R 1974 An inverse slope method for determining absolute resistivities. *Geophys.* **32** 1036-40.

IOP Conf. Series: Earth and Environmental Science **80** (2017) 012071 doi:10.1088/1755-1315/80/1/012071

[6] Selvam S and Sivasubramanian P 2012 Groundwater potential zone identification using geoelectrical survey: A case study from Medak district, Andhra Pradesh, India. *Int. J. Geomatics and Geosciences* **3** 55-62.
[7] Telford W M, Geldart L P and Sheriff R E 1990 A text book on Applied Geophysics, Cambridge University Press.
[8] Todd D K 1980 Groundwater Hydrology 2nd edn, *J. Wiley and Sons Inc.*, New York, 315.
[9] TWAD Handbook 2002.

IOP Conf. Series: Earth and Environmental Science **80** (2017) 012072 doi:10.1088/1755-1315/80/1/012072

Sustainable Irrigation Allocation Model for Dry and Wet Periods using Reservoir Storage and Inflow

S Surianarayanan*, C R Suribabu and K Ramakrishnan

School of Civil Engineering, SASTRA University, Thanjavur – 613 401. India.

*Email: suriyanarayanan@civil.sastra.edu

Abstract. The dry period agriculture is inevitable both for the farmers for their earning, and for the soil for its fertility by crop-rotation. In tropical countries like INDIA, dry period agriculture becomes difficult because of less (or) no rain fall . Hence a simple water balancing model for irrigation scheduling, using the measure "Volumetric Reliability" is prepared in this paper, with the storage and inflow of a reservoir both for the dry and wet periods.The case-study is done for a reservoir in INDIA with thirty one years of hydrological data(from 1982 to 2012). The objective of this paper is to prepare a simple water balance model taking 10 days periods of demand and supply for ID crop(Irrigated Dry crop, ground nut) with usage of volumetric reliability concept for the periods of deficiency and adoption of less water requirement crops to reduce the water-stress during critical periods of crop growth, and finally arrive at a feasible allocation schedule for the success of agriculture and the yield throughout the year both for wet and dry crops with the available storage on the start of irrigation for a particular year. The reservoir is divided for storages such as full, deficient and critical storages. The starting storage for the dry period from January is used after adequate allocation for wet crops, the quantity for riparian rights and for drinking water, for the sustainability. By the water-balancing, the time-series for thirty one years , it is found that for twenty two years the demand for the ID crops is satisfied with the storage in the reservoir, and in the remaining years of deficient inflows, for three years (1986,1996,2004)the demand is managed by using the safe reliability factor for demand which can nullify the deficit in demand for the whole supply period. But it is genuine to assure that the reduction in the amount of water for each 10 days periods should not exceed the survival limit of the crop. Necessary soil-moisture must be ensured in the crop in its root zone..A safe volumetric reliability factor of 0.75 is taken in this attempt to extend the success of dry period agriculture. The soil moisture balance in the clayey loam soil due to continuous irrigation can manage the deficiency, so that the soil-water does not go below the wilting point but must be checked with the soil-moisture sensors. The methods to reduce the evaporation from the soil , such as usage of drip irrigation and the mulching, which can reduce the crop water requirements in comparison with the conventional irrigation which is being followed at present in the spot of the case-study is suggested. In addition to these methods less water requirement crops like flowers , vegetables having less crop-period and provision of windbreak to reduce evapo-transpiration rates and other efficient methods of water management can make success of dry period agriculture even in the years critical storage level in the reservoir . The rainfall is not taken in this model which if in a year with ample quantity can be used for pre-sowing and for managing the water-stress. Augmentation of water with ground water, percolation tank can boost the level of success of agriculture. As the irrigation water allocation for the whole year, particularly for the ID crops is

prepared for the full, deficient and critical storages, this paper can help to achieve sustainable growth of crops throughout the year.
Keywords: ID crops, Volumetric Reliability and Sustainability.

1. Introduction

In the present scenario of decrease in rainfall ,untimely rainfall ,decrease in river- flow and increasing population in tropical countries like India where efficient conveyance and application methods in irrigation systems are seldom available, careful planning during allocation of water for irrigation is inevitable, to cater the water needs of the crops not only in the wet periods, but to save the crops from the deficiency of water during dry periods with the available water. Hence an irrigation allocation model for a multipurpose reservoir specifically for the irrigated-dry crops, (id crops) which also includes allocation for wet crops, drinking-water requirement, and quantity as per the riparian rights, is prepared in this paper. Volumetric reliability which is defined as the total volume of water supplied divided by the total demand during a specified period of time is the often used performance measure in rain water tank analysis. In deciding the optimal tank capacity, for the rain water tank system considered on a lot of single dwelling, 5kl tank is found to be adequate for a volumetric reliability of 0.8 than 10KL tank, however when replacing with the vegetables and grasses for garden with some water efficient species 3KL tank would fulfill the requirement according to C.H. Wang and J.M. Black more. The water consumption is optimized so as to get feasible size of water tank in the cost point of view also [1]. This performance measure of volumetric reliability is used in this paper to prepare irrigation water schedule for two things 1. To manage the water stress by reducing the volumetric reliability factor which is taken with in safe limit from 1 to 0.75 by which the crop can manage the water stress during the deficient inflow along with the available soil-moisture balance and 2. In addition to usage of volumetric reliability, replacement of water efficient species of grasses, vegetables and flowers instead of the regular dry period crop (ground nut) with efficient conveyance and irrigation methods like drip-irrigation is suggested as id crops which require more water during dry periods. The prime thought in this paper, is that when sufficient storage for the wet crops , water-supply etc, is reserved in a reservoir, a better water allocation schedule for the id crops is inevitable, genuinely for the deficient and critical storages, and can be used for sustainable irrigation during the whole year in a command area. By the success of ID crops the net benefit from agriculture can be ensured. The constraints are maximum, deficient, and critical reservoir storages which depend on the inflow in to the reservoir. The optimal operation of multi purpose reservoir for irrigation planning using surface and ground water is discussed by Nikam N.G et.al [2]. The critical water level for a soil below which the crops will be stressed is important. The reduced yield of crop is related to its growth stage and to the amount and duration that the soil water content is below this optimum value. Solutions to complex decision making problems are currently available for a variety of irrigation situation an integrated model is developed and is made of two modules one for intra seasonal allocation using lp and two for seasonal allocation to derive steady state reservoir operating policy as per martin M. Fogel et al. [3]. A short review of the mathematical models used in reservoir management and operation to present conclusions reached by previous state of art reviews, and to provide two ideas for closing the gap between theory and practice according to Slobodan P. Simonovic [4]. Water supply yield analysis techniques are developed to adequacy of wma water supply system to meet the escalating future demands for water. Historic yield values for the joint system yield model exceed 700mgd, indicating that the current water supply system is quite reliable according to. James A. Smith [5]. When design is applied in direct consultation with water users the system makes local water distribution more transparent and effective overall reliability = (actual volume delivered / planned volume) x actual supply duration / target supply duration according to Islander Abdulla et.al [6]. Reliability provides the inducement for farmers to invest in higher productivity to benefit individuals and society mapping of reliability and productivity through remote sensing is attempted sustainably, according to Chris Perry [7].The investigation strategies to alleviate the effects of drought on the probability and sustainability of irrigated agriculture this include conjunctive use of surface and ground water resources the simulation model consists of agronomic component and simulators for the system according to Gerritschoups et al [8]. According to S. Vedula, D. NageshKumar , the. Optimal seasonal multi crop irrigation water allocation and optimal stochastic intra seasonal in irrigation scheduling are carried out using two stage decomposition based on a stochastic dynamic programming method [9]. A optimal stochastic multi crop seasonal and intra seasonal irrigation control model is prepared by Judith A Sunantara et al [10]. The loss due to conveyance and the loss from soil surface can be reduced by adopting drip irrigation along with mulching. This will reduce the crop water requirement according to Battikhi and Hill [11]. Generally as the dry period agriculture is important for farmers for their earning, crop-rotation and food production which will generally have set-backs due to less availability of water and other reasons an attempt is made in this paper to analyze the possibility of success for dry crop by water scheduling using the operator "volumetric reliability".

1.1The need for water scheduling

Irrigation is the predominant sector where water requirement quantity is more than 85% in low and middle income group countries, where agriculture is the major economic sector (World Bank 1992). But the irrigated agriculture faces water stress due to shortages in demand. The demands of irrigation water have dependability on decisions for a particular period taken by the farmers which in turn depends on the type of crop and its period of growth. The water requirement decisions is generally based on 1.long term crops 2. Short -term crops. The food crops and the long term crops are generally given priority during the decisions for water allocations from a Reservoir. Hence, the crops sown during dry period, which is important for crop rotation and earning for farmers during dry season, is given secondary importance. Due to the fact that food crops like paddy, wheat are developed during wet season they have more chances of getting adequate water supply from a reservoir as well as from rainfall .But the farmers will be finding much difficult to do dry period agriculture because of the less amount of water available after allocation to wet crop irrigation and other components of demands for a multi-purpose reservoir. Also due to the stochastic nature of the inflow into the reservoir and rainfall in the agricultural field, there will be inadequate water for dry crop irrigation and sometimes it'll be severe to the extent that no agriculture can be carried using reservoir supply during dry period by the farmers. An attempt is made in this paper to manage the water stress during dry period using the reliability criteria, usage of less water requirement crops during the dry period to make the agriculture success to the possible extent. The standard water balancing equation and the simulation is used for the water allocation model in this paper. The Sathanur reservoir built in the year 1960,constructed across Poniyar river in India is taken as the case study The hydrological and hydraulic data with the history of the reservoir and all other details are collected for 31 years (1982—2012). This is a multi-purpose reservoir, which is satisfying the irrigation needs of the south Arcot district, in addition to the needs of drinking water supply for Tiruvannamalai city, hydro electricity power generation etc., This dam is mainly constructed to supply irrigation water during the dry season from December to April of every year in which there will be less rainfall and for the successful agriculture of Irrigated-Dry crops(ID crops).But due to the stochastic nature of the inflow which is the predominant factor for the ID-crop irrigation water, it is found that reliable water supply cannot be assured if the inflow in to the reservoir is very less (or) less than its demand. In this paper water balancing model is prepared for the irrigation water of ID crops, for managing the water-stress during periods of less inflow using the standard water balancing and the use of operator Reliability to the extent possible, during dry inflow periods. Out of 31 years , eight years were found to be deficient in satisfying the demand ,but by using the operators 3 years out of 8 years , the deficiency is managed .

1.2 Study area details

The area of case study is the Sathanur reservoir constructed during 1958, near Tiruvannamalai Town, North Arcot District, Tamilnadu, India. It is located across the river Ponnaiar. Its Catchment area is 10835 km2.The length of the Masonry Dam is 418 m and the length of the earth Dam is362 m. The height of Masonry Dam is 45 m above foundations. The Capacity of Reservoir is 229 million m 3and the live storage capacity is207 million m3.The spillway has 9 Vents of 12.2 m x 6.1 m. The capacity of Saddle Spillway is 2234 m3/s. Average inflow (1987-2007): 314.08 million m3 per year. The length of the river is7.2 km up to pickup weir. Period of wet season is September to January and the Period of dry season: December – March. Command area under wet period is is3020hectare and the crop is paddy. The Command area under dry period is15200 hectares and the ID crop is ground nut. There are two main canals such as LBC (left main canal), RBC (right main canal) supplying water for the crops for both the seasons. The salient features of dam (i) Capacity of the reservoir – 340 million cu.m (ii) No. of fillings – 20 out of 55 years (iii) Annual Storage – 6796 million cu.m (iv) Water spread area – 126.66 ha (v) Catchments area – 87.023 sq.m (vi) Area submerged – 102.935 ha (vii) Total dependable yield available at site – 10.26 million cu.m (viii) New command area – 740.57 ha (ix) Old command area – 109.67 ha (x) Number of spillway – 2 nos. The irrigated area after the constructions of the dam has been increased, additional area of irrigation was achieved in addition to the demands for the riparian rights.

Figure 1. Command area map of the Sathanur Reservoir

The scope of this paper is to arrive at an appropriate allocation decision model for dry crops by evaluating the reliability of water allocation model with the starting storage, inflow into the reservoir which is the predominant factors as sufficient, moderately deficient, and highly deficient supplies, by simulating the starting storage and the inflow on January 1 of the respective year until a positive allocation model is arrived by the usage of operator, " volumetric Reliability" ,change of crop having less water requirement, and using efficient irrigation methods.,

Table 1. Details of reservoir levels for 55 years.

S. No.	Year			Maximum Water Level in ft	Capacity in Mcft	Reservoir Filled	Reservoir Not Filled
1	1958	-	1959	102.80	5112	Filled	-
2	1959	-	1960	103.30	5188	Filled	-
3	1960	-	1961	102.10	5040	Filled	-
4	1961	-	1962	74.40	1947	-	Not filled
5	1962	-	1963	90.25	4739	-	Not filled
6	1963	-	1964	100.20	4732	-	Not filled
7	1964	-	1965	109.30	6178	-	Not filled
8	1965	-	1966	90.65	3502	-	Not filled
9	1966	-	1967	117.95	7859	-	Not filled
10	1967	-	1968	119.00	8083	Filled	-
11	1968	-	1969	118.55	7987	-	Not filled
12	1969	-	1970	118.95	8072	-	Not filled
13	1970	-	1971	118.35	7944	-	Not filled
14	1971	-	1972	118.90	8062	-	Not filled
15	1972	-	1973	122.50	8736	Filled	
16	1973	-	1974	117.35	7733	-	Not filled
17	1974	-	1975	117.00	7659	-	Not filled
18	1975	-	1976	118.40	7955	-	Not filled
19	1976	-	1977	78.85	2308	-	Not filled
20	1977	-	1978	119.00	8083	Filled	-
21	1978	-	1979	119.00	8083	Filled	-
22	1979	-	1980	119.00	8083	Filled	-
23	1980	-	1981	79.65	2378	-	Not filled
24	1981	-	1982	119.00	8083	Filled	-
25	1982	-	1983	73.75	1903	-	Not filled
26	1983	-	1984	94.20	3933	-	Not filled
27	1984	-	1985	95.05	4040	-	Not filled
28	1985	-	1986	87.40	2178	-	Not filled
29	1986	-	1987	118.10	7119	-	Not filled
30	1987	-	1988	101.80	4026	-	Not filled
31	1988	-	1989	114.95	6431	-	Not filled
32	1989	-	1990	83.25	1788	-	Not filled
33	1990	-	1991	68.90	794	-	Not filled
34	1991	-	1992	119.00	7321	Filled	-
35	1992	-	1993	119.00	7321	Filled	-
36	1993	-	1994	119.00	7321	Filled	-
37	1994	-	1995	115.20	6484	-	Not filled
38	1995	-	1996	108.10	4725	-	Not filled
39	1996	-	1997	119.00	7321	Filled	-
40	1997	-	1998	119.00	7321	Filled	-
41	1998	-	1999	119.00	7321	Filled	-
42	1999	-	2000	118.50	7209	-	Not filled

43	2000	-	2001	117.30	6942	-	Not filled
44	2001	-	2002	117.10	6897	-	Not filled
45	2002	-	2003	78.75	1415	-	Not filled
46	2003	-	2004	91.70	2648	-	Not filled
47	2004	-	2005	119.00	7321	Filled	-
48	2005	-	2006	119.00	7321	Filled	-
49	2006	-	2007	97.20	3358	-	Not filled
50	2007	-	2008	115.45	6538	-	Not filled
51	2008	-	2009	119.00	7321	Filled	
52	2009	-	2010	107.10	4901	-	Not filled
53	2010	-	2011	119.00	7321	Filled	-
54	2011	-	2012	119.00	7321	Filled	-
55	2012	-	2013	89.00	2342	-	Not filled

2. Objective

By adopting a safe volumetric reliability factor for the irrigation water supplied to the crop, during deficit inflow, the success of dry crop and the benefit can also achieved. The objective function is written as

1. Maximize Z1 = Maximize Reliability of irrigation = R (t)
2. Maximize Z2 = Maximize benefit B (t) / yield (t).

Rv= volumetric reliability is the measure used. = Volume of supply / Volume of Demand for 10 day periods. By increasing the reliability the level of success of agriculture as well as benefit can be achieved.

The standard water balancing equation and consecutive simulation is done for each 10 day periods. Water balancing equation used is,

$$S_t + I_t - L_t - D_t = S_{t+1}, \quad (1)$$

Where, t=1, 2, 3… n periods of crop each of 10 days period. And

S_t =Storage in the reservoir, at the start of irrigation in Mm3, I_t =Inflow during t (10days), D_t = Irrigation requirement during t, (0.75 times the inflow is only taken as I to account for the losses.)S_{t+1}= Storage at the start of t+1 period.
Where

$$R_v = \frac{\sum_{k=1}^{n} s}{\sum_{k=1}^{n} D}$$

$$0.75 \leq Rv \leq 1 \text{ for 9 periods from k=1 to n} \quad (2)$$

WhereR_v is the volumetric reliability.
By increasing the R_v ,the yield can be increased (or) maximized upto when $R_{v=1}$
Where,$\sum_{k=1}^{n}$ s = total supply during a period (10 days) for the dry crop

IOP Conf. Series: Earth and Environmental Science **80** (2017) 012072 doi:10.1088/1755-1315/80/1/012072

$\sum_{k=1}^{n}$ D = total demand during a period (10 days) for the dry crop
Y=Yield from the agriculture

The value of R=1 for full and medium storage. The value of R_v is reduced from 1 to 0.75for deficient and severe storages. The reliability of irrigation is assessed every year by simulating the time series of storage, inflow and demand using the water balance equation as in methodology.

3. Methodology

The reservoir is divided in to three zones like safe storage (207 to 100 M m^3), deficient storage (100 to 50 M m^3) and critical storage (below 50 M m^3). From the year 1982 to 2012, with the time series,the inflow, the irrigation requirement is simulated using The Standard Water Balancing Equation. For the years of deficient storage the Volumetric Reliability factor of 0.75 is used to come over the water stress and for the success of the dry crop. The flowchart for the run of the model with suggestions for the critical periods is also given below.

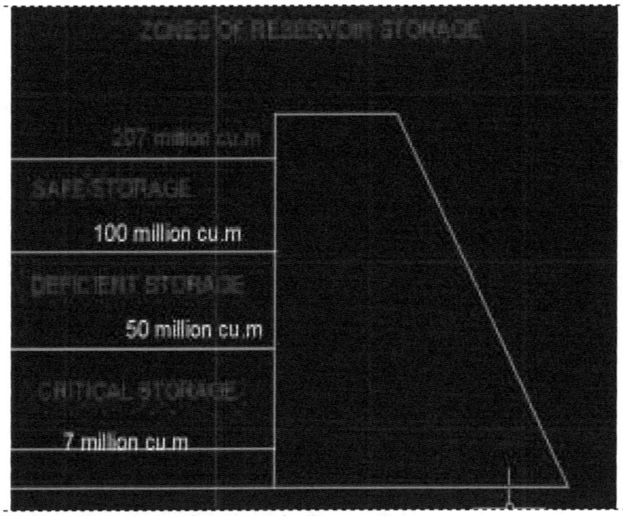

Figure 2. Zones of Reservoir Storage

IOP Conf. Series: Earth and Environmental Science **80** (2017) 012072 doi:10.1088/1755-1315/80/1/012072

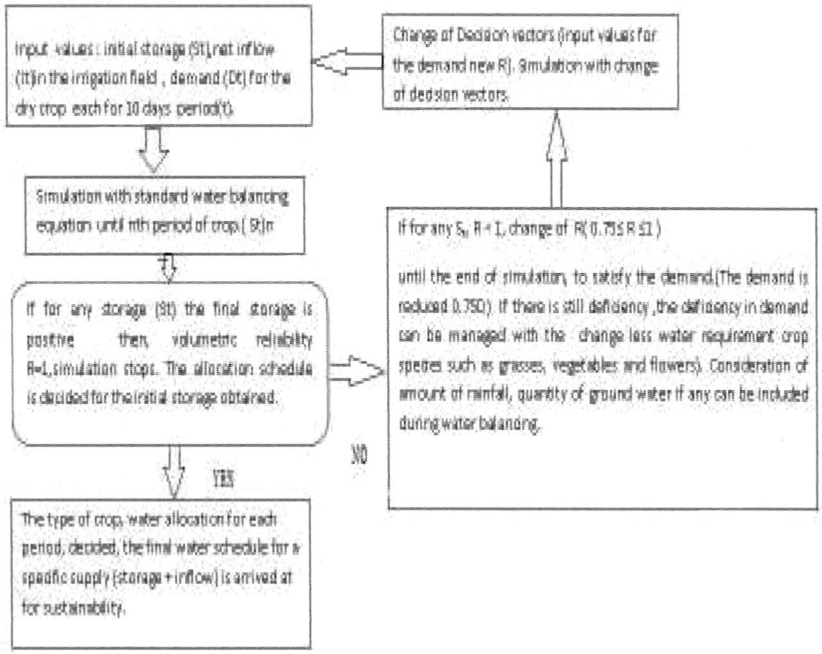

Figure 3. Methodology Flow chart

3.1 Steps of Calculations of demand and supply for water balancing

i. The demand and supply calculations are made for 10days periods and simulated starting from the storage on january 1 for 9 periods for the ID crop for all the 31 years (1982-2012)the demand for all 36 periods of the year is calculated taking the priorities for the riparian rights,drinking water requirement and wet crops.
ii. The demand is calculated by using Blaney-criddle method using the details of data obtained from the standard hydrological data for the area.
iii. The consumptive use for 10 days period for ground nut is calculated by
iv. Cu= kp/40 (1.8t +32), where, Cu= monthly consumptive use in 'mm'. The values for the k.p. t is obtained for the periods using standard reference book for the command area , prepared by the government. The total command area is 15, 045 hectares for two canals namely left bank canal and rigat bank canal.
v. The supply(intial storage inflow) on january 1 for each 31 years with inflows , is water balanced with the demand for all 9 periods for ID crops using standard water balancing andfor deficient and critical periods the deficiency of water is checked with the volumetric reliability factor of 0.75 to manage the water stress., and suggestive methods are given to relive from water-stress to the extent possible.

Table 2. Demand for dry crop (ID) Ground nut and others

Crop water requirement in mm (10days)	Area in hectares	Depth in m	Volume in cu.m	Volume in cu.m	Other allocations (dr.wtr+rap.ri) Million cum	Total req. for ID crop period M cum
12.87	15045	0.013	1955854	1.96	1	2.96
18	15045	0.018	2735460	2.74	5.28	8.02
22.84	15045	0.023	3470995	3.47	3.57	7.04
28.6	15045	0.029	4346342	4.35	1	5.35
38.84	15045	0.039	5902515	5.9	1	6.9
41.4	15045	0.041	6291558	6.29	4.4	10.69
35.45	15045	0.035	5387337	5.39	4.43	9.82
37	15045	0.037	5622890	5.62	3.83	9.45
33	15045	0.033	5015010	5.02	3.83	8.85
268				40.73	28.34	69.07

Table 3. Demand Calculations(CWR) for ID crop including priorities for other requirements

S. No.	AREA (in Hec)	Depth (in m)	Vol. (in m3)	Other allocations (Million cu.m)	ID crop (Million cu.m)	Total (Million cu.m)	Field requirement considering losses 25% (M cu.m)
1.	15197	0.013	1955854	1	1.96	2.96	3.94
2	15197	0.018	2735460	5.28	2.74	8.02	10.68
3	15197	0.023	3470995	3.57	3.47	7.04	9.38
4	15197	0.029	4346342	1	4.35	5.35	7.12
5	15197	0.039	5902515	1	5.90	6.90	9.20
6	15197	0.041	6291558	4.4	6.29	10.69	14.25
7	15197	0.035	5387337	4.43	5.39	9.82	13.08
8	15197	0.037	5622890	3.83	5.62	9.45	12.60
9	15197	0.033	5015010	3.83	5.02	8.85	11.79
			Total		40.73	69.07	92.09

Table 4. Table showing the actual demand in the field in cluding losses

Total CWR in the head of canal (M cu.m)	Total CWR in the field including losses taken as 25% (M cu.m)
2.96	3.94
8.02	10.69
7.04	9.39
5.35	7.13
6.90	9.20
10.69	14.26
9.82	13.09
9.45	12.60
8.85	11.79
69.07	92.09

Table 5. Allocations for wet crop and riparian rights etc., for 36 periods of the year

Month	Priorities With Periods(10 Days)	Drink.Wat 1	Tank Dem 2	ID (Gr.Nut) 3	Down Stream Command Demand(Thirukoil) 4
Jan	1	1	0	2.96	0
	2	1	4.28	8.02	0
	3	1	2.57	7.04	0
Feb	1	1	0	5.35	0
	2	1	0	6.90	0
	3	1	3.4	10.69	0
Mar	1	1	3.43	9.82	0
	2	1	2.83	9.45	5.67 During 2, 3 pds of March.
	3	1	2.83	8.85	0
Apr	1	1	0		0
	2	1	0		0
	3	1	0		0
May	1	1	0		0
	2	1	0		0
	3	1	0		0
June	1	1	0		0
	2	1	0		0
	3	1	0		0
July	1	1	0		0
	2	1	0		0
	3	1	0		0
Aug	1	1	0		0
	2	1	0		0
	3	1	0		0
Sep	1	1	9.5		0
	2	1	0		0
	3	1	0		0
Oct	1	1	0		0
	2	1	0		0
	3	1	0		0
Nov	1	1	0		0
	2	1	0		0
	3	1	0		0
Dec	1	1	0		0
	2	1	0		0
	3	1	13.61		0

Table 6. Water balancing for deficient storage for the year 2004

St	It	dt	Total
74.98524	0	3.94	71.04411
71.04411	0	10.69	60.35683
60.35683	0	9.39	50.96883
50.96883	1.25	7.13	45.09038
45.09038	0	9.20	35.88702
35.88702	0	14.26	21.63161
21.63161	0	13.09	8.541831
8.541831	0	12.60	-4.06202
-4.06202	0	11.79	-15.8554
		92.09	

Table 7. Water balancing for deficient storage for 2004 using volumetric reliability of (R_v)0.75 in demand

St	It	dt	Total
74.98524	0	2.96	72.02939
72.02939	0	8.02	64.01393
64.01393	0	7.04	56.97294
56.97294	1.25	5.35	52.87659
52.87659	0	6.90	45.97408
45.97408	0	10.69	35.28252
35.28252	0	9.82	25.46518
25.46518	0	9.45	16.01229
16.01229	0	8.85	7.167284

Table 8. Water balancing for storage (st=storage at time): it, dt(inflow, demand during t)-2002

St	It	dt	St+1
201.9379	21.13633	3.94	219.1331
219.1331	1.763135	10.69	210.209
210.209	0	9.39	200.821
204.6534	0	7.13	197.525
197.525	0	9.20	188.3216
188.3216	0	14.26	174.0662
174.1084	0	13.09	161.0186
161.0186	0	12.60	148.4148
148.4148	0	11.79	136.6214
		92.09	

Table 9. Water balancing for critical storage for the year 1986

St	It	dt	St+1
57.96	1.11	3.94	55.12
55.12	7.23	10.69	53.63
53.63	2.98	9.39	49.08
49.08	1.30	7.13	44.52
44.52	0.83	9.20	30.42
30.42	0.45	14.26	21.12
21.12	0.69	13.09	11.39
11.39	0.56	12.60	1.53
1.53	0.37	11.79	-8.52

Table 10. Water balancing for critical storage for 1986 using volumetric reliability of 0.75 in demand and water stress is managed with the inflow

St	It	dt	st+1
57.96	1.11	2.951295	56.11072
56.11	7.23	8.01546	55.33
55.33	2.98	7.037955	51.26
51.26	1.30	5.331145	47.23
47.23	0.83	14.59719	33.47
33.47	0.45	10.72156	23.20
23.20	0.69	5.984616	17.90
17.90	0.56	6.312871	12.15
12.15	0.37	6.62289	5.90
		67.57	

3.2 Design schedule for water allocations:
Normal season: Storage = 207 M cu.m

Table 11. Existing Schedule in operation by the government (Values in Million cu.m)

Case	Storage	To colony (ws)	Dead storage	II crop (riparian rights)	Drinking water supply	Total deduction	Wet crop	Dry crop	Remarks
Full storage	207	19	9	34	8	70	23	114	12 spells with summer irrigation for 15000 hectares,3164 hec of paddy and 2023 hec of second crop.
Deficient storage	100	17	9	-	8	34	13	53	8 spells of two third supply,10000 hectares of summer irrigation.
Scarcity season	70	13	9	-	8	30	13	27	8 spells of One third supply of 6755 hec of summer irrigation.

Table 12. Hypothetical schedule of allocations

Case	Storage	To colony (ws)	Dead storage	II crop (riparian rights)	Drinking water supply	Total deduction	Wet crop	Dry crop	Remarks
Full storage	207	19	9	34	8	70	23	114	12 spell considering full area both for summer, wet, and II crops.
Deficient storage	100	17	9	-	8	34	13	53	8 spells of two third supply considering the full area for summer crop, wet crop, and II crop if there is scope (supply with volumetric reliability of 0.75, and monitoring the minimum soil moisture)
Scarcity storage	70	13	9	-	8	30	13	27	8 spells of One third supply with much reduction in agricultural area.(Supply with volumetric reliability of 0.75 can increase the area of agriculture)

Table 13. Statement showing the average monthly rainfall of sathanur dam Rainfall Station

Jan	Feb	Mar	Apr	May	Jun	Jul	Aug	Sep	Oct	Nov	Dec	Total
11.688	9.135	13.9	18.125	86.12	51.88	72.79	109.95	162.19	193.52	166.97	82.658	978.923

Table 14. Command area for Sathanur dam in hectares

Canals	Dry Crop (Ha)	Wet Crop (Ha)
SLBC	8638	1130
SRBC	6559	1890
TOTAL	15197	3020

SLBC –sathanur left bank canal: SRBC sathanur right bank canal

Table 15. Water balancing with normal values for the year1996

St	it	dt	st+1
76.46	0.14	3.94	72.67
72.67	0.00	10.69	63.94
63.94	0.00	9.38	56.42
56.42	0.00	7.11	50.55
50.55	0.00	19.46	35.62
35.62	0.00	14.30	25.87
25.87	0.00	7.98	15.45
15.45	0.00	8.42	5.03
5.03	0.00	8.83	-5.39

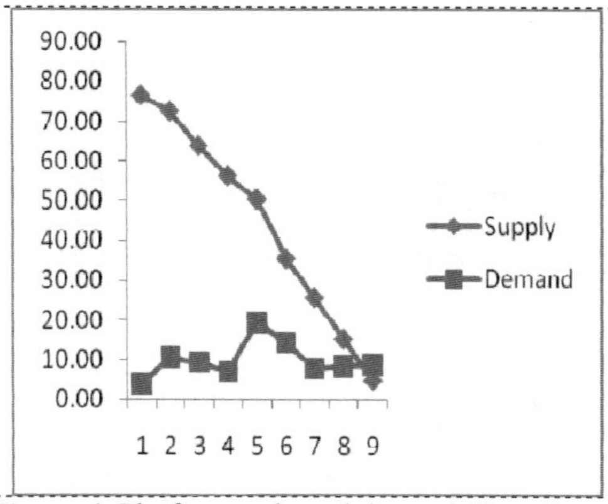

Figure 4. Plotfor supply and demand (Rv= 1)

Table 16. Water balancing with normal values for the year1996 using(R_v) =0.75

St	it	dt	st+1
76.46	0.14	2.95	73.65
73.65	0.00	8.02	65.63
65.63	0.00	7.04	58.60
58.60	0.00	5.33	53.27
53.27	0.00	14.60	38.67
38.67	0.00	10.72	27.95
27.95	0.00	5.98	21.96
21.96	0.00	6.31	15.65
15.65	0.00	6.62	9.03

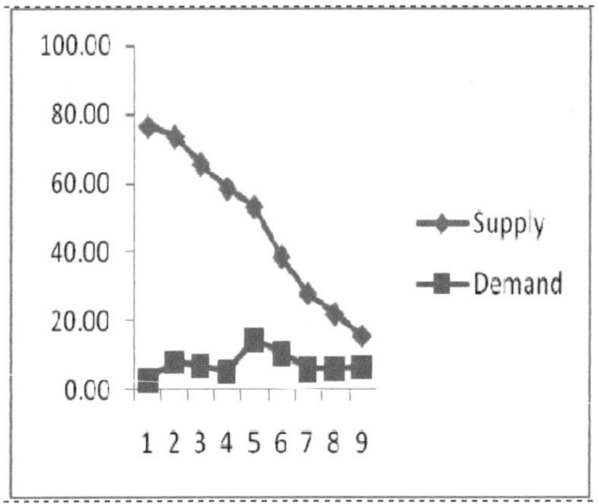

Figure 5. Plot for supply and demand (Rv=.0.75)

Table 17. Water balancing with normal values for the year 1986

St	it	dt	st+1
57.96	1.11	3.94	55.13
55.13	7.23	10.69	53.63
53.63	2.98	9.38	49.08
49.08	1.30	7.11	44.52
44.52	0.83	19.46	30.42
30.42	0.45	14.30	21.12
21.12	0.69	7.98	11.39
11.39	0.56	8.42	1.53
1.53	0.37	8.83	-8.52

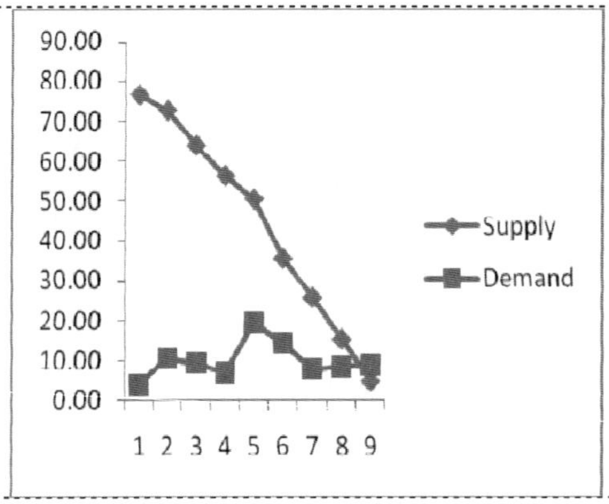

Figure 6. Plot for supply and demand (Rv=1)

4. Results and Discussion

The annul inflow in to the reservoir in the ascending order of lowest, middle and highest in quantity among each consecutive 5 years is taken and simulated using standard water balance equation for the set of data from 1982 to 2012 (31 years). Out of 31 years 8 years the demand was not satisfied or deficient and by using the volumetric reliability it was found that for 3 years the water deficiency canbe managed. For the remaining years of irrigation water deficiency the very less water requirement crop than groundnut with efficient methods of irrigation such as drip and sprinkler irrigation, with augmentation of groundwater can save the ID crops making sustainable irrigation through the year.

The rain fall is scarce and if there is rainfall in a year it can augment the irrigation requirement along with groundwater and saving in percolation ponds during excess year. In the allocation schedule prepared by the government, the area of irrigation for dry period is reduced according to the storage. But the total area of 15,045 hectares is taken for ID crops in this model for all the 31 years.

Table 18. Water balancing with normal values for the year 1986 using (R_v)=0.75

St	it	dt	st+1
57.96	1.11	2.95	55.92
55.92	7.23	8.02	55.13
55.13	2.98	7.04	51.07
51.07	1.30	5.33	47.04
47.04	0.83	14.60	33.27
33.27	0.45	10.72	23.00
23.00	0.69	5.98	17.71
17.71	0.56	6.31	11.96
11.96	0.37	6.62	5.70

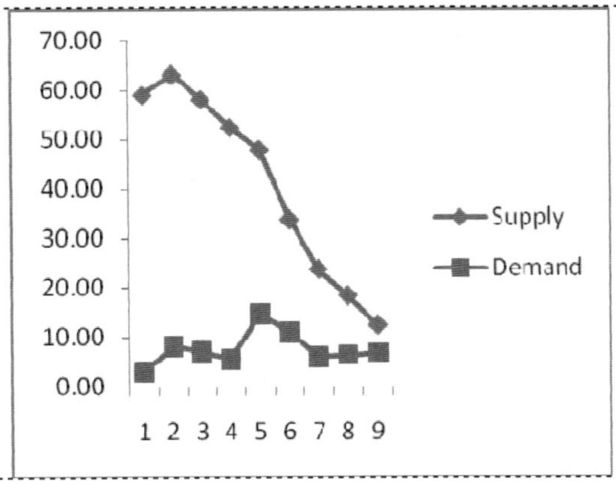

Figure 7. Plot for supply and demand (Rv=0.75)

Out of 55 years (from 1957 -2012) the dam got filled for 20 years and with in the remaining years 13 years the storage was critical making water stress for ID crops. The water balancing and usage of volumetric reliability is done for 31 years(1982-2012). (The table numbers 16,18 and the figures 4,5,6 and 7 shows how the reduced demand is satisfied using Rv) The model gives the water scheduling for three deficient years to make it success out of 8 years. For the remaining years of deficient and critical flows suggestions are given to over come the water-stress. Out of the hydrological parameters such as the reservoir inflow, rainfall and storage at the start of irrigation for the ID crops, the storage at the start is the only known parameter to decide the pattern of ID crops. As the period of ID crops starts from January coincides with opening of water for irrigation from reservoir storages up to December every year the ID crops can have probability of the propotionly ample amount of water for its allocation, with its storage. This model requires the prediction of inflow in to the reservoir which can be assessed by the rainfall in the catchment area.The rainfall which is not the reliable parameter during dry period is not taken in this model, if any can help for requirements during pre-sowing, mid or end periods of crop growth if in a particular year. Even though this work is hypothetical, it can be made practically possible by checking the soil moisture balance during each day of the crop period to ensure the minimum available moisture in the root zone of the crop, using soil moisture meters and sophisticated instruments which is a must for the adoption of this model in the field and finally the survival and success of ID crops can be achieved.

Table 19. Results and Discussions

Year	Storage (in Mm^3)	Inflow(in Mm^3)	Usage of volumetric Reliability criteria ,rainfall for managing the water stress	
			Deficiency in Demand, over coming using (R_v)	
1996	76.46	0.14	-5.39	Using R=0.75,the deficiency is reduced from - 5.39 to +9.02 , no rainfall, usage of limited ground water along efficient methods can increase the volumetric reliability from 0.75 to 1
1986	57.95	15.52	-8.52	Using R=0.75, the deficiency is reduced from - 8.52 to + 5.7, rainfall of 13.3 can augment. usage of limited ground water along efficient methods can increase the volumetric reliability from 0.75 to 1
2004	74.98	1.25	-15.85	Using R=0.75,the deficiency is reduced from - 15.85 to +7.16 , rainfall of 4.20 cm can augment usage of limited ground water along efficient methods can increase the volumetric reliability from 0.75 to 1
1989	56.87	2.05	-23.07	Using R=0.75, the deficiency is reduced -23.07 to -8.64 , rainfall of 6.00 can augment, the deficiency and the reliabity can be increased by other measures as given above.
1983	53.18	1.41	-27.39	No rainfall but suggestive methods of crops with less CWR and efficient methods of irrigation can be adopted. (Drip, sprinkler methods, mulching etc), with reduction in area for ID crops to meet out the water stress.
1990	31.47	3.55	-46.96	Even though a rainfall of 18 M m^3 can manage the water stress to some extent other measures can be carried out.
2003	38.62	0.00	-53.46	Severe water stress, no rainfall. Area of agriculture can be reduced so that ID crops can be developed using available water
1991	14.84	2.59	-64.56	Severe water stress , 0.45 rainfall . Area of agriculture can be reduced so that ID crops can be developed using available water

References

[1] Wang C H and Black J M 2012 Supply-demand Risk and Resilience Assessment. *Water Res. Manage* **26** 4381-96.
[2] Nikan N G and Regulwar D G 2015 Optimal Operation of Multipurpose Reservoir for Irrigation Planning with Conjunctive Use of Surface and Ground Water. *J. Water Res. Protection* **4** 34-42.
[3] Martin M F 1974 Optimal Control of Irrigation Water Application. *J. Hydrol.* **28** 343-58.
[4] Slobodan P S 1992.Reservoir System Analysis: closing gap between theory and practice. *ASCE J. Water Res. Plan. Manage.* **118** 262-80.
[5] James A S 1989 Water Supply Yield Analysis for Washington Metropolitan Area. *ASCE J. Water Res. Plan. Manage.* **115**(2) 230-42.
[6] Iskandar A 2006 The Reliability Improvement in Irrigation Services. Application of Rotational Water Distribution to Tertiary Canals in Central Asia. *IWMI Research Report* #100. 34 Colombo, Sri Lanka .
[7] Chris P 2005. Irrigation Reliability and Productivity of Water: A propsed methodology using evapo-transpiration mapping. Institute of Water and Environment, Cranfiled University, *Irrigation and Drainage System* **19** 211-21.
[8] Gerrit S, Adans L C, Joseluis M and Steven G M 2006 Sustainable Conjuctive Water Management in Irrigated Agriculture Model Formulation and Application to The Yaqui Valley, Mexico. *Water Res. Res.* **42**(10).
[9] Vedula S and Nagesh K D 1996 An Integrated Model for Optimal Reservoir Operation for Irrigation of Multiple Crops. *Water Res. Res.* **32**(4) 1101-8.
[10] Judith D S 1997 Optimal Stochastic Multi Crop Seasonal and Intra Seasonal Irrigation Control *ASCE J. Water Res. Plan. Manage.* **123** 39-48.
[11] Battikhi A M and Hill R W 1998 Irrigation scheduling and cantaloupe yield model for the Jordan Valley. *Agric. Water Manage.* **15** 177-87.